T0172025

Birkhäuser

Applied and Numerical Harmonic Analysis

Series Editor
John J. Benedetto
University of Maryland
College Park, MD, USA

Editorial Advisory Board

Akram Aldroubi
Vanderbilt University
Nashville, TN, USA

Jelena Kovačević
Carnegie Mellon University
Pittsburgh, PA, USA

Andrea Bertozzi
University of California
Los Angeles, CA, USA

Gitta Kutyniok
Technische Universität Berlin
Berlin, Germany

Douglas Cochran
Arizona State University
Phoenix, AZ, USA

Mauro Maggioni
Duke University
Durham, NC, USA

Hans G. Feichtinger
University of Vienna
Vienna, Austria

Zuowei Shen
National University of Singapore
Singapore, Singapore

Christopher Heil
Georgia Institute of Technology
Atlanta, GA, USA

Thomas Strohmer
University of California
Davis, CA, USA

Stéphane Jaffard
University of Paris XII
Paris, France

Yang Wang
Michigan State University
East Lansing, MI, USA

For further volumes:
http://www.springer.com/series/4968

Simon Foucart • Holger Rauhut

A Mathematical Introduction to Compressive Sensing

 Birkhäuser

Simon Foucart
Department of Mathematics
Drexel University
Philadelphia, PA, USA

Holger Rauhut
Lehrstuhl C für Mathematik (Analysis)
RWTH Aachen University
Aachen, Germany

ISBN 978-1-4939-0063-3   ISBN 978-0-8176-4948-7 (eBook)
DOI 10.1007/978-0-8176-4948-7
Springer New York Heidelberg Dordrecht London

Mathematics Subject Classification (2010): 46B09, 68P30, 90C90, 94A08, 94A12, 94A20

© Springer Science+Business Media New York 2013
Softcover re-print of the Hardcover 1st edition 2013
This work is subject to copyright. All rights are reserved by the Publisher, whether the whole or part of the material is concerned, specifically the rights of translation, reprinting, reuse of illustrations, recitation, broadcasting, reproduction on microfilms or in any other physical way, and transmission or information storage and retrieval, electronic adaptation, computer software, or by similar or dissimilar methodology now known or hereafter developed. Exempted from this legal reservation are brief excerpts in connection with reviews or scholarly analysis or material supplied specifically for the purpose of being entered and executed on a computer system, for exclusive use by the purchaser of the work. Duplication of this publication or parts thereof is permitted only under the provisions of the Copyright Law of the Publisher's location, in its current version, and permission for use must always be obtained from Springer. Permissions for use may be obtained through RightsLink at the Copyright Clearance Center. Violations are liable to prosecution under the respective Copyright Law.
The use of general descriptive names, registered names, trademarks, service marks, etc. in this publication does not imply, even in the absence of a specific statement, that such names are exempt from the relevant protective laws and regulations and therefore free for general use.
While the advice and information in this book are believed to be true and accurate at the date of publication, neither the authors nor the editors nor the publisher can accept any legal responsibility for any errors or omissions that may be made. The publisher makes no warranty, express or implied, with respect to the material contained herein.

Printed on acid-free paper

Springer is part of Springer Science+Business Media (www.birkhauser-science.com)

ANHA Series Preface

The *Applied and Numerical Harmonic Analysis* (ANHA) book series aims to provide the engineering, mathematical, and scientific communities with significant developments in harmonic analysis, ranging from abstract harmonic analysis to basic applications. The title of the series reflects the importance of applications and numerical implementation, but richness and relevance of applications and implementation depend fundamentally on the structure and depth of theoretical underpinnings. Thus, from our point of view, the interleaving of theory and applications and their creative symbiotic evolution is axiomatic.

Harmonic analysis is a wellspring of ideas and applicability that has flourished, developed, and deepened over time within many disciplines and by means of creative cross-fertilization with diverse areas. The intricate and fundamental relationship between harmonic analysis and fields such as signal processing, partial differential equations (PDEs), and image processing is reflected in our state-of-the-art *ANHA* series.

Our vision of modern harmonic analysis includes mathematical areas such as wavelet theory, Banach algebras, classical Fourier analysis, time–frequency analysis, and fractal geometry, as well as the diverse topics that impinge on them. For example, wavelet theory can be considered an appropriate tool to deal with some basic problems in digital signal processing, speech and image processing, geophysics, pattern recognition, biomedical engineering, and turbulence. These areas implement the latest technology from sampling methods on surfaces to fast algorithms and computer vision methods. The underlying mathematics of wavelet theory depends not only on classical Fourier analysis, but also on ideas from abstract harmonic analysis, including von Neumann algebras and the affine group. This leads to a study of the Heisenberg group and its relationship to Gabor systems, and of the metaplectic group for a meaningful interaction of signal decomposition methods. The unifying influence of wavelet theory in the aforementioned topics illustrates the justification for providing a means for centralizing and disseminating information from the broader, but still focused, area of harmonic analysis. This will be a key role of *ANHA*. We intend to publish the scope and interaction that such a host of issues demands.

Along with our commitment to publish mathematically significant works at the frontiers of harmonic analysis, we have a comparably strong commitment to publish major advances in the following applicable topics in which harmonic analysis plays a substantial role:

Antenna theory	*Prediction theory*
Biomedical signal processing	*Radar applications*
Digital signal processing	*Sampling theory*
Fast algorithms	*Spectral estimation*
Gabor theory and applications	*Speech processing*
Image processing	*Time–frequency and*
Numerical partial differential equations	*time-scale analysis*
	Wavelet theory

The above point of view for the *ANHA* book series is inspired by the history of Fourier analysis itself, whose tentacles reach into so many fields.

In the last two centuries, Fourier analysis has had a major impact on the development of mathematics, on the understanding of many engineering and scientific phenomena, and on the solution of some of the most important problems in mathematics and the sciences. Historically, Fourier series were developed in the analysis of some of the classical PDEs of mathematical physics; these series were used to solve such equations. In order to understand Fourier series and the kinds of solutions they could represent, some of the most basic notions of analysis were defined, e.g., the concept of "function". Since the coefficients of Fourier series are integrals, it is no surprise that Riemann integrals were conceived to deal with uniqueness properties of trigonometric series. Cantor's set theory was also developed because of such uniqueness questions.

A basic problem in Fourier analysis is to show how complicated phenomena, such as sound waves, can be described in terms of elementary harmonics. There are two aspects of this problem: first, to find, or even define properly, the harmonics or spectrum of a given phenomenon, e.g., the spectroscopy problem in optics; second, to determine which phenomena can be constructed from given classes of harmonics, as done, e.g., by the mechanical synthesizers in tidal analysis.

Fourier analysis is also the natural setting for many other problems in engineering, mathematics, and the sciences. For example, Wiener's Tauberian theorem in Fourier analysis not only characterizes the behavior of the prime numbers, but also provides the proper notion of spectrum for phenomena such as white light; this latter process leads to the Fourier analysis associated with correlation functions in filtering and prediction problems, and these problems, in turn, deal naturally with Hardy spaces in the theory of complex variables.

Nowadays, some of the theory of PDEs has given way to the study of Fourier integral operators. Problems in antenna theory are studied in terms of unimodular trigonometric polynomials. Applications of Fourier analysis abound in signal processing, whether with the fast Fourier transform (FFT), or filter design, or the adaptive modeling inherent in time–frequency-scale methods such as wavelet theory. The coherent states of mathematical physics are translated and modulated Fourier transforms, and these are used, in conjunction with the uncertainty principle, for dealing with signal reconstruction in communications theory. We are back to the raison d'être of the *ANHA* series!

University of Maryland   *John J. Benedetto*
College Park   Series Editor

Dedicated to our families

Pour Jeanne

Für Daniela, Niels, Paulina und Antonella

Preface

Recent years have seen an explosion of research activities in a fascinating area called compressed sensing, compressive sensing, or compressive sampling. A Google Scholar search for articles containing one of these three terms in their title returned about 4,400 hits at the time this preface was written. The area of compressive sensing, at the intersection of mathematics, electrical engineering, computer science, and physics, takes its name from the premise that data acquisition and compression can be performed simultaneously. This is possible because many real-world signals are sparse, and even though they are acquired with seemingly too few measurements, exploiting sparsity enables one to solve the resulting underdetermined systems of linear equations. The reconstruction of sparse signals is not only feasible in theory, but efficient algorithms also exist to perform the reconstruction in practice. Moreover, involving randomness in the acquisition step enables one to utilize the minimal number of measurements. These realizations, together with their potential applications, have triggered the interest of the scientific community since around 2004. Some of the ingredients are of course much older than the advent of compressive sensing itself, and the underlying theory builds on various branches of mathematics. These branches include linear algebra, approximation theory, convex analysis, optimization, probability theory (in particular, random matrices), Banach space geometry, harmonic analysis, and graph theory. This book is a detailed and self-contained introduction to the rich and elegant mathematical theory of compressive sensing. It presents all the necessary background material without assuming any special prior knowledge—just basic analysis, linear algebra, and probability theory.

The perspective adopted here is definitely a mathematical one, only complemented at the beginning with a teaser on the strong potential for applications. Our taste partly dictated the choice of topics, which was limited by the need to keep this volume an introduction rather than an exhaustive treatise. However, the exposition is complete in the sense that we wanted every result to be fully proved for the material to become accessible to graduate students in mathematics as well as to engineers, computer scientists, and physicists. We have also made efforts to produce short and natural proofs that are often simplified versions of the ones found in the literature.

We both, independently, went through the process of rendering the foundations of compressive sensing understandable to students when we prepared lecture notes for courses given at Vanderbilt University, Drexel University, the University of Bonn, and ETH Zurich. This monograph is a further attempt to clarify the theory even more. Lecturers wishing to prepare a course based on it will find some hints at the end of Chap. 1.

The overall organization follows a path from simple to more complicated (so does the organization within chapters). This results in the structure of the book outlined below. The first chapter gives a brief introduction to the essentials of compressive sensing, describes some motivations and applications, and provides a detailed overview of the whole book. Chapters 2–6 treat the deterministic theory of compressive sensing. There, we cover the notion of sparsity, introduce basic algorithms, and analyze their performance based on various properties. Since the major breakthroughs rely on random matrices, we present the required tools from probability theory in Chaps. 7 and 8. Then Chaps. 9–12 deal with sparse recovery based on random matrices and with related topics. Chapter 13 looks into the use of lossless expanders and Chap. 14 covers recovery of random sparse signals with deterministic matrices. Finally, Chap. 15 examines some algorithms for ℓ_1-minimization. The book concludes with three appendices which cover basic material from matrix analysis, convex analysis, and other miscellaneous topics.

Each chapter ends with a "Notes" section. This is the place where we provide useful tangential comments which would otherwise disrupt the flow of the text, such as relevant references, historical remarks, additional facts, or open questions. We have compiled a selection of exercises for each chapter. They give the reader an opportunity to work on the material and to establish further interesting results. For instance, snapshots on the related theory of low-rank matrix recovery appear as exercises throughout the book.

A variety of sparse recovery algorithms appear in this book, together with their theoretical analysis. A practitioner may wonder which algorithm to choose for a precise purpose. In general, all the algorithms should be relatively efficient, but determining which one performs best and/or fastest in the specific setup is a matter of numerical experiments. To avoid creating a bias towards any algorithm, we decided not to present numerical comparisons for the simple reason that running experiments in all possible setups is unfeasible. Nevertheless, some crude hints are given in the Notes section of Chap. 3.

It was a challenge to produce a monograph on a rapidly evolving field such as compressive sensing. Some developments in the area occurred during the writing process and forced us to make a number of revisions and additions. We believe that the current material represents a solid foundation for the mathematical theory of compressive sensing and that further developments will build on it rather than replace it. Of course, we cannot be totally confident in a prediction about a field moving so quickly, and maybe the material will require some update in some years.

Many researchers have influenced the picture of compressive sensing we paint in this book. We have tried to carefully cite their contributions. However, we are bound to have forgotten some important works, and we apologize to their authors for that. Our vision benefited from various collaborations and discussions, in particular with (in alphabetical order) Nir Ailon, Akram Aldroubi, Ulaş Ayaz, Sören Bartels, Helmut Boelcskei, Petros Boufounos, Emmanuel Candès, Volkan Cevher, Albert Cohen, Ingrid Daubechies, Ron DeVore, Sjoerd Dirksen, Yonina Eldar, Jalal Fadili, Maryam Fazel, Hans Feichtinger, Massimo Fornasier, Rémi Gribonval, Karlheinz Gröchenig, Jarvis Haupt, Pawel Hitczenko, Franz Hlawatsch, Max Hügel, Mark Iwen, Maryia Kabanava, Felix Krahmer, Stefan Kunis, Gitta Kutyniok, Ming-Jun Lai, Ignace Loris, Shahar Mendelson, Alain Pajor, Götz Pfander, Alexander Powell, Justin Romberg, Karin Schnass, Christoph Schwab, Željka Stojanac, Jared Tanner, Georg Tauböck, Vladimir Temlyakov, Joel Tropp, Tino Ullrich, Pierre Vandergheynst, Roman Vershynin, Jan Vybiral, Rachel Ward, Hugo Woerdeman, Przemysław Wojtaszczyk, and Stefan Worm. We greatly acknowledge the help of several colleagues for proofreading and commenting parts of the manuscript. They are (in alphabetical order) David Aschenbrücker, Ulaş Ayaz, Bubacarr Bah, Sören Bartels, Jean-Luc Bouchot, Volkan Cevher, Christine DeMol, Sjoerd Dirksen, Massimo Fornasier, Rémi Gribonval, Karlheinz Gröchenig, Anders Hansen, Aicke Hinrichs, Pawel Hitczenko, Max Hügel, Mark Iwen, Maryia Kabanava, Emily King, Felix Krahmer, Guillaume Lecué, Ignace Loris, Arian Maleki, Michael Minner, Deanna Needell, Yaniv Plan, Alexander Powell, Omar Rivasplata, Rayan Saab, Željka Stojanac, Thomas Strohmer, Joel Tropp, Tino Ullrich, Jan Vybiral, Rachel Ward, and Hugo Woerdeman. We are grateful to Richard Baraniuk, Michael Lustig, Jared Tanner, and Shreyas Vasanawala for generously supplying us with figures for our book. We thank our host institutions for their support and the excellent working environment they provided during the preparation of our project: Vanderbilt University, Université Pierre et Marie Curie, and Drexel University for Simon Foucart; the Hausdorff Center for Mathematics and the Institute for Numerical Simulation at the University of Bonn for Holger Rauhut. Parts of the book were written during research visits of Holger Rauhut at Université Pierre et Marie Curie, at ETH Zurich, and at the Institute for Mathematics and Its Applications at the University of Minnesota. Simon Foucart acknowledges the hospitality of the Hausdorff Center for Mathematics during his many visits to Bonn. We are grateful to the Numerical Harmonic Analysis Group (NuHAG) at the University of Vienna for allowing us to use their online BibTeX database for managing the references. Simon Foucart acknowledges the financial support from the NSF (National Science Foundation) under the grant DMS-1120622 and Holger Rauhut acknowledges the financial support from the WWTF (Wiener Wissenschafts-, Forschungs- und Technologie-Fonds) through the project SPORTS (MA07-004) as well as the European Research Council through the Starting Grant StG 258926.

Finally, we hope that the readers enjoy their time studying this book and that the efforts they invest in learning compressive sensing will be worthwhile.

Philadelphia, PA, USA   Simon Foucart
Bonn, Germany   Holger Rauhut

Contents

Chapter 1
An Invitation to Compressive Sensing

This first chapter introduces the standard compressive sensing problem and gives an overview of the content of this book. Since the mathematical theory is highly motivated by real-life problems, we also briefly describe some of the potential applications.

1.1 What is Compressive Sensing?

In many practical problems of science and technology, one encounters the task of inferring quantities of interest from measured information. For instance, in signal and image processing, one would like to reconstruct a signal from measured data. When the information acquisition process is linear, the problem reduces to solving a linear system of equations. In mathematical terms, the observed data $y \in \mathbb{C}^m$ is connected to the signal $x \in \mathbb{C}^N$ of interest via

$$\mathbf{A}\mathbf{x} = \mathbf{y}. \tag{1.1}$$

The matrix $\mathbf{A} \in \mathbb{C}^{m \times N}$ models the linear measurement (information) process. Then one tries to recover the vector $x \in \mathbb{C}^N$ by solving the above linear system. Traditional wisdom suggests that the number m of measurements, i.e., the amount of measured data, must be at least as large as the signal length N (the number of components of x). This principle is the basis for most devices used in current technology, such as analog-to-digital conversion, medical imaging, radar, and mobile communication. Indeed, if $m < N$, then classical linear algebra indicates that the linear system (1.1) is underdetermined and that there are infinitely many solutions (provided, of course, that there exists at least one). In other words, without additional information, it is impossible to recover x from y in the case $m < N$. This fact also relates to the Shannon sampling theorem, which states that the sampling rate of a continuous-time signal must be twice its highest frequency in order to ensure reconstruction.

S. Foucart and H. Rauhut, *A Mathematical Introduction to Compressive Sensing*, Applied and Numerical Harmonic Analysis, DOI 10.1007/978-0-8176-4948-7_1, © Springer Science+Business Media New York 2013

Fig. 1.1 Antonella, Niels, and Paulina. *Top*: Original Image. *Bottom*: Reconstruction using 1% of the largest absolute wavelet coefficients, i.e., 99 % of the coefficients are set to zero

Thus, it came as a surprise that under certain assumptions it is actually possible to reconstruct signals when the number m of available measurements is smaller than the signal length N. Even more surprisingly, efficient algorithms do exist for the reconstruction. The underlying assumption which makes all this possible is *sparsity*. The research area associated to this phenomenon has become known as *compressive sensing, compressed sensing, compressive sampling*, or *sparse recovery*. This whole book is devoted to the mathematics underlying this field.

Sparsity. A signal is called sparse if most of its components are zero. As empirically observed, many real-world signals are compressible in the sense that they are well approximated by sparse signals—often after an appropriate change of basis. This explains why compression techniques such as JPEG, MPEG, or MP3 work so well in practice. For instance, JPEG relies on the sparsity of images in the discrete cosine basis or wavelet basis and achieves compression by only storing the largest discrete cosine or wavelet coefficients. The other coefficients are simply set to zero. We refer to Fig. 1.1 for an illustration of the fact that natural images are sparse in the wavelet domain.

Let us consider again the acquisition of a signal and the resulting measured data. With the additional knowledge that the signal is sparse or compressible, the traditional approach of taking at least as many measurements as the signal length seems to waste resources: At first, substantial efforts are devoted to measuring all entries of the signal and then most coefficients are discarded in the compressed version. Instead, one would want to acquire the compressed version of a signal "directly" via significantly fewer measured data than the signal length—exploiting the sparsity or compressibility of the signal. In other words, we would like to compressively sense a compressible signal! This constitutes the basic goal of compressive sensing.

We emphasize that the main difficulty here lies in the locations of the nonzero entries of the vector $\mathbf{x}$ not being known beforehand. If they were, one would simply reduce the matrix $\mathbf{A}$ to the columns indexed by this location set. The resulting system of linear equations then becomes overdetermined and one can solve for the nonzero entries of the signal. Not knowing the nonzero locations of the vector to be reconstructed introduces some nonlinearity since s-sparse vectors (those having at most s nonzero coefficients) form a nonlinear set. Indeed, adding two s-sparse vectors gives a $2s$-sparse vector in general. Thus, any successful reconstruction method will necessarily be nonlinear.

Intuitively, the complexity or "intrinsic" information content of a compressible signal is much smaller than its signal length (otherwise compression would not be possible). So one may argue that the required amount of data (number of measurements) should be proportional to this intrinsic information content rather than the signal length. Nevertheless, it is not immediately clear how to achieve the reconstruction in this scenario.

Looking closer at the standard compressive sensing problem consisting in the reconstruction of a sparse vector $\mathbf{x} \in \mathbb{C}^N$ from underdetermined measurements $\mathbf{y} = \mathbf{Ax} \in \mathbb{C}^m$, $m < N$, one essentially identifies two questions:

- How should one design the linear measurement process? In other words, what matrices $\mathbf{A} \in \mathbb{C}^{m \times N}$ are suitable?
- How can one reconstruct $\mathbf{x}$ from $\mathbf{y} = \mathbf{Ax}$? In other words, what are efficient reconstruction algorithms?

These two questions are not entirely independent, as the reconstruction algorithm needs to take $\mathbf{A}$ into account, but we will see that one can often separate the analysis of the matrix $\mathbf{A}$ from the analysis of the algorithm.

Let us notice that the first question is by far not trivial. In fact, compressive sensing is not fitted for arbitrary matrices $\mathbf{A} \in \mathbb{C}^{m \times N}$. For instance, if $\mathbf{A}$ is made of rows of the identity matrix, then $\mathbf{y} = \mathbf{Ax}$ simply picks some entries of $\mathbf{x}$, and hence, it contains mostly zero entries. In particular, no information is obtained about the nonzero entries of $\mathbf{x}$ not caught in $\mathbf{y}$, and the reconstruction appears impossible for such a matrix $\mathbf{A}$. Therefore, compressive sensing is not only concerned with the

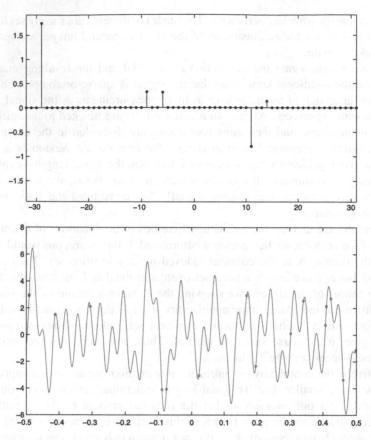

Fig. 1.2 *Top*: 5-sparse vector of Fourier coefficients of length 64. *Bottom*: real part of time-domain signal with 16 samples

recovery algorithm—the first question on the design of the measurement matrix is equally important and delicate. We also emphasize that the matrix $\mathbf{A}$ should ideally be designed for all signals $\mathbf{x}$ simultaneously, with a measurement process which is nonadaptive in the sense that the type of measurements for the datum y_j (i.e., the jth row of $\mathbf{A}$) does not depend on the previously observed data $y_1, \ldots, y_{j-1}$. As it turns out, adaptive measurements do not provide better theoretical performance in general (at least in a sense to be made precise in Chap. 10).

Algorithms. For practical purposes, the availability of reasonably fast reconstruction algorithms is essential. This feature is arguably the one which brought so much attention to compressive sensing. The first algorithmic approach coming to mind is probably ℓ_0-minimization. Introducing the notation $\|\mathbf{x}\|_0$ for the number of nonzero entries of a vector $\mathbf{x}$, it is natural to try to reconstruct $\mathbf{x}$ as a solution of the combinatorial optimization problem

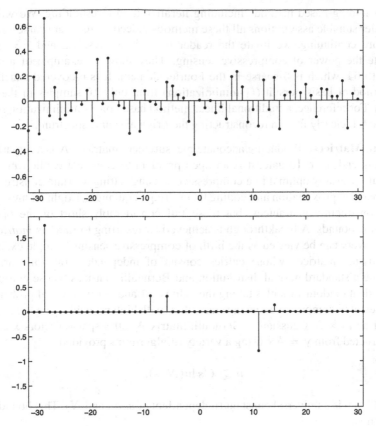

Fig. 1.3 *Top*: poor reconstruction via ℓ_2-minimization. *Bottom*: exact reconstruction via ℓ_1-minimization

$$\text{minimize } \|\mathbf{z}\|_0 \quad \text{subject to } \mathbf{A}\mathbf{z} = \mathbf{y}.$$

In words, we search for the sparsest vector consistent with the measured data $\mathbf{y} = \mathbf{A}\mathbf{x}$. Unfortunately, ℓ_0-minimization is NP-hard in general. Thus, it may seem quite surprising that fast and provably effective reconstruction algorithms do exist. A very popular and by now well-understood method is basis pursuit or ℓ_1-minimization, which consists in finding the minimizer of the problem

$$\text{minimize } \|\mathbf{z}\|_1 \quad \text{subject to } \mathbf{A}\mathbf{z} = \mathbf{y}. \tag{1.2}$$

Since the ℓ_1-norm $\| \cdot \|_1$ is a convex function, this optimization problem can be solved with efficient methods from convex optimization. Basis pursuit can be interpreted as the convex relaxation of ℓ_0-minimization. Alternative reconstruction methods include greedy-type methods such as orthogonal matching pursuit, as well as thresholding-based methods including iterative hard thresholding. We will see that under suitable assumptions all these methods indeed do recover sparse vectors.

Before continuing, we invite the reader to look at Figs. 1.2 and 1.3, which illustrate the power of compressive sensing. They show an example of a signal of length 64, which is 5-sparse in the Fourier domain. It is recovered exactly by the method of basis pursuit (ℓ_1-minimization) from only 16 samples in the time domain. For reference, a traditional linear method based on ℓ_2-minimization is also displayed. It clearly fails in reconstructing the original sparse spectrum.

Random Matrices. Producing adequate measurement matrices $\mathbf{A}$ is a remarkably intriguing endeavor. To date, it is an open problem to construct explicit matrices which are provably optimal in a compressive sensing setting. Certain constructions from sparse approximation and coding theory (e.g., equiangular tight frames) yield fair reconstruction guarantees, but these fall considerably short of the optimal achievable bounds. A breakthrough is achieved by resorting to *random matrices*— this discovery can be viewed as the birth of compressive sensing. Simple examples are Gaussian matrices whose entries consist of independent random variables following a standard normal distribution and Bernoulli matrices whose entries are independent random variables taking the values $+1$ and -1 with equal probability. A key result in compressive sensing states that, with high probability on the random draw of an $m \times N$ Gaussian or Bernoulli matrix $\mathbf{A}$, all s-sparse vectors $\mathbf{x}$ can be reconstructed from $\mathbf{y} = \mathbf{A}\mathbf{x}$ using a variety of algorithms provided

$$m \geq Cs \ln(N/s), \tag{1.3}$$

where $C > 0$ is a universal constant (independent of s, m, and N). This bound is in fact optimal.

According to (1.3), the amount m of data needed to recover s-sparse vectors scales linearly in s, while the signal length N only has a mild logarithmic influence. In particular, if the sparsity s is small compared to N, then the number m of measurements can also be chosen small in comparison to N, so that exact solutions of an underdetermined system of linear equations become plausible! This fascinating discovery impacts many potential applications.

We now invite the reader to examine Fig. 1.4. It compares the performance of two algorithms, namely, basis pursuit and hard thresholding pursuit, for the recovery of sparse vectors $\mathbf{x} \in \mathbb{C}^N$ from the measurement vectors $\mathbf{y} = \mathbf{A}\mathbf{x} \in \mathbb{C}^m$ based on simulations involving Gaussian random matrices $\mathbf{A}$ and randomly chosen s-sparse vectors $\mathbf{x}$. With a fixed sparsity s, the top plot shows the percentage of vectors $\mathbf{x}$ that were successfully recovered as a function of the number m of measurements. In particular, it indicates how large m has to be in comparison with s for the recovery to be guaranteed. With a fixed number m of measurements, the bottom plot shows the percentage of vectors $\mathbf{x}$ that were successfully recovered as a function of their sparsity s. In particular, it indicates how small s has to be in comparison with m for the recovery to be guaranteed. We note that the algorithm performing best is different for these two plots. This is due to the probability distribution chosen for the

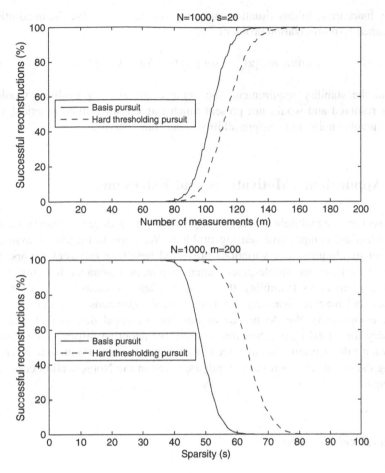

Fig. 1.4 *Top*: percentage of successful recoveries for Rademacher sparse vectors. *Bottom*: percentage of successful recoveries for Gaussian sparse vectors

nonzero entries of the sparse vectors: The top plot used a Rademacher distribution while the bottom plot used a Gaussian distribution.

The outlined recovery result extends from Gaussian random matrices to the more practical situation encountered in sampling theory. Here, assuming that a function of interest has a sparse expansion in a suitable orthogonal system (in trigonometric monomials, say), it can be recovered from a small number of randomly chosen samples (point evaluations) via ℓ_1-minimization or several other methods. This connection to sampling theory explains the alternative name compressive sampling.

Stability. Compressive sensing features another crucial aspect, namely, its reconstruction algorithms are stable. This means that the reconstruction error stays under control when the vectors are not exactly sparse and when the measurements **y** are slightly inaccurate. In this situation, one may, for instance, solve the quadratically constrained ℓ_1-minimization problem

$$\text{minimize } \|\mathbf{z}\|_1 \quad \text{subject to } \|\mathbf{A}\mathbf{z} - \mathbf{y}\|_2 \le \eta. \qquad (1.4)$$

Without the stability requirement, the compressive sensing problem would be swiftly resolved and would not present much interest since most practical applications involve noise and compressibility rather than sparsity.

1.2   Applications, Motivations, and Extensions

In this section, we highlight a selection of problems that reduce to or can be modeled as the standard compressive sensing problem. We hope to thereby convince the reader of its ubiquity. The variations presented here take different flavors: technological applications (single-pixel camera, magnetic resonance imaging, radar), scientific motivations (sampling theory, sparse approximation, error correction, statistics and machine learning), and theoretical extensions (low-rank recovery, matrix completion). We do not delve into the technical details that would be necessary for a total comprehension. Instead, we adopt an informal style and we focus on the description of an idealized mathematical model. Pointers to references treating the details in much more depth are given in the Notes section concluding the chapter.

Single-Pixel Camera

Compressive sensing techniques are implemented in a device called the single-pixel camera. The idea is to correlate in hardware a real-world image with independent realizations of Bernoulli random vectors and to measure these correlations (inner products) on a single pixel. It suffices to measure only a small number of such random inner products in order to reconstruct images via sparse recovery methods.

For the purpose of this exposition, images are represented via gray values of pixels collected in the vector $\mathbf{z} \in \mathbb{R}^N$, where $N = N_1 N_2$ and N_1, N_2 denote the width and height of the image in pixels. Images are not usually sparse in the canonical (pixel) basis, but they are often sparse after a suitable transformation, for instance, a wavelet transform or discrete cosine transform. This means that one can write $\mathbf{z} = \mathbf{W}\mathbf{x}$, where $\mathbf{x} \in \mathbb{R}^N$ is a sparse or compressible vector and $\mathbf{W} \in \mathbb{R}^{N \times N}$ is a unitary matrix representing the transform.

The crucial ingredient of the single-pixel camera is a microarray consisting of a large number of small mirrors that can be turned on or off individually. The light from the image is reflected on this microarray and a lens combines all the reflected beams in one sensor, the single pixel of the camera; see Fig. 1.5. Depending on

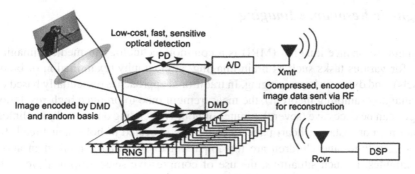

Fig. 1.5 Schematic representation of a single-pixel camera (Image courtesy of Rice University)

a small mirror being switched on or off, it contributes or not to the light intensity measured at the sensor. In this way, one realizes in hardware the inner product $\langle \mathbf{z}, \mathbf{b} \rangle$ of the image $\mathbf{z}$ with a vector $\mathbf{b}$ containing ones at the locations corresponding to switched-on mirrors and zeros elsewhere. In turn, one can also realize inner products with vectors $\mathbf{a}$ containing only $+1$ and -1 with equal probability by defining two auxiliary vectors $\mathbf{b}^1, \mathbf{b}^2 \in \{0, 1\}^N$ via

$$b_j^1 = \begin{cases} 1 & \text{if } a_j = 1, \\ 0 & \text{if } a_j = -1, \end{cases} \qquad b_j^2 = \begin{cases} 1 & \text{if } a_j = -1, \\ 0 & \text{if } a_j = 1, \end{cases}$$

so that $\langle \mathbf{z}, \mathbf{a} \rangle = \langle \mathbf{z}, \mathbf{b}^1 \rangle - \langle \mathbf{z}, \mathbf{b}^2 \rangle$. Choosing vectors $\mathbf{a}_1, \ldots, \mathbf{a}_m$ independently at random with entries taking the values ± 1 with equal probability, the measured intensities $y_\ell = \langle \mathbf{z}, \mathbf{a}_\ell \rangle$ are inner products with independent Bernoulli vectors. Therefore, we have $\mathbf{y} = \mathbf{A}\mathbf{z}$ for a (random) Bernoulli matrix $\mathbf{A} \in \mathbb{R}^{m \times N}$ whose action on the image $\mathbf{z}$ has been realized in hardware. Recalling that $\mathbf{z} = \mathbf{W}\mathbf{x}$ and writing $\mathbf{A}' = \mathbf{A}\mathbf{W}$ yield the system

$$\mathbf{y} = \mathbf{A}'\mathbf{x},$$

where the vector $\mathbf{x}$ is sparse or compressible. In this situation, the measurements are taken sequentially, and since this process may be time-consuming, it is desirable to use only few measurements. Thus, we have arrived at the standard compressive sensing problem. The latter allows for the reconstruction of $\mathbf{x}$ from $\mathbf{y}$ and finally the image is deduced as $\mathbf{z} = \mathbf{W}\mathbf{x}$. We will justify in Chap. 9 the validity of the accurate reconstruction from $m \geq Cs \ln(N/s)$ measurements for images that are (approximately) s-sparse in some transform domain.

Although the single-pixel camera is more a proof of concept than a new trend in camera design, it is quite conceivable that similar devices will be used for different imaging tasks. In particular, for certain wavelengths outside the visible spectrum, it is impossible or at least very expensive to build chips with millions of sensor pixels on an area of only several square millimeters. In such a context, the potential of a technology based on compressive sensing is expected to really pay off.

Magnetic Resonance Imaging

Magnetic resonance imaging (MRI) is a common technology in medical imaging used for various tasks such as brain imaging, angiography (examination of blood vessels), and dynamic heart imaging. In traditional approaches (essentially based on the Shannon sampling theorem), the measurement time to produce high-resolution images can be excessive (several minutes or hours depending on the task) in clinical situations. For instance, heart patients cannot be expected to hold their breath for too long a time, and children are too impatient to sit still for more than about two minutes. In such situations, the use of compressive sensing to achieve high-resolution images based on few samples appears promising.

MRI relies on the interaction of a strong magnetic field with the hydrogen nuclei (protons) contained in the body's water molecules. A static magnetic field polarizes the spin of the protons resulting in a magnetic moment. Applying an additional radio frequency excitation field produces a precessing magnetization transverse to the static field. The precession frequency depends linearly on the strength of the magnetic field. The generated electromagnetic field can be detected by sensors. Imposing further magnetic fields with a spatially dependent strength, the precession frequency depends on the spatial position as well. Exploiting the fact that the transverse magnetization depends on the physical properties of the tissue (for instance, proton density) allows one to reconstruct an image of the body from the measured signal.

In mathematical terms, we denote the transverse magnetization at position $\mathbf{z} \in \mathbb{R}^3$ by $X(\mathbf{z}) = |X(\mathbf{z})|e^{-i\phi(\mathbf{z})}$ where $|X(\mathbf{z})|$ is the magnitude and $\phi(\mathbf{z})$ is the phase. The additional possibly time-dependent magnetic field is designed to depend linearly on the position and is therefore called gradient field. Denoting by $\mathbf{G} \in \mathbb{R}^3$ the gradient of this magnetic field, the precession frequency (being a function of the position in $\mathbb{R}^3$) can be written as

$$\omega(\mathbf{z}) = \kappa(B + \langle \mathbf{G}, \mathbf{z} \rangle), \quad \mathbf{z} \in \mathbb{R}^3,$$

where B is the strength of the static field and κ is a physical constant. With a time-dependent gradient $\mathbf{G} : [0, T] \to \mathbb{R}^3$, the magnetization phase $\phi(\mathbf{z}) = \phi(\mathbf{z}, t)$ is the integral

$$\phi(\mathbf{z}, t) = 2\pi\kappa \int_0^t \langle \mathbf{G}(\tau), \mathbf{z} \rangle d\tau,$$

where $t = 0$ corresponds to the time of the radio frequency excitation. We introduce the function $\mathbf{k} : [0, T] \to \mathbb{R}^3$ defined by

$$\mathbf{k}(t) = \kappa \int_0^t \mathbf{G}(\tau) d\tau.$$

The receiver coil integrates over the whole spatial volume and measures the signal

$$f(t) = \int_{\mathbb{R}^3} |X(\mathbf{z})| e^{-2\pi i \langle \mathbf{k}(t), \mathbf{z} \rangle} \, d\mathbf{z} = \mathcal{F}(|X|)(\mathbf{k}(t)),$$

where $\mathcal{F}(|X|)(\boldsymbol{\xi})$ denotes the three-dimensional Fourier transform of the magnitude $|X|$ of the magnetization. It is also possible to measure slices of a body, in which case the three-dimensional Fourier transform is replaced by a two-dimensional Fourier transform.

In conclusion, the signal measured by the MRI system is the Fourier transform of the spatially dependent magnitude of the magnetization $|X|$ (the image), subsampled on the curve $\{\mathbf{k}(t) : t \in [0, T]\} \subset \mathbb{R}^3$. By repeating several radio frequency excitations with modified parameters, one obtains samples of the Fourier transform of $|X|$ along several curves $\mathbf{k}_1, \dots, \mathbf{k}_L$ in $\mathbb{R}^3$. The required measurement time is proportional to the number L of such curves, and we would like to minimize this number L.

A natural discretization represents each volume element (or area element in case of two-dimensional imaging of slices) by a single voxel (or pixel), so that the magnitude of the magnetization $|X|$ becomes a finite-dimensional vector $\mathbf{x} \in \mathbb{R}^N$ indexed by $Q := [N_1] \times [N_2] \times [N_3]$ with $N = \text{card}(Q) = N_1 N_2 N_3$ and $[N_i] := \{1, \dots, N_i\}$. After discretizing the curves $\mathbf{k}_1, \dots, \mathbf{k}_L$, too, the measured data become samples of the three-dimensional discrete Fourier transform of $\mathbf{x}$, i.e.,

$$(\mathcal{F}\mathbf{x})_{\mathbf{k}} = \sum_{\ell \in Q} x_\ell e^{-2\pi i \sum_{j=1}^{3} k_j \ell_j / N_j}, \quad \mathbf{k} \in Q.$$

Let $K \subset Q$ with $\text{card}(K) = m$ denote a subset of the discretized frequency space Q, which is covered by the trajectories $\mathbf{k}_1, \dots, \mathbf{k}_L$. Then the measured data vector $\mathbf{y}$ corresponds to

$$\mathbf{y} = \mathbf{R}_K \mathcal{F} \mathbf{x} = \mathbf{A}\mathbf{x},$$

where $\mathbf{R}_K$ is the linear map that restricts a vector indexed by Q to its indices in K. The measurement matrix $\mathbf{A} = \mathbf{R}_K \mathcal{F} \in \mathbb{C}^{m \times N}$ is a partial Fourier matrix. In words, the vector $\mathbf{y}$ collects the samples of the three-dimensional Fourier transform of the discretized image $\mathbf{x}$ on the set K. Since we would like to use a small number m of samples, we end up with an underdetermined system of equations.

In certain medical imaging applications such as angiography, it is realistic to assume that the image $\mathbf{x}$ is sparse with respect to the canonical basis, so that we immediately arrive at the standard compressive sensing problem. In the general scenario, the discretized image $\mathbf{x}$ will be sparse or compressible only after transforming into a suitable domain, using wavelets, for instance—in mathematical terms, we have $\mathbf{x} = \mathbf{W}\mathbf{x}'$ for some unitary matrix $\mathbf{W} \in \mathbb{C}^{N \times N}$ and some sparse vector $\mathbf{x}' \in \mathbb{C}^N$. This leads to the model

$$\mathbf{y} = \mathbf{A}'\mathbf{x}',$$

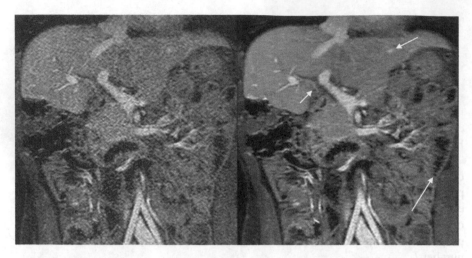

Fig. 1.6 Comparison of a traditional MRI reconstruction (*left*) and a compressive sensing reconstruction (*right*). The pictures show a coronal slice through an abdomen of a 3-year-old pediatric patient following an injection of a contrast agent. The image size was set to $320 \times 256 \times 160$ voxels. The data were acquired using a 32-channel pediatric coil. The acquisition was accelerated by a factor of 7.2 by random subsampling of the frequency domain. The *left image* is a traditional linear reconstruction showing severe artifacts. The *right image*, a (wavelet-based) compressive sensing reconstruction, exhibits diagnostic quality and significantly reduced artifacts. The subtle features indicated with *arrows* show well on the compressive sensing reconstruction, while almost disappearing in the traditional one (Image courtesy of Michael Lustig, Stanford University, and Shreyas Vasanawala, Lucile Packard Children's Hospital, Stanford University)

with the transformed measurement matrix $\mathbf{A}' = \mathbf{AW} = \mathbf{R}_K \mathcal{F} \mathbf{W} \in \mathbb{C}^{m \times N}$ and a sparse or compressible vector $\mathbf{x}' \in \mathbb{C}^N$. Again, we arrived at the standard compressive sensing problem.

The challenge is to determine good sampling sets K with small size that still ensure recovery of sparse images. The theory currently available predicts that sampling sets K chosen uniformly at random among all possible sets of cardinality m work well (at least when $\mathbf{W}$ is the identity matrix). Indeed, the results of Chap. 12 guarantee that an s-sparse $\mathbf{x}' \in \mathbb{C}^N$ can be reconstructed by ℓ_1-minimization if $m \geq Cs \ln N$.

Unfortunately, such random sets K are difficult to realize in practice due to the continuity constraints of the trajectories curves $\mathbf{k}_1, \ldots, \mathbf{k}_L$. Therefore, good realizable sets K are investigated empirically. One option that seems to work well takes the trajectories as parallel lines in $\mathbb{R}^3$ whose intersections with a coordinate plane are chosen uniformly at random. This gives some sort of approximation to the case where K is "completely" random. Other choices such as perturbed spirals are also possible.

Figure 1.6 shows a comparison of a traditional MRI reconstruction technique with reconstruction via compressive sensing. The compressive sensing reconstruction has much better visual quality and resolves some clinically important details, which are not visible in the traditional reconstruction at all.

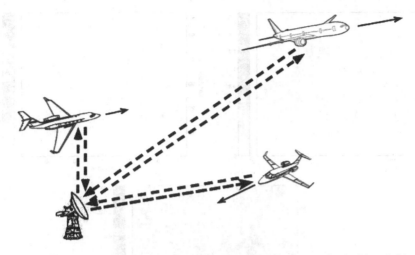

Fig. 1.7 Schematic illustration of a radar device measuring distances and velocities of objects

Radar

Compressive sensing can be applied to several radar frameworks. In the one presented here, an antenna sends out a properly designed electromagnetic wave—the radar pulse—which is scattered at objects in the surrounding environment, for instance, airplanes in the sky. A receive antenna then measures an electromagnetic signal resulting from the scattered waves. Based on the delay of the received signal, one can determine the distance of an object, and the Doppler effect allows one to deduce its speed with respect to the direction of view; see Fig. 1.7 for an illustration.

Let us describe a simple finite-dimensional model for this scenario. We denote by $(\mathbf{T}_k \mathbf{z})_j = z_{j-k \bmod m}$ the cyclic translation operator on $\mathbb{C}^m$ and by $(\mathbf{M}_\ell \mathbf{z})_j = e^{2\pi i \ell j/m} z_j$ the modulation operator on $\mathbb{C}^m$. The map transforming the sent signal to the received signal—also called channel—can be expressed as

$$\mathbf{B} = \sum_{(k,\ell)\in[m]^2} x_{k,\ell} \mathbf{T}_k \mathbf{M}_\ell,$$

where the translations correspond to delay and the modulations to Doppler effect. The vector $\mathbf{x} = (x_{k,\ell})$ characterizes the channel. A nonzero entry $x_{k,\ell}$ occurs if there is a scattering object present in the surroundings with distance and speed corresponding to the shift $\mathbf{T}_k$ and modulation $\mathbf{M}_\ell$. Only a limited number of scattering objects are usually present, which translates into the sparsity of the coefficient vector $\mathbf{x}$. The task is now to determine $\mathbf{x}$ and thereby to obtain information about scatterers in the surroundings by probing the channel with a suitable known radio pulse, modeled in this finite-dimensional setup by a vector $\mathbf{g} \in \mathbb{C}^m$. The received signal $\mathbf{y}$ is given by

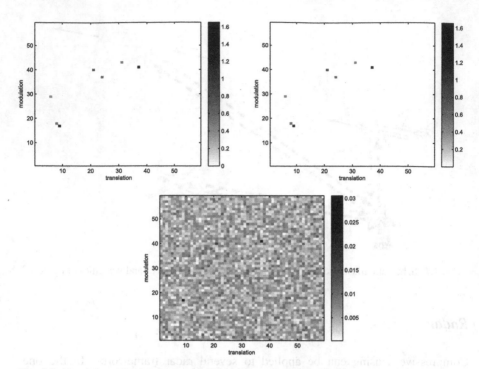

Fig. 1.8 *Top left*: original 7-sparse coefficient vector ($m = 59$) in the translation–modulation (delay-Doppler) plane. *Top right*: reconstruction by ℓ_1-minimization using the Alltop window. *Bottom*: for comparison, the reconstruction by traditional ℓ_2-minimization

$$\mathbf{y} = \mathbf{B}\mathbf{g} = \sum_{(k,\ell) \in [m]^2} x_{k,\ell} \mathbf{T}_k \mathbf{M}_\ell \mathbf{g} = \mathbf{A}_\mathbf{g}\mathbf{x},$$

where the m^2 columns of the measurement matrix $\mathbf{A}_\mathbf{g} \in \mathbb{C}^{m \times m^2}$ are equal to $\mathbf{T}_k \mathbf{M}_\ell \mathbf{g}$, $(k, \ell) \in [m]^2$. Recovering $\mathbf{x} \in \mathbb{C}^{m^2}$ from the measured signal $\mathbf{y}$ amounts to solving an underdetermined linear system. Taking the sparsity of $\mathbf{x}$ into consideration, we arrive at the standard compressive sensing problem. The associated reconstruction algorithms, including ℓ_1-minimization, apply.

It remains to find suitable radio pulse sequences $\mathbf{g} \in \mathbb{C}^m$ ensuring that $\mathbf{x}$ can be recovered from $\mathbf{y} = \mathbf{B}\mathbf{g}$. A popular choice of $\mathbf{g}$ is the so-called *Alltop* vector, which is defined for prime $m \geq 5$ as

$$g_\ell = e^{2\pi i \ell^3/m}, \quad \ell \in [m].$$

We refer to Chap. 5 for more details and to Fig. 1.8 for a numerical example.

Although the Alltop window works well in practice, the theoretical guarantees currently available are somewhat limited due to the fact that $\mathbf{g}$ is deterministic. As an alternative consistent with the general philosophy of compressive sensing, one can choose $\mathbf{g} \in \mathbb{C}^m$ at random, for instance, as a Bernoulli vector with independent ± 1 entries. In this case, it is known that an s-sparse vector $\mathbf{x} \in \mathbb{C}^{m^2}$ can be recovered from $\mathbf{y} = \mathbf{B}\mathbf{x} \in \mathbb{C}^m$ provided $s \leq Cm/\ln m$. More information can be found in the Notes section of Chap. 12.

Sampling Theory

Reconstructing a continuous-time signal from a discrete set of samples is an important task in many technological and scientific applications. Examples include image processing, sensor technology in general, and analog-to-digital conversion appearing, for instance, in audio entertainment systems or mobile communication devices. Currently, most sampling techniques rely on the Shannon sampling theorem, which states that a function of bandwidth B has to be sampled at the rate $2B$ in order to ensure reconstruction.

In mathematical terms, the Fourier transform of a continuous-time signal $f \in L^1(\mathbb{R})$ (meaning that $\int_{\mathbb{R}} |f(t)|dt < \infty$) is defined by

$$\hat{f}(\xi) = \int_{\mathbb{R}} f(t)e^{-2\pi it\xi}dt, \quad \xi \in \mathbb{R}.$$

We say that f is bandlimited with bandwidth B if $\hat{f}$ is supported in $[-B, B]$. The *Shannon sampling theorem* states that such f can be reconstructed from its discrete set of samples $\{f(k/(2B)), k \in \mathbb{Z}\}$ via the formula

$$f(t) = \sum_{k \in \mathbb{Z}} f\left(\frac{k}{2B}\right) \operatorname{sinc}(2\pi Bt - \pi k), \qquad (1.5)$$

where the sinc function is given by

$$\operatorname{sinc}(t) = \begin{cases} \dfrac{\sin t}{t} & \text{if } t \neq 0, \\ 1 & \text{if } t = 0. \end{cases}$$

To facilitate a comparison with compressive sensing, we also formulate the Shannon sampling theorem in a finite-dimensional setting. We consider trigonometric polynomials of maximal degree M, i.e., functions of the type

$$f(t) = \sum_{k=-M}^{M} x_k e^{2\pi ikt}, \quad t \in [0, 1]. \qquad (1.6)$$

The degree M serves as a substitute for the bandwidth B. Since the space of trigonometric polynomials of maximal degree M has dimension $N = 2M + 1$, it is expected that f can be reconstructed from $N = 2M + 1$ samples. Indeed, Theorem C.1 in the appendix states that

$$f(t) = \frac{1}{2M+1} \sum_{k=0}^{2M} f\left(\frac{k}{2M+1}\right) D_M\left(t - \frac{k}{2M+1}\right), \quad t \in [0,1],$$

where the Dirichlet kernel D_M is given by

$$D_M(t) = \sum_{k=-M}^{M} e^{2\pi i k t} = \begin{cases} \dfrac{\sin(\pi(2M+1)t)}{\sin(\pi t)} & \text{if } t \neq 0, \\ 2M+1 & \text{if } t = 0. \end{cases}$$

For dimensionality reasons, it is not possible to reconstruct trigonometric polynomials of maximal degree M from fewer than $N = 2M + 1$ samples. In practice, however, the required degree M may be large; hence, the number of samples must be large, too—sometimes significantly larger than realistic. So the question arises whether the required number of samples can be reduced by exploiting additional assumptions. Compressibility in the Fourier domain, for instance, is a reasonable assumption in many practical scenarios. In fact, if the vector $\mathbf{x} \in \mathbb{C}^N$ of Fourier coefficients of f in (1.6) is sparse (or compressible), then few samples do suffice for exact (or approximate) reconstruction.

Precisely, given a set $\{t_1, \ldots, t_m\} \subset [0,1]$ of m sampling points, we can write the vector $\mathbf{y} = (f(t_\ell))_{\ell=1}^m$ as

$$\mathbf{y} = \mathbf{A}\mathbf{x} \tag{1.7}$$

where $\mathbf{A} \in \mathbb{C}^{m \times N}$ is a Fourier-type matrix with entries

$$A_{\ell,k} = e^{2\pi i k t_\ell}, \quad \ell = 1, \ldots, m, \quad k = -M, \ldots, M.$$

The problem of recovering f from its vector $\mathbf{y}$ of m samples reduces to finding the coefficient vector $\mathbf{x}$. This amounts to solving the linear system (1.7), which is underdetermined when $m < N$. With the sparsity assumption, we arrive at the standard compressive sensing problem. A number of recovery algorithms, including ℓ_1-minimization, can then be applied. A crucial question now concerns the choice of sampling points. As indicated before, randomness helps. In fact, we will see in Chap. 12 that choosing the sampling points $t_1, \ldots, t_m$ independently and uniformly at random in $[0,1]$ allows one to reconstruct f with high probability from its m samples $f(t_1), \ldots, f(t_m)$ provided that $m \geq Cs \ln(N)$. Thus, few samples suffice if s is small. An illustrating example was already displayed in Figs. 1.2 and 1.3.

Sparse Approximation

Compressive sensing builds on the empirical observation that many types of signals can be approximated by sparse ones. In this sense, compressive sensing can be seen as a subfield of sparse approximation. There is a specific problem in sparse approximation similar to the standard compressive sensing problem of recovering a sparse vector $\mathbf{x} \in \mathbb{C}^N$ from the incomplete information $\mathbf{y} = \mathbf{A}\mathbf{x} \in \mathbb{C}^m$ with $m < N$.

Suppose that a vector $\mathbf{y} \in \mathbb{C}^m$ (usually a signal or an image in applications) is to be represented as a linear combination of prescribed elements $\mathbf{a}_1, \ldots, \mathbf{a}_N \in \mathbb{C}^m$ such that $\text{span}\{\mathbf{a}_1, \ldots, \mathbf{a}_N\} = \mathbb{C}^m$. The system $(\mathbf{a}_1, \ldots, \mathbf{a}_N)$ is often called a dictionary. Note that this system may be linearly dependent (redundant) since we allow $N > m$. Redundancy may be desired when linearly independence is too restrictive. For instance, in time–frequency analysis, bases of time–frequency shifts elements are only possible if the generator has poor time–frequency concentration— this is the Balian–Low theorem. Unions of several bases are also of interest. In such situations, a representation $\mathbf{y} = \sum_{j=1}^{N} x_j \mathbf{a}_j$ is not unique. Traditionally, one removes this drawback by considering a representation with the smallest number of terms, i.e., a sparsest representation.

Let us now form the matrix $\mathbf{A} \in \mathbb{C}^{m \times N}$ with columns $\mathbf{a}_1, \ldots, \mathbf{a}_N$. Finding the sparsest representation of $\mathbf{y}$ amounts to solving

$$\text{minimize } \|\mathbf{z}\|_0 \quad \text{subject to } \mathbf{A}\mathbf{z} = \mathbf{y}. \tag{P_0}$$

If we tolerate a representation error η, then one considers the slightly modified optimization problem

$$\text{minimize } \|\mathbf{z}\|_0 \quad \text{subject to } \|\mathbf{A}\mathbf{z} - \mathbf{y}\| \leq \eta. \tag{$P_{0,\eta}$}$$

The problem (P_0) is the same as the one encountered in the previous section. Both optimization problems (P_0) and $(P_{0,\eta})$ are NP-hard in general, but all the algorithmic approaches presented in this book for the standard compressive sensing problem, including ℓ_1-minimization, may be applied in this context to overcome the computational bottleneck. The conditions on $\mathbf{A}$ ensuring exact or approximate recovery of the sparsest vector $\mathbf{x}$, which will be derived in Chaps. 4, 5, and 6, remain valid.

There are, however, some differences in philosophy compared to the compressive sensing problem. In the latter, one is often free to design the matrix $\mathbf{A}$ with appropriate properties, while $\mathbf{A}$ is usually prescribed in the context of sparse approximation. In particular, it is not realistic to rely on randomness as in compressive sensing. Since it is hard to verify the conditions ensuring sparse recovery in the optimal parameter regime (m linear in s up to logarithmic factors), the theoretical guarantees fall short of the ones encountered for random matrices. An exception to this rule of thumb will be covered in Chap. 14 where recovery guarantees are obtained for randomly chosen signals.

The second difference between sparse approximation and compressive sensing appears in the targeted error estimates. In compressive sensing, one is interested in the error $\|\mathbf{x} - \mathbf{x}^\sharp\|$ at the coefficient level, where $\mathbf{x}$ and $\mathbf{x}^\sharp$ are the original and reconstructed coefficient vectors, respectively, while in sparse approximation, the goal is to approximate a given $\mathbf{y}$ with a sparse expansion $\mathbf{y}^\sharp = \sum_j x_j^\sharp \mathbf{a}_j$, so one is rather interested in $\|\mathbf{y} - \mathbf{y}^\sharp\|$. An estimate for $\|\mathbf{x} - \mathbf{x}^\sharp\|$ often yields an estimate for $\|\mathbf{y} - \mathbf{y}^\sharp\| = \|\mathbf{A}(\mathbf{x} - \mathbf{x}^\sharp)\|$, but the converse is not generally true.

Finally, we briefly describe some signal and image processing applications of sparse approximation.

- *Compression.* Suppose that we have found a sparse approximation $\hat{\mathbf{y}} = \mathbf{A}\hat{\mathbf{x}}$ of a signal $\mathbf{y}$ with a sparse vector $\hat{\mathbf{x}}$. Then storing $\hat{\mathbf{y}}$ amounts to storing only the nonzero coefficients of $\hat{\mathbf{x}}$. Since $\hat{\mathbf{x}}$ is sparse, significantly less memory is required than for storing the entries of the original signal $\mathbf{y}$.

- *Denoising.* Suppose that we observe a noisy version $\tilde{\mathbf{y}} = \mathbf{y} + \mathbf{e}$ of a signal $\mathbf{y}$, where $\mathbf{e}$ represents a noise vector with $\|\mathbf{e}\| \leq \eta$. The task is then to remove the noise and to recover a good approximation of the original signal $\mathbf{y}$. In general, if nothing is known about $\mathbf{y}$, this problem becomes ill-posed. However, assuming that $\mathbf{y}$ can be well represented by a sparse expansion, a reasonable approach consists in taking a sparse approximation of $\tilde{\mathbf{y}}$. More precisely, we ideally choose the solution $\hat{\mathbf{x}}$ of the ℓ_0-minimization problem $(\mathrm{P}_{0,\eta})$ with $\mathbf{y}$ replaced by the known signal $\tilde{\mathbf{y}}$. Then we form $\hat{\mathbf{y}} = \mathbf{A}\hat{\mathbf{x}}$ as the denoised version of $\mathbf{y}$. For a computationally tractable approach, one replaces the NP-hard problem $(\mathrm{P}_{0,\eta})$ by one of the compressive sensing (sparse approximation) algorithms, for instance, the ℓ_1-minimization variant (1.4) which takes noise into account, or the so-called basis pursuit denoising problem

$$\text{minimize } \lambda\|\mathbf{z}\|_1 + \|\mathbf{A}\mathbf{z} - \mathbf{y}\|_2^2.$$

- *Data Separation.* Suppose that a vector $\mathbf{y} \in \mathbb{C}^m$ is the composition of two (or more) components, say $\mathbf{y} = \mathbf{y}_1 + \mathbf{y}_2$. Given $\mathbf{y}$, we wish to extract the unknown vectors $\mathbf{y}_1, \mathbf{y}_2 \in \mathbb{C}^m$. This problem appears in several signal processing tasks. For instance, astronomers would like to separate point structures (stars, galaxy clusters) from filaments in their images. Similarly, an audio processing task consists in separating harmonic components (pure sinusoids) from short peaks. Without additional assumption, this separation problem is ill-posed. However, if both components $\mathbf{y}_1$ and $\mathbf{y}_2$ have sparse representations in dictionaries $(\mathbf{a}_1, \ldots, \mathbf{a}_{N_1})$ and $(\mathbf{b}_1, \ldots, \mathbf{b}_{N_2})$ of different nature (for instance, sinusoids and spikes), then the situation changes. We can then write

$$\mathbf{y} = \sum_{j=1}^{N_1} x_{1,j}\mathbf{a}_j + \sum_{j=1}^{N_2} x_{2,j}\mathbf{b}_j = \mathbf{A}\mathbf{x},$$

where the matrix $\mathbf{A} \in \mathbb{C}^{m \times (N_1 + N_2)}$ has columns $\mathbf{a}_1, \ldots, \mathbf{a}_{N_1}, \mathbf{b}_1, \ldots, \mathbf{b}_{N_2}$ and the vector $\mathbf{x} = [x_{1,1}, \ldots, x_{1,N_1}, x_{2,1}, \ldots, x_{2,N_2}]^\top$ is sparse. The compressive sensing methodology then allows one—under certain conditions—to determine the coefficient vector $\mathbf{x}$, hence to derive the two components $\mathbf{y}_1 = \sum_{j=1}^{N_1} x_{1,j} \mathbf{a}_j$ and $\mathbf{y}_2 = \sum_{j=1}^{N_2} x_{2,j} \mathbf{b}_j$.

Error Correction

In every realistic data transmission device, pieces of data are occasionally corrupted. To overcome this unavoidable issue, one designs schemes for the correction of such errors provided they do not occur too often.

Suppose that we have to transmit a vector $\mathbf{z} \in \mathbb{R}^n$. A standard strategy is to encode it into a vector $\mathbf{v} = \mathbf{B}\mathbf{z} \in \mathbb{R}^N$ of length $N = n + m$, where $\mathbf{B} \in \mathbb{R}^{N \times n}$. Intuitively, the redundancy in $\mathbf{B}$ (due to $N > n$) should help in identifying transmission errors. The number m reflects the amount of redundancy.

Assume that the receiver measures $\mathbf{w} = \mathbf{v} + \mathbf{x} \in \mathbb{R}^N$, where $\mathbf{x}$ represents transmission error. The assumption that transmission errors do not occur too often translates into the sparsity of $\mathbf{x}$, say $\|\mathbf{x}\|_0 \leq s$. For decoding, we construct a matrix $\mathbf{A} \in \mathbb{R}^{m \times N}$—called generalized checksum matrix—such that $\mathbf{A}\mathbf{B} = \mathbf{0}$, i.e., all rows of $\mathbf{A}$ are orthogonal to all columns of $\mathbf{B}$. We then form the generalized checksum

$$\mathbf{y} = \mathbf{A}\mathbf{w} = \mathbf{A}(\mathbf{v} + \mathbf{x}) = \mathbf{A}\mathbf{B}\mathbf{z} + \mathbf{A}\mathbf{x} = \mathbf{A}\mathbf{x}.$$

We arrived at the standard compressive sensing problem with the matrix $\mathbf{A}$ and the sparse error vector $\mathbf{x}$. Under suitable conditions, the methodology described in this book allows one to recover $\mathbf{x}$ and in turn the original transmit vector $\mathbf{v} = \mathbf{w} - \mathbf{x}$. Then one solves the overdetermined system $\mathbf{v} = \mathbf{B}\mathbf{z}$ to derive the data vector $\mathbf{z}$.

For concreteness of the scheme, we may choose a matrix $\mathbf{A} \in \mathbb{R}^{m \times N}$ as a suitable compressive sensing matrix, for instance, a Gaussian random matrix. Then we select the matrix $\mathbf{B} \in \mathbb{R}^{N \times n}$ with $n + m = N$ in such a way that its columns span the orthogonal complement of the row space of $\mathbf{A}$, thus guaranteeing that $\mathbf{A}\mathbf{B} = \mathbf{0}$. With these choices, we are able to correct a number s of transmission errors as large as $Cm / \ln(N/m)$.

Statistics and Machine Learning

The goal of statistical regression is to predict an outcome based on certain input data. It is common to choose the linear model

$$\mathbf{y} = \mathbf{A}\mathbf{x} + \mathbf{e},$$

where $\mathbf{A} \in \mathbb{R}^{m \times N}$—often called design or predictor matrix in this context—collects the input data and $\mathbf{y}$ the output data and $\mathbf{e}$ is a random noise vector. The vector $\mathbf{x}$ is a parameter that has to be estimated from the data. In a statistical framework, the notation (n, p) is generally used instead of (m, N), but we keep the latter for consistency. In a clinical study, e.g., the entries $A_{j,k}$ in the row associated to the jth patient may refer to blood pressure, weight, height, gene data, concentration of certain markers, etc. The corresponding output y_j would be another quantity of interest, for instance, the probability that jth patient suffers a certain disease. Having data for m patients, the regression task is to fit the model, i.e., to determine the parameter vector $\mathbf{x}$.

In practice, the number N of parameters is often much larger than the number m of observations, so even without noise, the problem of fitting the parameter $\mathbf{x}$ is ill-posed without further assumption. In many cases, however, only a small number of parameters contribute towards the effect to be predicted, but it is a priori unknown which of these parameters are influential. This leads to sparsity in the vector $\mathbf{x}$, and again we arrive at the standard compressive sensing problem. In statistical terms, determining a sparse parameter vector $\mathbf{x}$ corresponds to selecting the relevant explanatory variables, i.e., the support of $\mathbf{x}$. One also speaks of *model selection*.

The methods described in this book can be applied in this context, too. Still, there is a slight deviation from our usual setup due to the randomness of the noise vector $\mathbf{e}$. In particular, instead of the quadratically constrained ℓ_1-minimization problem (1.4), one commonly considers the so-called LASSO (least absolute shrinkage and selection operator)

$$\text{minimize } \|\mathbf{A}\mathbf{z} - \mathbf{y}\|_2^2 \quad \text{subject to } \|\mathbf{z}\|_1 \leq \tau \qquad (1.8)$$

for an appropriate regularization parameter τ depending on the variance of the noise. Further variants are the *Dantzig selector*

$$\text{minimize } \|\mathbf{z}\|_1 \quad \text{subject to } \|\mathbf{A}^*(\mathbf{A}\mathbf{z} - \mathbf{y})\|_\infty \leq \lambda, \qquad (1.9)$$

or the ℓ_1-regularized problem (sometimes also called LASSO or basis pursuit denoising in the literature)

$$\text{minimize } \lambda\|\mathbf{z}\|_1 + \|\mathbf{A}\mathbf{z} - \mathbf{y}\|_2^2,$$

again for appropriate choices of λ. We will not deal with the statistical context any further, but we simply mention that near-optimal statistical estimation properties can be shown for both the LASSO and the Dantzig selector under conditions on $\mathbf{A}$ that are similar to the ones of the following chapters.

A closely related regression problem arises in *machine learning*. Given random pairs of samples $(t_j, y_j)_{j=1}^m$, where t_j is some input parameter vector and y_j is a scalar output, one would like to predict the output y for a future input data t. The model relating the output y to the input t is

$$\mathbf{y} = f(t) + \mathbf{e},$$

where $\mathbf{e}$ is random noise. The task is to learn the function f based on training samples (t_j, y_j). Without further hypotheses on f, this is an impossible task. Therefore, we assume that f has a sparse expansion in a given dictionary of functions $\psi_1, \ldots, \psi_N$, i.e., that f is written as

$$f(t) = \sum_{\ell=1}^{N} x_\ell \psi_\ell(t),$$

where $\mathbf{x}$ is a sparse vector. Introducing the matrix $\mathbf{A} \in \mathbb{R}^{m \times N}$ with entries

$$A_{j,k} = \psi_k(t_j),$$

we arrive at the model

$$\mathbf{y} = \mathbf{A}\mathbf{x} + \mathbf{e},$$

and the task is to estimate the sparse coefficient vector $\mathbf{x}$. This has the same form as the problem described above, and the same estimation procedures including the LASSO and the Dantzig selector apply.

Low-Rank Matrix Recovery and Matrix Completion

Let us finally describe an extension of compressive sensing together with some of its applications. Rather than recovering a sparse vector $\mathbf{x} \in \mathbb{C}^N$, we now aim at recovering a matrix $\mathbf{X} \in \mathbb{C}^{n_1 \times n_2}$ from incomplete information. Sparsity is replaced by the assumption that $\mathbf{X}$ has low rank. Indeed, the small complexity of the set of matrices with a given low rank compared to the set of all matrices makes the recovery of such matrices plausible.

For a linear map $\mathcal{A} : \mathbb{C}^{n_1 \times n_2} \to \mathbb{C}^m$ with $m < n_1 n_2$, suppose that we are given the measurement vector

$$\mathbf{y} = \mathcal{A}(\mathbf{X}) \in \mathbb{C}^m.$$

The task is to reconstruct $\mathbf{X}$ from $\mathbf{y}$. To stand a chance of success, we assume that $\mathbf{X}$ has rank at most $r \ll \min\{n_1, n_2\}$. The naive approach of solving the optimization problem

$$\text{minimize } \text{rank}(\mathbf{Z}) \quad \text{subject to } \mathcal{A}(\mathbf{Z}) = \mathbf{y}$$

is NP-hard, but an analogy with the compressive sensing problem will help. To illustrate this analogy, we consider the singular value decomposition of $\mathbf{X}$, i.e.,

$$\mathbf{X} = \sum_{\ell=1}^{n} \sigma_\ell \mathbf{u}_\ell \mathbf{v}_\ell^*.$$

Here, $n = \min\{n_1, n_2\}$, $\sigma_1 \geq \sigma_2 \geq \cdots \sigma_n \geq 0$ are the singular values of $\mathbf{X}$, and $\mathbf{u}_\ell \in \mathbb{C}^{n_1}$, $\mathbf{v}_\ell \in \mathbb{C}^{n_2}$ are the left and right singular vectors, respectively. We refer to Appendix A.2 for details. The matrix $\mathbf{X}$ is of rank r if and only if the vector $\boldsymbol{\sigma} = \boldsymbol{\sigma}(\mathbf{X})$ of singular values is r-sparse, i.e., $\mathrm{rank}(\mathbf{X}) = \|\boldsymbol{\sigma}(\mathbf{X})\|_0$. Having the ℓ_1-minimization approach for compressive sensing in mind, it is natural to introduce the so-called *nuclear norm* as the ℓ_1-norm of the singular values, i.e.,

$$\|\mathbf{X}\|_* = \|\boldsymbol{\sigma}(\mathbf{X})\|_1 = \sum_{\ell=1}^{n} \sigma_\ell(\mathbf{X}).$$

Then we consider the nuclear norm minimization problem

$$\text{minimize } \|\mathbf{Z}\|_* \quad \text{subject to } \mathcal{A}(\mathbf{Z}) = \mathbf{y}. \tag{1.10}$$

This is a convex optimization problem which can be solved efficiently, for instance, after reformulation as a semidefinite program.

A theory very similar to the recovery of sparse vectors can be developed, and appropriate conditions on $\mathcal{A}$ ensure exact or approximate recovery via nuclear norm minimization (and other algorithms). Again, random maps $\mathcal{A}$ turn out to be optimal, and matrices $\mathbf{X}$ of rank at most r can be recovered from m measurements with high probability provided

$$m \geq Cr \max\{n_1, n_2\}.$$

This bound is optimal since the right-hand side corresponds to the number of degrees of freedom required to describe an $n_1 \times n_2$ matrix of rank r. In contrast to the vector case, there is remarkably no logarithmic factor involved.

As a popular special case, the *matrix completion* problem seeks to fill in missing entries of a low-rank matrix. Thus, the measurement map $\mathcal{A}$ samples the entries $\mathcal{A}(\mathbf{X})_\ell = X_{j,k}$ for some indices j, k depending on ℓ. This setup appears, for example, in consumer taste prediction. Assume that an (online) store sells products indexed by the rows of the matrix and consumers—indexed by the columns—are able to rate these products. Not every consumer will rate every product, so only a limited number of entries of this matrix are available. For purposes of individualized advertisement, the store is interested in predicting the whole matrix of consumer ratings. Often, if two customers both like some subset of products, then they will also both like or dislike other subsets of products (the "types" of customers are essentially limited). For this reason, it can be assumed that the matrix of ratings has (at least approximately) low rank, which is confirmed empirically. Therefore, methods from low-rank matrix recovery, including the nuclear norm minimization approach, apply in this setup.

Although certainly interesting, we will not treat low-rank recovery extensively in this book. Nevertheless, due to the close analogy with sparse recovery, the main results are covered in exercises, and the reader is invited to work through them.

1.3 Overview of the Book

Before studying the standard compressive sensing problem on a technical level, it is beneficial to draw a road map of the basic results and solving strategies presented in this book.

As previously revealed, the notions of *sparsity* and *compressibility* are at the core of compressive sensing. A vector $\mathbf{x} \in \mathbb{C}^N$ is called s-sparse if it has at most s nonzero entries, in other words, if $\|\mathbf{x}\|_0 := \mathrm{card}(\{j : x_j \neq 0\})$ is smaller than or equal to s. The notation $\|\mathbf{x}\|_0$ has become customary, even though it does not represent a norm. In practice, one encounters vectors that are not exactly s-sparse but compressible in the sense that they are well approximated by sparse ones. This is quantified by the error of best s-term approximation to $\mathbf{x}$ given by

$$\sigma_s(\mathbf{x})_p := \inf_{\|\mathbf{z}\|_0 \leq s} \|\mathbf{x} - \mathbf{z}\|_p.$$

Chapter 2 introduces these notions formally, establishes relations to weak ℓ_p-quasinorms, and shows elementary estimates for the error of best s-term approximation, including

$$\sigma_s(\mathbf{x})_2 \leq \frac{1}{s^{1/p-1/2}} \|\mathbf{x}\|_p, \quad p \leq 2. \tag{1.11}$$

This suggests that unit balls in the ℓ_p-quasinorm for small $p \leq 1$ are good models for compressible vectors. We further study the problem of determining the minimal number m of measurements—namely, $m = 2s$—required (at least in principle) to recover all s-sparse vectors $\mathbf{x}$ from $\mathbf{y} = \mathbf{A}\mathbf{x}$ with a matrix $\mathbf{A} \in \mathbb{C}^{m \times N}$. It is remarkable that the actual length N of the vectors $\mathbf{x}$ does not play any role. The basic recovery procedure associated to this first recovery guarantee is the ℓ_0-minimization, i.e.,

$$\text{minimize } \|\mathbf{z}\|_0 \quad \text{subject to } \mathbf{A}\mathbf{z} = \mathbf{y}.$$

We will show in Sect. 2.3 that the ℓ_0-minimization is NP-hard by relating it to the *exact cover by 3-sets problem*, which is known to be NP-complete. Thus, ℓ_0-minimization is intractable in general, hence useless for practical purposes.

In order to circumvent the computational bottleneck of ℓ_0-minimization, we introduce several tractable alternatives in *Chap. 3*. Here, rather than a detailed analysis, we only present some intuitive justification and elementary results for these recovery algorithms. They can be subsumed under roughly three categories: optimization methods, greedy methods, and thresholding-based methods. The optimization approaches include the ℓ_1-minimization (1.2) (also called *basis pursuit*) and the quadratically constrained ℓ_1-minimization (1.4) (sometimes also called basis pursuit denoising in the literature), which takes potential measurement error into account. These minimization problems can be solved with various methods from convex optimization such as interior-point methods. We will also present specialized numerical methods for ℓ_1-minimization later in Chap. 15.

Orthogonal matching pursuit is a greedy method that builds up the support set of the reconstructed sparse vector iteratively by adding one index to the current support set at each iteration. The selection process is greedy because the index is chosen to minimize the residual at each iteration. Another greedy method is *compressive sampling matching pursuit* (CoSaMP). At each iteration, it selects several elements of the support set and then refines this selection.

The simple recovery procedure known as *basic thresholding* determines the support set in one step by choosing the s indices maximizing the correlations $|\langle \mathbf{x}, \mathbf{a}_j \rangle|$ of the vector $\mathbf{x}$ with the columns of $\mathbf{A}$. The reconstructed vector is obtained after an orthogonal projection on the span of the corresponding columns. Although this method is very fast, its performance is limited. A more powerful method is *iterative hard thresholding*. Starting with $\mathbf{x}^0 = \mathbf{0}$, say, it iteratively computes

$$\mathbf{x}^{n+1} = H_s \left(\mathbf{x}^n + \mathbf{A}^* (\mathbf{y} - \mathbf{A} \mathbf{x}^n) \right),$$

where H_s denotes the hard thresholding operator that keeps the s largest absolute entries of a vector and sets the other entries to zero. In the absence of the operator H_s, this is well known in the area of inverse problems as Landweber iterations. Applying H_s ensures sparsity of $\mathbf{x}^n$ at each iteration. We will finally present the *hard thresholding pursuit* algorithm which combines iterative hard thresholding with an orthogonal projection step.

Chapter 4 is devoted to the analysis of basis pursuit (ℓ_1-minimization). First, we derive conditions for the exact recovery of sparse vectors. The null space property of order s is a necessary and sufficient condition (on the matrix $\mathbf{A}$) for the success of exact recovery of all s-sparse vectors $\mathbf{x}$ from $\mathbf{y} = \mathbf{A}\mathbf{x}$ via ℓ_1-minimization. It basically requires that every vector in the null space of $\mathbf{A}$ is far from being sparse. This is natural, since a nonzero vector $\mathbf{x} \in \ker \mathbf{A}$ cannot be distinguished from the zero vector using $\mathbf{y} = \mathbf{A}\mathbf{x} = \mathbf{0}$. Next, we refine the null space property— introducing the stable null space property and the robust null space property—to ensure that ℓ_1-recovery is stable under sparsity defect and robust under measurement error. We also derive conditions that ensure the ℓ_1-recovery of an individual sparse vector. These conditions (on the vector $\mathbf{x}$ and the matrix $\mathbf{A}$) are useful in later chapters to establish so-called nonuniform recovery results for randomly chosen measurement matrices. The chapter is brought to an end with two small detours.

The first one is a geometric interpretation of conditions for exact recovery. The second one considers low-rank recovery and the nuclear norm minimization (1.10). The success of the latter is shown to be equivalent to a suitable adaptation of the null space property. Further results concerning low-rank recovery are treated in exercises spread throughout the book.

The null space property is not easily verifiable by a direct computation. The *coherence*, introduced in *Chap. 5*, is a much simpler concept to assess the quality of a measurement matrix. For $\mathbf{A} \in \mathbb{C}^{m \times N}$ with ℓ_2-normalized columns $\mathbf{a}_1, \ldots, \mathbf{a}_N$, it is defined as

$$\mu := \max_{j \neq k} |\langle \mathbf{a}_j, \mathbf{a}_k \rangle|.$$

We also introduce the ℓ_1-coherence function μ_1 as a slight refinement of the coherence. Ideally, the coherence μ of a measurement matrix should be small. A fundamental lower bound on μ (a related bound on μ_1 holds, too) is

$$\mu \geq \sqrt{\frac{N - m}{m(N - 1)}}.$$

For large N, the right-hand side scales like $1/\sqrt{m}$. The matrices achieving this lower bound are equiangular tight frames. We investigate conditions on m and N for the existence of equiangular tight frames and provide an explicit example of an $m \times m^2$ matrix (m being prime) with near-minimal coherence. Finally, based on the coherence, we analyze several recovery algorithms, in particular ℓ_1-minimization and orthogonal matching pursuit. For both of them, we obtain a verifiable sufficient condition for the recovery of all s-sparse vectors $\mathbf{x}$ from $\mathbf{y} = \mathbf{Ax}$, namely,

$$(2s - 1)\mu < 1.$$

Consequently, for a small enough sparsity, the algorithms are able to recover sparse vectors from incomplete information. Choosing a matrix $\mathbf{A} \in \mathbb{C}^{m \times N}$ with near-minimal coherence of order $c/\sqrt{m}$ (which imposes some mild conditions on N), s-sparse recovery is achievable with m of order s^2. In particular, s-sparse recovery is achievable from incomplete information ($m \ll N$) when s is small ($s \ll \sqrt{N}$). As already outlined, this can be significantly improved. In fact, we will see in later chapters that the optimal order for m is $s \ln(N/s)$. But for now the lower bound $\mu \geq c/\sqrt{m}$ implies that the coherence-based approach relying on $(2s - 1)\mu < 1$ necessitates

$$m \geq Cs^2. \tag{1.12}$$

This yields a number of measurements that scale quadratically in the sparsity rather than linearly (up to logarithmic factors). However, the coherence-based approach has the advantage of simplicity (the analysis of various recovery algorithms is relatively short) and of availability of explicit (deterministic) constructions of measurement matrices.

The concept of *restricted isometric property* (RIP) proves very powerful to overcome the quadratic bottleneck (1.12). The restricted isometry constant δ_s of a matrix $\mathbf{A} \in \mathbb{C}^{m \times N}$ is defined as the smallest $\delta \geq 0$ such that

$$(1 - \delta)\|\mathbf{x}\|_2^2 \leq \|\mathbf{A}\mathbf{x}\|_2^2 \leq (1 + \delta)\|\mathbf{x}\|_2^2 \quad \text{for all } s\text{-sparse } \mathbf{x}.$$

Informally, the matrix $\mathbf{A}$ is said to possess the RIP if δ_s is small for sufficiently large s. The RIP requires all submatrices formed by s columns of $\mathbf{A}$ to be well conditioned, since $\mathbf{A}\mathbf{x} = \mathbf{A}_S\mathbf{x}_S$ whenever $\mathbf{x} \in \mathbb{C}^N$ is supported on a set S of size s. Here, $\mathbf{A}_S \in \mathbb{C}^{m \times s}$ denotes the submatrix formed with columns of $\mathbf{A}$ indexed by S and $\mathbf{x}_S \in \mathbb{C}^s$ denotes the restriction of $\mathbf{x}$ to S.

Chapter 6 starts with basic results on the restricted isometry constants. For instance, there is the relation $\delta_2 = \mu$ with the coherence when the columns of $\mathbf{A}$ are ℓ_2-normalized. In this sense, restricted isometry constants generalize the coherence by considering all s-tuples rather than all pairs of columns. Other relations include the simple (and quite pessimistic) bound $\delta_s \leq (s-1)\mu$, which can be derived directly from Gershgorin's disk theorem.

We then turn to the analysis of the various recovery algorithms based on the restricted isometry property of $\mathbf{A}$. Typically, under conditions of the type

$$\delta_{\kappa s} \leq \delta_* \tag{1.13}$$

for some integer κ and some threshold $\delta_* < 1$ (both depending only on the algorithm), every s-sparse vector $\mathbf{x}$ is recoverable from $\mathbf{y} = \mathbf{A}\mathbf{x}$. The table below summarizes the sufficient conditions for basis pursuit, iterative hard thresholding, hard thresholding pursuit, orthogonal matching pursuit, and compressive sampling matching pursuit.

BP	IHT	HTP	OMP	CoSaMP
$\delta_{2s} < 0.6248$	$\delta_{3s} < 0.5773$	$\delta_{3s} < 0.5773$	$\delta_{13s} < 0.1666$	$\delta_{4s} < 0.4782$

Moreover, the reconstructions are stable when sparsity is replaced by compressibility and robust when measurement error occurs. More precisely, denoting by $\mathbf{x}^\sharp$ the output of the above algorithms run with $\mathbf{y} = \mathbf{A}\mathbf{x} + \mathbf{e}$ and $\|\mathbf{e}\|_2 \leq \eta$, the error estimates

$$\|\mathbf{x} - \mathbf{x}^\sharp\|_2 \leq C \frac{\sigma_s(\mathbf{x})_1}{\sqrt{s}} + D\eta, \tag{1.14}$$

$$\|\mathbf{x} - \mathbf{x}^\sharp\|_1 \leq C\sigma_s(\mathbf{x})_1 + D\sqrt{s}\eta, \tag{1.15}$$

hold for all $\mathbf{x} \in \mathbb{C}^N$ with absolute constants $C, D > 0$.

At the time of writing, finding explicit (deterministic) constructions of matrices satisfying (1.13) in the regime where m scales linearly in s up to logarithmic factors is an open problem. The reason lies in the fact that usual tools (such as Gershgorin's theorem) to estimate condition numbers essentially involve the coherence (or ℓ_1-coherence function), as in $\delta_{\kappa s} \leq (\kappa s - 1)\mu$. Bounding the latter by a fixed δ_* still faces the quadratic bottleneck (1.12).

We resolve this issue by passing to random matrices. Then a whole new set of tools from probability theory becomes available. When the matrix $\mathbf{A}$ is drawn at random, these tools enable to show that the restricted isometry property or other conditions ensuring recovery hold with high probability provided $m \geq Cs \ln(N/s)$. *Chapters 7 and 8* introduce all the necessary background on probability theory.

We start in *Chap. 7* by recalling basic concepts such as expectation, moments, Gaussian random variables and vectors, and Jensen's inequality. Next, we treat the relation between the moments of a random variable and its tails. Bounds on the tails of sums of independent random variables will be essential later, and Cramér's theorem provides general estimates involving the moment generating functions of the random variables. Hoeffding's inequality specializes to the sum of independent bounded mean-zero random variables. Gaussian and Rademacher/Bernoulli variables (the latter taking the values $+1$ or -1 with equal probability) fall into the larger class of subgaussian random variables, for which we also present basic results. Finally, Bernstein inequalities refine Hoeffding's inequality by taking into account the variance of the random variables. Furthermore, they extend to possibly unbounded subexponential random variables.

For many compressive sensing results with Gaussian or Bernoulli random matrices—that is, for large parts of Chaps. 9 and 11, including bounds for the restricted isometry constants—the relatively simple tools of Chap. 7 are already sufficient. Several topics in compressive sensing, however, notably the analysis of random partial Fourier matrices, build on more advanced tools from probability theory. *Chapter 8* presents the required material. For instance, we cover Rademacher sums of the form $\sum_j \epsilon_j a_j$ where the $\epsilon_j = \pm 1$ are independent Rademacher variables and the symmetrization technique leading to such sums. Khintchine inequalities bound the moments of Rademacher sums. The noncommutative Bernstein inequality provides a tail bound for the operator norm of independent mean-zero random matrices. Dudley's inequality bounds the expected supremum over a family of random variables by a geometric quantity of the set indexing the family. Concentration of measure describes the high-dimensional phenomenon which sees functions of random vectors concentrating around their means. Such a result is presented for Lipschitz functions of Gaussian random vectors.

With the probabilistic tools at hand, we are prepared to study Gaussian, Bernoulli, and more generally subgaussian random matrices in *Chap. 9*. A crucial ingredient for the proof of the restricted isometry property is the concentration inequality

$$\mathbb{P}(|\|\mathbf{Ax}\|_2^2 - \|\mathbf{x}\|_2^2| \geq t\|\mathbf{x}\|_2^2) \leq 2\exp(-cmt^2), \tag{1.16}$$

valid for any fixed $\mathbf{x} \in \mathbb{R}^N$ and $t \in (0, 1)$ with a random draw of a properly scaled $m \times N$ subgaussian random matrix $\mathbf{A}$. Using covering arguments—in particular, exploiting bounds on covering numbers from Appendix C.2—we deduce that the restricted isometry constants satisfy $\delta_s \leq \delta$ with high probability provided

$$m \geq C\delta^{-2}s\ln(eN/s). \tag{1.17}$$

The invariance of the concentration inequality under orthogonal transformations implies that subgaussian random matrices are *universal* in the sense that they allow for the recovery of vectors that are sparse not only in the canonical basis but also in an arbitrary (but fixed) orthonormal basis.

In the special case of Gaussian random matrices, one can exploit refined methods not available in the subgaussian case, such as Gordon's lemma and concentration of measure. We will deduce good explicit constants in the nonuniform setting where we only target recovery of a fixed s-sparse vector using a random draw of an $m \times N$ Gaussian matrix. For large dimensions, we roughly obtain that

$$m > 2s\ln(N/s)$$

is sufficient to recover an s-sparse vector using ℓ_1-minimization; see Chap. 9 for precise statements. This is the general rule of thumb reflecting the outcome of empirical tests, even for nongaussian random matrices—although the proof applies only to the Gaussian case.

We close Chap. 9 with a detour to the Johnson–Lindenstrauss lemma which states that a finite set of points in a large dimensional space can be mapped to a significantly lower-dimensional space while almost preserving all mutual distances (no sparsity assumption is involved here). This is somewhat equivalent to the concentration inequality (1.16). In this sense, the Johnson–Lindenstrauss lemma implies the RIP. We will conversely show that if a matrix satisfies the RIP, then randomizing the signs of its column yields a Johnson–Lindenstrauss embedding with high probability.

In *Chap. 10*, we show that the number of measurements (1.3) for sparse recovery using subgaussian random matrices is optimal. This is done by relating the standard compressive sensing problem to Gelfand widths of ℓ_1-balls. More precisely, for a subset K of a normed space $X = (\mathbb{R}^N, \|\cdot\|)$ and for $m < N$, we introduce the quantity

$$E^m(K, X) := \inf\left\{\sup_{\mathbf{x} \in K} \|\mathbf{x} - \Delta(\mathbf{A}\mathbf{x})\|, \ \mathbf{A} \in \mathbb{R}^{m \times N}, \ \Delta : \mathbb{R}^m \to \mathbb{R}^N\right\}.$$

It quantifies the worst-case reconstruction error over K of optimal measurement/reconstruction schemes in compressive sensing. The Gelfand width of K is defined as

$$d^m(K, X) := \inf\left\{\sup_{\mathbf{x} \in K \cap \ker \mathbf{A}} \|\mathbf{x}\|, \ \mathbf{A} \in \mathbb{R}^{m \times N}\right\}.$$

If $K = -K$ and $K + K \subset aK$ for some constant a, as it is the case with $a = 2$ for the unit ball of some norm, then

$$d^m(K, X) \leq E^m(K, X) \leq a d^m(K, X).$$

Since by (1.11) unit balls $K = B_p^N$ in the N-dimensional ℓ_p-space, $p \leq 1$, are good models for compressible vectors, we are led to study their Gelfand widths. For ease of exposition, we only cover the case $p = 1$. An upper bound for $E^m(B_1^N, \ell_2^N)$, and thereby for $d^m(B_1^N, \ell_2^N)$, can be easily derived from the error estimate (1.14) combined with the number of measurements that ensure the RIP for subgaussian random matrices. This gives

$$d^m(B_1^N, \ell_2^N) \leq C \min \left\{ 1, \frac{\ln(eN/m)}{m} \right\}^{1/2}.$$

We derive the matching lower bound

$$d^m(B_1^N, \ell_2^N) \geq c \min \left\{ 1, \frac{\ln(eN/m)}{m} \right\}^{1/2},$$

and we deduce that the bound (1.17) is necessary to guarantee the existence of a stable scheme for s-sparse recovery. An intermediate step in the proof of this lower bound is of independent interest. It states that a necessary condition on the number of measurements to guarantee that every s-sparse vector $\mathbf{x}$ is recoverable from $\mathbf{y} = \mathbf{A}\mathbf{x}$ via ℓ_1-minimization (stability is not required) is

$$m \geq Cs \ln(eN/s). \tag{1.18}$$

The error bound (1.14) includes the term $\sigma_s(\mathbf{x})_1/\sqrt{s}$, although the error is measured in ℓ_2-norm. This raises the question of the possibility of an error bound with the term $\sigma_s(\mathbf{x})_2$ on the right-hand side. *Chapter 11* investigates this question and the more general question of the existence of pairs of measurement matrix $\mathbf{A} \in \mathbb{R}^{m \times N}$ and reconstruction map $\Delta : \mathbb{R}^m \to \mathbb{R}^N$ satisfying

$$\|\mathbf{x} - \Delta(\mathbf{A}\mathbf{x})\|_q \leq \frac{C}{s^{1/p - 1/q}} \sigma_s(\mathbf{x})_p \quad \text{for all } \mathbf{x} \in \mathbb{R}^N.$$

This bound is referred to as *mixed (ℓ_q, ℓ_p)-instance optimality* and simply as ℓ_p-*instance optimality* when $q = p$. The ℓ_1-instance optimality implies the familiar bound $m \geq Cs \ln(eN/s)$. However, ℓ_2-instance optimality necessarily leads to

$$m \geq cN.$$

This regime of parameters is not interesting in compressive sensing. However, we may ask for less, namely, that the error bound in ℓ_2 holds in a nonuniform setting, i.e., for fixed $\mathbf{x}$ with high probability on a draw of a subgaussian random matrix $\mathbf{A}$. As it turns out, with Δ_1 denoting the ℓ_1-minimization map, the error bound

$$\|\mathbf{x} - \Delta_1(\mathbf{A}\mathbf{x})\|_2 \leq C\sigma_s(\mathbf{x})_2$$

does hold with high probability under the condition $m \geq Cs\ln(eN/s)$. The analysis necessitates the notion of ℓ_1-quotient property. It is proved for Gaussian random matrices, and a slight variation is proved for subgaussian random matrices.

In addition, Chap. 11 investigates a question about measurement error. When it is present, one may use the quadratically constrained ℓ_1-minimization

$$\text{minimize } \|\mathbf{z}\|_1 \quad \text{subject to } \|\mathbf{A}\mathbf{z} - \mathbf{y}\|_2 \leq \eta,$$

yet this requires an estimation of the noise level η (other algorithms do not require an estimation of η, but they require an estimation of the sparsity level s instead). Only then are the error bounds (1.14) and (1.15) valid under RIP. We will establish that, somewhat unexpectedly, the equality-constrained ℓ_1-minimization (1.2) can also be performed in the presence of measurement error using Gaussian measurement matrices. Indeed, the ℓ_1-quotient property implies the same reconstruction bounds (1.14) and (1.15) even without knowledge of the noise level η.

Subgaussian random matrices are of limited practical use, because specific applications may impose a structure on the measurement matrix that totally random matrices lack. As mentioned earlier, deterministic measurement matrices providing provable recovery guarantees are missing from the current theory. This motivates the study of *structured random matrices*. In *Chap. 12*, we investigate a particular class of structured random matrices arising in sampling problems. This includes random partial Fourier matrices.

Let $(\psi_1, \ldots, \psi_N)$ be a system of complex-valued functions which are orthonormal with respect to some probability measure ν on a set $\mathcal{D}$, i.e.,

$$\int_{\mathcal{D}} \psi_j(t)\overline{\psi_k(t)}d\nu(t) = \delta_{j,k}.$$

We call this system a *bounded orthonormal system* if there exists a constant $K \geq 1$ (ideally independent of N) such that

$$\sup_{1 \leq j \leq N} \sup_{t \in \mathcal{D}} |\psi_j(t)| \leq K.$$

A particular example is the trigonometric system where $\psi_j(t) = e^{2\pi i j t}$ for $j \in \Gamma \subset \mathbb{Z}$ with $\text{card}(\Gamma) = N$, in which case $K = 1$. We consider functions in the span of a bounded orthonormal system, i.e.,

$$f(t) = \sum_{j=1}^{N} x_j \psi_j(t),$$

and we assume that the coefficient vector $\mathbf{x} \in \mathbb{C}^N$ is sparse. The task is to reconstruct f (or equivalently $\mathbf{x}$) from sample values at locations $t_1, \ldots, t_m$, namely,

$$y_k = f(t_k) = \sum_{j=1}^{N} x_j \psi_j(t_k).$$

Introducing the *sampling matrix* $\mathbf{A} \in \mathbb{C}^{m \times N}$ with entries

$$A_{j,k} = \psi_j(t_k), \qquad (1.19)$$

the vector of samples is given by $\mathbf{y} = \mathbf{A}\mathbf{x}$. We are back to the standard compressive sensing problem with a matrix $\mathbf{A}$ taking this particular form. Randomness enters the picture by way of the sampling locations $t_1, \ldots, t_m$ which are chosen independently at random according to the probability measure ν. This makes $\mathbf{A}$ a structured random matrix. Before studying its performance, we relate this sampling setup with discrete uncertainty principles and establish performance limitations. In the context of the Hadamard transform, in slight contrast to (1.18), we show that now at least $m \geq Cs \ln N$ measurements are necessary.

Deriving recovery guarantees for the random sampling matrix $\mathbf{A}$ in (1.19) is more involved than for subgaussian random matrices where all the entries are independent. In fact, the matrix $\mathbf{A}$ has mN entries, but it is generated only by m independent random variables. We proceed by increasing level of difficulty and start by showing nonuniform sparse recovery guarantees for ℓ_1-minimization. The number of samples allowing one to recover a fixed s-sparse coefficient vector $\mathbf{x}$ with high probability is then $m \geq CK^2 s \ln N$.

The bound for the restricted isometry constants of the random sampling matrix $\mathbf{A}$ in (1.19) is a highlight of the theory of compressive sensing. It states that $\delta_s \leq \delta$ with high probability provided

$$m \geq CK^2 \delta^{-2} s \ln^4(N).$$

We close Chap. 12 by illustrating some connections to the Λ_1-problem from harmonic analysis.

A further type of measurement matrix used in compressive sensing is considered in *Chap. 13*. It arises as the adjacency matrix of certain bipartite graphs called lossless expanders. Hence, its entries take only the values 0 and 1. The existence of lossless expanders with optimal parameters is shown via probabilistic (combinatorial) arguments. We then show that the $m \times N$ adjacency matrix of a lossless expander allows for uniform recovery of all s-sparse vectors via ℓ_1-minimization provided that

$$m \geq Cs \ln(N/s).$$

Moreover, we present two iterative reconstruction algorithms. One of them has the remarkable feature that its runtime is *sublinear* in the signal length N; more precisely, its execution requires $\mathcal{O}(s^2 \ln^3 N)$ operations. Since only the locations and the values of s nonzero entries need to be identified, such superfast algorithms are not implausible. In fact, sublinear algorithms are possible in other contexts, too, but they are always designed together with the measurement matrix $\mathbf{A}$.

In *Chap. 14*, we follow a different approach to sparse recovery guarantees by considering a fixed (deterministic) matrix $\mathbf{A}$ and choosing the s-sparse vector $\mathbf{x}$ at random. More precisely, we select its support set S uniformly at random among all subsets of $[N] = \{1, 2, \ldots, N\}$ with cardinality s. The signs of the nonzero coefficients of $\mathbf{x}$ are chosen at random as well, but their magnitudes are kept arbitrary. Under a very mild condition on the coherence μ of $\mathbf{A} \in \mathbb{C}^{m \times N}$, namely,

$$\mu \leq \frac{c}{\ln N}, \tag{1.20}$$

and under the condition

$$\frac{s\|\mathbf{A}\|_{2 \to 2}}{N} \leq \frac{c}{\ln N}, \tag{1.21}$$

the vector $\mathbf{x}$ is recoverable from $\mathbf{y} = \mathbf{Ax}$ via ℓ_1-minimization with high probability. The (deterministic or random) matrices $\mathbf{A}$ usually used in compressive sensing and signal processing, for instance, tight frames, obey (1.21) provided

$$m \geq Cs \ln N. \tag{1.22}$$

Since (1.20) is also satisfied for these matrices, we again obtain sparse recovery in the familiar parameter regime (1.22). The analysis relies on the crucial fact that a random column submatrix of $\mathbf{A}$ is well conditioned under (1.20) and (1.21). We note that this random signal model may not always reflect the type of signals encountered in practice, so the theory for random matrices remains important. Nevertheless, the result for random signals explains the outcome of numerical experiments where the signals are often constructed at random.

The ℓ_1-minimization principle (basis pursuit) is one of the most powerful sparse recovery methods—as should have become clear by now. *Chapter 15* presents a selection of efficient algorithms to perform this optimization task in practice (the selection is nonexhaustive, and the algorithms have been chosen not only for their efficiency but also for their simplicity and diversity). First, the homotopy method applies to the real-valued case $\mathbf{A} \in \mathbb{R}^{m \times N}$, $\mathbf{y} \in \mathbb{R}^m$. For a parameter $\lambda > 0$, we consider the functional

$$F_\lambda(\mathbf{x}) = \frac{1}{2}\|\mathbf{Ax} - \mathbf{y}\|_2^2 + \lambda\|\mathbf{x}\|_1.$$

Its minimizer $\mathbf{x}_\lambda$ converges to the minimizer $\mathbf{x}^\sharp$ of the equality-constrained ℓ_1-minimization problem (1.2). The map $\lambda \mapsto \mathbf{x}_\lambda$ turns out to be piecewise linear. The homotopy method starts with a sufficiently large λ, for which $\mathbf{x}_\lambda = \mathbf{0}$, and traces the endpoints of the linear pieces until $\lambda = 0^+$, for which $\mathbf{x}_\lambda = \mathbf{x}^\sharp$. At each step of the algorithm, an element is added or removed from the support set of the current minimizer. Since one mostly adds elements to the support, this algorithm is usually very efficient for small sparsity.

As a second method, we treat Chambolle and Pock's primal–dual algorithm. This algorithm applies to a large class of optimization problems including ℓ_1-minimization. It consists of a simple iterative procedure which updates a primal, a dual, and an auxiliary variable at each step. All of the computations are easy to perform. We show convergence of the sequence of primal variables generated by the algorithm to the minimizer of the given functional and outline its specific form for three types of ℓ_1-minimization problems. In contrast to the homotopy method, it applies also in the complex-valued case.

Finally, we discuss a method that iteratively solves weighted ℓ_2-minimization problems. The weights are suitably updated in each iteration based on the solution of the previous iteration. Since weighted ℓ_2-minimization can be performed efficiently (in fact, this is a linear problem), each step of the algorithm can be computed quickly. Although this algorithm is strongly motivated by ℓ_1-minimization, its convergence to the ℓ_1-minimizer is not guaranteed. Nevertheless, under the null space property of the matrix $\mathbf{A}$ (equivalent to sparse recovery via ℓ_1-minimization), we show that the iteratively reweighted least squares algorithm recovers every s-sparse vector from $\mathbf{y} = \mathbf{A}\mathbf{x}$. Recovery is stable when passing to compressible vectors. Moreover, we give an estimate of the convergence rate in the exactly sparse case.

The book is concluded with three appendices. *Appendix A* covers background material from linear algebra and matrix analysis, including vector and matrix norms, eigenvalues and singular values, and matrix functions. Basic concepts and results from convex analysis and convex optimization are presented in *Appendix B*. We also treat matrix convexity and present a proof of Lieb's theorem on the concavity of the matrix function $\mathbf{X} \mapsto \operatorname{tr} \exp(\mathbf{H} + \ln \mathbf{X})$ on the set of positive definite matrices. *Appendix C* presents miscellaneous material including covering numbers, Fourier transforms, elementary estimates on binomial coefficients, the Gamma function and Stirling's formula, smoothing of Lipschitz functions via convolution, distributional derivatives, and differential inequalities.

Notation is usually introduced when it first appears. Additionally, a collection of symbols used in the text can be found on pp. 589. All the constants in this book are universal unless stated otherwise. This means that they do not depend on any other quantity. Often, the value of a constant is given explicitly or it can be deduced from the proof.

Notes

The field of compressive sensing was initiated with the papers [94] by Candès, Romberg, and Tao and [152] by Donoho who coined the term *compressed sensing*. Even though there have been predecessors on various aspects of the field, these papers seem to be the first ones to combine the ideas of ℓ_1-minimization with a random choice of measurement matrix and to realize the effectiveness of this combination for solving underdetermined systems of equations. Also, they emphasized the potential of compressive sensing for many signal processing tasks.

We now list some of the highlights from preceding works and earlier developments connected to compressive sensing. Details and references on the advances of compressive sensing itself will be given in the Notes sections at the end of each subsequent chapter. References [29, 84, 100, 182, 204, 411, 427] provide overview articles on compressive sensing.

Arguably, the first contribution connected to sparse recovery was made by de Prony [402] as far back as 1795. He developed a method for identifying the frequencies $\omega_j \in \mathbb{R}$ and the amplitudes $x_j \in \mathbb{C}$ in a nonharmonic trigonometric sum of the form $f(t) = \sum_{j=1}^{s} x_j e^{2\pi i \omega_j t}$. His method takes equidistant samples and solves an eigenvalue problem to compute the ω_j. This method is related to Reed–Solomon decoding covered in the next chapter; see Theorem 2.15. For more information on the Prony method, we refer to [344, 401].

The use of ℓ_1-minimization appeared in the 1965 Ph.D. thesis [332] of Logan in the context of sparse frequency estimation, and an early theoretical work on L_1-minimization is the paper [161] by Donoho and Logan. Geophysicists observed in the late 1970s that ℓ_1-minimization can be successfully used to compute a sparse reflection function indicating changes between subsurface layers [441, 469]. The use of total-variation minimization, which is closely connected to ℓ_1-minimization, appeared in the 1990s in the work on image processing by Rudin, Osher, and Fatemi [436]. The use of ℓ_1-minimization and related greedy methods in statistics was greatly popularized by the work of Tibshirani [473] on the LASSO (Least Absolute Shrinkage and Selection Operator).

The theory of sparse approximation and associated algorithms began in the 1990s with the papers [114, 342, 359]. The theoretical understanding of conditions allowing greedy methods and ℓ_1-minimization to recover the sparsest solution developed with the work in [155, 158, 181, 215, 224, 239, 476, 479].

Compressive sensing has connections with the area of information-based complexity which considers the general question of how well functions f from a class $\mathcal{F}$ can be approximated from m sample values or more generally from the evaluation of m linear or nonlinear functionals applied to f; see [474]. The optimal recovery error defined as the maximal reconstruction error for the best sampling and recovery methods over all functions in the class $\mathcal{F}$ is closely related to the so-called *Gelfand width* of $\mathcal{F}$ [370]; see also Chap. 10. Of particular interest in compressive sensing is the ℓ_1-ball B_1^N in $\mathbb{R}^N$. Famous results due to Kashin [299] and Gluskin and Garnaev [219, 227] sharply bound the Gelfand widths of B_1^N from above and below; see also Chap. 10. Although the original interest of Kashin was to estimate m-widths of Sobolev classes, these results give precise performance bounds on how well any method may recover (approximately) sparse vectors from linear measurements. It is remarkable that [219, 299] already employed Bernoulli and Gaussian random matrices in ways similar to their use in compressive sensing (see Chap. 9).

In computer science, too, sparsity appeared before the advent of compressive sensing through the area of sketching. Here, one is not only interested in recovering huge data sets (such as data streams on the Internet) from vastly undersampled data, but one requires in addition that the associated algorithms have sublinear runtime in the signal length. There is no a priori contradiction in this desideratum because one only needs to report locations and values of nonzero entries. Such algorithms often use ideas from *group testing* [173], which dates back to World War II, when Dorfman [171] devised an efficient method for detecting draftees with syphilis. One usually designs the matrix and the fast algorithm simultaneously [131, 225] in this setup. Lossless expanders as studied in Chap. 13 play a key role in some of the constructions [41]. Quite remarkably, sublinear algorithms are also available for sparse Fourier transforms [223, 261, 262, 287, 288, 519].

Applications of Compressive Sensing. We next provide comments and references on the applications and motivations described in Sect. 1.2.

Single-pixel camera. The single-pixel camera was developed by Baraniuk and coworkers [174] as an elegant proof of concept that the ideas of compressive sensing can be implemented in hardware.

Magnetic resonance imaging. The initial paper [94] on compressive sensing was motivated by medical imaging—although Candès et al. have in fact treated the very similar problem of computerized tomography. The application of compressive sensing techniques to magnetic resonance imaging (MRI) was investigated in [255, 338, 358, 497]. Background on the theoretical foundations of MRI can be found, for instance, in [252, 267, 512]. Applications of compressive sensing to the related problem of *nuclear magnetic resonance spectroscopy* are contained in [278, 447]. Background on the methods related to Fig. 1.6 is described in the work of Lustig, Vasanawala and coworkers [358, 497].

Radar. The particular radar application outlined in Sect. 1.2 is described in more detail in [268]. The same mathematical model appears also in sonar and in the channel estimation problem of wireless communications [384, 385, 412]. The application of compressive sensing to other radar scenarios can be found, for instance, in [185, 189, 283, 397, 455].

Sampling theory. The classical sampling theorem (1.5) can be associated with the names of Shannon, Nyquist, Whittaker, and Kotelnikov. Sampling theory is a broad and well-developed area. We refer to [39, 195, 271, 272, 294] for further information on the classical aspects. The use of sparse recovery techniques in sampling problems appeared early in the development of the compressive sensing theory [94, 97, 408, 409, 411, 416]. In fact, the alternative name *compressive sampling* indicates that compressive sensing can be viewed as a part of sampling theory—although it draws from quite different mathematical tools than classical sampling theory itself.

Sparse approximation. The theory of compressive sensing can also be viewed as a part of sparse approximation with roots in signal processing, harmonic analysis

[170], and numerical analysis [122]. A general source for background on sparse approximation and its applications are the books [179,451,472] as well as the survey paper [73].

The principle of representing a signal by a small number of terms in a suitable basis in order to achieve compression is realized, for instance, in the ubiquitous compression standards JPEG, MPEG, and MP3. Wavelets [137] are known to provide a good basis for images, and the analysis of the best (nonlinear) approximation reaches into the area of function spaces, more precisely Besov spaces [508]. Similarly, Gabor expansions [244] may compress audio signals. Since good Gabor systems are always redundant systems (frames) and never bases, computational tools to compute the sparsest representation of a signal are essential. It was realized in [342, 359] that this problem is in general NP-hard. The greedy approach via orthogonal matching pursuit was then introduced in [342] (although it had appeared earlier in different contexts), while basis pursuit (ℓ_1-minimization) was introduced in [114].

The use of the uncertainty principle for deducing a positive statement on the data separation problem with respect to the Fourier and canonical bases appeared in [163,164]. For further information on the separation problem, we refer the reader to [92,158,160,181,238,331,482]. Background on denoising via sparse representations can be found in [105,150,159,180,407,450].

The analysis of conditions allowing algorithms such as ℓ_1-minimization or orthogonal matching pursuit to recover the sparsest representation has started with the contributions [155–158, 224, 476, 479], and these early results are the basis for the advances in compressive sensing.

Error correction. The idealized setup of error correction and the compressive sensing approach described in Sect. 1.2 appeared in [96, 167, 431]. For more background on error correction, we refer to [282].

Statistics and machine learning. Sparsity has a long history in statistics and in linear regression models in particular. The corresponding area is sometimes referred to as high-dimensional statistics or model selection because the support set of the coefficient vector $\mathbf{x}$ determines the relevant explanatory variables and thereby selects a model. Stepwise forward regression methods are closely related to greedy algorithms such as (orthogonal) matching pursuit. The LASSO, i.e., the minimization problem (1.8), was introduced by Tibshirani in [473]. Candès and Tao have introduced the Dantzig selector (1.9) in [98] and realized that methods of compressive sensing (the restricted isometry property) are useful for the analysis of sparse regression methods. We refer to [48] and the monograph [76] for details. For more information on machine learning, we direct the reader to [18, 133, 134, 444]. Connections between sparsity and machine learning can be found, for instance, in [23, 147, 513].

Low-rank matrix recovery. The extension of compressive sensing to the recovery of low-rank matrices from incomplete information emerged with the papers [90, 99, 418]. The idea of replacing the rank minimization problem by the nuclear norm minimization appeared in the Ph.D. thesis of Fazel [190]. The matrix completion problem is treated in [90, 99, 417] and the more general problem of quantum state tomography in [245, 246, 330].

Let us briefly mention further applications and relations to other fields.

In *inverse problems*, sparsity has also become an important concept for regularization methods. Instead of Tikhonov regularization with a Hilbert space norm [186], one uses an ℓ_1-norm regularization approach [138, 406]. In many practical applications, this improves the recovered solutions. Ill-posed inverse problems appear, for instance, in geophysics where ℓ_1-norm regularization was already used in [441, 469] but without rigorous mathematical theory at that time. We refer to the survey papers [269, 270] dealing with compressive sensing in seismic exploration.

Total-variation minimization is a classical and successful approach for image denoising and other tasks in image processing [104, 106, 436]. Since the total variation is the ℓ_1-norm of the gradient, the minimization problem is closely related to basis pursuit. In fact, the motivating example for the first contribution [94] of Candès, Romberg, and Tao to compressive sensing came from total-variation minimization in computer tomography. The restricted isometry property can be used to analyze image recovery via total-variation minimization [364]. The primal–dual algorithm of Chambolle and Pock to be presented in Chap. 15 was originally motivated by total-variation minimization as well [107].

Further applications of compressive sensing and sparsity in general include imaging (tomography, ultrasound, photoacoustic imaging, hyperspectral imaging, etc.), analog-to-digital conversion [353, 488], DNA microarray processing, astronomy [507], and wireless communications [27, 468].

Topics not Covered in this Book. It is impossible to give a detailed account of all the directions that have so far cropped up around compressive sensing. This book certainly makes a selection, but we believe that we cover the most important aspects and mathematical techniques. With this basis, the reader should be well equipped to read the original references on further directions, generalizations, and applications. Let us only give a brief account of additional topics together with the relevant references. Again, no claim about completeness of the list is made.

Structured sparsity models. One often has additional a priori knowledge than just pure sparsity in the sense that the support set of the sparse vector to be recovered possesses a certain structure, i.e., only specific support sets are allowed. Let us briefly describe the *joint-sparsity* and *block-sparsity* model.

Suppose that we take measurements not only of a single signal but of a collection of signals that are somewhat coupled. Rather than assuming that each signal is sparse (or compressible) on its own, we assume that the unknown support set is the same for all signals in the collection. In this case, we speak of joint sparsity. A motivating example is color images where each signal corresponds to a color channel of the image, say red, green, and blue. Since edges usually appear at the same location for all channels, the gradient features some joint sparsity. Instead of the usual ℓ_1-minimization problem, one considers mixed ℓ_1/ℓ_2-norm minimization or greedy algorithms exploiting the joint-sparsity structure. A similar setup is described by the block-sparsity (or group-sparsity) model, where certain indices of the sparse vector are grouped together. Then a signal is block sparse if most groups (blocks) of coefficients are zero. In other words, nonzero coefficients appear in groups. Recovery algorithms may exploit this prior knowledge to improve the recovery performance. A theory can be developed along similar lines as usual sparsity [143, 183, 184, 203, 241, 478, 487]. The so-called *model-based compressive sensing* [30] provides a further, very general structured sparsity setup.

Sublinear algorithms. This type of algorithms have been developed in computer science for a longer time. The fact that only the locations and values of nonzero entries of a sparse vector have to be reported enables one to design recovery algorithms whose runtime is sublinear in the vector length. Recovery methods are also called streaming algorithms or heavy hitters. We will only cover a toy sublinear algorithm in Chap. 13, and we refer to [41, 131, 222, 223, 225, 261, 285, 289] for more information.

Connection with the geometry of random polytopes. Donoho and Tanner [154, 165–167] approached the analysis of sparse recovery via ℓ_1-minimization through polytope geometry. In fact, the recovery of s-sparse vectors via ℓ_1-minimization is equivalent to a geometric property—called neighborliness—of the projected ℓ_1-ball under the action of the measurement matrix; see also Corollary 4.39. When the measurement matrix is a Gaussian random matrix, Donoho and Tanner give a precise analysis of so-called phase transitions that predict in which ranges of (s, m, N) sparse recovery is successful and unsuccessful with high probability. In particular, their analysis provides the value of the optimal constant C such that $m \approx C s \ln(N/s)$ allows for s-sparse recovery via ℓ_1-minimization. We only give a brief account of their work in the Notes of Chap. 9.

Compressive sensing and quantization. If compressive sensing is used for signal acquisition, then a realistic sensor must quantize the measured data. This means that only a finite number of values for the measurements y_ℓ are possible. For instance, 8 bits provide $2^8 = 256$ values for an approximation of y_ℓ to be stored. If the quantization is coarse, then this additional source of error cannot be ignored and a revised theoretical analysis becomes necessary. We refer to [249, 316, 520] for background information. We also mention the extreme case of 1-bit compressed sensing where only the signs of the measurements are available via $\mathbf{y} = \mathrm{sgn}(\mathbf{Ax})$ [290, 393, 394].

Dictionary learning. Sparsity usually occurs in a specific basis or redundant dictionary. In certain applications, it may not be immediately clear which dictionary is suitable to sparsify the signals of interest. Dictionary learning tries to identify a good dictionary using training signals. Algorithmic approaches include the K-SVD algorithm [5, 429] and optimization methods [242]. Optimizing over both the dictionary and the coefficients in the expansions results in a nonconvex program, even when using ℓ_1-minimization. Therefore, it is notoriously hard to establish a rigorous mathematical theory of dictionary learning despite the fact that the algorithms perform well in practice. Nevertheless, there are a few interesting mathematical results available in the spirit of compressive sensing [221, 242].

Recovery of functions of many variables. Techniques from compressive sensing can be exploited for the reconstruction of functions on a high-dimensional space from point samples. Traditional approaches suffer the curse of dimensionality, which predicts that the number of samples required to achieve a certain reconstruction accuracy scales exponentially with the spatial dimension even for classes of infinitely differentiable functions [371, 474]. It is often a reasonable assumption in practice that the function to be reconstructed depends only on a small number of (a priori unknown) variables. This model is investigated in [125, 149], and ideas of compressive sensing allow one to dramatically reduce the number of required samples. A more general model considers functions of the form $f(\mathbf{x}) = g(\mathbf{Ax})$, where $\mathbf{x}$ belongs to a subset $\mathcal{D} \subset \mathbb{R}^N$ with N being large, $\mathbf{A} \in \mathbb{R}^{m \times N}$ with $m \ll N$, and g is a smooth function on an m-dimensional domain. Both g and $\mathbf{A}$ are unknown a priori and are to be reconstructed from suitable samples of f. Again, under suitable assumptions on g and on $\mathbf{A}$, one can build on methods from compressive sensing to recover f from a relatively small number of samples. We refer to [124, 206, 266] for details.

Hints for Preparing a Course. This book can be used for a course on compressive sensing at the graduate level. Although the whole material exceeds what can be reasonably covered in a one-semester class, properly selected topics do convert into self-contained components. We suggest the following possibilities:

- For a comprehensive treatment of the deterministic issues, Chaps. 2–6 complemented by Chap. 10 are appropriate. If a proof of the restricted isometry property for random matrices is desired, one can add the simple arguments of Sect. 9.1, which only rely on a few tools from Chap. 7. In a class lasting only one quarter rather than one semester, one can remove Sect. 4.5 and mention only briefly the stability and robustness results of Chaps. 4 and 6. One can also concentrate only on ℓ_1-minimization and discard Chap. 3 as well as Sects. 5.3, 5.5, 6.3, and 6.4 if the variety of algorithms is not a priority.
- On the other hand, for a course focusing on algorithmic aspects, Chaps. 2–6 as well as (parts of) Chap. 15 are appropriate, possibly replacing Chap. 5 by Chap. 13 and including (parts of) Appendix B.
- For a course focusing on probabilistic issues, we recommend Chaps. 7–9 and Chaps. 11, 12, and 14. This can represent a second one-semester class. However, if this material has to be delivered as a first course, Chap. 4 (especially Sects. 4.1 and 4.4) and Chap. 6 (especially Sects. 6.1 and 6.2) need to be included.

Of course, parts of particular chapters may also be dropped depending on the desired emphasis.

We will be happy to receive feedback on these suggestions from instructors using this book in their class. They may also contact us to obtain typed-out solutions for some of the exercises.

Chapter 2
Sparse Solutions of Underdetermined Systems

In this chapter, we define the notions of vector sparsity and compressibility, and we establish some related inequalities used throughout the book. We will use basic results on vector and matrix norms, which can be found in Appendix A. We then investigate, in two different settings, the minimal number of linear measurements required to recover sparse vectors. We finally prove that ℓ_0-minimization, the ideal recovery scheme, is NP-hard in general.

2.1 Sparsity and Compressibility

We start by defining the ideal notion of *sparsity*. We first introduce the notations $[N]$ for the set $\{1, 2, \ldots, N\}$ and $\mathrm{card}(S)$ for the cardinality of a set S. Furthermore, we write $\overline{S}$ for the complement $[N] \setminus S$ of a set S in $[N]$.

Definition 2.1. The *support* of a vector $\mathbf{x} \in \mathbb{C}^N$ is the index set of its nonzero entries, i.e.,

$$\mathrm{supp}(\mathbf{x}) := \{j \in [N] : x_j \neq 0\}.$$

The vector $\mathbf{x} \in \mathbb{C}^N$ is called *s-sparse* if at most s of its entries are nonzero, i.e., if

$$\|\mathbf{x}\|_0 := \mathrm{card}(\mathrm{supp}(\mathbf{x})) \leq s.$$

The customary notation $\|\mathbf{x}\|_0$—the notation $\|\mathbf{x}\|_0^0$ would in fact be more appropriate—comes from the observation that

$$\|\mathbf{x}\|_p^p := \sum_{j=1}^N |x_j|^p \xrightarrow[p \to 0]{} \sum_{j=1}^N \mathbf{1}_{\{x_j \neq 0\}} = \mathrm{card}(\{j \in [N] : x_j \neq 0\}).$$

S. Foucart and H. Rauhut, *A Mathematical Introduction to Compressive Sensing*, Applied and Numerical Harmonic Analysis, DOI 10.1007/978-0-8176-4948-7_2, © Springer Science+Business Media New York 2013

Here, we used the notations $1_{\{x_j \neq 0\}} = 1$ if $x_j \neq 0$ and $1_{\{x_j \neq 0\}} = 0$ if $x_j = 0$. In other words the quantity $\|\mathbf{x}\|_0$ is the limit as p decreases to zero of the pth power of the ℓ_p-quasinorm of $\mathbf{x}$. It is abusively called the ℓ_0-norm of $\mathbf{x}$, although it is neither a norm nor a quasinorm—see Appendix A for precise definitions of these notions. In practice, sparsity can be a strong constraint to impose, and we may prefer the weaker concept of *compressibility*. For instance, we may consider vectors that are nearly s-sparse, as measured by the *error of best s-term approximation*.

Definition 2.2. For $p > 0$, the ℓ_p-*error of best s-term approximation* to a vector $\mathbf{x} \in \mathbb{C}^N$ is defined by

$$\sigma_s(\mathbf{x})_p := \inf \left\{ \|\mathbf{x} - \mathbf{z}\|_p, \ \mathbf{z} \in \mathbb{C}^N \text{ is } s\text{-sparse} \right\}.$$

In the definition of $\sigma_s(\mathbf{x})_p$, the infimum is achieved by an s-sparse vector $\mathbf{z} \in \mathbb{C}^N$ whose nonzero entries equal the s largest absolute entries of $\mathbf{x}$. Hence, although such a vector $\mathbf{z} \in \mathbb{C}^N$ may not be unique, it achieves the infimum independently of $p > 0$.

Informally, we may call $\mathbf{x} \in \mathbb{C}^N$ a *compressible* vector if the error of its best s-term approximation decays quickly in s. According to the following proposition, this happens in particular if $\mathbf{x}$ belongs to the unit ℓ_p-ball for some small $p > 0$, where the unit ℓ_p-ball is defined by

$$B_p^N := \{ \mathbf{z} \in \mathbb{C}^N : \|\mathbf{z}\|_p \leq 1 \}.$$

Consequently, the nonconvex balls B_p^N for $p < 1$ serve as good models for compressible vectors.

Proposition 2.3. *For any $q > p > 0$ and any $\mathbf{x} \in \mathbb{C}^N$,*

$$\sigma_s(\mathbf{x})_q \leq \frac{1}{s^{1/p-1/q}} \|\mathbf{x}\|_p.$$

Before proving this proposition, it is useful to introduce the notion of *nonincreasing rearrangement*.

Definition 2.4. The *nonincreasing rearrangement* of the vector $\mathbf{x} \in \mathbb{C}^N$ is the vector $\mathbf{x}^* \in \mathbb{R}^N$ for which

$$x_1^* \geq x_2^* \geq \ldots \geq x_N^* \geq 0$$

and there is a permutation $\pi : [N] \to [N]$ with $x_j^* = |x_{\pi(j)}|$ for all $j \in [N]$.

Proof (of Proposition 2.3). If $\mathbf{x}^* \in \mathbb{R}_+^N$ is the nonincreasing rearrangement of $\mathbf{x} \in \mathbb{C}^N$, we have

$$\sigma_s(\mathbf{x})_q^q = \sum_{j=s+1}^{N} (x_j^*)^q \leq (x_s^*)^{q-p} \sum_{j=s+1}^{N} (x_j^*)^p \leq \left(\frac{1}{s}\sum_{j=1}^{s}(x_j^*)^p\right)^{\frac{q-p}{p}} \left(\sum_{j=s+1}^{N}(x_j^*)^p\right)$$

$$\leq \left(\frac{1}{s}\|\mathbf{x}\|_p^p\right)^{\frac{q-p}{p}}\|\mathbf{x}\|_p^p = \frac{1}{s^{q/p-1}}\|\mathbf{x}\|_p^q.$$

The result follows by taking the power $1/q$ in both sides of this inequality. $\qquad\square$

We strengthen the previous proposition by finding the smallest possible constant $c_{p,q}$ in the inequality $\sigma_s(\mathbf{x})_q \leq c_{p,q}s^{-1/p+1/q}\|\mathbf{x}\|_p$. This can be skipped on first reading, but it is nonetheless informative because the proof technique, which consists in solving a convex optimization problem by hand, will reappear in Theorem 5.8 and Lemma 6.14.

Theorem 2.5. *For any $q > p > 0$ and any $\mathbf{x} \in \mathbb{C}^N$, the inequality*

$$\sigma_s(\mathbf{x})_q \leq \frac{c_{p,q}}{s^{1/p-1/q}}\|\mathbf{x}\|_p$$

holds with

$$c_{p,q} := \left[\left(\frac{p}{q}\right)^{p/q}\left(1-\frac{p}{q}\right)^{1-p/q}\right]^{1/p} \leq 1.$$

Let us point out that the frequent choice $p = 1$ and $q = 2$ gives

$$\sigma_s(\mathbf{x})_2 \leq \frac{1}{2\sqrt{s}}\|\mathbf{x}\|_1.$$

Proof. Let $\mathbf{x}^* \in \mathbb{R}_+^N$ be the nonincreasing rearrangement of $\mathbf{x} \in \mathbb{C}^N$. Setting $\alpha_j := (x_j^*)^p$, we will prove the equivalent statement

$$\left.\begin{array}{r}\alpha_1 \geq \alpha_2 \geq \cdots \geq \alpha_N \geq 0 \\ \alpha_1 + \alpha_2 + \cdots + \alpha_N \leq 1\end{array}\right\} \implies \alpha_{s+1}^{q/p} + \alpha_{s+2}^{q/p} + \cdots + \alpha_N^{q/p} \leq \frac{c_{p,q}^q}{s^{q/p-1}}.$$

Thus, with $r := q/p > 1$, we aim at maximizing the convex function

$$f(\alpha_1, \alpha_2, \ldots, \alpha_N) := \alpha_{s+1}^r + \alpha_{s+2}^r + \cdots + \alpha_N^r$$

over the convex polygon

$$\mathcal{C} := \{(\alpha_1, \ldots, \alpha_N) \in \mathbb{R}^N : \alpha_1 \geq \cdots \geq \alpha_N \geq 0 \text{ and } \alpha_1 + \cdots + \alpha_N \leq 1\}.$$

According to Theorem B.16, the maximum of f is attained at a vertex of $\mathcal{C}$. The vertices of $\mathcal{C}$ are obtained as intersections of N hyperplanes arising by turning N

of the $(N + 1)$ inequality constraints into equalities. Thus, we have the following possibilities:

- If $\alpha_1 = \cdots = \alpha_N = 0$, then $f(\alpha_1, \alpha_2, \ldots, \alpha_N) = 0$.
- If $\alpha_1 + \cdots + \alpha_N = 1$ and $\alpha_1 = \cdots = \alpha_k > \alpha_{k+1} = \cdots = \alpha_N = 0$ for some $1 \leq k \leq s$, then $f(\alpha_1, \alpha_2, \ldots, \alpha_N) = 0$.
- If $\alpha_1 + \cdots + \alpha_N = 1$ and $\alpha_1 = \cdots = \alpha_k > \alpha_{k+1} = \cdots = \alpha_N = 0$ for some $s + 1 \leq k \leq N$, then $\alpha_1 = \cdots = \alpha_k = 1/k$, and consequently $f(\alpha_1, \alpha_2, \ldots, \alpha_N) = (k - s)/k^r$.

It follows that

$$\max_{(\alpha_1, \ldots, \alpha_N) \in \mathcal{C}} f(\alpha_1, \alpha_2, \ldots, \alpha_N) = \max_{s+1 \leq k \leq N} \frac{k - s}{k^r}.$$

Considering k as a continuous variable, we now observe that the function $g(k) := (k - s)/k^r$ is increasing until the critical point $k^* = (r/(r - 1))s$ and decreasing thereafter. We obtain

$$\max_{(\alpha_1, \ldots, \alpha_N) \in \mathcal{C}} f(\alpha_1, \alpha_2, \ldots, \alpha_N) \leq g(k^*) = \frac{1}{r}\left(1 - \frac{1}{r}\right)^{r-1} \frac{1}{s^{r-1}} = c_{p,q}^q \frac{1}{s^{q/p-1}}.$$

This is the desired result.   □

Another possibility to define *compressibility* is to call a vector $\mathbf{x} \in \mathbb{C}^N$ *compressible* if the number

$$\mathrm{card}(\{j \in [N] : |x_j| \geq t\})$$

of its significant—rather than nonzero—components is small. This naturally leads to the introduction of weak ℓ_p-spaces.

Definition 2.6. For $p > 0$, the weak ℓ_p space $w\ell_p^N$ denotes the space $\mathbb{C}^N$ equipped with the quasinorm

$$\|\mathbf{x}\|_{p,\infty} := \inf\left\{M \geq 0 : \mathrm{card}(\{j \in [N] : |x_j| \geq t\}) \leq \frac{M^p}{t^p} \text{ for all } t > 0\right\}.$$

To verify that the previous quantity indeed defines a quasinorm, we check, for any $\mathbf{x}, \mathbf{y} \in \mathbb{C}^N$ and any $\lambda \in \mathbb{C}$, that $\|\mathbf{x}\| = 0 \Rightarrow \mathbf{x} = 0$, $\|\lambda\mathbf{x}\| = |\lambda|\|\mathbf{x}\|$, and $\|\mathbf{x} + \mathbf{y}\|_{p,\infty} \leq 2^{\max\{1,1/p\}}(\|\mathbf{x}\|_{p,\infty} + \|\mathbf{y}\|_{p,\infty})$. The first two properties are easy, while the third property is a consequence of the more general statement below.

Proposition 2.7. *Let* $\mathbf{x}^1, \ldots, \mathbf{x}^k \in \mathbb{C}^N$. *Then, for* $p > 0$,

$$\|\mathbf{x}^1 + \cdots + \mathbf{x}^k\|_{p,\infty} \leq k^{\max\{1,1/p\}}(\|\mathbf{x}^1\|_{p,\infty} + \cdots + \|\mathbf{x}^k\|_{p,\infty}).$$

Proof. Let $t > 0$. If $|x_j^1 + \cdots + x_j^k| \geq t$ for some $j \in [N]$, then we have $|x_j^i| \geq t/k$ for some $i \in [k]$. This means that

$$\{j \in [N] : |x_j^1 + \cdots + x_j^k| \geq t\} \subset \bigcup_{i \in [k]} \{j \in [N] : |x_j^i| \geq t/k\} .$$

We derive

$$\text{card}(\{j \in [N] : |x_j^1 + \cdots + x_j^k| \geq t\}) \leq \sum_{i \in [k]} \frac{\|\mathbf{x}^i\|_{p,\infty}^p}{(t/k)^p}$$

$$= \frac{k^p(\|\mathbf{x}^1\|_{p,\infty}^p + \cdots + \|\mathbf{x}^k\|_{p,\infty}^p)}{t^p} .$$

According to the definition of the weak ℓ_p-quasinorm of $\mathbf{x}^1 + \cdots + \mathbf{x}^k$, we obtain

$$\|\mathbf{x}^1 + \cdots + \mathbf{x}^k\|_{p,\infty} \leq k\big(\|\mathbf{x}^1\|_{p,\infty}^p + \cdots + \|\mathbf{x}^k\|_{p,\infty}^p\big)^{1/p} .$$

Now, if $p \leq 1$, comparing the ℓ_p and ℓ_1 norms in $\mathbb{R}^k$ gives

$$\big(\|\mathbf{x}^1\|_{p,\infty}^p + \cdots + \|\mathbf{x}^k\|_{p,\infty}^p\big)^{1/p} \leq k^{1/p-1}(\|\mathbf{x}^1\|_{p,\infty} + \cdots + \|\mathbf{x}^k\|_{p,\infty}),$$

and if $p \geq 1$, comparing the ℓ_p and ℓ_1 norms in $\mathbb{R}^k$ gives

$$\big(\|\mathbf{x}^1\|_{p,\infty}^p + \cdots + \|\mathbf{x}^k\|_{p,\infty}^p\big)^{1/p} \leq \|\mathbf{x}^1\|_{p,\infty} + \cdots + \|\mathbf{x}^k\|_{p,\infty} .$$

The result immediately follows. □

Remark 2.8. The constant $k^{\max\{1,1/p\}}$ in Proposition 2.7 is sharp; see Exercise 2.2.

It is sometimes preferable to invoke the following alternative expression for the weak ℓ_p-quasinorm of a vector $\mathbf{x} \in \mathbb{C}^N$.

Proposition 2.9. *For $p > 0$, the weak ℓ_p-quasinorm of a vector $\mathbf{x} \in \mathbb{C}^N$ can be expressed as*

$$\|\mathbf{x}\|_{p,\infty} = \max_{k \in [N]} k^{1/p} x_k^* ,$$

where $\mathbf{x}^* \in \mathbb{R}_+^N$ denotes the nonincreasing rearrangement of $\mathbf{x} \in \mathbb{C}^N$.

Proof. Given $\mathbf{x} \in \mathbb{C}^N$, in view of $\|\mathbf{x}\|_{p,\infty} = \|\mathbf{x}^*\|_{p,\infty}$, we need to establish that $\|\mathbf{x}\| := \max_{k \in [N]} k^{1/p} x_k^*$ equals $\|\mathbf{x}^*\|_{p,\infty}$. For $t > 0$, we first note that either $\{j \in [N] : x_j^* \geq t\} = [k]$ for some $k \in [N]$ or $\{j \in [N] : x_j^* \geq t\} = \emptyset$. In the former case, $t \leq x_k^* \leq \|\mathbf{x}\|/k^{1/p}$, and hence, $\text{card}(\{j \in [N] : x_j^* \geq t\}) = k \leq \|\mathbf{x}\|^p/t^p$. This inequality holds trivially in the case that $\{j \in [N] : x_j^* \geq t\} = \emptyset$.

According to the definition of the weak ℓ_p-quasinorm, we obtain $\|\mathbf{x}^*\|_{p,\infty} \le \|\mathbf{x}\|$. Let us now suppose that $\|\mathbf{x}\| > \|\mathbf{x}^*\|_{p,\infty}$, so that $\|\mathbf{x}\| \ge (1+\epsilon)\|\mathbf{x}^*\|_{p,\infty}$ for some $\epsilon > 0$. This means that $k^{1/p}x_k^* \ge (1+\epsilon)\|\mathbf{x}^*\|_{p,\infty}$ for some $k \in [N]$. Therefore, the set

$$\{j \in [N] : x_j^* \ge (1+\epsilon)\|\mathbf{x}^*\|_{p,\infty}/k^{1/p}\}$$

contains the set $[k]$. The definition of the weak ℓ_p-quasinorm yields

$$k \le \frac{\|\mathbf{x}^*\|_{p,\infty}^p}{\left((1+\epsilon)\|\mathbf{x}^*\|_{p,\infty}/k^{1/p}\right)^p} = \frac{k}{(1+\epsilon)^p} ,$$

which is a contradiction. We conclude that $\|\mathbf{x}\| = \|\mathbf{x}^*\|_{p,\infty}$.   □

This alternative expression of the weak ℓ_p-quasinorm provides a slightly easier way to compare it to the ℓ_p-(quasi)norm, as follows.

Proposition 2.10. *For any $p > 0$ and any $\mathbf{x} \in \mathbb{C}^N$,*

$$\|\mathbf{x}\|_{p,\infty} \le \|\mathbf{x}\|_p .$$

Proof. For $k \in [N]$, we write

$$\|\mathbf{x}\|_p^p = \sum_{j=1}^{N}(x_j^*)^p \ge \sum_{j=1}^{k}(x_j^*)^p \ge k(x_k^*)^p .$$

Raising to the power $1/p$ and taking the maximum over k gives the result.   □

The alternative expression of the weak ℓ_p-quasinorm also enables us to easily establish a variation of Proposition 2.3 where weak ℓ_p replaces ℓ_p.

Proposition 2.11. *For any $q > p > 0$ and $\mathbf{x} \in \mathbb{C}^N$, the inequality*

$$\sigma_s(\mathbf{x})_q \le \frac{d_{p,q}}{s^{1/p-1/q}}\|\mathbf{x}\|_{p,\infty}$$

holds with

$$d_{p,q} := \left(\frac{p}{q-p}\right)^{1/q} .$$

Proof. We may assume without loss of generality that $\|\mathbf{x}\|_{p,\infty} \le 1$, so that $x_k^* \le 1/k^{1/p}$ for all $k \in [N]$. We then have

$$\sigma_s(\mathbf{x})_q^q = \sum_{k=s+1}^{N} (x_k^*)^q \leq \sum_{k=s+1}^{N} \frac{1}{k^{q/p}} \leq \int_s^N \frac{1}{t^{q/p}} dt = -\frac{1}{q/p-1} \frac{1}{t^{q/p-1}} \Big|_{t=s}^{t=N}$$

$$\leq \frac{p}{q-p} \frac{1}{s^{q/p-1}} .$$

Taking the power $1/q$ yields the desired result. □

Proposition 2.11 shows that vectors $\mathbf{x} \in \mathbb{C}^N$ which are compressible in the sense that $\|\mathbf{x}\|_{p,\infty} \leq 1$ for small $p > 0$ are also compressible in the sense that their errors of best s-term approximation decay quickly with s.

We close this section with a technical result on the nonincreasing rearrangement.

Lemma 2.12. *The nonincreasing rearrangement satisfies, for* $\mathbf{x}, \mathbf{z} \in \mathbb{C}^N$,

$$\|\mathbf{x}^* - \mathbf{z}^*\|_\infty \leq \|\mathbf{x} - \mathbf{z}\|_\infty . \tag{2.1}$$

Moreover, for $s \in [N]$,

$$|\sigma_s(\mathbf{x})_1 - \sigma_s(\mathbf{z})_1| \leq \|\mathbf{x} - \mathbf{z}\|_1 , \tag{2.2}$$

and for $k > s$,

$$(k-s)x_k^* \leq \|\mathbf{x} - \mathbf{z}\|_1 + \sigma_s(\mathbf{z})_1 . \tag{2.3}$$

Proof. For $j \in [N]$, the index set of j largest absolute entries of $\mathbf{x}$ intersects the index set of $N - j + 1$ smallest absolute entries of $\mathbf{z}$. Picking an index ℓ in this intersection, we obtain

$$x_j^* \leq |x_\ell| \leq |z_\ell| + \|\mathbf{x} - \mathbf{z}\|_\infty \leq z_j^* + \|\mathbf{x} - \mathbf{z}\|_\infty.$$

Reversing the roles of $\mathbf{x}$ and $\mathbf{z}$ shows (2.1).

Next, let $\mathbf{v} \in \mathbb{C}^N$ be a best s-term approximation to $\mathbf{z}$. Then

$$\sigma_s(\mathbf{x})_1 \leq \|\mathbf{x} - \mathbf{v}\|_1 \leq \|\mathbf{x} - \mathbf{z}\|_1 + \|\mathbf{z} - \mathbf{v}\|_1 = \|\mathbf{x} - \mathbf{z}\|_1 + \sigma_s(\mathbf{z})_1 ,$$

and again by symmetry this establishes (2.2). The inequality (2.3) follows from (2.2) by noting that

$$(k-s)x_k^* \leq \sum_{j=s+1}^{k} x_j^* \leq \sum_{j \geq s+1} x_j^* = \sigma_s(\mathbf{x})_1 .$$

This completes the proof. □

2.2 Minimal Number of Measurements

The compressive sensing problem consists in reconstructing an s-sparse vector $\mathbf{x} \in \mathbb{C}^N$ from

$$\mathbf{y} = \mathbf{A}\mathbf{x}$$

where $\mathbf{A} \in \mathbb{C}^{m \times N}$ is the so-called measurement matrix. With $m < N$, this system of linear equations is underdetermined, but the sparsity assumption hopefully helps in identifying the original vector $\mathbf{x}$.

In this section, we examine the question of the minimal number of linear measurements needed to reconstruct s-sparse vectors from these measurements, regardless of the practicality of the reconstruction scheme. This question can in fact take two meanings, depending on whether we require that the measurement scheme allows for the reconstruction of all s-sparse vectors $\mathbf{x} \in \mathbb{C}^N$ simultaneously or whether we require that, given an s-sparse vector $\mathbf{x} \in \mathbb{C}^N$, the measurement scheme allows for the reconstruction of this specific vector. While the second scenario seems to be unnatural at first sight because the vector $\mathbf{x}$ is unknown a priori, it will become important later when aiming at recovery guarantees when the matrix $\mathbf{A}$ is chosen at random and the sparse vector $\mathbf{x}$ is fixed (so-called nonuniform recovery guarantees).

The minimal number m of measurements depends on the setting considered, namely, it equals $2s$ in the first case and $s + 1$ in the second case. However, we will see in Chap. 11 that if we also require the reconstruction scheme to be stable (the meaning will be made precise later), then the minimal number of required measurements additionally involves a factor of $\ln(N/s)$, so that recovery will never be stable with only $2s$ measurements.

Before separating the two settings discussed above, it is worth pointing out the equivalence of the following properties for given sparsity s, matrix $\mathbf{A} \in \mathbb{C}^{m \times N}$, and s-sparse $\mathbf{x} \in \mathbb{C}^N$:

(a) The vector $\mathbf{x}$ is the unique s-sparse solution of $\mathbf{A}\mathbf{z} = \mathbf{y}$ with $\mathbf{y} = \mathbf{A}\mathbf{x}$, that is, $\{\mathbf{z} \in \mathbb{C}^N : \mathbf{A}\mathbf{z} = \mathbf{A}\mathbf{x}, \|\mathbf{z}\|_0 \leq s\} = \{\mathbf{x}\}$.

(b) The vector $\mathbf{x}$ can be reconstructed as the unique solution of

$$\underset{\mathbf{z} \in \mathbb{C}^N}{\text{minimize}} \ \|\mathbf{z}\|_0 \quad \text{subject to } \mathbf{A}\mathbf{z} = \mathbf{y}. \tag{P_0}$$

Indeed, if an s-sparse $\mathbf{x} \in \mathbb{C}^N$ is the unique s-sparse solution of $\mathbf{A}\mathbf{z} = \mathbf{y}$ with $\mathbf{y} = \mathbf{A}\mathbf{x}$, then a solution $\mathbf{x}^\sharp$ of (P_0) is s-sparse and satisfies $\mathbf{A}\mathbf{x}^\sharp = \mathbf{y}$, so that $\mathbf{x}^\sharp = \mathbf{x}$. This shows $(a) \Rightarrow (b)$. The implication $(b) \Rightarrow (a)$ is clear.

Recovery of All Sparse Vectors

Before stating the main result for this case, we observe that the uniqueness of sparse solutions of underdetermined linear systems can be reformulated in several ways.

For a matrix $\mathbf{A} \in \mathbb{C}^{m \times N}$ and a subset $S \subset [N]$, we use the notation $\mathbf{A}_S$ to indicate the column submatrix of $\mathbf{A}$ consisting of the columns indexed by S. Similarly, for $\mathbf{x} \in \mathbb{C}^N$ we denote by $\mathbf{x}_S$ either the subvector in $\mathbb{C}^S$ consisting of the entries indexed by S, that is, $(\mathbf{x}_S)_\ell = x_\ell$ for $\ell \in S$, or the vector in $\mathbb{C}^N$ which coincides with $\mathbf{x}$ on the entries in S and is zero on the entries outside S, that is,

$$(\mathbf{x}_S)_\ell = \begin{cases} x_\ell & \text{if } \ell \in S \,, \\ 0 & \text{if } \ell \notin S \,. \end{cases} \tag{2.4}$$

It should always be clear from the context which of the two options applies.

Theorem 2.13. *Given $\mathbf{A} \in \mathbb{C}^{m \times N}$, the following properties are equivalent:*

(a) *Every s-sparse vector $\mathbf{x} \in \mathbb{C}^N$ is the unique s-sparse solution of $\mathbf{A}\mathbf{z} = \mathbf{A}\mathbf{x}$, that is, if $\mathbf{A}\mathbf{x} = \mathbf{A}\mathbf{z}$ and both $\mathbf{x}$ and $\mathbf{z}$ are s-sparse, then $\mathbf{x} = \mathbf{z}$.*
(b) *The null space $\ker \mathbf{A}$ does not contain any $2s$-sparse vector other than the zero vector, that is, $\ker \mathbf{A} \cap \{\mathbf{z} \in \mathbb{C}^N : \|\mathbf{z}\|_0 \leq 2s\} = \{\mathbf{0}\}$.*
(c) *For every $S \subset [N]$ with $\operatorname{card}(S) \leq 2s$, the submatrix $\mathbf{A}_S$ is injective as a map from $\mathbb{C}^S$ to $\mathbb{C}^m$.*
(d) *Every set of $2s$ columns of $\mathbf{A}$ is linearly independent.*

Proof. (b)$\Rightarrow$(a) Let $\mathbf{x}$ and $\mathbf{z}$ be s-sparse with $\mathbf{A}\mathbf{x} = \mathbf{A}\mathbf{z}$. Then $\mathbf{x} - \mathbf{z}$ is $2s$-sparse and $\mathbf{A}(\mathbf{x} - \mathbf{z}) = \mathbf{0}$. If the kernel does not contain any $2s$-sparse vector different from the zero vector, then $\mathbf{x} = \mathbf{z}$.

(a)$\Rightarrow$(b) Conversely, assume that for every s-sparse vector $\mathbf{x} \in \mathbb{C}^N$, we have $\{\mathbf{z} \in \mathbb{C}^N : \mathbf{A}\mathbf{z} = \mathbf{A}\mathbf{x}, \|\mathbf{z}\|_0 \leq s\} = \{\mathbf{x}\}$. Let $\mathbf{v} \in \ker \mathbf{A}$ be $2s$-sparse. We can write $\mathbf{v} = \mathbf{x} - \mathbf{z}$ for s-sparse vectors $\mathbf{x}, \mathbf{z}$ with $\operatorname{supp} \mathbf{x} \cap \operatorname{supp} \mathbf{z} = \emptyset$. Then $\mathbf{A}\mathbf{x} = \mathbf{A}\mathbf{z}$ and by assumption $\mathbf{x} = \mathbf{z}$. Since the supports of $\mathbf{x}$ and $\mathbf{z}$ are disjoint, it follows that $\mathbf{x} = \mathbf{z} = \mathbf{0}$ and $\mathbf{v} = \mathbf{0}$.

For the equivalence of (b), (c), and (d), we observe that for a $2s$-sparse vector $\mathbf{v}$ with $S = \operatorname{supp} \mathbf{v}$, we have $\mathbf{A}\mathbf{v} = \mathbf{A}_S \mathbf{v}_S$. Noting that $S = \operatorname{supp} \mathbf{v}$ ranges through all possible subsets of $[N]$ of cardinality $\operatorname{card}(S) \leq 2s$ when $\mathbf{v}$ ranges through all possible $2s$-sparse vectors completes the proof by basic linear algebra. $\qquad\square$

We observe, in particular, that if it is possible to reconstruct every s-sparse vector $\mathbf{x} \in \mathbb{C}^N$ from the knowledge of its measurement vector $\mathbf{y} = \mathbf{A}\mathbf{x} \in \mathbb{C}^m$, then (a) holds and consequently so does (d). This implies $\operatorname{rank}(\mathbf{A}) \geq 2s$. We also have $\operatorname{rank}(\mathbf{A}) \leq m$, because the rank is at most equal to the number of rows. Therefore, the number of measurements needed to reconstruct every s-sparse vector always satisfies

$$m \geq 2s.$$

We are now going to see that $m = 2s$ measurements suffice to reconstruct every s-sparse vector—at least in theory.

Theorem 2.14. *For any integer* $N \geq 2s$, *there exists a measurement matrix* $\mathbf{A} \in \mathbb{C}^{m \times N}$ *with* $m = 2s$ *rows such that every s-sparse vector* $\mathbf{x} \in \mathbb{C}^N$ *can be recovered from its measurement vector* $\mathbf{y} = \mathbf{Ax} \in \mathbb{C}^m$ *as a solution of* (P$_0$).

Proof. Let us fix $t_N > \cdots > t_2 > t_1 > 0$ and consider the matrix $\mathbf{A} \in \mathbb{C}^{m \times N}$ with $m = 2s$ defined by

$$\mathbf{A} = \begin{bmatrix} 1 & 1 & \cdots & 1 \\ t_1 & t_2 & \cdots & t_N \\ \vdots & \vdots & \cdots & \vdots \\ t_1^{2s-1} & t_2^{2s-1} & \cdots & t_N^{2s-1} \end{bmatrix}. \tag{2.5}$$

Let $S = \{j_1 < \cdots < j_{2s}\}$ be an index set of cardinality $2s$. The square matrix $\mathbf{A}_S \in \mathbb{C}^{2s \times 2s}$ is (the transpose of) a *Vandermonde matrix*. Theorem A.24 yields

$$\det(\mathbf{A}_S) = \begin{vmatrix} 1 & 1 & \cdots & 1 \\ t_{j_1} & t_{j_2} & \cdots & t_{j_{2s}} \\ \vdots & \vdots & \cdots & \vdots \\ t_{j_1}^{2s-1} & t_{j_2}^{2s-1} & \cdots & t_{j_{2s}}^{2s-1} \end{vmatrix} = \prod_{k < \ell} (t_{j_\ell} - t_{j_k}) > 0.$$

This shows that $\mathbf{A}_S$ is invertible, in particular injective. Since the condition (c) of Theorem 2.13 is fulfilled, every s-sparse vector $\mathbf{x} \in \mathbb{C}^N$ is the unique s-sparse vector satisfying $\mathbf{Az} = \mathbf{Ax}$, so it can be recovered as the unique solution of (P$_0$).
$\square$

Many other matrices meet the condition (c) of Theorem 2.13. As an example, the integer powers of $t_1, \ldots, t_N$ in the matrix of (2.5) do not need to be the consecutive integers $0, 1, \ldots, 2s - 1$. Instead of the $N \times N$ Vandermonde matrix associated with $t_N > \cdots > t_1 > 0$, we can start with any matrix $\mathbf{M} \in \mathbb{R}^{N \times N}$ that is *totally positive*, i.e., that satisfies $\det \mathbf{M}_{I,J} > 0$ for any sets $I, J \subset [N]$ of the same cardinality, where $\mathbf{M}_{I,J}$ represents the submatrix of $\mathbf{M}$ with rows indexed by I and columns indexed by J. We then select any $m = 2s$ rows of $\mathbf{M}$, indexed by a set I, say, to form the matrix $\mathbf{A}$. For an index $S \subset [N]$ of cardinality $2s$, the matrix $\mathbf{A}_S$ reduces to $\mathbf{M}_{I,S}$; hence, it is invertible. As another example, the numbers $t_N, \ldots, t_1$ do not need to be positive nor real, as long as $\det(\mathbf{A}_S) \neq 0$ instead of $\det(\mathbf{A}_S) > 0$. In particular, with $t_\ell = e^{2\pi i (\ell-1)/N}$ for $\ell \in [N]$, Theorem A.24 guarantees that the (rescaled) partial Fourier matrix

$$\mathbf{A} = \begin{bmatrix} 1 & 1 & 1 & \cdots & 1 \\ 1 & e^{2\pi i/N} & e^{2\pi i 2/N} & \cdots & e^{2\pi i(N-1)/N} \\ \vdots & \vdots & \vdots & \vdots & \vdots \\ 1 & e^{2\pi i(2s-1)/N} & e^{2\pi i(2s-1)2/N} & \cdots & e^{2\pi i(2s-1)(N-1)/N} \end{bmatrix}$$

allows for the reconstruction of every s-sparse vector $\mathbf{x} \in \mathbb{C}^N$ from $\mathbf{y} = \mathbf{A}\mathbf{x} \in \mathbb{C}^{2s}$. In fact, an argument similar to the one we will use for Theorem 2.16 below shows that the set of $(2s) \times N$ matrices such that $\det(\mathbf{A}_S) = 0$ for some $S \subset [N]$ with $\mathrm{card}(S) \leq 2s$ has Lebesgue measure zero; hence, most $(2s) \times N$ matrices allow the reconstruction of every s-sparse vector $\mathbf{x} \in \mathbb{C}^N$ from $\mathbf{y} = \mathbf{A}\mathbf{x} \in \mathbb{C}^{2s}$. In general, the reconstruction procedure consisting of solving (P$_0$) is not feasible in practice, as will be shown in Sect. 2.3. However, in the case of Fourier measurements, a better reconstruction scheme based on the Prony method can be used.

Theorem 2.15. *For any $N \geq 2s$, there exists a practical procedure for the reconstruction of every $2s$-sparse vector from its first $m = 2s$ discrete Fourier measurements.*

Proof. Let $\mathbf{x} \in \mathbb{C}^N$ be an s-sparse vector, which we interpret as a function x from $\{0, 1, \ldots, N-1\}$ into $\mathbb{C}$ supported on an index set $S \subset \{0, 1, \ldots, N-1\}$ of size s. We suppose that this vector is observed via its first $2s$ discrete Fourier coefficients $\hat{x}(0), \ldots, \hat{x}(2s-1)$, where

$$\hat{x}(j) := \sum_{k=0}^{N-1} x(k) e^{-2\pi i j k / N}, \qquad 0 \leq j \leq N-1.$$

We consider the trigonometric polynomial of degree s defined by

$$p(t) := \frac{1}{N} \prod_{k \in S} \left(1 - e^{-2\pi i k / N} e^{2\pi i t / N}\right).$$

This polynomial vanishes exactly for $t \in S$, so we aim at finding the unknown set S by determining p or equivalently its Fourier transform $\hat{p}$. We note that, since x vanishes on the complementary set $\overline{S}$ of S in $\{0, 1, \ldots, N-1\}$, we have $p(t)x(t) = 0$ for all $0 \leq t \leq N-1$. By discrete convolution, we obtain $\hat{p} * \hat{x} = \widehat{p \cdot x} = 0$, that is to say,

$$(\hat{p} * \hat{x})(j) := \sum_{k=0}^{N-1} \hat{p}(k) \cdot \hat{x}(j - k \mod N) = 0 \qquad \text{for all } 0 \leq j \leq N-1. \quad (2.6)$$

We also note that, since $\frac{1}{N}\hat{p}(k)$ is the coefficient of $p(t)$ on the monomial $e^{2\pi i k t / N}$ and since p has degree s, we have $\hat{p}(0) = 1$ and $\hat{p}(k) = 0$ for all $k > s$. It remains to determine the s discrete Fourier coefficients $\hat{p}(1), \ldots, \hat{p}(s)$. For this purpose, we write the s equations (2.6) in the range $s \leq j \leq 2s - 1$ in the form

$$
\begin{aligned}
\hat{x}(s) &+ \hat{p}(1)\hat{x}(s-1) + \cdots + \hat{p}(s)\hat{x}(0) &= 0, \\
\hat{x}(s+1) &+ \hat{p}(1)\hat{x}(s) + \cdots + \hat{p}(s)\hat{x}(1) &= 0, \\
&\;\;\vdots \qquad\qquad \vdots \qquad\quad \ddots \qquad \vdots &\vdots \\
\hat{x}(2s-1) &+ \hat{p}(1)\hat{x}(2s-2) + \cdots + \hat{p}(s)\hat{x}(s-1) &= 0.
\end{aligned}
$$

This translates into the system

$$\begin{bmatrix} \hat{x}(s-1) & \hat{x}(s-2) & \cdots & \hat{x}(0) \\ \hat{x}(s) & \hat{x}(s-1) & \cdots & \hat{x}(1) \\ \vdots & \vdots & \ddots & \vdots \\ \hat{x}(2s-2) & \hat{x}(2s-3) & \cdots & \hat{x}(s-1) \end{bmatrix} \begin{bmatrix} \hat{p}(1) \\ \hat{p}(2) \\ \vdots \\ \hat{p}(s) \end{bmatrix} = - \begin{bmatrix} \hat{x}(s) \\ \hat{x}(s+1) \\ \vdots \\ \hat{x}(2s-1) \end{bmatrix}.$$

Because $\hat{x}(0), \ldots, \hat{x}(2s-1)$ are known, we solve for $\hat{p}(1), \ldots, \hat{p}(s)$. Since the Toeplitz matrix above is not always invertible—take, e.g., $x = [1, 0, \ldots, 0]^\top$, so that $\hat{x} = [1, 1, \ldots, 1]^\top$—we obtain a solution $\hat{q}(1), \ldots, \hat{q}(s)$ not guaranteed to be $\hat{p}(1), \ldots, \hat{p}(s)$. Appending the values $\hat{q}(0) = 1$ and $\hat{q}(k) = 0$ for all $k > s$, the linear system reads

$$(\hat{q} * \hat{x})(j) = 0 \qquad \text{for all } s \leq j \leq 2s - 1.$$

Therefore, the s-sparse vector $q \cdot x$ has a Fourier transform $\widehat{q \cdot x} = \hat{q} * \hat{x}$ vanishing on a set of s consecutive indices. Writing this in matrix form and using Theorem A.24, we derive that $q \cdot x = 0$, so that the trigonometric polynomial q vanishes on S. Since the degree of q is at most s, the set of zeros of q coincide with the set S, which can thus be found by solving a polynomial equation—or simply by identifying the s smallest values of $|q(j)|$, $0 \leq j \leq N - 1$. Finally, the values of $x(j)$, $j \in S$, are obtained by solving the overdetermined system of $2s$ linear equations imposed by the knowledge of $\hat{x}(0), \ldots, \hat{x}(2s-1)$.   □

Despite its appeal, the reconstruction procedure just described hides some important drawbacks. Namely, it is not stable with respect to sparsity defects nor is it robust with respect to measurement errors. The reader is invited to verify this statement numerically in Exercise 2.8. In fact, we will prove in Chap. 11 that any stable scheme for s-sparse reconstruction requires at least $m \approx c \, s \ln(eN/s)$ linear measurements, where $c > 0$ is a constant depending on the stability requirement.

Recovery of Individual Sparse Vectors

In the next setting, the s-sparse vector $\mathbf{x} \in \mathbb{C}^N$ is fixed before the measurement matrix $\mathbf{A} \in \mathbb{C}^{m \times N}$ is chosen. The conditions for the vector $\mathbf{x}$ to be the unique s-sparse vector consistent with the measurements depend on $\mathbf{A}$ as well as on $\mathbf{x}$ itself. While this seems unnatural at first sight because $\mathbf{x}$ is unknown a priori, the philosophy is that the conditions will be met for *most* $(s + 1) \times N$ matrices. This setting is relevant since the measurement matrices are often chosen at random.

Theorem 2.16. *For any $N \geq s + 1$, given an s-sparse vector $\mathbf{x} \in \mathbb{C}^N$, there exists a measurement matrix $\mathbf{A} \in \mathbb{C}^{m \times N}$ with $m = s + 1$ rows such that the vector $\mathbf{x}$ can be reconstructed from its measurement vector $\mathbf{y} = \mathbf{A}\mathbf{x} \in \mathbb{C}^m$ as a solution of* (P_0).

Proof. Let $\mathbf{A} \in \mathbb{C}^{(s+1) \times N}$ be a matrix for which the s-sparse vector $\mathbf{x}$ cannot be recovered from $\mathbf{y} = \mathbf{A}\mathbf{x}$ (via ℓ_0-minimization). This means that there exists a vector $\mathbf{z} \in \mathbb{C}^N$ distinct from $\mathbf{x}$, supported on a set $S = \text{supp}(\mathbf{z}) = \{j_1, \ldots, j_s\}$ of size at most s (if $\|\mathbf{z}\|_0 < s$, we fill up S with arbitrary elements $j_\ell \in [N]$), such that $\mathbf{A}\mathbf{z} = \mathbf{A}\mathbf{x}$. If $\text{supp}(\mathbf{x}) \subset S$, then the equality $(\mathbf{A}(\mathbf{z} - \mathbf{x}))_{[s]} = 0$ shows that the square matrix $\mathbf{A}_{[s],S}$ is noninvertible; hence,

$$f(a_{1,1}, \ldots, a_{1,N}, \ldots, a_{m,1}, \ldots, a_{m,N}) := \det(A_{[s],S}) = 0.$$

If $\text{supp}(\mathbf{x}) \not\subset S$, then the space $V := \{\mathbf{u} \in \mathbb{C}^N : \text{supp}(\mathbf{u}) \subset S\} + \mathbb{C}\mathbf{x}$ has dimension $s + 1$, and the linear map $G : V \to \mathbb{C}^{s+1}, \mathbf{v} \mapsto \mathbf{A}\mathbf{v}$ is noninvertible, since $G(\mathbf{z} - \mathbf{x}) = 0$. The matrix of the linear map G in the basis $(\mathbf{e}_{j_1}, \ldots, \mathbf{e}_{j_s}, \mathbf{x})$ of V takes the form

$$B_{\mathbf{x},S} := \begin{bmatrix} a_{1,j_1} & \cdots & a_{1,j_s} & \sum_{j \in \text{supp}(\mathbf{x})} x_j a_{1,j} \\ \vdots & \ddots & \vdots & \vdots \\ a_{s+1,j_1} & \cdots & a_{s+1,j_s} & \sum_{j \in \text{supp}(\mathbf{x})} x_j a_{s+1,j} \end{bmatrix},$$

and we have

$$g_S(a_{1,1}, \ldots, a_{1,N}, \ldots, a_{m,1}, \ldots, a_{m,N}) := \det(B_{\mathbf{x},S}) = 0.$$

This shows that the entries of the matrix $\mathbf{A}$ satisfy

$$(a_{1,1}, \ldots, a_{1,N}, \ldots, a_{m,1}, \ldots, a_{m,N}) \in f^{-1}(\{0\}) \cup \bigcup_{\text{card}(S)=s} g_S^{-1}(\{0\}).$$

But since f and all g_S, $\text{card}(S) = s$, are nonzero polynomial functions of the variables $(a_{1,1}, \ldots, a_{1,N}, \ldots, a_{m,1}, \ldots, a_{m,N})$, the sets $f^{-1}(\{0\})$ and $g_S^{-1}(\{0\})$, $\text{card}(S) = s$, have Lebesgue measure zero and so does their union. It remains to choose the entries of the matrix $\mathbf{A}$ outside of this union of measure zero to ensure that the vector $\mathbf{x}$ can be recovered from $\mathbf{y} = \mathbf{A}\mathbf{x}$. $\qquad\square$

2.3 NP-Hardness of ℓ_0-Minimization

As mentioned in Sect. 2.2, reconstructing an s-sparse vector $\mathbf{x} \in \mathbb{C}^N$ from its measurement vector $\mathbf{y} \in \mathbb{C}^m$ amounts to solving the ℓ_0-minimization problem

$$\underset{\mathbf{z} \in \mathbb{C}^N}{\text{minimize}} \ \|\mathbf{z}\|_0 \quad \text{subject to } \mathbf{A}\mathbf{z} = \mathbf{y}. \tag{P_0}$$

Since a minimizer has sparsity at most s, the straightforward approach for finding it consists in solving every rectangular system $\mathbf{A}_S \mathbf{u} = \mathbf{y}$, or rather every square system $\mathbf{A}_S^* \mathbf{A}_S \mathbf{u} = \mathbf{A}_S^* \mathbf{y}$, for $\mathbf{u} \in \mathbb{C}^S$ where S runs through all the possible subsets of $[N]$ with size s. However, since the number $\binom{N}{s}$ of these subsets is prohibitively large, such a straightforward approach is completely unpractical. By way of illustration, for small problem sizes $N = 1000$ and $s = 10$, we would have to solve $\binom{1000}{10} \geq \left(\frac{1000}{10}\right)^{10} = 10^{20}$ linear systems of size 10×10. Even if each such system could be solved in 10^{-10} seconds, the time required to solve (P$_0$) with this approach would still be 10^{10} seconds, i.e., more than 300 years. We are going to show that solving (P$_0$) in fact is intractable for any possible approach. Precisely, for any fixed $\eta \geq 0$, we are going to show that the more general problem

$$\underset{\mathbf{z} \in \mathbb{C}^N}{\text{minimize}} \, \|\mathbf{z}\|_0 \quad \text{subject to } \|\mathbf{A}\mathbf{z} - \mathbf{y}\|_2 \leq \eta \qquad \qquad (\text{P}_{0,\eta})$$

is NP-hard.

We start by introducing the necessary terminology from computational complexity. First, a polynomial-time algorithm is an algorithm performing its task in a number of steps bounded by a polynomial expression in the size of the input. Next, let us describe in a rather informal way a few classes of decision problems:

- The class $\mathfrak{P}$ of P-problems consists of all decision problems for which there exists a polynomial-time algorithm finding a solution.
- The class $\mathfrak{NP}$ of NP-problems consists of all decision problems for which there exists a polynomial-time algorithm certifying a solution. Note that the class $\mathfrak{P}$ is clearly contained in the class $\mathfrak{NP}$.
- The class $\mathfrak{NP}$-hard of NP-hard problems consist of all problems (not necessarily decision problems) for which a solving algorithm could be transformed in polynomial time into a solving algorithm for any NP-problem. Roughly speaking, this is the class of problems at least as hard as any NP-problem. Note that the class $\mathfrak{NP}$-hard is not contained in the class $\mathfrak{NP}$.
- The class $\mathfrak{NP}$-complete of NP-complete problems consist of all problems that are both NP and NP-hard; in other words, it consists of all the NP-problems at least as hard as any other NP-problem.

The situation can be summarized visually as in Fig. 2.1. It is a common belief that $\mathfrak{P}$ is strictly contained in $\mathfrak{NP}$, that is to say, that there are problems for which potential solutions can be certified, but for which a solution cannot be found in polynomial time. However, this remains a major open question to this day. There is a vast catalog of NP-complete problems, the most famous of which being perhaps the traveling salesman problem. The one we are going to use is *exact cover by 3-sets*.

Exact cover by 3-sets problem

Given a collection $\{\mathcal{C}_i, i \in [N]\}$ of 3-element subsets of $[m]$, does there exist an exact cover (a partition) of $[m]$, i.e., a set $J \subset [N]$ such that $\cup_{j \in J} \mathcal{C}_j = [m]$ and $\mathcal{C}_j \cap \mathcal{C}_{j'} = \emptyset$ for all $j, j' \in J$ with $j \neq j'$?

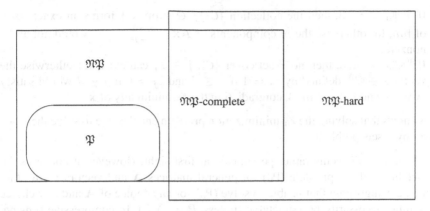

Fig. 2.1 Schematic representation of P, NP, NP-complete, and NP-hard problems

Taking for granted that this problem is NP-complete, we can now prove the main result of this section.

Theorem 2.17. *For any $\eta \geq 0$, the ℓ_0-minimization problem* $(\mathrm{P}_{0,\eta})$ *for general* $\mathbf{A} \in \mathbb{C}^{m \times N}$ *and* $\mathbf{y} \in \mathbb{C}^m$ *is NP-hard.*

Proof. By rescaling, we may and do assume that $\eta < 1$. According to the previous considerations, it is enough to show that the exact cover by 3-sets problem can be reduced in polynomial time to the ℓ_0-minimization problem. Let then $\{\mathcal{C}_i, i \in [N]\}$ be a collection of 3-element subsets of $[m]$. We define vectors $\mathbf{a}_1, \mathbf{a}_2, \ldots, \mathbf{a}_N \in \mathbb{C}^m$ by

$$(\mathbf{a}_i)_j = \begin{cases} 1 & \text{if } j \in \mathcal{C}_i, \\ 0 & \text{if } j \notin \mathcal{C}_i. \end{cases}$$

We then define a matrix $\mathbf{A} \in \mathbb{C}^{m \times N}$ and a vector $\mathbf{y} \in \mathbb{C}^m$ by

$$\mathbf{A} = \left[\mathbf{a}_1 \middle| \mathbf{a}_2 \middle| \cdots \middle| \mathbf{a}_N \right], \qquad \mathbf{y} = [1, 1, \ldots, 1]^\top.$$

Since $N \leq \binom{m}{3}$, this construction can be done in polynomial time. If a vector $\mathbf{z} \in \mathbb{C}^N$ obeys $\|\mathbf{A}\mathbf{z} - \mathbf{y}\|_2 \leq \eta$, then all the m components of the vector $\mathbf{A}\mathbf{z}$ are distant to 1 by at most η, so they are nonzero and $\|\mathbf{A}\mathbf{z}\|_0 = m$. But since each vector $\mathbf{a}_i$ has exactly 3 nonzero components, the vector $\mathbf{A}\mathbf{z} = \sum_{j=1}^N z_j \mathbf{a}_j$ has at most $3\|\mathbf{z}\|_0$ nonzero components, $\|\mathbf{A}\mathbf{z}\|_0 \leq 3\|\mathbf{z}\|_0$. Therefore, a vector $\mathbf{z} \in \mathbb{C}^N$ obeying $\|\mathbf{A}\mathbf{z} - \mathbf{y}\|_2 \leq \eta$ must satisfy $\|\mathbf{z}\|_0 \geq m/3$. Let us now run the ℓ_0-minimization problem, and let $\mathbf{x} \in \mathbb{C}^N$ denote the output. We separate two cases:

1. If $\|\mathbf{x}\|_0 = m/3$, then the collection $\{\mathcal{C}_j, j \in \text{supp}(\mathbf{x})\}$ forms an exact cover of $[m]$, for otherwise the m components of $\mathbf{Ax} = \sum_{j=1}^{N} x_j \mathbf{a}_j$ would not all be nonzero.

2. If $\|\mathbf{x}\|_0 > m/3$, then no exact cover $\{\mathcal{C}_j, j \in J\}$ can exist, for otherwise the vector $\mathbf{z} \in \mathbb{C}^N$ defined by $z_j = 1$ if $j \in J$ and $z_j = 0$ if $j \notin J$ would satisfy $\mathbf{Az} = \mathbf{y}$ and $\|\mathbf{z}\|_0 = m/3$, contradicting the ℓ_0-minimality of $\mathbf{x}$.

This shows that solving the ℓ_0-minimization problem enables one to solve the exact cover by 3-sets problem.   □

Theorem 2.17 seems rather pessimistic at first sight. However, it concerns the intractability of the problem (P$_0$) for general matrices $\mathbf{A}$ and vectors $\mathbf{y}$. In other words, any algorithm that is able to solve (P$_0$) for *any* choice of $\mathbf{A}$ and *any* choice of $\mathbf{y}$ must necessarily be intractable (unless $P = NP$). In compressive sensing, we will rather consider special choices of $\mathbf{A}$ and choose $\mathbf{y} = \mathbf{Ax}$ for some sparse $\mathbf{x}$. We will see that a variety of tractable algorithms will then provably recover $\mathbf{x}$ from $\mathbf{y}$ and thereby solve (P$_0$) for such specifically designed matrices $\mathbf{A}$. However, to emphasize this point once more, such algorithms will *not* successfully solve the ℓ_0-minimization problem for *all* possible choices of $\mathbf{A}$ and $\mathbf{y}$ due to NP-hardness. A selection of tractable algorithms is introduced in the coming chapter.

Notes

Proposition 2.3 is an observation due to Stechkin. In the case $p = 1$ and $q = 2$, the optimal constant $c_{1,2} = 1/2$ was obtained by Gilbert, Strauss, Tropp, and Vershynin in [225]. Theorem 2.5 with optimal constants $c_{p,q}$ for all $q > p > 0$ is a particular instance of a more general result, which also contains the *shifting inequality* of Exercise 6.15; see [209].

The weak ℓ_p-spaces are weak L_p-spaces for purely atomic measures. The weak L_p-spaces are also denoted $L_{p,\infty}$ and generalize to Lorentz spaces $L_{p,q}$ [284]. Thus, weak ℓ_p-spaces are a particular instance of more general spaces equipped with the quasinorm

$$\|\mathbf{x}\|_{p,q} = \left(\sum_{k=1}^{N} k^{q/p-1} (x_k^*)^q \right)^{1/q}.$$

The result of Theorem 2.16 is due to Wakin in [503]. Theorem 2.13 can be found in the article by Cohen, Dahmen, and DeVore [123]. One can also add an equivalent proposition expressed in terms of *spark* or in terms of *Kruskal rank*. The spark $\text{sp}(\mathbf{A})$ of a matrix $\mathbf{A}$ was defined by Donoho and Elad in [155] as the minimal size of a linearly dependent set of columns of $\mathbf{A}$. It is related to the Kruskal rank $\text{kr}(\mathbf{A})$ of $\mathbf{A}$, defined in [313] as the maximal integer k such that any k columns of $\mathbf{A}$ are linearly independent, via $\mathrm{sp}(\mathbf{A}) = \mathrm{kr}(\mathbf{A}) + 1$. Thus, according to Theorem 2.13, every s-sparse vector $\mathbf{x} \in \mathbb{C}^N$ is the unique s-sparse solution of $\mathbf{Az} = \mathbf{Ax}$ if and only if $\mathrm{kr}(\mathbf{A}) \geq 2s$ or if $\mathrm{sp}(\mathbf{A}) > 2s$.

Totally positive matrices were extensively studied by Karlin in [298]. One can also consult the more recent book [389] by Pinkus.

The reconstruction procedure of Theorem 2.15 based on a discrete version of the Prony method was known long before the development of compressive sensing. It is also related to Reed–Solomon decoding [52, 232]. The general Prony method [402] is designed for recovering a nonharmonic Fourier series of the form

$$f(t) = \sum_{k=1}^{s} x_k e^{2\pi i \omega_k t}$$

from equidistant samples $f(0), f(k/\alpha), f(2k/\alpha), \ldots, f(2s/\alpha)$. Here both the $\omega_k \in \mathbb{R}$ and the x_k are unknown. First the ω_k are found by solving an eigenvalue problem for a Hankel matrix associated to the samples of f. In the second step, the x_k are found by solving a linear system of equations. The difference to the method of Theorem 2.15 is due to the fact that the ω_k are not assumed to lie on a grid anymore. We refer to [344, 357] for more details. The Prony method has the disadvantage of being unstable. Several approaches have been proposed to stabilize it [14, 15, 45, 46, 401], although there seems to be a limit of how stable it can get when the number s of terms gets larger. The recovery methods in the so-called theory of *finite rate of innovation* are also related to the Prony method [55].

For an introduction to computational complexity, one can consult [19]. The NP-hardness of the ℓ_0-minimization problem was proved by Natarajan in [359]. It was later proved by Ge, Jiang, and Ye in [220] that the ℓ_p-minimization problem is NP-hard also for any $p < 1$; see Exercise 2.10.

Exercises

2.1. For $0 < p < 1$, prove that the pth power of the ℓ_p-quasinorm satisfies the triangle inequality

$$\|\mathbf{x} + \mathbf{y}\|_p^p \leq \|\mathbf{x}\|_p^p + \|\mathbf{y}\|_p^p, \qquad \mathbf{x}, \mathbf{y} \in \mathbb{C}^N.$$

For $0 < p < \infty$, deduce the inequality

$$\|\mathbf{x}_1 + \cdots + \mathbf{x}_k\|_p \leq k^{\max\{0, 1/p-1\}} \big(\|\mathbf{x}_1\|_p + \cdots + \|\mathbf{x}_k\|_p\big), \qquad \mathbf{x}_1, \ldots, \mathbf{x}_k \in \mathbb{C}^N.$$

2.2. Show that the constant $k^{\max\{1, 1/p\}}$ in Proposition 2.7 is sharp.

2.3. If $\mathbf{u}, \mathbf{v} \in \mathbb{C}^N$ are disjointly supported, prove that

$$\max(\|\mathbf{u}\|_{1,\infty}, \|\mathbf{v}\|_{1,\infty}) \leq \|\mathbf{u} + \mathbf{v}\|_{1,\infty} \leq \|\mathbf{u}\|_{1,\infty} + \|\mathbf{v}\|_{1,\infty}$$

and show that these inequalities are sharp.

2.4. As a converse to Proposition 2.10, prove that for any $p > 0$ and any $\mathbf{x} \in \mathbb{C}^N$,

$$\|\mathbf{x}\|_p \leq \ln(eN)^{1/p} \|\mathbf{x}\|_{p,\infty}.$$

2.5. Given $q > p > 0$ and $\mathbf{x} \in \mathbb{C}^N$, modify the proof of Proposition 2.3 to obtain

$$\sigma_s(\mathbf{x})_q \leq \frac{1}{s^{1/p-1/q}} \|\mathbf{x}\|_{p,\infty}^{1-p/q} \|\mathbf{x}\|_p^{p/q}.$$

2.6. Let $(B_0^n, B_1^n, \ldots, B_n^n)$ be the *Bernstein polynomials* of degree n defined by

$$B_i^n(x) := \binom{n}{i} x^i (1-x)^{n-i}.$$

For $0 < x_0 < x_1 < \cdots < x_n < 1$, prove that the matrix $[B_i^n(x_j)]_{i,j=0}^n$ is totally positive.

2.7. Prove that the product of two totally positive matrices is totally positive.

2.8. Implement the reconstruction procedure based on $2s$ discrete Fourier measurements as described in Sect. 2.2. Test it on a few random examples. Then pass from sparse to compressible vectors $\mathbf{x}$ having small $\sigma_s(\mathbf{x})_1$ and test on perturbed measurements $\mathbf{y} = \mathbf{A}\mathbf{x} + \mathbf{e}$ with small $\|\mathbf{e}\|_2$.

2.9. Let us assume that the vectors $\mathbf{x} \in \mathbb{R}^N$ are no longer observed via linear measurements $\mathbf{y} = \mathbf{A}\mathbf{x} \in \mathbb{R}^m$, but rather via measurements $\mathbf{y} = f(\mathbf{x})$, where $f : \mathbb{R}^N \to \mathbb{R}^m$ is a continuous map satisfying $f(-\mathbf{x}) = -f(\mathbf{x})$ for all $\mathbf{x} \in \mathbb{R}^N$. Prove that the minimal number of measurements needed to reconstruct every s-sparse vector equals $2s$. You may use the *Borsuk–Ulam theorem*:
If a continuous map F from the sphere S^n—relative to an arbitrary norm—of $\mathbb{R}^{n+1}$ into $\mathbb{R}^n$ is antipodal, i.e.,

$$F(-\mathbf{x}) = -F(\mathbf{x}) \qquad \text{for all } \mathbf{x} \in S^n,$$

then it vanishes at least once, i.e.,

$$F(\mathbf{x}) = \mathbf{0} \qquad \text{for some } \mathbf{x} \in S^n.$$

2.10. *NP-Hardness of ℓ_p-minimization for $0 < p < 1$*
Given $\mathbf{A} \in \mathbb{C}^{m \times N}$ and $\mathbf{y} \in \mathbb{C}^m$, the ℓ_p-minimization problem consists in computing a vector $\mathbf{x} \in \mathbb{C}^N$ with minimal ℓ_p-quasinorm subject to $\mathbf{A}\mathbf{x} = \mathbf{y}$.

The *partition problem* consists, given integers $a_1, \ldots, a_n$, in deciding whether there exist two sets $I, J \subset [n]$ such that $I \cap J = \emptyset$, $I \cup J = [n]$, and $\sum_{i \in I} a_i = \sum_{j \in J} a_j$. Assuming the NP-completeness of the partition problem, prove that the ℓ_p-minimization problem is NP-hard. It will be helpful to introduce the matrix $\mathbf{A}$ and the vector $\mathbf{y}$ defined by

$$
\mathbf{A} := \begin{bmatrix} a_1 & a_2 & \cdots & a_n & -a_1 & -a_2 & \cdots & -a_n \\ 1 & 0 & \cdots & 0 & 1 & 0 & \cdots & 0 \\ 0 & 1 & \cdots & 0 & 0 & 1 & \cdots & 0 \\ \vdots & & \ddots & 0 & \vdots & & \ddots & 0 \\ 0 & \cdots & 0 & 1 & 0 & \cdots & 0 & 1 \end{bmatrix} \quad \text{and} \quad \mathbf{y} = [0, 1, 1, \ldots, 1]^{\top}.
$$

2.11. *NP-Hardness of rank minimization*
Show that the rank-minimization problem

$$
\underset{\mathbf{Z} \in \mathbb{C}^{n_1 \times n_2}}{\text{minimize}} \ \text{rank}(\mathbf{Z}) \quad \text{subject to} \ \mathcal{A}(\mathbf{Z}) = \mathbf{y} \ .
$$

is NP-hard on the set of linear measurement maps $\mathcal{A} : \mathbb{C}^{n_1 \times n_2} \to \mathbb{C}^m$ and vectors $\mathbf{y} \in \mathbb{C}^m$.

Chapter 3
Basic Algorithms

In this chapter, a selection of popular algorithms used in compressive sensing is presented. The algorithms are divided into three categories: optimization methods, greedy methods, and thresholding-based methods. Their rigorous analyses are postponed until later, when appropriate tools such as coherence and restricted isometry constants become available. Only intuitive justification is given for now.

3.1 Optimization Methods

An *optimization problem* is a problem of the type

$$\underset{\mathbf{x}\in\mathbb{R}^N}{\text{minimize}}\ F_0(\mathbf{x}) \quad \text{subject to } F_i(\mathbf{x}) \le b_i,\ i \in [n],$$

where the function $F_0 : \mathbb{R}^N \to \mathbb{R}$ is called an *objective function* and the functions $F_1, \ldots, F_n : \mathbb{R}^N \to \mathbb{R}$ are called *constraint functions*. This general framework also encompasses equality constraints of the type $G_i(\mathbf{x}) = c_i$, since the equality $G_i(\mathbf{x}) = c_i$ is equivalent to the inequalities $G_i(\mathbf{x}) \le c_i$ and $-G_i(\mathbf{x}) \le -c_i$. If $F_0, F_1, \ldots, F_n$ are all convex functions, then the problem is called a *convex optimization problem*—see Appendix B.5 for more information. If $F_0, F_1, \ldots, F_n$ are all linear functions, then the problem is called a *linear program*. Our sparse recovery problem is in fact an optimization problem, since it translates into

$$\text{minimize } \|\mathbf{z}\|_0 \quad \text{subject to } \mathbf{Az} = \mathbf{y}. \tag{P_0}$$

This is a nonconvex problem, and we even have seen in Theorem 2.17 that it is NP-hard in general. However, keeping in mind that $\|\mathbf{z}\|_q^q$ approaches $\|\mathbf{z}\|_0$ as $q > 0$ tends to zero, we can approximate (P_0) by the problem

$$\text{minimize } \|\mathbf{z}\|_q \quad \text{subject to } \mathbf{Az} = \mathbf{y}. \tag{P_q}$$

S. Foucart and H. Rauhut, *A Mathematical Introduction to Compressive Sensing*, Applied and Numerical Harmonic Analysis, DOI 10.1007/978-0-8176-4948-7_3, © Springer Science+Business Media New York 2013

For $q > 1$, even 1-sparse vectors are not solutions of (P_q)—see Exercise 3.1. For $0 < q < 1$, (P_q) is again a nonconvex problem, which is also NP-hard in general—see Exercise 2.10. But for the critical value $q = 1$, it becomes the following convex problem (interpreted as the convex relaxation of (P_0); see Sect. B.3 for the definition of convex relaxation):

$$\text{minimize } \|\mathbf{z}\|_1 \quad \text{subject to } \mathbf{Az} = \mathbf{y}. \tag{P_1}$$

This principle is usually called ℓ_1-*minimization* or *basis pursuit*. There are several specific algorithms to solve the optimization problem, and some of them are presented in Chap. 15.

Basis pursuit

Input: measurement matrix $\mathbf{A}$, measurement vector $\mathbf{y}$.
Instruction:

$$\mathbf{x}^\sharp = \text{argmin } \|\mathbf{z}\|_1 \quad \text{subject to } \mathbf{Az} = \mathbf{y}. \tag{BP}$$

Output: the vector $\mathbf{x}^\sharp$.

Let us complement the previous intuitive justification by the observation that ℓ_1-minimizers are sparse, at least in the real setting. In the complex setting, this is not necessarily true; see Exercise 3.2.

Theorem 3.1. *Let $\mathbf{A} \in \mathbb{R}^{m \times N}$ be a measurement matrix with columns $\mathbf{a}_1, \ldots, \mathbf{a}_N$. Assuming the uniqueness of a minimizer $\mathbf{x}^\sharp$ of*

$$\underset{\mathbf{z} \in \mathbb{R}^N}{\text{minimize }} \|\mathbf{z}\|_1 \quad \text{subject to } \mathbf{Az} = \mathbf{y},$$

the system $\{\mathbf{a}_j, j \in \text{supp}(\mathbf{x}^\sharp)\}$ is linearly independent, and in particular

$$\|\mathbf{x}^\sharp\|_0 = \text{card}(\text{supp}(\mathbf{x}^\sharp)) \leq m.$$

Proof. By way of contradiction, let us assume that the system $\{\mathbf{a}_j, j \in S\}$ is linearly dependent, where $S = \text{supp}(\mathbf{x}^\sharp)$. This means that there exists a nonzero vector $\mathbf{v} \in \mathbb{R}^N$ supported on S such that $\mathbf{Av} = \mathbf{0}$. Then, for any $t \neq 0$,

$$\|\mathbf{x}^\sharp\|_1 < \|\mathbf{x}^\sharp + t\mathbf{v}\|_1 = \sum_{j \in S} |x_j^\sharp + tv_j| = \sum_{j \in S} \text{sgn}(x_j^\sharp + tv_j)(x_j^\sharp + tv_j).$$

If $|t|$ is small enough, namely, $|t| < \min_{j \in S} |x_j^\sharp|/\|\mathbf{v}\|_\infty$, we have

$$\text{sgn}(x_j^\sharp + tv_j) = \text{sgn}(x_j^\sharp) \quad \text{for all } j \in S.$$

It follows that, for $t \neq 0$ with $|t| < \min_{j \in S} |x_j^\sharp|/\|\mathbf{v}\|_\infty$,

$$\|\mathbf{x}^\sharp\|_1 < \sum_{j \in S} \mathrm{sgn}(x_j^\sharp)(x_j^\sharp + t v_j) = \sum_{j \in S} \mathrm{sgn}(x_j^\sharp) x_j^\sharp + t \sum_{j \in S} \mathrm{sgn}(x_j^\sharp) v_j$$

$$= \|\mathbf{x}^\sharp\|_1 + t \sum_{j \in S} \mathrm{sgn}(x_j^\sharp) v_j.$$

This is a contradiction, because we can always choose a small $t \neq 0$ such that $t \sum_{j \in S} \mathrm{sgn}(x_j^\sharp) v_j \leq 0$. $\qquad\qquad\square$

In the real setting, it is also worth pointing out that (P_1) can be recast as a linear program by introducing slack variables $\mathbf{z}^+, \mathbf{z}^- \in \mathbb{R}^N$. Given $\mathbf{z} \in \mathbb{R}^N$, these are defined, for $j \in [N]$, by

$$z_j^+ = \begin{cases} z_j & \text{if } z_j > 0, \\ 0 & \text{if } z_j \leq 0, \end{cases} \qquad z_j^- = \begin{cases} 0 & \text{if } z_j > 0, \\ -z_j & \text{if } z_j \leq 0. \end{cases}$$

The problem (P_1) is thus equivalent to a linear program with optimization variables $\mathbf{z}^+, \mathbf{z}^- \in \mathbb{R}^N$, namely, to

$$\underset{\mathbf{z}^+, \mathbf{z}^- \in \mathbb{R}^N}{\text{minimize}} \sum_{j=1}^N (z_j^+ + z_j^-) \quad \text{subject to} \quad \left[\mathbf{A} \,\middle|\, -\mathbf{A}\right] \begin{bmatrix} \mathbf{z}^+ \\ \mathbf{z}^- \end{bmatrix} = \mathbf{y}, \quad \begin{bmatrix} \mathbf{z}^+ \\ \mathbf{z}^- \end{bmatrix} \geq 0. \quad (\mathrm{P}_1')$$

Given the solution $(\mathbf{x}^+)^\sharp, (\mathbf{x}^-)^\sharp$ of this program, the solution of (P_1) is recovered by $\mathbf{x}^\sharp = (\mathbf{x}^+)^\sharp - (\mathbf{x}^-)^\sharp$.

These considerations do not carry over to the complex setting. In this case, we present alternative considerations that directly extend to a more general ℓ_1-minimization taking measurement error into account, namely,

$$\text{minimize} \ \|\mathbf{z}\|_1 \quad \text{subject to} \ \|\mathbf{A}\mathbf{z} - \mathbf{y}\|_2 \leq \eta. \qquad\qquad (\mathrm{P}_{1,\eta})$$

This variation is natural because in general the measurement vector $\mathbf{y} \in \mathbb{C}^m$ is not exactly equal to $\mathbf{A}\mathbf{x} \in \mathbb{C}^m$, but rather to $\mathbf{A}\mathbf{x} + \mathbf{e}$ for some measurement error $\mathbf{e} \in \mathbb{C}^m$ that can be estimated in ℓ_2-norm, say, by $\|\mathbf{e}\|_2 \leq \eta$ for some $\eta \geq 0$. Then, given a vector $\mathbf{z} \in \mathbb{C}^N$, we introduce its real and imaginary parts $\mathbf{u}, \mathbf{v} \in \mathbb{R}^N$ and a vector $\mathbf{c} \in \mathbb{R}^N$ such that $c_j \geq |z_j| = \sqrt{u_j^2 + v_j^2}$ for all $j \in [N]$. The problem $(\mathrm{P}_{1,\eta})$ is then equivalent to the following problem with optimization variables $\mathbf{c}, \mathbf{u}, \mathbf{v} \in \mathbb{R}^N$:

$$\underset{\mathbf{c},\mathbf{u},\mathbf{v} \in \mathbb{R}^N}{\text{minimize}} \sum_{j=1}^N c_j \quad \text{subject to} \ \left\| \begin{bmatrix} \mathrm{Re}(\mathbf{A}) & -\mathrm{Im}(\mathbf{A}) \\ \mathrm{Im}(\mathbf{A}) & \mathrm{Re}(\mathbf{A}) \end{bmatrix} \begin{bmatrix} \mathbf{u} \\ \mathbf{v} \end{bmatrix} - \begin{bmatrix} \mathrm{Re}(\mathbf{y}) \\ \mathrm{Im}(\mathbf{y}) \end{bmatrix} \right\|_2 \leq \eta, \quad (\mathrm{P}_{1,\eta}')$$

$$\sqrt{u_1^2 + v_1^2} \leq c_1,$$

$$\vdots$$

$$\sqrt{u_N^2 + v_N^2} \leq c_N.$$

This is an instance of a *second-order cone problem*; see Appendix B.5 for more details. Given its solution $(\mathbf{c}^\sharp, \mathbf{u}^\sharp, \mathbf{v}^\sharp)$, the solution to $(P_{1,\eta})$ is given by $\mathbf{x}^\sharp = \mathbf{u}^\sharp + i\mathbf{v}^\sharp$. Note that the choice $\eta = 0$ yields the second-order cone formulation of (P_1) in the complex case.

The principle of solving $(P_{1,\eta})$ is called *quadratically constrained basis pursuit* (or sometimes noise-aware ℓ_1-minimization). Again, there is a choice of algorithms to perform this task.

Quadratically constrained basis pursuit

Input: measurement matrix $\mathbf{A}$, measurement vector $\mathbf{y}$, noise level η.
Instruction:

$$\mathbf{x}^\sharp = \operatorname{argmin} \|\mathbf{z}\|_1 \qquad \text{subject to } \|\mathbf{A}\mathbf{z} - \mathbf{y}\|_2 \le \eta. \qquad (\mathrm{BP}_\eta)$$

Output: the vector $\mathbf{x}^\sharp$.

The solution $\mathbf{x}^\sharp$ of

$$\underset{\mathbf{z}\in\mathbb{C}^N}{\text{minimize}} \ \|\mathbf{z}\|_1 \qquad \text{subject to } \|\mathbf{A}\mathbf{z} - \mathbf{y}\|_2 \le \eta \qquad (3.1)$$

is strongly linked to the output of the *basis pursuit denoising*, which consists in solving, for some parameter $\lambda \ge 0$,

$$\underset{\mathbf{z}\in\mathbb{C}^N}{\text{minimize}} \ \lambda\|\mathbf{z}\|_1 + \|\mathbf{A}\mathbf{z} - \mathbf{y}\|_2^2 . \qquad (3.2)$$

The solution of (3.1) is also related to the output of the *LASSO*, which consists in solving, for some parameter $\tau \ge 0$,

$$\underset{\mathbf{z}\in\mathbb{C}^N}{\text{minimize}} \ \|\mathbf{A}\mathbf{z} - \mathbf{y}\|_2 \qquad \text{subject to } \|\mathbf{z}\|_1 \le \tau. \qquad (3.3)$$

Precisely, some links between the three approaches are given below.

Proposition 3.2. *(a) If* $\mathbf{x}$ *is a minimizer of the basis pursuit denoising (3.2) with* $\lambda > 0$, *then there exists* $\eta = \eta_\mathbf{x} \ge 0$ *such that* $\mathbf{x}$ *is a minimizer of the quadratically constrained basis pursuit (3.1).*
(b) If $\mathbf{x}$ *is a unique minimizer of the quadratically constrained basis pursuit (3.1) with* $\eta \ge 0$, *then there exists* $\tau = \tau_\mathbf{x} \ge 0$ *such that* $\mathbf{x}$ *is a unique minimizer of the LASSO (3.3).*
(c) If $\mathbf{x}$ *is a minimizer of the LASSO (3.3) with* $\tau > 0$, *then there exists* $\lambda = \lambda_\mathbf{x} \ge 0$ *such that* $\mathbf{x}$ *is a minimizer of the basis pursuit denoising (3.2).*

Proof. (a) We set $\eta := \|\mathbf{A}\mathbf{x} - \mathbf{y}\|_2$ and consider $\mathbf{z} \in \mathbb{C}^N$ such that $\|\mathbf{A}\mathbf{z} - \mathbf{y}\|_2 \le \eta$. Together with the fact that $\mathbf{x}$ is a minimizer of (3.2), this yields

$$\lambda\|\mathbf{x}\|_1 + \|\mathbf{Ax} - \mathbf{y}\|_2^2 \le \lambda\|\mathbf{z}\|_1 + \|\mathbf{Az} - \mathbf{y}\|_2^2 \le \lambda\|\mathbf{z}\|_1 + \|\mathbf{Ax} - \mathbf{y}\|_2^2.$$

After simplification, we obtain $\|\mathbf{x}\|_1 \le \|\mathbf{z}\|_1$; hence, $\mathbf{x}$ is a minimizer of (3.1).

(b) We set $\tau := \|\mathbf{x}\|_1$ and consider $\mathbf{z} \in \mathbb{C}^N$, $\mathbf{z} \ne \mathbf{x}$, such that $\|\mathbf{z}\|_1 \le \tau$. Since $\mathbf{x}$ is a unique minimizer of (3.1), this implies that $\mathbf{z}$ cannot satisfy the constraint of (3.1); hence, $\|\mathbf{Az} - \mathbf{y}\|_2 > \eta \ge \|\mathbf{Ax} - \mathbf{y}\|_2$. This shows that $\mathbf{x}$ is a unique minimizer of (3.3).

(c) This part uses tools from convex analysis, and we refer to Theorem B.28 for a proof. □

Another type of ℓ_1-minimization problem is the Dantzig selector,

$$\underset{\mathbf{z}\in\mathbb{C}^N}{\text{minimize}} \ \|\mathbf{z}\|_1 \qquad \text{subject to } \|\mathbf{A}^*(\mathbf{Az} - \mathbf{y})\|_\infty \le \tau. \tag{3.4}$$

This is again a convex optimization problem. The intuition for the constraint is that the residual $\mathbf{r} = \mathbf{Az} - \mathbf{y}$ should have small correlation with all columns $\mathbf{a}_j$ of the matrix $\mathbf{A}$—indeed, $\|\mathbf{A}^*(\mathbf{Az} - \mathbf{y})\|_\infty = \max_{j\in[N]} |\langle \mathbf{r}, \mathbf{a}_j \rangle|$. A similar theory developed for the ℓ_1-minimization problems (BP) and (BP$_\eta$) later in the book is valid for the Dantzig selector as well. Some aspects are covered in Exercises 4.11, 6.18, 9.11, and 15.6.

3.2 Greedy Methods

In this section, we introduce two iterative greedy algorithms commonly used in compressive sensing. The first algorithm, called *orthogonal matching pursuit*, adds one index to a target support S^n at each iteration and updates a target vector $\mathbf{x}^n$ as the vector supported on the target support S^n that best fits the measurements. The algorithm is formally described as follows.

Orthogonal matching pursuit (OMP)

Input: measurement matrix $\mathbf{A}$, measurement vector $\mathbf{y}$.
Initialization: $S^0 = \emptyset$, $\mathbf{x}^0 = 0$.
Iteration: repeat until a stopping criterion is met at $n = \bar{n}$:

$$S^{n+1} = S^n \cup \{j_{n+1}\}, \quad j_{n+1} := \underset{j\in[N]}{\text{argmax}}\{|(\mathbf{A}^*(\mathbf{y} - \mathbf{Ax}^n))_j|\}, \qquad (\text{OMP}_1)$$

$$\mathbf{x}^{n+1} = \underset{\mathbf{z}\in\mathbb{C}^N}{\text{argmin}}\ \{\|\mathbf{y} - \mathbf{Az}\|_2, \text{supp}(\mathbf{z}) \subset S^{n+1}\}. \qquad (\text{OMP}_2)$$

Output: the $\bar{n}$-sparse vector $\mathbf{x}^\sharp = \mathbf{x}^{\bar{n}}$.

The projection step (OMP$_2$) is the most costly part of the orthogonal matching pursuit algorithm. It can be accelerated by using the QR-decomposition of $\mathbf{A}_{S_n}$. In fact, efficient methods exist for updating the QR-decomposition when a column is added to the matrix. If available one may alternatively exploit fast matrix–vector multiplications for $\mathbf{A}$ (like the fast Fourier transform; see Sect. C.1). We refer to the discussion at the end of Sect. A.3 for details. In the case that fast matrix–vector multiplication routines are available for $\mathbf{A}$ and $\mathbf{A}^*$, they should also be used for speed up of the computation of $\mathbf{A}^*(\mathbf{y} - \mathbf{A}\mathbf{x}^n)$.

The choice of the index j_{n+1} is dictated by a greedy strategy where one aims to reduce the ℓ_2-norm of the residual $\mathbf{y} - \mathbf{A}\mathbf{x}^n$ as much as possible at each iteration. The following lemma (refined in Exercise 3.10) applied with $S = S^n$ and $\mathbf{v} = \mathbf{x}^n$ gives some insight as to why an index j maximizing $|(\mathbf{A}^*(\mathbf{y} - \mathbf{A}\mathbf{x}^n))_j|$ is a good candidate for a large decrease of the ℓ_2-norm of the residual.

Lemma 3.3. *Let* $\mathbf{A} \in \mathbb{C}^{m \times N}$ *be a matrix with* ℓ_2-*normalized columns. Given* $S \subset [N]$, $\mathbf{v}$ *supported on* S, *and* $j \in [N]$, *if*

$$\mathbf{w} := \operatorname*{argmin}_{\mathbf{z} \in \mathbb{C}^N} \left\{ \|\mathbf{y} - \mathbf{A}\mathbf{z}\|_2, \operatorname{supp}(\mathbf{z}) \subset S \cup \{j\} \right\},$$

then

$$\|\mathbf{y} - \mathbf{A}\mathbf{w}\|_2^2 \leq \|\mathbf{y} - \mathbf{A}\mathbf{v}\|_2^2 - |(\mathbf{A}^*(\mathbf{y} - \mathbf{A}\mathbf{v}))_j|^2.$$

Proof. Since any vector of the form $\mathbf{v} + t\mathbf{e}_j$ with $t \in \mathbb{C}$ is supported on $S \cup \{j\}$, we have

$$\|\mathbf{y} - \mathbf{A}\mathbf{w}\|_2^2 \leq \min_{t \in \mathbb{C}} \|\mathbf{y} - \mathbf{A}(\mathbf{v} + t\mathbf{e}_j)\|_2^2.$$

Writing $t = \rho e^{i\theta}$ with $\rho \geq 0$ and $\theta \in [0, 2\pi)$, we compute

$$
\begin{aligned}
\|\mathbf{y} - \mathbf{A}(\mathbf{v} + t\mathbf{e}_j)\|_2^2 &= \|\mathbf{y} - \mathbf{A}\mathbf{v} - t\mathbf{A}\mathbf{e}_j\|_2^2 \\
&= \|\mathbf{y} - \mathbf{A}\mathbf{v}\|_2^2 + |t|^2 \|\mathbf{A}\mathbf{e}_j\|_2^2 - 2\operatorname{Re}(\bar{t}\langle \mathbf{y} - \mathbf{A}\mathbf{v}, \mathbf{A}\mathbf{e}_j \rangle) \\
&= \|\mathbf{y} - \mathbf{A}\mathbf{v}\|_2^2 + \rho^2 - 2\operatorname{Re}(\rho e^{-i\theta}(\mathbf{A}^*(\mathbf{y} - \mathbf{A}\mathbf{v}))_j) \\
&\geq \|\mathbf{y} - \mathbf{A}\mathbf{v}\|_2^2 + \rho^2 - 2\rho |(\mathbf{A}^*(\mathbf{y} - \mathbf{A}\mathbf{v}))_j|,
\end{aligned}
$$

with equality for a properly chosen θ. As a quadratic polynomial in ρ, the latter expression is minimized when $\rho = |(\mathbf{A}^*(\mathbf{y} - \mathbf{A}\mathbf{u}))_j|$. This shows that

$$\min_{t \in \mathbb{C}} \|\mathbf{y} - \mathbf{A}(\mathbf{v} + t\mathbf{e}_j)\|_2^2 = \|\mathbf{y} - \mathbf{A}\mathbf{v}\|_2^2 - |(\mathbf{A}^*(\mathbf{y} - \mathbf{A}\mathbf{u}))_j|^2,$$

which concludes the proof. $\qquad\square$

We point out that the step (OMP$_2$) also reads as

$$\mathbf{x}_{S^{n+1}}^{n+1} = \mathbf{A}_{S^{n+1}}^\dagger \mathbf{y},$$

where $\mathbf{x}_{S^{n+1}}^{n+1}$ denotes the restriction of $\mathbf{x}^{n+1}$ to its support set S^{n+1} and where $\mathbf{A}_{S^{n+1}}^\dagger$ is the pseudo-inverse of $\mathbf{A}_{S^{n+1}}$; see Sect. A.2 for details. This simply says that $\mathbf{z} = \mathbf{x}_{S^{n+1}}^{n+1}$ is a solution of $\mathbf{A}_{S^{n+1}}^* \mathbf{A}_{S^{n+1}} \mathbf{z} = \mathbf{A}_{S^{n+1}}^* \mathbf{y}$. This fact is justified by the following lemma, which will also be useful for other algorithms containing a step similar to (OMP$_2$).

Lemma 3.4. *Given an index set $S \subset [N]$, if*

$$\mathbf{v} := \underset{\mathbf{z} \in \mathbb{C}^N}{\operatorname{argmin}} \left\{ \|\mathbf{y} - \mathbf{A}\mathbf{z}\|_2, \operatorname{supp}(\mathbf{z}) \subset S \right\},$$

then

$$(\mathbf{A}^*(\mathbf{y} - \mathbf{A}\mathbf{v}))_S = \mathbf{0}. \tag{3.5}$$

Proof. According the definition of $\mathbf{v}$, the vector $\mathbf{A}\mathbf{v}$ is the orthogonal projection of $\mathbf{y}$ onto the space $\{\mathbf{A}\mathbf{z}, \operatorname{supp}(\mathbf{z}) \subset S\}$; hence, it is characterized by the orthogonality condition

$$\langle \mathbf{y} - \mathbf{A}\mathbf{v}, \mathbf{A}\mathbf{z} \rangle = 0 \qquad \text{for all } \mathbf{z} \in \mathbb{C}^N \text{ with } \operatorname{supp}(\mathbf{z}) \subset S.$$

This means that $\langle \mathbf{A}^*(\mathbf{y} - \mathbf{A}\mathbf{v}), \mathbf{z} \rangle = 0$ for all $\mathbf{z} \in \mathbb{C}^N$ with $\operatorname{supp}(\mathbf{z}) \subset S$, which holds if and only if (3.5) is satisfied. $\qquad\square$

A natural stopping criterion for the orthogonal matching pursuit algorithm is $\mathbf{A}\mathbf{x}^{\bar{n}} = \mathbf{y}$. However, to account for measurement and computation errors, we use instead $\|\mathbf{y} - \mathbf{A}\mathbf{x}^{\bar{n}}\|_2 \leq \varepsilon$ or $\|\mathbf{A}^*(\mathbf{y} - \mathbf{A}\mathbf{x}^{\bar{n}})\|_\infty \leq \varepsilon$ for some chosen tolerance $\epsilon > 0$. If there is an estimate for the sparsity s of the vector $\mathbf{x} \in \mathbb{C}^N$ to be recovered, another possible stopping criterion can simply be $\bar{n} = s$, since then the target vector $\mathbf{x}^{\bar{n}}$ is s-sparse. For instance, if $\mathbf{A}$ is a square orthogonal matrix, then the algorithm with this stopping criterion successfully recovers an s-sparse vector $\mathbf{x} \in \mathbb{C}^N$ from $\mathbf{y} = \mathbf{A}\mathbf{x}$, since it can be seen that the vector $\mathbf{x}^n$ produced at the nth iteration equals the n-sparse vector consisting of n largest entries of $\mathbf{x}$. More generally, the success of recovery of s-sparse vectors via s iterations of the orthogonal matching pursuit algorithm is determined by the following result.

Proposition 3.5. *Given a matrix $\mathbf{A} \in \mathbb{C}^{m \times N}$, every nonzero vector $\mathbf{x} \in \mathbb{C}^N$ supported on a set S of size s is recovered from $\mathbf{y} = \mathbf{A}\mathbf{x}$ after at most s iterations of orthogonal matching pursuit if and only if the matrix $\mathbf{A}_S$ is injective and*

$$\max_{j \in S} |(\mathbf{A}^*\mathbf{r})_j| > \max_{\ell \in \overline{S}} |(\mathbf{A}^*\mathbf{r})_\ell| \tag{3.6}$$

for all nonzero $\mathbf{r} \in \{\mathbf{A}\mathbf{z}, \operatorname{supp}(\mathbf{z}) \subset S\}$.

Proof. Let us assume that the orthogonal matching pursuit algorithm recovers all vectors supported on a set S in at most $s = \mathrm{card}(S)$ iterations. Then, since two vectors supported on S which have the same measurement vector must be equal, the matrix $\mathbf{A}_S$ is injective. Moreover, since the index chosen at the first iteration always stays in the target support, if $\mathbf{y} = \mathbf{A}\mathbf{x}$ for some $\mathbf{x} \in \mathbb{C}^N$ exactly supported on S, then an index $\ell \in \overline{S}$ cannot be chosen at the first iteration, i.e., $\max_{j \in S} |(\mathbf{A}^*\mathbf{y})_j| > |(\mathbf{A}^*\mathbf{y})_\ell|$. Therefore, we have $\max_{j \in S} |(\mathbf{A}^*\mathbf{y})_j| > \max_{\ell \in \overline{S}} |(\mathbf{A}^*\mathbf{y})_\ell|$ for all nonzero $\mathbf{y} \in \{\mathbf{A}\mathbf{z}, \mathrm{supp}(\mathbf{z}) \subset S\}$. This shows the necessity of the two conditions given in the proposition.

To prove their sufficiency, assuming that $\mathbf{A}\mathbf{x}^1 \neq \mathbf{y}, \dots, \mathbf{A}\mathbf{x}^{s-1} \neq \mathbf{y}$ (otherwise there is nothing to do), we are going to prove that S^n is a subset of S of size n for any $0 \leq n \leq s$. This will imply $S^s = S$; hence, $\mathbf{A}\mathbf{x}^s = \mathbf{y}$ by (OMP$_2$), and finally $\mathbf{x}^s = \mathbf{x}$ by the injectivity of $\mathbf{A}_S$. To establish our claim, given $0 \leq n \leq s - 1$, we first notice that $S^n \subset S$ yields $\mathbf{r}^n := \mathbf{y} - \mathbf{A}\mathbf{x}^n \in \{\mathbf{A}\mathbf{z}, \mathrm{supp}(\mathbf{z}) \subset S\}$, so that the index j_{n+1} lies in S by (3.6) and $S^{n+1} = S^n \cup \{j_{n+1}\} \subset S$ by (OMP$_1$). This inductively proves that S^n is a subset of S for any $0 \leq n \leq s$. Next, given $1 \leq n \leq s - 1$, Lemma 3.4 implies that $(\mathbf{A}^*\mathbf{r}^n)_{S^n} = \mathbf{0}$. Therefore, according to its definition in (OMP$_1$), the index j_{n+1} does not lie in S^n, since this would mean that $\mathbf{A}^*\mathbf{r}^n = \mathbf{0}$ and in turn that $\mathbf{r}^n = \mathbf{0}$ by (3.6). This inductively proves that S^n is a set of size n. The proof is now complete.   □

Remark 3.6. A more concise way to formulate the necessary and sufficient conditions of Proposition 3.5 is the *exact recovery condition*, which reads

$$\|\mathbf{A}_S^\dagger \mathbf{A}_{\overline{S}}\|_{1 \to 1} < 1; \tag{3.7}$$

see Sect. A.1 for the definition of matrix norms. Implicitly, the existence of the pseudo-inverse $\mathbf{A}_S^\dagger = (\mathbf{A}_S^*\mathbf{A}_S)^{-1}\mathbf{A}_S^*$ is equivalent to the injectivity of $\mathbf{A}_S$. Moreover, (3.6) is then equivalent to

$$\|\mathbf{A}_S^*\mathbf{A}_S\mathbf{u}\|_\infty > \|\mathbf{A}_{\overline{S}}^*\mathbf{A}_S\mathbf{u}\|_\infty \quad \text{for all } \mathbf{u} \in \mathbb{C}^s \setminus \{\mathbf{0}\}.$$

Making the change $\mathbf{v} = \mathbf{A}_S^*\mathbf{A}_S\mathbf{u}$, this can be written as

$$\|\mathbf{v}\|_\infty > \|\mathbf{A}_{\overline{S}}^*\mathbf{A}_S(\mathbf{A}_S^*\mathbf{A}_S)^{-1}\mathbf{v}\|_\infty = \|\mathbf{A}_{\overline{S}}^*(\mathbf{A}_S^\dagger)^*\mathbf{v}\|_\infty \quad \text{for all } \mathbf{v} \in \mathbb{C}^s \setminus \{\mathbf{0}\}.$$

The latter reads $\|\mathbf{A}_{\overline{S}}^*(\mathbf{A}_S^\dagger)^*\|_{\infty \to \infty} < 1$, that is to say, $\|\mathbf{A}_S^\dagger \mathbf{A}_{\overline{S}}\|_{1 \to 1} < 1$; see also Remark A.6(a).

A weakness of the orthogonal matching pursuit algorithm is that, once an incorrect index has been selected in a target support S^n, it remains in all the subsequent target supports $S^{n'}$ for $n' \geq n$—see Sect. 6.4 where this issue is illustrated on a detailed example. Hence, if an incorrect index has been selected, s iterations of the orthogonal matching pursuit are not enough to recover a vector with sparsity s.

A possible solution is to increase the number of iterations. The following algorithm, called *compressive sampling matching pursuit algorithm*, proposes another strategy when an estimation of the sparsity s is available. To describe it, it is convenient to introduce the notations $H_s(\mathbf{z})$ for the best s-term approximation to $\mathbf{z} \in \mathbb{C}^N$ and $L_s(\mathbf{z})$ for the support of the latter, i.e.,

$$L_s(\mathbf{z}) := \text{index set of } s \text{ largest absolute entries of } \mathbf{z} \in \mathbb{C}^N, \quad (3.8)$$

$$H_s(\mathbf{z}) := \mathbf{z}_{L_s(\mathbf{z})}. \quad (3.9)$$

The nonlinear operator H_s is called *hard thresholding operator* of order s. Given the vector $\mathbf{z} \in \mathbb{C}^N$, the operator H_s keeps its s largest absolute entries and sets the other ones to zero. Note that it may not be uniquely defined. To resolve this issue, we choose the index set $L_s(\mathbf{z})$ out of all possible candidates according to a predefined rule, for instance, the lexicographic order.

Compressive sampling matching pursuit (CoSaMP)

Input: measurement matrix $\mathbf{A}$, measurement vector $\mathbf{y}$, sparsity level s.
Initialization: s-sparse vector $\mathbf{x}^0$, typically $\mathbf{x}^0 = \mathbf{0}$.
Iteration: repeat until a stopping criterion is met at $n = \bar{n}$:

$$U^{n+1} = \text{supp}(\mathbf{x}^n) \cup L_{2s}(\mathbf{A}^*(\mathbf{y} - \mathbf{A}\mathbf{x}^n)), \quad (\text{CoSaMP}_1)$$

$$\mathbf{u}^{n+1} = \underset{\mathbf{z} \in \mathbb{C}^N}{\text{argmin}} \left\{ \|\mathbf{y} - \mathbf{A}\mathbf{z}\|_2, \text{supp}(\mathbf{z}) \subset U^{n+1} \right\}, \quad (\text{CoSaMP}_2)$$

$$\mathbf{x}^{n+1} = H_s(\mathbf{u}^{n+1}). \quad (\text{CoSaMP}_3)$$

Output: the s-sparse vector $\mathbf{x}^\sharp = \mathbf{x}^{\bar{n}}$.

3.3 Thresholding-Based Methods

In this section, we describe further algorithms involving the hard thresholding operator H_s. The intuition for these algorithms, which justifies categorizing them in a different family, relies on the approximate inversion of the action on sparse vectors of the measurement matrix $\mathbf{A}$ by the action of its adjoint $\mathbf{A}^*$. Thus, the *basic thresholding algorithm* consists in determining the support of the s-sparse vector $\mathbf{x} \in \mathbb{C}^N$ to be recovered from the measurement vector $\mathbf{y} = \mathbf{A}\mathbf{x} \in \mathbb{C}^m$ as the indices of s largest absolute entries of $\mathbf{A}^*\mathbf{y}$ and then in finding the vector with this support that best fits the measurement. Formally, the algorithm reads as follows.

Basic thresholding

Input: measurement matrix $\mathbf{A}$, measurement vector $\mathbf{y}$, sparsity level s.
Instruction:

$$S^{\sharp} = L_s(\mathbf{A}^*\mathbf{y}), \tag{BT_1}$$

$$\mathbf{x}^{\sharp} = \underset{\mathbf{z}\in\mathbb{C}^N}{\operatorname{argmin}}\left\{\|\mathbf{y} - \mathbf{Az}\|_2, \operatorname{supp}(\mathbf{z}) \subset S^{\sharp}\right\}. \tag{BT_2}$$

Output: the s-sparse vector $\mathbf{x}^{\sharp}$.

A necessary and sufficient condition resembling (3.6) can be given for the success of s-sparse recovery using this simple algorithm.

Proposition 3.7. *A vector $\mathbf{x} \in \mathbb{C}^N$ supported on a set S is recovered from $\mathbf{y} = \mathbf{Ax}$ via basic thresholding if and only if*

$$\min_{j\in S}|(\mathbf{A}^*\mathbf{y})_j| > \max_{\ell\in\overline{S}}|(\mathbf{A}^*\mathbf{y})_\ell|. \tag{3.10}$$

Proof. It is clear that the vector $\mathbf{x}$ is recovered if and only if the index set $S^{\sharp}$ defined in (BT$_1$) coincides with the set S, that is to say, if and only if any entry of $\mathbf{A}^*\mathbf{y}$ on S is greater than any entry of $\mathbf{A}^*\mathbf{y}$ on $\overline{S}$. This is property (3.10). □

The more elaborate *iterative hard thresholding algorithm* is an iterative algorithm to solve the rectangular system $\mathbf{Az} = \mathbf{y}$, knowing that the solution is s-sparse. We shall solve the square system $\mathbf{A}^*\mathbf{Az} = \mathbf{A}^*\mathbf{y}$ instead, which can be interpreted as the fixed-point equation $\mathbf{z} = (\mathbf{Id} - \mathbf{A}^*\mathbf{A})\mathbf{z} + \mathbf{A}^*\mathbf{y}$. Classical iterative methods suggest the fixed-point iteration $\mathbf{x}^{n+1} = (\mathbf{Id} - \mathbf{A}^*\mathbf{A})\mathbf{x}^n + \mathbf{A}^*\mathbf{y}$. Since we target s-sparse vectors, we only keep the s largest absolute entries of $(\mathbf{Id} - \mathbf{A}^*\mathbf{A})\mathbf{x}^n + \mathbf{A}^*\mathbf{y} = \mathbf{x}^n + \mathbf{A}^*(\mathbf{y} - \mathbf{Ax}^n)$ at each iteration. The resulting algorithm reads as follows.

Iterative hard thresholding (IHT)

Input: measurement matrix $\mathbf{A}$, measurement vector $\mathbf{y}$, sparsity level s.
Initialization: s-sparse vector $\mathbf{x}^0$, typically $\mathbf{x}^0 = \mathbf{0}$.
Iteration: repeat until a stopping criterion is met at $n = \bar{n}$:

$$\mathbf{x}^{n+1} = H_s(\mathbf{x}^n + \mathbf{A}^*(\mathbf{y} - \mathbf{Ax}^n)). \tag{IHT}$$

Output: the s-sparse vector $\mathbf{x}^{\sharp} = \mathbf{x}^{\bar{n}}$.

The iterative hard thresholding algorithm does not require the computation of any orthogonal projection. If we are willing to pay the price of the orthogonal projections, like in the greedy methods, it makes sense to look at the vector with the same support as $\mathbf{x}^{n+1}$ that best fits the measurements. This leads to the *hard thresholding pursuit algorithm* defined below.

Hard thresholding pursuit (HTP)

Input: measurement matrix $\mathbf{A}$, measurement vector $\mathbf{y}$, sparsity level s.
Initialization: s-sparse vector $\mathbf{x}^0$, typically $\mathbf{x}^0 = \mathbf{0}$.
Iteration: repeat until a stopping criterion is met at $n = \bar{n}$:

$$S^{n+1} = L_s(\mathbf{x}^n + \mathbf{A}^*(\mathbf{y} - \mathbf{A}\mathbf{x}^n)), \tag{HTP$_1$}$$

$$\mathbf{x}^{n+1} = \underset{\mathbf{z} \in \mathbb{C}^N}{\operatorname{argmin}} \left\{ \|\mathbf{y} - \mathbf{A}\mathbf{z}\|_2, \operatorname{supp}(\mathbf{z}) \subset S^{n+1} \right\}. \tag{HTP$_2$}$$

Output: the s-sparse vector $\mathbf{x}^\sharp = \mathbf{x}^{\bar{n}}$.

Notes

More background on convex optimization can be found in Appendix B and in the books [70, 369] by Boyd and Vandenberghe and by Nocedal and Wright, respectively.

Basis pursuit was introduced by Chen, Donoho, and Saunders in [114]. The LASSO (Least Absolute Shrinkage and Selection Operator) algorithm is more popular in the statistics literature than the quadratically constrained basis pursuit or basis pursuit denoising algorithms. It was introduced by Tibshirani in [473]. The Dantzig selector (3.4) was introduced by Candès and Tao in [98]. Like the LASSO, it is more popular in statistics than in signal processing.

A greedy strategy that does not involve any orthogonal projection consists in updating $\mathbf{x}^n$ as $\mathbf{x}^{n+1} = \mathbf{x}^n + t\mathbf{e}_j$, where $t \in \mathbb{C}$ and $j \in [N]$ are chosen to minimize $\|\mathbf{y} - \mathbf{A}\mathbf{x}^{n+1}\|$. The argument of Lemma 3.3 imposes the choice of j as a maximizer of $|(\mathbf{A}^*(\mathbf{y} - \mathbf{A}\mathbf{x}^n))_j|$ and then $t = (\mathbf{A}^*(\mathbf{y} - \mathbf{A}\mathbf{x}^n))_j$. This corresponds to the *matching pursuit* algorithm, introduced in signal processing by Mallat and Zhang in [342] and by Qian and Chen in [404] and in statistics as the projection pursuit regression by Friedman and Stuetzle in [214]. In approximation theory, it is known as pure greedy algorithm; see, for instance, the surveys [470, 471] and the monograph [472] by Temlyakov. There, the orthogonal matching pursuit algorithm is also known as orthogonal greedy algorithm. Just like matching pursuit, it was introduced independently by several researchers in different fields, e.g., by Davis, Mallat, and Zhang in [145], by Pati, Rezaiifar, and Krishnaprasad in [378], by Chen, Billings, and Luo in [113], or in [277] by Högborn, where it was called CLEAN in the context of astronomical data processing. The orthogonal matching pursuit algorithm was analyzed in terms of sparse recovery by Tropp in [476].

The *compressive sampling matching pursuit* algorithm was devised by Needell and Tropp in [361]. It was inspired by the earlier *regularized orthogonal matching pursuit* developed and analyzed by Needell and Vershynin in [362, 363].

The *subspace pursuit* algorithm, introduced by Dai and Milenkovic in [135], is another algorithm in the greedy family, but it will not be examined in this book (see [135] or [472] for its analysis). It bears some resemblance with compressive sampling matching pursuit, except that, instead of $2s$, only s indices of largest absolute entries of the residual vector are selected and that an additional orthogonal projection step is performed at each iteration. Its description is given below.

Subspace pursuit

Input: measurement matrix $\mathbf{A}$, measurement vector $\mathbf{y}$, sparsity level s.
Initialization: s-sparse vector $\mathbf{x}^0$, typically $\mathbf{x}^0 = \mathbf{0}$, $S^0 = \mathrm{supp}(\mathbf{x}^0)$.
Iteration: repeat until a stopping criterion is met at $n = \bar{n}$:

$$U^{n+1} = S^n \cup L_s(\mathbf{A}^*(\mathbf{y} - \mathbf{A}\mathbf{x}^n)), \qquad (\mathrm{SP}_1)$$

$$\mathbf{u}^{n+1} = \operatorname*{argmin}_{\mathbf{z} \in \mathbb{C}^N} \left\{ \|\mathbf{y} - \mathbf{A}\mathbf{z}\|_2, \mathrm{supp}(\mathbf{z}) \subset U^{n+1} \right\}, \qquad (\mathrm{SP}_2)$$

$$S^{n+1} = L_s(\mathbf{u}^{n+1}), \qquad (\mathrm{SP}_3)$$

$$\mathbf{x}^{n+1} = \operatorname*{argmin}_{\mathbf{z} \in \mathbb{C}^N} \left\{ \|\mathbf{y} - \mathbf{A}\mathbf{z}\|_2, \mathrm{supp}(\mathbf{z}) \subset S^{n+1} \right\}. \qquad (\mathrm{SP}_4)$$

Output: the s-sparse vector $\mathbf{x}^\sharp = \mathbf{x}^{\bar{n}}$.

The thresholding-based family also contains algorithms that do not require an estimation of the sparsity s. In such algorithms, the hard thresholding operator gives way to a *soft thresholding operator* with threshold $\tau > 0$. This operator, also encountered in (15.22) and (B.18), acts componentwise on a vector $\mathbf{z} \in \mathbb{C}^N$ by sending the entry z_j to

$$S_\tau(z_j) = \begin{cases} \mathrm{sgn}(z_j)(|z_j| - \tau) & \text{if } |z_j| \geq \tau, \\ 0 & \text{otherwise.} \end{cases}$$

Another important method for sparse recovery is the message-passing algorithm studied by Donoho, Maleki, and Montanari in [162]. The soft thresholding algorithms will not be analyzed in this book (but see the Notes section of Chap. 15).

Which Algorithm Should One Choose? In principle, all the algorithms introduced in this chapter work reasonably well in practice (with the possible exception of basic thresholding). Given a set of requirements, which algorithm should be favored depends on the precise situation, e.g., via the specific measurement matrix $\mathbf{A}$ and via the values of the parameters s, m, N.

As a first criterion, the minimal number m of measurements for a sparsity s and a signal length N may vary with each algorithm. Comparing the recovery rates and identifying the best algorithm is then a matter of numerical tests. For this criterion, the recovery performance of basic thresholding is significantly worse than the one of other algorithms, although it is the fastest algorithms since it identifies the support in only one step.

The speed of the algorithm is a second criterion, and it is also a matter of numerical tests. However, we can give at least the following rough guidelines. If the sparsity s is quite small, then orthogonal matching pursuit is extremely fast because the speed essentially depends on the number of iterations, which typically equals s when the algorithm succeeds. Compressive sampling matching pursuit and hard thresholding pursuit are fast for small s, too, because each step involves the computation of an orthogonal projection relative to $\mathbf{A}_S$ with small $S \subset [N]$. But if the sparsity s is not that small compared to N, then orthogonal matching pursuit may nonetheless require a significant time. The same applies to the homotopy method (see Chap. 15) which builds the support of an ℓ_1-minimizer iteratively. The runtime of iterative hard thresholding is almost not influenced by the sparsity s at all.

Basis pursuit, per se, is not an algorithm, so the runtime depends on the actual algorithm that is used for the minimization. For Chambolle and Pock's primal dual algorithm (see Chap. 15), which constructs a sequence converging to an ℓ_1-minimizer, the sparsity s has no serious influence on the speed. Hence, for mildly large s, it can be significantly faster than orthogonal matching pursuit. We emphasize this point to dispute the common misconception that greedy algorithms are always faster than ℓ_1-minimization—this is only true for small sparsity. The iteratively reweighted least squares method (see Chap. 15) may also be a good alternative for mildly large sparsity.

Additionally, another important feature of an algorithm is the possibility to exploit fast matrix–vector multiplication routines that are available for $\mathbf{A}$ and $\mathbf{A}^*$. In principle, any of the proposed methods can be sped up in this case, but the task is complicated if orthogonal projection steps are involved. Fast matrix–vector multiplications are easily integrated in the iterative hard thresholding algorithm and in Chambolle and Pock's primal dual algorithm for ℓ_1-minimization. The acceleration achieved in this context depends on the algorithm, and the fastest algorithm should again be determined by numerical tests in the precise situation.

Exercises

3.1. Let $q > 1$ and let $\mathbf{A}$ be an $m \times N$ matrix with $m < N$. Prove that there exists a 1-sparse vector which is not a minimizer of (P_q).

3.2. Given $\mathbf{A} = \begin{bmatrix} 1 & 0 & -1 \\ 0 & 1 & -1 \end{bmatrix}$, prove that the vector $\mathbf{x} = [1, e^{i2\pi/3}, e^{i4\pi/3}]^\top$ is the unique minimizer of $\|\mathbf{z}\|_1$ subject to $\mathbf{Az} = \mathbf{Ax}$. This shows that, in the complex setting, a unique ℓ_1-minimizer is not necessarily m-sparse, m being the number of rows of the measurement matrix.

3.3. Let $\mathbf{A} \in \mathbb{R}^{m \times N}$ and $\mathbf{y} \in \mathbb{R}^m$. Assuming the uniqueness of the minimizer $\mathbf{x}^\sharp$ of

$$\underset{\mathbf{z} \in \mathbb{R}^N}{\text{minimize}} \, \|\mathbf{z}\|_1 \quad \text{subject to } \|\mathbf{Az} - \mathbf{y}\| \leq \eta,$$

where $\eta \geq 0$ and $\|\cdot\|$ is an arbitrary norm on $\mathbb{R}^m$, prove that $\mathbf{x}^\sharp$ is necessarily m-sparse.

3.4. Given $\mathbf{A} \in \mathbb{R}^{m \times N}$, suppose that every $m \times m$ submatrix of $\mathbf{A}$ is invertible. For $\mathbf{x} \in \mathbb{R}^N$, let $\mathbf{x}^\sharp$ be the unique minimizer of $\|\mathbf{z}\|_1$ subject to $\mathbf{Az} = \mathbf{Ax}$. Prove that either $\mathbf{x}^\sharp = \mathbf{x}$ or $\text{supp}(\mathbf{x}) \not\subset \text{supp}(\mathbf{x}^\sharp)$.

3.5. For $\mathbf{A} \in \mathbb{R}^{m \times N}$ and $\mathbf{x} \in \mathbb{R}^N$, prove that there is no ambiguity between $\mathbf{z} \in \mathbb{R}^N$ and $\mathbf{z} \in \mathbb{C}^N$ when one says that the vector $\mathbf{x}$ is the unique minimizer of $\|\mathbf{z}\|_1$ subject to $\mathbf{Az} = \mathbf{Ax}$.

3.6. Carefully check the equivalences of (P_1) with (P_1') and $(P_{1,\eta})$ with $(P_{1,\eta}')$.

3.7. Given $\mathbf{A} \in \mathbb{C}^{m \times N}$ and $\tau > 0$, show that the solution of

$$\underset{\mathbf{z} \in \mathbb{C}^N}{\text{minimize}} \, \|\mathbf{Az} - \mathbf{y}\|_2^2 + \tau \|\mathbf{z}\|_2^2$$

is given by

$$\mathbf{z}^\sharp = (\mathbf{A}^* \mathbf{A} + \tau \mathbf{Id})^{-1} \mathbf{A}^* \mathbf{y}.$$

3.8. Given $\mathbf{A} \in \mathbb{C}^{m \times N}$, suppose that there is a unique minimizer $f(\mathbf{y}) \in \mathbb{C}^N$ of $\|\mathbf{z}\|_1$ subject to $\|\mathbf{Az} - \mathbf{y}\|_2 \leq \eta$ whenever $\mathbf{y}$ belongs to some set $\mathcal{S}$. Prove that the map f is continuous on $\mathcal{S}$.

3.9. Prove that any 1-sparse vector $\mathbf{x} \in \mathbb{C}^3$ is recovered with one iteration of the orthogonal matching pursuit algorithm for the measurement matrix

$$\mathbf{A} = \begin{bmatrix} 1 & -1/2 & -1/2 \\ 0 & \sqrt{3}/2 & -\sqrt{3}/2 \end{bmatrix}.$$

We now add a measurement by appending the row $[1 \ 3 \ 3]$ to $\mathbf{A}$, thus forming the matrix

$$\hat{\mathbf{A}} = \begin{bmatrix} 1 & -1/2 & -1/2 \\ 0 & \sqrt{3}/2 & -\sqrt{3}/2 \\ 1 & 3 & 3 \end{bmatrix}.$$

Prove that the 1-sparse vector $\mathbf{x} = [1 \ 0 \ 0]^\top$ cannot be recovered via the orthogonal matching pursuit algorithm with the measurement matrix $\hat{\mathbf{A}}$.

3.10. Given a matrix $\mathbf{A} \in \mathbb{C}^{m \times N}$ with ℓ_2-normalized columns $\mathbf{a}_1, \ldots, \mathbf{a}_N$ and given a vector $\mathbf{y} \in \mathbb{C}^m$, we consider an iterative algorithm where the index set S^n is updated via $S^{n+1} = S^n \cup \{j^{n+1}\}$ for an unspecified index j^{n+1} and where the output vector is updated via $\mathbf{x}^{n+1} = \mathrm{argmin}\{\|\mathbf{y} - \mathbf{A}\mathbf{z}\|_2, \mathrm{supp}(\mathbf{z}) \subset S^{n+1}\}$. Prove that the ℓ_2-norm of the residual decreases according to

$$\|\mathbf{y} - \mathbf{A}\mathbf{x}^{n+1}\|_2^2 = \|\mathbf{y} - \mathbf{A}\mathbf{x}^n\|_2^2 - \Delta_n,$$

where the quantity Δ_n satisfies

$$\Delta_n = \|\mathbf{A}(\mathbf{x}^{n+1} - \mathbf{x}^n)\|_2^2 = x_{j^{n+1}}^{n+1}(\mathbf{A}^*(\mathbf{y} - \mathbf{A}\mathbf{x}^n))_{j^{n+1}}$$

$$= \frac{|(\mathbf{A}^*(\mathbf{y} - \mathbf{A}\mathbf{x}^n))_{j^{n+1}}|^2}{\mathrm{dist}(\mathbf{a}_{j^{n+1}}, \mathrm{span}\{\mathbf{a}_j, j \in S^n\})^2}$$

$$\geq |(\mathbf{A}^*(\mathbf{y} - \mathbf{A}\mathbf{x}^n))_{j^{n+1}}|^2.$$

3.11. Implement the algorithms of this chapter. Choose $\mathbf{A} \in \mathbb{R}^{m \times N}$ with independent random entries equal to $1/\sqrt{m}$ or $-1/\sqrt{m}$, each with probability $1/2$. Test the algorithms on randomly generated s-sparse signals, where first the support is chosen at random and then the nonzero coefficients. By varying N, m, s, evaluate the empirical success probability of recovery.

Chapter 4
Basis Pursuit

Let us recall that the intuitive approach to the compressive sensing problem of recovering a sparse vector $\mathbf{x} \in \mathbb{C}^N$ from its measurement vector $\mathbf{y} = \mathbf{A}\mathbf{x} \in \mathbb{C}^m$, where $m < N$, consists in the ℓ_0-minimization problem

$$\underset{\mathbf{z} \in \mathbb{C}^N}{\text{minimize}} \, \|\mathbf{z}\|_0 \quad \text{subject to } \mathbf{A}\mathbf{z} = \mathbf{y}. \tag{P_0}$$

We have seen in Chap. 2 that this problem is unfortunately NP-hard in general. Chapter 3 has therefore outlined several tractable strategies to solve the standard compressive sensing problem. In the current chapter, we focus on the basis pursuit (ℓ_1-minimization) strategy, which consists in solving the convex optimization problem

$$\underset{\mathbf{z} \in \mathbb{C}^N}{\text{minimize}} \, \|\mathbf{z}\|_1 \quad \text{subject to } \mathbf{A}\mathbf{z} = \mathbf{y}. \tag{P_1}$$

We investigate conditions on the matrix $\mathbf{A}$ which ensure exact or approximate reconstruction of the original sparse or compressible vector $\mathbf{x}$. In Sect. 4.1, we start with a necessary and sufficient condition for the exact reconstruction of every sparse vector $\mathbf{x} \in \mathbb{C}^N$ as a solution of (P_1) with the vector $\mathbf{y} \in \mathbb{C}^m$ obtained as $\mathbf{y} = \mathbf{A}\mathbf{x}$. This condition is called the null space property. In Sects. 4.2 and 4.3, we strengthen this null space property to make the reconstruction via basis pursuit stable with respect to sparsity defect and robust with respect to measurement error. In Sect. 4.4, we discuss other types of necessary and sufficient conditions for the success of basis pursuit to reconstruct an individual sparse vector $\mathbf{x}$. Although such conditions do not seem useful at first sight because they involve the a priori unknown vector $\mathbf{x}$, they will nevertheless be essential to establish the so-called nonuniform recovery guarantees when the measurement matrix is random. In Sect. 4.5, we interpret the recovery of sparse vectors via basis pursuit in terms of polytope geometry. Section 4.6 closes the chapter with a short digression to the low-rank recovery

S. Foucart and H. Rauhut, *A Mathematical Introduction to Compressive Sensing*, Applied and Numerical Harmonic Analysis, DOI 10.1007/978-0-8176-4948-7_4, © Springer Science+Business Media New York 2013

problem and its approach via nuclear norm minimization. Again, a version of the null space property is equivalent to the successful recovery of every low-rank matrix.

4.1 Null Space Property

In this section, we introduce the null space property, and we prove that it is a necessary and sufficient condition for exact recovery of sparse vectors via basis pursuit. The arguments are valid in the real and complex settings alike, so we first state the results for a field $\mathbb{K}$ that can either be $\mathbb{R}$ or $\mathbb{C}$. Then we establish the equivalence of the real and complex null space properties. We recall that for a vector $\mathbf{v} \in \mathbb{C}^N$ and a set $S \subset [N]$, we denote by $\mathbf{v}_S$ either the vector in $\mathbb{C}^S$, which is the restriction of $\mathbf{v}$ to the indices in S, or the vector in $\mathbb{C}^N$ which coincides with $\mathbf{v}$ on the indices in S and is extended to zero outside S; see also (2.4). It should always become clear from the context which variant of $\mathbf{v}_S$ is meant (and sometimes both variants lead to the same quantity, such as in expressions like $\|\mathbf{v}_S\|_1$).

Definition 4.1. A matrix $\mathbf{A} \in \mathbb{K}^{m \times N}$ is said to satisfy the *null space property* relative to a set $S \subset [N]$ if

$$\|\mathbf{v}_S\|_1 < \|\mathbf{v}_{\overline{S}}\|_1 \quad \text{for all } \mathbf{v} \in \ker \mathbf{A} \setminus \{\mathbf{0}\}. \tag{4.1}$$

It is said to satisfy the null space property of order s if it satisfies the null space property relative to any set $S \subset [N]$ with $\mathrm{card}(S) \leq s$.

Remark 4.2. It is important to observe that, for a given $\mathbf{v} \in \ker \mathbf{A} \setminus \{\mathbf{0}\}$, the condition $\|\mathbf{v}_S\|_1 < \|\mathbf{v}_{\overline{S}}\|_1$ holds for any set $S \subset [N]$ with $\mathrm{card}(S) \leq s$ as soon as it holds for an index set of s largest (in modulus) entries of $\mathbf{v}$.

Remark 4.3. There are two convenient reformulations of the null space property. The first one is obtained by adding $\|\mathbf{v}_S\|_1$ to both sides of the inequality $\|\mathbf{v}_S\|_1 < \|\mathbf{v}_{\overline{S}}\|_1$. Thus, the null space property relative to S reads

$$2 \|\mathbf{v}_S\|_1 < \|\mathbf{v}\|_1 \quad \text{for all } \mathbf{v} \in \ker \mathbf{A} \setminus \{\mathbf{0}\}. \tag{4.2}$$

The second one is obtained by choosing S as an index set of s largest (in modulus) entries of $\mathbf{v}$ and this time by adding $\|\mathbf{v}_{\overline{S}}\|_1$ to both sides of the inequality. Thus, the null space property of order s reads

$$\|\mathbf{v}\|_1 < 2\,\sigma_s(\mathbf{v})_1 \quad \text{for all } \mathbf{v} \in \ker \mathbf{A} \setminus \{\mathbf{0}\}, \tag{4.3}$$

where we recall from Definition 2.2 that, for $p > 0$, the ℓ_p-error of *best s-term approximation* to $\mathbf{x} \in \mathbb{K}^N$ is defined by

$$\sigma_s(\mathbf{x})_p = \inf_{\|\mathbf{z}\|_0 \leq s} \|\mathbf{x} - \mathbf{z}\|_p.$$

We now indicate the link between the null space property and exact recovery of sparse vectors via basis pursuit.

Theorem 4.4. *Given a matrix* $\mathbf{A} \in \mathbb{K}^{m \times N}$, *every vector* $\mathbf{x} \in \mathbb{K}^N$ *supported on a set* S *is the unique solution of* (P_1) *with* $\mathbf{y} = \mathbf{A}\mathbf{x}$ *if and only if* $\mathbf{A}$ *satisfies the null space property relative to* S.

Proof. Given a fixed index set S, let us first assume that every vector $\mathbf{x} \in \mathbb{K}^N$ supported on S is the unique minimizer of $\|\mathbf{z}\|_1$ subject to $\mathbf{A}\mathbf{z} = \mathbf{A}\mathbf{x}$. Thus, for any $\mathbf{v} \in \ker \mathbf{A} \setminus \{\mathbf{0}\}$, the vector $\mathbf{v}_S$ is the unique minimizer of $\|\mathbf{z}\|_1$ subject to $\mathbf{A}\mathbf{z} = \mathbf{A}\mathbf{v}_S$. But we have $\mathbf{A}(-\mathbf{v}_{\overline{S}}) = \mathbf{A}\mathbf{v}_S$ and $-\mathbf{v}_{\overline{S}} \neq \mathbf{v}_S$, because $\mathbf{A}(\mathbf{v}_{\overline{S}} + \mathbf{v}_S) = \mathbf{A}\mathbf{v} = \mathbf{0}$ and $\mathbf{v} \neq \mathbf{0}$. We conclude that $\|\mathbf{v}_S\|_1 < \|\mathbf{v}_{\overline{S}}\|_1$. This establishes the null space property relative to S.

Conversely, let us assume that the null space property relative to S holds. Then, given a vector $\mathbf{x} \in \mathbb{K}^N$ supported on S and a vector $\mathbf{z} \in \mathbb{K}^N$, $\mathbf{z} \neq \mathbf{x}$, satisfying $\mathbf{A}\mathbf{z} = \mathbf{A}\mathbf{x}$, we consider the vector $\mathbf{v} := \mathbf{x} - \mathbf{z} \in \ker \mathbf{A} \setminus \{\mathbf{0}\}$. In view of the null space property, we obtain

$$\|\mathbf{x}\|_1 \leq \|\mathbf{x} - \mathbf{z}_S\|_1 + \|\mathbf{z}_S\|_1 = \|\mathbf{v}_S\|_1 + \|\mathbf{z}_S\|_1$$
$$< \|\mathbf{v}_{\overline{S}}\|_1 + \|\mathbf{z}_S\|_1 = \|-\mathbf{z}_{\overline{S}}\|_1 + \|\mathbf{z}_S\|_1 = \|\mathbf{z}\|_1.$$

This establishes the required minimality of $\|\mathbf{x}\|_1$.   □

Letting the set S vary, we immediately obtain the following result as a consequence of Theorem 4.4.

Theorem 4.5. *Given a matrix* $\mathbf{A} \in \mathbb{K}^{m \times N}$, *every* s-*sparse vector* $\mathbf{x} \in \mathbb{K}^N$ *is the unique solution of* (P_1) *with* $\mathbf{y} = \mathbf{A}\mathbf{x}$ *if and only if* $\mathbf{A}$ *satisfies the null space property of order* s.

Remark 4.6. (a) This theorem shows that for every $\mathbf{y} = \mathbf{A}\mathbf{x}$ with s-sparse $\mathbf{x}$ the ℓ_1-minimization strategy (P_1) actually solves the ℓ_0-minimization problem (P_0) when the null space property of order s holds. Indeed, assume that every s-sparse vector $\mathbf{x}$ is recovered via ℓ_1-minimization from $\mathbf{y} = \mathbf{A}\mathbf{x}$. Let $\mathbf{z}$ be the minimizer of the ℓ_0-minimization problem (P_0) with $\mathbf{y} = \mathbf{A}\mathbf{x}$ then $\|\mathbf{z}\|_0 \leq \|\mathbf{x}\|_0$ so that also $\mathbf{z}$ is s-sparse. But since every s-sparse vector is the unique ℓ_1-minimizer, it follows that $\mathbf{x} = \mathbf{z}$.

(b) It is desirable for any reconstruction scheme to preserve sparse recovery if some measurements are rescaled, reshuffled, or added. Basis pursuit actually features such properties. Indeed, mathematically speaking, these operations consist in replacing the original measurement matrix $\mathbf{A}$ by new measurement matrices $\hat{\mathbf{A}}$ and $\tilde{\mathbf{A}}$ defined by

$$\hat{\mathbf{A}} := \mathbf{G}\mathbf{A}, \quad \text{where } \mathbf{G} \text{ is some invertible } m \times m \text{ matrix,}$$

$$\tilde{\mathbf{A}} := \begin{bmatrix} \mathbf{A} \\ \mathbf{B} \end{bmatrix}, \quad \text{where } \mathbf{B} \text{ is some } m' \times N \text{ matrix.}$$

We observe that $\ker \hat{\mathbf{A}} = \ker \mathbf{A}$ and $\ker \tilde{\mathbf{A}} \subset \ker \mathbf{A}$, hence the null space property for the matrices $\hat{\mathbf{A}}$ and $\tilde{\mathbf{A}}$ remains fulfilled if it is satisfied for the matrix $\mathbf{A}$. It is not true that the null space property remains valid if we multiply on the right by an invertible matrix—see Exercise 4.2.

We close this section by inspecting the influence of the underlying field. Unifying the arguments by using $\mathbb{K}$ for either $\mathbb{R}$ or $\mathbb{C}$ had the advantage of brevity, but it results in a potential ambiguity about null space properties. Indeed, we often encounter real-valued measurement matrices, and they can also be regarded as complex-valued matrices. Thus, for such $\mathbf{A} \in \mathbb{R}^{m \times N}$, the distinction between the real null space $\ker_{\mathbb{R}} \mathbf{A}$ and the complex null space $\ker_{\mathbb{C}} \mathbf{A} = \ker_{\mathbb{R}} \mathbf{A} + i \ker_{\mathbb{R}} \mathbf{A}$ leads, on the one hand, to the real null space property relative to a set S, namely,

$$\sum_{j \in S} |v_j| < \sum_{\ell \in \overline{S}} |v_\ell| \quad \text{for all } \mathbf{v} \in \ker_{\mathbb{R}} \mathbf{A}, \mathbf{v} \neq \mathbf{0}, \tag{4.4}$$

and, on the other hand, to the complex null space property relative to S, namely,

$$\sum_{j \in S} \sqrt{v_j^2 + w_j^2} < \sum_{\ell \in \overline{S}} \sqrt{v_\ell^2 + w_\ell^2} \quad \text{for all } \mathbf{v}, \mathbf{w} \in \ker_{\mathbb{R}} \mathbf{A}, (\mathbf{v}, \mathbf{w}) \neq (\mathbf{0}, \mathbf{0}). \tag{4.5}$$

We are going to show below that the real and complex versions are in fact equivalent. Therefore, there is no ambiguity when we say that a real measurement matrix allows the exact recovery of all sparse vectors via basis pursuit: These vectors can be interpreted as real or as complex vectors. This explains why we usually work in the complex setting.

Theorem 4.7. *Given a matrix $\mathbf{A} \in \mathbb{R}^{m \times N}$, the real null space property (4.4) relative to a set S is equivalent to the complex null space property (4.5) relative to this set S.*

In particular, the real null space property of order s is equivalent to the complex null space property of order s.

Proof. We notice first that (4.4) immediately follows from (4.5) by setting $\mathbf{w} = \mathbf{0}$. So let us assume that (4.4) holds. We consider $\mathbf{v}, \mathbf{w} \in \ker_{\mathbb{R}} \mathbf{A}$ with $(\mathbf{v}, \mathbf{w}) \neq (\mathbf{0}, \mathbf{0})$. If $\mathbf{v}$ and $\mathbf{w}$ are linearly dependent, then the inequality $\sum_{j \in S} \sqrt{v_j^2 + w_j^2} < \sum_{\ell \in \overline{S}} \sqrt{v_\ell^2 + w_\ell^2}$ is clear, so we may suppose that they are linearly independent. Then $\mathbf{u} := \cos \theta \, \mathbf{v} + \sin \theta \, \mathbf{w} \in \ker_{\mathbb{R}} \mathbf{A}$ is nonzero, and (4.4) yields, for any $\theta \in \mathbb{R}$,

$$\sum_{j \in S} |\cos \theta \, v_j + \sin \theta \, w_j| < \sum_{\ell \in \overline{S}} |\cos \theta \, v_\ell + \sin \theta \, w_\ell|. \tag{4.6}$$

For each $k \in [N]$, we define $\theta_k \in [-\pi, \pi]$ by the equalities

$$v_k = \sqrt{v_k^2 + w_k^2} \, \cos \theta_k, \quad w_k = \sqrt{v_k^2 + w_k^2} \, \sin \theta_k,$$

so that (4.6) reads

$$\sum_{j \in S} \sqrt{v_j^2 + w_j^2} \, |\cos(\theta - \theta_j)| < \sum_{\ell \in \overline{S}} \sqrt{v_\ell^2 + w_\ell^2} \, |\cos(\theta - \theta_\ell)|.$$

We now integrate over $\theta \in [-\pi, \pi]$ to obtain

$$\sum_{j \in S} \sqrt{v_j^2 + w_j^2} \int_{-\pi}^{\pi} |\cos(\theta - \theta_j)| d\theta < \sum_{\ell \in \overline{S}} \sqrt{v_\ell^2 + w_\ell^2} \int_{-\pi}^{\pi} |\cos(\theta - \theta_\ell)| d\theta.$$

For the inequality $\sum_{j \in S} \sqrt{v_j^2 + w_j^2} < \sum_{\ell \in \overline{S}} \sqrt{v_\ell^2 + w_\ell^2}$, it remains to observe that

$$\int_{-\pi}^{\pi} |\cos(\theta - \theta')| d\theta$$

is a positive constant independent of $\theta' \in [-\pi, \pi]$—namely, 4. The proof is now complete.   □

Remark 4.8. The equivalence of Theorem 4.7 extends to the stable and robust null space properties introduced later in this chapter; see Exercise 4.5.

Nonconvex Minimization

Recall that the number of nonzero entries of a vector $z \in \mathbb{C}^N$ is approximated by the qth power of its ℓ_q-quasinorm,

$$\sum_{j=1}^{N} |z_j|^q \xrightarrow[q \to 0]{} \sum_{j=1}^{N} 1_{\{z_j \neq 0\}} = \|z\|_0.$$

This observation suggests to replace the ℓ_0-minimization problem (P_0) by the optimization problem

$$\underset{z \in \mathbb{C}^N}{\text{minimize}} \, \|z\|_q \quad \text{subject to } Az = y. \tag{P_q}$$

This optimization problem fails to recover even 1-sparse vectors for $q > 1$; see Exercise 3.1. For $0 < q < 1$, on the other hand, the optimization problem becomes nonconvex and is even NP-hard; see Exercise 2.10. Thus, the case $q = 1$ might appear as the only important one. Nonetheless, the properties of the ℓ_q-minimization for $0 < q < 1$ can prove useful on theoretical questions. Our goal here is merely to justify the intuitive prediction that the problem (P_q) does not provide a worse approximation of the original problem (P_0) when q gets smaller. For this purpose, we need an analog of the null space property for $0 < q < 1$. The proof of our next result, left as Exercise 4.12, duplicates the proof of Theorem 4.4. It relies on the fact that the qth power of the ℓ_q-quasinorm satisfies the triangle inequality; see Exercise 2.1.

Theorem 4.9. *Given a matrix* $\mathbf{A} \in \mathbb{C}^{m \times N}$ *and* $0 < q \leq 1$, *every* s-sparse vector $\mathbf{x} \in \mathbb{C}^N$ *is the unique solution of* (P_q) *with* $\mathbf{y} = \mathbf{A}\mathbf{x}$ *if and only if, for any set* $S \subset [N]$ *with* $\mathrm{card}(S) \leq s$,

$$\|\mathbf{v}_S\|_q < \|\mathbf{v}_{\overline{S}}\|_q \quad \text{for all } \mathbf{v} \in \ker \mathbf{A} \setminus \{\mathbf{0}\}.$$

We can now prove that sparse recovery via ℓ_q-minimization implies sparse recovery via ℓ_p-minimization whenever $0 < p < q \leq 1$.

Theorem 4.10. *Given a matrix* $\mathbf{A} \in \mathbb{C}^{m \times N}$ *and* $0 < p < q \leq 1$, *if every* s-sparse vector $\mathbf{x} \in \mathbb{C}^N$ *is the unique solution of* (P_q) *with* $\mathbf{y} = \mathbf{A}\mathbf{x}$, *then every* s-sparse vector $\mathbf{x} \in \mathbb{C}^N$ *is also the unique solution of* (P_p) *with* $\mathbf{y} = \mathbf{A}\mathbf{x}$.

Proof. According to Theorem 4.9, it is enough to prove that, if $\mathbf{v} \in \ker \mathbf{A} \setminus \{\mathbf{0}\}$ and if S is an index set of s largest absolute entries of $\mathbf{v}$, then

$$\sum_{j \in S} |v_j|^p < \sum_{\ell \in \overline{S}} |v_\ell|^p, \tag{4.7}$$

as soon as (4.7) holds with q in place of p. Indeed, if (4.7) holds for q, then necessarily $\mathbf{v}_{\overline{S}} \neq \mathbf{0}$ since S is an index of largest absolute entries and $\mathbf{v} \neq \mathbf{0}$. The desired inequality (4.7) can therefore be rewritten as

$$\sum_{j \in S} \frac{1}{\sum_{\ell \in \overline{S}} (|v_\ell|/|v_j|)^p} < 1. \tag{4.8}$$

Now observe that $|v_\ell|/|v_j| \leq 1$ for $\ell \in \overline{S}$ and $j \in S$. This makes the left-hand side of (4.8) a nondecreasing function of $0 < p \leq 1$. Hence, its value at $p < q$ does not exceed its value at q, which is less than one by hypothesis. This shows the validity of (4.7) and concludes the proof.   □

4.2   Stability

The vectors we aim to recover via basis pursuit—or other schemes, for that matter— are sparse only in idealized situations. In more realistic scenarios, we can only claim that they are close to sparse vectors. In such cases, we would like to recover a vector $\mathbf{x} \in \mathbb{C}^N$ with an error controlled by its distance to s-sparse vectors. This property is usually referred to as the *stability* of the reconstruction scheme with respect to sparsity defect. We shall prove that the basis pursuit is stable under a slightly strengthened version of the null space property.

Definition 4.11. A matrix $\mathbf{A} \in \mathbb{C}^{m \times N}$ is said to satisfy the *stable null space property* with constant $0 < \rho < 1$ relative to a set $S \subset [N]$ if

$$\|\mathbf{v}_S\|_1 \leq \rho \|\mathbf{v}_{\overline{S}}\|_1 \quad \text{for all } \mathbf{v} \in \ker \mathbf{A}.$$

It is said to satisfy the stable null space property of order s with constant $0 < \rho < 1$ if it satisfies the stable null space property with constant $0 < \rho < 1$ relative to any set $S \subset [N]$ with $\operatorname{card}(S) \leq s$.

The main stability result of this section reads as follows.

Theorem 4.12. *Suppose that a matrix $\mathbf{A} \in \mathbb{C}^{m \times N}$ satisfies the stable null space property of order s with constant $0 < \rho < 1$. Then, for any $\mathbf{x} \in \mathbb{C}^N$, a solution $\mathbf{x}^\sharp$ of (P_1) with $\mathbf{y} = \mathbf{A}\mathbf{x}$ approximates the vector $\mathbf{x}$ with ℓ_1-error*

$$\|\mathbf{x} - \mathbf{x}^\sharp\|_1 \leq \frac{2(1 + \rho)}{(1 - \rho)} \sigma_s(\mathbf{x})_1. \tag{4.9}$$

Remark 4.13. In contrast to Theorem 4.4 we cannot guarantee uniqueness of the ℓ_1-minimizer anymore—although nonuniqueness is rather pathological. In any case, even when the ℓ_1-minimizer is not unique, the theorem above states that *every* solution $\mathbf{x}^\sharp$ of (P_1) with $\mathbf{y} = \mathbf{A}\mathbf{x}$ satisfies (4.9).

We are actually going to prove a stronger "if and only if" theorem below. The result is a statement valid for any index set S in which the vector $\mathbf{x}^\sharp \in \mathbb{C}^N$ is replaced by any vector $\mathbf{z} \in \mathbb{C}^N$ satisfying $\mathbf{A}\mathbf{z} = \mathbf{A}\mathbf{x}$. Apart from improving Theorem 4.12, the result also says that, under the stable null space property relative to S, the distance between a vector $\mathbf{x} \in \mathbb{C}^N$ supported on S and a vector $\mathbf{z} \in \mathbb{C}^N$ satisfying $\mathbf{A}\mathbf{z} = \mathbf{A}\mathbf{x}$ is controlled by the difference between their norms.

Theorem 4.14. *The matrix $\mathbf{A} \in \mathbb{C}^{m \times N}$ satisfies the stable null space property with constant $0 < \rho < 1$ relative to S if and only if*

$$\|\mathbf{z} - \mathbf{x}\|_1 \leq \frac{1 + \rho}{1 - \rho} \left(\|\mathbf{z}\|_1 - \|\mathbf{x}\|_1 + 2 \|\mathbf{x}_{\overline{S}}\|_1 \right) \tag{4.10}$$

for all vectors $\mathbf{x}, \mathbf{z} \in \mathbb{C}^N$ with $\mathbf{A}\mathbf{z} = \mathbf{A}\mathbf{x}$.

The error bound (4.9) follows from Theorem 4.14 as follows: Take S to be a set of s largest absolute coefficients of $\mathbf{x}$, so that $\|\mathbf{x}_{\overline{S}}\|_1 = \sigma_s(\mathbf{x})_1$. If $\mathbf{x}^\sharp$ is a minimizer of (P_1), then $\|\mathbf{x}^\sharp\|_1 \leq \|\mathbf{x}\|_1$ and $\mathbf{A}\mathbf{x}^\sharp = \mathbf{A}\mathbf{x}$. The right-hand side of inequality (4.10) with $\mathbf{z} = \mathbf{x}^\sharp$ can therefore be estimated by the right hand of (4.9).

Before turning to the proof of Theorem 4.14, we isolate the following observation, as it will also be needed later.

Lemma 4.15. *Given a set* $S \subset [N]$ *and vectors* $\mathbf{x}, \mathbf{z} \in \mathbb{C}^N$,

$$\|(\mathbf{x} - \mathbf{z})_{\overline{S}}\|_1 \leq \|\mathbf{z}\|_1 - \|\mathbf{x}\|_1 + \|(\mathbf{x} - \mathbf{z})_S\|_1 + 2\|\mathbf{x}_{\overline{S}}\|_1.$$

Proof. The result simply follows from

$$\|\mathbf{x}\|_1 = \|\mathbf{x}_{\overline{S}}\|_1 + \|\mathbf{x}_S\|_1 \leq \|\mathbf{x}_{\overline{S}}\|_1 + \|(\mathbf{x} - \mathbf{z})_S\|_1 + \|\mathbf{z}_S\|_1,$$

$$\|(\mathbf{x} - \mathbf{z})_{\overline{S}}\|_1 \leq \|\mathbf{x}_{\overline{S}}\|_1 + \|\mathbf{z}_{\overline{S}}\|_1.$$

These two inequalities sum up to give

$$\|\mathbf{x}\|_1 + \|(\mathbf{x} - \mathbf{z})_{\overline{S}}\|_1 \leq 2\|\mathbf{x}_{\overline{S}}\|_1 + \|(\mathbf{x} - \mathbf{z})_S\|_1 + \|\mathbf{z}\|_1.$$

This is the desired inequality.   □

Proof (of Theorem 4.14). Let us first assume that the matrix $\mathbf{A}$ satisfies (4.10) for all vectors $\mathbf{x}, \mathbf{z} \in \mathbb{C}^N$ with $\mathbf{Az} = \mathbf{Ax}$. Given a vector $\mathbf{v} \in \ker \mathbf{A}$, since $\mathbf{Av}_{\overline{S}} = \mathbf{A}(-\mathbf{v}_S)$, we can apply (4.10) with $\mathbf{x} = -\mathbf{v}_S$ and $\mathbf{z} = \mathbf{v}_{\overline{S}}$. It yields

$$\|\mathbf{v}\|_1 \leq \frac{1 + \rho}{1 - \rho}(\|\mathbf{v}_{\overline{S}}\|_1 - \|\mathbf{v}_S\|_1).$$

This can be written as

$$(1 - \rho)(\|\mathbf{v}_S\|_1 + \|\mathbf{v}_{\overline{S}}\|_1) \leq (1 + \rho)(\|\mathbf{v}_{\overline{S}}\|_1 - \|\mathbf{v}_S\|_1).$$

After rearranging the terms, we obtain

$$\|\mathbf{v}_S\|_1 \leq \rho\|\mathbf{v}_{\overline{S}}\|_1,$$

and we recognize the stable null space property with constant $0 < \rho < 1$ relative to S.

Conversely, let us now assume that the matrix $\mathbf{A}$ satisfies the stable null space property with constant $0 < \rho < 1$ relative to S. For $\mathbf{x}, \mathbf{z} \in \mathbb{C}^N$ with $\mathbf{Az} = \mathbf{Ax}$, since $\mathbf{v} := \mathbf{z} - \mathbf{x} \in \ker \mathbf{A}$, the stable null space property yields

$$\|\mathbf{v}_S\|_1 \leq \rho\|\mathbf{v}_{\overline{S}}\|_1. \tag{4.11}$$

Moreover, Lemma 4.15 gives

$$\|\mathbf{v}_{\overline{S}}\|_1 \leq \|\mathbf{z}\|_1 - \|\mathbf{x}\|_1 + \|\mathbf{v}_S\|_1 + 2\|\mathbf{x}_{\overline{S}}\|_1. \tag{4.12}$$

Substituting (4.11) into (4.12), we obtain

$$\|\mathbf{v}_{\overline{S}}\|_1 \leq \|\mathbf{z}\|_1 - \|\mathbf{x}\|_1 + \rho\|\mathbf{v}_{\overline{S}}\|_1 + 2\|\mathbf{x}_{\overline{S}}\|_1.$$

Since $\rho < 1$, this can be rewritten as

$$\|v_{\overline{S}}\|_1 \le \frac{1}{1-\rho}\left(\|z\|_1 - \|x\|_1 + 2\|x_{\overline{S}}\|_1\right).$$

Using (4.11) once again, we derive

$$\|v\|_1 = \|v_{\overline{S}}\|_1 + \|v_S\|_1 \le (1+\rho)\|v_{\overline{S}}\|_1 \le \frac{1+\rho}{1-\rho}\left(\|z\|_1 - \|x\|_1 + 2\|x_{\overline{S}}\|_1\right),$$

which is the desired inequality.   □

Remark 4.16. Given the matrix $A \in \mathbb{C}^{m \times N}$, let us consider, for each index set $S \subset [N]$ with $\mathrm{card}(S) \le s$, the operator R_S defined on $\ker A$ by $R_S(v) = v_S$. The formulation (4.2) of the null space property says that

$$\mu := \max\{\|R_S\|_{1 \to 1} : S \subset [N], \mathrm{card}(S) \le s\} < 1/2.$$

It then follows that A satisfies the stable null space property with constant $\rho := \mu/(1-\mu) < 1$. Thus, the stability of the basis pursuit comes for free if sparse vectors are exactly recovered. However, the constant $2(1+\rho)/(1-\rho)$ in (4.9) may be very large if ρ is close to one.

4.3 Robustness

In realistic situations, it is also inconceivable to measure a signal $x \in \mathbb{C}^N$ with infinite precision. This means that the measurement vector $y \in \mathbb{C}^m$ is only an approximation of the vector $Ax \in \mathbb{C}^m$, with

$$\|Ax - y\| \le \eta$$

for some $\eta \ge 0$ and for some norm $\| \cdot \|$ on $\mathbb{C}^m$—usually the ℓ_2-norm, but the ℓ_1-norm will also be considered in Chap. 13. In this case, the reconstruction scheme should be required to output a vector $x^\star \in \mathbb{C}^N$ whose distance to the original vector $x \in \mathbb{C}^N$ is controlled by the measurement error $\eta \ge 0$. This property is usually referred to as the *robustness* of the reconstruction scheme with respect to measurement error. We are going to show that if the problem (P_1) is replaced by the convex optimization problem

$$\underset{z \in \mathbb{C}^N}{\text{minimize}} \|z\|_1 \quad \text{subject to} \|Az - y\| \le \eta, \qquad (P_{1,\eta})$$

then the robustness of the basis pursuit algorithm is guaranteed by the following additional strengthening of the null space property.

Definition 4.17. The matrix $\mathbf{A} \in \mathbb{C}^{m \times N}$ is said to satisfy the *robust null space property* (with respect to $\|\cdot\|$) with constants $0 < \rho < 1$ and $\tau > 0$ relative to a set $S \subset [N]$ if

$$\|\mathbf{v}_S\|_1 \leq \rho \|\mathbf{v}_{\overline{S}}\|_1 + \tau \|\mathbf{A}\mathbf{v}\| \quad \text{for all } \mathbf{v} \in \mathbb{C}^N. \tag{4.13}$$

It is said to satisfy the robust null space property of order s with constants $0 < \rho < 1$ and $\tau > 0$ if it satisfies the robust null space property with constants ρ, τ relative to any set $S \subset [N]$ with $\mathrm{card}(S) \leq s$.

Remark 4.18. Observe that the above definition does not require that $\mathbf{v}$ is contained in $\ker \mathbf{A}$. In fact, if $\mathbf{v} \in \ker \mathbf{A}$, then the term $\|\mathbf{A}\mathbf{v}\|$ in (4.13) vanishes, and we see that the robust null space property implies the stable null space property in Definition 4.11.

The following theorem constitutes the first main result of this section. It incorporates the conclusion of Theorem 4.12 as the special case $\eta = 0$. The special case of an s-sparse vector $\mathbf{x} \in \mathbb{C}^N$ is also worth a separate look.

Theorem 4.19. *Suppose that a matrix $\mathbf{A} \in \mathbb{C}^{m \times N}$ satisfies the robust null space property of order s with constants $0 < \rho < 1$ and $\tau > 0$. Then, for any $\mathbf{x} \in \mathbb{C}^N$, a solution $\mathbf{x}^\sharp$ of $(\mathrm{P}_{1,\eta})$ with $\mathbf{y} = \mathbf{A}\mathbf{x} + \mathbf{e}$ and $\|\mathbf{e}\| \leq \eta$ approximates the vector $\mathbf{x}$ with ℓ_1-error*

$$\|\mathbf{x} - \mathbf{x}^\sharp\|_1 \leq \frac{2(1+\rho)}{(1-\rho)} \sigma_s(\mathbf{x})_1 + \frac{4\tau}{1-\rho} \eta.$$

In the spirit of Theorem 4.14, we are going to prove a stronger "if and only if" statement valid for any index set S.

Theorem 4.20. *The matrix $\mathbf{A} \in \mathbb{C}^{m \times N}$ satisfies the robust null space property with constants $0 < \rho < 1$ and $\tau > 0$ relative to S if and only if*

$$\|\mathbf{z} - \mathbf{x}\|_1 \leq \frac{1+\rho}{1-\rho} \left(\|\mathbf{z}\|_1 - \|\mathbf{x}\|_1 + 2 \|\mathbf{x}_{\overline{S}}\|_1 \right) + \frac{2\tau}{1-\rho} \|\mathbf{A}(\mathbf{z} - \mathbf{x})\| \tag{4.14}$$

for all vectors $\mathbf{x}, \mathbf{z} \in \mathbb{C}^N$.

Proof. We basically follow the same steps as in the proof of Theorem 4.14. First, we assume that the matrix $\mathbf{A}$ satisfies (4.14) for all vectors $\mathbf{x}, \mathbf{z} \in \mathbb{C}^N$. Thus, for $\mathbf{v} \in \mathbb{C}^N$, taking $\mathbf{x} = -\mathbf{v}_S$ and $\mathbf{z} = \mathbf{v}_{\overline{S}}$ yields

$$\|\mathbf{v}\|_1 \leq \frac{1+\rho}{1-\rho} (\|\mathbf{v}_{\overline{S}}\|_1 - \|\mathbf{v}_S\|_1) + \frac{2\tau}{1-\rho} \|\mathbf{A}\mathbf{v}\|.$$

Rearranging the terms gives

$$(1-\rho)(\|\mathbf{v}_S\|_1 + \|\mathbf{v}_{\overline{S}}\|_1) \leq (1+\rho)(\|\mathbf{v}_{\overline{S}}\|_1 - \|\mathbf{v}_S\|_1) + 2\tau\|\mathbf{A}\mathbf{v}\|,$$

that is to say

$$\|\mathbf{v}_S\|_1 \leq \rho \|\mathbf{v}_{\overline{S}}\|_1 + \tau \|\mathbf{A}\mathbf{v}\|.$$

This is the robust null space property with constants $0 < \rho < 1$ and $\tau > 0$ relative to S.

Conversely, we assume that the matrix $\mathbf{A}$ satisfies the robust null space property with constant $0 < \rho < 1$ and $\tau > 0$ relative to S. For $\mathbf{x}, \mathbf{z} \in \mathbb{C}^N$, setting $\mathbf{v} := \mathbf{z} - \mathbf{x}$, the robust null space property and Lemma 4.15 yield

$$\|\mathbf{v}_S\|_1 \leq \rho \|\mathbf{v}_{\overline{S}}\|_1 + \tau \|\mathbf{A}\mathbf{v}\|,$$

$$\|\mathbf{v}_{\overline{S}}\|_1 \leq \|\mathbf{z}\|_1 - \|\mathbf{x}\|_1 + \|\mathbf{v}_S\|_1 + 2\|\mathbf{x}_{\overline{S}}\|_1.$$

Combining these two inequalities gives

$$\|\mathbf{v}_{\overline{S}}\|_1 \leq \frac{1}{1-\rho} \big(\|\mathbf{z}\|_1 - \|\mathbf{x}\|_1 + 2\|\mathbf{x}_{\overline{S}}\|_1 + \tau \|\mathbf{A}\mathbf{v}\|\big).$$

Using the robust null space property once again, we derive

$$\begin{aligned}
\|\mathbf{v}\|_1 &= \|\mathbf{v}_{\overline{S}}\|_1 + \|\mathbf{v}_S\|_1 \leq (1+\rho)\|\mathbf{v}_{\overline{S}}\|_1 + \tau \|\mathbf{A}\mathbf{v}\| \\
&\leq \frac{1+\rho}{1-\rho}\big(\|\mathbf{z}\|_1 - \|\mathbf{x}\|_1 + 2\|\mathbf{x}_{\overline{S}}\|_1\big) + \frac{2\tau}{1-\rho}\|\mathbf{A}\mathbf{v}\|,
\end{aligned}$$

which is the desired inequality. $\qquad\square$

We now turn to the second main result of this section. It enhances the previous robustness result by replacing the ℓ_1-error estimate by an ℓ_p-error estimate for $p \geq 1$. A final strengthening of the null space property is required. The corresponding property could be defined relative to any fixed set $S \subset [N]$, but it is not introduced as such because this will not be needed later.

Definition 4.21. Given $q \geq 1$, the matrix $\mathbf{A} \in \mathbb{C}^{m \times N}$ is said to satisfy the ℓ_q-robust null space property of order s (with respect to $\|\cdot\|$) with constants $0 < \rho < 1$ and $\tau > 0$ if, for any set $S \subset [N]$ with $\mathrm{card}(S) \leq s$,

$$\|\mathbf{v}_S\|_q \leq \frac{\rho}{s^{1-1/q}} \|\mathbf{v}_{\overline{S}}\|_1 + \tau \|\mathbf{A}\mathbf{v}\| \quad \text{for all } \mathbf{v} \in \mathbb{C}^N.$$

In view of the inequality $\|\mathbf{v}_S\|_p \leq s^{1/p-1/q}\|\mathbf{v}_S\|_q$ for $1 \leq p \leq q$, we observe that the ℓ_q-robust null space property with constants $0 < \rho < 1$ and $\tau > 0$ implies that, for any set $S \subset [N]$ with $\mathrm{card}(S) \leq s$,

$$\|\mathbf{v}_S\|_p \leq \frac{\rho}{s^{1-1/p}} \|\mathbf{v}_{\overline{S}}\|_1 + \tau \, s^{1/p-1/q} \|\mathbf{A}\mathbf{v}\| \quad \text{for all } \mathbf{v} \in \mathbb{C}^N.$$

Thus, for $1 \leq p \leq q$, the ℓ_q-robust null space property implies the ℓ_p-robust null space property with identical constants, modulo the change of norms $\|\cdot\| \leftarrow s^{1/p-1/q}\|\cdot\|$. This justifies in particular that the ℓ_q-robust null space property is a strengthening of the previous robust null space property. In Sect. 6.2, we will establish the ℓ_2-robust null space property for measurement matrices with small restricted isometry constants. The robustness of the quadratically constrained basis pursuit algorithm is then deduced according to the following theorem.

Theorem 4.22. *Suppose that the matrix* $\mathbf{A} \in \mathbb{C}^{m \times N}$ *satisfies the* ℓ_2-*robust null space property of order* s *with constants* $0 < \rho < 1$ *and* $\tau > 0$. *Then, for any* $\mathbf{x} \in \mathbb{C}^N$, *a solution* $\mathbf{x}^\sharp$ *of* $(P_{1,\eta})$ *with* $\| \cdot \| = \| \cdot \|_2$, $\mathbf{y} = \mathbf{A}\mathbf{x} + \mathbf{e}$, *and* $\|\mathbf{e}\|_2 \leq \eta$ *approximates the vector* $\mathbf{x}$ *with* ℓ_p-*error*

$$\|\mathbf{x} - \mathbf{x}^\sharp\|_p \leq \frac{C}{s^{1-1/p}} \sigma_s(\mathbf{x})_1 + D s^{1/p-1/2} \eta, \qquad 1 \leq p \leq 2, \tag{4.15}$$

for some constants $C, D > 0$ *depending only on* ρ *and* τ.

The estimates for the extremal values $p = 1$ and $p = 2$ are the most familiar. They read

$$\|\mathbf{x} - \mathbf{x}^\sharp\|_1 \leq C \sigma_s(\mathbf{x})_1 + D \sqrt{s} \eta,$$

$$\|\mathbf{x} - \mathbf{x}^\sharp\|_2 \leq \frac{C}{\sqrt{s}} \sigma_s(\mathbf{x})_1 + D \eta. \tag{4.16}$$

One should remember that the coefficient of $\sigma_s(\mathbf{x})_1$ is a constant for $p = 1$ and scales like $1/\sqrt{s}$ for $p = 2$, while the coefficient of η scales like $\sqrt{s}$ for $p = 1$ and is a constant for $p = 2$. We then retrieve the correct powers of s appearing in Theorem 4.22 for any $1 \leq p \leq 2$ via interpolating the powers of s with linear functions in $1/p$.

Remark 4.23. Let us comment on the fact that, regardless of the ℓ_p-space in which the error is estimated, the best s-term approximation error $\sigma_s(\mathbf{x})_1$ with respect to the ℓ_1-norm always appears on the right-hand side. One may wonder why the error estimate in ℓ_2 does not involve $\sigma_s(\mathbf{x})_2$ instead of $\sigma_s(\mathbf{x})_1/\sqrt{s}$. In fact, we will see in Theorem 11.5 that such an estimate is impossible in parameter regimes of (m, N) that are interesting for compressive sensing. Besides, we have seen that unit ℓ_q-balls with $q < 1$ provide good models for compressible vectors by virtue of Theorem 2.3 and its refinement Theorem 2.5. Indeed, if $\|\mathbf{x}\|_q \leq 1$ for $q < 1$, then, for $p \geq 1$,

$$\sigma_s(\mathbf{x})_p \leq s^{1/p-1/q}.$$

Assuming perfect measurements (that is, $\eta = 0$), the error bound (4.15) yields

$$\|\mathbf{x} - \mathbf{x}^\sharp\|_p \leq \frac{C}{s^{1-1/p}} \sigma_s(\mathbf{x})_1 \leq C s^{1/p-1/q}, \qquad 1 \leq p \leq 2.$$

Therefore, the reconstruction error in ℓ_p obeys the same decay rate in s as the error of best s-term approximation in ℓ_p for all $p \in [1, 2]$. From this point of view, the term $\sigma_s(\mathbf{x})_1 / s^{1-1/p}$ is not significantly worse than $\sigma_s(\mathbf{x})_p$.

Remark 4.24. The ℓ_q-robust null space property may seem mysterious at first sight, but it is necessary—save for the condition $\rho < 1$—to obtain estimates of the type

$$\|\mathbf{x} - \mathbf{x}^\sharp\|_q \leq \frac{C}{s^{1-1/q}} \sigma_s(\mathbf{x})_1 + D\,\eta, \tag{4.17}$$

where $\mathbf{x}^\sharp$ is a minimizer of $(P_{1,\eta})$ with $\mathbf{y} = \mathbf{A}\mathbf{x} + \mathbf{e}$ and $\|\mathbf{e}\| \leq \eta$. Indeed, given $\mathbf{v} \in \mathbb{C}^N$ and $S \subset [N]$ with $\operatorname{card}(S) \leq s$, we apply (4.17) with $\mathbf{x} = \mathbf{v}$, $\mathbf{e} = -\mathbf{A}\mathbf{v}$, and $\eta = \|\mathbf{A}\mathbf{v}\|$, so that $\mathbf{x}^\sharp = \mathbf{0}$, to obtain

$$\|\mathbf{v}\|_q \leq \frac{C}{s^{1-1/q}} \|\mathbf{v}_{\overline{S}}\|_1 + D\,\|\mathbf{A}\mathbf{v}\|,$$

and in particular

$$\|\mathbf{v}_S\|_q \leq \frac{C}{s^{1-1/q}} \|\mathbf{v}_{\overline{S}}\|_1 + D\,\|\mathbf{A}\mathbf{v}\|.$$

For the proof of Theorem 4.22, we establish the stronger result of Theorem 4.25 below. The result of Theorem 4.22 follows by choosing $q = 2$ and specifying $\mathbf{z}$ to be $\mathbf{x}^\sharp$.

Theorem 4.25. *Given $1 \leq p \leq q$, suppose that the matrix $\mathbf{A} \in \mathbb{C}^{m \times N}$ satisfies the ℓ_q-robust null space property of order s with constants $0 < \rho < 1$ and $\tau > 0$. Then, for any $\mathbf{x}, \mathbf{z} \in \mathbb{C}^N$,*

$$\|\mathbf{z} - \mathbf{x}\|_p \leq \frac{C}{s^{1-1/p}} \left(\|\mathbf{z}\|_1 - \|\mathbf{x}\|_1 + 2\sigma_s(\mathbf{x})_1 \right) + D\,s^{1/p-1/q} \|\mathbf{A}(\mathbf{z} - \mathbf{x})\|,$$

where $C := (1 + \rho)^2 / (1 - \rho)$ and $D := (3 + \rho)\tau / (1 - \rho)$.

Proof. Let us first remark that the ℓ_q-robust null space properties imply the ℓ_1-robust and ℓ_p-robust null space property $(p \leq q)$ in the forms

$$\|\mathbf{v}_S\|_1 \leq \rho \|\mathbf{v}_{\overline{S}}\|_1 + \tau\,s^{1-1/q} \|\mathbf{A}\mathbf{v}\|, \tag{4.18}$$

$$\|\mathbf{v}_S\|_p \leq \frac{\rho}{s^{1-1/p}} \|\mathbf{v}_{\overline{S}}\|_1 + \tau\,s^{1/p-1/q} \|\mathbf{A}\mathbf{v}\|, \tag{4.19}$$

for all $\mathbf{v} \in \mathbb{C}^N$ and all $S \subset [N]$ with $\operatorname{card}(S) \leq s$. Thus, in view of (4.18), applying Theorem 4.20 with S chosen as an index set of s largest (in modulus) entries of $\mathbf{x}$ leads to

$$\|\mathbf{z} - \mathbf{x}\|_1 \leq \frac{1 + \rho}{1 - \rho} \left(\|\mathbf{z}\|_1 - \|\mathbf{x}\|_1 + 2\,\sigma_s(\mathbf{x})_1 \right) + \frac{2\tau}{1 - \rho}\,s^{1-1/q} \|\mathbf{A}(\mathbf{z} - \mathbf{x})\|. \tag{4.20}$$

Then, choosing S as an index set of s largest (in modulus) entries of $\mathbf{z} - \mathbf{x}$, we use Theorem 2.5 to notice that

$$\|\mathbf{z} - \mathbf{x}\|_p \le \|(\mathbf{z} - \mathbf{x})_{\overline{S}}\|_p + \|(\mathbf{z} - \mathbf{x})_S\|_p \le \frac{1}{s^{1-1/p}} \|\mathbf{z} - \mathbf{x}\|_1 + \|(\mathbf{z} - \mathbf{x})_S\|_p.$$

In view of (4.19), we derive

$$\|\mathbf{z} - \mathbf{x}\|_p \le \frac{1}{s^{1-1/p}} \|\mathbf{z} - \mathbf{x}\|_1 + \frac{\rho}{s^{1-1/p}} \|(\mathbf{z} - \mathbf{x})_{\overline{S}}\|_1 + \tau \, s^{1/p-1/q} \|\mathbf{A}(\mathbf{z} - \mathbf{x})\|$$

$$\le \frac{1+\rho}{s^{1-1/p}} \|\mathbf{z} - \mathbf{x}\|_1 + \tau \, s^{1/p-1/q} \|\mathbf{A}(\mathbf{z} - \mathbf{x})\|. \tag{4.21}$$

It remains to substitute (4.20) into the latter to obtain the desired result.   □

4.4   Recovery of Individual Vectors

In some cases, we deal with specific sparse vectors rather than with all vectors supported on a given set or all vectors with a given sparsity. We then require some recovery conditions that are finer than the null space property. This section provides such conditions, with a subtle difference between the real and the complex settings, due to the fact that the *sign* of a number z, defined as

$$\mathrm{sgn}(z) := \begin{cases} \dfrac{z}{|z|} & \text{if } z \neq 0, \\[2mm] 0 & \text{if } z = 0, \end{cases}$$

is a discrete quantity when z is real, but it is not when z is complex. For a vector $\mathbf{x} \in \mathbb{C}^N$, we denote by $\mathrm{sgn}(\mathbf{x}) \in \mathbb{C}^N$ the vector with components $\mathrm{sgn}(x_j), j \in [N]$. Let us start with the complex version of a recovery condition valid for individual sparse vectors.

Theorem 4.26. *Given a matrix $\mathbf{A} \in \mathbb{C}^{m \times N}$, a vector $\mathbf{x} \in \mathbb{C}^N$ with support S is the unique minimizer of $\|\mathbf{z}\|_1$ subject to $\mathbf{Az} = \mathbf{Ax}$ if one of the following equivalent conditions holds:*

(a) $\left| \sum_{j \in S} \overline{\mathrm{sgn}(x_j)} v_j \right| < \|\mathbf{v}_{\overline{S}}\|_1$ *for all $\mathbf{v} \in \ker \mathbf{A} \setminus \{\mathbf{0}\}$,*

(b) $\mathbf{A}_S$ is injective, and there exists a vector $\mathbf{h} \in \mathbb{C}^m$ such that

$$(\mathbf{A}^*\mathbf{h})_j = \mathrm{sgn}(x_j), \quad j \in S, \qquad |(\mathbf{A}^*\mathbf{h})_\ell| < 1, \quad \ell \in \overline{S}.$$

Proof. Let us start by proving that (a) implies that $\mathbf{x}$ is the unique minimizer of $\|\mathbf{z}\|_1$ subject to $\mathbf{Az} = \mathbf{Ax}$. For a vector $\mathbf{z} \neq \mathbf{x}$ such that $\mathbf{Az} = \mathbf{Ax}$, we just have to write, with $\mathbf{v} := \mathbf{x} - \mathbf{z} \in \ker \mathbf{A} \setminus \{\mathbf{0}\}$,

$$\|\mathbf{z}\|_1 = \|\mathbf{z}_S\|_1 + \|\mathbf{z}_{\overline{S}}\|_1 = \|(\mathbf{x} - \mathbf{v})_S\|_1 + \|\mathbf{v}_{\overline{S}}\|_1$$
$$> |\langle \mathbf{x} - \mathbf{v}, \operatorname{sgn}(\mathbf{x})_S \rangle| + |\langle \mathbf{v}, \operatorname{sgn}(\mathbf{x})_S \rangle| \geq |\langle \mathbf{x}, \operatorname{sgn}(\mathbf{x})_S \rangle| = \|\mathbf{x}\|_1.$$

The implication $(b) \Rightarrow (a)$ is also simple. Indeed, observing that $\mathbf{A}\mathbf{v}_S = -\mathbf{A}\mathbf{v}_{\overline{S}}$ for $\mathbf{v} \in \ker \mathbf{A} \setminus \{\mathbf{0}\}$, we write

$$\left| \sum_{j \in S} \overline{\operatorname{sgn}(x_j)} v_j \right| = |\langle \mathbf{v}_S, \mathbf{A}^*\mathbf{h} \rangle| = |\langle \mathbf{A}\mathbf{v}_S, \mathbf{h} \rangle| = |\langle \mathbf{A}\mathbf{v}_{\overline{S}}, \mathbf{h} \rangle|$$

$$= |\langle \mathbf{v}_{\overline{S}}, \mathbf{A}^*\mathbf{h} \rangle| \leq \max_{\ell \in \overline{S}} |(\mathbf{A}^*\mathbf{h})_\ell| \, \|\mathbf{v}_{\overline{S}}\|_1 < \|\mathbf{v}_{\overline{S}}\|_1.$$

The strict inequality holds since $\|\mathbf{v}_{\overline{S}}\|_1 > 0$; otherwise, the nonzero vector $\mathbf{v} \in \ker \mathbf{A}$ would be supported on S, contradicting the injectivity of $\mathbf{A}_S$.

The remaining implication $(a) \Rightarrow (b)$ requires more work. We start by noticing that (a) implies $\|\mathbf{v}_{\overline{S}}\|_1 > 0$ for all $\mathbf{v} \in \ker \mathbf{A} \setminus \{\mathbf{0}\}$. It follows that the matrix $\mathbf{A}_S$ is injective. Indeed, assume $\mathbf{A}_S \mathbf{v}_S = \mathbf{0}$ for some $\mathbf{v}_S \neq \mathbf{0}$ and complete $\mathbf{v}_S$ to a vector $\mathbf{v} \in \mathbb{C}^N$ by setting $\mathbf{v}_{\overline{S}} = \mathbf{0}$. Then $\mathbf{v}$ is contained in $\ker \mathbf{A} \setminus \{\mathbf{0}\}$, which is in contradiction with $\|\mathbf{v}_{\overline{S}}\|_1 > 0$ for all $\mathbf{v} \in \ker \mathbf{A} \setminus \{\mathbf{0}\}$. Next, since the continuous function $\mathbf{v} \mapsto |\langle \mathbf{v}, \operatorname{sgn}(\mathbf{x})_S \rangle| / \|\mathbf{v}_{\overline{S}}\|_1$ takes values less than one on the unit sphere of $\ker \mathbf{A}$, which is compact, its maximum μ satisfies $\mu < 1$. By homogeneity, we deduce

$$|\langle \mathbf{v}, \operatorname{sgn}(\mathbf{x})_S \rangle| \leq \mu \|\mathbf{v}_{\overline{S}}\|_1 \quad \text{for all } \mathbf{v} \in \ker \mathbf{A}.$$

We then define, for $\mu < \nu < 1$, the convex set $\mathcal{C}$ and the affine set $\mathcal{D}$ by

$$\mathcal{C} := \left\{ \mathbf{z} \in \mathbb{C}^N : \|\mathbf{z}_S\|_1 + \nu \|\mathbf{z}_{\overline{S}}\|_1 \leq \|\mathbf{x}\|_1 \right\},$$
$$\mathcal{D} := \left\{ \mathbf{z} \in \mathbb{C}^N : \mathbf{A}\mathbf{z} = \mathbf{A}\mathbf{x} \right\}.$$

The intersection $\mathcal{C} \cap \mathcal{D}$ reduces to $\{\mathbf{x}\}$. Indeed, we observe that $\mathbf{x} \in \mathcal{C} \cap \mathcal{D}$, and if $\mathbf{z} \neq \mathbf{x}$ belongs to $\mathcal{C} \cap \mathcal{D}$, setting $\mathbf{v} := \mathbf{x} - \mathbf{z} \in \ker \mathbf{A} \setminus \{\mathbf{0}\}$, we obtain a contradiction from

$$\|\mathbf{x}\|_1 \geq \|\mathbf{z}_S\|_1 + \nu \|\mathbf{z}_{\overline{S}}\|_1 = \|(\mathbf{x} - \mathbf{v})_S\|_1 + \nu \|\mathbf{v}_{\overline{S}}\|_1$$
$$> \|(\mathbf{x} - \mathbf{v})_S\|_1 + \mu \|\mathbf{v}_{\overline{S}}\|_1 \geq |\langle \mathbf{x} - \mathbf{v}, \operatorname{sgn}(\mathbf{x})_S \rangle| + |\langle \mathbf{v}, \operatorname{sgn}(\mathbf{x})_S \rangle|$$
$$\geq |\langle \mathbf{x}, \operatorname{sgn}(\mathbf{x})_S \rangle| = \|\mathbf{x}\|_1.$$

Thus, by the separation of convex sets via hyperplanes (see Theorem B.4 and Remark B.5) there exists a vector $\mathbf{w} \in \mathbb{C}^N$ such that

$$\mathcal{C} \subset \left\{ \mathbf{z} \in \mathbb{C}^N : \operatorname{Re}\langle \mathbf{z}, \mathbf{w} \rangle \leq \|\mathbf{x}\|_1 \right\}, \tag{4.22}$$

$$\mathcal{D} \subset \left\{ \mathbf{z} \in \mathbb{C}^N : \operatorname{Re}\langle \mathbf{z}, \mathbf{w} \rangle = \|\mathbf{x}\|_1 \right\}. \tag{4.23}$$

In view of (4.22), we have

$$\|\mathbf{x}\|_1 \geq \max_{\|\mathbf{z}_S + \nu \mathbf{z}_{\overline{S}}\|_1 \leq \|\mathbf{x}\|_1} \mathrm{Re}\,\langle \mathbf{z}, \mathbf{w} \rangle$$

$$= \max_{\|\mathbf{z}_S + \nu \mathbf{z}_{\overline{S}}\|_1 \leq \|\mathbf{x}\|_1} \mathrm{Re}\left(\sum_{j \in S} z_j \overline{w_j} + \sum_{j \in \overline{S}} \nu z_j \overline{w_j}/\nu \right)$$

$$= \max_{\|\mathbf{z}_S + \nu \mathbf{z}_{\overline{S}}\|_1 \leq \|\mathbf{x}\|_1} \mathrm{Re}\,\langle \mathbf{z}_S + \nu \mathbf{z}_{\overline{S}}, \mathbf{w}_S + (1/\nu)\mathbf{w}_{\overline{S}} \rangle$$

$$= \|\mathbf{x}\|_1 \,\|\mathbf{w}_S + (1/\nu)\mathbf{w}_{\overline{S}}\|_\infty = \|\mathbf{x}\|_1 \max\left\{ \|\mathbf{w}_S\|_\infty, (1/\nu)\|\mathbf{w}_{\overline{S}}\|_\infty \right\}.$$

Setting aside the case $\mathbf{x} \neq \mathbf{0}$ (where the choice $\mathbf{h} = \mathbf{0}$ would do), we obtain $\|\mathbf{w}_S\|_\infty \leq 1$ and $\|\mathbf{w}_{\overline{S}}\|_\infty \leq \nu < 1$. From (4.23), we derive $\mathrm{Re}\,\langle \mathbf{x}, \mathbf{w} \rangle = \|\mathbf{x}\|_1$, i.e., $w_j = \mathrm{sgn}(x_j)$ for all $j \in S$, and also $\mathrm{Re}\,\langle \mathbf{v}, \mathbf{w} \rangle = 0$ for all $\mathbf{v} \in \ker \mathbf{A}$, i.e., $\mathbf{w} \in (\ker \mathbf{A})^\perp$. Since $(\ker \mathbf{A})^\perp = \mathrm{ran}\,\mathbf{A}^*$, we write $\mathbf{w} = \mathbf{A}^* \mathbf{h}$ for some $\mathbf{h} \in \mathbb{C}^m$. This establishes (b). $\square$

Remark 4.27. (a) If a vector $\mathbf{x} \in \mathbb{C}^N$ with support S satisfies condition (a) of the previous theorem, then all vectors $\mathbf{x}' \in \mathbb{C}^N$ with support $S' \subset S$ and $\mathrm{sgn}(\mathbf{x}')_{S'} = \mathrm{sgn}(\mathbf{x})_{S'}$ are also exactly recovered via basis pursuit. Indeed, for $\mathbf{v} \in \ker \mathbf{A} \setminus \{\mathbf{0}\}$,

$$\left| \sum_{j \in S'} \mathrm{sgn}(x'_j) v_j \right| = \left| \sum_{j \in S} \mathrm{sgn}(x_j) v_j - \sum_{j \in S \setminus S'} \mathrm{sgn}(x_j) v_j \right|$$

$$\leq \left| \sum_{j \in S} \mathrm{sgn}(x_j) v_j \right| + \sum_{j \in S \setminus S'} |v_j| < \|\mathbf{v}_{\overline{S}}\|_1 + \|\mathbf{v}_{S \setminus S'}\|_1 = \|\mathbf{v}_{\overline{S'}}\|_1.$$

(b) Theorem 4.26 can be made stable under noise on the measurements and under passing to compressible vectors; see Exercise 4.17 and also compare with Theorem 4.33 below. However, the resulting error bounds are slightly weaker than the ones of Theorem 4.25 under the ℓ_2-robust null space property.

The equalities $(\mathbf{A}^* \mathbf{h})_j = \mathrm{sgn}(x_j)$, $j \in S$, considered in (ii) translate into $\mathbf{A}_S^* \mathbf{h} = \mathrm{sgn}(\mathbf{x}_S)$. This is satisfied for the choice $\mathbf{h} = \left(\mathbf{A}_S^\dagger\right)^* \mathrm{sgn}(\mathbf{x}_S)$, where the expression $\mathbf{A}_S^\dagger := (\mathbf{A}_S^* \mathbf{A}_S)^{-1} \mathbf{A}_S^*$ of the *Moore–Penrose pseudo-inverse* of $\mathbf{A}_S$ is justified by its injectivity; see (A.28). Since the conditions $|(\mathbf{A}^* \mathbf{h})_\ell| < 1$, $\ell \in \overline{S}$, then read $|\langle \mathbf{a}_\ell, \mathbf{h} \rangle| < 1$, $\ell \in \overline{S}$, where $\mathbf{a}_1, \ldots, \mathbf{a}_N$ are the columns of $\mathbf{A}$, we can state the following result.

Corollary 4.28. *Let $\mathbf{a}_1, \ldots, \mathbf{a}_N$ be the columns of $\mathbf{A} \in \mathbb{C}^{m \times N}$. For $\mathbf{x} \in \mathbb{C}^N$ with support S, if the matrix $\mathbf{A}_S$ is injective and if*

$$|\langle \mathbf{A}_S^\dagger \mathbf{a}_\ell, \mathrm{sgn}(\mathbf{x}_S) \rangle| < 1 \quad \text{for all } \ell \in \overline{S}, \tag{4.24}$$

then the vector $\mathbf{x}$ is the unique solution of (P_1) with $\mathbf{y} = \mathbf{Ax}$.

Remark 4.29. In general, there is no converse to Theorem 4.26. Let us consider, for instance,

$$\mathbf{A} := \begin{bmatrix} 1 & 0 & -1 \\ 0 & 1 & -1 \end{bmatrix}, \qquad \mathbf{x} = \begin{bmatrix} e^{-\pi i/3} \\ e^{\pi i/3} \\ 0 \end{bmatrix}.$$

We can verify that $\mathbf{x}$ is the unique minimizer of $\|\mathbf{z}\|_1$ subject to $\mathbf{Az} = \mathbf{Ax}$; see Exercise 4.14. However, (a) fails. Indeed, for a vector $\mathbf{v} = [\zeta, \zeta, \zeta] \in \ker \mathbf{A} \setminus \{\mathbf{0}\}$, we have $|\overline{\text{sgn}(x_1)}v_1 + \overline{\text{sgn}(x_2)}v_2| = |(e^{\pi i/3} + e^{-\pi i/3})\zeta| = |\zeta|$, while $\|\mathbf{v}_{\{3\}}\|_1 = |\zeta|$. In contrast, a converse to Theorem 4.26 holds in the real setting.

Theorem 4.30. *Given a matrix $\mathbf{A} \in \mathbb{R}^{m \times N}$, a vector $\mathbf{x} \in \mathbb{R}^N$ with support S is the unique minimizer of $\|\mathbf{z}\|_1$ subject to $\mathbf{Az} = \mathbf{Ax}$ if and only if one of the following equivalent conditions holds:*

(a) $\left| \sum_{j \in S} \text{sgn}(x_j)v_j \right| < \|\mathbf{v}_{\overline{S}}\|_1$ *for all $\mathbf{v} \in \ker \mathbf{A} \setminus \{\mathbf{0}\}$.*

(b) $\mathbf{A}_S$ *is injective, and there exists a vector $\mathbf{h} \in \mathbb{R}^m$ such that*

$$(\mathbf{A}^\top \mathbf{h})_j = \text{sgn}(x_j), \quad j \in S, \qquad |(\mathbf{A}^\top \mathbf{h})_\ell| < 1, \quad \ell \in \overline{S}.$$

Proof. The arguments given in the proof of Theorem 4.26 still hold in the real setting; hence, it is enough to show that (a) holds as soon as $\mathbf{x}$ is the unique minimizer of $\|\mathbf{z}\|_1$ subject to $\mathbf{Az} = \mathbf{Ax}$. In this situation, for $\mathbf{v} \in \ker \mathbf{A} \setminus \{\mathbf{0}\}$, the vector $\mathbf{z} := \mathbf{x} - \mathbf{v}$ satisfies $\mathbf{z} \neq \mathbf{x}$ and $\mathbf{Az} = \mathbf{Ax}$, so that

$$\|\mathbf{x}\|_1 < \|\mathbf{z}\|_1 = \|\mathbf{z}_S\|_1 + \|\mathbf{z}_{\overline{S}}\|_1 = \langle \mathbf{z}, \text{sgn}(\mathbf{z})_S \rangle + \|\mathbf{z}_{\overline{S}}\|_1.$$

Taking $\|\mathbf{x}\|_1 \geq \langle \mathbf{x}, \text{sgn}(\mathbf{z})_S \rangle$ into account, we derive $\langle \mathbf{x} - \mathbf{z}, \text{sgn}(\mathbf{z})_S \rangle < \|\mathbf{z}_{\overline{S}}\|_1$. Hence, we have

$$\langle \mathbf{v}, \text{sgn}(\mathbf{x} - \mathbf{v})_S \rangle < \|\mathbf{v}_{\overline{S}}\|_1 \quad \text{for all } \mathbf{v} \in \ker \mathbf{A} \setminus \{\mathbf{0}\}.$$

Writing the latter with $\mathbf{v} \in \ker \mathbf{A} \setminus \{\mathbf{0}\}$ replaced by $t\mathbf{v}$, $t > 0$, and simplifying by t, we obtain

$$\langle \mathbf{v}, \text{sgn}(\mathbf{x} - t\mathbf{v})_S \rangle < \|\mathbf{v}_{\overline{S}}\|_1 \quad \text{for all } \mathbf{v} \in \ker \mathbf{A} \setminus \{\mathbf{0}\} \text{ and all } t > 0.$$

Taking $t > 0$ small enough so that $\text{sgn}(x_j - tv_j) = \text{sgn}(x_j)$—note that it is essential for $\mathbf{x}$ to be *exactly* supported on S—we conclude

$$\langle \mathbf{v}, \text{sgn}(\mathbf{x})_S \rangle < \|\mathbf{v}_{\overline{S}}\|_1 \quad \text{for all } \mathbf{v} \in \ker \mathbf{A} \setminus \{\mathbf{0}\},$$

which implies (a) by replacing $\mathbf{v}$ by $-\mathbf{v}$ if necessary. $\qquad \square$

Remark 4.31. Theorem 4.30 shows that in the real setting the recovery of a given vector via basis pursuit depends only on its sign pattern, but not on the magnitude of its entries. Moreover, by Remark 4.27(a), if a vector $\mathbf{x} \in \mathbb{R}^N$ with support S is exactly recovered via basis pursuit, then all vectors $x' \in \mathbb{R}^N$ with support $S' \subset S$ and $\mathrm{sgn}(\mathbf{x}')_{S'} = \mathrm{sgn}(\mathbf{x})_{S'}$ are also exactly recovered via basis pursuit.

The construction of the "dual vector" $\mathbf{h}$ described in property (b) of Theorems 4.26 and 4.30 is not always straightforward. The following condition involving an "inexact dual vector" is sometimes easier to verify.

Theorem 4.32. *Let $\mathbf{a}_1, \dots, \mathbf{a}_N$ be the columns of $\mathbf{A} \in \mathbb{C}^{m \times N}$ and let $\mathbf{x} \in \mathbb{C}^N$ with support S. For $\alpha, \beta, \gamma, \theta \geq 0$, assume that*

$$\|(\mathbf{A}_S^* \mathbf{A}_S)^{-1}\|_{2 \to 2} \leq \alpha, \qquad \max_{\ell \in \overline{S}} \|\mathbf{A}_S^* \mathbf{a}_\ell\|_2 \leq \beta, \qquad (4.25)$$

and that there exists a vector $\mathbf{u} = \mathbf{A}^* \mathbf{h} \in \mathbb{C}^N$ with $\mathbf{h} \in \mathbb{C}^m$ such that

$$\|\mathbf{u}_S - \mathrm{sgn}(\mathbf{x}_S)\|_2 \leq \gamma \quad and \quad \|\mathbf{u}_{\overline{S}}\|_\infty \leq \theta. \qquad (4.26)$$

If $\theta + \alpha\beta\gamma < 1$, then $\mathbf{x}$ is the unique minimizer of $\|\mathbf{z}\|_1$ subject to $\mathbf{Az} = \mathbf{Ax}$.

Proof. According to Theorem 4.26, it is enough to prove that

$$|\langle \mathbf{v}, \mathrm{sgn}(\mathbf{x}_S) \rangle| < \|\mathbf{v}_{\overline{S}}\|_1$$

for all $\mathbf{v} \in \ker \mathbf{A} \setminus \{\mathbf{0}\}$. For such a $\mathbf{v} \in \ker \mathbf{A} \setminus \{\mathbf{0}\}$, since $\mathbf{u} \in \mathrm{ran}\, \mathbf{A}^* = (\ker \mathbf{A})^{\perp}$,

$$|\langle \mathbf{v}, \mathrm{sgn}(\mathbf{x}_S) \rangle| = |\langle \mathbf{v}, \mathrm{sgn}(\mathbf{x}_S) - \mathbf{u} \rangle| \leq |\langle \mathbf{v}_{\overline{S}}, \mathbf{u}_{\overline{S}} \rangle| + |\langle \mathbf{v}_S, \mathrm{sgn}(\mathbf{x}_S) - \mathbf{u}_S \rangle|$$

$$\leq \theta \|\mathbf{v}_{\overline{S}}\|_1 + \gamma \|\mathbf{v}_S\|_2,$$

where the last step uses both inequalities of (4.26). From (4.25), we derive

$$\|\mathbf{v}_S\|_2 = \|(\mathbf{A}_S^* \mathbf{A}_S)^{-1} \mathbf{A}_S^* \mathbf{A}_S \mathbf{v}_S\|_2 \leq \alpha \|\mathbf{A}_S^* \mathbf{A}_S \mathbf{v}_S\|_2 = \alpha \|\mathbf{A}_S^* \mathbf{A}_{\overline{S}} \mathbf{v}_{\overline{S}}\|_2$$

$$= \alpha \Big\| \sum_{\ell \in \overline{S}} v_\ell \mathbf{A}_S^* \mathbf{a}_\ell \Big\|_2 \leq \alpha \sum_{\ell \in \overline{S}} |v_\ell| \|\mathbf{A}_S^* \mathbf{a}_\ell\|_2 \leq \alpha\beta \|\mathbf{v}_{\overline{S}}\|_1.$$

This implies that

$$|\langle \mathbf{v}, \mathrm{sgn}(\mathbf{x}_S) \rangle| \leq (\theta + \alpha\beta\gamma) \|\mathbf{v}_{\overline{S}}\|_1.$$

The conclusion follows from $\theta + \alpha\beta\gamma < 1$. $\qquad \qquad \square$

The next statement makes the result stable and robust with respect to sparsity defect and measurement error. However, due to the appearance of an extra factor $\sqrt{s}$, the error bound is not as sharp as (4.16), which was obtained under the ℓ_2-robust null space property. Nevertheless, since it applies under weaker conditions on $\mathbf{A}$, it proves useful in certain situations, especially when the null space property (or the restricted isometry property to be studied in Chap. 6) is not known to hold or harder to establish; see, for instance, Chap. 12.

Theorem 4.33. *Let* $\mathbf{a}_1, \ldots, \mathbf{a}_N$ *be the columns of* $\mathbf{A} \in \mathbb{C}^{m \times N}$, *let* $\mathbf{x} \in \mathbb{C}^N$ *with* s *largest absolute entries supported on* S, *and let* $\mathbf{y} = \mathbf{A}\mathbf{x} + \mathbf{e}$ *with* $\|\mathbf{e}\|_2 \le \eta$. *For* $\delta, \beta, \gamma, \theta, \tau \ge 0$ *with* $\delta < 1$, *assume that*

$$\|\mathbf{A}_S^* \mathbf{A}_S - \mathbf{Id}\|_{2 \to 2} \le \delta, \qquad \max_{\ell \in \overline{S}} \|\mathbf{A}_S^* \mathbf{a}_\ell\|_2 \le \beta, \qquad (4.27)$$

and that there exists a vector $\mathbf{u} = \mathbf{A}^* \mathbf{h} \in \mathbb{C}^N$ *with* $\mathbf{h} \in \mathbb{C}^m$ *such that*

$$\|\mathbf{u}_S - \mathrm{sgn}(\mathbf{x}_S)\|_2 \le \gamma, \quad \|\mathbf{u}_{\overline{S}}\|_\infty \le \theta, \quad \text{and} \quad \|\mathbf{h}\|_2 \le \tau\sqrt{s}. \qquad (4.28)$$

If $\rho := \theta + \beta\gamma/(1 - \delta) < 1$, *then a minimizer* $\mathbf{x}^\sharp$ *of* $\|\mathbf{z}\|_1$ *subject to* $\|\mathbf{A}\mathbf{z} - \mathbf{y}\|_2 \le \eta$ *satisfies*

$$\|\mathbf{x} - \mathbf{x}^\sharp\|_2 \le C_1 \sigma_s(\mathbf{x})_1 + (C_2 + C_3\sqrt{s})\eta$$

for some constants $C_1, C_2, C_3 > 0$ *depending only on* $\delta, \beta, \gamma, \theta, \tau$.

Remark 4.34. The proof reveals explicit values of the constants, namely,

$$C_1 = \frac{2}{1 - \rho}\left(1 + \frac{\beta}{1 - \delta}\right), \qquad C_2 = \frac{2\sqrt{1 + \delta}}{1 - \delta}\mu\left(\frac{\gamma}{1 - \rho}\left(1 + \frac{\beta}{1 - \delta}\right) + 1\right),$$

$$C_3 = \frac{2\tau}{1 - \rho}\left(1 + \frac{\beta}{1 - \delta}\right).$$

For instance, the specific choice $\delta = \beta = \gamma = 1/2$, $\theta = 1/4$, and $\tau = 2$, for which $\rho = 3/4$, results in $C_1 \approx 16$, $C_2 = 10\sqrt{6} \approx 24.49$, and $C_3 \approx 32$.

Proof. Observe that $\mathbf{x}$ is feasible for the quadratically constrained ℓ_1-minimization problem due to the assumed ℓ_2-bound of the perturbation $\mathbf{e}$. Setting $\mathbf{v} := \mathbf{x}^\sharp - \mathbf{x}$, the minimality of $\|\mathbf{x}^\sharp\|_1$ implies

$$\|\mathbf{x}\|_1 \ge \|\mathbf{x}^\sharp\|_1 = \|\mathbf{x} + \mathbf{v}\|_1 = \|(\mathbf{x} + \mathbf{v})_S\|_1 + \|(\mathbf{x} + \mathbf{v})_{\overline{S}}\|_1$$

$$\ge \mathrm{Re}\langle(\mathbf{x} + \mathbf{v})_S, \mathrm{sgn}(\mathbf{x}_S)\rangle + \|\mathbf{v}_{\overline{S}}\|_1 - \|\mathbf{x}_{\overline{S}}\|_1$$

$$= \|\mathbf{x}_S\|_1 + \mathrm{Re}\langle\mathbf{v}_S, \mathrm{sgn}(\mathbf{x}_S)\rangle + \|\mathbf{v}_{\overline{S}}\|_1 - \|\mathbf{x}_{\overline{S}}\|_1.$$

Rearranging and using the fact that $\|\mathbf{x}\|_1 = \|\mathbf{x}_S\|_1 + \|\mathbf{x}_{\overline{S}}\|_1$ yields

$$\|\mathbf{v}_{\overline{S}}\|_1 \le 2\|\mathbf{x}_{\overline{S}}\|_1 + |\langle\mathbf{v}_S, \mathrm{sgn}(\mathbf{x}_S)\rangle|. \qquad (4.29)$$

In view of (4.28), we have

$$|\langle \mathbf{v}_S, \operatorname{sgn}(\mathbf{x}_S) \rangle| \leq |\langle \mathbf{v}_S, \operatorname{sgn}(\mathbf{x}_S) - \mathbf{u}_S \rangle| + |\langle \mathbf{v}_S, \mathbf{u}_S \rangle|$$

$$\leq \gamma \|\mathbf{v}_S\|_2 + |\langle \mathbf{v}, \mathbf{u} \rangle| + |\langle \mathbf{v}_{\overline{S}}, \mathbf{u}_{\overline{S}} \rangle|. \tag{4.30}$$

The first inequality of (4.27) guarantees (see, for instance, Lemma A.12 and Proposition A.15) that $\|(\mathbf{A}_S^* \mathbf{A}_S)^{-1}\|_{2 \to 2} \leq 1/(1 - \delta)$ and $\|\mathbf{A}_S^*\|_{2 \to 2} \leq \sqrt{1 + \delta}$. Hence,

$$\|\mathbf{v}_S\|_2 \leq \frac{1}{1 - \delta} \|\mathbf{A}_S^* \mathbf{A}_S \mathbf{v}_S\|_2 \leq \frac{1}{1 - \delta} \|\mathbf{A}_S^* \mathbf{A}_{\overline{S}} \mathbf{v}_{\overline{S}}\|_2 + \frac{1}{1 - \delta} \|\mathbf{A}_S^* \mathbf{A} \mathbf{v}\|_2$$

$$\leq \frac{1}{1 - \delta} \sum_{\ell \in \overline{S}} |v_\ell| \|\mathbf{A}_S^* \mathbf{a}_\ell\|_2 + \frac{\sqrt{1 + \delta}}{1 - \delta} \|\mathbf{A} \mathbf{v}\|_2$$

$$\leq \frac{\beta}{1 - \delta} \|\mathbf{v}_{\overline{S}}\|_1 + \frac{2\sqrt{1 + \delta}}{1 - \delta} \eta. \tag{4.31}$$

The last step involved the inequality $\|\mathbf{A}\mathbf{v}\|_2 \leq 2\eta$, which follows from the optimization constraint as

$$\|\mathbf{A}\mathbf{v}\|_2 = \|\mathbf{A}(\mathbf{x}^\sharp - \mathbf{x})\|_2 \leq \|\mathbf{A}\mathbf{x}^\sharp - \mathbf{y}\|_2 + \|\mathbf{y} - \mathbf{A}\mathbf{x}\|_2 \leq 2\eta.$$

The latter inequality combined with $\|\mathbf{h}\|_2 \leq \tau \sqrt{s}$ also gives

$$|\langle \mathbf{v}, \mathbf{u} \rangle| = |\langle \mathbf{v}, \mathbf{A}^* \mathbf{h} \rangle| = |\langle \mathbf{A}\mathbf{v}, \mathbf{h} \rangle| \leq \|\mathbf{A}\mathbf{v}\|_2 \|\mathbf{h}\|_2 \leq 2\tau \eta \sqrt{s},$$

while $\|\mathbf{u}_{\overline{S}}\|_\infty \leq \theta$ implies $|\langle \mathbf{v}_{\overline{S}}, \mathbf{u}_{\overline{S}} \rangle| \leq \theta \|\mathbf{v}_{\overline{S}}\|_1$. Substituting these estimates in (4.30) and in turn in (4.29) yields

$$\|\mathbf{v}_{\overline{S}}\|_1 \leq 2\|\mathbf{x}_{\overline{S}}\|_1 + \left(\theta + \frac{\beta \gamma}{1 - \delta} \right) \|\mathbf{v}_{\overline{S}}\|_1 + \left(2\gamma \frac{\sqrt{1 + \delta}}{1 - \delta} + 2\tau \sqrt{s} \right) \eta.$$

Since $\rho = \theta + \beta \gamma/(1 - \delta) < 1$, this can be rearranged as

$$\|\mathbf{v}_{\overline{S}}\|_1 \leq \frac{2}{1 - \rho} \|\mathbf{x}_{\overline{S}}\|_1 + \frac{2(\mu \gamma + \tau \sqrt{s})}{1 - \rho} \eta, \tag{4.32}$$

where $\mu := \sqrt{1 + \delta}/(1 - \delta)$. Using (4.31) once again, we derive

$$\|\mathbf{v}_S\|_2 \leq \frac{2\beta}{(1 - \rho)(1 - \delta)} \|\mathbf{x}_{\overline{S}}\|_1 + \left(\frac{2\beta(\mu \gamma + \tau \sqrt{s})}{(1 - \rho)(1 - \delta)} + 2\mu \right) \eta. \tag{4.33}$$

Finally, combining (4.32) and (4.33), we obtain

$$\|\mathbf{v}\|_2 \le \|\mathbf{v}_{\overline{S}}\|_2 + \|\mathbf{v}_S\|_2 \le \|\mathbf{v}_{\overline{S}}\|_1 + \|\mathbf{v}_S\|_2$$

$$\le \frac{2}{1-\rho}\left(1 + \frac{\beta}{1-\delta}\right)\|\mathbf{x}_{\overline{S}}\|_1 + \left(\frac{2(\mu\gamma + \tau\sqrt{s})}{1-\rho}\left(1 + \frac{\beta}{1-\delta}\right) + 2\mu\right)\eta.$$

Taking $\|\mathbf{x}_{\overline{S}}\|_1 = \sigma_s(\mathbf{x})_1$ into account, we arrive at the desired result. $\square$

The next characterization of exact recovery via ℓ_1-minimization involves tangent cones to the ℓ_1-ball. For a vector $\mathbf{x} \in \mathbb{R}^N$, we introduce the convex cone

$$T(\mathbf{x}) = \text{cone}\{\mathbf{z} - \mathbf{x} : \mathbf{z} \in \mathbb{R}^N, \|\mathbf{z}\|_1 \le \|\mathbf{x}\|_1\}, \tag{4.34}$$

where the notation cone represents the conic hull; see (B.4).

Theorem 4.35. *For $\mathbf{A} \in \mathbb{R}^{m \times N}$, a vector $\mathbf{x} \in \mathbb{R}^N$ is the unique minimizer of $\|\mathbf{z}\|_1$ subject to $\mathbf{A}\mathbf{z} = \mathbf{A}\mathbf{x}$ if and only if $\ker \mathbf{A} \cap T(\mathbf{x}) = \{\mathbf{0}\}$.*

Proof. Assume that $\ker \mathbf{A} \cap T(\mathbf{x}) = \{\mathbf{0}\}$. Let $\mathbf{x}^\sharp$ be an ℓ_1-minimizer. We have $\|\mathbf{x}^\sharp\|_1 \le \|\mathbf{x}\|_1$ and $\mathbf{A}\mathbf{x}^\sharp = \mathbf{A}\mathbf{x}$, so that $\mathbf{v} := \mathbf{x}^\sharp - \mathbf{x} \in T(\mathbf{x}) \cap \ker \mathbf{A} = \{\mathbf{0}\}$; hence, $\mathbf{x}^\sharp = \mathbf{x}$. This means that $\mathbf{x}$ is the unique ℓ_1-minimizer. Conversely, assume that $\mathbf{x}$ is the unique ℓ_1-minimizer. A vector $\mathbf{v} \in T(\mathbf{x}) \setminus \{\mathbf{0}\}$ can be written as $\mathbf{v} = \sum t_j(\mathbf{z}_j - \mathbf{x})$ with $t_j \ge 0$ and $\|\mathbf{z}_j\|_1 \le \|\mathbf{x}\|_1$. Note that $\mathbf{v} \ne \mathbf{0}$ implies $\sum t_i > 0$, and we can consider $t'_j := t_j/(\sum t_i)$. If $\mathbf{v} \in \ker \mathbf{A}$, we would have $\mathbf{A}(\sum t'_j \mathbf{z}_j) = \mathbf{A}\mathbf{x}$, while $\|\sum t'_j \mathbf{z}_j\|_1 \le \sum t'_j \|\mathbf{z}_j\|_1 \le \|\mathbf{x}\|_1$. By uniqueness of an ℓ_1-minimizer, this would imply $\sum t'_j \mathbf{z}_j = \mathbf{x}$, so that $\mathbf{v} = \mathbf{0}$, which is a contradiction. We conclude that $(T(\mathbf{x}) \setminus \{\mathbf{0}\}) \cap \ker \mathbf{A} = \emptyset$, i.e., $T(\mathbf{x}) \cap \ker \mathbf{A} = \{\mathbf{0}\}$, as desired. $\square$

The above theorem extends to robust recovery as follows.

Theorem 4.36. *For $\mathbf{A} \in \mathbb{R}^{m \times N}$, let $\mathbf{x} \in \mathbb{R}^N$ and $\mathbf{y} = \mathbf{A}\mathbf{x} + \mathbf{e} \in \mathbb{R}^m$ with $\|\mathbf{e}\|_2 \le \eta$. If*

$$\inf_{\mathbf{v} \in T(\mathbf{x}), \|\mathbf{v}\|_2 = 1} \|\mathbf{A}\mathbf{v}\|_2 \ge \tau$$

for some $\tau > 0$, then a minimizer $\mathbf{x}^\sharp$ of $\|\mathbf{z}\|_1$ subject to $\|\mathbf{A}\mathbf{z} - \mathbf{y}\|_2 \le \eta$ satisfies

$$\|\mathbf{x} - \mathbf{x}^\sharp\|_2 \le \frac{2\eta}{\tau}. \tag{4.35}$$

Proof. The inequality $\|\mathbf{x}^\sharp\|_1 \le \|\mathbf{x}\|_1$ yields $\mathbf{v} := (\mathbf{x}^\sharp - \mathbf{x})/\|\mathbf{x}^\sharp - \mathbf{x}\|_2 \in T(\mathbf{x})$—note that $\mathbf{x}^\sharp - \mathbf{x} \ne \mathbf{0}$ can be safely assumed. Since $\|\mathbf{v}\|_2 = 1$, the assumption implies $\|\mathbf{A}\mathbf{v}\|_2 \ge \tau$, i.e., $\|\mathbf{A}(\mathbf{x}^\sharp - \mathbf{x})\|_2 \ge \tau\|\mathbf{x}^\sharp - \mathbf{x}\|_2$. It remains to remark that

$$\|\mathbf{A}(\mathbf{x}^\sharp - \mathbf{x})\|_2 \le \|\mathbf{A}\mathbf{x}^\sharp - \mathbf{y}\|_2 + \|\mathbf{A}\mathbf{x} - \mathbf{y}\|_2 \le 2\eta$$

in order to obtain the desired result. $\square$

Remark 4.37. Theorems 4.35 and 4.36 both remain valid in the complex case by considering the complex cone $T(\mathbf{x}) = \text{cone}\{\mathbf{z} - \mathbf{x} : \mathbf{z} \in \mathbb{C}^N, \|\mathbf{z}\|_1 \leq \|\mathbf{x}\|_1\}$.

4.5   The Projected Cross-Polytope

In this section, we build on Theorem 4.30 to characterize geometrically the success of basis pursuit in recovering an individual sparse vector or all sparse vectors simultaneously. We first recall that a *convex polytope* P in $\mathbb{R}^n$ can be viewed either as the convex hull of a finite set of points or as a bounded intersection of finitely many half-spaces. For instance, with $(\mathbf{e}_1, \ldots, \mathbf{e}_N)$ denoting the canonical basis of $\mathbb{R}^N$, the unit ball of ℓ_1^N described as

$$ B_1^N := \text{conv}\{\mathbf{e}_1, -\mathbf{e}_1, \ldots, \mathbf{e}_N, -\mathbf{e}_N\} = \bigcap_{\varepsilon \in \{-1,1\}^N} \left\{ \mathbf{z} \in \mathbb{R}^N : \sum_{i=1}^N \varepsilon_i z_i \leq 1 \right\} $$

is a convex polytope which is sometimes called *cross-polytope*. Its image under a matrix $\mathbf{A} \in \mathbb{R}^{m \times N}$ is also a convex polytope, since it is the convex hull of $\{\mathbf{A}\mathbf{e}_1, -\mathbf{A}\mathbf{e}_1, \ldots, \mathbf{A}\mathbf{e}_N, -\mathbf{A}\mathbf{e}_N\}$. A *face* of a convex polytope P in $\mathbb{R}^n$ is a set of the form

$$ F = \{\mathbf{z} \in P : \langle \mathbf{z}, \mathbf{h} \rangle = c\} $$

for some $\mathbf{h} \in \mathbb{R}^n$ and $c \in \mathbb{R}$, given that $\langle \mathbf{z}, \mathbf{h} \rangle \leq c$ holds for all $\mathbf{z} \in P$. Note that $c > 0$ if F is a proper face of a symmetric convex polytope P, so we may always assume $c = 1$ in this case. A face F of P is called a *k-face* if its affine hull has dimension k. The 0-, 1-, $(n-2)$-, and $(n-1)$-faces are called vertices, edges, ridges, and facets, respectively. For $0 \leq k \leq N-1$, it can be verified that the k-faces of B_1^N are the $2^{k+1} \binom{N}{k+1}$ sets

$$ \left\{ \mathbf{z} \in B_1^N : \sum_{j \in J} \varepsilon_j z_j = 1 \right\} = \text{conv}\{\varepsilon_j \mathbf{e}_j, j \in J\}, $$

where $J \subset [N]$ has size $k+1$ and $\varepsilon \in \{-1, 1\}^J$. Thus, if a vector $\mathbf{x} \in \mathbb{R}^N$ with $\|\mathbf{x}\|_1 = 1$ is exactly s-sparse, it is contained in exactly one $(s-1)$-face of B_1^N, namely,

$$ F_\mathbf{x} := \text{conv}\{\text{sgn}(x_j)\mathbf{e}_j, j \in \text{supp}(\mathbf{x})\}. $$

We are now in a position to state the necessary and sufficient condition for the recovery of individual vectors via basis pursuit.

Theorem 4.38. *For a matrix* $\mathbf{A} \in \mathbb{R}^{m \times N}$, *a vector* $\mathbf{x} \in \mathbb{R}^N$ *with support* S *of size* $s \geq 1$ *is the unique minimizer of* $\|\mathbf{z}\|_1$ *subject to* $\mathbf{Az} = \mathbf{Ax}$ *if and only if the* $(s-1)$*-face* $F_{\mathbf{x}/\|\mathbf{x}\|_1}$ *of* B_1^N *maps to an* $(s-1)$*-face of* $\mathbf{A}B_1^N$ *not containing any* $\varepsilon \mathbf{A e}_\ell, \varepsilon = \pm 1, \ell \in \overline{S}$.

Proof. We assume without loss of generality that $\|\mathbf{x}\|_1 = 1$. Before proving both implications, we notice that if $\mathbf{A}F_{\mathbf{x}}$ is a face of $\mathbf{A}B_1^N$, then it is an $(s-1)$-face if and only if its affine hull $\mathbf{Ax} + V$,

$$V := \Big\{ \sum_{j \in S} t_j \mathrm{sgn}(x_j) \mathbf{A e}_j, \ \mathbf{t} \in \mathbb{R}^S \ \text{with} \ \sum_{j \in S} t_j = 0 \Big\},$$

has dimension $s - 1$. This is equivalent to saying that the range of the linear map $f : \mathbf{t} \mapsto \sum_{j \in S} t_j \mathrm{sgn}(x_j) \mathbf{A e}_j$ defined on $T := \{ \mathbf{t} \in \mathbb{R}^S : \sum_{j \in S} t_j = 0 \}$ has dimension $s - 1$. Since $\dim(T) = s - 1$, this is also equivalent to saying that f is injective (in other words, that $\sum_{j \in S} t_j \mathrm{sgn}(x_j) \mathbf{A e}_j = 0$ and $\sum_{j \in S} t_j = 0$ can both occur only when $\mathbf{t} = 0$ or that the system $\{ \mathrm{sgn}(x_j) \mathbf{A e}_j, j \in S \}$ is affinely independent).

Let us assume on the one hand that $\mathbf{A}F_{\mathbf{x}}$ is an $(s-1)$-face of $\mathbf{A}B_1^N$ not containing any $\varepsilon \mathbf{A e}_\ell, \varepsilon = \pm 1, \ell \in \overline{S}$. Since it is a face of $\mathbf{A}B_1^N$, there exists $\mathbf{h} \in \mathbb{R}^m$ such that if $\mathbf{z} \in B_1^N$, then

$$\langle \mathbf{Az}, \mathbf{h} \rangle \leq 1 \qquad \text{and} \qquad [\langle \mathbf{Az}, \mathbf{h} \rangle = 1 \Leftrightarrow \mathbf{Az} \in \mathbf{A}F_{\mathbf{x}}].$$

For $j \in S$, since $\mathrm{sgn}(x_j) \mathbf{A e}_j \in \mathbf{A}F_{\mathbf{x}}$, we have

$$(\mathbf{A}^\top \mathbf{h})_j = \langle \mathbf{e}_j, \mathbf{A}^\top \mathbf{h} \rangle = \langle \mathbf{A e}_j, \mathbf{h} \rangle = \mathrm{sgn}(x_j) \langle \mathrm{sgn}(x_j) \mathbf{A e}_j, \mathbf{h} \rangle = \mathrm{sgn}(x_j).$$

For $\ell \in \overline{S}$, since $\mathbf{A e}_\ell \notin \mathbf{A}F_{\mathbf{x}}$ and $-\mathbf{A e}_\ell \notin \mathbf{A}F_{\mathbf{x}}$, we have, for some $\varepsilon = \pm 1$,

$$|(\mathbf{A}^\top \mathbf{h})_\ell| = \varepsilon (\mathbf{A}^\top \mathbf{h})_\ell = \varepsilon \langle \mathbf{e}_\ell, \mathbf{A}^\top \mathbf{h} \rangle = \langle \varepsilon \mathbf{A e}_\ell, \mathbf{h} \rangle < 1.$$

Let us now consider $\mathbf{z} \in \mathbb{R}^N$ satisfying $\mathbf{Az} = \mathbf{Ax}$ and $\|\mathbf{z}\|_1 \leq \|\mathbf{x}\|_1 = 1$. From

$$1 = \langle \mathbf{Ax}, \mathbf{h} \rangle = \langle \mathbf{Az}, \mathbf{h} \rangle = \langle \mathbf{z}, \mathbf{A}^\top \mathbf{h} \rangle = \sum_{j \in S} z_j \mathrm{sgn}(x_j) + \sum_{\ell \in \overline{S}} z_\ell (\mathbf{A}^\top \mathbf{h})_\ell$$

$$\leq \sum_{j \in S} |z_j| + \sum_{\ell \in \overline{S}} |z_\ell| = \|\mathbf{z}\|_1 \leq 1, \tag{4.36}$$

we derive that all inequalities above are equalities, i.e., $z_j \mathrm{sgn}(x_j) = |z_j|$ for all $j \in S$, $z_\ell = 0$ for all $\ell \in \overline{S}$ (otherwise there is at least one strict inequality $z_\ell (\mathbf{A}^\top \mathbf{h})_\ell < |z_\ell|$), and $\|\mathbf{z}\|_1 = 1$. With $t_j := |z_j| - |x_j|, j \in S$, we then obtain $\sum_{j \in S} t_j = \|\mathbf{z}\|_1 - \|\mathbf{x}\|_1 = 0$ and

$$\sum_{j \in S} t_j \mathrm{sgn}(x_j) \mathbf{A} \mathbf{e}_j = \sum_{j \in S} z_j \mathbf{A} \mathbf{e}_j - \sum_{j \in S} x_j \mathbf{A} \mathbf{e}_j = \mathbf{A} \mathbf{z} - \mathbf{A} \mathbf{x} = 0.$$

The injectivity of f (due to the fact that $\mathbf{A} B_1^N$ is an $(s-1)$-face) implies $|z_j| = |x_j|$ for all $j \in S$ and in turn $\mathbf{z} = \mathbf{x}$. This proves that $\mathbf{x}$ is the unique minimizer of $\|\mathbf{z}\|_1$ subject to $\mathbf{A} \mathbf{z} = \mathbf{A} \mathbf{x}$.

On the other hand, let us assume that $\mathbf{x}$ is the unique minimizer of $\|\mathbf{z}\|_1$ subject to $\mathbf{A} \mathbf{z} = \mathbf{A} \mathbf{x}$. By Theorem 4.30, the matrix $\mathbf{A}_S$ is injective, and there exists $\mathbf{h} \in \mathbb{R}^m$ such that $(\mathbf{A}^\top \mathbf{h})_j = \mathrm{sgn}(x_j)$ for all $j \in S$ and $|(\mathbf{A}^\top \mathbf{h})_\ell| < 1$ for all $\ell \in \overline{S}$. We observe that, for $\mathbf{z} \in B_1^N$,

$$\langle \mathbf{A} \mathbf{z}, \mathbf{h} \rangle = \langle \mathbf{z}, \mathbf{A}^\top \mathbf{h} \rangle \leq \|\mathbf{z}\|_1 \|\mathbf{A}^\top \mathbf{h}\|_\infty \leq 1. \tag{4.37}$$

Moreover, if $\mathbf{A} \mathbf{z} \in \mathbf{A} F_\mathbf{x}$, then $\langle \mathbf{A} \mathbf{z}, \mathbf{h} \rangle = 1$. Conversely, if $\langle \mathbf{A} \mathbf{z}, \mathbf{h} \rangle = 1$, equality throughout (4.37) yields (as with (4.36)) $z_j = |z_j| \mathrm{sgn}(x_j)$ for $j \in S$, $z_\ell = 0$ for $\ell \in \overline{S}$, and $\|\mathbf{z}\|_1 = 1$, i.e., $\mathbf{z} \in F_\mathbf{x}$ and in turn $\mathbf{A} \mathbf{z} \in \mathbf{A} F_\mathbf{x}$. This shows that $\mathbf{A} F_\mathbf{x}$ is a face of $\mathbf{A} B_1^N$. Next, the injectivity of $\mathbf{A}_S$ immediately implies the injectivity of f, so that $\mathbf{A} F_\mathbf{x}$ is an $(s-1)$-face. Finally, if $\varepsilon \mathbf{A} \mathbf{e}_\ell \in \mathbf{A} F_\mathbf{x}$ for some $\varepsilon = \pm 1$ and $\ell \in \overline{S}$, then $\varepsilon (\mathbf{A}^\top \mathbf{h})_\ell = \langle \varepsilon \mathbf{A} \mathbf{e}_\ell, \mathbf{h} \rangle = 1$, which is impossible. We have therefore established that $\mathbf{A} F_\mathbf{x}$ is an $(s-1)$-face of $\mathbf{A} B_1^N$ not containing any $\varepsilon \mathbf{A} \mathbf{e}_\ell, \varepsilon = \pm 1$, $\ell \in \overline{S}$. $\square$

We finally use Theorem 4.38 to characterize the success of sparse recovery via basis pursuit as a geometric property about the projected cross-polytope $\mathbf{A} B_1^N$. The characterization basically says that all low-dimensional faces of B_1^N remain faces after decreasing the dimension through the action of $\mathbf{A}$. This phenomenon goes against the geometric intuition in small dimension, as some edges of the three-dimensional cross-polytope are necessarily swallowed up when it is projected to be drawn in two dimensions. The precise statement below involves the number $f_k(P)$ of k-faces of a convex polytope P.

Theorem 4.39. *For a matrix $\mathbf{A} \in \mathbb{R}^{m \times N}$ and an integer $s \geq 1$, the following properties are equivalent.*

(a) *Every s-sparse vector $\mathbf{x} \in \mathbb{R}^N$ is the unique minimizer of $\|\mathbf{z}\|_1$ subject to $\mathbf{A} \mathbf{z} = \mathbf{A} \mathbf{x}$.*
(b) *$f_k(\mathbf{A} B_1^N) = f_k(B_1^N)$ whenever $0 \leq k \leq s-1$.*
(c) *$f_k(\mathbf{A} B_1^N) = f_k(B_1^N)$ for $k = 0$ and $k = s-1$.*
(d) *$\mathbf{A} B_1^N$ has $2N$ vertices, and every $(s-1)$-face of B_1^N maps to an $(s-1)$-face of $\mathbf{A} B_1^N$.*

Proof. Before proving the chain of implications $(a) \Rightarrow (b) \Rightarrow (c) \Rightarrow (d) \Rightarrow (a)$, we point out that every k-face of $\mathbf{A} B_1^N$ can be written as $\mathbf{A} F$, where F is a k-face of B_1^N. In other words, if $\mathcal{F}_k(P)$ denotes the set of k-faces of a convex polytope P, then

$$\mathcal{F}_k(\mathbf{A} B_1^N) \subset \mathbf{A} \mathcal{F}_k(B_1^N). \tag{4.38}$$

To see this, considering a k-face G of $\mathbf{A}B_1^N$, there exists $\mathbf{h} \in \mathbb{R}^m$ such that if $\mathbf{z} \in B_1^N$, then

$$\langle \mathbf{A}\mathbf{z}, \mathbf{h} \rangle \leq 1 \qquad \text{and} \qquad [\langle \mathbf{A}\mathbf{z}, \mathbf{h} \rangle = 1 \Leftrightarrow \mathbf{A}\mathbf{z} \in G].$$

Setting $F := \{\mathbf{z} \in B_1^N : \mathbf{A}\mathbf{z} \in G\}$, we see that $\mathbf{A}F = G$ and that if $\mathbf{z} \in B_1^N$, then

$$\langle \mathbf{z}, \mathbf{A}^\top \mathbf{h} \rangle \leq 1 \qquad \text{and} \qquad [\langle \mathbf{z}, \mathbf{A}^\top \mathbf{h} \rangle = 1 \Leftrightarrow \mathbf{z} \in F],$$

so that F is a face of B_1^N. Thus, we can write $F = \operatorname{conv}\{\varepsilon_j \mathbf{e}_j, j \in J\}$ for some $J \subset [N]$ and $\varepsilon \in \{-1, 1\}^J$. As derived in the beginning of the proof of Theorem 4.38, the fact that $G = \mathbf{A}F$ is a k-face is equivalent to the bijectivity of the map $f :$ $\mathbf{t} \mapsto \sum_{j \in J} t_j \varepsilon_j \mathbf{A}\mathbf{e}_j$ from $T := \{\mathbf{t} \in \mathbb{R}^J : \sum_{j \in J} t_j = 0\}$ onto the k-dimensional vector space directing the affine hull of G. Hence, it follows that $\dim(T) = k$, i.e., $\operatorname{card}(J) = k + 1$, so that F is indeed a k-face.

Let us now turn to the implication $(a) \Rightarrow (b)$. For $0 \leq k \leq s - 1$, since every k-face of B_1^N can be written as some $F_\mathbf{x}$ for an exactly $(k+1)$-sparse vector $\mathbf{x} \in \mathbb{R}^N$ with $\|\mathbf{x}\|_1 = 1$, Theorem 4.38 implies that every k-face of B_1^N maps to a k-face of $\mathbf{A}B_1^N$. Thus, $F \mapsto \mathbf{A}F$ is a well-defined mapping from $\mathcal{F}_k(B_1^N)$ to $\mathcal{F}_k(\mathbf{A}B_1^N)$, and (4.38) says that it is surjective. Let us now assume that $\mathbf{A}F = \mathbf{A}F'$ for two distinct k-faces F and F' of B_1^N. We write $F = F_\mathbf{x}$ and $F' = F_{\mathbf{x}'}$ for some $\mathbf{x}, \mathbf{x}' \in \mathbb{R}^N$ with $\|\mathbf{x}\|_1 = \|\mathbf{x}'\|_1 = 1$ and with supports S, S' of size $k + 1$. If $S = S'$, then $\operatorname{sgn}(x_j) \neq \operatorname{sgn}(x_j')$ for some $j \in S$, and both $\operatorname{sgn}(x_j)\mathbf{A}\mathbf{e}_j$ and $\operatorname{sgn}(x_j')\mathbf{A}\mathbf{e}_j$ would belong to the k-face $\mathbf{A}F = \mathbf{A}F'$, which is absurd since otherwise this face would contain the zero vector as their midpoint. If $S \neq S'$, then we pick $\ell \in S' \setminus S$ and $\varepsilon = \pm 1$ with $\varepsilon \mathbf{e}_\ell \in F'$; hence, $\varepsilon \mathbf{A}\mathbf{e}_\ell \in \mathbf{A}F'$, while $\ell \in \overline{S}$ yields $\varepsilon \mathbf{A}\mathbf{e}_\ell \notin \mathbf{A}F = \mathbf{A}F'$, which is also absurd. This shows that the mapping $F \mapsto \mathbf{A}F$ is injective, hence bijective from $\mathcal{F}_k(B_1^N)$ to $\mathcal{F}_k(\mathbf{A}B_1^N)$. We conclude that the cardinality of these two sets are identical, i.e., that $f_k(\mathbf{A}B_1^N) = f_k(B_1^N)$, as desired.

The implication $(b) \Rightarrow (c)$ is straightforward, so we focus on $(c) \Rightarrow (d)$. Since vertices are just 0-faces, $f_0(\mathbf{A}B_1^N) = f_0(B_1^N)$ directly implies that $\mathbf{A}B_1^N$ has $2N$ vertices. Next, according to (4.38), for each $(s - 1)$-face G of $\mathbf{A}B_1^N$, we can find an $(s - 1)$-face F of B_1^N such that $G = \mathbf{A}F$. Thus, $G \mapsto F$ defines an injective mapping from $\mathcal{F}_{s-1}(\mathbf{A}B_1^N)$ into $\mathcal{F}_{s-1}(B_1^N)$. Since these two sets have the same cardinality, we deduce that the mapping is surjective, too. This means that, for every $(s - 1)$-face F of B_1^N, there is an $(s - 1)$-face G of $\mathbf{A}B_1^N$ with $G = \mathbf{A}F$; hence, $\mathbf{A}F$ is indeed an $(s - 1)$-face of $\mathbf{A}B_1^N$. This proves (d).

Finally, let us show the implication $(d) \Rightarrow (a)$. In view of Remark 4.31, we only need to prove that every exactly s-sparse vector $\mathbf{x} \in \mathbb{R}^N$ is the unique minimizer of $\|\mathbf{z}\|_1$ subject to $\mathbf{A}\mathbf{z} = \mathbf{A}\mathbf{x}$. Using Theorem 4.38, we need to prove that if $\mathbf{x} \in \mathbb{R}^N$ has support S of size s, then $\mathbf{A}F_\mathbf{x}$ is an $(s - 1)$-face of $\mathbf{A}B_1^N$ and $\mathbf{A}F_\mathbf{x}$ does not contain any $\varepsilon \mathbf{A}\mathbf{e}_\ell$ with $\varepsilon = \pm 1$ and $\ell \in \overline{S}$. Since $F_\mathbf{x}$ is an $(s-1)$-face of B_1^N, the first part of this statement follows directly from (d). For the second part of the statement, given $\varepsilon = \pm 1$ and $\ell \in \overline{S}$, if we had $\varepsilon \mathbf{A}\mathbf{e}_\ell \in \mathbf{A}F_\mathbf{x} = \operatorname{conv}\{\operatorname{sgn}(x_j)\mathbf{A}\mathbf{e}_j, j \in S\}$, then either $\varepsilon \mathbf{A} \mathbf{e}_\ell = \mathrm{sgn}(x_j) \mathbf{A} \mathbf{e}_j$ for some $j \in S$ or $\varepsilon \mathbf{A} \mathbf{e}_\ell$ is not a vertex of $\mathbf{A} B_1^N$. In any case, this yields $\mathrm{card}(\mathbf{A} \mathcal{F}_0(B_1^N)) < 2N$ and then $\mathrm{card}(\mathcal{F}_0(\mathbf{A} B_1^N)) < 2N$ by (4.38), which contradicts the fact that $\mathbf{A} B_1^N$ has $2N$ vertices. We have now proved (a).   □

4.6 Low-Rank Matrix Recovery

In this section, we shortly digress on the problem of recovering matrices of low rank from incomplete linear measurements, which was already mentioned in Sect. 1.2 (p. 21). In this context, the number of nonzero singular values—the rank of a matrix—replaces the number of nonzero entries: the sparsity of a vector.

We suppose that a matrix $\mathbf{X} \in \mathbb{C}^{n_1 \times n_2}$ of rank at most r is observed via the measurement vector $\mathbf{y} = \mathcal{A}(\mathbf{X}) \in \mathbb{C}^m$ where $\mathcal{A}$ is a linear map from $\mathbb{C}^{n_1 \times n_2}$ to $\mathbb{C}^m$. As in the vector case the first approach to this problem that probably comes to mind is to solve the rank-minimization problem

$$\underset{\mathbf{Z} \in \mathbb{C}^{n_1 \times n_2}}{\text{minimize}} \ \mathrm{rank}(\mathbf{Z}) \quad \text{subject to } \mathcal{A}(\mathbf{X}) = \mathbf{y}.$$

Unfortunately, like ℓ_0-minimization this problem is NP-hard; see Exercise 2.11. In fact, the rank of $\mathbf{Z}$ equals the ℓ_0-norm of the vector $[\sigma_1(\mathbf{Z}), \dots, \sigma_n(\mathbf{Z})]^\top$ of singular values of $\mathbf{Z}$. Motivated by the vector case where ℓ_1-minimization is a good strategy, we relax the minimization of the rank to the nuclear norm minimization problem

$$\underset{\mathbf{Z} \in \mathbb{C}^{n_1 \times n_2}}{\text{minimize}} \ \|\mathbf{Z}\|_* \quad \text{subject to } \mathcal{A}(\mathbf{Z}) = \mathbf{y}. \tag{4.39}$$

Here, the *nuclear norm*, defined by

$$\|\mathbf{Z}\|_* := \sum_{j=1}^n \sigma_j(\mathbf{Z}), \qquad n := \min\{n_1, n_2\},$$

is the ℓ_1-norm of the vector of singular values of $\mathbf{Z}$. We refer to (A.25) and Appendix A.2 in general for the fact that $\|\cdot\|_*$ is indeed a norm. The problem (4.39) is a convex optimization problem, and it is actually equivalent to a semidefinite program; see Exercise 4.18.

The analysis of the nuclear norm minimization strategy (4.39) is analogous to the vector case. In particular, the success of the strategy is equivalent to a null space property.

Theorem 4.40. *Given a linear map $\mathcal{A}$ from $\mathbb{C}^{n_1 \times n_2}$ to $\mathbb{C}^m$, every matrix $\mathbf{X} \in \mathbb{C}^{n_1 \times n_2}$ of rank at most r is the unique solution of (4.39) with $\mathbf{y} = \mathcal{A}(\mathbf{X})$ if and* *only if, for all* $\mathbf{M} \in \ker \mathcal{A} \setminus \{\mathbf{0}\}$ *with singular values* $\sigma_1(\mathbf{M}) \geq \cdots \geq \sigma_n(\mathbf{M}) \geq 0$, $n := \min\{n_1, n_2\}$,

$$\sum_{j=1}^{r} \sigma_j(\mathbf{M}) < \sum_{j=r+1}^{n} \sigma_j(\mathbf{M}). \tag{4.40}$$

Proof. Let us first assume that every matrix $\mathbf{X} \in \mathbb{C}^{n_1 \times n_2}$ of rank at most r is the unique solution of (4.39) with $\mathbf{y} = \mathcal{A}(\mathbf{X})$. We consider the singular value decomposition of a matrix $\mathbf{M} \in \ker \mathcal{A} \setminus \{\mathbf{0}\}$ and write $\mathbf{M} = \mathbf{U}\mathrm{diag}(\sigma_1, \ldots, \sigma_n)\mathbf{V}^*$ for $\sigma_1 \geq \cdots \geq \sigma_n \geq 0$ and $\mathbf{U} \in \mathbb{C}^{n_1 \times n_1}, \mathbf{V} \in \mathbb{C}^{n_2 \times n_2}$ unitary. Setting $\mathbf{M}_1 = \mathbf{U}\mathrm{diag}(\sigma_1, \ldots, \sigma_r, 0, \ldots, 0)\mathbf{V}^*$ and $\mathbf{M}_2 = \mathbf{U}\mathrm{diag}(0, \ldots, 0, -\sigma_{r+1}, \ldots, -\sigma_n)\mathbf{V}^*$, we have $\mathbf{M} = \mathbf{M}_1 - \mathbf{M}_2$. Thus, $\mathcal{A}(\mathbf{M}) = \mathbf{0}$ translates into $\mathcal{A}(\mathbf{M}_1) = \mathcal{A}(\mathbf{M}_2)$. Since the rank of $\mathbf{M}_1$ is at most r, its nuclear norm must be smaller than the nuclear norm of $\mathbf{M}_2$. This means that $\sigma_1 + \cdots + \sigma_r < \sigma_{r+1} + \cdots + \sigma_n$, as desired.

Conversely, let us now assume that $\sum_{j=1}^{r} \sigma_j(\mathbf{M}) < \sum_{j=r+1}^{n} \sigma_j(\mathbf{M})$ for every $\mathbf{M} \in \ker \mathcal{A} \setminus \{\mathbf{0}\}$ with singular values $\sigma_1(\mathbf{M}) \geq \cdots \geq \sigma_n(\mathbf{M}) \geq 0$. Consider a matrix $\mathbf{X} \in \mathbb{C}^{n_1 \times n_2}$ of rank at most r and a matrix $\mathbf{Z} \in \mathbb{C}^{n_1 \times n_2}, \mathbf{Z} \neq \mathbf{X}$, satisfying $\mathcal{A}(\mathbf{Z}) = \mathcal{A}(\mathbf{X})$. We aim at proving that $\|\mathbf{Z}\|_* > \|\mathbf{X}\|_*$. Let us set $\mathbf{M} := \mathbf{X} - \mathbf{Z} \in \ker \mathcal{A} \setminus \{\mathbf{0}\}$. Lemma A.18 ensures that the singular values $\sigma_j(\mathbf{M}), \sigma_j(\mathbf{Z}), \sigma_j(\mathbf{M})$ satisfy

$$\|\mathbf{Z}\|_* = \sum_{j=1}^{n} \sigma_j(\mathbf{X} - \mathbf{M}) \geq \sum_{j=1}^{n} |\sigma_j(\mathbf{X}) - \sigma_j(\mathbf{M})|.$$

For $j \in [r]$, we have $|\sigma_j(\mathbf{X}) - \sigma_j(\mathbf{M})| \geq \sigma_j(\mathbf{X}) - \sigma_j(\mathbf{M})$, and for $r + 1 \leq j \leq n$, it holds $|\sigma_j(\mathbf{X}) - \sigma_j(\mathbf{M})| = \sigma_j(\mathbf{M})$. In view of our hypothesis, we derive

$$\|\mathbf{Z}\|_* \geq \sum_{j=1}^{r} \sigma_j(\mathbf{X}) - \sum_{j=1}^{r} \sigma_j(\mathbf{M}) + \sum_{j=r+1}^{n} \sigma_j(\mathbf{M}) > \sum_{j=1}^{r} \sigma_j(\mathbf{X}) = \|\mathbf{X}\|_*.$$

This establishes the desired inequality.   □

Like in the vector case, one can introduce stable and robust versions of the rank null space property (4.40) and show corresponding error estimates for reconstruction via nuclear norm minimization; see Exercises 4.19 and 4.20. Also, recovery conditions for individual low-rank matrices can be shown, analogously to the results for the vector case in Sect. 4.4; see Exercise 4.21.

In the remainder of the book, the low-rank recovery problem will only be treated via exercises; see, e.g., Exercises 6.25 and 9.12. The reader is, of course, very welcome to work through them.

Notes

Throughout the chapter, we have insisted on sparse vectors to be unique solutions of (P_1). If we dropped the uniqueness requirement, then a necessary and sufficient condition for every s-sparse vector to be a solution of (P_1) would be a weak null space property where the strict inequality is replaced by a weak inequality sign.

The null space property is somewhat folklore in the compressive sensing literature. It appeared implicitly in works of Donoho and Elad [155], of Donoho and Huo [158], and of Elad and Bruckstein [181]. Gribonval and Nielsen also isolated the notion in [239]. The name was first used by Cohen et al. in [123], albeit for a property slightly more general than (4.3), namely, $\|\mathbf{v}\|_1 \leq C\sigma_s(\mathbf{v})_1$ for all $\mathbf{v} \in \ker \mathbf{A}$, where $C \geq 1$ is an unspecified constant. We have coined the terms stable and robust null space properties for some notions that are implicit in the literature.

The equivalence between the real and complex null space properties was established in [211] using a different argument than the one of Theorem 4.7. The result was generalized by Lai and Liu in [315]. The proof of Theorem 4.7 follows their argument.

Given a measurement matrix $\mathbf{A} \in \mathbb{K}^{m \times N}$ and a vector $\mathbf{x} \in \mathbb{K}^N$, one can rephrase the optimization problem (P_1) with $\mathbf{y} = \mathbf{A}\mathbf{x}$ as the problem of best approximation to $\mathbf{x} \in \mathbb{K}^N$ from the subspace $\ker \mathbf{A}$ of $\mathbb{K}^N$ in the ℓ_1-norm. Some of the results of this chapter can be derived using known characterizations of best ℓ_1-approximation. The book [388] by Pinkus is a good source on the subject, although it does not touch the complex setting.

The term *instance optimality* is sometimes also used for what we called stability in this chapter. Chapter 11 gives more details on this topic.

The stability and robustness of sparse reconstruction via basis pursuit, as stated after Theorem 4.22, were established by Candès et al. in [95] under a restricted isometry property—see Chap. 6—condition on the measurement matrix.

The fact that sparse recovery via ℓ_q-minimization implies sparse recovery via ℓ_p-minimization whenever $0<p<q\leq1$ was proved by Gribonval and Nielsen in [240].

The sufficient condition (b) of Theorems 4.26 and 4.30, as well as Corollary 4.28, can be found in works of Fuchs [215] and of Tropp [477]. In [94], Candès et al. stated that condition (b) is also necessary for partial Fourier matrices. The argument is slightly imprecise, since it would generalize to other measurement matrices and contradict Remark 4.29, but it is correct for real matrices; see Theorem 4.30. The discrepancy between real and complex settings disappears if uniqueness of minimizers and strict inequalities are not required; see Exercise 4.15.

The recovery conditions of Theorems 4.35 and 4.36 appeared in more general form in [108].

The success of sparse recovery via basis pursuit was characterized in terms of faces of polytopes by Donoho in [151], where the condition of Corollary 4.39 was also interpreted in terms of *neighborliness* of the polytope $\mathbf{A}B_1^N$—see Exercise 4.16.

Exercises

4.1. Suppose that $\mathbf{A} \in \mathbb{C}^{m \times N}$ satisfies the null space property of order s. Theorems 4.5 and 2.13 guarantee that $\ker \mathbf{A}$ does not contain any $2s$-sparse vectors other than the zero vector. Give a direct proof of this fact.

4.2. Find a 2×3 matrix $\mathbf{A}$ and a nonsingular 3×3 diagonal matrix $\mathbf{D}$ such that $\mathbf{A}$ has the null space property of order 1 but $\mathbf{AD}$ does not.

4.3. Prove that there exist a matrix $\mathbf{A}$ and an individual vector $\mathbf{x}$ such that $\mathbf{x}$ can be recovered from $\mathbf{Ax}$ via basis pursuit with a number of measurements $m < 2s$.

4.4. Suppose that the null space of a real matrix $\mathbf{A}$ is a two-dimensional space with basis $(\mathbf{v}, \mathbf{w})$. Prove that $\mathbf{A}$ has the null space property of order s if and only if

$$\sum_{j \in S} |v_j| < \sum_{\ell \in \overline{S}} |v_\ell|, \quad \sum_{j \in S} |w_j| < \sum_{\ell \in \overline{S}} |w_\ell|, \quad \sum_{j \in S} |v_i w_j - v_j w_i| < \sum_{\ell \in \overline{S}} |v_i w_\ell - v_\ell w_i|,$$

for all $i \in [N]$ and all $S \subset [N]$ with $\mathrm{card}(S) \leq s$.

4.5. Prove the equivalence between the real and complex stable null space properties with constant $0 < \rho < 1$ relative to a set S. More generally, prove the equivalence between the real and complex robust null space properties, in the following sense: Given $\mathbf{A} \in \mathbb{R}^{m \times N}$, a constant $\rho > 0$, a set S, and a norm on $\mathbb{C}^m$ invariant by complex conjugation (i.e., $\|\overline{\mathbf{y}}\| = \|\mathbf{y}\|$ for all $\mathbf{y} \in \mathbb{C}^m$), there exists a constant $\tau > 0$ such that

$$\|\mathbf{v}_S\|_1 \leq \rho \|\mathbf{v}_{\overline{S}}\|_1 + \tau \|\mathbf{Av}\| \qquad \text{for all } \mathbf{v} \in \mathbb{R}^N$$

if and only if there exists a constant $\tau' > 0$ such that

$$\|\mathbf{v}_S\|_1 \leq \rho \|\mathbf{v}_{\overline{S}}\|_1 + \tau' \|\mathbf{Av}\| \qquad \text{for all } \mathbf{v} \in \mathbb{C}^N.$$

4.6. Given $\mathbf{A} \in \mathbb{C}^{m \times N}$ and $0 < c < 1$, prove the equivalence of the properties:

(i) $\|\mathbf{v}_S\|_1 \leq \|\mathbf{v}_{\overline{S}}\|_1 - c \|\mathbf{v}\|_1$ for all $\mathbf{v} \in \ker \mathbf{A}$ and $S \subset [N]$ with $\mathrm{card}(S) \leq s$.
(ii) $\|\mathbf{x}\|_1 \leq \|\mathbf{z}\|_1 - c \|\mathbf{x} - \mathbf{z}\|_1$ for all s-sparse $\mathbf{x} \in \mathbb{C}^N$ and $\mathbf{z} \in \mathbb{C}^N$ with $\mathbf{Az} = \mathbf{Ax}$.

4.7. Given $S \subset [N]$, prove that a minimizer $\mathbf{x}^\sharp$ of $\|\mathbf{z}\|_1$ subject to a constraint met by $\mathbf{x} \in \mathbb{C}^N$ satisfies

$$\left\| (\mathbf{x} - \mathbf{x}^\sharp)_{\overline{S}} \right\|_1 \leq \left\| (\mathbf{x} - \mathbf{x}^\sharp)_S \right\|_1 + 2 \left\| \mathbf{x}_{\overline{S}} \right\|_1.$$

4.8. Given $\mathbf{A} \in \mathbb{R}^{m \times N}$, prove that every nonnegative s-sparse vector $\mathbf{x} \in \mathbb{R}^N$ is the unique solution of

$$\underset{\mathbf{z} \in \mathbb{R}^N}{\text{minimize}} \ \|\mathbf{z}\|_1 \quad \text{subject to } \mathbf{Az} = \mathbf{Ax} \text{ and } \mathbf{z} \geq 0$$

if and only if

$$\mathbf{v}_{\overline{S}} \geq \mathbf{0} \Longrightarrow \sum_{j=1}^{N} v_j > 0$$

for all $\mathbf{v} \in \ker \mathbf{A} \setminus \{\mathbf{0}\}$ and all $S \subset [N]$ with $\mathrm{card}(S) \leq s$.

4.9. Let $\mathbf{A} \in \mathbb{R}^{m \times N}$ be a matrix for which $\sum_{j=1}^{N} v_j = 0$ whenever $\mathbf{v} \in \ker \mathbf{A}$, and let $S \subset [N]$ be a fixed index set. Suppose that every nonnegative vector supported on S is uniquely recovered by ℓ_1-minimization. Prove that every nonnegative vector $\mathbf{x}$ supported on S is in fact the unique vector in the set $\{\mathbf{z} \in \mathbb{R}^N : \mathbf{z} \geq \mathbf{0}, \mathbf{A}\mathbf{z} = \mathbf{A}\mathbf{x}\}$.

4.10. Given matrices $\mathbf{A} \in \mathbb{C}^{m \times N}$ and $\mathbf{M} \in \mathbb{C}^{m \times m}$, suppose that $\mathbf{M}\mathbf{A}$ satisfies the ℓ_2-robust null space property of order s with constants $0 < \rho < 1$ and $\tau > 0$. Prove that there exist constants $C, D > 0$ depending only on ρ, τ, and $\|\mathbf{M}\|_{2 \to 2}$ such that, for any $\mathbf{x} \in \mathbb{C}^N$,

$$\|\mathbf{x} - \mathbf{x}^\sharp\|_p \leq \frac{C}{s^{1-1/p}} \sigma_s(\mathbf{x})_1 + D \, s^{1/p-1/2} \, \eta, \qquad 1 \leq p \leq 2,$$

where $\mathbf{x}^\sharp \in \mathbb{C}^N$ is a solution of $(\mathrm{P}_{1,\eta})$ with $\|\cdot\| = \|\cdot\|_2$, $\mathbf{y} = \mathbf{A}\mathbf{x} + \mathbf{e}$, and $\|\mathbf{e}\|_2 \leq \eta$.

4.11. Let $\mathbf{A} \in \mathbb{C}^{m \times N}$ be such that, for some constants $0 < \rho < 1$ and $\tau \geq 0$,

$$\|\mathbf{v}_S\|_2 \leq \frac{\rho}{\sqrt{s}} \|\mathbf{v}_{\overline{S}}\|_1 + \tau \|\mathbf{A}^* \mathbf{A}\mathbf{v}\|_\infty \quad \text{for all } S \subset [N] \text{ and all } \mathbf{v} \in \mathbb{C}^N.$$

For $\mathbf{x} \in \mathbb{C}^N$, let $\mathbf{y} = \mathbf{A}\mathbf{x} + \mathbf{e}$ for some $\mathbf{e} \in \mathbb{C}^m$ with $\|\mathbf{A}^* \mathbf{e}\|_\infty \leq \eta$. Let $\mathbf{x}^\sharp$ be a minimizer of the Dantzig selector

$$\underset{\mathbf{z} \in \mathbb{C}^N}{\text{minimize}} \, \|\mathbf{z}\|_1 \quad \text{subject to } \|\mathbf{A}^*(\mathbf{A}\mathbf{z} - \mathbf{y})\|_\infty \leq \eta.$$

Show that

$$\|\mathbf{x} - \mathbf{x}^\sharp\|_2 \leq \frac{C\sigma_s(\mathbf{x})_1}{\sqrt{s}} + D\eta,$$

for constants $C, D > 0$ depending only on ρ and τ.

4.12. Prove Theorem 4.9 and generalize other results from Sects. 4.1 to 4.3 when the ℓ_1-norm is replaced by the ℓ_q-quasinorm for $0 < q < 1$.

4.13. Given an integer $s \geq 1$ and an exponent $q \in (0, 1)$, find a measurement matrix that allows reconstruction of s-sparse vectors via ℓ_p-minimization for $p < q$ but not for $p > q$.

4.14. Verify the statement of Remark 4.29 that $\mathbf{x} = [e^{-\pi i/3}, e^{\pi i/3}, 0]^\top$ is the unique minimizer of $\|\mathbf{z}\|_1$ subject to $\mathbf{Az} = \mathbf{Ax}$, where $\mathbf{A} = \begin{bmatrix} 1 & 0 & -1 \\ 0 & 1 & -1 \end{bmatrix}$.

4.15. Given a matrix $\mathbf{A} \in \mathbb{C}^{m \times N}$ and a vector $\mathbf{x} \in \mathbb{C}^N$ with support S, prove that $\mathbf{x}$ is a minimizer of $\|\mathbf{z}\|_1$ subject to $\mathbf{Az} = \mathbf{Ax}$ if one of the following equivalent conditions holds:

(i) $\left| \sum_{j \in S} \overline{\mathrm{sgn}(x_j)} v_j \right| \le \|\mathbf{v}_{\overline{S}}\|_1$ for all $\mathbf{v} \in \ker \mathbf{A}$.

(ii) There exists a vector $\mathbf{h} \in \mathbb{C}^m$ such that

$$(\mathbf{A}^* \mathbf{h})_j = \mathrm{sgn}(x_j), \quad j \in S, \qquad |(\mathbf{A}^* \mathbf{h})_\ell| \le 1, \quad \ell \in \overline{S}.$$

4.16. A symmetric convex polytope P is called *centrally k-neighborly* if any set of k of its vertices, not containing a pair $\{\mathbf{v}, -\mathbf{v}\}$, spans a $(k-1)$-face of P. Given a matrix $\mathbf{A} \in \mathbb{R}^{m \times N}$, prove that every s-sparse vector $\mathbf{x} \in \mathbb{R}^N$ is the unique solution of (P_1) with $\mathbf{y} = \mathbf{Ax}$ if and only if the convex polytope $\mathbf{A}B_1^N$ has $2N$ vertices and is s-neighborly.

4.17. Stable and robust recovery via dual certificate

Let $\mathbf{A} \in \mathbb{C}^{m \times N}$ be a matrix with ℓ_2-normalized columns. Let $\mathbf{x} \in \mathbb{C}^N$ and let $S \subset [N]$ be an index set of s largest absolute entries of $\mathbf{x}$. Assume that

$$\|\mathbf{A}_S^* \mathbf{A}_S - \mathbf{Id}\|_{2 \to 2} \le \alpha$$

for some $\alpha \in (0, 1)$ and that there exists a dual certificate $\mathbf{u} = \mathbf{A}^* \mathbf{h} \in \mathbb{C}^N$ with $\mathbf{h} \in \mathbb{C}^m$ such that

$$\mathbf{u}_S = \mathrm{sgn}(\mathbf{x}_S), \quad \|\mathbf{u}_{\overline{S}}\|_\infty \le \beta, \quad \|\mathbf{h}\|_2 \le \gamma\sqrt{s},$$

for some constants $0 < \beta < 1$ and $\gamma > 0$. Suppose that we are given corrupted measurements $\mathbf{y} = \mathbf{Ax} + \mathbf{e}$ with $\|\mathbf{e}\|_2 \le \eta$. Show that a solution $\mathbf{x}^\sharp \in \mathbb{C}^N$ of the ℓ_1-minimization problem

$$\underset{\mathbf{z} \in \mathbb{C}^N}{\text{minimize}} \ \|\mathbf{z}\|_1 \quad \text{subject to} \quad \|\mathbf{Az} - \mathbf{y}\|_2 \le \eta$$

satisfies

$$\|\mathbf{x} - \mathbf{x}^\sharp\|_2 \le C\sigma_s(\mathbf{x})_1 + D\sqrt{s}\eta$$

for appropriate constants $C, D > 0$ depending only on α, β, and γ.

4.18. Nuclear norm minimization via semidefinite programming

Let $\| \cdot \|_*$ denote the nuclear norm. Given a linear map $\mathcal{A} : \mathbb{C}^{n_1 \times n_2} \to \mathbb{C}^m$ and $\mathbf{y} \in \mathbb{C}^m$, show that the nuclear norm minimization problem

$$\underset{\mathbf{X} \in \mathbb{C}^{n_1 \times n_2}}{\text{minimize}} \|\mathbf{X}\|_* \quad \text{subject to } \mathcal{A}(\mathbf{X}) = \mathbf{y}$$

is equivalent to the semidefinite program

$$\underset{\mathbf{X} \in \mathbb{C}^{n_1 \times n_2}, \mathbf{Y} \in \mathbb{C}^{n_1 \times n_1}, \mathbf{Z} \in \mathbb{C}^{n_2 \times n_2}}{\text{minimize}} \quad \text{tr } \mathbf{Y} + \text{tr } \mathbf{Z} \quad \text{subject to } \mathcal{A}(\mathbf{X}) = \mathbf{y}$$

$$\text{and } \begin{bmatrix} \mathbf{Y} & \mathbf{X} \\ \mathbf{X}^* & \mathbf{Z} \end{bmatrix} \succcurlyeq \mathbf{0}.$$

4.19. Stable rank null space property

Let $\mathcal{A} : \mathbb{C}^{n_1 \times n_2} \to \mathbb{C}^m$ be a linear measurement map and let $n = \min\{n_1, n_2\}$. Assume that $\mathcal{A}$ satisfies the stable rank null space property of order r with constant $0 < \rho < 1$, i.e., that for all $\mathbf{M} \in \ker \mathcal{A} \setminus \{\mathbf{0}\}$, the singular values of $\mathbf{M}$ satisfy

$$\sum_{\ell=1}^{r} \sigma_\ell(\mathbf{M}) \leq \rho \sum_{\ell=r+1}^{n} \sigma_\ell(\mathbf{M}).$$

Show that, for all $\mathbf{X}, \mathbf{Z} \in \mathbb{C}^{n_1 \times n_2}$ with $\mathcal{A}(\mathbf{X}) = \mathcal{A}(\mathbf{Z})$,

$$\|\mathbf{X} - \mathbf{Z}\|_* \leq \frac{1+\rho}{1-\rho} \left(\|\mathbf{Z}\|_* - \|\mathbf{X}\|_* + 2 \sum_{\ell=r+1}^{n} \sigma_\ell(\mathbf{X}) \right). \tag{4.41}$$

For $\mathbf{X} \in \mathbb{C}^{n_1 \times n_2}$, let $\mathbf{X}^\sharp$ be a solution of the nuclear norm minimization problem

$$\underset{\mathbf{Z} \in \mathbb{C}^{n_1 \times n_2}}{\text{minimize}} \|\mathbf{Z}\|_* \quad \text{subject to } \mathcal{A}(\mathbf{Z}) = \mathcal{A}(\mathbf{X}).$$

Deduce the error estimate

$$\|\mathbf{X} - \mathbf{X}^\sharp\|_* \leq \frac{2(1+\rho)}{1-\rho} \sum_{\ell=r+1}^{n} \sigma_\ell(\mathbf{X}).$$

Conversely, show that if (4.41) holds for all $\mathbf{X}, \mathbf{Z}$ such that $\mathcal{A}(\mathbf{X}) = \mathcal{A}(\mathbf{Z})$, then $\mathcal{A}$ satisfies the stable rank null space property of order r with constant $0 < \rho < 1$.

4.20. Robust rank null space property

Let $\mathcal{A} : \mathbb{C}^{n_1 \times n_2} \to \mathbb{C}^m$ be a linear measurement map, let $n = \min\{n_1, n_2\}$, and let $\|\cdot\|$ be some norm on $\mathbb{C}^m$.

(a) We say that $\mathcal{A}$ satisfies the robust rank null space property of order r (with respect to $\|\cdot\|$) with constants $0 < \rho < 1$ and $\tau > 0$ if, for all $\mathbf{M} \in \mathbb{C}^{n_1 \times n_2}$, the singular values of $\mathbf{M}$ satisfy

$$\sum_{\ell=1}^{r} \sigma_\ell(\mathbf{M}) \leq \rho \sum_{\ell=r+1}^{n} \sigma_\ell(\mathbf{M}) + \tau \|\mathcal{A}(\mathbf{M})\|.$$

Show that

$$\|\mathbf{X} - \mathbf{Z}\|_* \leq \frac{1+\rho}{1-\rho} \left(\|\mathbf{Z}\|_* - \|\mathbf{X}\|_* + 2 \sum_{\ell=r+1}^{n} \sigma_\ell(\mathbf{X}) \right) + \frac{2\tau}{1-\rho} \|\mathcal{A}(\mathbf{Z} - \mathbf{X})\|$$

holds for all $\mathbf{X}, \mathbf{Z} \in \mathbb{C}^{n_1 \times n_2}$ if and only if $\mathcal{A}$ satisfies the robust rank null space property of order r with constants ρ and τ.

(b) Assume that $\mathcal{A}$ satisfies the Frobenius robust rank null space property of order r (with respect to $\|\cdot\|$) with constants $0 < \rho < 1$ and $\tau > 0$, i.e., that for all $\mathbf{M} \in \ker \mathcal{A} \setminus \{\mathbf{0}\}$,

$$\left(\sum_{\ell=1}^{r} \sigma_\ell(\mathbf{M})^2 \right)^{1/2} \leq \frac{\rho}{\sqrt{r}} \sum_{\ell=r+1}^{n} \sigma_\ell(\mathbf{M}) + \tau \|\mathcal{A}(\mathbf{M})\|.$$

For $\mathbf{X} \in \mathbb{C}^{n_1 \times n_2}$, assume that $\mathbf{y} = \mathcal{A}(\mathbf{X}) + \mathbf{e}$ with $\|\mathbf{e}\|_2 \leq \eta$ for some $\eta \geq 0$. Let $\mathbf{X}^\sharp$ be a solution of the quadratically constrained nuclear norm minimization problem

$$\underset{\mathbf{Z} \in \mathbb{C}^{n_1 \times n_2}}{\text{minimize}} \|\mathbf{Z}\|_* \quad \text{subject to } \|\mathcal{A}(\mathbf{Z}) - \mathbf{y}\|_2 \leq \eta.$$

Show that

$$\|\mathbf{X} - \mathbf{X}^\sharp\|_F \leq \frac{C}{\sqrt{r}} \sum_{\ell=r+1}^{n} \sigma_\ell(\mathbf{X}) + D\eta$$

for constants $C, D > 0$ only depending on ρ and τ.

4.21. Low-rank matrix recovery via dual certificate

Let $\|\mathbf{X}\|_*$ denote the nuclear norm of a matrix $\mathbf{X} \in \mathbb{C}^{n_1 \times n_2}$ and $\langle \mathbf{X}, \mathbf{Y} \rangle_F = \operatorname{tr}(\mathbf{X}\mathbf{Y}^*)$ the Frobenius inner product of two matrices $\mathbf{X}, \mathbf{Y} \in \mathbb{C}^{n_1 \times n_2}$.

(a) Show that the nuclear norm is the dual norm of the operator norm, i.e.,

$$\|\mathbf{X}\|_* = \sup_{\mathbf{Y} \in \mathbb{C}^{n_1 \times n_2}, \|\mathbf{Y}\|_{2 \to 2} \leq 1} |\langle \mathbf{X}, \mathbf{Y} \rangle_F|, \qquad \mathbf{X} \in \mathbb{C}^{n_1 \times n_2}.$$

(b) Let $\mathbf{X}, \mathbf{Y} \in \mathbb{C}^{n_1 \times n_2}$ be such that $\mathbf{X}\mathbf{Y}^* = \mathbf{0}$ and $\mathbf{X}^*\mathbf{Y} = \mathbf{0}$. Show that

$$\|\mathbf{X}\|_* + \|\mathbf{Y}\|_* = \|\mathbf{X} + \mathbf{Y}\|_*.$$

(c) Let $\mathbf{X} \in \mathbb{C}^{n_1 \times n_2}$ be a matrix of rank r with singular value decomposition $\mathbf{X} = \sum_{\ell=1}^{r} \sigma_\ell \mathbf{u}_\ell \mathbf{v}_\ell^*$, where both $\{\mathbf{u}_\ell, \ell \in [r]\}$ and $\{\mathbf{v}_\ell, \ell \in [r]\}$ are orthonormal systems. Let $T \subset \mathbb{C}^{n_1 \times n_2}$ be the linear space spanned by

$$\{\mathbf{u}_\ell \mathbf{x}_\ell^*, \mathbf{x}_\ell \in \mathbb{C}^{n_2}, \ell \in [r]\} \cup \{\mathbf{y}_\ell \mathbf{v}_\ell^*, \mathbf{y}_\ell \in \mathbb{C}^{n_1}, \ell \in [r]\},$$

and let $T^{\perp}$ be its orthogonal complement relative to the Frobenius inner product. Denote by $\mathbf{P}_U = \sum_{\ell=1}^{r} \mathbf{u}_\ell \mathbf{u}_\ell^* \in \mathbb{C}^{n_1 \times n_1}$ the matrix of the orthogonal projection onto the span of $\{\mathbf{u}_\ell, \ell \in [r]\}$ and by $\mathbf{P}_V \in \mathbb{C}^{n_2 \times n_2}$ the matrix of the orthogonal projection onto the span of $\{\mathbf{v}_\ell, \ell \in [r]\}$. Show that the orthogonal projections $\mathcal{P}_T : \mathbb{C}^{n_1 \times n_2} \to T$ and $\mathcal{P}_{T^{\perp}} : \mathbb{C}^{n_1 \times n_2} \to T^{\perp}$ are given, for $\mathbf{Z} \in \mathbb{C}^{n_1 \times n_2}$, by

$$\mathcal{P}_T(\mathbf{Z}) = \mathbf{P}_U \mathbf{Z} + \mathbf{Z} \mathbf{P}_V - \mathbf{P}_U \mathbf{Z} \mathbf{P}_V,$$

$$\mathcal{P}_{T^{\perp}}(\mathbf{Z}) = (\mathbf{Id} - \mathbf{P}_U)\mathbf{Z}(\mathbf{Id} - \mathbf{P}_V).$$

(d) Given a linear map $\mathcal{A} : \mathbb{C}^{n_1 \times n_2} \to \mathbb{C}^m$, show that $\mathbf{X}$ is the unique solution of the nuclear norm minimization problem

$$\underset{\mathbf{Z} \in \mathbb{C}^{n_1 \times n_2}}{\text{minimize}} \|\mathbf{Z}\|_* \quad \text{subject to} \quad \mathcal{A}(\mathbf{Z}) = \mathcal{A}(\mathbf{X})$$

if $\mathcal{A}$ restricted to T is injective and if there exists a dual certificate $\mathbf{M} = \mathcal{A}^* \mathbf{h} \in \mathbb{C}^{n_1 \times n_2}$ with $\mathbf{h} \in \mathbb{C}^m$ satisfying

$$\mathcal{P}_T(\mathbf{M}) = \sum_{\ell=1}^{r} \mathbf{u}_\ell \mathbf{v}_\ell^* \quad \text{and} \quad \|\mathcal{P}_{T^{\perp}}(\mathbf{M})\|_{2 \to 2} < 1.$$

Chapter 5
Coherence

In compressive sensing, the analysis of recovery algorithms usually involves a quantity that measures the suitability of the measurement matrix. The coherence is a very simple such measure of quality. In general, the smaller the coherence, the better the recovery algorithms perform. In Sect. 5.1, we introduce the notion of coherence of a matrix and some of its generalizations. In Sect. 5.2, we examine how small the coherence can be and we point out some matrices with small coherence. In Sects. 5.3, 5.4, and 5.5, we give some sufficient conditions expressed in terms of the coherence that guarantee the success of orthogonal matching pursuit, basis pursuit, and thresholding algorithms.

5.1 Definitions and Basic Properties

We start with the definition of the coherence of a matrix. We stress that the columns of the matrix are always implicitly understood to be ℓ_2-normalized.

Definition 5.1. Let $\mathbf{A} \in \mathbb{C}^{m \times N}$ be a matrix with ℓ_2-normalized columns $\mathbf{a}_1, \ldots, \mathbf{a}_N$, i.e., $\|\mathbf{a}_i\|_2 = 1$ for all $i \in [N]$. The *coherence* $\mu = \mu(A)$ of the matrix $\mathbf{A}$ is defined as

$$\mu := \max_{1 \leq i \neq j \leq N} |\langle \mathbf{a}_i, \mathbf{a}_j \rangle|. \tag{5.1}$$

Next we introduce the more general concept of ℓ_1-coherence function, which incorporates the usual coherence as the particular value $s = 1$ of its argument.

Definition 5.2. Let $\mathbf{A} \in \mathbb{C}^{m \times N}$ be a matrix with ℓ_2-normalized columns $\mathbf{a}_1, \ldots, \mathbf{a}_N$. The *$\ell_1$-coherence function* μ_1 of the matrix $\mathbf{A}$ is defined for $s \in [N-1]$ by

$$\mu_1(s) := \max_{i \in [N]} \max \left\{ \sum_{j \in S} |\langle \mathbf{a}_i, \mathbf{a}_j \rangle|, \; S \subset [N], \, \mathrm{card}(S) = s, \, i \notin S \right\}.$$

S. Foucart and H. Rauhut, *A Mathematical Introduction to Compressive Sensing*, Applied and Numerical Harmonic Analysis, DOI 10.1007/978-0-8176-4948-7_5, © Springer Science+Business Media New York 2013

It is straightforward to observe that, for $1 \leq s \leq N - 1$,

$$\mu \leq \mu_1(s) \leq s\,\mu, \qquad (5.2)$$

and more generally that, for $1 \leq s, t \leq N - 1$ with $s + t \leq N - 1$,

$$\max\{\mu_1(s), \mu_1(t)\} \leq \mu_1(s + t) \leq \mu_1(s) + \mu_1(t). \qquad (5.3)$$

We remark that the coherence, and more generally the ℓ_1-coherence function, is invariant under multiplication on the left by a unitary matrix $\mathbf{U}$, for the columns of $\mathbf{UA}$ are the ℓ_2-normalized vectors $\mathbf{Ua}_1, \ldots, \mathbf{Ua}_N$ and they satisfy $\langle \mathbf{Ua}_i, \mathbf{Ua}_j \rangle = \langle \mathbf{a}_i, \mathbf{a}_j \rangle$. Moreover, because of the Cauchy–Schwarz inequality $|\langle \mathbf{a}_i, \mathbf{a}_j \rangle| \leq \|\mathbf{a}_i\|_2 \|\mathbf{a}_j\|_2$, it is clear that the coherence of a matrix is bounded above by one, i.e.,

$$\mu \leq 1.$$

Let us consider for a moment a matrix $\mathbf{A} \in \mathbb{C}^{m \times N}$ with $m \geq N$. We observe that $\mu = 0$ if and only if the columns of $\mathbf{A}$ form an orthonormal system. In particular, in the case of a square matrix, we have $\mu = 0$ if and only if $\mathbf{A}$ is a unitary matrix. From now on, we concentrate on the situation occurring in compressive sensing, i.e., we only consider matrices $\mathbf{A} \in \mathbb{C}^{m \times N}$ with $m < N$. In this case, there are limitations on how small the coherence can be. These limitations are given in Sect. 5.2. For the moment, we simply point out that a small coherence implies that column submatrices of moderate size are well conditioned. Let us recall that the notation $\mathbf{A}_S$ denotes the matrix formed by the columns of $\mathbf{A} \in \mathbb{C}^{m \times N}$ indexed by a subset S of $[N]$.

Theorem 5.3. *Let $\mathbf{A} \in \mathbb{C}^{m \times N}$ be a matrix with ℓ_2-normalized columns and let $s \in [N]$. For all s-sparse vectors $\mathbf{x} \in \mathbb{C}^N$,*

$$\left(1 - \mu_1(s - 1)\right) \|\mathbf{x}\|_2^2 \leq \|\mathbf{Ax}\|_2^2 \leq \left(1 + \mu_1(s - 1)\right) \|\mathbf{x}\|_2^2,$$

or equivalently, for each set $S \subset [N]$ with $\mathrm{card}(S) \leq s$, the eigenvalues of the matrix $\mathbf{A}_S^* \mathbf{A}_S$ lie in the interval $\left[1 - \mu_1(s - 1), 1 + \mu_1(s - 1)\right]$. In particular, if $\mu_1(s - 1) < 1$, then $\mathbf{A}_S^* \mathbf{A}_S$ is invertible.

Proof. For a set $S \subset [N]$ with $\mathrm{card}(S) \leq s$, since the matrix $\mathbf{A}_S^* \mathbf{A}_S$ is positive semidefinite, it has an orthonormal basis of eigenvectors associated with real, positive eigenvalues. We denote the minimal eigenvalue by $\lambda_{\min}$ and the maximal eigenvalue by $\lambda_{\max}$. Then, since $\mathbf{Ax} = \mathbf{A}_S \mathbf{x}_S$ for any $\mathbf{x} \in \mathbb{C}^N$ supported on S, it is easy to see that the maximum of

$$\|\mathbf{Ax}\|_2^2 = \langle \mathbf{A}_S \mathbf{x}_S, \mathbf{A}_S \mathbf{x}_S \rangle = \langle \mathbf{A}_S^* \mathbf{A}_S \mathbf{x}_S, \mathbf{x}_S \rangle$$

over the set $\{\mathbf{x} \in \mathbb{C}^N, \mathrm{supp}\,\mathbf{x} \subset S, \|\mathbf{x}\|_2 = 1\}$ is $\lambda_{\max}$ and that its minimum is $\lambda_{\min}$. This explains the equivalence mentioned in the theorem. Now, due to the normalizations $\|a_j\|_2 = 1$ for all $j \in [N]$, the diagonal entries of $A_S^* A_S$ all equal one. By Gershgorin's disk theorem (see Theorem A.11), the eigenvalues of $A_S^* A_S$ are contained in the union of the disks centered at 1 with radii

$$r_j := \sum_{\ell \in S, \ell \neq j} |(A_S^* A_S)_{j,\ell}| = \sum_{\ell \in S, \ell \neq j} |\langle a_\ell, a_j \rangle| \leq \mu_1(s-1), \qquad j \in S.$$

Since these eigenvalues are real, they must lie in $\left[1 - \mu_1(s-1), 1 + \mu_1(s-1) \right]$, as announced. □

Corollary 5.4. *Given a matrix* $A \in \mathbb{C}^{m \times N}$ *with* ℓ_2*-normalized columns and an integer* $s \geq 1$, *if*

$$\mu_1(s) + \mu_1(s-1) < 1,$$

then, for each set $S \subset [N]$ *with* $\text{card}(S) \leq 2s$, *the matrix* $A_S^* A_S$ *is invertible and the matrix* A_S *injective. In particular, the conclusion holds if*

$$\mu < \frac{1}{2s-1}.$$

Proof. In view of (5.3), the condition $\mu_1(s) + \mu_1(s-1) < 1$ implies that $\mu_1(2s-1) < 1$. For a set $S \subset [N]$ with $\text{card}(S) \leq 2s$, according to Theorem 5.3, the smallest eigenvalue of the matrix $A_S^* A_S$ satisfies $\lambda_{\min} \geq 1 - \mu_1(2s-1) > 0$, which shows that $A_S^* A_S$ is invertible. To see that A_S is injective, we simply observe that $A_S z = 0$ yields $A_S^* A_S z = 0$, so that $z = 0$. This proves the first statement. The second one simply follows from $\mu_1(s) + \mu_1(s-1) \leq (2s-1)\mu < 1$ if $\mu < 1/(2s-1)$. □

5.2 Matrices with Small Coherence

In this section, we give lower bounds for the coherence and for the ℓ_1-coherence function of a matrix $A \in \mathbb{C}^{m \times N}$ with $m < N$. We also study the feasibility of achieving these lower bounds. We then give an example of a matrix with an almost minimal coherence. The analysis is carried out for matrices $A \in \mathbb{K}^{m \times N}$, where the field $\mathbb{K}$ can either be $\mathbb{R}$ or $\mathbb{C}$, because the matrices achieving the lower bounds have different features in the real and complex settings. In both cases, however, their columns are equiangular tight frames, which are defined below.

Definition 5.5. A system of ℓ_2-normalized vectors $(a_1, \ldots, a_N)$ in $\mathbb{K}^m$ is called *equiangular* if there is a constant $c \geq 0$ such that

$$|\langle a_i, a_j \rangle| = c \qquad \text{for all } i, j \in [N], \, i \neq j.$$

Tight frames can be defined by several conditions, whose equivalence is left as Exercise 5.2.

Definition 5.6. A system of vectors $(a_1, \ldots, a_N)$ in $\mathbb{K}^m$ is called a *tight frame* if there exists a constant $\lambda > 0$ such that one of the following equivalent conditions holds:

(a) $\|x\|_2^2 = \lambda \sum_{j=1}^{N} |\langle x, a_j \rangle|^2$ for all $x \in \mathbb{K}^m$,

(b) $x = \lambda \sum_{j=1}^{N} \langle x, a_j \rangle a_j$ for all $x \in \mathbb{K}^m$,

(c) $AA^* = \dfrac{1}{\lambda} \mathrm{Id}_m$, where A is the matrix with columns $a_1, \ldots, a_N$.

Unsurprisingly, a system of ℓ_2-normalized vectors is called an equiangular tight frame if it is both an equiangular system and a tight frame. Such systems are the ones achieving the lower bound given below and known as the *Welch bound*.

Theorem 5.7. *The coherence of a matrix* $A \in \mathbb{K}^{m \times N}$ *with* ℓ_2-*normalized columns satisfies*

$$\mu \geq \sqrt{\frac{N-m}{m(N-1)}} . \tag{5.4}$$

Equality holds if and only if the columns $a_1, \ldots, a_N$ *of the matrix* A *form an equiangular tight frame.*

Proof. Let us introduce the *Gram matrix* $G := A^*A \in \mathbb{K}^{N \times N}$ of the system $(a_1, \ldots, a_N)$, which has entries

$$G_{i,j} = \overline{\langle a_i, a_j \rangle} = \langle a_j, a_i \rangle , \qquad i, j \in [N] ,$$

and the matrix $H := AA^* \in \mathbb{K}^{m \times m}$. On the one hand, since the system $(a_1, \ldots, a_N)$ is ℓ_2-normalized, we have

$$\mathrm{tr}\,(G) = \sum_{i=1}^{N} \|a_i\|_2^2 = N. \tag{5.5}$$

On the other hand, since the inner product

$$\langle U, V \rangle_F := \mathrm{tr}\,(UV^*) = \sum_{i,j=1}^{n} U_{i,j} \overline{V_{i,j}}$$

induces the so-called *Froebenius norm* $\| \cdot \|_F$ on $\mathbb{K}^{n \times n}$ (see (A.16)), the Cauchy–Schwarz inequality yields

$$\operatorname{tr}(\mathbf{H}) = \langle \mathbf{H}, \mathbf{Id}_m \rangle_F \leq \|\mathbf{H}\|_F \|\mathbf{Id}_m\|_F = \sqrt{m} \ \sqrt{\operatorname{tr}(\mathbf{HH}^*)}. \tag{5.6}$$

Let us now observe that

$$\operatorname{tr}(\mathbf{HH}^*) = \operatorname{tr}(\mathbf{AA}^*\mathbf{AA}^*) = \operatorname{tr}(\mathbf{A}^*\mathbf{AA}^*\mathbf{A}) = \operatorname{tr}(\mathbf{GG}^*) = \sum_{i,j=1}^{N} |\langle \mathbf{a}_i, \mathbf{a}_j \rangle|^2$$

$$= \sum_{i=1}^{N} \|\mathbf{a}_i\|_2^2 + \sum_{i,j=1, i \neq j}^{N} |\langle \mathbf{a}_i, \mathbf{a}_j \rangle|^2 = N + \sum_{i,j=1, i \neq j}^{N} |\langle \mathbf{a}_i, \mathbf{a}_j \rangle|^2. \tag{5.7}$$

In view of $\operatorname{tr}(\mathbf{G}) = \operatorname{tr}(\mathbf{H})$, combining (5.5), (5.6), and (5.7) yields

$$N^2 \leq m \left(N + \sum_{i,j=1, i \neq j}^{N} |\langle \mathbf{a}_i, \mathbf{a}_j \rangle|^2 \right). \tag{5.8}$$

Taking into account that

$$|\langle \mathbf{a}_i, \mathbf{a}_j \rangle| \leq \mu \qquad \text{for all } i, j \in [N], \ i \neq j, \tag{5.9}$$

we obtain

$$N^2 \leq m \left(N + (N^2 - N)\mu^2 \right),$$

which is a simple rearrangement of (5.4). Moreover, equality holds in (5.4) exactly when equalities hold in (5.6) and in (5.9). Equality in (5.6) says that $\mathbf{H} = \lambda \mathbf{Id}_m$ for some—necessarily nonnegative—constant λ, i.e., that the system $(\mathbf{a}_1, \ldots, \mathbf{a}_N)$ is a tight frame. Equality in (5.9) says that this system is equiangular. □

The Welch bound can be extended to the ℓ_1-coherence function for small values of its argument.

Theorem 5.8. *The ℓ_1-coherence function of a matrix $\mathbf{A} \in \mathbb{K}^{m \times N}$ with ℓ_2-normalized columns satisfies*

$$\mu_1(s) \geq s \sqrt{\frac{N - m}{m(N - 1)}} \qquad \text{whenever} \quad s < \sqrt{N - 1}. \tag{5.10}$$

Equality holds if and only if the columns $\mathbf{a}_1, \ldots, \mathbf{a}_N$ of the matrix $\mathbf{A}$ form an equiangular tight frame.

The proof is based on the following lemma.

Lemma 5.9. *For $k < \sqrt{n}$, if the finite sequence $(\alpha_1, \alpha_2, \ldots, \alpha_n)$ satisfies*

$$\alpha_1 \geq \alpha_2 \geq \cdots \geq \alpha_n \geq 0 \qquad \text{and} \qquad \alpha_1^2 + \alpha_2^2 + \cdots + \alpha_n^2 \geq \frac{n}{k^2},$$

then

$$\alpha_1 + \alpha_2 + \cdots + \alpha_k \geq 1,$$

with equality if and only if $\alpha_1 = \alpha_2 = \cdots = \alpha_n = 1/k$.

Proof. We are going to show the equivalent statement

$$\left.\begin{array}{r} \alpha_1 \geq \alpha_2 \geq \cdots \geq \alpha_n \geq 0 \\ \alpha_1 + \alpha_2 + \cdots + \alpha_k \leq 1 \end{array}\right\} \implies \alpha_1^2 + \alpha_2^2 + \cdots + \alpha_n^2 \leq \frac{n}{k^2},$$

with equality if and only if $\alpha_1 = \alpha_2 = \cdots = \alpha_n = 1/k$. This is the problem of maximizing the convex function

$$f(\alpha_1, \alpha_2, \ldots, \alpha_n) := \alpha_1^2 + \alpha_2^2 + \cdots + \alpha_n^2$$

over the convex polygon

$$\mathcal{C} := \{(\alpha_1, \ldots, \alpha_n) \in \mathbb{R}^n : \alpha_1 \geq \cdots \geq \alpha_n \geq 0 \text{ and } \alpha_1 + \cdots + \alpha_k \leq 1\}.$$

Because any point in $\mathcal{C}$ is a convex combination of its vertices (so that the extreme points of $\mathcal{C}$ are vertices) and because the function f is convex, the maximum is attained at a vertex of $\mathcal{C}$ by Theorem B.16. The vertices of $\mathcal{C}$ are obtained as intersections of n hyperplanes arising by turning n of the $(n + 1)$ inequality constraints into equalities. Thus, we have the following possibilities:

- If $\alpha_1 = \alpha_2 = \cdots = \alpha_n = 0$, then $f(\alpha_1, \alpha_2, \ldots, \alpha_n) = 0$.
- If $\alpha_1 + \cdots + \alpha_k = 1$ and $\alpha_1 = \cdots = \alpha_\ell > \alpha_{\ell+1} = \cdots = \alpha_n = 0$ for $1 \leq \ell \leq k$, then $\alpha_1 = \cdots = \alpha_\ell = 1/\ell$, and consequently $f(\alpha_1, \alpha_2, \ldots, \alpha_n) = 1/\ell$.
- If $\alpha_1 + \cdots + \alpha_k = 1$ and $\alpha_1 = \cdots = \alpha_\ell > \alpha_{\ell+1} = \cdots = \alpha_n = 0$ for $k < \ell \leq n$, then $\alpha_1 = \cdots = \alpha_\ell = 1/k$, and consequently $f(\alpha_1, \alpha_2, \ldots, \alpha_n) = \ell/k^2$.

Taking $k < \sqrt{n}$ into account, it follows that

$$\max_{(\alpha_1, \ldots, \alpha_n) \in \mathcal{C}} f(\alpha_1, \ldots, \alpha_n) = \max\left\{\max_{1 \leq \ell \leq k} \frac{1}{\ell}, \max_{k < \ell \leq n} \frac{\ell}{k^2}\right\} = \max\left\{1, \frac{n}{k^2}\right\} = \frac{n}{k^2},$$

with equality only in the case $\ell = n$ where $\alpha_1 = \alpha_2 = \cdots = \alpha_n = 1/k$.   □

Proof (of Theorem 5.8). Let us recall from (5.8) that we have

$$\sum_{i,j=1, i \neq j}^{N} |\langle \mathbf{a}_i, \mathbf{a}_j \rangle|^2 \geq \frac{N^2}{m} - N = \frac{N(N - m)}{m},$$

which yields

$$\max_{i \in [N]} \sum_{j=1, j \neq i}^{N} |\langle \mathbf{a}_i, \mathbf{a}_j \rangle|^2 \geq \frac{1}{N} \sum_{i,j=1, i \neq j}^{N} |\langle \mathbf{a}_i, \mathbf{a}_j \rangle|^2 \geq \frac{N - m}{m}.$$

For an index $i^* \in [N]$ achieving the latter maximum, we reorder the sequence $(|\langle \mathbf{a}_{i^*}, \mathbf{a}_j \rangle|)_{j=1, j \neq i^*}^N$ as $\beta_1 \geq \beta_2 \geq \cdots \geq \beta_{N-1} \geq 0$, so that

$$\beta_1^2 + \beta_2^2 + \cdots + \beta_{N-1}^2 \geq \frac{N-m}{m}.$$

Lemma 5.9 with $n = N-1$, $k = s$, and $\alpha_\ell := \left(\sqrt{m(N-1)/(N-m)}/s \right) \beta_\ell$ gives $\alpha_1 + \alpha_2 + \cdots + \alpha_s \geq 1$. It follows that

$$\mu_1(s) \geq \beta_1 + \beta_2 + \cdots + \beta_s \geq s \sqrt{\frac{N-m}{m(N-1)}},$$

as announced. Let us now assume that equality holds in (5.10), so that all the previous inequalities are in fact equalities. As in the proof of Theorem 5.7, equality in (5.8) implies that the system $(\mathbf{a}_1, \ldots, \mathbf{a}_N)$ is a tight frame. The case of equality in Lemma 5.9 implies that $|\langle \mathbf{a}_{i^*}, \mathbf{a}_j \rangle| = \sqrt{(N-m)/(m(N-1))}$ for all $j \in [N]$, $j \neq i^*$. Since the index i^* can be arbitrarily chosen in $[N]$, the system $(\mathbf{a}_1, \ldots, \mathbf{a}_N)$ is also equiangular. Conversely, the proof that equiangular tight frames yield equality in (5.10) follows easily from Theorem 5.7 and (5.2). $\qquad \square$

In compressive sensing, we are interested not only in small coherence but also in $m \times N$ matrices where N is much larger than m. This restriction makes it impossible to meet the Welch bound. Indeed, the next theorem shows that the number of vectors in an equiangular tight frame—or in an equiangular system, for that matter—cannot be arbitrarily large.

Theorem 5.10. *The cardinality N of an equiangular system $(\mathbf{a}_1, \ldots, \mathbf{a}_N)$ of ℓ_2-normalized vectors in $\mathbb{K}^m$ satisfies*

$$N \leq \frac{m(m+1)}{2} \qquad \qquad \text{when } \mathbb{K} = \mathbb{R},$$

$$N \leq m^2 \qquad \qquad \text{when } \mathbb{K} = \mathbb{C}.$$

If equality is achieved, then the system $(\mathbf{a}_1, \ldots, \mathbf{a}_N)$ is also a tight frame.

We will use the following simple lemma twice in the proof of this theorem.

Lemma 5.11. *For any $z \in \mathbb{C}$, the $n \times n$ matrix*

$$\begin{bmatrix} 1 & z & z & \cdots & z \\ z & 1 & z & \cdots & z \\ \vdots & \ddots & \ddots & \ddots & \vdots \\ z & \cdots & z & 1 & z \\ z & \cdots & z & z & 1 \end{bmatrix}$$

admits $1 + (n - 1)z$ *as a single eigenvalue and* $1 - z$ *as a multiple eigenvalue of multiplicity* $n - 1$.

Proof. Summing the columns of the matrix, we see that the vector $[1, \ldots, 1]^\top$ is an eigenvector for the eigenvalue $1 + (n - 1)z$. Then, subtracting from the first column each subsequent column, we also see that the $(n - 1)$ linearly independent vectors $[1, -1, 0, \ldots, 0]^\top$, $[1, 0, -1, 0, \ldots, 0]^\top$, $\ldots$, $[1, 0 \ldots, 0, -1]^\top$ are eigenvectors for the eigenvalue $1 - z$. The proof is now complete.   □

Proof (of Theorem 5.10). The key point is to lift our considerations from the space $\mathbb{K}^m$ to a subspace $\mathcal{S}_m$ of operators on $\mathbb{K}^m$. In the case $\mathbb{K} = \mathbb{R}$, $\mathcal{S}_m$ is the space of symmetric operators on $\mathbb{R}^m$, and in the case $\mathbb{K} = \mathbb{C}$, $\mathcal{S}_m$ is simply the space of operators on $\mathbb{C}^m$. (It is tempting to consider Hermitian operators, but they do not form a linear space.) These spaces are endowed with the *Frobenius inner product*

$$\langle \mathbf{P}, \mathbf{Q} \rangle_F = \operatorname{tr}(\mathbf{P}\mathbf{Q}^*), \qquad \mathbf{P}, \mathbf{Q} \in \mathcal{S}_m.$$

Let us introduce the orthogonal projectors $\mathbf{P}_1, \ldots, \mathbf{P}_N \in \mathcal{S}_m$ onto the lines spanned by $\mathbf{a}_1, \ldots, \mathbf{a}_N$. These operators are defined, for $i \in [N]$, by

$$\mathbf{P}_i(\mathbf{v}) = \langle \mathbf{v}, \mathbf{a}_i \rangle \, \mathbf{a}_i, \qquad \mathbf{v} \in \mathbb{K}^m.$$

Let us denote by c the common magnitude of the inner products $\langle \mathbf{a}_i, \mathbf{a}_j \rangle$, $i \neq j$, and by $(\mathbf{e}_1, \ldots, \mathbf{e}_m)$ the canonical basis of $\mathbb{K}^m$. Using the fact that $\mathbf{P}_i^2 = \mathbf{P}_i = \mathbf{P}_i^*$, we calculate, for $i, j \in [N], i \neq j$,

$$\langle \mathbf{P}_i, \mathbf{P}_i \rangle_F = \operatorname{tr}(\mathbf{P}_i \mathbf{P}_i^*) = \operatorname{tr}(\mathbf{P}_i) = \sum_{k=1}^m \langle \mathbf{P}_i(\mathbf{e}_k), \mathbf{e}_k \rangle = \sum_{k=1}^m \langle \mathbf{e}_k, \mathbf{a}_i \rangle \langle \mathbf{a}_i, \mathbf{e}_k \rangle$$

$$= \sum_{k=1}^m |\langle \mathbf{a}_i, \mathbf{e}_k \rangle|^2 = \|\mathbf{a}_i\|_2^2 = 1,$$

$$\langle \mathbf{P}_i, \mathbf{P}_j \rangle_F = \operatorname{tr}(\mathbf{P}_i \mathbf{P}_j^*) = \operatorname{tr}(\mathbf{P}_i \mathbf{P}_j) = \sum_{k=1}^m \langle \mathbf{P}_i \mathbf{P}_j(\mathbf{e}_k), \mathbf{e}_k \rangle = \sum_{k=1}^m \langle \mathbf{P}_j(\mathbf{e}_k), \mathbf{P}_i(\mathbf{e}_k) \rangle$$

$$= \sum_{k=1}^m \langle \mathbf{e}_k, \mathbf{a}_j \rangle \overline{\langle \mathbf{e}_k, \mathbf{a}_i \rangle} \langle \mathbf{a}_j, \mathbf{a}_i \rangle = \overline{\langle \mathbf{a}_i, \mathbf{a}_j \rangle} \Big\langle \sum_{k=1}^m \langle \mathbf{a}_i, \mathbf{e}_k \rangle \mathbf{e}_k, \mathbf{a}_j \Big\rangle$$

$$= \overline{\langle \mathbf{a}_i, \mathbf{a}_j \rangle} \langle \mathbf{a}_i, \mathbf{a}_j \rangle = |\langle \mathbf{a}_i, \mathbf{a}_j \rangle|^2 = c^2.$$

Thus, the Gram matrix of the system $(\mathbf{P}_1, \ldots, \mathbf{P}_N)$ is the $N \times N$ matrix

$$\begin{bmatrix} 1 & c^2 & c^2 & \cdots & c^2 \\ c^2 & 1 & c^2 & \cdots & c^2 \\ \vdots & \ddots & \ddots & \ddots & \vdots \\ c^2 & \cdots & c^2 & 1 & c^2 \\ c^2 & \cdots & c^2 & c^2 & 1 \end{bmatrix}.$$

In view of $0 \le c^2 < 1$, Lemma 5.11 implies that this Gram matrix is invertible, which means that the system $(\mathbf{P}_1, \ldots, \mathbf{P}_N)$ is linearly independent. But this system lies in the space $\mathcal{S}_m$, which has dimension $m(m+1)/2$ when $\mathbb{K} = \mathbb{R}$ and dimension m^2 when $\mathbb{K} = \mathbb{C}$. Therefore, we obtain

$$N \le \frac{m(m+1)}{2} \qquad \text{when } \mathbb{K} = \mathbb{R},$$

$$N \le m^2 \qquad \text{when } \mathbb{K} = \mathbb{C}.$$

Let us now assume that equality holds. Then the system $(\mathbf{Id}_m, \mathbf{P}_1, \ldots, \mathbf{P}_N)$ is linearly dependent; hence, the determinant of its Gram matrix vanishes. This translates into

$$\begin{vmatrix} m & 1 & 1 & 1 & \cdots & 1 \\ 1 & 1 & c^2 & c^2 & \cdots & c^2 \\ 1 & c^2 & 1 & c^2 & \cdots & c^2 \\ \vdots & \vdots & \ddots & \ddots & \ddots & \vdots \\ 1 & c^2 & \cdots & c^2 & 1 & c^2 \\ 1 & c^2 & \cdots & c^2 & c^2 & 1 \end{vmatrix} = 0.$$

Subtracting the first row divided by m from all the other rows and expanding with respect to the first column, we derive the $N \times N$ identity

$$\begin{vmatrix} 1 & b & b & \cdots & b \\ b & 1 & b & \cdots & b \\ \vdots & \ddots & \ddots & \ddots & \vdots \\ b & \cdots & b & 1 & b \\ b & \cdots & b & b & 1 \end{vmatrix} = 0, \qquad \text{where} \quad b := \frac{mc^2 - 1}{m - 1}.$$

Since $1 - b = m(1 - c^2)/(m - 1) \neq 0$, Lemma 5.11 implies that $1 + (N - 1)b = 0$, which reads after simplification

$$c^2 = \frac{N - m}{m(N - 1)}.$$

This shows that the ℓ_2-normalized system $(\mathbf{a}_1, \ldots, \mathbf{a}_N)$ meets the Welch bound. Thus, according to Theorem 5.7, it is an equiangular tight frame. □

The upper bounds on the number of vectors in an equiangular system are sharp. For instance, equiangular systems of 6 vectors in $\mathbb{R}^3$ and of 28 vectors in $\mathbb{R}^7$ are given in Exercise 5.5, while equiangular systems of 4 vectors in $\mathbb{C}^2$ and of 9 vectors in $\mathbb{C}^3$ are given in Exercise 5.6. In contrast with $\mathbb{C}^m$, where systems of m^2 equiangular vectors in $\mathbb{C}^m$ seem to exist for all m, systems of $m(m + 1)/2$ equiangular vectors in $\mathbb{R}^m$ do not exist for all m, as shown below. They are known to exist when m is equal to 2, 3, 7, and 23, but the cases of other allowed values are not settled.

Theorem 5.12. *For $m \geq 3$, if there is an equiangular system of $m(m+1)/2$ vectors in $\mathbb{R}^m$, then $m + 2$ is necessarily the square of an odd integer.*

Proof. Let $(\mathbf{a}_1, \ldots, \mathbf{a}_N)$ be a system of $N = m(m + 1)/2$ equiangular ℓ_2-normalized vectors. According to Theorem 5.10, this system is a tight frame; hence, the matrix $\mathbf{A}$ with columns $\mathbf{a}_1, \ldots, \mathbf{a}_N$ satisfies $\mathbf{AA}^* = \lambda \mathrm{Id}_m$ for some $\lambda > 0$. Since the matrix $\mathbf{G} := \mathbf{A}^*\mathbf{A}$ has the same nonzero eigenvalues as $\mathbf{AA}^*$, i.e., λ with multiplicity m, it also has zero as an eigenvalue of multiplicity $N - m$. Moreover, since $\mathbf{G}$ is the Gram matrix of the system $(\mathbf{a}_1, \ldots, \mathbf{a}_N)$, its diagonal entries all equal one, while its off-diagonal entries all have the same absolute value c. Consequently, the matrix $\mathbf{B} := (\mathbf{G} - \mathrm{Id}_N)/c$ has the form

$$\mathbf{B} = \begin{bmatrix} 0 & b_{1,2} & \cdots & & b_{1,N} \\ b_{2,1} & 0 & \ddots & & \vdots \\ \vdots & \ddots & \ddots & & b_{N-1,N} \\ b_{N,1} & \cdots & b_{N,N-1} & & 0 \end{bmatrix}, \quad \text{where } b_{i,j} = \pm 1,$$

and has $-1/c$ as an eigenvalue of multiplicity $N - m$. Thus, its characteristic polynomial $P_\mathbf{B}(x) := \sum_{k=0}^{N} \beta_k(-x)^k$, $\beta_N = 1$, has integer coefficients β_k and vanishes at $x = -1/c$. Given that

$$c = \sqrt{\frac{N - m}{m(N - 1)}} = \sqrt{\frac{(m+1)/2 - 1}{m(m+1)/2 - 1}} = \sqrt{\frac{m - 1}{m^2 + m - 2}} = \frac{1}{\sqrt{m + 2}},$$

we have $P_\mathbf{B}(-\sqrt{m + 2}) = 0$, i.e,

$$\left(\sum_{0 \leq k \leq N/2} b_{2k}(m + 2)^k \right) + \sqrt{m + 2} \left(\sum_{0 \leq k \leq (N-1)/2} b_{2k+1}(m + 2)^k \right) = 0.$$

Noticing that the two sums above, denoted by Σ_1 and Σ_2, are both integers, we obtain the equality $\Sigma_1^2 = (m + 2)\Sigma_2^2$, which shows that $m + 2$ is a square, since any prime factor of $m + 2$ must appear an even number of times in its prime factor decomposition. We now need to show that $n := \sqrt{m + 2}$ is odd. Let us introduce the

$N \times N$ matrix $\mathbf{J}_N$ whose entries are all equal to one. Its null space has dimension $N - 1$, so it intersects the $(N - m)$-dimensional eigenspace of $\mathbf{B}$ corresponding to the eigenvalue $-1/c = -n$, since $N - 1 + N - m > N$ for $m \geq 3$, i.e., $N = m(m + 1)/2 > m + 1$. Consequently, the matrix $\mathbf{C} := (\mathbf{B} - \mathbf{Id}_N + \mathbf{J}_N)/2$ admits $-(n + 1)/2$ as an eigenvalue. Its diagonal entries are all equal to zero, while its off-diagonal entries are all equal to zero or one. Thus, its characteristic polynomial $P_\mathbf{C}(x) := \sum_{k=0}^{N} \gamma_k(-x)^k$, $\gamma_N = 1$, has integer coefficients γ_k and vanishes at $x = -(n+1)/2$. The equality $P_\mathbf{C}(-(n+1)/2) = 0$ can be rewritten as

$$(n + 1)^N = - \sum_{k=0}^{N-1} 2^{N-k} \gamma_k (n + 1)^k.$$

This shows that $(n + 1)^N$ is an even integer, hence so is $n + 1$. This completes the proof that $n = \sqrt{m + 2}$ is an odd integer.   □

In the complex setting, it seems plausible that equiangular systems of $N = m^2$ vectors exist for all values of m. This would yield $m \times m^2$ matrices with coherence equal to $1/\sqrt{m + 1}$, but no construction of such systems is known at the moment. We present below an explicit $m \times m^2$ matrix with coherence equal to $1/\sqrt{m}$ instead. Let us incidentally notice that $1/\sqrt{m}$ is the limit of the Welch bound when N goes to infinity.

Proposition 5.13. *For each prime number $m \geq 5$, there is an explicit $m \times m^2$ complex matrix with coherence $\mu = 1/\sqrt{m}$.*

Proof. Throughout the proof, we identify the set $[m]$ with $\mathbb{Z}/m\mathbb{Z} =: \mathbb{Z}_m$. For $k, \ell \in \mathbb{Z}_m$, we introduce the *translation* and *modulation* operators $\mathbf{T}_k$ and $\mathbf{M}_\ell$ defined, for $\mathbf{z} \in \mathbb{C}^{\mathbb{Z}_m}$ and $j \in \mathbb{Z}_m$, by

$$(\mathbf{T}_k \mathbf{z})_j = z_{j-k}, \qquad (\mathbf{M}_\ell \mathbf{z})_j = e^{2\pi i \ell j/m} z_j.$$

These operators are isometries of $\ell_2(\mathbb{Z}_m)$. We also introduce the so-called *Alltop* vector, which is the ℓ_2-normalized vector $\mathbf{x} \in \mathbb{C}^{\mathbb{Z}_m}$ defined by

$$x_j := \frac{1}{\sqrt{m}} e^{2\pi i j^3/m}, \qquad j \in \mathbb{Z}_m.$$

The explicit $m \times m^2$ matrix of the proposition is the one with columns $\mathbf{M}_\ell \mathbf{T}_k \mathbf{x}$, $k, \ell \in \mathbb{Z}_m$, i.e., the matrix

$$\left[\mathbf{M}_1 \mathbf{T}_1 \mathbf{x} \big| \cdots \big| \mathbf{M}_1 \mathbf{T}_m \mathbf{x} \big| \mathbf{M}_2 \mathbf{T}_1 \mathbf{x} \big| \cdots \big| \cdots \big| \mathbf{M}_m \mathbf{T}_1 \mathbf{x} \big| \cdots \big| \mathbf{M}_m \mathbf{T}_m \mathbf{x} \right].$$

The inner product of two different columns indexed by (k, ℓ) and (k', ℓ') is

$$\langle \mathbf{M}_\ell \mathbf{T}_k \mathbf{x}, \mathbf{M}_{\ell'} \mathbf{T}_{k'} \mathbf{x} \rangle = \sum_{j \in \mathbb{Z}_m} (\mathbf{M}_\ell \mathbf{T}_k \mathbf{x})_j \overline{(\mathbf{M}_{\ell'} \mathbf{T}_{k'} \mathbf{x})_j}$$

$$= \sum_{j \in \mathbb{Z}_m} e^{2\pi i \ell j/m} x_{j-k} e^{-2\pi i \ell' j/m} \overline{x_{j-k'}}$$

$$= \frac{1}{m} \sum_{j \in \mathbb{Z}_m} e^{2\pi i (\ell - \ell') j/m} e^{2\pi i ((j-k)^3 - (j-k')^3)/m}.$$

Setting $a := \ell - \ell'$ and $b := k - k'$, so that $(a, b) \neq (0, 0)$, we make the change of summation index $h = j - k'$ to obtain

$$\left| \langle \mathbf{M}_\ell \mathbf{T}_k \mathbf{x}, \mathbf{M}_{\ell'} \mathbf{T}_{k'} \mathbf{x} \rangle \right| = \frac{1}{m} \left| e^{2\pi i a k'/m} \sum_{h \in \mathbb{Z}_m} e^{2\pi i a h/m} e^{2\pi i ((h-b)^3 - h^3)/m} \right|$$

$$= \frac{1}{m} \left| \sum_{h \in \mathbb{Z}_m} e^{2\pi i a h/m} e^{2\pi i (-3bh^2 + 3b^2 h - b^3)/m} \right|$$

$$= \frac{1}{m} \left| \sum_{h \in \mathbb{Z}_m} e^{2\pi i (-3bh^2 + (a+3b^2)h)/m} \right|.$$

We now set $c := -3b$ and $d := a + 3b^2$, and we look at the previous modulus squared. We have

$$\left| \langle \mathbf{M}_\ell \mathbf{T}_k \mathbf{x}, \mathbf{M}_{\ell'} \mathbf{T}_{k'} \mathbf{x} \rangle \right|^2 = \frac{1}{m^2} \sum_{h \in \mathbb{Z}_m} e^{2\pi i (ch^2 + dh)/m} \sum_{h' \in \mathbb{Z}_m} e^{-2\pi i (ch'^2 + dh')/m}$$

$$= \frac{1}{m^2} \sum_{h, h' \in \mathbb{Z}_m} e^{2\pi i (h-h')(c(h+h')+d)/m}$$

$$= \frac{1}{m^2} \sum_{h', h'' \in \mathbb{Z}_m} e^{2\pi i h''(c(h''+2h')+d)/m}$$

$$= \frac{1}{m^2} \sum_{h'' \in \mathbb{Z}_m} e^{2\pi i h''(ch''+d)/m} \left(\sum_{h' \in \mathbb{Z}_m} e^{4\pi i ch'' h'/m} \right).$$

For each $h'' \in \mathbb{Z}_m$, we observe that

$$\sum_{h' \in \mathbb{Z}_m} e^{4\pi i ch'' h'/m} = \begin{cases} m & \text{if } 2ch'' = 0 \mod m, \\ 0 & \text{if } 2ch'' \neq 0 \mod m. \end{cases}$$

Let us separate two cases:

1. $c = 0 \mod m$:
   Since $c = -3b$ and $3 \neq 0 \mod m$, we have $b = 0$; hence, $d = a + 3b^2 \neq 0$

mod m, so that

$$\left|\langle \mathbf{M}_\ell \mathbf{T}_k \mathbf{x}, \mathbf{M}_{\ell'} \mathbf{T}_{k'} \mathbf{x} \rangle\right|^2 = \frac{1}{m} \sum_{h'' \in \mathbb{Z}_m} e^{2\pi i d h''/m} = 0.$$

2. $c \neq 0 \mod m$:
   Since $2 \neq 0 \mod m$, the equality $2ch'' = 0$ only occurs when $h'' = 0 \mod m$, so that

$$\left|\langle \mathbf{M}_\ell \mathbf{T}_k \mathbf{x}, \mathbf{M}_{\ell'} \mathbf{T}_{k'} \mathbf{x} \rangle\right|^2 = \frac{1}{m}.$$

This allows to conclude that the coherence of the matrix is equal to $1/\sqrt{m}$. □

5.3 Analysis of Orthogonal Matching Pursuit

We claimed at the beginning of this chapter that the performance of sparse recovery algorithms is enhanced by a small coherence. We justify this claim in the remaining sections. For instance, in view of (5.3), Theorems 5.14 and 5.15 guarantee the exact recovery of every s-sparse vector via orthogonal matching pursuit and via basis pursuit when the measurement matrix has a coherence $\mu < 1/(2s - 1)$. We focus on the orthogonal matching pursuit algorithm in this section.

Theorem 5.14. *Let $\mathbf{A} \in \mathbb{C}^{m \times N}$ be a matrix with ℓ_2-normalized columns. If*

$$\mu_1(s) + \mu_1(s - 1) < 1, \tag{5.11}$$

then every s-sparse vector $\mathbf{x} \in \mathbb{C}^N$ is exactly recovered from the measurement vector $\mathbf{y} = \mathbf{Ax}$ after at most s iterations of orthogonal matching pursuit.

Proof. Let $\mathbf{a}_1, \ldots, \mathbf{a}_N$ denote the ℓ_2-normalized columns of $\mathbf{A}$. According to Proposition 3.5, we need to prove that, for any $S \subset [N]$ with $\mathrm{card}(S) = s$, the matrix $\mathbf{A}_S$ is injective and that

$$\max_{j \in S} |\langle \mathbf{r}, \mathbf{a}_j \rangle| > \max_{\ell \in \overline{S}} |\langle \mathbf{r}, \mathbf{a}_\ell \rangle| \tag{5.12}$$

for all nonzero $\mathbf{r} \in \{\mathbf{Az}, \mathrm{supp}(\mathbf{z}) \subset S\}$. Let then $\mathbf{r} := \sum_{i \in S} r_i \mathbf{a}_i$ be such a vector, and let $k \in S$ be chosen so that $|r_k| = \max_{i \in S} |r_i| > 0$. On the one hand, for $\ell \in \overline{S}$, we have

$$|\langle \mathbf{r}, \mathbf{a}_\ell \rangle| = \left| \sum_{i \in S} r_i \langle \mathbf{a}_i, \mathbf{a}_\ell \rangle \right| \leq \sum_{i \in S} |r_i| |\langle \mathbf{a}_i, \mathbf{a}_\ell \rangle| \leq |r_k| \mu_1(s).$$

On the other hand, we have

$$\left| \langle \mathbf{r}, \mathbf{a}_k \rangle \right| = \left| \sum_{i \in S} r_i \langle \mathbf{a}_i, \mathbf{a}_k \rangle \right| \geq |r_k| \, |\langle \mathbf{a}_k, \mathbf{a}_k \rangle| - \sum_{i \in S, i \neq k} |r_i| \, |\langle \mathbf{a}_i, \mathbf{a}_k \rangle|$$

$$\geq |r_k| - |r_k| \mu_1(s - 1).$$

Thus, (5.12) is fulfilled because $1 - \mu_1(s - 1) > \mu_1(s)$ according to (5.11). Finally, the injectivity of $\mathbf{A}_S$ follows from Corollary 5.4.   □

5.4 Analysis of Basis Pursuit

In this section, we show that a small coherence also guarantees the success of basis pursuit. As a matter of fact, any condition guaranteeing the success of the recovery of all vectors supported on a set S via card(S) iterations of orthogonal matching pursuit also guarantees the success of the recovery of all vectors supported on S via basis pursuit. This follows from the fact that the exact recovery condition (3.7) implies the null space property (4.1). Indeed, given $\mathbf{v} \in \ker \mathbf{A} \setminus \{\mathbf{0}\}$, we have $\mathbf{A}_S \mathbf{v}_S = -\mathbf{A}_{\overline{S}} \mathbf{v}_{\overline{S}}$, and

$$\|\mathbf{v}_S\|_1 = \|\mathbf{A}_S^\dagger \mathbf{A}_S \mathbf{v}_S\|_1 = \|\mathbf{A}_S^\dagger \mathbf{A}_{\overline{S}} \mathbf{v}_{\overline{S}}\|_1 \leq \|\mathbf{A}_S^\dagger \mathbf{A}_{\overline{S}}\|_{1 \to 1} \|\mathbf{v}_{\overline{S}}\|_1 < \|\mathbf{v}_{\overline{S}}\|_1.$$

Thus, the following result is immediate. We nonetheless give an alternative self-contained proof.

Theorem 5.15. *Let $\mathbf{A} \in \mathbb{C}^{m \times N}$ be a matrix with ℓ_2-normalized columns. If*

$$\mu_1(s) + \mu_1(s - 1) < 1, \tag{5.13}$$

then every s-sparse vector $\mathbf{x} \in \mathbb{C}^N$ is exactly recovered from the measurement vector $\mathbf{y} = \mathbf{A}\mathbf{x}$ via basis pursuit.

Proof. According to Theorem 4.5, it is necessary and sufficient to prove that the matrix $\mathbf{A}$ satisfies the null space property of order s, i.e., that

$$\|\mathbf{v}_S\|_1 < \|\mathbf{v}_{\overline{S}}\|_1 \tag{5.14}$$

for any nonzero vector $\mathbf{v} \in \ker \mathbf{A}$ and any index set $S \subset [N]$ with card$(S) = s$. If $\mathbf{a}_1, \ldots, \mathbf{a}_N$ denote the columns of $\mathbf{A}$, then the condition $\mathbf{v} \in \ker \mathbf{A}$ translates into $\sum_{j=1}^N v_j \mathbf{a}_j = 0$. Thus, taking the inner product with a particular $\mathbf{a}_i$, $i \in S$, and isolating the term in v_i, we obtain

$$v_i = v_i \langle \mathbf{a}_i, \mathbf{a}_i \rangle = - \sum_{j=1, j \neq i}^{N} v_j \langle \mathbf{a}_j, \mathbf{a}_i \rangle = - \sum_{\ell \in \overline{S}} v_\ell \langle \mathbf{a}_\ell, \mathbf{a}_i \rangle - \sum_{j \in S, j \neq i} v_j \langle \mathbf{a}_j, \mathbf{a}_i \rangle.$$

It follows that

$$|v_i| \leq \sum_{\ell \in \overline{S}} |v_\ell| |\langle \mathbf{a}_\ell, \mathbf{a}_i \rangle| + \sum_{j \in S, j \neq i} |v_j| |\langle \mathbf{a}_j, \mathbf{a}_i \rangle|.$$

Summing over all $i \in S$ and interchanging the summations, we derive

$$\|\mathbf{v}_S\|_1 = \sum_{i \in S} |v_i| \leq \sum_{\ell \in \overline{S}} |v_\ell| \sum_{i \in S} |\langle \mathbf{a}_\ell, \mathbf{a}_i \rangle| + \sum_{j \in S} |v_j| \sum_{i \in S, i \neq j} |\langle \mathbf{a}_j, \mathbf{a}_i \rangle|$$

$$\leq \sum_{\ell \in \overline{S}} |v_\ell| \, \mu_1(s) + \sum_{j \in S} |v_j| \, \mu_1(s-1) = \mu_1(s) \|\mathbf{v}_{\overline{S}}\|_1 + \mu_1(s-1) \|\mathbf{v}_S\|_1 .$$

After rearrangement, this reads $(1 - \mu_1(s-1)) \|\mathbf{v}_S\|_1 \leq \mu_1(s) \|\mathbf{v}_{\overline{S}}\|_1$, and (5.14) is fulfilled because $1 - \mu_1(s-1) > \mu_1(s)$, which is a rewriting of (5.13). $\square$

Choosing a matrix $\mathbf{A} \in \mathbb{C}^{m \times N}$ with small coherence $\mu \leq c/\sqrt{m}$ (for instance, the one of Theorem 5.13), we see that the condition $(2s-1)\mu < 1$ ensuring recovery of s-sparse vectors via orthogonal matching pursuit as well as via ℓ_1-minimization is satisfied once

$$m \geq Cs^2 . \tag{5.15}$$

This gives a first estimate in terms of sparsity for the required number of measurements using practical recovery algorithms and specific matrices $\mathbf{A}$. This result can sometimes be satisfying, but an estimate where the sparsity s enters quadratically is often too pessimistic, especially for mildly large s. We will later see that a linear scaling of m in s is possible up to logarithmic factors.

Let us point out that it is not possible to overcome the quadratic bottleneck in (5.15) using Theorems 5.14 and 5.15. Indeed, let us assume on the contrary that the sufficient condition $\mu_1(s) + \mu_1(s-1) < 1$ holds with $m \leq (2s-1)^2/2$ and $s < \sqrt{N-1}$, say. Provided N is large, say $N \geq 2m$, we apply Theorem 5.8 to derive a contradiction from

$$1 > \mu_1(s) + \mu_1(s-1) \geq (2s-1) \sqrt{\frac{N-m}{m(N-1)}} \geq \sqrt{\frac{2(N-m)}{N-1}} \geq \sqrt{\frac{N}{N-1}}.$$

In the following chapters, we will reduce the number of required measurements below the order s^2 by introducing new tools for the analysis of sparse recovery algorithms.

5.5 Analysis of Thresholding Algorithms

In this final section, we show that thresholding algorithms can also be analyzed using the coherence. For instance, under the same condition as before, even the basic thresholding algorithm will successfully recover sparse vectors with nonzero entries of constant magnitude.

Theorem 5.16. *Let* $\mathbf{A} \in \mathbb{C}^{m \times N}$ *be a matrix with* ℓ_2*-normalized columns and let* $\mathbf{x} \in \mathbb{C}^N$ *be a vector supported on a set* S *of size* s. *If*

$$\mu_1(s) + \mu_1(s-1) < \frac{\min_{i \in S} |x_i|}{\max_{i \in S} |x_i|} \,, \tag{5.16}$$

then the vector $\mathbf{x} \in \mathbb{C}^N$ *is exactly recovered from the measurement vector* $\mathbf{y} = \mathbf{A}\mathbf{x}$ *via basic thresholding.*

Proof. Let $\mathbf{a}_1, \dots, \mathbf{a}_N$ denote the ℓ_2-normalized columns of $\mathbf{A}$. According to Proposition 3.7, we need to prove that, for any $j \in S$ and any $\ell \in \overline{S}$,

$$|\langle \mathbf{A}\mathbf{x}, \mathbf{a}_j \rangle| > |\langle \mathbf{A}\mathbf{x}, \mathbf{a}_\ell \rangle| \,. \tag{5.17}$$

We observe that

$$|\langle \mathbf{A}\mathbf{x}, \mathbf{a}_\ell \rangle| = |\sum_{i \in S} x_i \langle \mathbf{a}_i, \mathbf{a}_\ell \rangle| \leq \sum_{i \in S} |x_i||\langle \mathbf{a}_i, \mathbf{a}_\ell \rangle| \leq \mu_1(s) \max_{i \in S} |x_i|,$$

$$|\langle \mathbf{A}\mathbf{x}, \mathbf{a}_j \rangle| = |\sum_{i \in S} x_i \langle \mathbf{a}_i, \mathbf{a}_j \rangle| \geq |x_j| - \sum_{i \in S, i \neq j} |x_i||\langle \mathbf{a}_i, \mathbf{a}_j \rangle|$$

$$\geq \min_{i \in S} |x_i| - \mu_1(s-1) \max_{i \in S} |x_i|.$$

Thus, taking (5.16) into account, we obtain

$$|\langle \mathbf{A}\mathbf{x}, \mathbf{a}_j \rangle| - |\langle \mathbf{A}\mathbf{x}, \mathbf{a}_\ell \rangle| \geq \min_{i \in S} |x_i| - (\mu(s) + \mu_1(s-1)) \max_{i \in S} |x_i| > 0.$$

This shows (5.17) and concludes the proof. $\square$

It is possible to prove the success of sparse recovery via the iterative hard thresholding algorithm under some coherence conditions—see Exercise 5.10. We now turn directly to the more involved hard thresholding pursuit algorithm. Just as for orthogonal matching pursuit, we show that s iterations are enough for the recovery of s-sparse vectors under a condition quite similar to (5.11). In view of (5.2), we observe that the condition in question is met when the coherence of the measurement matrix satisfies $\mu < 1/(3s - 1)$.

Theorem 5.17. *Let* $\mathbf{A} \in \mathbb{C}^{m \times N}$ *be a matrix with* ℓ_2*-normalized columns. If*

$$2\mu_1(s) + \mu_1(s-1) < 1,$$

then every s-sparse vector $\mathbf{x} \in \mathbb{C}^N$ *is exactly recovered from the measurement vector* $\mathbf{y} = \mathbf{A}\mathbf{x}$ *after at most s iterations of hard thresholding pursuit.*

Proof. Let us consider indices $j_1, j_2, \ldots, j_N$ such that

$$|x_{j_1}| \geq |x_{j_2}| \geq \cdots \geq |x_{j_s}| > |x_{j_{s+1}}| = \cdots = |x_{j_N}| = 0.$$

We are going to prove that, for $0 \leq n \leq s - 1$, the set $\{j_1, \ldots, j_{n+1}\}$ is included in S^{n+1} defined by (HTP$_1$) with $\mathbf{y} = \mathbf{A}\mathbf{x}$ as the set of largest absolute entries of

$$\mathbf{z}^{n+1} := \mathbf{x}^n + \mathbf{A}^* \mathbf{A}(\mathbf{x} - \mathbf{x}^n). \tag{5.18}$$

This will imply that $S^s = S = \mathrm{supp}(\mathbf{x})$ and consequently that $\mathbf{x}^s = \mathbf{x}$ by (HTP$_2$). Note that it is sufficient to prove that

$$\min_{1 \leq k \leq n+1} |z_{j_k}^{n+1}| > \max_{\ell \in \overline{S}} |z_\ell^{n+1}|. \tag{5.19}$$

We notice that, for every $j \in [N]$,

$$z_j^{n+1} = x_j^n + \sum_{i=1}^N (x_i - x_i^n)\langle \mathbf{a}_i, \mathbf{a}_j \rangle = x_j + \sum_{i \neq j} (x_i - x_i^n)\langle \mathbf{a}_i, \mathbf{a}_j \rangle.$$

Therefore, we have

$$|z_j^{n+1} - x_j| \leq \sum_{i \in S^n, i \neq j} |x_i - x_i^n||\langle \mathbf{a}_i, \mathbf{a}_j \rangle| + \sum_{i \in S \backslash S^n, i \neq j} |x_i||\langle \mathbf{a}_i, \mathbf{a}_j \rangle|. \tag{5.20}$$

We derive, for $1 \leq k \leq n + 1$ and $\ell \in \overline{S}$, that

$$|z_{j_k}^{n+1}| \geq |x_{j_k}| - \mu_1(s)\|(\mathbf{x} - \mathbf{x}^n)_{S^n}\|_\infty - \mu_1(s)\|\mathbf{x}_{S \backslash S^n}\|_\infty, \tag{5.21}$$

$$|z_\ell^{n+1}| \leq \mu_1(s)\|(\mathbf{x} - \mathbf{x}^n)_{S^n}\|_\infty + \mu_1(s)\|\mathbf{x}_{S \backslash S^n}\|_\infty. \tag{5.22}$$

In particular, for $n = 0$, substituting $\|(\mathbf{x} - \mathbf{x}^n)_{S^n}\|_\infty = 0$ into (5.21) and (5.22) gives

$$|z_{j_1}^1| \geq (1 - \mu_1(s))\|\mathbf{x}\|_\infty > \mu_1(s)\|\mathbf{x}\|_\infty \geq |z_\ell^1| \quad \text{for all } \ell \in \overline{S},$$

by virtue of $2\mu_1(s) < 1$. Therefore, the base case of the inductive hypothesis (5.19) holds for $n = 0$. Let us now assume that this hypothesis holds for $n - 1$ with $n \geq 1$. This implies that $\{j_1, \ldots, j_n\} \subset S^n$. We notice that (HTP$_2$) with n replaced by $n - 1$ gives, in view of Lemma 3.4,

$$(\mathbf{A}^*\mathbf{A}(\mathbf{x} - \mathbf{x}^n))_{S^n} = 0 .$$

Hence, for any $j \in S^n$, the definition (5.18) of $\mathbf{z}^{n+1}$ implies that $z_j^{n+1} = x_j^n$, and then (5.20) yields

$$|x_j^n - x_j| \le \mu_1(s-1)\|(\mathbf{x} - \mathbf{x}^n)_{S^n}\|_\infty + \mu_1(s-1)\|\mathbf{x}_{S\setminus S^n}\|_\infty .$$

Taking the maximum over $j \in S^n$ and rearranging gives

$$\|(\mathbf{x} - \mathbf{x}^n)_{S^n}\|_\infty \le \frac{\mu_1(s-1)}{1 - \mu_1(s-1)}\|\mathbf{x}_{S\setminus S^n}\|_\infty .$$

Substituting the latter into (5.21) and (5.22), we obtain, for $1 \le k \le n+1$ and $\ell \in \overline{S}$,

$$|z_{j_k}^{n+1}| \ge \left(1 - \frac{\mu_1(s)}{1 - \mu_1(s-1)}\right)|x_{j_{n+1}}| ,$$

$$|z_\ell^{n+1}| \le \frac{\mu_1(s)}{1 - \mu_1(s-1)}|x_{j_{n+1}}| .$$

Since $\mu_1(s)/(1 - \mu_1(s-1)) < 1/2$, this shows that (5.19) holds for n, too. The proof by induction is now complete.   □

Notes

The analysis of sparse recovery algorithms could be carried out using merely the coherence. For instance, the conclusion of Theorem 5.17 can be achieved under the sufficient condition $\mu < 1/(3s - 1)$, as obtained by Maleki in [340]. Similarly, the conclusion of Theorem 5.15 can be achieved under the sufficient condition $\mu < 1/(2s-1)$, as obtained earlier by Gribonval and Nielsen in [239] and also by Donoho and Elad in [155]. This followed previous work [158] on ℓ_1-minimization by Donoho and Huo. They considered matrices formed by the union of two orthonormal bases and introduced the concept of mutual (in)coherence; see Exercise 5.1.

Theorems 5.14 and 5.15 in their present form were established by Tropp in [476]. What we call ℓ_1-coherence function here is called *cumulative coherence function* there. This concept also appears under the name *Babel function*. A straightforward extension to any $p > 0$ would be the ℓ_p-coherence function of a matrix $\mathbf{A}$ with ℓ_2-normalized columns $\mathbf{a}_1, \ldots, \mathbf{a}_N$ defined by

$$\mu_p(s) := \max_{i \in [N]} \max\left\{\left(\sum_{j \in S}|\langle \mathbf{a}_i, \mathbf{a}_j\rangle|^p\right)^{1/p}, S \subseteq [N], \operatorname{card}(S) = s, i \notin S\right\} .$$

Theorem 5.8 on the Welch-type lower bound for the ℓ_1-coherence function appeared in [443]. The $m \times m^2$ matrix with coherence equal to $1/\sqrt{m}$ of Proposition 5.13 is taken from [12, 456]. One can observe that its columns are the union of m orthonormal bases. In [250], Gurevich, Hadani, and Sochen uncovered a matrix with p rows (p being a prime number), roughly p^5 columns, and coherence bounded above by $4/\sqrt{p}$. Another number theoretic construction of $p \times p^k$ matrices ($p > k$ being a prime number) with coherence bounded above by $(k-1)/\sqrt{p}$ can be found, for instance, in [472, Chap. 5.7.4].

There is a vast literature dedicated to frames. The notion is not restricted to the finite-dimensional setting, although this is the only one we considered. Good starting places to learn about the subject are Christensen's books [119] and [120]. As mentioned in the text, not everything is known about equiangular tight frames. In particular, whether equiangular systems of m^2 vectors in $\mathbb{C}^m$ exist for all values of m is not known—the numerical experiments performed for $m \leq 45$ by Renes, Blume-Kohout, Scott, and Caves in [420] seem to indicate that they do. More details on the subject of equiangular tight frames, and more generally tight frames in finite dimension, can be found in Waldron's book [504].

Exercises

5.1. The *mutual coherence* between two orthonormal bases $\mathbf{U} = (\mathbf{u}_1, \ldots, \mathbf{u}_m)$ and $\mathbf{V} = (\mathbf{v}_1, \ldots, \mathbf{v}_m)$ of $\mathbb{C}^m$ is defined as

$$\mu(\mathbf{U}, \mathbf{V}) := \max_{1 \leq i, j \leq m} |\langle \mathbf{u}_i, \mathbf{v}_j \rangle|.$$

Establish the inequalities

$$\frac{1}{\sqrt{m}} \leq \mu(\mathbf{U}, \mathbf{V}) \leq 1$$

and prove that they are sharp.

5.2. Prove the equivalence of the three conditions of Definition 5.6 and find the value of the constant λ when the vectors $\mathbf{a}_1, \ldots, \mathbf{a}_N$ are ℓ_2-normalized.

5.3. Establish the alternative expressions for the ℓ_1-coherence function

$$\mu_1(s) = \max_{\mathrm{card}(S) \leq s+1} \|\mathbf{A}_S^* \mathbf{A}_S - \mathbf{Id}\|_{1 \to 1} = \max_{\mathrm{card}(S) \leq s+1} \|\mathbf{A}_S^* \mathbf{A}_S - \mathbf{Id}\|_{\infty \to \infty}.$$

5.4. Prove that the $m+1$ vertices of a regular simplex in $\mathbb{R}^m$ centered at the origin form an equiangular tight frame.

5.5. With $c := (\sqrt{5} - 1)/2$, prove that the columns of the matrix

$$\begin{bmatrix} 1 & 0 & c & 1 & 0 & -c \\ c & 1 & 0 & -c & 1 & 0 \\ 0 & c & 1 & 0 & -c & 1 \end{bmatrix}$$

form an equiangular system of 6 vectors in $\mathbb{R}^3$. Prove also that the vectors obtained by unit cyclic shifts on four vectors $[1, \pm 1, 0, \pm 1, 0, 0, 0]^\top$ form an equiangular system of 28 vectors in $\mathbb{R}^7$.

5.6. With $c := e^{i\pi/4}\sqrt{2 - \sqrt{3}}$, prove that the columns of the matrix

$$\begin{bmatrix} 1 & c & 1 & -c \\ c & 1 & -c & 1 \end{bmatrix}$$

form an equiangular system of 4 vectors in $\mathbb{C}^2$. With $\omega := e^{i2\pi/3}$, prove also that the columns of the matrix

$$\begin{bmatrix} -2 & 1 & 1 & -2 & \omega^2 & \omega & -2 & \omega & \omega^2 \\ 1 & -2 & 1 & \omega & -2 & \omega^2 & \omega^2 & -2 & \omega \\ 1 & 1 & -2 & \omega^2 & \omega & -2 & \omega & \omega^2 & -2 \end{bmatrix}$$

form an equiangular system of 9 vectors in $\mathbb{C}^3$.

5.7. Prove that the columns of the matrix considered in Proposition 5.13 form a tight frame.

5.8. Suppose that a known vector is an s-sparse linear combination of vectors from the canonical and Fourier bases $\mathbf{E} = (\mathbf{e}_1, \ldots, \mathbf{e}_m)$ and $\mathbf{F} = (\mathbf{f}_1, \ldots, \mathbf{f}_m)$, defined as

$$\mathbf{e}_k = [0, \ldots, 0, \underbrace{1}_{\text{index } k}, 0, \ldots, 0]^\top, \qquad \mathbf{f}_k = \frac{1}{\sqrt{m}}[1, e^{2\pi i k/m}, \ldots, e^{2\pi i k(m-1)/m}]^\top.$$

Prove that the unknown coefficients can be found by orthogonal matching pursuit or basis pursuit if $s < (\sqrt{m} + 1)/2$.

5.9. Given $\nu < 1/2$, suppose that a matrix $\mathbf{A} \in \mathbb{C}^{m \times N}$ satisfies

$$\mu_1(s) \leq \nu.$$

Prove that, for any $\mathbf{x} \in \mathbb{C}^N$ and $\mathbf{y} = \mathbf{Ax} + \mathbf{e}$ with $\|\mathbf{e}\|_2 \leq \eta$, a minimizer $\mathbf{x}^\sharp$ of $\|\mathbf{z}\|_1$ subject to $\|\mathbf{Az} - \mathbf{y}\|_2 \leq \eta$ approximates the vector $\mathbf{x}$ with ℓ_1-error

$$\|\mathbf{x} - \mathbf{x}^\sharp\|_1 \le C \, \sigma_s(\mathbf{x})_1 + D \, s \, \eta,$$

for some positive constants C and D depending only on ν.

5.10. Let $\mathbf{A} \in \mathbb{C}^{m \times N}$ be a matrix with ℓ_2-normalized columns. Prove that, if $\mu_1(2s) < 1/2$ (in particular if $\mu < 1/(4s)$), then every s-sparse vector $\mathbf{x} \in \mathbb{C}^N$ is recovered from $\mathbf{y} = \mathbf{A}\mathbf{x}$ via iterative hard thresholding.

Chapter 6
Restricted Isometry Property

The coherence is a simple and useful measure of the quality of a measurement matrix. However, the lower bound on the coherence in Theorem 5.7 limits the performance analysis of recovery algorithms to rather small sparsity levels. A finer measure of the quality of a measurement matrix is needed to overcome this limitation. This is provided by the concept of *restricted isometry property*, also known as *uniform uncertainty principle*. It ensures the success of the sparse recovery algorithms presented in this book. Restricted isometry constants are introduced in Sect. 6.1. The success of sparse recovery is then established under some conditions on these constants for basis pursuit in Sect. 6.2, for thresholding-based algorithms in Sect. 6.3, and for greedy algorithms in Sect. 6.4.

6.1 Definitions and Basic Properties

Unlike the coherence, which only takes pairs of columns of a matrix into account, the restricted isometry constant of order s involves all s-tuples of columns and is therefore more suited to assess the quality of the matrix. As with the coherence, small restricted isometry constants are desired. Here is their formal definition.

Definition 6.1. The sth *restricted isometry constant* $\delta_s = \delta_s(\mathbf{A})$ of a matrix $\mathbf{A} \in \mathbb{C}^{m \times N}$ is the smallest $\delta \geq 0$ such that

$$(1 - \delta)\|\mathbf{x}\|_2^2 \leq \|\mathbf{A}\mathbf{x}\|_2^2 \leq (1 + \delta)\|\mathbf{x}\|_2^2 \tag{6.1}$$

for all s-sparse vectors $\mathbf{x} \in \mathbb{C}^N$. Equivalently, it is given by

$$\delta_s = \max_{S \subset [N], \operatorname{card}(S) \leq s} \|\mathbf{A}_S^* \mathbf{A}_S - \mathbf{Id}\|_{2 \to 2}. \tag{6.2}$$

We say that $\mathbf{A}$ satisfies the *restricted isometry property* if δ_s is small for reasonably large s—the meaning of small δ_s and large s will be made precise later.

S. Foucart and H. Rauhut, *A Mathematical Introduction to Compressive Sensing*,
Applied and Numerical Harmonic Analysis, DOI 10.1007/978-0-8176-4948-7_6,
© Springer Science+Business Media New York 2013

We make a few remarks before establishing the equivalence of these two definitions. The first one is that the sequence of restricted isometry constants is nondecreasing, i.e.,

$$\delta_1 \leq \delta_2 \leq \cdots \leq \delta_s \leq \delta_{s+1} \leq \cdots \leq \delta_N.$$

The second one is that, although $\delta_s \geq 1$ is not forbidden, the relevant situation occurs for $\delta_s < 1$. Indeed, (6.2) says that each column submatrix $\mathbf{A}_S$, $S \subset [N]$ with $\operatorname{card}(S) \leq s$, has all its singular values in the interval $[1 - \delta_s, 1 + \delta_s]$ and is therefore injective when $\delta_s < 1$. In fact, $\delta_{2s} < 1$ is more relevant, since the inequality (6.1) yields $\|\mathbf{A}(\mathbf{x} - \mathbf{x}')\|_2^2 > 0$ for all distinct s-sparse vectors $\mathbf{x}, \mathbf{x}' \in \mathbb{C}^N$; hence, distinct s-sparse vectors have distinct measurement vectors. The third and final remark is that, if the entries of the measurement matrix $\mathbf{A}$ are real, then δ_s could also be defined as the smallest $\delta \geq 0$ such that (6.1) holds for all real s-sparse vectors $\mathbf{x} \in \mathbb{R}^N$. This is because the operator norms of the real symmetric matrix $\mathbf{A}_S^* \mathbf{A}_S - \mathbf{Id}$ relative $\ell_2(\mathbb{R})$ and to $\ell_2(\mathbb{C})$ are equal—both to its largest eigenvalues in modulus—and because the two definitions of restricted isometry constants would be equivalent in the real setting, too.

For the equivalence of (6.1) and (6.2) in the complex setting, we start by noticing that (6.1) is equivalent to

$$\left| \|\mathbf{A}_S \mathbf{x}\|_2^2 - \|\mathbf{x}\|_2^2 \right| \leq \delta \|\mathbf{x}\|_2^2 \quad \text{for all } S \subset [N], \operatorname{card}(S) \leq s, \text{ and all } \mathbf{x} \in \mathbb{C}^S.$$

We then observe that, for $\mathbf{x} \in \mathbb{C}^S$,

$$\|\mathbf{A}_S \mathbf{x}\|_2^2 - \|\mathbf{x}\|_2^2 = \langle \mathbf{A}_S \mathbf{x}, \mathbf{A}_S \mathbf{x} \rangle - \langle \mathbf{x}, \mathbf{x} \rangle = \langle (\mathbf{A}_S^* \mathbf{A}_S - \mathbf{Id}) \mathbf{x}, \mathbf{x} \rangle.$$

Since the matrix $\mathbf{A}_S^* \mathbf{A}_S - \mathbf{Id}$ is Hermitian, we have

$$\max_{\mathbf{x} \in \mathbb{C}^S \setminus \{0\}} \frac{\langle (\mathbf{A}_S^* \mathbf{A}_S - \mathbf{Id}) \mathbf{x}, \mathbf{x} \rangle}{\|\mathbf{x}\|_2^2} = \|\mathbf{A}_S^* \mathbf{A}_S - \mathbf{Id}\|_{2 \to 2},$$

so that (6.1) is equivalent to

$$\max_{S \subset [N], \operatorname{card}(S) \leq s} \|\mathbf{A}_S^* \mathbf{A}_S - \mathbf{Id}\|_{2 \to 2} \leq \delta.$$

This proves the identity (6.2), as δ_s is the smallest such δ.

It is now possible to compare the restricted isometry constants of a matrix with its coherence μ and coherence function μ_1; see Definitions 5.1 and 5.2.

Proposition 6.2. *If the matrix* $\mathbf{A}$ *has* ℓ_2*-normalized columns* $\mathbf{a}_1, \ldots, \mathbf{a}_N$, *i.e.,* $\|\mathbf{a}_j\|_2 = 1$ *for all* $j \in [N]$, *then*

$$\delta_1 = 0, \qquad \delta_2 = \mu, \qquad \delta_s \leq \mu_1(s - 1) \leq (s - 1)\mu, \quad s \geq 2.$$

Proof. The ℓ_2-normalization of the columns means that $\|\mathbf{A}\mathbf{e}_j\|_2^2 = \|\mathbf{e}_j\|_2^2$ for all $j \in [N]$, that is to say $\delta_1 = 0$. Next, with $\mathbf{a}_1, \ldots, \mathbf{a}_N$ denoting the columns of the matrix $\mathbf{A}$, we have

$$\delta_2 = \max_{1 \leq i \neq j \leq N} \|\mathbf{A}^*_{\{i,j\}}\mathbf{A}_{\{i,j\}} - \mathbf{Id}\|_{2 \to 2}, \qquad \mathbf{A}^*_{\{i,j\}}\mathbf{A}_{\{i,j\}} = \begin{bmatrix} 1 & \langle \mathbf{a}_j, \mathbf{a}_i \rangle \\ \langle \mathbf{a}_i, \mathbf{a}_j \rangle & 1 \end{bmatrix}.$$

The eigenvalues of the matrix $\mathbf{A}^*_{\{i,j\}}\mathbf{A}_{\{i,j\}} - \mathbf{Id}$ are $|\langle \mathbf{a}_i, \mathbf{a}_j \rangle|$ and $-|\langle \mathbf{a}_i, \mathbf{a}_j \rangle|$, so its operator norm is $|\langle \mathbf{a}_i, \mathbf{a}_j \rangle|$. Taking the maximum over $1 \leq i \neq j \leq N$ yields the equality $\delta_2 = \mu$. The inequality $\delta_s \leq \mu_1(s-1) \leq (s-1)\mu$ follows from Theorem 5.3. $\qquad \square$

In view of the existence of $m \times m^2$ matrices with coherence μ equal to $1/\sqrt{m}$ (see Chap. 5), this already shows the existence of $m \times m^2$ matrices with restricted isometry constant $\delta_s < 1$ for $s \leq \sqrt{m}$. We will establish that, given $\delta < 1$, there exist $m \times N$ matrices with restricted isometry constant $\delta_s \leq \delta$ for $s \leq cm/\ln(eN/m)$, where c is a constant depending only on δ; see Chap. 9. This is essentially the largest range possible; see Chap. 10. Matrices with a small restricted isometry constant of this optimal order are informally said to satisfy the *restricted isometry property* or *uniform uncertainty principle*.

We now make a simple but essential observation, which motivates the related notion of restricted orthogonality constant.

Proposition 6.3. *Let* $\mathbf{u}, \mathbf{v} \in \mathbb{C}^N$ *be vectors with* $\|\mathbf{u}\|_0 \leq s$ *and* $\|\mathbf{v}\|_0 \leq t$. *If* $\text{supp}(\mathbf{u}) \cap \text{supp}(\mathbf{v}) = \emptyset$, *then*

$$|\langle \mathbf{A}\mathbf{u}, \mathbf{A}\mathbf{v} \rangle| \leq \delta_{s+t}\|\mathbf{u}\|_2\|\mathbf{v}\|_2. \tag{6.3}$$

Proof. Let $S := \text{supp}(\mathbf{u}) \cup \text{supp}(\mathbf{v})$, and let $\mathbf{u}_S, \mathbf{v}_S \in \mathbb{C}^S$ be the restrictions of $\mathbf{u}, \mathbf{v} \in \mathbb{C}^N$ to S. Since $\mathbf{u}$ and $\mathbf{v}$ have disjoint supports, we have $\langle \mathbf{u}_S, \mathbf{v}_S \rangle = 0$. We derive

$$|\langle \mathbf{A}\mathbf{u}, \mathbf{A}\mathbf{v} \rangle| = |\langle \mathbf{A}_S\mathbf{u}_S, \mathbf{A}_S\mathbf{v}_S \rangle - \langle \mathbf{u}_S, \mathbf{v}_S \rangle| = |\langle (\mathbf{A}^*_S\mathbf{A}_S - \mathbf{Id})\mathbf{u}_S, \mathbf{v}_S \rangle|$$

$$\leq \|(\mathbf{A}^*_S\mathbf{A}_S - \mathbf{Id})\mathbf{u}_S\|_2\|\mathbf{v}_S\|_2 \leq \|\mathbf{A}^*_S\mathbf{A}_S - \mathbf{Id}\|_{2 \to 2}\|\mathbf{u}_S\|_2\|\mathbf{v}_S\|_2,$$

and the conclusion follows from (6.2), $\|\mathbf{u}_S\|_2 = \|\mathbf{u}\|_2$, and $\|\mathbf{v}_S\|_2 = \|\mathbf{v}\|_2$. $\qquad \square$

Definition 6.4. The (s,t)-*restricted orthogonality constant* $\theta_{s,t} = \theta_{s,t}(\mathbf{A})$ of a matrix $\mathbf{A} \in \mathbb{C}^{m \times N}$ is the smallest $\theta \geq 0$ such that

$$|\langle \mathbf{A}\mathbf{u}, \mathbf{A}\mathbf{v} \rangle| \leq \theta \|\mathbf{u}\|_2\|\mathbf{v}\|_2 \tag{6.4}$$

for all disjointly supported s-sparse and t-sparse vectors $\mathbf{u}, \mathbf{v} \in \mathbb{C}^N$. Equivalently, it is given by

$$\theta_{s,t} = \max \{ \|\mathbf{A}^*_T\mathbf{A}_S\|_{2 \to 2}, \ S \cap T = \emptyset, \ \text{card}(S) \leq s, \ \text{card}(T) \leq t \}. \tag{6.5}$$

The justification of the equivalence between the two definitions is left as Exercise 6.4. We now give a comparison result between restricted isometry constants and restricted orthogonality constants.

Proposition 6.5. *Restricted isometry constants and restricted orthogonality constants are related by*

$$\theta_{s,t} \le \delta_{s+t} \le \frac{1}{s+t}\left(s\,\delta_s + t\,\delta_t + 2\sqrt{st}\,\theta_{s,t}\right).$$

The special case $t = s$ gives the inequalities

$$\theta_{s,s} \le \delta_{2s} \quad \text{and} \quad \delta_{2s} \le \delta_s + \theta_{s,s}.$$

Proof. The first inequality is Proposition 6.3. For the second inequality, given an $(s+t)$-sparse vector $\mathbf{x} \in \mathbb{C}^N$ with $\|\mathbf{x}\|_2 = 1$, we need to show that

$$\left|\|\mathbf{A}\mathbf{x}\|_2^2 - \|\mathbf{x}\|_2^2\right| \le \frac{1}{s+t}\left(s\,\delta_s + t\,\delta_t + 2\sqrt{st}\,\theta_{s,t}\right).$$

Let $\mathbf{u}, \mathbf{v} \in \mathbb{C}^N$ be two disjointly supported vectors such that $\mathbf{u} + \mathbf{v} = \mathbf{x}$, where $\mathbf{u}$ is s-sparse and $\mathbf{v}$ is t-sparse, respectively. We have

$$\|\mathbf{A}\mathbf{x}\|_2^2 = \langle \mathbf{A}(\mathbf{u}+\mathbf{v}), \mathbf{A}(\mathbf{u}+\mathbf{v})\rangle = \|\mathbf{A}\mathbf{u}\|_2^2 + \|\mathbf{A}\mathbf{v}\|_2^2 + 2\,\mathrm{Re}\langle \mathbf{A}\mathbf{u}, \mathbf{A}\mathbf{v}\rangle.$$

Taking $\|\mathbf{x}\|_2^2 = \|\mathbf{u}\|_2^2 + \|\mathbf{v}\|_2^2$ into account, we derive

$$\left|\|\mathbf{A}\mathbf{x}\|_2^2 - \|\mathbf{x}\|_2^2\right| \le \left|\|\mathbf{A}\mathbf{u}\|_2^2 - \|\mathbf{u}\|_2^2\right| + \left|\|\mathbf{A}\mathbf{v}\|_2^2 - \|\mathbf{v}\|_2^2\right| + 2\left|\langle \mathbf{A}\mathbf{u}, \mathbf{A}\mathbf{v}\rangle\right|$$
$$\le \delta_s\|\mathbf{u}\|_2^2 + \delta_t\|\mathbf{v}\|_2^2 + 2\theta_{s,t}\|\mathbf{u}\|_2\|\mathbf{v}\|_2 = f\left(\|\mathbf{u}\|_2^2\right),$$

where we have set, for $\alpha \in [0, 1]$,

$$f(\alpha) := \delta_s\alpha + \delta_t(1 - \alpha) + 2\theta_{s,t}\sqrt{\alpha(1 - \alpha)}. \tag{6.6}$$

It can be shown that there is an $\alpha^* \in [0, 1]$ such that this function is nondecreasing on $[0, \alpha^*]$ and then nonincreasing on $[\alpha^*, 1]$—see Exercise 6.5. Depending on the location of this α^* with respect to $s/(s+t)$, the function f is either nondecreasing on $[0, s/(s+t)]$ or nonincreasing on $[s/(s+t), 1]$. By properly choosing the vector $\mathbf{u}$, we can always assume that $\|\mathbf{u}\|_2^2$ is in one of these intervals. Indeed, if $\mathbf{u}$ is made of s smallest absolute entries of $\mathbf{x}$ while $\mathbf{v}$ is made of t largest absolute entries of $\mathbf{x}$, then we have

$$\frac{\|\mathbf{u}\|_2^2}{s} \le \frac{\|\mathbf{v}\|_2^2}{t} = \frac{1 - \|\mathbf{u}\|_2^2}{t}, \quad \text{so that } \|\mathbf{u}\|_2^2 \le \frac{s}{s+t},$$

and if $\mathbf{u}$ was made of s largest absolute entries of $\mathbf{x}$, then we would likewise have $\|\mathbf{u}\|_2^2 \geq s/(s+t)$. This implies

$$\left|\|\mathbf{Ax}\|_2^2 - \|\mathbf{x}\|_2^2\right| \leq f\left(\frac{s}{s+t}\right) = \delta_s \frac{s}{s+t} + \delta_t \frac{t}{s+t} + 2\theta_{s,t} \frac{\sqrt{st}}{s+t}.$$

The proof is complete. □

We continue by proving that restricted isometry constants and restricted orthogonality constants of high order can be controlled by those of lower order.

Proposition 6.6. *For integers $r, s, t \geq 1$ with $t \geq s$,*

$$\theta_{t,r} \leq \sqrt{\frac{t}{s}}\, \theta_{s,r},$$

$$\delta_t \leq \frac{t-d}{s} \delta_{2s} + \frac{d}{s} \delta_s \qquad \text{where } d := \gcd(s,t).$$

The special case $t = cs$ gives

$$\delta_{cs} \leq c\,\delta_{2s}.$$

Remark 6.7. There are other relations enabling to control constants of higher order by constants of lower order; see Exercise 6.10.

Proof. Given a t-sparse vector $\mathbf{u} \in \mathbb{C}^N$ and an r-sparse vector $\mathbf{v} \in \mathbb{C}^N$ that are disjointly supported, we need to show that

$$|\langle \mathbf{Au}, \mathbf{Av}\rangle| \leq \sqrt{\frac{t}{s}}\, \theta_{s,r} \|\mathbf{u}\|_2 \|\mathbf{v}\|_2, \tag{6.7}$$

$$\left|\|\mathbf{Au}\|_2^2 - \|\mathbf{u}\|_2^2\right| \leq \left(\frac{t-d}{s}\delta_{2s} + \frac{d}{s}\delta_s\right)\|\mathbf{u}\|_2^2. \tag{6.8}$$

Let d be a common divisor of s and t. We introduce integers k, n such that

$$s = kd, \qquad t = nd.$$

Let $T = \{j_1, j_2, \ldots, j_t\}$ denote the support of $\mathbf{u}$. We consider the n subsets $S_1, S_2, \ldots, S_n \subset T$ of size s defined by

$$S_i = \{j_{(i-1)d+1}, j_{(i-1)d+2}, \ldots, j_{(i-1)d+s}\},$$

where indices are meant modulo t. In this way, each $j \in T$ belongs to exactly $s/d = k$ sets S_i, so that

$$\mathbf{u} = \frac{1}{k} \sum_{i=1}^{n} \mathbf{u}_{S_i}, \qquad \|\mathbf{u}\|_2^2 = \frac{1}{k} \sum_{i=1}^{n} \|\mathbf{u}_{S_i}\|_2^2.$$

We now derive (6.7) from

$$|\langle \mathbf{Au}, \mathbf{Av} \rangle| \le \frac{1}{k} \sum_{i=1}^{n} |\langle \mathbf{Au}_{S_i}, \mathbf{Av} \rangle| \le \frac{1}{k} \sum_{i=1}^{n} \theta_{s,r} \|\mathbf{u}_{S_i}\|_2 \|\mathbf{v}\|_2$$

$$\le \theta_{s,r} \frac{\sqrt{n}}{k} \left(\sum_{i=1}^{n} \|\mathbf{u}_{S_i}\|_2^2 \right)^{1/2} \|\mathbf{v}\|_2 = \theta_{s,r} \left(\frac{n}{k} \right)^{1/2} \|\mathbf{u}\|_2 \|\mathbf{v}\|_2.$$

Inequality (6.8) follows from

$$\left| \|\mathbf{Au}\|_2^2 - \|\mathbf{u}\|_2^2 \right| = \left| \langle (\mathbf{A}^*\mathbf{A} - \mathbf{Id})\mathbf{u}, \mathbf{u} \rangle \right|$$

$$\le \frac{1}{k^2} \sum_{i=1}^{n} \sum_{j=1}^{n} \left| \langle (\mathbf{A}^*\mathbf{A} - \mathbf{Id})\mathbf{u}_{S_i}, \mathbf{u}_{S_j} \rangle \right|$$

$$= \frac{1}{k^2} \left(\sum_{1 \le i \ne j \le n} \left| \langle (\mathbf{A}_{S_i \cup S_j}^* \mathbf{A}_{S_i \cup S_j} - \mathbf{Id})\mathbf{u}_{S_i}, \mathbf{u}_{S_j} \rangle \right| \right.$$

$$\left. + \sum_{i=1}^{n} \left| \langle (\mathbf{A}_{S_i}^* \mathbf{A}_{S_i} - \mathbf{Id})\mathbf{u}_{S_i}, \mathbf{u}_{S_i} \rangle \right| \right)$$

$$\le \frac{1}{k^2} \left(\sum_{1 \le i \ne j \le n} \delta_{2s} \|\mathbf{u}_{S_i}\|_2 \|\mathbf{u}_{S_j}\|_2 + \sum_{i=1}^{n} \delta_s \|\mathbf{u}_{S_i}\|_2^2 \right)$$

$$= \frac{\delta_{2s}}{k^2} \left(\sum_{i=1}^{n} \|\mathbf{u}_{S_i}\|_2 \right)^2 - \frac{\delta_{2s} - \delta_s}{k^2} \sum_{i=1}^{n} \|\mathbf{u}_{S_i}\|_2^2$$

$$\le \left(\frac{\delta_{2s} n}{k^2} - \frac{\delta_{2s} - \delta_s}{k^2} \right) \sum_{i=1}^{n} \|\mathbf{u}_{S_i}\|_2^2 = \left(\frac{n}{k} \delta_{2s} - \frac{1}{k}(\delta_{2s} - \delta_s) \right) \|\mathbf{u}\|_2^2$$

$$= \left(\frac{t}{s} \delta_{2s} - \frac{1}{k}(\delta_{2s} - \delta_s) \right) \|\mathbf{u}\|_2^2.$$

To make the latter as small as possible, we take k as small as possible, i.e., we choose d as the greatest common divisor of s and t. This completes the proof. □

Just like for the coherence, it is important to know how small the sth restricted isometry constant of a matrix $\mathbf{A} \in \mathbb{C}^{m \times N}$ can be. In the case $N \ge Cm$ of relevance in compressive sensing, Theorem 6.8 below states that the restricted isometry constant must satisfy $\delta_s \ge c \sqrt{s/m}$. For $s = 2$, this reads $\mu \ge c'/\sqrt{m}$, which is reminiscent of the Welch bound of Theorem 5.7. In fact, the proof is an adaptation of the proof of this theorem.

Theorem 6.8. *For* $\mathbf{A} \in \mathbb{C}^{m \times N}$ *and* $2 \leq s \leq N$, *one has*

$$m \geq c \frac{s}{\delta_s^2}, \tag{6.9}$$

provided $N \geq C m$ *and* $\delta_s \leq \delta_*$, *where the constants* c, C, *and* δ_* *depend only on each other. For instance, the choices* $c = 1/162$, $C = 30$, *and* $\delta_* = 2/3$ *are valid.*

Proof. We first notice that the statement cannot hold for $s = 1$, as $\delta_1 = 0$ if all the columns of $\mathbf{A}$ have ℓ_2-norm equal to 1. Let us set $t := \lfloor s/2 \rfloor \geq 1$, and let us decompose the matrix $\mathbf{A}$ in blocks of size $m \times t$—except possibly the last one which may have less columns—as

$$\mathbf{A} = [\ \mathbf{A}_1 \mid \mathbf{A}_2 \mid \cdots \mid \mathbf{A}_n\ ], \qquad N \leq nt.$$

From (6.2) and (6.5), we recall that, for all $i, j \in [n], i \neq j$,

$$\|\mathbf{A}_i^* \mathbf{A}_i - \mathbf{Id}\|_{2\to 2} \leq \delta_t \leq \delta_s, \qquad \|\mathbf{A}_i^* \mathbf{A}_j\|_{2\to 2} \leq \theta_{t,t} \leq \delta_{2t} \leq \delta_s,$$

so that the eigenvalues of $\mathbf{A}_i^* \mathbf{A}_i$ and the singular values of $\mathbf{A}_i^* \mathbf{A}_j$ satisfy

$$1 - \delta_s \leq \lambda_k(\mathbf{A}_i^* \mathbf{A}_i) \leq 1 + \delta_s, \qquad \sigma_k(\mathbf{A}_i^* \mathbf{A}_j) \leq \delta_s.$$

Let us introduce the matrices

$$\mathbf{H} := \mathbf{A}\mathbf{A}^* \in \mathbb{C}^{m \times m}, \qquad \mathbf{G} := \mathbf{A}^* \mathbf{A} = [\mathbf{A}_i^* \mathbf{A}_j]_{1 \leq i,j \leq n} \in \mathbb{C}^{N \times N}.$$

On the one hand, we have the lower bound

$$\mathrm{tr}(\mathbf{H}) = \mathrm{tr}(\mathbf{G}) = \sum_{i=1}^{n} \mathrm{tr}(\mathbf{A}_i^* \mathbf{A}_i) = \sum_{i=1}^{n} \sum_{k=1}^{t} \lambda_k(\mathbf{A}_i^* \mathbf{A}_i) \geq n\,t\,(1 - \delta_s). \tag{6.10}$$

On the other hand, writing $\langle \mathbf{M}_1, \mathbf{M}_2 \rangle_F = \mathrm{tr}(\mathbf{M}_2^* \mathbf{M}_1)$ for the Frobenius inner product of two matrices $\mathbf{M}_1$ and $\mathbf{M}_2$ (see (A.15)), we have

$$\mathrm{tr}(\mathbf{H})^2 = \langle \mathbf{Id}_m, \mathbf{H} \rangle_F^2 \leq \|\mathbf{Id}_m\|_F^2 \|\mathbf{H}\|_F^2 = m\,\mathrm{tr}(\mathbf{H}^* \mathbf{H}).$$

Then, by cyclicity of the trace,

$$\mathrm{tr}(\mathbf{H}^* \mathbf{H}) = \mathrm{tr}(\mathbf{A}\mathbf{A}^* \mathbf{A}\mathbf{A}^*) = \mathrm{tr}(\mathbf{A}^* \mathbf{A}\mathbf{A}^* \mathbf{A}) = \mathrm{tr}(\mathbf{G}\mathbf{G}^*)$$

$$= \sum_{i=1}^{n} \mathrm{tr}\left(\sum_{j=1}^{m} \mathbf{A}_i^* \mathbf{A}_j \mathbf{A}_j^* \mathbf{A}_i \right)$$

$$= \sum_{1 \leq i \neq j \leq n} \sum_{k=1}^{t} \sigma_k(\mathbf{A}_i^* \mathbf{A}_j)^2 + \sum_{i=1}^{n} \sum_{k=1}^{t} \lambda_k(\mathbf{A}_i^* \mathbf{A}_i)^2$$

$$\leq n\,(n-1)\,t\,\delta_s^2 + n\,t\,(1 + \delta_s)^2.$$

We derive the upper bound

$$\text{tr}(\mathbf{H})^2 \leq m\,n\,t\left((n-1)\delta_s^2 + (1+\delta_s)^2\right). \tag{6.11}$$

Combining the bounds (6.10) and (6.11) yields

$$m \geq \frac{n\,t\,(1-\delta_s)^2}{(n-1)\,\delta_s^2 + (1+\delta_s)^2}.$$

If $(n-1)\delta_s^2 < (1+\delta_s)^2/5$, we would obtain, using $\delta_s \leq 2/3$,

$$m > \frac{n\,t\,(1-\delta_s)^2}{6(1+\delta_s)^2/5} \geq \frac{5(1-\delta_s)^2}{6(1+\delta_s)^2}\,N \geq \frac{1}{30}\,N,$$

which contradicts our assumption. We therefore have $(n-1)\,\delta_s^2 \geq (1+\delta_s)^2/5$, which yields, using $\delta_s \leq 2/3$ again and $s \leq 3t$,

$$m \geq \frac{n\,t\,(1-\delta_s)^2}{6(n-1)\,\delta_s^2} \geq \frac{1}{54}\frac{t}{\delta_s^2} \geq \frac{1}{162}\frac{s}{\delta_s^2}.$$

This is the desired result.   □

Let us compare the previous lower bound on restricted isometry constants, namely,

$$\delta_s \geq \sqrt{cs/m}, \tag{6.12}$$

with upper bounds available so far. Precisely, choosing a matrix $\mathbf{A} \in \mathbb{C}^{m \times N}$ with a coherence of optimal order $\mu \leq c/\sqrt{m}$, Proposition 6.2 implies that

$$\delta_s \leq (s-1)\mu \leq cs/\sqrt{m}. \tag{6.13}$$

There is a significant gap between (6.12) and (6.13). In particular, (6.13) with the quadratic scaling

$$m \geq c's^2 \tag{6.14}$$

allows δ_s to be small, while this requires under (6.12) that $m \geq c's$. However, whether such a condition can be sufficient is unknown at this point. We will see later in Chap. 9 that certain random matrices $\mathbf{A} \in \mathbb{R}^{m \times N}$ satisfy $\delta_s \leq \delta$ with high probability for some $\delta > 0$ provided

$$m \geq C\delta^{-2}s\ln(eN/s). \tag{6.15}$$

We will also see in Corollary 10.8 that $\delta_s \leq \delta$ requires $m \geq C_\delta s \ln(eN/s)$. Therefore, the lower bound $m \geq c's$ is optimal up to logarithmic factors, and (6.9) is optimal regarding the scaling $C_\delta = C\delta^{-2}$.

Difficulty of Deterministic Constructions of Matrices with RIP. As just mentioned, random matrices will be used to obtain the restricted isometry property $\delta_s \leq \delta$ (abbreviated RIP) in the optimal regime (6.15) for the number m of measurements in terms of the sparsity s and the vector length N. To date, finding deterministic (i.e., explicit or at least constructible in polynomial time) matrices satisfying $\delta_s \leq \delta$ in this regime is a major open problem. Essentially all available estimations of δ_s for deterministic matrices combine a coherence estimation and Proposition 6.2 in one form or another (with one exception commented on in the Notes section). This leads to bounds of the type (6.13) and in turn to the quadratic bottleneck (6.14). Thus, the lower bound of Theorem 5.7 in principle prevents such a proof technique to generate improved results. The intrinsic difficulty in bounding the restricted isometry constants of explicit matrices $\mathbf{A}$ lies in the basic tool for estimating the eigenvalues of $\mathbf{A}_S^* \mathbf{A}_S - \mathbf{Id}$, namely, Gershgorin's disk theorem (Theorem A.11). Indeed, assuming ℓ_2-normalization of the columns of $\mathbf{A}$ and taking the supremum over all $S \subset [N]$ with $\mathrm{card}(S) = s$ leads then to the ℓ_1-coherence function $\mu_1(s-1)$—this is how we showed the bound $\delta_s \leq \mu_1(s-1)$ of Proposition 6.2; see also Theorem 5.3. Then the lower bound for the ℓ_1-coherence function from Theorem 5.8 tells us that the quadratic bottleneck is unavoidable when using Gershgorin's theorem to estimate restricted isometry constants. It seems that not only the magnitude of the entries of the Gramian $\mathbf{A}^*\mathbf{A}$ but also their signs should be taken into account in order to improve estimates for deterministic matrices, but which tools to be used for this purpose remains unclear (a slight improvement over the quadratic bottleneck is discussed in the Notes section). One may conjecture that some of the matrices with coherence of optimal order, for instance, the one of Theorem 5.13, also satisfy the restricted isometry property when m scales linearly in s up to logarithmic factors. All the same, when passing to random matrices, a powerful set of tools becomes available for the estimation of the restricted isometry constants in the optimal regime.

6.2 Analysis of Basis Pursuit

In this section, we establish the success of sparse recovery via basis pursuit for measurement matrices with small restricted isometry constants. We give two proofs of this fact. The first proof is simple and quite natural. It shows that the condition $\delta_{2s} < 1/3$ is sufficient to guarantee exact recovery of all s-sparse vectors via ℓ_1-minimization. The second proof is more involved. It shows that the weaker condition $\delta_{2s} < 0.6246$ is actually sufficient to guarantee stable and robust recovery of all s-sparse vectors via ℓ_1-minimization. We start by presenting the simple argument which ignores stability and robustness issues (although such issues can be treated with only a slight extension of the argument).

Theorem 6.9. *Suppose that the $2s$th restricted isometry constant of the matrix $\mathbf{A} \in \mathbb{C}^{m \times N}$ satisfies*

$$\delta_{2s} < \frac{1}{3}. \tag{6.16}$$

Then every s-sparse vector $\mathbf{x} \in \mathbb{C}^N$ *is the unique solution of*

$$\underset{\mathbf{z} \in \mathbb{C}^N}{\text{minimize}} \ \|\mathbf{z}\|_1 \quad \text{subject to } \mathbf{Az} = \mathbf{Ax}.$$

The following observation is recurring in our argument, so we isolate it from the proof.

Lemma 6.10. *Given* $q > p > 0$, *if* $\mathbf{u} \in \mathbb{C}^s$ *and* $\mathbf{v} \in \mathbb{C}^t$ *satisfy*

$$\max_{i \in [s]} |u_i| \le \min_{j \in [t]} |v_j|, \tag{6.17}$$

then

$$\|\mathbf{u}\|_q \le \frac{s^{1/q}}{t^{1/p}} \|\mathbf{v}\|_p.$$

The special case $p = 1$, $q = 2$, *and* $t = s$ *gives*

$$\|\mathbf{u}\|_2 \le \frac{1}{\sqrt{s}} \|\mathbf{v}\|_1.$$

Proof. For the first statement, we only need to notice that

$$\frac{\|\mathbf{u}\|_q}{s^{1/q}} = \left[\frac{1}{s} \sum_{i=1}^{s} |u_i|^q \right]^{1/q} \le \max_{i \in [s]} |u_i|,$$

$$\frac{\|\mathbf{v}\|_p}{t^{1/p}} = \left[\frac{1}{t} \sum_{j=1}^{t} |v_j|^p \right]^{1/p} \ge \min_{j \in [t]} |v_j|,$$

and to use (6.17). The second statement is an immediate consequence. $\square$

Proof (of Theorem 6.9). According to Corollary 4.5, it is enough to establish the null space property of order s in the form

$$\|\mathbf{v}_S\|_1 < \frac{1}{2} \|\mathbf{v}\|_1 \quad \text{for all } \mathbf{v} \in \ker \mathbf{A} \setminus \{\mathbf{0}\} \text{ and all } S \subset [N] \text{ with } \mathrm{card}(S) = s.$$

This will follow from the stronger statement

$$\|\mathbf{v}_S\|_2 \le \frac{\rho}{2\sqrt{s}} \|\mathbf{v}\|_1 \quad \text{for all } \mathbf{v} \in \ker \mathbf{A} \text{ and all } S \subset [N] \text{ with } \mathrm{card}(S) = s,$$

where

$$\rho := \frac{2\delta_{2s}}{1 - \delta_{2s}}$$

satisfies $\rho < 1$ whenever $\delta_{2s} < 1/3$. Given $\mathbf{v} \in \ker \mathbf{A}$, we notice that it is enough to consider an index set $S =: S_0$ of s largest absolute entries of the vector $\mathbf{v}$. We partition the complement $\overline{S_0}$ of S_0 in $[N]$ as $\overline{S_0} = S_1 \cup S_2 \cup \ldots$, where

$$S_1 : \text{index set of } s \text{ largest absolute entries of } \mathbf{v} \text{ in } \overline{S_0},$$

$$S_2 : \text{index set of } s \text{ largest absolute entries of } \mathbf{v} \text{ in } \overline{S_0 \cup S_1},$$

etc. In view of $\mathbf{v} \in \ker \mathbf{A}$, we have $\mathbf{A}(\mathbf{v}_{S_0}) = \mathbf{A}(-\mathbf{v}_{S_1} - \mathbf{v}_{S_2} - \cdots)$, so that

$$\|\mathbf{v}_{S_0}\|_2^2 \leq \frac{1}{1 - \delta_{2s}} \|\mathbf{A}(\mathbf{v}_{S_0})\|_2^2 = \frac{1}{1 - \delta_{2s}} \langle \mathbf{A}(\mathbf{v}_{S_0}), \mathbf{A}(-\mathbf{v}_{S_1}) + \mathbf{A}(-\mathbf{v}_{S_2}) + \cdots \rangle$$

$$= \frac{1}{1 - \delta_{2s}} \sum_{k \geq 1} \langle \mathbf{A}(\mathbf{v}_{S_0}), \mathbf{A}(-\mathbf{v}_{S_k}) \rangle. \tag{6.18}$$

According to Proposition 6.3, we also have

$$\langle \mathbf{A}(\mathbf{v}_{S_0}), \mathbf{A}(-\mathbf{v}_{S_k}) \rangle \leq \delta_{2s} \|\mathbf{v}_{S_0}\|_2 \|\mathbf{v}_{S_k}\|_2. \tag{6.19}$$

Substituting (6.19) into (6.18) and dividing by $\|\mathbf{v}_{S_0}\|_2 > 0$, we obtain

$$\|\mathbf{v}_{S_0}\|_2 \leq \frac{\delta_{2s}}{1 - \delta_{2s}} \sum_{k \geq 1} \|\mathbf{v}_{S_k}\|_2 = \frac{\rho}{2} \sum_{k \geq 1} \|\mathbf{v}_{S_k}\|_2.$$

For $k \geq 1$, the s absolute entries of $\mathbf{v}_{S_k}$ do not exceed the s absolute entries of $\mathbf{v}_{S_{k-1}}$, so that Lemma 6.10 yields

$$\|\mathbf{v}_{S_k}\|_2 \leq \frac{1}{\sqrt{s}} \|\mathbf{v}_{S_{k-1}}\|_1.$$

We then derive

$$\|\mathbf{v}_{S_0}\|_2 \leq \frac{\rho}{2\sqrt{s}} \sum_{k \geq 1} \|\mathbf{v}_{S_{k-1}}\|_1 \leq \frac{\rho}{2\sqrt{s}} \|\mathbf{v}\|_1.$$

This is the desired inequality.   □

Remark 6.11. In (6.18), the vector $\mathbf{v}_{S_0}$ was interpreted as being $2s$-sparse, although it is in fact s-sparse. The better bound $\|\mathbf{v}_{S_0}\|_2^2 \leq \|\mathbf{A}(\mathbf{v}_{S_0})\|_2^2/(1 - \delta_s)$ could therefore be invoked. In (6.19), the restricted orthogonality constant $\theta_{s,s}$ could also have been used instead of δ_{2s}. This would yield the sufficient condition $\delta_s + 2\theta_{s,s} < 1$ instead of (6.16).

It is instructive to refine the above proof by establishing stability and robustness. The reader is invited to do so in Exercise 6.12. Here, stability and robustness are incorporated in Theorem 6.12 below, which also improves on Theorem 6.9 by relaxing the sufficient condition (6.16).

Theorem 6.12. *Suppose that the* 2*sth restricted isometry constant of the matrix* $\mathbf{A} \in \mathbb{C}^{m \times N}$ *satisfies*

$$\delta_{2s} < \frac{4}{\sqrt{41}} \approx 0.6246. \tag{6.20}$$

Then, for any $\mathbf{x} \in \mathbb{C}^N$ *and* $\mathbf{y} \in \mathbb{C}^m$ *with* $\|\mathbf{A}\mathbf{x} - \mathbf{y}\|_2 \leq \eta$, *a solution* $\mathbf{x}^\sharp$ *of*

$$\underset{\mathbf{z} \in \mathbb{C}^N}{\text{minimize}} \|\mathbf{z}\|_1 \qquad \text{subject to } \|\mathbf{A}\mathbf{z} - \mathbf{y}\|_2 \leq \eta$$

approximates the vector $\mathbf{x}$ *with errors*

$$\|\mathbf{x} - \mathbf{x}^\sharp\|_1 \leq C \sigma_s(\mathbf{x})_1 + D \sqrt{s}\, \eta,$$

$$\|\mathbf{x} - \mathbf{x}^\sharp\|_2 \leq \frac{C}{\sqrt{s}} \sigma_s(\mathbf{x})_1 + D \eta,$$

where the constants $C, D > 0$ *depend only on* δ_{2s}.

These error estimates—in fact, ℓ_p-error estimates for any $1 \leq p \leq 2$—are immediately deduced from Theorem 4.22 and the following result.

Theorem 6.13. *If the* 2*sth restricted isometry constant of* $\mathbf{A} \in \mathbb{C}^{m \times N}$ *obeys* (6.20), *then the matrix* $\mathbf{A}$ *satisfies the* ℓ_2-*robust null space property of order s with constants* $0 < \rho < 1$ *and* $\tau > 0$ *depending only on* δ_{2s}.

The argument makes use of the following lemma, called *square root lifting* inequality. It can be viewed as a counterpart of the inequality $\|\mathbf{a}\|_1 \leq \sqrt{s}\|\mathbf{a}\|_2$ for $\mathbf{a} \in \mathbb{C}^s$.

Lemma 6.14. *For* $a_1 \geq a_2 \geq \cdots \geq a_s \geq 0$,

$$\sqrt{a_1^2 + \cdots + a_s^2} \leq \frac{a_1 + \cdots + a_s}{\sqrt{s}} + \frac{\sqrt{s}}{4}(a_1 - a_s).$$

Proof. We prove the equivalent statement

$$\left. \begin{array}{r} a_1 \geq a_2 \geq \cdots \geq a_s \geq 0 \\ \dfrac{a_1 + a_2 + \cdots + a_s}{\sqrt{s}} + \dfrac{\sqrt{s}}{4} a_1 \leq 1 \end{array} \right\} \implies \sqrt{a_1^2 + \cdots + a_s^2} + \frac{\sqrt{s}}{4} a_s \leq 1.$$

Thus, we aim at maximizing the convex function

$$f(a_1, a_2, \ldots, a_s) := \sqrt{a_1^2 + \cdots + a_s^2} + \frac{\sqrt{s}}{4} a_s$$

over the convex polytope

$$\mathcal{C} := \{(a_1, \ldots, a_s) \in \mathbb{R}^s : a_1 \geq \cdots \geq a_s \geq 0 \text{ and } \frac{a_1 + \cdots + a_s}{\sqrt{s}} + \frac{\sqrt{s}}{4} a_1 \leq 1\}.$$

Because any point in $\mathcal{C}$ is a convex combination of its vertices and because the function f is convex, the maximum is attained at a vertex of $\mathcal{C}$; see Theorem B.16. The vertices of $\mathcal{C}$ are obtained as intersections of s hyperplanes arising by turning s of the $(s + 1)$ inequality constraints into equalities. We have the following possibilities:

- If $a_1 = \cdots = a_s = 0$, then $f(a_1, a_2, \ldots, a_s) = 0$.
- If $(a_1 + \cdots + a_s)/\sqrt{s} + \sqrt{s}\, a_1/4 = 1$ and $a_1 = \cdots = a_k > a_{k+1} = \cdots = a_s = 0$ for some $1 \leq k \leq s - 1$, then one has $a_1 = \cdots = a_k = 4\sqrt{s}/(4k + s)$, and consequently $f(a_1, \ldots, a_s) = 4\sqrt{ks}/(4k + s) \leq 1$.
- If $(a_1 + \cdots + a_s)/\sqrt{s} + \sqrt{s}\, a_1/4 = 1$ and $a_1 = \cdots = a_s > 0$, then one has $a_1 = \cdots = a_s = 4/(5\sqrt{s})$, and consequently $f(a_1, \ldots, a_s) = 4/5 + 1/5 = 1$.

We have obtained

$$\max_{(a_1, \ldots, a_s) \in \mathcal{C}} f(a_1, a_2, \ldots, a_s) = 1,$$

which is the desired result.   □

We are now ready to establish the robust null space property stated in Theorem 6.13. To simplify the initial reading of the proof, the reader may consider only the stable null space property by specifying $\mathbf{v} \in \ker \mathbf{A}$ in the following argument.

Proof (of Theorem 6.13). We need to find constants $0 < \rho < 1$ and $\tau > 0$ such that, for any $\mathbf{v} \in \mathbb{C}^N$ and any $S \subset [N]$ with $\mathrm{card}(S) = s$,

$$\|\mathbf{v}_S\|_2 \leq \frac{\rho}{\sqrt{s}} \|\mathbf{v}_{\overline{S}}\|_1 + \tau \|\mathbf{A}\mathbf{v}\|_2. \tag{6.21}$$

Given $\mathbf{v} \in \mathbb{C}^N$, it is enough to consider an index set $S =: S_0$ of s largest absolute entries of $\mathbf{v}$. As before, we partition the complement of S_0 as $\overline{S_0} = S_1 \cup S_2 \cup \ldots$, where

$$S_1 : \text{index set of } s \text{ largest absolute entries of } \mathbf{v} \text{ in } \overline{S_0},$$

$$S_2 : \text{index set of } s \text{ largest absolute entries of } \mathbf{v} \text{ in } \overline{S_0 \cup S_1},$$

etc. Since the vector $\mathbf{v}_{S_0}$ is s-sparse, we can write

$$\|\mathbf{A}\mathbf{v}_{S_0}\|_2^2 = (1 + t)\|\mathbf{v}_{S_0}\|_2^2 \qquad \text{with } |t| \leq \delta_s.$$

We are going to establish that, for any $k \geq 1$,

$$|\langle \mathbf{A}\mathbf{v}_{S_0}, \mathbf{A}\mathbf{v}_{S_k} \rangle| \leq \sqrt{\delta_{2s}^2 - t^2} \, \|\mathbf{v}_{S_0}\|_2 \|\mathbf{v}_{S_k}\|_2. \tag{6.22}$$

To do so, we normalize the vectors $\mathbf{v}_{S_0}$ and $\mathbf{v}_{S_k}$ by setting $\mathbf{u} := \mathbf{v}_{S_0}/\|\mathbf{v}_{S_0}\|_2$ and $\mathbf{w} := e^{i\theta}\mathbf{v}_{S_k}/\|\mathbf{v}_{S_k}\|_2$, θ being chosen so that $|\langle \mathbf{A}\mathbf{u}, \mathbf{A}\mathbf{w}\rangle| = \mathrm{Re}\langle \mathbf{A}\mathbf{u}, \mathbf{A}\mathbf{w}\rangle$. Then, for real numbers $\alpha, \beta \geq 0$ to be chosen later, we write

$$2|\langle \mathbf{A}\mathbf{u}, \mathbf{A}\mathbf{w}\rangle| = \frac{1}{\alpha+\beta}\left[\|\mathbf{A}(\alpha\mathbf{u}+\mathbf{w})\|_2^2 - \|\mathbf{A}(\beta\mathbf{u}-\mathbf{w})\|_2^2 - (\alpha^2-\beta^2)\|\mathbf{A}\mathbf{u}\|_2^2\right]$$

$$\leq \frac{1}{\alpha+\beta}\left[(1+\delta_{2s})\|\alpha\mathbf{u}+\mathbf{w}\|_2^2 - (1-\delta_{2s})\|\beta\mathbf{u}-\mathbf{w}\|_2^2 - (\alpha^2-\beta^2)(1+t)\|\mathbf{u}\|_2^2\right]$$

$$= \frac{1}{\alpha+\beta}\left[(1+\delta_{2s})(\alpha^2+1) - (1-\delta_{2s})(\beta^2+1) - (\alpha^2-\beta^2)(1+t)\right]$$

$$= \frac{1}{\alpha+\beta}\left[\alpha^2(\delta_{2s}-t) + \beta^2(\delta_{2s}+t) + 2\delta_{2s}\right].$$

Making the choice $\alpha = (\delta_{2s}+t)/\sqrt{\delta_{2s}^2-t^2}$ and $\beta = (\delta_{2s}-t)/\sqrt{\delta_{2s}^2-t^2}$, we derive

$$2|\langle \mathbf{A}\mathbf{u}, \mathbf{A}\mathbf{w}\rangle| \leq \frac{\sqrt{\delta_{2s}^2-t^2}}{2\delta_{2s}}\left[\delta_{2s}+t+\delta_{2s}-t+2\delta_{2s}\right] = 2\sqrt{\delta_{2s}^2-t^2},$$

which is a reformulation of the desired inequality (6.22). Next, we observe that

$$\|\mathbf{A}\mathbf{v}_{S_0}\|_2^2 = \left\langle \mathbf{A}\mathbf{v}_{S_0}, \mathbf{A}\left(\mathbf{v} - \sum_{k\geq1}\mathbf{v}_{S_k}\right)\right\rangle = \langle \mathbf{A}\mathbf{v}_{S_0}, \mathbf{A}\mathbf{v}\rangle - \sum_{k\geq1}\langle \mathbf{A}\mathbf{v}_{S_0}, \mathbf{A}\mathbf{v}_{S_k}\rangle$$

$$\leq \|\mathbf{A}\mathbf{v}_{S_0}\|_2\|\mathbf{A}\mathbf{v}\|_2 + \sum_{k\geq1}\sqrt{\delta_{2s}^2-t^2}\|\mathbf{v}_{S_0}\|_2\|\mathbf{v}_{S_k}\|_2$$

$$= \|\mathbf{v}_{S_0}\|\left(\sqrt{1+t}\,\|\mathbf{A}\mathbf{v}\|_2 + \sqrt{\delta_{2s}^2-t^2}\sum_{k\geq1}\|\mathbf{v}_{S_k}\|_2\right). \tag{6.23}$$

For each $k \geq 1$, we denote by v_k^- and v_k^+ the smallest and largest absolute entries of $\mathbf{v}$ on S_k, and we use Lemma 6.14 to obtain

$$\sum_{k\geq1}\|\mathbf{v}_{S_k}\|_2 \leq \sum_{k\geq1}\left(\frac{1}{\sqrt{s}}\|\mathbf{v}_{S_k}\|_1 + \frac{\sqrt{s}}{4}(v_k^+ - v_k^-)\right)$$

$$\leq \frac{1}{\sqrt{s}}\|\mathbf{v}_{\overline{S_0}}\|_1 + \frac{\sqrt{s}}{4}v_1^+ \leq \frac{1}{\sqrt{s}}\|\mathbf{v}_{\overline{S_0}}\|_1 + \frac{1}{4}\|\mathbf{v}_{S_0}\|_2.$$

Substituting the latter in the right-hand side of (6.23), while replacing $\|\mathbf{A}\mathbf{v}_{S_0}\|_2^2$ by $(1+t)\|\mathbf{v}_{S_0}\|_2^2$ in the left-hand side and dividing through by $\|\mathbf{v}_{S_0}\|_2$, yields

$$(1+t)\|\mathbf{v}_{S_0}\|_2 \leq \sqrt{1+t}\,\|\mathbf{Av}\|_2 + \frac{\sqrt{\delta_{2s}^2 - t^2}}{\sqrt{s}}\|\mathbf{v}_{\overline{S_0}}\|_1 + \frac{\sqrt{\delta_{2s}^2 - t^2}}{4}\|\mathbf{v}_{S_0}\|_2$$

$$\leq (1+t)\left(\frac{1}{\sqrt{1+t}}\|\mathbf{Av}\|_2 + \frac{\delta_{2s}}{\sqrt{s}\sqrt{1-\delta_{2s}^2}}\|\mathbf{v}_{\overline{S_0}}\|_1 + \frac{\delta_{2s}}{4\sqrt{1-\delta_{2s}^2}}\|\mathbf{v}_{S_0}\|_2\right),$$

where we used the easily verified inequality $\sqrt{\delta_{2s}^2 - t^2}/(1+t) \leq \delta_{2s}/\sqrt{1-\delta_{2s}}$. Dividing through by $1+t$, using $1/\sqrt{1+t} \leq 1/\sqrt{1-\delta_{2s}}$, and rearranging gives

$$\|\mathbf{v}_{S_0}\|_2 \leq \frac{\delta_{2s}}{\sqrt{1-\delta_{2s}^2} - \delta_{2s}/4}\frac{\|\mathbf{v}_{\overline{S_0}}\|_1}{\sqrt{s}} + \frac{\sqrt{1+\delta_{2s}}}{\sqrt{1-\delta_{2s}^2} - \delta_{2s}/4}\|\mathbf{Av}\|_2.$$

This takes the form of the desired inequality (6.21) as soon as

$$\rho := \frac{\delta_{2s}}{\sqrt{1-\delta_{2s}^2} - \delta_{2s}/4} < 1,$$

i.e., $5\delta_{2s}/4 < \sqrt{1-\delta_{2s}^2}$, or $41\delta_{2s}^2 < 16$, which reduces to Condition (6.20). □

We close this section by highlighting some limitations of the restricted isometry property in the context of basis pursuit. We recall from Remark 4.6 that s-sparse recovery via basis pursuit is preserved if some measurements are rescaled, reshuffled, or added. However, these operations may deteriorate the restricted isometry constants. Reshuffling measurements corresponds to replacing the measurement matrix $\mathbf{A} \in \mathbb{C}^{m \times N}$ by $\mathbf{PA}$, where $\mathbf{P} \in \mathbb{C}^{m \times m}$ is a permutation matrix. This operation leaves the restricted isometry constants unchanged, since in fact $\delta_s(\mathbf{UA}) = \delta_s(\mathbf{A})$ for any unitary matrix $\mathbf{U} \in \mathbb{C}^{m \times m}$. Adding a measurement, however, which corresponds to appending a row to the measurement matrix, may increase the restricted isometry constant. Consider, for instance, a matrix $\mathbf{A} \in \mathbb{C}^{m \times N}$ with sth order restricted isometry constant $\delta_s(\mathbf{A}) < 1$, and let $\delta > \delta_s(\mathbf{A})$. We construct a matrix $\tilde{\mathbf{A}}$ by appending the row $[0 \ldots 0 \ \sqrt{1+\delta}]$. With $\mathbf{x} := [0 \ldots 0 \ 1]^\top$, it is easy to see that $\|\mathbf{Ax}\|_2^2 \geq 1 + \delta$. This implies that $\delta_1(\tilde{\mathbf{A}}) \geq \delta$ and consequently that $\delta_s(\tilde{\mathbf{A}}) > \delta_s(\mathbf{A})$. Likewise, rescaling the measurements, which corresponds to replacing the measurement matrix $\mathbf{A} \in \mathbb{C}^{m \times N}$ by $\mathbf{DA}$, where $\mathbf{D} \in \mathbb{C}^{m \times m}$ is a diagonal matrix, may also increase the restricted isometry constant. This is even the case for scalar rescaling, i.e., replacing $\mathbf{A}$ by $d\mathbf{A}$ for $d \in \mathbb{C}$. For instance, if $\mathbf{A} \in \mathbb{C}^{m \times N}$ has an sth order restricted isometry constant $\delta_s(\mathbf{A}) < 3/5$, then the sth order restricted isometry constant of $2\mathbf{A}$ satisfies $\delta_s(2\mathbf{A}) \geq 3 - 4\delta_s(\mathbf{A}) > \delta_s(\mathbf{A})$. In order to circumvent the issue of scalar rescaling, one can work instead with the sth *restricted isometry ratio* $\gamma_s = \gamma_s(\mathbf{A})$, defined as

$$\gamma_s := \frac{\beta_s}{\alpha_s} \geq 1,$$

where α_s and β_s are the largest and smallest constants $\alpha, \beta \geq 0$ such that

$$\alpha \|\mathbf{x}\|_2^2 \leq \|\mathbf{A}\mathbf{x}\|_2^2 \leq \beta \|\mathbf{x}\|_2^2$$

for all s-sparse vectors $\mathbf{x} \in \mathbb{C}^N$. Note that this does not settle the issue of general rescaling. Consider indeed the $(2s) \times (2s+1)$ matrix $\mathbf{A}$ and the $(2s) \times (2s)$ diagonal matrix $\mathbf{D}_\varepsilon$ defined by

$$\mathbf{A} = \begin{bmatrix} 1 & 0 & \cdots & 0 & -1 \\ 0 & 1 & \ddots & 0 & -1 \\ \vdots & \ddots & \ddots & 0 & \vdots \\ 0 & \cdots & 0 & 1 & -1 \end{bmatrix}, \quad \mathbf{D}_\varepsilon = \mathrm{diag}[\varepsilon, 1/\varepsilon, 1, \ldots, 1].$$

Since $\ker \mathbf{D}_\varepsilon \mathbf{A} = \ker \mathbf{A}$ is spanned by $[1, 1, \ldots, 1]^\top$, the matrices $\mathbf{D}_\varepsilon \mathbf{A}$ and $\mathbf{A}$ both satisfy the sth order null space property, hence allow s-sparse recovery via basis pursuit. However, the sth order restricted isometry ratio of $\mathbf{D}_\varepsilon \mathbf{A}$ can be made arbitrarily large, since $\gamma_s(\mathbf{D}_\varepsilon \mathbf{A}) \geq 1/\varepsilon^4$. Incidentally, this shows that there are matrices allowing s-sparse recovery via basis pursuit but whose sth order restricted isometry constant is arbitrarily close to 1—even after scalar renormalization; see Exercise 6.2.

6.3   Analysis of Thresholding Algorithms

In this section, we establish the success of sparse recovery via iterative hard thresholding and via hard thresholding pursuit for measurement matrices with small restricted isometry constants. Again, we start with a simple and quite natural proof of the success of s-sparse recovery via iterative hard thresholding under the condition $\delta_{3s} < 0.5$. This is done in the ideal situation of exactly sparse vectors acquired with perfect accuracy. We then cover the more realistic situation of approximately sparse vectors measured with some errors. The improved result only requires the weaker condition $\delta_{3s} < 0.5773$. It applies to both iterative hard thresholding and hard thresholding pursuit, but its proof is more involved. Before all this, we recall from Sect. 3 that the iterative hard thresholding algorithm starts with an initial s-sparse vector $\mathbf{x}^0 \in \mathbb{C}^N$, typically $\mathbf{x}^0 = \mathbf{0}$, and produces a sequence $(\mathbf{x}^n)$ defined inductively by

$$\mathbf{x}^{n+1} = H_s(\mathbf{x}^n + \mathbf{A}^*(\mathbf{y} - \mathbf{A}\mathbf{x}^n)). \tag{IHT}$$

The hard thresholding operator H_s keeps the s largest absolute entries of a vector, so that $H_s(\mathbf{z})$ is a (not necessarily unique) best s-term approximation to $\mathbf{z} \in \mathbb{C}^N$. For small restricted isometry constants, the success of iterative hard thresholding is intuitively justified by the fact that $\mathbf{A}^*\mathbf{A}$ behaves like the identity when its domain and range are restricted to small support sets. Thus, if $\mathbf{y} = \mathbf{A}\mathbf{x}$ for some sparse $\mathbf{x} \in \mathbb{C}^N$, the contribution to $\mathbf{x}^{n+1}$ of $\mathbf{A}^*(\mathbf{y} - \mathbf{A}\mathbf{x}^n) = \mathbf{A}^*\mathbf{A}(\mathbf{x} - \mathbf{x}^n)$ is roughly $\mathbf{x} - \mathbf{x}^n$, which sums with $\mathbf{x}^n$ to the desired $\mathbf{x}$. Here is a formal statement of the success of iterative hard thresholding.

Theorem 6.15. *Suppose that the 3sth restricted isometry constant of the matrix* $\mathbf{A} \in \mathbb{C}^{m \times N}$ *satisfies*

$$\delta_{3s} < \frac{1}{2}. \tag{6.24}$$

Then, for every s-sparse vector $\mathbf{x} \in \mathbb{C}^N$, *the sequence* $(\mathbf{x}^n)$ *defined by* (IHT) *with* $\mathbf{y} = \mathbf{A}\mathbf{x}$ *converges to* $\mathbf{x}$.

The following observation is recurring in our arguments, so we isolate it from the proof.

Lemma 6.16. *Given vectors* $\mathbf{u}, \mathbf{v} \in \mathbb{C}^N$ *and an index set* $S \subset [N]$,

$$|\langle \mathbf{u}, (\mathbf{Id} - \mathbf{A}^*\mathbf{A})\mathbf{v}\rangle| \le \delta_t \|\mathbf{u}\|_2 \|\mathbf{v}\|_2 \quad \textit{if } \mathrm{card}(\mathrm{supp}(\mathbf{u}) \cup \mathrm{supp}(\mathbf{v})) \le t,$$

$$\|((\mathbf{Id} - \mathbf{A}^*\mathbf{A})\mathbf{v})_S\|_2 \le \delta_t \|\mathbf{v}\|_2 \quad \textit{if } \mathrm{card}(S \cup \mathrm{supp}(\mathbf{v})) \le t.$$

Proof. For the first inequality, let $T := \mathrm{supp}(\mathbf{u}) \cup \mathrm{supp}(\mathbf{v})$, and let $\mathbf{u}_T$ and $\mathbf{v}_T$ denote the subvectors of $\mathbf{u}$ and $\mathbf{v}$ obtained by only keeping the entries indexed by T. We write

$$|\langle \mathbf{u}, (\mathbf{Id} - \mathbf{A}^*\mathbf{A})\mathbf{v}\rangle| = |\langle \mathbf{u}, \mathbf{v}\rangle - \langle \mathbf{A}\mathbf{u}, \mathbf{A}\mathbf{v}\rangle| = |\langle \mathbf{u}_T, \mathbf{v}_T\rangle - \langle \mathbf{A}_T\mathbf{u}_T, \mathbf{A}_T\mathbf{v}_T\rangle|$$

$$= |\langle \mathbf{u}_T, (\mathbf{Id} - \mathbf{A}_T^*\mathbf{A}_T)\mathbf{v}_T\rangle| \le \|\mathbf{u}_T\|_2 \|(\mathbf{Id} - \mathbf{A}_T^*\mathbf{A}_T)\mathbf{v}_T\|_2$$

$$\le \|\mathbf{u}_T\|_2 \|\mathbf{Id} - \mathbf{A}_T^*\mathbf{A}_T\|_{2\to 2} \|\mathbf{v}_T\|_2 \le \delta_t \|\mathbf{u}\|_2 \|\mathbf{v}\|_2.$$

The second inequality follows from the first one by observing that

$$\|((\mathbf{Id} - \mathbf{A}^*\mathbf{A})\mathbf{v})_S\|_2^2 = \langle ((\mathbf{Id} - \mathbf{A}^*\mathbf{A})\mathbf{v})_S, (\mathbf{Id} - \mathbf{A}^*\mathbf{A})\mathbf{v}\rangle \le \delta_t \|((\mathbf{Id} - \mathbf{A}^*\mathbf{A})\mathbf{v})_S\|_2 \|\mathbf{v}\|_2.$$

We divide through by $\|((\mathbf{Id} - \mathbf{A}^*\mathbf{A})\mathbf{v})_S\|_2$ to complete the proof.   □

Proof (of Theorem 6.15). It is enough to find a constant $0 \le \rho < 1$ such that

$$\|\mathbf{x}^{n+1} - \mathbf{x}\|_2 \le \rho \|\mathbf{x}^n - \mathbf{x}\|_2, \qquad n \ge 0, \tag{6.25}$$

since this implies by induction that

$$\|\mathbf{x}^n - \mathbf{x}\|_2 \le \rho^n \|\mathbf{x}^0 - \mathbf{x}\|_2 \xrightarrow[n\to\infty]{} 0.$$

By definition, the s-sparse vector $\mathbf{x}^{n+1}$ is a better (or at least equally good) approximation to

$$\mathbf{u}^n := \mathbf{x}^n + \mathbf{A}^*(\mathbf{y} - \mathbf{A}\mathbf{x}^n) = \mathbf{x}^n + \mathbf{A}^*\mathbf{A}(\mathbf{x} - \mathbf{x}^n)$$

than the s-sparse vector $\mathbf{x}$. This implies

$$\|\mathbf{u}^n - \mathbf{x}^{n+1}\|_2^2 \le \|\mathbf{u}^n - \mathbf{x}\|_2^2.$$

Expanding $\|\mathbf{u}^n - \mathbf{x}^{n+1}\|_2^2 = \|(\mathbf{u}^n - \mathbf{x}) - (\mathbf{x}^{n+1} - \mathbf{x})\|_2^2$ and rearranging yields

$$\|\mathbf{x}^{n+1} - \mathbf{x}\|_2^2 \le 2\,\mathrm{Re}\langle \mathbf{u}^n - \mathbf{x}, \mathbf{x}^{n+1} - \mathbf{x}\rangle. \tag{6.26}$$

We now use Lemma 6.16 to obtain

$$\mathrm{Re}\langle \mathbf{u}^n - \mathbf{x}, \mathbf{x}^{n+1} - \mathbf{x}\rangle = \mathrm{Re}\langle (\mathbf{Id} - \mathbf{A}^*\mathbf{A})(\mathbf{x}^n - \mathbf{x}), \mathbf{x}^{n+1} - \mathbf{x}\rangle$$

$$\le \delta_{3s}\|\mathbf{x}^n - \mathbf{x}\|_2\|\mathbf{x}^{n+1} - \mathbf{x}\|_2. \tag{6.27}$$

If $\|\mathbf{x}^{n+1} - \mathbf{x}\|_2 > 0$, we derive from (6.26) and (6.27) that

$$\|\mathbf{x}^{n+1} - \mathbf{x}\|_2 \le 2\delta_{3s}\|\mathbf{x}^n - \mathbf{x}\|_2,$$

which is obviously true if $\|\mathbf{x}^{n+1} - \mathbf{x}\|_2 = 0$. Thus, the desired inequality (6.25) holds with $\rho = 2\delta_{3s} < 1$.   □

Remark 6.17. Sufficient conditions for the success of s-sparse recovery via basis pursuit were previously given in terms of δ_{2s}. Such sufficient conditions can also be given for iterative hard thresholding. For instance, since $\delta_{3s} \le 2\delta_{2s} + \delta_s \le 3\delta_{2s}$ by Proposition 6.6, it is enough to assume $\delta_{2s} < 1/6$ to guarantee $\delta_{3s} < 1/2$, hence the success of s-sparse recovery via iterative hard thresholding. This condition may be weakened to $\delta_{2s} < 1/4$ by refining the argument in the proof of Theorem 6.15—see Exercise 6.20. It can be further weakened to $\delta_{2s} < 1/3$ with a slight modification of the algorithm—see Exercise 6.21.

It is again instructive to refine the proof above for approximately sparse vectors measured with some errors, and the reader is invited to do so in Exercise 6.19. Theorem 6.18 below covers this case while improving on Theorem 6.15 by relaxing the sufficient condition (6.24). As a consequence, we will obtain in Theorem 6.21 error estimates similar to the ones for basis pursuit. We underline that the arguments are valid for both iterative hard thresholding and hard thresholding pursuit. As a reminder, this latter algorithm starts with an initial s-sparse vector $\mathbf{x}^0 \in \mathbb{C}^N$, typically $\mathbf{x}^0 = 0$, and produces a sequence $(\mathbf{x}^n)$ defined inductively by

$$S^{n+1} = L_s(\mathbf{x}^n + \mathbf{A}^*(\mathbf{y} - \mathbf{A}\mathbf{x}^n)), \tag{HTP$_1$}$$

$$\mathbf{x}^{n+1} = \arg\min \left\{ \|\mathbf{y} - \mathbf{A}\mathbf{z}\|_2, \mathrm{supp}(\mathbf{z}) \subset S^{n+1} \right\}. \tag{HTP$_2$}$$

We recall that $L_s(\mathbf{z})$ denotes an index set of s largest absolute entries of a vector $\mathbf{z} \in \mathbb{C}^N$; see (3.8).

Theorem 6.18. *Suppose that the $3s$th restricted isometry constant of the matrix $\mathbf{A} \in \mathbb{C}^{m \times N}$ satisfies*

$$\delta_{3s} < \frac{1}{\sqrt{3}} \approx 0.5773. \tag{6.28}$$

Then, for $\mathbf{x} \in \mathbb{C}^N$, $\mathbf{e} \in \mathbb{C}^m$, and $S \subset [N]$ with $\mathrm{card}(S) = s$, the sequence $(\mathbf{x}^n)$ defined by (IHT) or by (HTP$_1$), (HTP$_2$) with $\mathbf{y} = \mathbf{Ax} + \mathbf{e}$ satisfies, for any $n \geq 0$,

$$\|\mathbf{x}^n - \mathbf{x}_S\|_2 \leq \rho^n \|\mathbf{x}^0 - \mathbf{x}_S\|_2 + \tau \|\mathbf{A}\mathbf{x}_{\overline{S}} + \mathbf{e}\|_2, \tag{6.29}$$

where $\rho = \sqrt{3}\,\delta_{3s} < 1$, $\tau \leq 2.18/(1 - \rho)$ for (IHT), $\rho = \sqrt{2\delta_{3s}^2/(1 - \delta_{2s}^2)} < 1$, $\tau \leq 5.15/(1 - \rho)$ for (HTP$_1$), (HTP$_2$).

Remark 6.19. The intuitive superiority of the hard thresholding pursuit algorithm over the iterative hard thresholding algorithm is not reflected in a weaker sufficient condition in terms of restricted isometry constants but rather in a faster rate of convergence justified by $\sqrt{2\delta_{3s}^2/(1 - \delta_{2s}^2)} < \sqrt{3}\,\delta_{3s}$ when $\delta_{3s} < 1/\sqrt{3}$.

We isolate the following observation from the proof of the theorem.

Lemma 6.20. *Given $\mathbf{e} \in \mathbb{C}^m$ and $S \in [N]$ with $\mathrm{card}(S) \leq s$,*

$$\|(\mathbf{A}^*\mathbf{e})_S\|_2 \leq \sqrt{1 + \delta_s}\,\|\mathbf{e}\|_2.$$

Proof. We only need to write

$$\|(\mathbf{A}^*\mathbf{e})_S\|_2^2 = \langle \mathbf{A}^*\mathbf{e}, (\mathbf{A}^*\mathbf{e})_S \rangle = \langle \mathbf{e}, \mathbf{A}((\mathbf{A}^*\mathbf{e})_S) \rangle \leq \|\mathbf{e}\|_2 \|\mathbf{A}((\mathbf{A}^*\mathbf{e})_S)\|_2$$

$$\leq \|\mathbf{e}\|_2 \sqrt{1 + \delta_s}\,\|(\mathbf{A}^*\mathbf{e})_S\|_2,$$

and to divide through by $\|(\mathbf{A}^*\mathbf{e})_S\|_2$. $\square$

Proof (of Theorem 6.18). Given $\mathbf{x} \in \mathbb{C}^N$, $\mathbf{e} \in \mathbb{C}^m$, $S \subset [N]$ with $\mathrm{card}(S) = s$, our aim is to prove that, for any $n \geq 0$,

$$\|\mathbf{x}^{n+1} - \mathbf{x}_S\|_2 \leq \rho\|\mathbf{x}^n - \mathbf{x}_S\|_2 + (1 - \rho)\tau\|\mathbf{A}\mathbf{x}_{\overline{S}} + \mathbf{e}\|_2. \tag{6.30}$$

The estimate (6.29) then follows by induction. For both iterative hard thresholding and hard thresholding pursuit, the index set $S^{n+1} := \mathrm{supp}(\mathbf{x}^{n+1})$ consists of s largest absolute entries of $\mathbf{x}^n + \mathbf{A}^*(\mathbf{y} - \mathbf{Ax}^n)$, so we have

$$\|(\mathbf{x}^n + \mathbf{A}^*(\mathbf{y} - \mathbf{Ax}^n))_S\|_2^2 \leq \|(\mathbf{x}^n + \mathbf{A}^*(\mathbf{y} - \mathbf{Ax}^n))_{S^{n+1}}\|_2^2.$$

Eliminating the contribution on $S \cap S^{n+1}$, we derive

$$\|(\mathbf{x}^n + \mathbf{A}^*(\mathbf{y} - \mathbf{A}\mathbf{x}^n))_{S \setminus S^{n+1}}\|_2 \leq \|(\mathbf{x}^n + \mathbf{A}^*(\mathbf{y} - \mathbf{A}\mathbf{x}^n))_{S^{n+1} \setminus S}\|_2.$$

The right-hand side may be written as

$$\|(\mathbf{x}^n + \mathbf{A}^*(\mathbf{y} - \mathbf{A}\mathbf{x}^n))_{S^{n+1} \setminus S}\|_2 = \|(\mathbf{x}^n - \mathbf{x}_S + \mathbf{A}^*(\mathbf{y} - \mathbf{A}\mathbf{x}^n))_{S^{n+1} \setminus S}\|_2.$$

The left-hand side satisfies

$$\|(\mathbf{x}^n + \mathbf{A}^*(\mathbf{y} - \mathbf{A}\mathbf{x}^n))_{S \setminus S^{n+1}}\|_2$$
$$= \|(\mathbf{x}_S - \mathbf{x}^{n+1} + \mathbf{x}^n - \mathbf{x}_S + \mathbf{A}^*(\mathbf{y} - \mathbf{A}\mathbf{x}^n))_{S \setminus S^{n+1}}\|_2$$
$$\geq \|(\mathbf{x}_S - \mathbf{x}^{n+1})_{S \setminus S^{n+1}}\|_2 - \|(\mathbf{x}^n - \mathbf{x}_S + \mathbf{A}^*(\mathbf{y} - \mathbf{A}\mathbf{x}^n))_{S \setminus S^{n+1}}\|_2.$$

With $S \Delta S^{n+1} = (S \setminus S^{n+1}) \cup (S^{n+1} \setminus S)$ denoting the symmetric difference of the sets S and S^{n+1}, we conclude that

$$\|(\mathbf{x}_S - \mathbf{x}^{n+1})_{S \setminus S^{n+1}}\|_2 \leq \|(\mathbf{x}^n - \mathbf{x}_S + \mathbf{A}^*(\mathbf{y} - \mathbf{A}\mathbf{x}^n))_{S \setminus S^{n+1}}\|_2$$
$$+ \|(\mathbf{x}^n - \mathbf{x}_S + \mathbf{A}^*(\mathbf{y} - \mathbf{A}\mathbf{x}^n))_{S^{n+1} \setminus S}\|_2$$
$$\leq \sqrt{2}\,\|(\mathbf{x}^n - \mathbf{x}_S + \mathbf{A}^*(\mathbf{y} - \mathbf{A}\mathbf{x}^n))_{S \Delta S^{n+1}}\|_2. \quad (6.31)$$

Let us first concentrate on iterative hard thresholding. In this case,

$$\mathbf{x}^{n+1} = \left(\mathbf{x}^n + \mathbf{A}^*(\mathbf{y} - \mathbf{A}\mathbf{x}^n)\right)_{S^{n+1}}.$$

It then follows that

$$\|\mathbf{x}^{n+1} - \mathbf{x}_S\|_2^2 = \|(\mathbf{x}^{n+1} - \mathbf{x}_S)_{S^{n+1}}\|_2^2 + \|(\mathbf{x}^{n+1} - \mathbf{x}_S)_{\overline{S^{n+1}}}\|_2^2$$
$$= \|(\mathbf{x}^n - \mathbf{x}_S + \mathbf{A}^*(\mathbf{y} - \mathbf{A}\mathbf{x}^n))_{S^{n+1}}\|_2^2 + \|(\mathbf{x}^{n+1} - \mathbf{x}_S)_{S \setminus S^{n+1}}\|_2^2.$$

Together with (6.31), we obtain

$$\|\mathbf{x}^{n+1} - \mathbf{x}_S\|_2^2 \leq \|(\mathbf{x}^n - \mathbf{x}_S + \mathbf{A}^*(\mathbf{y} - \mathbf{A}\mathbf{x}^n))_{S^{n+1}}\|_2^2$$
$$+ 2\,\|(\mathbf{x}^n - \mathbf{x}_S + \mathbf{A}^*(\mathbf{y} - \mathbf{A}\mathbf{x}^n))_{S \Delta S^{n+1}}\|_2^2$$
$$\leq 3\,\|(\mathbf{x}^n - \mathbf{x}_S + \mathbf{A}^*(\mathbf{y} - \mathbf{A}\mathbf{x}^n))_{S \cup S^{n+1}}\|_2^2.$$

We now write $\mathbf{y} = \mathbf{A}\mathbf{x} + \mathbf{e} = \mathbf{A}\mathbf{x}_S + \mathbf{e}'$ with $\mathbf{e}' := \mathbf{A}\mathbf{x}_{\overline{S}} + \mathbf{e}$, and we call upon Lemma 6.16 (noticing that $\text{card}(S \cup S^{n+1} \cup \text{supp}(\mathbf{x}^n - \mathbf{x}_S)) \leq 3s$) and Lemma 6.20 to deduce

$$\|\mathbf{x}^{n+1} - \mathbf{x}_S\|_2 \le \sqrt{3}\,\|(\mathbf{x}^n - \mathbf{x}_S + \mathbf{A}^*\mathbf{A}(\mathbf{x}_S - \mathbf{x}^n) + \mathbf{A}^*\mathbf{e}')_{S \cup S^{n+1}}\|_2$$

$$\le \sqrt{3}\,\left[\|((\mathbf{Id} - \mathbf{A}^*\mathbf{A})(\mathbf{x}^n - \mathbf{x}_S))_{S \cup S^{n+1}}\|_2 + \|(\mathbf{A}^*\mathbf{e}')_{S \cup S^{n+1}}\|_2\right]$$

$$\le \sqrt{3}\,\left[\delta_{3s}\|\mathbf{x}^n - \mathbf{x}_S\|_2 + \sqrt{1 + \delta_{2s}}\,\|\mathbf{e}'\|_2\right].$$

This is the desired inequality (6.30) for iterative hard thresholding. We notice that $\rho = \sqrt{3}\,\delta_{3s}$ is indeed smaller than one as soon as $\delta_{3s} < 1/\sqrt{3}$ and that $(1 - \rho)\tau = \sqrt{3}\sqrt{1 + \delta_{2s}} \le \sqrt{3 + \sqrt{3}} \le 2.18$.

Let us now concentrate on hard thresholding pursuit. In this case,

$$\mathbf{x}^{n+1} = \operatorname{argmin}\left\{\|\mathbf{y} - \mathbf{A}\mathbf{z}\|_2, \operatorname{supp}(\mathbf{z}) \subset S^{n+1}\right\}.$$

As the best ℓ_2-approximation to $\mathbf{y}$ from the space $\{\mathbf{A}\mathbf{z}, \operatorname{supp}(\mathbf{z}) \subset S^{n+1}\}$, the vector $\mathbf{A}\mathbf{x}^{n+1}$ is characterized by

$$\langle \mathbf{y} - \mathbf{A}\mathbf{x}^{n+1}, \mathbf{A}\mathbf{z}\rangle = 0 \qquad \text{whenever}\operatorname{supp}(\mathbf{z}) \subset S^{n+1},$$

that is to say, by $\langle \mathbf{A}^*(\mathbf{y} - \mathbf{A}\mathbf{x}^{n+1}), \mathbf{z}\rangle = 0$ whenever $\operatorname{supp}(\mathbf{z}) \subset S^{n+1}$ or

$$(\mathbf{A}^*(\mathbf{y} - \mathbf{A}\mathbf{x}^{n+1}))_{S^{n+1}} = \mathbf{0}.$$

Taking this and (6.31) into consideration, we write

$$\|\mathbf{x}^{n+1} - \mathbf{x}_S\|_2^2 = \|(\mathbf{x}^{n+1} - \mathbf{x}_S)_{S^{n+1}}\|_2^2 + \|(\mathbf{x}^{n+1} - \mathbf{x}_S)_{S \setminus S^{n+1}}\|_2^2$$

$$\le \|(\mathbf{x}^{n+1} - \mathbf{x}_S + \mathbf{A}^*(\mathbf{y} - \mathbf{A}\mathbf{x}^{n+1}))_{S^{n+1}}\|_2^2$$

$$+ 2\,\|(\mathbf{x}^n - \mathbf{x}_S + \mathbf{A}^*(\mathbf{y} - \mathbf{A}\mathbf{x}^n))_{S \triangle S^{n+1}}\|_2^2$$

$$\le \left[\|((\mathbf{Id} - \mathbf{A}^*\mathbf{A})(\mathbf{x}^{n+1} - \mathbf{x}_S))_{S^{n+1}}\|_2 + \|(\mathbf{A}^*\mathbf{e}')_{S^{n+1}}\|_2\right]^2$$

$$+ 2\left[\|((\mathbf{Id} - \mathbf{A}^*\mathbf{A})(\mathbf{x}^n - \mathbf{x}_S))_{S \triangle S^{n+1}}\|_2 + \|(\mathbf{A}^*\mathbf{e}')_{S \triangle S^{n+1}}\|_2\right]^2.$$

Applying Lemma 6.16 and Lemma 6.20 yields

$$\|\mathbf{x}^{n+1} - \mathbf{x}_S\|_2^2 \le \left[\delta_{2s}\|\mathbf{x}^{n+1} - \mathbf{x}_S\|_2 + \sqrt{1 + \delta_s}\,\|\mathbf{e}'\|_2\right]^2$$

$$+ 2\left[\delta_{3s}\|\mathbf{x}^n - \mathbf{x}_S\|_2 + \sqrt{1 + \delta_{2s}}\,\|\mathbf{e}'\|_2\right]^2.$$

After rearrangement, this reads

$$2\left[\delta_{3s}\|\mathbf{x}^n - \mathbf{x}_S\|_2 + \sqrt{1 + \delta_{2s}}\,\|\mathbf{e}'\|_2\right]^2$$

$$\ge (1 - \delta_{2s}^2)\left(\|\mathbf{x}^{n+1} - \mathbf{x}_S\|_2 + \frac{\sqrt{1 + \delta_s}}{1 + \delta_{2s}}\,\|\mathbf{e}'\|_2\right)\left(\|\mathbf{x}^{n+1} - \mathbf{x}_S\|_2 - \frac{\sqrt{1 + \delta_s}}{1 - \delta_{2s}}\,\|\mathbf{e}'\|_2\right).$$

Since we may assume $\|\mathbf{x}^{n+1} - \mathbf{x}_S\|_2 \geq \sqrt{1+\delta_s}\,\|\mathbf{e}'\|_2/(1-\delta_{2s})$ to make the latter expression in parentheses nonnegative—otherwise (6.30) already holds for the value of $(1-\rho)\tau$ given below—we obtain

$$2\left[\delta_{3s}\|\mathbf{x}^n - \mathbf{x}_S\|_2 + \sqrt{1+\delta_{2s}}\,\|\mathbf{e}'\|_2\right]^2 \geq (1-\delta_{2s}^2)\left(\|\mathbf{x}^{n+1} - \mathbf{x}_S\|_2 - \frac{\sqrt{1+\delta_s}}{1-\delta_{2s}}\|\mathbf{e}'\|_2\right)^2.$$

From here, taking the square root and rearranging gives

$$\|\mathbf{x}^{n+1} - \mathbf{x}_S\|_2 \leq \frac{\sqrt{2}\,\delta_{3s}}{\sqrt{1-\delta_{2s}^2}}\|\mathbf{x}^n - \mathbf{x}_S\|_2 + \left(\frac{\sqrt{2}}{\sqrt{1-\delta_{2s}}} + \frac{\sqrt{1+\delta_s}}{1-\delta_{2s}}\right)\|\mathbf{e}'\|_2.$$

This is the desired inequality (6.30) for hard thresholding pursuit. We notice that $\rho := \sqrt{2}\,\delta_{3s}/\sqrt{1-\delta_{2s}^2} \leq \sqrt{2}\,\delta_{3s}/\sqrt{1-\delta_{3s}^2}$ is smaller than one as soon as $\delta_{3s} < 1/\sqrt{3}$ and that $(1-\rho)\tau = \sqrt{2}/\sqrt{1-\delta_{2s}} + \sqrt{1+\delta_s}/(1-\delta_{2s}) \leq 5.15$. $\square$

Taking the limit as $n \to \infty$ in (6.29) yields $\|\mathbf{x}^\sharp - \mathbf{x}_S\|_2 \leq \tau\|\mathbf{A}\mathbf{x}_{\overline{S}} + \mathbf{e}\|_2$ if $\mathbf{x}^\sharp \in \mathbb{C}^N$ is the limit of the sequence $(\mathbf{x}^n)$ or at least one of its cluster points. Note that the existence of this limit is not at all guaranteed by our argument, but at least the existence of cluster points is guaranteed by the boundedness of $\|\mathbf{x}^n\|$ which follows from (6.29). In any case, we have $\|\mathbf{x} - \mathbf{x}^\sharp\|_2 \leq \|\mathbf{x}_{\overline{S}}\|_2 + \|\mathbf{x}_S - \mathbf{x}^\sharp\|_2$ by the triangle inequality, so choosing S as an index set of s largest absolute entries of $\mathbf{x}$ gives

$$\|\mathbf{x} - \mathbf{x}^\sharp\|_2 \leq \sigma_s(\mathbf{x})_2 + \tau\|\mathbf{A}\mathbf{x}_{\overline{S}} + \mathbf{e}\|_2. \tag{6.32}$$

This estimate does not resemble the basis pursuit estimates of Theorem 6.12. However, such estimates are available for thresholding algorithms, too, provided we replace the parameter s in (IHT) and (HTP$_1$), (HTP$_2$) by $2s$, say. The precise statement is as follows.

Theorem 6.21. *Suppose that the $6s$th order restricted isometry constant of the matrix $\mathbf{A} \in \mathbb{C}^{m \times N}$ satisfies $\delta_{6s} < 1/\sqrt{3}$. Then, for all $\mathbf{x} \in \mathbb{C}^N$ and $\mathbf{e} \in \mathbb{C}^m$, the sequence $(\mathbf{x}^n)$ defined by (IHT) or by (HTP$_1$), (HTP$_2$) with $\mathbf{y} = \mathbf{A}\mathbf{x} + \mathbf{e}$, $\mathbf{x}^0 = \mathbf{0}$, and s replaced by $2s$ satisfies, for any $n \geq 0$,*

$$\|\mathbf{x} - \mathbf{x}^n\|_1 \leq C\,\sigma_s(\mathbf{x})_1 + D\,\sqrt{s}\,\|\mathbf{e}\|_2 + 2\,\rho^n\sqrt{s}\,\|\mathbf{x}\|_2,$$

$$\|\mathbf{x} - \mathbf{x}^n\|_2 \leq \frac{C}{\sqrt{s}}\,\sigma_s(\mathbf{x})_1 + D\,\|\mathbf{e}\|_2 + 2\,\rho^n\,\|\mathbf{x}\|_2,$$

where the constants $C, D > 0$ and $0 < \rho < 1$ depend only on δ_{6s}. In particular, if the sequence $(\mathbf{x}^n)$ clusters around some $\mathbf{x}^\sharp \in \mathbb{C}^N$, then

$$\|\mathbf{x} - \mathbf{x}^\sharp\|_1 \leq C\,\sigma_s(\mathbf{x})_1 + D\,\sqrt{s}\,\|\mathbf{e}\|_2, \tag{6.33}$$

$$\|\mathbf{x} - \mathbf{x}^\sharp\|_2 \leq \frac{C}{\sqrt{s}}\,\sigma_s(\mathbf{x})_1 + D\,\|\mathbf{e}\|_2. \tag{6.34}$$

Remark 6.22. (a) Error estimates of the type (6.33) and (6.34) are not only valid for cluster points $\mathbf{x}^\sharp$ but also for $\mathbf{x}^n$ with n large enough when $C\sigma_s(\mathbf{x})_1 + D\sqrt{s}\|\mathbf{e}\|_2 > 0$. Indeed, in this case, if $n \geq n_0$ where n_0 is large enough, then

$$2\rho^n\sqrt{s}\|\mathbf{x}\|_2 \leq C\sigma_s(\mathbf{x})_1 + D\sqrt{s}\|\mathbf{e}\|_2 .$$

Therefore, the general error estimates above imply that, for all $n \geq n_0$,

$$\|\mathbf{x} - \mathbf{x}^n\|_1 \leq 2C\,\sigma_s(\mathbf{x})_1 + 2D\sqrt{s}\,\|\mathbf{e}\|_2,$$

$$\|\mathbf{x} - \mathbf{x}^n\|_2 \leq \frac{2C}{\sqrt{s}}\,\sigma_s(\mathbf{x})_1 + 2D\,\|\mathbf{e}\|_2.$$

(b) A major drawback when running hard thresholding algorithms is that an estimation of the targeted sparsity s is needed. This estimation is not needed for the inequality-constrained ℓ_1-minimization, but an estimation of the measurement error η is (a priori) needed instead. We will see in Chap. 11 that running the equality-constrained ℓ_1-minimization (P_1) on corrupted measurements may in some cases still have the benefit of stable and robust estimates (6.33) and (6.34).

The auxiliary result below plays a central role when proving statements such as Theorem 6.21.

Lemma 6.23. *Suppose* $\mathbf{A} \in \mathbb{C}^{m \times N}$ *has restricted isometry constant* $\delta_s < 1$. *Given* $\kappa, \tau > 0$, $\xi \geq 0$, *and* $\mathbf{e} \in \mathbb{C}^m$, *assume that two vectors* $\mathbf{x}, \mathbf{x}' \in \mathbb{C}^N$ *satisfy* $\|\mathbf{x}'\|_0 \leq \kappa s$ *and*

$$\|\mathbf{x}_T - \mathbf{x}'\|_2 \leq \tau\|\mathbf{A}\mathbf{x}_{\overline{T}} + \mathbf{e}\|_2 + \xi,$$

where T *denotes an index set of* $2s$ *largest absolute entries of* $\mathbf{x}$. *Then, for any* $1 \leq p \leq 2$,

$$\|\mathbf{x} - \mathbf{x}'\|_p \leq \frac{1 + c_\kappa \tau}{s^{1-1/p}}\,\sigma_s(\mathbf{x})_1 + d_\kappa\,\tau\,s^{1/p-1/2}\|\mathbf{e}\|_2 + d_\kappa\,s^{1/p-1/2}\xi, \qquad (6.35)$$

where the constants $c_\kappa, d_\kappa > 0$ *depend only on* κ.

Proof. We first use the fact that the vector $\mathbf{x}_T - \mathbf{x}'$ is $(2 + \kappa)s$-sparse to write

$$\|\mathbf{x} - \mathbf{x}'\|_p \leq \|\mathbf{x}_{\overline{T}}\|_p + \|\mathbf{x}_T - \mathbf{x}'\|_p \leq \|\mathbf{x}_{\overline{T}}\|_p + ((2+\kappa)s)^{1/p-1/2}\|\mathbf{x}_T - \mathbf{x}'\|_2$$

$$\leq \|\mathbf{x}_{\overline{T}}\|_p + \sqrt{2 + \kappa}\,s^{1/p-1/2}(\tau\|\mathbf{A}\mathbf{x}_{\overline{T}} + \mathbf{e}\|_2 + \xi). \qquad (6.36)$$

Let now $S \subset T$ denote an index set of s largest absolute entries of $\mathbf{x}$. We observe that, according to Proposition 2.3,

$$\|\mathbf{x}_{\overline{T}}\|_p = \sigma_s(\mathbf{x}_{\overline{S}})_p \leq \frac{1}{s^{1-1/p}}\|\mathbf{x}_{\overline{S}}\|_1 = \frac{1}{s^{1-1/p}}\sigma_s(\mathbf{x})_1. \qquad (6.37)$$

Let us partition the complement of T as $\overline{T} = S_2 \cup S_3 \cup \ldots$, where

S_2 : index set of s largest absolute entries of $\mathbf{x}$ in $\overline{T}$,

S_3 : index set of s largest absolute entries of $\mathbf{x}$ in $\overline{T \cup S_2}$,

etc. In this way, we have

$$\|\mathbf{A}\mathbf{x}_{\overline{T}} + \mathbf{e}\|_2 \leq \sum_{k \geq 2} \|\mathbf{A}\mathbf{x}_{S_k}\|_2 + \|\mathbf{e}\|_2 \leq \sum_{k \geq 2} \sqrt{1 + \delta_s} \|\mathbf{x}_{S_k}\|_2 + \|\mathbf{e}\|_2$$

$$\leq \sqrt{2} \sum_{k \geq 2} \|\mathbf{x}_{S_k}\|_2 + \|\mathbf{e}\|_2.$$

Using Lemma 6.10, it has become usual to derive

$$\sum_{k \geq 2} \|\mathbf{x}_{S_k}\|_2 \leq \frac{1}{s^{1/2}} \|\mathbf{x}_{\overline{S}}\|_1 = \frac{1}{s^{1/2}} \sigma_s(\mathbf{x})_1;$$

hence, we obtain

$$\|\mathbf{A}\mathbf{x}_{\overline{T}} + \mathbf{e}\|_2 \leq \frac{\sqrt{2}}{s^{1/2}} \sigma_s(\mathbf{x})_1 + \|\mathbf{e}\|_2. \tag{6.38}$$

Substituting (6.37) and (6.38) into (6.36), we obtain the estimate (6.35) with $c_\kappa = \sqrt{4 + 2\kappa}$ and $d_\kappa = \sqrt{2 + \kappa}$.   □

Proof (of Theorem 6.21). Given $\mathbf{x} \in \mathbb{C}^N$ and $\mathbf{e} \in \mathbb{C}^m$, under the present hypotheses, Theorem 6.18 implies that there exist $0 < \rho < 1$ and $\tau > 0$ depending only on δ_{6s} such that, for any $n \geq 0$,

$$\|\mathbf{x}_T - \mathbf{x}^n\|_2 \leq \tau \|\mathbf{A}\mathbf{x}_{\overline{T}} + \mathbf{e}\|_2 + \rho^n \|\mathbf{x}_T\|_2,$$

where T denotes an index set of $2s$ largest absolute entries of $\mathbf{x}$. Then Lemma 6.23 with $\mathbf{x}' = \mathbf{x}^n$ and $\xi = \rho^n \|\mathbf{x}_T\|_2 \leq \rho^n \|\mathbf{x}\|_2$ implies that, for any $1 \leq p \leq 2$,

$$\|\mathbf{x} - \mathbf{x}^n\|_p \leq \frac{C}{s^{1-1/p}} \sigma_s(\mathbf{x})_1 + D s^{1/p-1/2} \|\mathbf{e}\|_2 + 2\rho^n s^{1/p-1/2} \|\mathbf{x}\|_2,$$

where $C, D > 0$ depend only on τ, hence only on δ_{6s}. The desired estimates are the particular cases $p = 1$ and $p = 2$.   □

6.4   Analysis of Greedy Algorithms

In this final section, we establish the success of sparse recovery via the greedy algorithms presented in Sect. 3.2, namely, orthogonal matching pursuit and compressive sampling matching pursuit. For the orthogonal matching pursuit algorithm, we first remark that standard restricted isometry conditions are not enough to guarantee the recovery of all s-sparse vectors in at most s iterations. Indeed, for a fixed $1 < \eta < \sqrt{s}$, consider the $(s + 1) \times (s + 1)$ matrix with ℓ_2-normalized columns defined by

$$\mathbf{A} := \left[\begin{array}{c|c} \mathbf{Id} & \begin{matrix} \frac{\eta}{s} \\ \vdots \\ \frac{\eta}{s} \end{matrix} \\ \hline 0 \cdots 0 & \sqrt{\frac{s-\eta^2}{s}} \end{array} \right]. \tag{6.39}$$

We calculate

$$\mathbf{A}^*\mathbf{A} - \mathbf{Id} = \left[\begin{array}{c|c} \mathbf{0} & \begin{matrix} \frac{\eta}{s} \\ \vdots \\ \frac{\eta}{s} \end{matrix} \\ \hline \frac{\eta}{s} \cdots \frac{\eta}{s} & 0 \end{array} \right].$$

This matrix has eigenvalues $-\eta/\sqrt{s}$, $\eta/\sqrt{s}$ and 0 with multiplicity $s - 1$. Thus,

$$\delta_{s+1} = \|\mathbf{A}^*\mathbf{A} - \mathbf{Id}\|_{2 \to 2} = \frac{\eta}{\sqrt{s}}.$$

However, the s-sparse vector $\mathbf{x} = [1, \ldots, 1, 0]^\top$ is not recovered from $\mathbf{y} = \mathbf{A}\mathbf{x}$ after s iterations, since the wrong index $s + 1$ is picked at the first iteration. Indeed,

$$\mathbf{A}^*(\mathbf{y} - \mathbf{A}\mathbf{x}^0) = \mathbf{A}^*\mathbf{A}\mathbf{x} = \left[\begin{array}{c|c} \mathbf{Id} & \begin{matrix} \frac{\eta}{s} \\ \vdots \\ \frac{\eta}{s} \end{matrix} \\ \hline \frac{\eta}{s} \cdots \frac{\eta}{s} & 1 \end{array} \right] \begin{bmatrix} 1 \\ \vdots \\ 1 \\ 0 \end{bmatrix} = \begin{bmatrix} 1 \\ \vdots \\ 1 \\ \eta \end{bmatrix}.$$

There are two possibilities to bypass this issue: either perform more than s iterations or find a way to reject the wrong indices by modifying the orthogonal matching pursuit, which is the rationale behind compressive sampling matching pursuit. In both cases, sparse recovery will be established under restricted isometry conditions. In what follows, we do not separate the ideal situation of exactly sparse vectors measured with perfect accuracy, and we directly give the more cumbersome proofs for stable and robust s-sparse recovery under the condition $\delta_{13s} < 0.1666$ for $12s$ iterations of orthogonal matching pursuit and $\delta_{4s} < 0.4782$ for compressive sampling matching pursuit. Although the argument for the compressive sampling matching pursuit algorithm is closer to the argument used in the previous section, we start with the orthogonal matching pursuit algorithm.

Orthogonal Matching Pursuit

For the purpose of proving the main result, we consider the slightly more general algorithm starting with an index set S^0 and with

$$\mathbf{x}^0 := \text{argmin}\{\|\mathbf{y} - \mathbf{A}\mathbf{z}\|, \text{supp}(\mathbf{z}) \subset S^0\}, \tag{6.40}$$

and iterating the scheme

$$S^{n+1} = S^n \cup L_1(\mathbf{A}^*(\mathbf{y} - \mathbf{A}\mathbf{x}^n)), \tag{OMP'_1}$$

$$\mathbf{x}^{n+1} = \text{argmin}\left\{\|\mathbf{y} - \mathbf{A}\mathbf{z}\|_2, \text{supp}(\mathbf{z}) \subset S^{n+1}\right\}. \tag{OMP'_2}$$

The usual orthogonal matching pursuit algorithm corresponds to the default choice of $S^0 = \emptyset$ and $\mathbf{x}^0 = \mathbf{0}$. The following proposition is the key.

Proposition 6.24. *Given* $\mathbf{A} \in \mathbb{C}^{m \times N}$, *let* $\mathbf{y} = \mathbf{A}\mathbf{x} + \mathbf{e}$ *for some s-sparse* $\mathbf{x} \in \mathbb{C}^N$ *with* $S = \text{supp}(\mathbf{x})$ *and some* $\mathbf{e} \in \mathbb{C}^m$. *Let* $(\mathbf{x}^n)$ *denote the sequence defined by* (OMP'_1), (OMP'_2) *started at an index set* S^0. *With* $s^0 = \text{card}(S^0)$ *and* $s' = \text{card}(S \setminus S^0)$, *if* $\delta_{s+s^0+12s'} < 1/6$, *then there is a constant* $C > 0$ *depending only on* $\delta_{s+s^0+12s'}$ *such that*

$$\|\mathbf{y} - \mathbf{A}\mathbf{x}^{\bar{n}}\|_2 \le C\|\mathbf{e}\|_2, \qquad \bar{n} = 12s'.$$

Note that if $\mathbf{e} = \mathbf{0}$ and $S^0 = \emptyset$, this proposition implies exact s-sparse recovery via (OMP'_1), (OMP'_2) in $12s$ iterations. Indeed, we have $\mathbf{A}(\mathbf{x} - \mathbf{x}^{12s}) = \mathbf{0}$, which implies $\mathbf{x} - \mathbf{x}^{12s} = \mathbf{0}$ since $\|\mathbf{x} - \mathbf{x}^{12s}\|_0 \le 13s$ and $\delta_{13s} < 1$. Proposition 6.24 also implies some stability and robustness results stated in a familiar form.

Theorem 6.25. *Suppose that* $\mathbf{A} \in \mathbb{C}^{m \times N}$ *has restricted isometry constant*

$$\delta_{13s} < \frac{1}{6}.$$

Then there is a constant $C > 0$ *depending only on* δ_{13s} *such that, for all* $\mathbf{x} \in \mathbb{C}^N$ *and* $\mathbf{e} \in \mathbb{C}^m$, *the sequence* $(\mathbf{x}^n)$ *defined by* (OMP'_1), (OMP'_2) *with* $\mathbf{y} = \mathbf{A}\mathbf{x} + \mathbf{e}$ *satisfies*

$$\|\mathbf{y} - \mathbf{A}\mathbf{x}^{12s}\|_2 \le C\|\mathbf{A}\mathbf{x}_{\overline{S}} + \mathbf{e}\|_2$$

for any $S \subset [N]$ *with* $\text{card}(S) = s$. *Furthermore, if* $\delta_{26s} < 1/6$, *then there are constants* $C, D > 0$ *depending only on* δ_{26s} *such that, for all* $\mathbf{x} \in \mathbb{C}^N$ *and* $\mathbf{e} \in \mathbb{C}^m$, *the sequence* $(\mathbf{x}^n)$ *defined by* (OMP'_1), (OMP'_2) *with* $\mathbf{y} = \mathbf{A}\mathbf{x} + \mathbf{e}$ *satisfies, for any* $1 \le p \le 2$,

$$\|\mathbf{x} - \mathbf{x}^{24s}\|_p \le \frac{C}{s^{1-1/p}}\sigma_s(\mathbf{x})_1 + Ds^{1/p-1/2}\|\mathbf{e}\|_2.$$

Proof. Given $S \subset [N]$ with $\text{card}(S) = s$, we can write $\mathbf{y} = \mathbf{A}\mathbf{x}_S + \mathbf{e}'$ where $\mathbf{e}' := \mathbf{A}\mathbf{x}_{\overline{S}} + \mathbf{e}$. Applying Proposition 6.24 with $S^0 = \emptyset$ then gives the desired inequality

$$\|\mathbf{y} - \mathbf{A}\mathbf{x}^{12s}\|_2 \leq C\|\mathbf{e}'\|_2 = C\|\mathbf{A}\mathbf{x}_{\overline{S}} + \mathbf{e}\|_2$$

for some constant $C > 0$ depending only on $\delta_{12s} < 1/6$. For the second inequality, we choose T to be an index set of $2s$ largest absolute entries of $\mathbf{x}$, so the previous argument yields

$$\|\mathbf{y} - \mathbf{A}\mathbf{x}^{24s}\|_2 \leq C'\|\mathbf{A}\mathbf{x}_{\overline{T}} + \mathbf{e}\|_2$$

for some constant $C' > 0$ depending only on $\delta_{26s} < 1/6$. Now, in view of

$$\|\mathbf{y} - \mathbf{A}\mathbf{x}^{24s}\|_2 = \|\mathbf{A}(\mathbf{x}_T - \mathbf{x}^{24s}) + \mathbf{A}\mathbf{x}_{\overline{T}} + \mathbf{e}\|_2$$

$$\geq \|\mathbf{A}(\mathbf{x}_T - \mathbf{x}^{24s})\|_2 - \|\mathbf{A}\mathbf{x}_{\overline{T}} + \mathbf{e}\|_2$$

$$\geq \sqrt{1 - \delta_{26s}}\|\mathbf{x}^{24s} - \mathbf{x}_T\|_2 - \|\mathbf{A}\mathbf{x}_{\overline{T}} + \mathbf{e}\|_2,$$

we derive

$$\|\mathbf{x}^{24s} - \mathbf{x}_T\|_2 \leq \frac{C' + 1}{\sqrt{1 - \delta_{26s}}}\|\mathbf{A}\mathbf{x}_{\overline{T}} + \mathbf{e}\|_2.$$

An application of Lemma 6.23 with $\xi = 0$ gives the desired result. $\square$

It remains to establish the crucial Proposition 6.24. It is proved with the help of the following lemma.

Lemma 6.26. *Let $(\mathbf{x}^n)$ be the sequence defined by* (OMP'_1), (OMP'_2) *with $\mathbf{y} = \mathbf{A}\mathbf{x} + \mathbf{e}$ for some s-sparse $\mathbf{x} \in \mathbb{C}^N$ and for some $\mathbf{e} \in \mathbb{C}^m$. Then, for $n \geq 0$, $T \subset [N]$ not included in S^n, and $\mathbf{z} \in \mathbb{C}^N$ supported on T,*

$$\|\mathbf{y} - \mathbf{A}\mathbf{x}^{n+1}\|_2^2$$

$$\leq \|\mathbf{y} - \mathbf{A}\mathbf{x}^n\|_2^2 - \frac{\|\mathbf{A}(\mathbf{z} - \mathbf{x}^n)\|_2^2}{\|\mathbf{z}_{T \setminus S^n}\|_1^2} \max\{0, \|\mathbf{y} - \mathbf{A}\mathbf{x}^n\|_2^2 - \|\mathbf{y} - \mathbf{A}\mathbf{z}\|_2^2\}$$

$$\leq \|\mathbf{y} - \mathbf{A}\mathbf{x}^n\|_2^2 - \frac{1 - \delta}{\text{card}(T \setminus S^n)} \max\{0, \|\mathbf{y} - \mathbf{A}\mathbf{x}^n\|_2^2 - \|\mathbf{y} - \mathbf{A}\mathbf{z}\|_2^2\},$$

where $\delta := \delta_{\text{card}(T \cup S^n)}$.

Proof. The second inequality follows from the first one by noticing that

$$\|\mathbf{A}(\mathbf{x}^n - \mathbf{z})\|_2^2 \geq (1 - \delta)\|\mathbf{x}^n - \mathbf{z}\|_2^2 \geq (1 - \delta)\|(\mathbf{x}^n - \mathbf{z})_{T \setminus S^n}\|_2^2,$$

$$\|\mathbf{z}_{T \setminus S^n}\|_1^2 \leq \text{card}(T \setminus S^n)\|\mathbf{z}_{T \setminus S^n}\|_2^2 = \text{card}(T \setminus S^n)\|(\mathbf{x}^n - \mathbf{z})_{T \setminus S^n}\|_2^2.$$

We now recall from Lemma 3.3 that the decrease in the squared ℓ_2-norm of the residual is at least $|(\mathbf{A}^*(\mathbf{y} - \mathbf{A}\mathbf{x}^n))_{j^{n+1}}|^2$, where j^{n+1} denotes the index of a largest absolute entry of $\mathbf{A}^*(\mathbf{y} - \mathbf{A}\mathbf{x}^n)$. Thus, the first inequality follows from

$$|(\mathbf{A}^*(\mathbf{y} - \mathbf{A}\mathbf{x}^n))_{j^{n+1}}|^2 \geq \frac{\|\mathbf{A}(\mathbf{z} - \mathbf{x}^n)\|_2^2}{\|\mathbf{z}_{T \setminus S^n}\|_1^2} \left(\|\mathbf{y} - \mathbf{A}\mathbf{x}^n\|_2^2 - \|\mathbf{y} - \mathbf{A}\mathbf{z}\|_2^2\right) \quad (6.41)$$

when $\|\mathbf{y} - \mathbf{A}\mathbf{x}^n\|_2^2 \geq \|\mathbf{y} - \mathbf{A}\mathbf{z}\|_2^2$. Let us also recall from Lemma 3.4 that $(\mathbf{A}^*(\mathbf{y} - \mathbf{A}\mathbf{x}^n))_{S^n} = \mathbf{0}$ to observe on the one hand that

$$\text{Re}\langle \mathbf{A}(\mathbf{z} - \mathbf{x}^n), \mathbf{y} - \mathbf{A}\mathbf{x}^n \rangle$$
$$= \text{Re}\langle \mathbf{z} - \mathbf{x}^n, \mathbf{A}^*(\mathbf{y} - \mathbf{A}\mathbf{x}^n) \rangle = \text{Re}\langle \mathbf{z} - \mathbf{x}^n, (\mathbf{A}^*(\mathbf{y} - \mathbf{A}\mathbf{x}^n))_{\overline{S^n}} \rangle$$
$$= \text{Re}\langle (\mathbf{z} - \mathbf{x}^n)_{T \setminus S^n}, (\mathbf{A}^*(\mathbf{y} - \mathbf{A}\mathbf{x}^n))_{T \setminus S^n} \rangle$$
$$\leq \|(\mathbf{z} - \mathbf{x}^n)_{T \setminus S^n}\|_1 \|\mathbf{A}^*(\mathbf{y} - \mathbf{A}\mathbf{x}^n)\|_\infty$$
$$= \|\mathbf{z}_{T \setminus S^n}\|_1 |(\mathbf{A}^*(\mathbf{y} - \mathbf{A}\mathbf{x}^n))_{j^{n+1}}|. \quad (6.42)$$

On the other hand, we have

$$2\,\text{Re}\langle \mathbf{A}(\mathbf{z} - \mathbf{x}^n), \mathbf{y} - \mathbf{A}\mathbf{x}^n \rangle$$
$$= \|\mathbf{A}(\mathbf{z} - \mathbf{x}^n)\|_2^2 + \|\mathbf{y} - \mathbf{A}\mathbf{x}^n\|_2^2 - \|\mathbf{A}(\mathbf{z} - \mathbf{x}^n) - (\mathbf{y} - \mathbf{A}\mathbf{x}^n)\|_2^2$$
$$= \|\mathbf{A}(\mathbf{z} - \mathbf{x}^n)\|_2^2 + \left(\|\mathbf{y} - \mathbf{A}\mathbf{x}^n\|_2^2 - \|\mathbf{y} - \mathbf{A}\mathbf{z}\|_2^2\right)$$
$$\geq 2\|\mathbf{A}(\mathbf{z} - \mathbf{x}^n)\|_2 \sqrt{\|\mathbf{y} - \mathbf{A}\mathbf{x}^n\|_2^2 - \|\mathbf{y} - \mathbf{A}\mathbf{z}\|_2^2}, \quad (6.43)$$

where the inequality between the arithmetic and the geometric mean was used in the last step. Combining the squared versions of (6.42) and (6.43), we arrive at

$$\|\mathbf{A}(\mathbf{z} - \mathbf{x}^n)\|_2^2 \left(\|\mathbf{y} - \mathbf{A}\mathbf{x}^n\|_2^2 - \|\mathbf{y} - \mathbf{A}\mathbf{z}\|_2^2\right) \leq \|\mathbf{z}_{T \setminus S^n}\|_1^2 |(\mathbf{A}^*(\mathbf{y} - \mathbf{A}\mathbf{x}^n))_{j^{n+1}}|^2.$$

The desired inequality (6.41) follows from here. $\square$

We are now ready for the proof of the main proposition.

Proof (of Proposition 6.24). The proof proceeds by induction on $\text{card}(S \setminus S^0)$. If it is zero, i.e., if $S \subset S^0$, then the definition of $\mathbf{x}^0$ implies

$$\|\mathbf{y} - \mathbf{A}\mathbf{x}^0\|_2 \leq \|\mathbf{y} - \mathbf{A}\mathbf{x}\|_2 = \|\mathbf{e}\|_2,$$

and the result holds with $C = 1$. Let us now assume that the result holds for all S and S^0 such that $\text{card}(S \setminus S^0) \leq s' - 1$, $s' \geq 1$, and let us show that it holds when $\text{card}(S \setminus S^0) = s'$. We consider subsets of $S \setminus S^0$ defined by $T^0 = \emptyset$ and

$$T^\ell = \{\text{indices of } 2^{\ell-1} \text{ largest absolute entries of } \mathbf{x}_{\overline{S^0}} \} \text{ for } \ell \geq 1,$$

to which we associate the vectors

$$\tilde{\mathbf{x}}^\ell := \mathbf{x}_{\overline{S^0 \cup T^\ell}}, \quad \ell \geq 0.$$

Note that the last T^ℓ, namely, $T^{\lceil \log_2(s') \rceil + 1}$, is taken to be the whole set $S \setminus S^0$ (and may have less than $2^{\ell-1}$ elements), so that $\tilde{\mathbf{x}}^\ell = \mathbf{0}$. For a constant $\mu > 0$ to be chosen later, since $\|\tilde{\mathbf{x}}^{\ell-1}\|_2^2 \geq \mu \|\tilde{\mathbf{x}}^\ell\|_2^2 = 0$ for this last index, we can consider the smallest integer $1 \leq L \leq \lceil \log_2(s') \rceil + 1$ such that

$$\|\tilde{\mathbf{x}}^{L-1}\|_2^2 \geq \mu \|\tilde{\mathbf{x}}^L\|_2^2.$$

This definition implies the (possibly empty) list of inequalities

$$\|\tilde{\mathbf{x}}^0\|_2^2 < \mu \|\tilde{\mathbf{x}}^1\|_2^2, \ldots, \|\tilde{\mathbf{x}}^{L-2}\|_2^2 < \mu \|\tilde{\mathbf{x}}^{L-1}\|_2^2.$$

For each $\ell \in [L]$, we apply Lemma 6.26 to the vector $\mathbf{z} = \mathbf{x} - \tilde{\mathbf{x}}^\ell$, which is supported on $S^0 \cup T^\ell$. Taking into account that $(S^0 \cup T^\ell) \cup S^n \subset S \cup S^n$ and that $(S^0 \cup T^\ell) \setminus S^n \subset (S^0 \cup T^\ell) \setminus S^0 = T^\ell$, we obtain, after subtracting $\|\mathbf{y} - \mathbf{A}\mathbf{z}\|_2^2 = \|\mathbf{A}\tilde{\mathbf{x}}^\ell + \mathbf{e}\|_2^2$ from both sides,

$$\max\{0, \|\mathbf{y} - \mathbf{A}\mathbf{x}^{n+1}\|_2^2 - \|\mathbf{A}\tilde{\mathbf{x}}^\ell + \mathbf{e}\|_2^2\}$$
$$\leq \left(1 - \frac{1 - \delta_{s+n}}{\mathrm{card}(T^\ell)}\right) \max\{0, \|\mathbf{y} - \mathbf{A}\mathbf{x}^n\|_2^2 - \|\mathbf{A}\tilde{\mathbf{x}}^\ell + \mathbf{e}\|_2^2\}$$
$$\leq \exp\left(-\frac{1 - \delta_{s+n}}{\mathrm{card}(T^\ell)}\right) \max\{0, \|\mathbf{y} - \mathbf{A}\mathbf{x}^n\|_2^2 - \|\mathbf{A}\tilde{\mathbf{x}}^\ell + \mathbf{e}\|_2^2\}.$$

For any $K \geq 0$ and any $n, k \geq 0$ satisfying $n + k \leq K$, we derive by induction that

$$\max\{0, \|\mathbf{y} - \mathbf{A}\mathbf{x}^{n+k}\|_2^2 - \|\mathbf{A}\tilde{\mathbf{x}}^\ell + \mathbf{e}\|_2^2\}$$
$$\leq \exp\left(-\frac{k(1 - \delta_{s+K})}{\mathrm{card}(T^\ell)}\right) \max\{0, \|\mathbf{y} - \mathbf{A}\mathbf{x}^n\|_2^2 - \|\mathbf{A}\tilde{\mathbf{x}}^\ell + \mathbf{e}\|_2^2\}.$$

By separating cases in the rightmost maximum, we easily deduce

$$\|\mathbf{y} - \mathbf{A}\mathbf{x}^{n+k}\|_2^2 \leq \exp\left(-\frac{k(1 - \delta_{s+K})}{\mathrm{card}(T^\ell)}\right) \|\mathbf{y} - \mathbf{A}\mathbf{x}^n\|_2^2 + \|\mathbf{A}\tilde{\mathbf{x}}^\ell + \mathbf{e}\|_2^2.$$

For some positive integer κ to be chosen later, applying this successively with

$$k_1 := \kappa \, \mathrm{card}(T^1), \ldots, \ k_L := \kappa \, \mathrm{card}(T^L), \quad \text{and} \quad K := k_1 + \cdots + k_L,$$

yields, with $\nu := \exp(\kappa(1 - \delta_{s+K}))$,

$$\|\mathbf{y} - \mathbf{A}\mathbf{x}^{k_1}\|_2^2 \leq \frac{1}{\nu}\|\mathbf{y} - \mathbf{A}\mathbf{x}^0\|_2^2 + \|\mathbf{A}\tilde{\mathbf{x}}^1 + \mathbf{e}\|_2^2,$$

$$\|\mathbf{y} - \mathbf{A}\mathbf{x}^{k_1+k_2}\|_2^2 \leq \frac{1}{\nu}\|\mathbf{y} - \mathbf{A}\mathbf{x}^{k_1}\|_2^2 + \|\mathbf{A}\tilde{\mathbf{x}}^2 + \mathbf{e}\|_2^2,$$

$$\vdots$$

$$\|\mathbf{y} - \mathbf{A}\mathbf{x}^{k_1+\cdots+k_{L-1}+k_L}\|_2^2 \leq \frac{1}{\nu}\|\mathbf{y} - \mathbf{A}\mathbf{x}^{k_1+\cdots+k_{L-1}}\|_2^2 + \|\mathbf{A}\tilde{\mathbf{x}}^L + \mathbf{e}\|_2^2.$$

By combining these inequalities, we obtain

$$\|\mathbf{y}-\mathbf{A}\mathbf{x}^K\|_2^2 \leq \frac{\|\mathbf{y} - \mathbf{A}\mathbf{x}^0\|_2^2}{\nu^L} + \frac{\|\mathbf{A}\tilde{\mathbf{x}}^1 + \mathbf{e}\|_2^2}{\nu^{L-1}} + \cdots + \frac{\|\mathbf{A}\tilde{\mathbf{x}}^{L-1} + \mathbf{e}\|_2^2}{\nu} + \|\mathbf{A}\tilde{\mathbf{x}}^L + \mathbf{e}\|_2^2.$$

Taking into account that $\mathbf{x} - \tilde{\mathbf{x}}^0$ is supported on $S^0 \cup T^0 = S^0$, the definition (6.40) of $\mathbf{x}^0$ implies that $\|\mathbf{y} - \mathbf{A}\mathbf{x}^0\|_2^2 \leq \|\mathbf{y} - \mathbf{A}(\mathbf{x} - \tilde{\mathbf{x}}^0)\|_2^2 = \|\mathbf{A}\tilde{\mathbf{x}}^0 + \mathbf{e}\|_2^2$; hence

$$\|\mathbf{y} - \mathbf{A}\mathbf{x}^K\|_2^2 \leq \sum_{\ell=0}^{L} \frac{\|\mathbf{A}\tilde{\mathbf{x}}^\ell + \mathbf{e}\|_2^2}{\nu^{L-\ell}} \leq \sum_{\ell=0}^{L} \frac{2(\|\mathbf{A}\tilde{\mathbf{x}}^\ell\|_2^2 + \|\mathbf{e}\|_2^2)}{\nu^{L-\ell}}.$$

Let us remark that, for $\ell \leq L - 1$ and also for $\ell = L$,

$$\|\mathbf{A}\tilde{\mathbf{x}}^\ell\|_2^2 \leq (1 + \delta_s)\|\tilde{\mathbf{x}}^\ell\|_2^2 \leq (1 + \delta_s)\mu^{L-1-\ell}\|\tilde{\mathbf{x}}^{L-1}\|_2^2.$$

As a result, we have

$$\|\mathbf{y} - \mathbf{A}\mathbf{x}^K\|_2^2 \leq \frac{2(1 + \delta_s)\|\tilde{\mathbf{x}}^{L-1}\|_2^2}{\mu} \sum_{\ell=0}^{L} \left(\frac{\mu}{\nu}\right)^{L-\ell} + 2\|\mathbf{e}\|_2^2 \sum_{\ell=0}^{L} \frac{1}{\nu^{L-\ell}}$$

$$\leq \frac{2(1 + \delta_s)\|\tilde{\mathbf{x}}^{L-1}\|_2^2}{\mu(1 - \mu/\nu)} + \frac{2\|\mathbf{e}\|_2^2}{1 - \nu}.$$

We choose $\mu = \nu/2$ so that $\mu(1 - \mu/\nu)$ takes its maximal value $\nu/4$. It follows that, with $\alpha := \sqrt{8(1 + \delta_s)/\nu}$ and $\beta := \sqrt{2/(1 - \nu)}$,

$$\|\mathbf{y} - \mathbf{A}\mathbf{x}^K\|_2 \leq \alpha\|\tilde{\mathbf{x}}^{L-1}\|_2 + \beta\|\mathbf{e}\|_2. \qquad (6.44)$$

On the other hand, with $\gamma := \sqrt{1 - \delta_{s+s^0+K}}$, we have

$$\|\mathbf{y} - \mathbf{A}\mathbf{x}^K\|_2 = \|\mathbf{A}(\mathbf{x} - \mathbf{x}^K) + \mathbf{e}\|_2 \geq \|\mathbf{A}(\mathbf{x} - \mathbf{x}^K)\|_2 - \|\mathbf{e}\|_2$$

$$\geq \gamma\|\mathbf{x} - \mathbf{x}^K\|_2 - \|\mathbf{e}\|_2 \geq \gamma\|\mathbf{x}_{\overline{SK}}\|_2 - \|\mathbf{e}\|_2.$$

We deduce that

$$\|\mathbf{x}_{\overline{S^K}}\|_2 \leq \frac{\alpha}{\gamma}\|\tilde{\mathbf{x}}^{L-1}\|_2 + \frac{\beta+1}{\gamma}\|\mathbf{e}\|_2. \tag{6.45}$$

Let us now choose $\kappa = 3$, which guarantees that

$$\frac{\alpha}{\gamma} = \sqrt{\frac{8(1+\delta_s)}{(1-\delta_{s+s^0+K})\exp(\kappa(1-\delta_{s+K}))}} \leq 0.92 < 1,$$

since $\delta_s \leq \delta_{s+K} \leq \delta_{s+s^0+K} \leq \delta_{s+s^0+12s'} < 1/6$. Hereby, we have used the fact that $L \leq \lceil \log_2(s')\rceil + 1$ to derive

$$K = \kappa(1 + \cdots + 2^{L-2} + \mathrm{card}(T^L)) < \kappa(2^{L-1} + s') \leq 3\kappa s' = 9s'.$$

Thus, in the case $((\beta+1)/\gamma)\|\mathbf{e}\|_2 < (1-\alpha/\gamma)\|\tilde{\mathbf{x}}^{L-1}\|_2$, we derive from (6.45) that

$$\|\mathbf{x}_{\overline{S^K}}\|_2 < \|\tilde{\mathbf{x}}^{L-1}\|_2, \quad \text{i.e.,} \quad \|(\mathbf{x}_{\overline{S^0}})_{S\setminus S^K}\|_2 < \|(\mathbf{x}_{\overline{S^0}})_{(S\setminus S^0)\setminus T^{L-1}}\|_2.$$

But since T^{L-1} lists the 2^{L-2} largest absolute entries of $\mathbf{x}_{\overline{S^0}}$, this yields

$$\mathrm{card}(S \setminus S^K) < \mathrm{card}((S \setminus S^0) \setminus T^{L-1}) = s' - 2^{L-2}.$$

Note that continuing the algorithm from iteration K amounts to starting it again with S^0 replaced by S^K. In view of $K \leq \kappa(1 + \cdots + 2^{L-2} + 2^{L-1}) < 3 \cdot 2^L$, we have

$$s + \mathrm{card}(S^K) + 12\,\mathrm{card}(S \setminus S^K) \leq s + s^0 + K + 12(s' - 2^{L-2}) \leq s + s^0 + 12s'$$

and consequently $\mathrm{card}(S \setminus S^K) < s'$. Therefore, the induction hypothesis applies to give

$$\|\mathbf{y} - \mathbf{A}\mathbf{x}^{K+n}\|_2 \leq C\|\mathbf{e}\|_2, \quad \text{for } n = 12\,\mathrm{card}(S \setminus S^K).$$

Thus, the number of required iterations satisfies $K + n \leq 12s'$, as desired. In the alternative case where $((\beta+1)/\gamma)\|\mathbf{e}\|_2 \geq (1-\alpha/\gamma)\|\tilde{\mathbf{x}}^{L-1}\|_2$, the situation is easier, since (6.44) yields

$$\|\mathbf{y} - \mathbf{A}\mathbf{x}^K\|_2 \leq \frac{\alpha(\beta+1)}{\gamma-\alpha}\|\mathbf{e}\|_2 + \beta\|\mathbf{e}\|_2 =: C\|\mathbf{e}\|_2,$$

where the constant $C \geq 1$ depends only on $\delta_{s+s^0+12s'}$. This shows that the induction hypothesis holds when $\mathrm{card}(S \setminus S^0) = s'$. $\square$

Compressive Sampling Matching Pursuit

As a reminder, we recall that the compressive sampling matching pursuit algorithm (CoSaMP) starts with an initial s-sparse vector $\mathbf{x}^0 \in \mathbb{C}^N$, typically $\mathbf{x}^0 = \mathbf{0}$, and produces a sequence $(\mathbf{x}^n)$ defined inductively by

$$U^{n+1} = \operatorname{supp}(\mathbf{x}^n) \cup L_{2s}(\mathbf{A}^*(\mathbf{y} - \mathbf{A}\mathbf{x}^n)), \qquad (\text{CoSaMP}_1)$$

$$\mathbf{u}^{n+1} = \operatorname{argmin}\left\{\|\mathbf{y} - \mathbf{A}\mathbf{z}\|_2, \operatorname{supp}(\mathbf{z}) \subset U^{n+1}\right\}, \qquad (\text{CoSaMP}_2)$$

$$\mathbf{x}^{n+1} = H_s(\mathbf{u}^{n+1}). \qquad (\text{CoSaMP}_3)$$

Here are the main results for this algorithm.

Theorem 6.27. *Suppose that the $4s$th restricted isometry constant of the matrix $\mathbf{A} \in \mathbb{C}^{m \times N}$ satisfies*

$$\delta_{4s} < \frac{\sqrt{\sqrt{11/3} - 1}}{2} \approx 0.4782. \qquad (6.46)$$

Then, for $\mathbf{x} \in \mathbb{C}^N$, $\mathbf{e} \in \mathbb{C}^m$, and $S \subset [N]$ with $\operatorname{card}(S) = s$, the sequence $(\mathbf{x}^n)$ defined by (CoSaMP_1), (CoSaMP_2), (CoSaMP_3) with $\mathbf{y} = \mathbf{A}\mathbf{x} + \mathbf{e}$ satisfies

$$\|\mathbf{x}^n - \mathbf{x}_S\|_2 \le \rho^n \|\mathbf{x}^0 - \mathbf{x}_S\|_2 + \tau \|\mathbf{A}\mathbf{x}_{\overline{S}} + \mathbf{e}\|_2, \qquad (6.47)$$

where the constant $0 < \rho < 1$ and $\tau > 0$ depend only on δ_{4s}.

Note that, if $\mathbf{x}$ is s-sparse and if $\mathbf{e} = \mathbf{0}$, then $\mathbf{x}$ is recovered as the limit of the sequence $(\mathbf{x}^n)$. In a more general situation, there is no guarantee that the sequence $(\mathbf{x}^n)$ converges. But (6.47) implies at least boundedness of the sequence $(\|\mathbf{x}^n\|_2)$ so that existence of cluster points is guaranteed. Stability and robustness results can then be stated as follows.

Theorem 6.28. *Suppose that the $8s$th restricted isometry constant of the matrix $\mathbf{A} \in \mathbb{C}^{m \times N}$ satisfies $\delta_{8s} < 0.4782$. Then, for $\mathbf{x} \in \mathbb{C}^N$ and $\mathbf{e} \in \mathbb{C}^m$, the sequence $(\mathbf{x}^n)$ defined by (CoSaMP_1), (CoSaMP_2), (CoSaMP_3) with $\mathbf{y} = \mathbf{A}\mathbf{x} + \mathbf{e}$, $\mathbf{x}^0 = \mathbf{0}$, and s replaced by $2s$ satisfies, for any $n \ge 0$,*

$$\|\mathbf{x} - \mathbf{x}^n\|_1 \le C \sigma_s(\mathbf{x})_1 + D \sqrt{s} \|\mathbf{e}\|_2 + 2\rho^n \sqrt{s} \|\mathbf{x}\|_2,$$

$$\|\mathbf{x} - \mathbf{x}^n\|_2 \le \frac{C}{\sqrt{s}} \sigma_s(\mathbf{x})_1 + D \|\mathbf{e}\|_2 + 2\rho^n \|\mathbf{x}\|_2,$$

where the constants $C, D > 0$ and $0 < \rho < 1$ depend only on δ_{8s}. In particular, if $\mathbf{x}^\sharp \in \mathbb{C}^N$ denotes a cluster point of the sequence $(\mathbf{x}^n)$, then

$$\|\mathbf{x} - \mathbf{x}^\sharp\|_1 \leq C \sigma_s(\mathbf{x})_1 + D \sqrt{s} \|\mathbf{e}\|_2,$$

$$\|\mathbf{x} - \mathbf{x}^\sharp\|_2 \leq \frac{C}{\sqrt{s}} \sigma_s(\mathbf{x})_1 + D \|\mathbf{e}\|_2.$$

Remark 6.29. Similarly to Remark 6.22(a), the error estimates are also valid if $\mathbf{x}^\sharp$ is replaced by $\mathbf{x}^n$ for n large enough, provided the right-hand side is nonzero.

Theorem 6.28 follows from Theorem 6.27 via Lemma 6.23 in the same way as Theorem 6.21 follows from Theorem 6.18 for thresholding algorithms. We therefore only concentrate on establishing Theorem 6.27.

Proof (of Theorem 6.27). As in the proof of Theorem 6.18, we establish that for any $n \geq 0$,

$$\|\mathbf{x}^{n+1} - \mathbf{x}_S\|_2 \leq \rho \|\mathbf{x}^n - \mathbf{x}_S\|_2 + (1 - \rho)\tau \|\mathbf{A}\mathbf{x}_{\overline{S}} + \mathbf{e}\|_2 \qquad (6.48)$$

with $0 < \rho < 1$ and $\tau > 0$ to be determined below. This implies the estimate (6.47) by induction. The strategy for proving (6.48) consists in inferring a consequence of each (CoSaMP) step: discarding $\mathbf{A}\mathbf{x}_{\overline{S}} + \mathbf{e}$ here for simplicity, (CoSaMP$_1$) yields an estimate for $\|(\mathbf{x}_S - \mathbf{u}^{n+1})_{\overline{U^{n+1}}}\|_2$ in terms of $\|\mathbf{x}^n - \mathbf{x}_S\|_2$, (CoSaMP$_2$) yields an estimate for $\|(\mathbf{x}_S - \mathbf{u}^{n+1})_{U^{n+1}}\|_2$ in terms of $\|(\mathbf{x}_S - \mathbf{u}^{n+1})_{\overline{U^{n+1}}}\|_2$, and (CoSaMP$_3$) yields an estimate for $\|\mathbf{x}^{n+1} - \mathbf{x}_S\|_2$ in terms of $\|(\mathbf{x}_S - \mathbf{u}^{n+1})_{U^{n+1}}\|_2$ and $\|(\mathbf{x}_S - \mathbf{u}^{n+1})_{\overline{U^{n+1}}}\|_2$, so overall an estimate for $\|\mathbf{x}^{n+1} - \mathbf{x}_S\|_2$ in terms of $\|\mathbf{x}^n - \mathbf{x}_S\|_2$ is deduced.

We start with (CoSaMP$_3$). Specifically, we observe that $\mathbf{x}^{n+1}$ is a better (or at least equally good) s-term approximation to $\mathbf{u}^{n+1}$ than $\mathbf{x}_{S \cup U^{n+1}}$. Denoting $S^{n+1} = \text{supp}(\mathbf{x}^{n+1})$ and observing that $S^{n+1} \subset U^{n+1}$, we conclude that

$$\|(\mathbf{x}_S - \mathbf{x}^{n+1})_{U^{n+1}}\|_2 = \|\mathbf{x}_{S \cap U^{n+1}} - \mathbf{x}^{n+1}\|_2$$
$$\leq \|\mathbf{u}^{n+1} - \mathbf{x}^{n+1}\|_2 + \|\mathbf{u}^{n+1} - \mathbf{x}_{S \cap U^{n+1}}\|_2$$
$$\leq 2\|\mathbf{u}^{n+1} - \mathbf{x}_{S \cap U^{n+1}}\|_2 = 2\|(\mathbf{x}_S - \mathbf{u}^{n+1})_{U^{n+1}}\|_2.$$

Then, using $(\mathbf{x}^{n+1})_{\overline{U^{n+1}}} = \mathbf{0}$ and $(\mathbf{u}^{n+1})_{\overline{U^{n+1}}} = \mathbf{0}$, it follows that

$$\|\mathbf{x}_S - \mathbf{x}^{n+1}\|_2^2 = \|(\mathbf{x}_S - \mathbf{x}^{n+1})_{\overline{U^{n+1}}}\|_2^2 + \|(\mathbf{x}_S - \mathbf{x}^{n+1})_{U^{n+1}}\|_2^2$$
$$\leq \|(\mathbf{x}_S - \mathbf{u}^{n+1})_{\overline{U^{n+1}}}\|_2^2 + 4\|(\mathbf{x}_S - \mathbf{u}^{n+1})_{U^{n+1}}\|_2^2. \qquad (6.49)$$

Now, as a consequence of (CoSaMP$_2$), the vector $\mathbf{A}\mathbf{u}^{n+1}$ is characterized by

$$\langle \mathbf{y} - \mathbf{A}\mathbf{u}^{n+1}, \mathbf{A}\mathbf{z} \rangle = 0 \qquad \text{whenever } \text{supp}(\mathbf{z}) \subset U^{n+1}.$$

This is equivalent to $\langle \mathbf{A}^*(\mathbf{y} - \mathbf{A}\mathbf{u}^{n+1}), \mathbf{z}\rangle = 0$ whenever $\mathrm{supp}(\mathbf{z}) \subset U^{n+1}$ or to $(\mathbf{A}^*(\mathbf{y} - \mathbf{A}\mathbf{u}^{n+1}))_{U^{n+1}} = 0$. Since $\mathbf{y} = \mathbf{A}\mathbf{x}_S + \mathbf{e}'$ with $\mathbf{e}' := \mathbf{A}\mathbf{x}_{\overline{S}} + \mathbf{e}$, this means

$$(\mathbf{A}^*\mathbf{A}(\mathbf{x}_S - \mathbf{u}^{n+1}))_{U^{n+1}} = -(\mathbf{A}^*\mathbf{e}')_{U^{n+1}}.$$

We make use of this fact to obtain

$$\|(\mathbf{x}_S - \mathbf{u}^{n+1})_{U^{n+1}}\|_2 \le \|((\mathbf{Id} - \mathbf{A}^*\mathbf{A})(\mathbf{x}_S - \mathbf{u}^{n+1}))_{U^{n+1}}\|_2 + \|(\mathbf{A}^*\mathbf{e}')_{U^{n+1}}\|_2$$
$$\le \delta_{4s}\|\mathbf{x}_S - \mathbf{u}^{n+1}\|_2 + \|(\mathbf{A}^*\mathbf{e}')_{U^{n+1}}\|_2,$$

where the last inequality follows from Lemma 6.16. Note that we may assume $\|(\mathbf{x}_S - \mathbf{u}^{n+1})_{U^{n+1}}\|_2 > \|(\mathbf{A}^*\mathbf{e}')_{U^{n+1}}\|_2/(1 - \delta_{4s})$, otherwise (6.50) below is immediate. In this case, the previous inequality is equivalent, by virtue of $\|(\mathbf{x}_S - \mathbf{u}^{n+1})_{U^{n+1}}\|_2 > \|(\mathbf{A}^*\mathbf{e}')_{U^{n+1}}\|_2$, to

$$\left[\|(\mathbf{x}_S - \mathbf{u}^{n+1})_{U^{n+1}}\|_2 - \|(\mathbf{A}^*\mathbf{e}')_{U^{n+1}}\|_2\right]^2$$
$$\le \delta_{4s}^2\|(\mathbf{x}_S - \mathbf{u}^{n+1})_{U^{n+1}}\|_2^2 + \delta_{4s}^2\|(\mathbf{x}_S - \mathbf{u}^{n+1})_{\overline{U^{n+1}}}\|_2^2.$$

Using the identity $a^2 - b^2 = (a + b)(a - b)$, we derive

$$\delta_{4s}^2\|(\mathbf{x}_S - \mathbf{u}^{n+1})_{\overline{U^{n+1}}}\|_2^2 \ge (1 - \delta_{4s}^2)$$
$$\times \left(\|(\mathbf{x}_S - \mathbf{u}^{n+1})_{U^{n+1}}\|_2 - \frac{1}{1 + \delta_{4s}}\|(\mathbf{A}^*\mathbf{e}')_{U^{n+1}}\|_2\right)$$
$$\times \left(\|(\mathbf{x}_S - \mathbf{u}^{n+1})_{U^{n+1}}\|_2 - \frac{1}{1 - \delta_{4s}}\|(\mathbf{A}^*\mathbf{e}')_{U^{n+1}}\|_2\right).$$

Bounding the middle term from below by the bottom term, we obtain

$$\frac{\delta_{4s}^2}{1 - \delta_{4s}^2}\|(\mathbf{x}_S - \mathbf{u}^{n+1})_{\overline{U^{n+1}}}\|_2^2 \ge \left(\|(\mathbf{x}_S - \mathbf{u}^{n+1})_{U^{n+1}}\|_2 - \frac{1}{1 - \delta_{4s}}\|(\mathbf{A}^*\mathbf{e}')_{U^{n+1}}\|_2\right)^2.$$

Taking the square root and rearranging gives

$$\|(\mathbf{x}_S - \mathbf{u}^{n+1})_{U^{n+1}}\|_2 \le \frac{\delta_{4s}}{\sqrt{1 - \delta_{4s}^2}}\|(\mathbf{x}_S - \mathbf{u}^{n+1})_{\overline{U^{n+1}}}\|_2$$
$$+ \frac{1}{1 - \delta_{4s}}\|(\mathbf{A}^*\mathbf{e}')_{U^{n+1}}\|_2. \qquad (6.50)$$

Next, as a consequence of (CoSaMP$_1$), if S^n denotes the support of $\mathbf{x}^n$ and if T^{n+1} denotes a set of $2s$ largest absolute entries of $\mathbf{A}^*(\mathbf{y} - \mathbf{A}\mathbf{x}^n)$, we have

$$\left\| \left(\mathbf{A}^*(\mathbf{y} - \mathbf{A}\mathbf{x}^n)\right)_{S \cup S^n} \right\|_2^2 \leq \left\| \left(\mathbf{A}^*(\mathbf{y} - \mathbf{A}\mathbf{x}^n)\right)_{T^{n+1}} \right\|_2^2.$$

Eliminating the contribution on $(S \cup S^n) \cap T^{n+1}$, we derive

$$\left\| \left(\mathbf{A}^*(\mathbf{y} - \mathbf{A}\mathbf{x}^n)\right)_{(S \cup S^n) \setminus T^{n+1}} \right\|_2 \leq \left\| \left(\mathbf{A}^*(\mathbf{y} - \mathbf{A}\mathbf{x}^n)\right)_{T^{n+1} \setminus (S \cup S^n)} \right\|_2.$$

The right-hand side may be written as

$$\left\| \left(\mathbf{A}^*(\mathbf{y} - \mathbf{A}\mathbf{x}^n)\right)_{T^{n+1} \setminus (S \cup S^n)} \right\|_2 = \left\| \left(\mathbf{x}^n - \mathbf{x}_S + \mathbf{A}^*(\mathbf{y} - \mathbf{A}\mathbf{x}^n)\right)_{T^{n+1} \setminus (S \cup S^n)} \right\|_2.$$

The left-hand side satisfies

$$\left\| \left(\mathbf{A}^*(\mathbf{y} - \mathbf{A}\mathbf{x}^n)\right)_{(S \cup S^n) \setminus T^{n+1}} \right\|_2 \geq \left\| \left(\mathbf{x}_S - \mathbf{x}^n\right)_{\overline{T^{n+1}}} \right\|_2$$
$$- \left\| \left(\mathbf{x}^n - \mathbf{x}_S + \mathbf{A}^*(\mathbf{y} - \mathbf{A}\mathbf{x}^n)\right)_{(S \cup S^n) \setminus T^{n+1}} \right\|_2.$$

These observations imply that

$$\left\| \left(\mathbf{x}_S - \mathbf{x}^n\right)_{\overline{T^{n+1}}} \right\|_2 \leq \left\| \left(\mathbf{x}^n - \mathbf{x}_S + \mathbf{A}^*(\mathbf{y} - \mathbf{A}\mathbf{x}^n)\right)_{(S \cup S^n) \setminus T^{n+1}} \right\|_2$$
$$+ \left\| \left(\mathbf{x}^n - \mathbf{x}_S + \mathbf{A}^*(\mathbf{y} - \mathbf{A}\mathbf{x}^n)\right)_{T^{n+1} \setminus (S \cup S^n)} \right\|_2$$
$$\leq \sqrt{2} \left\| \left(\mathbf{x}^n - \mathbf{x}_S + \mathbf{A}^*(\mathbf{y} - \mathbf{A}\mathbf{x}^n)\right)_{(S \cup S^n) \Delta T^{n+1}} \right\|_2$$
$$\leq \sqrt{2} \left\| \left((\mathbf{Id} - \mathbf{A}^*\mathbf{A})(\mathbf{x}^n - \mathbf{x}_S)\right)_{(S \cup S^n) \Delta T^{n+1}} \right\|_2$$
$$+ \sqrt{2} \left\| \left(\mathbf{A}^*\mathbf{e}'\right)_{(S \cup S^n) \Delta T^{n+1}} \right\|_2,$$

where $(S \cup S^n) \Delta T^{n+1}$ denotes the symmetric difference of the sets $S \cup S^n$ and T^{n+1} and where $\mathbf{y} = \mathbf{A}\mathbf{x}_S + \mathbf{e}'$ has been used. Since $T^{n+1} \subset U^{n+1}$ by (CoSaMP$_1$) and $S^n \subset U^{n+1}$ by (CoSaMP$_3$), the left-hand side can be bounded from below as

$$\left\| \left(\mathbf{x}_S - \mathbf{x}^n\right)_{\overline{T^{n+1}}} \right\|_2 \geq \left\| \left(\mathbf{x}_S - \mathbf{x}^n\right)_{\overline{U^{n+1}}} \right\|_2 = \left\| \left(\mathbf{x}_S\right)_{\overline{U^{n+1}}} \right\|_2 = \left\| \left(\mathbf{x}_S - \mathbf{u}^{n+1}\right)_{\overline{U^{n+1}}} \right\|_2.$$

Since the right-hand side can be bounded from above using Lemma 6.16, we derive accordingly

$$\left\| \left(\mathbf{x}_S - \mathbf{u}^{n+1}\right)_{\overline{U^{n+1}}} \right\|_2 \leq \sqrt{2}\,\delta_{4s} \left\| \mathbf{x}^n - \mathbf{x}_S \right\|_2$$
$$+ \sqrt{2} \left\| \left(\mathbf{A}^*\mathbf{e}'\right)_{(S \cup S^n) \Delta T^{n+1}} \right\|_2. \tag{6.51}$$

It remains to put (6.49)–(6.51) together. First combining (6.49) and (6.50), and using the inequality $a^2 + (b + c)^2 \leq (\sqrt{a^2 + b^2} + c)^2$, gives

$$\|\mathbf{x}_S - \mathbf{x}^{n+1}\|_2^2 \le \|(\mathbf{x}_S - \mathbf{u}^{n+1})_{\overline{U^{n+1}}}\|_2^2$$

$$+ 4\left(\frac{\delta_{4s}}{\sqrt{1 - \delta_{4s}^2}} \|(\mathbf{x}_S - \mathbf{u}^{n+1})_{\overline{U^{n+1}}}\|_2 + \frac{1}{1 - \delta_{4s}} \|(\mathbf{A}^*\mathbf{e}')_{U^{n+1}}\|_2\right)^2$$

$$\le \left(\sqrt{\frac{1 + 3\delta_{4s}^2}{1 - \delta_{4s}^2}} \|(\mathbf{x}_S - \mathbf{u}^{n+1})_{\overline{U^{n+1}}}\|_2 + \frac{2}{1 - \delta_{4s}} \|(\mathbf{A}^*\mathbf{e}')_{U^{n+1}}\|_2\right)^2.$$

Next, taking (6.51) into account we obtain

$$\|\mathbf{x}_S - \mathbf{x}^{n+1}\|_2 \le \sqrt{\frac{2\delta_{4s}^2(1 + 3\delta_{4s}^2)}{1 - \delta_{4s}^2}} \|\mathbf{x}^n - \mathbf{x}_S\|_2$$

$$+ \sqrt{\frac{2(1 + 3\delta_{4s}^2)}{1 - \delta_{4s}^2}} \|(\mathbf{A}^*\mathbf{e}')_{(S \cup S^n)\Delta T^{n+1}}\|_2 + \frac{2}{1 - \delta_{4s}} \|(\mathbf{A}^*\mathbf{e}')_{U^{n+1}}\|_2.$$

In view of Lemma 6.20, we conclude that the desired inequality (6.48) holds with

$$\rho = \sqrt{\frac{2\delta_{4s}^2(1 + 3\delta_{4s}^2)}{1 - \delta_{4s}^2}}, \quad (1 - \rho)\tau = \sqrt{\frac{2(1 + 3\delta_{4s}^2)}{1 - \delta_{4s}^2}} + \frac{2\sqrt{1 + \delta_{4s}}}{1 - \delta_{4s}}.$$

The constant ρ is less than 1 if and only if $6\delta_{4s}^4 + 3\delta_{4s}^2 - 1 < 0$. This occurs as soon as δ_{4s}^2 is smaller than the largest root of $6t^2 + 3t - 1$, i.e., as soon as $\delta_{4s}^2 < (\sqrt{11/3} - 1)/4$, which is Condition (6.46).   □

Notes

Candès and Tao introduced the concept of uniform uncertainty principle in [97], which they refined by defining the restricted isometry constants and the restricted orthogonality constants in [96]. In the latter, they proved the inequality $\delta_{s+t} \le \max(\delta_s, \delta_t) + \theta_{s,t}$, which we slightly improved in Proposition 6.5. Some authors define the restricted isometry constants "without squares." For instance, Cohen et al. considered in [123] the smallest $\delta \ge 0$ such that the inequality

$$(1 - \delta)\|\mathbf{x}\|_2 \le \|\mathbf{A}\mathbf{x}\|_2 \le (1 + \delta)\|\mathbf{x}\|_2$$

holds for all s-sparse vectors $\mathbf{x} \in \mathbb{C}^N$. Up to transformation of the constants, this is essentially equivalent to our definition.

Candès and Tao showed in [96] that the condition $\delta_{2s} + \delta_{3s} < 1$ guarantees exact s-sparse recovery via ℓ_1-minimization. Candès et al. further showed in [95] that the condition $\delta_{3s} + 3\delta_{4s} < 2$ guarantees stable and robust s-sparse recovery via ℓ_1-minimization. Later, a sufficient condition for stable and robust s-sparse recovery involving only δ_{2s} was obtained by Candès in [85], namely, $\delta_{2s} < \sqrt{2} - 1 \approx 0.414$. This sufficient condition was improved several times; see [16,81,82,207,212,354]. Exercises 6.13–6.16 retrace some of these improvements. A crucial tool is the *shifting inequality* (see Exercise 6.15) put forward by Cai et al. in [82]. These authors also introduced the square root lifting inequality of Lemma 6.14 in [81]. We obtained the condition $\delta_{2s} < 0.6246$ of Theorem 6.12 by invoking this inequality in the approach of [16]. On the other hand, Davies and Gribonval constructed in [144] matrices with restricted isometry constant δ_{2s} arbitrarily close to $1/\sqrt{2} \approx 0.7071$ for which some s-sparse vectors are not recovered via ℓ_1-minimization. Cai and Zhang proved in [83] that $\delta_s < 1/3$ is another sufficient condition for s-sparse recovery via ℓ_1-minimization and uncovered matrices with $\delta_s = 1/3$ such that some s-sparse vectors are not recovered via ℓ_1-minimization. We point out that other sufficient conditions involving δ_k with $k \neq s$ and $k \neq 2s$ can also be found; see, for instance, Exercises 6.14 and 6.16. As a matter of fact, Blanchard and Thompson argue that the parameters s and $2s$ are not the best choices for Gaussian random matrices; see [54]. Theorem 6.8, which appeared in [208], has to be kept in mind when assessing such conditions.

The use of the iterative hard thresholding algorithm in the context of compressive sensing was initiated by Blumensath and Davies in [56]. In [57], they established stable and robust estimates under the sufficient condition $\delta_{3s} < 1/\sqrt{8}$. The weaker condition $\delta_{3s} < 1/2$ of Theorem 6.15 appeared in [209]. The improved condition $\delta_{3s} < 1/\sqrt{3}$ of Theorem 6.18 was established in the paper [208] dedicated to the analysis of the hard thresholding pursuit algorithm. There, Theorem 6.18 was in fact established for a family of thresholding algorithms indexed by an integer k, with iterative hard thresholding and hard thresholding pursuit corresponding to the cases $k = 0$ and $k = \infty$, respectively. Exercise 6.21, which considers a variation of the iterative hard thresholding algorithm where a factor $\mu \neq 1$ is introduced in front of $\mathbf{A}^*(\mathbf{y} - \mathbf{A}\mathbf{x}^n)$, is inspired by the paper [218] by Garg and Khandekar. This factor μ is allowed to depend on n in some algorithms, notably in the normalized iterative hard thresholding algorithm of Blumensath and Davies [58].

The impossibility of s-sparse recovery via s iterations of orthogonal matching pursuit under a standard restricted isometry condition was first observed in [153, Sect. 7]; see also [409]. The example given at the beginning of Sect. 6.4 is taken from the article [355] by Mo and Shen, who also established the result of Exercise 6.23. The possibility of s-sparse recovery via a number of iterations of orthogonal matching pursuit that is proportional to s was shown in [515] by Zhang, who also proved the stability and robustness of the recovery by establishing Proposition 6.24 with $\bar{n} = 30\,\mathrm{card}(\mathrm{supp}(\mathbf{x}) \setminus S^0)$ under the condition $\delta_{31s} < 1/3$. Our proof follows his argument, which is also valid in more general settings.

In the original article [361] of Needell and Tropp introducing the compressive sampling matching pursuit algorithm, stability and robustness were stated under the condition $\delta_{4s} \leq 0.1$, although the arguments actually yield the condition $\delta_{4s} < 0.17157$. Theorem 6.27, which gives the condition $\delta_{4s} < 0.4782$, appears here for the first time. The first analysis of a greedy algorithm under the restricted isometry property appeared in [362, 363] for the regularized orthogonal matching pursuit algorithm where, however, an additional logarithmic factor appeared in the condition on the restricted isometry constant. The subspace pursuit algorithm of Dai and Milenkovic was also proved to be stable and robust under some restricted isometry conditions. We refer to the original paper [135] for details.

We mentioned at the end of Sect. 6.1 that most explicit (deterministic) constructions of matrices with the restricted isometry property are based on the coherence, hence the number m of measurement scales quadratically in the sparsity s as in (6.14). The sophisticated construction by Bourgain et al. in [64] (see also [65]) is a notable exception. The authors showed that their matrix $\mathbf{A} \in \mathbb{C}^{m \times N}$ has small restricted isometry constant δ_s once $m \geq Cs^{2-\varepsilon}$ and when $s^{2-\varepsilon} \leq N \leq s^{2+\varepsilon}$ for some $\varepsilon > 0$. While this slightly overcomes the quadratic bottleneck, which is certainly an important contribution to the theory, the limited range of s, m, and N is less relevant for practical purposes.

It was shown in [386] that for a given matrix $\mathbf{A} \in \mathbb{C}^{m \times N}$ and a sparsity level s, computing δ_s is an NP-hard problem; see also [28] for an earlier result in this direction.

Exercises

6.1. Suppose that $\mathbf{A} \in \mathbb{C}^{m \times N}$ has an sth order restricted isometry constant satisfying $\delta_s < 1$. Prove that, for any $S \subset [N]$ with card$(S) \leq s$,

$$\frac{1}{1 + \delta_s} \leq \left\| (\mathbf{A}_S^* \mathbf{A}_S)^{-1} \right\|_{2 \to 2} \leq \frac{1}{1 - \delta_s} \quad \text{and} \quad \frac{1}{\sqrt{1 + \delta_s}} \leq \left\| \mathbf{A}_S^\dagger \right\|_{2 \to 2} \leq \frac{1}{\sqrt{1 - \delta_s}}.$$

6.2. Given $\mathbf{A} \in \mathbb{C}^{m \times N}$, let α_s and β_s be the largest and smallest positive constants α and β such that

$$\alpha \left\| \mathbf{x} \right\|_2^2 \leq \left\| \mathbf{A} \mathbf{x} \right\|_2^2 \leq \beta \left\| \mathbf{x} \right\|_2^2$$

for all s-sparse vectors $\mathbf{x} \in \mathbb{C}^N$. Find the scaling factor $t > 0$ for which $\delta_s(t\mathbf{A})$ takes its minimal value, and prove that this value equals $(\beta - \alpha)/(\beta + \alpha)$.

6.3. Find a matrix $\mathbf{A} \in \mathbb{R}^{2 \times 3}$ with minimal 2nd order restricted isometry constant.

6.4. Prove the equivalence of the two definitions (6.4) and (6.5) of restricted orthogonality constants.

6.5. Verify in details that the function f defined on $[0, 1]$ as in (6.6) is first nondecreasing and then nonincreasing.

6.6. Given $\mathbf{x} \in \mathbb{C}^N$ and $\mathbf{A} \in \mathbb{C}^{m \times N}$ with sth restricted isometry constant δ_s, prove that

$$\|\mathbf{A}\mathbf{x}\|_2 \le \sqrt{1+\delta_s}\left(\|\mathbf{x}\|_2 + \frac{\|\mathbf{x}\|_1}{\sqrt{s}}\right).$$

6.7. Let $D_{s,N} = \{\mathbf{x} \in \mathbb{C}^N : \|\mathbf{x}\|_2 \le 1, \|\mathbf{x}\|_0 \le s\}$ be the Euclidean unit ball restricted to the s-sparse vectors. Show that

$$D_{s,N} \subset \text{conv}(D_{s,N}) \subset \sqrt{s}B_1^N \cap B_2^N \subset 2\,\text{conv}(D_{s,N}),$$

where $B_p^N = \{\mathbf{x} \in \mathbb{C}^N : \|\mathbf{x}\|_p \le 1\}$ is the unit ball in ℓ_p and conv denotes the convex hull; see Definition B.2.

6.8. Prove Proposition 6.3 directly from (6.1), without using (6.2) but rather with the help of the *polarization formula*

$$\text{Re}\langle \mathbf{x}, \mathbf{y} \rangle = \frac{1}{4}\left(\|\mathbf{x}+\mathbf{y}\|_2^2 - \|\mathbf{x}-\mathbf{y}\|_2^2\right).$$

6.9. In the case $t = ns$ where t is a multiple of s, improve the second inequality of Proposition 6.6 by showing that

$$\delta_{ns} \le (n-1)\theta_{s,s} + \delta_s.$$

6.10. Let $\mathbf{A} \in \mathbb{C}^{m \times N}$ be a matrix with ℓ_2-normalized columns. Given a vector $\mathbf{x} \in \mathbb{C}^N$ supported on T with $\text{card}(T) = t > s > 1$, prove that

$$\|\mathbf{A}\mathbf{x}\|_2^2 - \|\mathbf{x}\|_2^2 = \frac{1}{\binom{t-2}{s-2}} \sum_{S \subset T, \text{card}(S)=s} \left(\|\mathbf{A}\mathbf{x}_S\|_2^2 - \|\mathbf{x}_S\|_2^2\right)$$

$$= \frac{1}{\binom{t-2}{s-1}} \sum_{S \subset T, \text{card}(S)=s} \langle \mathbf{A}\mathbf{x}_S, \mathbf{A}\mathbf{x}_{T \setminus S} \rangle.$$

Deduce that

$$\delta_t \le \frac{t-1}{s-1}\delta_s, \qquad \delta_t \le \frac{t(t-1)}{2s(t-s)}\theta_{s,t-s}, \qquad \text{and in particular } \delta_{2s} \le 2\,\theta_{s,s}.$$

Extend these inequalities by incorporating δ_1 when the columns of $\mathbf{A}$ are not necessarily ℓ_2-normalized.

6.11. Suppose that the columns of the matrix $\mathbf{A} \in \mathbb{C}^{m \times N}$ are ℓ_2-normalized. Under the assumption $N > s^2+1$, derive the result of Theorem 6.8 with constants $c = 1/2$, $C = 2$, and without restriction on δ_*. Use Theorem 5.8, Exercise 5.3, and compare the matrix norms induced by the ℓ_2 and the ℓ_1 norms.

6.12. Let $\mathbf{A} \in \mathbb{C}^{m \times N}$ with $\delta_{2s} < 1/3$. For $\mathbf{x} \in \mathbb{C}^N$, let $\mathbf{y} = \mathbf{A}\mathbf{x} + \mathbf{e}$ for some $\mathbf{e} \in \mathbb{C}^m$ satisfying $\|\mathbf{e}\|_2 \le \eta$. Refine the proof of Theorem 6.9 in order to establish the stability and robustness of s-sparse recovery via quadratically constrained basis pursuit in the form

$$\|\mathbf{x} - \mathbf{x}^\sharp\|_2 \leq \frac{C}{\sqrt{s}}\sigma_s(\mathbf{x})_1 + D\eta,$$

where $\mathbf{x}^\sharp$ is a minimizer of $\|\mathbf{z}\|_1$ subject to $\|\mathbf{Az} - \mathbf{y}\|_2 \leq \eta$.

6.13. Let $\mathbf{A} \in \mathbb{C}^{m \times N}$, and let $S_0, S_1, S_2, \ldots$ denote index sets of size s ordered by decreasing magnitude of entries of a vector $\mathbf{v} \in \ker \mathbf{A}$. Prove that

$$\|\mathbf{v}_{S_0}\|_2^2 + \|\mathbf{v}_{S_1}\|_2^2 \leq \frac{2\delta_{2s}}{1 - \delta_{2s}} \sum_{k \geq 2} \|\mathbf{v}_{S_k}\|_2 \, (\|\mathbf{v}_{S_0}\|_2 + \|\mathbf{v}_{S_1}\|_2).$$

By interpreting this as the equation of a disk or by completing squares, deduce that

$$\|\mathbf{v}_{S_0}\|_2 \leq \frac{\rho}{\sqrt{s}}\|\mathbf{v}_{\overline{S_0}}\|_1, \quad \text{where } \rho := \frac{1 + \sqrt{2}}{2}\frac{\delta_{2s}}{1 - \delta_{2s}}.$$

Conclude that s-sparse recovery via basis pursuit is guaranteed if $\delta_{2s} < 0.453$.

6.14. For an integer $k \geq 1$, suppose that $\mathbf{A} \in \mathbb{C}^{m \times N}$ has restricted isometry constant $\delta_{(2k+1)s} < 1 - 1/\sqrt{2k}$. Prove that every s-sparse vector $\mathbf{x} \in \mathbb{C}^N$ can be recovered from $\mathbf{y} = \mathbf{Ax} \in \mathbb{C}^m$ via ℓ_1-minimization. [Hint: to establish the null space property, partition $[N]$ as $S \cup T_1 \cup T_2 \cup \ldots$, where S has size s and $T_1, T_2, \ldots$ have size ks.]

6.15. Given $a_1 \geq a_2 \geq \cdots \geq a_{k+\ell} \geq 0$, prove the *shifting inequality*

$$\sqrt{a_{\ell+1}^2 + \cdots + a_{\ell+k}^2} \leq c_{k,\ell}(a_1 + \cdots + a_k), \quad \text{where } c_{k,\ell} := \max\left(\frac{1}{\sqrt{k}}, \frac{1}{\sqrt{4\ell}}\right).$$

6.16. Suppose that $s =: 4r$ is a multiple of 4. For a matrix $\mathbf{A} \in \mathbb{C}^{m \times N}$, establish the success of s-sparse recovery via basis pursuit if $\delta_{5r} + \theta_{5r,s} < 1$. Show in particular that this holds if $\delta_{9s/4} < 0.5$, $\delta_{2s} < 1/(1 + \sqrt{5/4}) \approx 0.472$, or $\delta_{5s/4} < 1/(1 + \sqrt{10/3}) \approx 0.353$.

6.17. Using the square root lifting inequality of Lemma 6.14, find a condition on δ_s that guarantees the exact recovery of every s-sparse vector via basis pursuit.

6.18. Let $\mathbf{A} \in \mathbb{C}^{m \times N}$ with $\delta_{2s} < 1/3$. For $\mathbf{x} \in \mathbb{C}^N$, let $\mathbf{y} = \mathbf{Ax} + \mathbf{e}$ for some $\mathbf{e} \in \mathbb{C}^m$ satisfying $\|\mathbf{A}^*\mathbf{e}\|_\infty \leq \eta$. Let $\mathbf{x}^\sharp$ be a minimizer of the Dantzig selector

$$\underset{\mathbf{z} \in \mathbb{C}^N}{\text{minimize}} \|\mathbf{z}\|_1 \quad \text{subject to } \|\mathbf{A}^*(\mathbf{Az} - \mathbf{y})\|_\infty \leq \eta.$$

Show that

$$\|\mathbf{x} - \mathbf{x}^\#\|_2 \leq \frac{C\sigma_s(\mathbf{x})_1}{\sqrt{s}} + D\sqrt{s}\eta$$

for constants $C, D > 0$ depending only on δ_{2s}.

6.19. Refine the proof of Theorem 6.15 in order to establish the stability and robustness of s-sparse recovery via iterative hard thresholding when $\delta_{3s} < 1/3$.

6.20. Given $\mathbf{A} \in \mathbb{C}^{m \times N}$, prove that every s-sparse vector $\mathbf{x} \in \mathbb{C}^N$ is exactly recovered from $\mathbf{y} = \mathbf{A}\mathbf{x} \in \mathbb{C}^m$ via iterative hard thresholding if $\delta_{2s} < 1/4$. To do so, return to the proof of Theorem 6.15, precisely to (6.27), and separate the contributions to the inner product from the index sets of size $2s$ given by

$$(S \cup S^n) \cap (S \cup S^{n+1}), \quad (S \cup S^n) \setminus (S \cup S^{n+1}), \quad (S \cup S^{n+1}) \setminus (S \cup S^n),$$

where $S := \mathrm{supp}(\mathbf{x})$, $S^n := \mathrm{supp}(\mathbf{x}^n)$, and $S^{n+1} := \mathrm{supp}(\mathbf{x}^{n+1})$.

6.21. Given $\mathbf{A} \in \mathbb{C}^{m \times N}$ and $\mathbf{y} = \mathbf{A}\mathbf{x} \in \mathbb{C}^m$ for some s-sparse $\mathbf{x} \in \mathbb{C}^N$, we define a sequence $(\mathbf{x}^n)$ inductively, starting with an initial s-sparse vector $\mathbf{x}^0 \in \mathbb{C}^N$, by

$$\mathbf{x}^{n+1} = H_s(\mathbf{x}^n + \mu\, \mathbf{A}^*(\mathbf{y} - \mathbf{A}\mathbf{x}^n)), \qquad n \geq 0,$$

where the constant μ is to be determined later. Establish the identity

$$\|\mathbf{A}(\mathbf{x}^{n+1} - \mathbf{x})\|_2^2 - \|\mathbf{A}(\mathbf{x}^n - \mathbf{x})\|_2^2$$
$$= \|\mathbf{A}(\mathbf{x}^{n+1} - \mathbf{x}^n)\|_2^2 + 2\langle \mathbf{x}^n - \mathbf{x}^{n+1}, \mathbf{A}^*\mathbf{A}(\mathbf{x} - \mathbf{x}^n)\rangle.$$

Prove also the inequality

$$2\mu\langle \mathbf{x}^n - \mathbf{x}^{n+1}, \mathbf{A}^*\mathbf{A}(\mathbf{x} - \mathbf{x}^n)\rangle$$
$$\leq \|\mathbf{x}^n - \mathbf{x}\|_2^2 - 2\mu\|\mathbf{A}(\mathbf{x}^n - \mathbf{x})\|_2^2 - \|\mathbf{x}^{n+1} - \mathbf{x}^n\|_2^2.$$

With δ_{2s} denoting the $2s$th order restricted isometry constant of $\mathbf{A}$, derive the inequality

$$\|\mathbf{A}(\mathbf{x}^{n+1} - \mathbf{x})\|_2^2 \leq \left(1 - \frac{1}{\mu(1 + \delta_{2s})}\right)\|\mathbf{A}(\mathbf{x}^{n+1} - \mathbf{x}^n)\|_2^2$$
$$+ \left(\frac{1}{\mu(1 - \delta_{2s})} - 1\right)\|\mathbf{A}(\mathbf{x}^n - \mathbf{x})\|_2^2.$$

Deduce that the sequence $(\mathbf{x}^n)$ converges to $\mathbf{x}$ when $1 + \delta_{2s} < 1/\mu < 2(1 - \delta_{2s})$. Conclude by justifying the choice $\mu = 3/4$ under the condition $\delta_{2s} < 1/3$.

6.22. Verify the claims made at the start of Sect. 6.4 about the matrix $\mathbf{A}$ defined in (6.39).

6.23. Prove that every s-sparse vector $\mathbf{x} \in \mathbb{C}^N$ can be recovered from $\mathbf{y} = \mathbf{A}\mathbf{x} \in \mathbb{C}^m$ via s iterations of orthogonal matching pursuit provided the restricted isometry constant of $\mathbf{A}$ satisfies

$$\delta_{s+1} < \frac{1}{\sqrt{s+1}}.$$

6.24. Improve Proposition 6.24 in the case $\mathbf{e} = \mathbf{0}$ by reducing the number of required iterations and by weakening the restricted isometry condition.

6.25. Rank restricted isometry property
For a linear map $\mathcal{A} : \mathbb{C}^{n_1 \times n_2} \to \mathbb{C}^m$ and for $r \leq n := \min\{n_1, n_2\}$, the *rank restricted isometry constant* $\delta_r = \delta_r(\mathcal{A})$ is the defined as the smallest $\delta \geq 0$ such that

$$(1 - \delta)\|\mathbf{X}\|_F^2 \leq \|\mathcal{A}(\mathbf{X})\|_2^2 \leq (1 + \delta)\|\mathbf{X}\|_F^2$$

for all matrices $\mathbf{X} \in \mathbb{C}^{n_1 \times n_2}$ of rank at most r.

(a) Let $\mathbf{X}, \mathbf{Z} \in \mathbb{C}^{n_1 \times n_2}$ with $\langle \mathbf{X}, \mathbf{Z} \rangle_F = \mathrm{tr}\,(\mathbf{X}\mathbf{Z}^*) = 0$ and $\mathrm{rank}(\mathbf{X}) + \mathrm{rank}(\mathbf{Z}) \leq r$. Show that

$$|\langle \mathcal{A}(\mathbf{X}), \mathcal{A}(\mathbf{Z}) \rangle| \leq \delta_r \|\mathbf{X}\|_F \|\mathbf{Z}\|_F.$$

(b) Assume that $\delta_{2r} < 1/3$. Show that $\mathcal{A}$ possesses the rank null space property of order r defined by (4.40). In particular, every $\mathbf{X} \in \mathbb{C}^{n_1 \times n_2}$ of rank at most r is the unique solution to the nuclear norm minimization problem (see Sect. 4.6)

$$\underset{\mathbf{Z} \in \mathbb{C}^{n_1 \times n_2}}{\text{minimize}} \|\mathbf{Z}\|_* \quad \text{subject to } \mathcal{A}(\mathbf{Z}) = \mathcal{A}(\mathbf{X}).$$

(c) Assume that $\delta_{2r} < 0.6246$. Let $\mathbf{X} \in \mathbb{C}^{n_1 \times n_2}$ and $\mathbf{y} = \mathcal{A}(\mathbf{X}) + \mathbf{e}$ with $\|\mathbf{e}\|_2 \leq \eta$. Let $\mathbf{X}^\sharp$ be the solution to the quadratically constrained nuclear norm minimization problem

$$\underset{\mathbf{Z} \in \mathbb{C}^{n_1 \times n_2}}{\text{minimize}} \|\mathbf{Z}\|_* \quad \text{subject to } \|\mathcal{A}(\mathbf{Z}) - \mathbf{y}\|_2 \leq \eta.$$

Show that

$$\|\mathbf{X} - \mathbf{X}^\sharp\|_F \leq \frac{C}{\sqrt{r}} \sum_{\ell=r+1}^{n} \sigma_\ell(\mathbf{X}) + D\eta$$

for appropriate constants $C, D > 0$ depending only on δ_{2r}.

Chapter 7
Basic Tools from Probability Theory

The major breakthrough in proving recovery results in compressive sensing is obtained using random matrices. Most parts of the remainder of this book indeed require tools from probability theory. This and the next chapter are therefore somewhat exceptional in the sense that they do not deal directly with compressive sensing. Instead, we collect the necessary background material from probability theory. In this chapter, we introduce a first set of tools that will be sufficient to understand a large part of the theory in connection with sparse recovery and random matrices, in particular, the material covered in Sect. 9.1 and the first part of Sect. 9.2, Chap. 11, and Chap. 13. More advanced tools that will be used only in parts of the remainder of the book (remaining parts of Chap. 9 as well as Chaps. 12 and 14) are postponed to Chap. 8.

We only assume that the reader has basic knowledge of probability theory as can be found in most introductory textbooks on the subject. We recall the most basic facts of probability in Sect. 7.1. The relation of moments of random variables to their tails is presented in Sect. 7.2. Then in Sect. 7.3 we study deviation inequalities for sums of independent random variables by means of moment-generating functions. Cramér's theorem gives a very general estimate from which we deduce Hoeffding's inequality and later in Sect. 7.5 Bernstein inequality for bounded and subgaussian random variables. We introduce the latter in Sect. 7.4.

7.1 Essentials from Probability

In this section, we recall some important facts from basic probability theory and prove simple statements that might not be found in all basic textbooks.

Let $(\Omega, \Sigma, \mathbb{P})$ be a probability space, where Σ denotes a σ-algebra on the sample space Ω and $\mathbb{P}$ a probability measure on (Ω, Σ). The probability of an event $B \in \Sigma$ is denoted by

S. Foucart and H. Rauhut, *A Mathematical Introduction to Compressive Sensing*, Applied and Numerical Harmonic Analysis, DOI 10.1007/978-0-8176-4948-7_7, © Springer Science+Business Media New York 2013

$$\mathbb{P}(B) = \int_B d\mathbb{P}(\omega) = \int_\Omega I_B(\omega) d\mathbb{P}(\omega),$$

where the characteristic function $I_B(\omega)$ takes the value 1 if $\omega \in B$ and 0 otherwise. We say that an event B occurs almost surely if $\mathbb{P}(B) = 1$.

The *union bound* (or Bonferroni's inequality or Boole's inequality) states that for a collection of events $B_\ell \in \Sigma$, $\ell \in [n]$, we have

$$\mathbb{P}\left(\bigcup_{\ell=1}^n B_\ell\right) \le \sum_{\ell=1}^n \mathbb{P}(B_\ell). \tag{7.1}$$

A *random variable* X is a real-valued *measurable function* on (Ω, Σ). Recall that X is called measurable if the preimage $X^{-1}(A) = \{\omega \in \Omega : X(\omega) \in A\}$ belongs to Σ for all Borel-measurable subsets $A \subset \mathbb{R}$. Usually, every reasonable function X will be measurable, in particular, all functions appearing in this book. In what follows, we will usually not mention the underlying probability space $(\Omega, \Sigma, \mathbb{P})$ when speaking about random variables. The *distribution function* $F = F_X$ of X is defined as

$$F(t) = \mathbb{P}(X \le t), \quad t \in \mathbb{R}.$$

A random variable X possesses a *probability density function* $\phi : \mathbb{R} \to \mathbb{R}_+$ if

$$\mathbb{P}(a < X \le b) = \int_a^b \phi(t) dt \quad \text{for all } a < b \in \mathbb{R}. \tag{7.2}$$

In this case, $\phi(t) = \dfrac{d}{dt} F(t)$. The *expectation* or mean of a random variable will be denoted by

$$\mathbb{E}X = \int_\Omega X(\omega) d\mathbb{P}(\omega)$$

whenever the integral exists. If X has probability density function ϕ, then for a function $g : \mathbb{R} \to \mathbb{R}$,

$$\mathbb{E}g(X) = \int_{-\infty}^\infty g(t)\phi(t) dt. \tag{7.3}$$

The quantities $\mathbb{E}X^p$ for integer p are called *moments* of X, while $\mathbb{E}|X|^p$, for real-valued $p > 0$, are called *absolute moments*. (Sometimes we may omit "absolute.") The quantity $\mathbb{E}(X - \mathbb{E}X)^2 = \mathbb{E}X^2 - (\mathbb{E}X)^2$ is called *variance*. For $1 \le p < \infty$, $(\mathbb{E}|X|^p)^{1/p}$ defines a norm on the $L_p(\Omega, \mathbb{P})$-space of all p-integrable random variables; in particular, the triangle inequality

$$(\mathbb{E}|X + Y|^p)^{1/p} \le (\mathbb{E}|X|^p)^{1/p} + (\mathbb{E}|Y|^p)^{1/p} \tag{7.4}$$

holds for all p-integrable random variables X, Y on $(\Omega, \Sigma, \mathbb{P})$.

Hölder's inequality states that, for random variables X, Y on a common probability space and $p, q \ge 1$ with $1/p + 1/q = 1$, we have

$$|\mathbb{E}XY| \le (\mathbb{E}|X|^p)^{1/q} (\mathbb{E}|Y|^q)^{1/q}.$$

In the case $p = 1, q = \infty$, the last term is replaced by $\operatorname{ess\,sup}_\omega |Y(\omega)|$, i.e., the smallest constant K such that $|Y| \le K$ almost surely.

The special case $p = q = 2$ is the *Cauchy–Schwarz inequality*

$$|\mathbb{E}XY| \le \sqrt{\mathbb{E}|X|^2 \mathbb{E}|Y|^2}.$$

Since the constant (deterministic) random variable 1 has expectation $\mathbb{E}1 = 1$, Hölder's inequality shows that $\mathbb{E}|X|^p = \mathbb{E}[1 \cdot |X|^p] \le (\mathbb{E}|X|^{pr})^{1/r}$ for all $p > 0, r \ge 1$, and therefore, for all $0 < p \le q < \infty$,

$$(\mathbb{E}|X|^p)^{1/p} \le (\mathbb{E}|X|^q)^{1/q}. \tag{7.5}$$

Let $X_n, n \in \mathbb{N}$, be a sequence of random variables such that X_n converges to X as $n \to \infty$ in the sense that $\lim_{n\to\infty} X_n(\omega) = X(\omega)$ for almost all ω. *Lebesgue's dominated convergence theorem* states that if there exists a random variable Y with $\mathbb{E}|Y| < \infty$ such that $|X_n| \le |Y|$ almost surely, then $\lim_{n\to\infty} \mathbb{E}X_n = \mathbb{E}X$. Lebesgue's dominated convergence theorem has an obvious counterpart for integrals of sequences of functions.

Fubini's theorem on the integration of functions of two variables can be formulated as follows. Let $f : A \times B \to \mathbb{C}$ be measurable, where (A, ν) and (B, μ) are measurable spaces. If $\int_{A\times B} |f(x, y)| d(\nu \otimes \mu)(x, y) < \infty$, then

$$\int_A \left(\int_B f(x, y) d\mu(y) \right) d\nu(x) = \int_B \left(\int_A f(x, y) d\nu(x) \right) d\mu(y).$$

A formulation for expectations of functions of independent random vectors is provided below in (7.14).

Absolute moments can be computed by means of the following formula.

Proposition 7.1. *The absolute moments of a random variable X can be expressed as*

$$\mathbb{E}|X|^p = p \int_0^\infty \mathbb{P}(|X| \ge t) t^{p-1} dt, \quad p > 0.$$

Proof. Recall that $I_{\{|X|^p \ge x\}}$ is the random variable that takes the value 1 on the event $|X|^p \ge x$ and 0 otherwise. Using Fubini's theorem, we derive

$$\mathbb{E}|X|^p = \int_\Omega |X|^p d\mathbb{P} = \int_\Omega \int_0^{|X|^p} 1 dx d\mathbb{P} = \int_\Omega \int_0^\infty I_{\{|X|^p \geq x\}} dx d\mathbb{P}$$

$$= \int_0^\infty \int_\Omega I_{\{|X|^p \geq x\}} d\mathbb{P} dx = \int_0^\infty \mathbb{P}(|X|^p \geq x) dx$$

$$= p \int_0^\infty \mathbb{P}(|X|^p \geq t^p) t^{p-1} dt = p \int_0^\infty \mathbb{P}(|X| \geq t) t^{p-1} dt,$$

where we also applied a change of variables.   □

Corollary 7.2. *For a random variable X, the expectation satisfies*

$$\mathbb{E}X = \int_0^\infty \mathbb{P}(X \geq t) dt - \int_0^\infty \mathbb{P}(X \leq -t) dt.$$

Proof. We can write $X = X I_{\{X \in [0,\infty)\}} + X I_{\{X \in (-\infty,0)\}}$, so that

$$\mathbb{E}X = \mathbb{E}X I_{\{X \in [0,\infty)\}} - \mathbb{E}(-X I_{\{-X \in (0,\infty)\}}).$$

Both $X I_{\{X \in [0,\infty)\}}$ and $-X I_{\{-X \in (0,\infty)\}}$ are nonnegative random variables, so that an application of Proposition 7.1 shows the statement.

The function $t \mapsto \mathbb{P}(|X| \geq t)$ is called the *tail* of X. A simple but often effective tool to estimate the tail by expectations and moments is *Markov's inequality*.

Theorem 7.3. *Let X be a random variable. Then*

$$\mathbb{P}(|X| \geq t) \leq \frac{\mathbb{E}|X|}{t} \quad \text{for all } t > 0.$$

Proof. Note that $\mathbb{P}(|X| \geq t) = \mathbb{E}I_{\{|X| \geq t\}}$ and that $t I_{\{|X| \geq t\}} \leq |X|$. Hence, $t \mathbb{P}(|X| \geq t) = \mathbb{E}t I_{\{|X| \geq t\}} \leq \mathbb{E}|X|$ and the proof is complete.   □

Remark 7.4. As an important consequence we note that for, $p > 0$,

$$\mathbb{P}(|X| \geq t) = \mathbb{P}(|X|^p \geq t^p) \leq t^{-p} \mathbb{E}|X|^p \quad \text{for all } t > 0.$$

The special case $p = 2$ with X replaced by $X - \mathbb{E}X$ is referred to as the Chebyshev inequality. Similarly, for $\theta > 0$, we obtain

$$\mathbb{P}(X \geq t) = \mathbb{P}(\exp(\theta X) \geq \exp(\theta t)) \leq \exp(-\theta t) \mathbb{E}\exp(\theta X) \quad \text{for all } t \in \mathbb{R}.$$

The function $\theta \mapsto \mathbb{E}\exp(\theta X)$ is usually called the *Laplace transform* or the *moment-generating function* of X.

A *median* of a random variable X is a number M such that

$$\mathbb{P}(X \geq M) \geq 1/2 \quad \text{and} \quad \mathbb{P}(X \leq M) \geq 1/2.$$

The *binomial distribution* is the discrete probability distribution counting the number of successes in a sequence of N independent experiments where the probability of each individual success is p. If X has the binomial distribution, then

$$\mathbb{P}(X = k) = \binom{N}{k} p^k (1 - p)^{N-k}.$$

The expectation of X is given by $\mathbb{E}X = pN$. If pN is an integer, then the median $M = M(X)$ coincides with the expectation:

$$M(X) = pN. \tag{7.6}$$

A *normally distributed* random variable or *Gaussian random variable* X has probability density function

$$\psi(t) = \frac{1}{\sqrt{2\pi\sigma^2}} \exp\left(-\frac{(t - \mu)^2}{2\sigma^2}\right). \tag{7.7}$$

It has mean $\mathbb{E}X = \mu$ and variance $\mathbb{E}(X - \mu)^2 = \sigma^2$. A Gaussian random variable g with $\mathbb{E}g = 0$ and $\mathbb{E}g^2 = 1$ is called a *standard Gaussian* random variable, a standard normal variable, or simply a standard Gaussian. Its tail satisfies the following simple estimates.

Proposition 7.5. *Let g be a standard Gaussian random variable. Then, for all $u > 0$,*

$$\mathbb{P}(|g| \geq u) \leq \exp\left(-\frac{u^2}{2}\right), \tag{7.8}$$

$$\mathbb{P}(|g| \geq u) \leq \sqrt{\frac{2}{\pi}} \frac{1}{u} \exp\left(-\frac{u^2}{2}\right), \tag{7.9}$$

$$\mathbb{P}(|g| \geq u) \geq \sqrt{\frac{2}{\pi}} \frac{1}{u} \left(1 - \frac{1}{u^2}\right) \exp\left(-\frac{u^2}{2}\right),$$

$$\mathbb{P}(|g| \geq u) \geq \left(1 - \sqrt{\frac{2}{\pi}} u\right) \exp\left(-\frac{u^2}{2}\right).$$

Proof. By (7.7), we have

$$\mathbb{P}(|g| \geq u) = \frac{2}{\sqrt{2\pi}} \int_u^\infty e^{-t^2/2} dt. \tag{7.10}$$

Therefore, the stated estimates follow from Lemmas C.7 and C.8. $\qquad \square$

Let us compute the moment-generating function of a standard Gaussian.

Lemma 7.6. *Let g be a standard Gaussian random variable. Then, for $\theta \in \mathbb{R}$,*

$$\mathbb{E}\exp(\theta g) = \exp\left(\theta^2/2\right), \tag{7.11}$$

and more generally, for $\theta \in \mathbb{R}$ and $a < 1/2$,

$$\mathbb{E}\exp(ag^2 + \theta g) = \frac{1}{\sqrt{1 - 2a}}\exp\left(\frac{\theta^2}{2(1 - 2a)}\right).$$

Proof. For $\theta, a \in \mathbb{R}$, we have

$$\mathbb{E}\left(\exp(ag^2 + \theta g)\right) = \frac{1}{\sqrt{2\pi}}\int_{-\infty}^{\infty}\exp(ax^2 + \theta x)\exp\left(-x^2/2\right)\,dx$$

Noting the identity

$$ax^2 - x^2/2 + \theta x = -\frac{1 - 2a}{2}\left(x - \frac{\theta}{1 - 2a}\right)^2 + \frac{\theta^2}{2(1 - 2a)}$$

and applying a change of variable, we obtain

$$\mathbb{E}\exp(ag^2 + \theta g) = \exp\left(\frac{\theta^2}{2(1 - 2a)}\right)\frac{1}{\sqrt{2\pi}}\int_{-\infty}^{\infty}\exp\left(-\frac{1 - 2a}{2}x^2\right)\,dx$$

$$= \exp\left(\frac{\theta^2}{2(1 - 2a)}\right)\frac{1}{\sqrt{1 - 2a}}.$$

Hereby, we have used the fact that the probability density function (7.7) with $\mu = 1$ and $\sigma = 1/\sqrt{1 - 2a}$ integrates to 1. $\qquad\square$

The proof of the next result should explain the terminology *moment-generating function*.

Corollary 7.7. *The even moments of a standard Gaussian random variable g are given by*

$$\mathbb{E}g^{2n} = \frac{(2n)!}{2^n n!}, \quad n \in \mathbb{N}.$$

Proof. On the one hand, by Taylor expansion, we can write the moment-generating function as

$$\mathbb{E}\exp(\theta g) = \sum_{j=0}^{\infty}\frac{\theta^j \mathbb{E}[g^j]}{j!} = \sum_{n=0}^{\infty}\frac{\theta^{2n}\mathbb{E}g^{2n}}{(2n)!},$$

where we have used the fact that $\mathbb{E}g^j = 0$ for all odd j. On the other hand, Lemma 7.6 gives

$$\mathbb{E}\exp(\theta g) = \exp(\theta^2/2) = \sum_{n=0}^{\infty} \frac{\theta^{2n}}{2^n n!}.$$

Comparing coefficients gives

$$\frac{\mathbb{E}g^{2n}}{(2n)!} = \frac{1}{2^n n!},$$

which is equivalent to the claim.   □

A *random vector* $\mathbf{X} = [X_1, \ldots, X_n]^\top \in \mathbb{R}^n$ is a collection of n random variables on a common probability space $(\Omega, \Sigma, \mathbb{P})$. Its expectation is the vector $\mathbb{E}\mathbf{X} = [\mathbb{E}X_1, \ldots, \mathbb{E}X_n]^\top \in \mathbb{R}^n$, while its *joint distribution function* is defined as

$$F(t_1, \ldots, t_n) = \mathbb{P}(X_1 \leq t_1, \ldots, X_n \leq t_n), \quad t_1, \ldots, t_n \in \mathbb{R}.$$

Similarly to the univariate case, the random vector $\mathbf{X}$ has a *joint probability density* if there exists a function $\phi : \mathbb{R}^n \to \mathbb{R}_+$ such that, for any measurable domain $D \subset \mathbb{R}^n$,

$$\mathbb{P}(\mathbf{X} \in D) = \int_D \phi(t_1, \ldots, t_n) dt_1 \cdots dt_n.$$

A *complex random vector* $\mathbf{Z} = \mathbf{X} + i\mathbf{Y} \in \mathbb{C}^n$ is a special case of a $2n$-dimensional real random vector $(\mathbf{X}, \mathbf{Y}) \in \mathbb{R}^{2n}$.

A collection of random variables $X_1, \ldots, X_n$ is (stochastically) independent if for all $t_1, \ldots, t_n \in \mathbb{R}$,

$$\mathbb{P}(X_1 \leq t_1, \ldots, X_n \leq t_n) = \prod_{\ell=1}^{n} \mathbb{P}(X_\ell \leq t_\ell).$$

For independent random variables, we have

$$\mathbb{E}\left[\prod_{\ell=1}^{n} X_\ell\right] = \prod_{\ell=1}^{n} \mathbb{E}[X_\ell]. \tag{7.12}$$

If they have a joint probability density function ϕ, then the latter factorizes as

$$\phi(t_1, \ldots, t_n) = \phi_1(t_1) \times \cdots \times \phi_n(t_n)$$

where the $\phi_1, \ldots, \phi_n$ are the probability density functions of $X_1, \ldots, X_n$.

In generalization, a collection $\mathbf{X}_1 \in \mathbb{R}^{n_1}, \ldots, \mathbf{X}_m \in \mathbb{R}^{n_m}$ of random vectors are independent if for any collection of measurable sets $A_\ell \subset \mathbb{R}^{n_\ell}$, $\ell \in [m]$,

$$\mathbb{P}(\mathbf{X}_1 \in A_1, \ldots, \mathbf{X}_m \in A_m) = \prod_{\ell=1}^{m} \mathbb{P}(\mathbf{X}_\ell \in A_\ell).$$

If furthermore $f_\ell : \mathbb{R}^{n_\ell} \to \mathbb{R}^{N_\ell}$, $\ell \in [m]$, are measurable functions, then also the random vectors $f_1(\mathbf{X}_1), \ldots, f_m(\mathbf{X}_m)$ are independent. A collection $\mathbf{X}_1, \ldots, \mathbf{X}_m \in \mathbb{R}^n$ of independent random vectors that all have the same distribution is called *independent identically distributed* (i.i.d.).

A random vector $\mathbf{X}'$ will be called an independent copy of $\mathbf{X}$ if $\mathbf{X}$ and $\mathbf{X}'$ are independent and have the same distribution.

The sum $X + Y$ of two independent random variables X, Y having probability density functions ϕ_X, ϕ_Y has probability density function ϕ_{X+Y} given by the convolution

$$\phi_{X+Y}(t) = (\phi_X * \phi_Y)(t) = \int_{-\infty}^{\infty} \phi_X(u)\phi_Y(t-u)du. \qquad (7.13)$$

Fubini's theorem for expectations takes the following form. Let $\mathbf{X}, \mathbf{Y} \in \mathbb{R}^n$ be two independent random vectors (or simply random variables) and $f : \mathbb{R}^n \times \mathbb{R}^n \to \mathbb{R}$ be a measurable function such that $\mathbb{E}|f(\mathbf{X}, \mathbf{Y})| < \infty$. Then the functions

$$f_1 : \mathbb{R}^n \to \mathbb{R}, \ f_1(\mathbf{x}) = \mathbb{E}f(\mathbf{x}, \mathbf{Y}), \qquad f_2 : \mathbb{R}^n \to \mathbb{R}, \ f_2(\mathbf{y}) = \mathbb{E}f(\mathbf{X}, \mathbf{y})$$

are measurable ($\mathbb{E}|f_1(\mathbf{X})| < \infty$ and $\mathbb{E}|f_2(\mathbf{Y})| < \infty$) and

$$\mathbb{E}f_1(\mathbf{X}) = \mathbb{E}f_2(\mathbf{Y}) = \mathbb{E}f(\mathbf{X}, \mathbf{Y}). \qquad (7.14)$$

The random variable $f_1(\mathbf{X})$ is also called conditional expectation or expectation conditional on $\mathbf{X}$ and will sometimes be denoted by $\mathbb{E}_Y f(\mathbf{X}, \mathbf{Y})$.

A random vector $\mathbf{g} \in \mathbb{R}^n$ is called a *standard Gaussian vector* if its components are independent standard normally distributed random variables. More generally, a random vector $\mathbf{X} \in \mathbb{R}^n$ is said to be a Gaussian vector or multivariate normally distributed if there exists a matrix $\mathbf{A} \in \mathbb{R}^{n \times k}$ such that $\mathbf{X} = \mathbf{A}\mathbf{g} + \boldsymbol{\mu}$, where $\mathbf{g} \in \mathbb{R}^k$ is a standard Gaussian vector and $\boldsymbol{\mu} \in \mathbb{R}^n$ is the mean of $\mathbf{X}$. The matrix $\boldsymbol{\Sigma} = \mathbf{A}\mathbf{A}^*$ is then the covariance matrix of $\mathbf{X}$, i.e., $\boldsymbol{\Sigma} = \mathbb{E}(\mathbf{X} - \boldsymbol{\mu})(\mathbf{X} - \boldsymbol{\mu})^*$. If $\boldsymbol{\Sigma}$ is nondegenerate, i.e., $\boldsymbol{\Sigma}$ is positive definite, then $\mathbf{X}$ has a joint probability density function of the form

$$\psi(\mathbf{x}) = \frac{1}{(2\pi)^{n/2}\sqrt{\det(\boldsymbol{\Sigma})}} \exp\left(-\frac{1}{2}\langle \mathbf{x} - \boldsymbol{\mu}, \boldsymbol{\Sigma}^{-1}(\mathbf{x} - \boldsymbol{\mu})\rangle\right), \quad \mathbf{x} \in \mathbb{R}^n.$$

In the degenerate case when $\boldsymbol{\Sigma}$ is not invertible, $\mathbf{X}$ does not have a density. It is easily deduced from the density that a rotated standard Gaussian $\mathbf{U}\mathbf{g}$, where $\mathbf{U}$ is an orthogonal matrix, has the same distribution as $\mathbf{g}$ itself.

If $X_1, \ldots, X_n$ are independent and normally distributed random variables with means $\mu_1, \ldots, \mu_n$ and variances $\sigma_1^2, \ldots, \sigma_n^2$, then $\mathbf{X} = [X_1, \ldots, X_n]^\top$ has a multivariate normal distribution and the sum $Z = \sum_{\ell=1}^{n} X_\ell$ has the univariate normal distribution with mean $\mu = \sum_{\ell=1}^{n} \mu_\ell$ and variance $\sigma^2 = \sum_{\ell=1}^{n} \sigma_\ell^2$, as can be calculated from (7.13).

The central limit theorem recalled below highlights the importance of the normal distribution.

Theorem 7.8. *Let $(X_j)_{j \in \mathbb{N}}$ be a sequence of independent and identically distributed random variables with mean $\mathbb{E}X_j = \mu$ and variance $\mathbb{E}(X_j - \mu)^2 = \sigma^2$. Then the sequence of random variables*

$$Y_n := \frac{\sum_{j=1}^{n} (X_j - \mu)}{\sigma \sqrt{n}}$$

converges in distribution to a standard Gaussian random variable g, i.e., $\lim_{n \to \infty} \mathbb{E}f(Y_n) = \mathbb{E}f(g)$ for all bounded continuous functions $f : \mathbb{R} \to \mathbb{R}$.

The next statement is concerned with another important distribution derived from the normal distribution.

Lemma 7.9. *Let $\mathbf{g} = [g_1, \ldots, g_n]^\top$ be a standard Gaussian vector. Then the random variable*

$$Z = \sum_{\ell=1}^{n} g_\ell^2$$

has the $\chi^2(n)$-distribution whose probability density function ϕ_n is given by

$$\phi_n(u) = \frac{1}{2^{n/2} \Gamma(n/2)} u^{(n/2)-1} \exp(-u/2) I_{(0,\infty)}(u), \quad \text{for all } u \in \mathbb{R}, \quad (7.15)$$

where Γ is the Gamma function (see Appendix C.3).

Proof. We proceed by induction on n. The distribution function of a scalar squared standard Gaussian g^2 is given by $\mathbb{P}(g^2 \leq u) = 0$ for $u < 0$ and $\mathbb{P}(g^2 \leq u) = \mathbb{P}(-\sqrt{u} \leq g \leq \sqrt{u}) = F(\sqrt{u}) - F(-\sqrt{u})$ for $u \geq 0$, where F is the distribution function of g. If ψ denotes the corresponding probability density function, it follows that the probability density ϕ_1 of the random variable g^2 is given for $u < 0$ by $\phi_1 = 0$ and for $u \geq 0$ by

$$\phi_1(u) = \frac{d}{du} \left(F(\sqrt{u}) - F(-\sqrt{u}) \right) = \frac{1}{2} u^{-1/2} \psi(\sqrt{u}) + \frac{1}{2} u^{-1/2} \psi(-\sqrt{u})$$

$$= \frac{1}{\sqrt{2\pi}} u^{-1/2} e^{-u/2}.$$

Hence, for $n = 1$, (7.15) is established since $\Gamma(1/2) = \sqrt{\pi}$.

Now assume that the formula (7.15) has already been established for $n \geq 1$. For $u \leq 0$, we have $\phi_{n+1}(u) = 0$, and for $u > 0$, since by (7.13), the probability

density function of the sum of independent random variables is the convolution of
their probability density functions, we have

$$
\begin{aligned}
\phi_{n+1}(u) = (\phi_n * \phi_1)(u) &= \int_{-\infty}^{\infty} \phi_n(t)\phi_1(u-t)dt \\
&= \frac{1}{2^{n/2+1/2}\Gamma(n/2)\Gamma(1/2)} \int_0^{\infty} t^{(n/2)-1}e^{-t/2}(u-t)^{-1/2}e^{-(u-t)/2}I_{(0,\infty)}(u-t)dt \\
&= \frac{1}{2^{(n+1)/2}\Gamma(1/2)\Gamma(n/2)}e^{-u/2}\int_0^u t^{(n/2)-1}(u-t)^{-1/2}dt \\
&= \frac{1}{2^{(n+1)/2}\Gamma(1/2)\Gamma(n/2)}e^{-u/2}u^{(n/2)-1/2}\int_0^1 t^{(n/2)-1}(1-t)^{-1/2}dt \\
&= \frac{1}{2^{(n+1)/2}\Gamma(1/2)\Gamma(n/2)}e^{-u/2}u^{(n+1)/2-1}B(n/2,1/2) \\
&= \frac{1}{2^{(n+1)/2}\Gamma((n+1)/2)}u^{(n+1)/2-1}e^{-u/2},
\end{aligned}
$$

where we used that the Beta function B satisfies

$$
B(x,y) := \int_0^1 u^{x-1}(1-u)^{y-1}du = \frac{\Gamma(x)\Gamma(y)}{\Gamma(x+y)}, \qquad x, y > 0, \qquad (7.16)
$$

see Exercise 7.1. Thus, we proved the formula (7.15) for $n+1$. This completes the
proof by induction. □

Jensen's inequality reads as follows.

Theorem 7.10. *Let $f : \mathbb{R}^n \to \mathbb{R}$ be a convex function and let $\mathbf{X} \in \mathbb{R}^n$ be a random
vector. Then*

$$
f(\mathbb{E}\mathbf{X}) \le \mathbb{E}f(\mathbf{X}).
$$

Proof. Let $\mathbf{v}$ be an element of the subdifferential $\partial f(\mathbb{E}\mathbf{X})$; see Definition B.20.
(Note that the subdifferential of a convex function is always nonempty at every
point.) By definition of ∂f, we have, for any realization of $\mathbf{X}$,

$$
f(\mathbb{E}\mathbf{X}) \le f(\mathbf{X}) + \langle \mathbf{v}, \mathbb{E}\mathbf{X} - \mathbf{X} \rangle.
$$

Taking expectations on both sides of this inequality gives the statement by noting
that $\mathbb{E}[\mathbb{E}\mathbf{X} - \mathbf{X}] = 0$. □

Note that $-f$ is convex if f is concave, so that for concave functions f, Jensen's
inequality reads

$$
\mathbb{E}f(\mathbf{X}) \le f(\mathbb{E}\mathbf{X}). \qquad (7.17)
$$

This concludes our outline of basic facts of probability theory.

7.2 Moments and Tails

Moment and tail estimates of random variables are intimately related. We start with a simple statement in this direction.

Proposition 7.11. *Suppose that Z is a random variable satisfying*

$$(\mathbb{E}|Z|^p)^{1/p} \le \alpha \beta^{1/p} p^{1/\gamma} \quad \text{for all } p \in [p_0, p_1] \tag{7.18}$$

for some constants $\alpha, \beta, \gamma, p_1 > p_0 > 0$. Then

$$\mathbb{P}(|Z| \ge e^{1/\gamma} \alpha u) \le \beta e^{-u^\gamma/\gamma} \text{ for all } u \in [p_0^{1/\gamma}, p_1^{1/\gamma}].$$

Proof. By Markov's inequality (Theorem 7.3), we obtain for an arbitrary $\kappa > 0$

$$\mathbb{P}(|Z| \ge e^\kappa \alpha u) \le \frac{\mathbb{E}|Z|^p}{(e^\kappa \alpha u)^p} \le \beta \left(\frac{\alpha p^{1/\gamma}}{e^\kappa \alpha u} \right)^p.$$

Choosing $p = u^\gamma$ yields $\mathbb{P}(|Z| \ge e^\kappa \alpha u) \le \beta e^{-\kappa u^\gamma}$ and further setting $\kappa = 1/\gamma$ yields the claim. □

Remark 7.12. Important special cases of Proposition 7.11 are $\gamma = 1, 2$. Indeed, if $(\mathbb{E}|Z|^p)^{1/p} \le \alpha \beta^{1/p} \sqrt{p}$ for all $p \ge 2$, then

$$\mathbb{P}(|Z| \ge e^{1/2} \alpha u) \le \beta e^{-u^2/2} \quad \text{for all } u \ge \sqrt{2} ; \tag{7.19}$$

while if $(\mathbb{E}|Z|^p)^{1/p} \le \alpha \beta^{1/p} p$ for all $p \ge 2$, then

$$\mathbb{P}(|Z| \ge e \alpha u) \le \beta e^{-u} \quad \text{for all } u \ge 2. \tag{7.20}$$

If one replaces β by $\beta' = \max\{\beta, e^{2/\gamma}\}$, $\gamma = 1, 2$, on the right-hand sides of (7.19) and (7.20), then the inequalities hold for all $u \ge 0$, since for $u < 2^{1/\gamma}$, they become trivial, i.e., the right-hand sides become larger than 1.

A converse to Proposition 7.11 involving the Gamma function Γ (see Appendix C.3) also holds.

Proposition 7.13. *Suppose that a random variable Z satisfies, for some $\gamma > 0$,*

$$\mathbb{P}(|Z| \ge e^{1/\gamma} \alpha u) \le \beta e^{-u^\gamma/\gamma} \quad \text{for all } u > 0.$$

Then, for $p > 0$,

$$\mathbb{E}|Z|^p \le \beta \alpha^p (e\gamma)^{p/\gamma} \Gamma\left(\frac{p}{\gamma} + 1 \right). \tag{7.21}$$

As a consequence, for $p \geq 1$,

$$(\mathbb{E}|Z|^p)^{1/p} \leq C_1 \alpha (C_{2,\gamma}\beta)^{1/p} p^{1/\gamma} \quad \text{for all } p \geq 1, \tag{7.22}$$

where $C_1 = e^{1/(2e)} \approx 1.2019$ and $C_{2,\gamma} = \sqrt{2\pi/\gamma}e^{\gamma/12}$. In particular, one has $C_{2,1} \approx 2.7245$, $C_{2,2} \approx 2.0939$.

Proof. Using Proposition 7.1 and two changes of variables, we obtain

$$\mathbb{E}|Z|^p = p \int_0^\infty \mathbb{P}(|Z| > t) t^{p-1} dt = p\alpha^p e^{p/\gamma} \int_0^\infty \mathbb{P}(|Z| \geq e^{1/\gamma}\alpha u) u^{p-1} du$$

$$\leq p\alpha^p e^{p/\gamma} \int_0^\infty \beta e^{-u^\gamma/\gamma} u^{p-1} du = p\beta\alpha^p e^{p/\gamma} \int_0^\infty e^{-v}(\gamma v)^{p/\gamma-1} dv$$

$$= \beta\alpha^p (e\gamma)^{p/\gamma} \frac{p}{\gamma} \Gamma\left(\frac{p}{\gamma}\right). \tag{7.23}$$

This shows (7.21) taking into account the functional equation for the Gamma function. Applying Stirling's formula (C.12) yields

$$\mathbb{E}|Z|^p \leq \beta\alpha^p (e\gamma)^{p/\gamma} \sqrt{2\pi} \left(\frac{p}{\gamma}\right)^{p/\gamma+1/2} e^{-p/\gamma} e^{\gamma/(12p)}$$

$$= \sqrt{2\pi}\beta\alpha^p e^{\gamma/(12p)} p^{p/\gamma+1/2} \gamma^{-1/2}.$$

Using the assumption $p \geq 1$, we obtain

$$(\mathbb{E}|Z|^p)^{1/p} \leq \left(\frac{\sqrt{2\pi}e^{\gamma/12}}{\sqrt{\gamma}}\beta\right)^{1/p} \alpha p^{1/\gamma} p^{1/(2p)}.$$

Finally, $p^{1/(2p)}$ takes its maximum value for $p = e$, i.e., $p^{1/(2p)} \leq e^{1/(2e)}$. This yields the statement of the proposition.   □

Next we consider the expectation $\mathbb{E}|Z|$ of a random variable Z satisfying a subgaussian tail estimate (see (7.31) below) and improve on the general estimate (7.22) for $p = 1$.

Proposition 7.14. *Let Z be a random variable satisfying*

$$\mathbb{P}(|Z| \geq \alpha u) \leq \beta e^{-u^2/2} \quad \text{for all } u \geq \sqrt{2\ln(\beta)},$$

for some constants $\alpha > 0, \beta \geq 2$. Then

$$\mathbb{E}|Z| \leq C_\beta \alpha \sqrt{\ln(4\beta)}$$

with $C_\beta = \sqrt{2} + 1/(4\sqrt{2}\ln(4\beta)) \leq \sqrt{2} + 1/(4\sqrt{2}\ln(8)) \approx 1.499 < 3/2$.

Proof. Let $\kappa \geq \sqrt{2\ln(\beta)}$ be some number to be chosen later. By Proposition 7.11, the expectation can be expressed as

$$\mathbb{E}|Z| = \int_0^\infty \mathbb{P}(|Z| \geq u)du = \alpha \int_0^\infty \mathbb{P}(|Z| \geq \alpha u)du$$

$$\leq \alpha \left(\int_0^\kappa 1 du + \beta \int_\kappa^\infty e^{-u^2/2} du \right) \leq \alpha \left(\kappa + \frac{\beta}{\kappa} e^{-\kappa^2/2} \right).$$

In the second line, we used the fact that any probability is bounded by 1, and in the last step, we applied Lemma C.7. Choosing $\kappa = \sqrt{2\ln(4\beta)}$ completes the proof. $\qquad \square$

Let us also provide a slight variation on Proposition 7.11.

Proposition 7.15. *Suppose Z is a random variable satisfying*

$$(\mathbb{E}|Z|^p)^{1/p} \leq \beta^{1/p}(\alpha_1 p + \alpha_2\sqrt{p} + \alpha_3) \quad \text{for all } p \geq p_0.$$

Then, for $u \geq p_0$,

$$\mathbb{P}\big(|Z| \geq e(\alpha_1 u + \alpha_2\sqrt{u} + \alpha_3)\big) \leq \beta e^{-u}.$$

Proof. The proof is basically the same as the one of Proposition 7.11 and left as Exercise 7.15. $\qquad \square$

Tail probabilities can also be bounded from below using moments. We start with the classical *Paley–Zygmund* inequality.

Lemma 7.16. *If a nonnegative random variable Z has finite second moment, then*

$$\mathbb{P}(Z > t) \geq \frac{(\mathbb{E}Z - t)^2}{\mathbb{E}Z^2}, \qquad 0 \leq t \leq \mathbb{E}Z.$$

Proof. For $t \geq 0$, the Cauchy–Schwarz inequality yields

$$\mathbb{E}Z = \mathbb{E}[ZI_{\{Z>t\}}] + \mathbb{E}[ZI_{\{Z\leq t\}}]$$

$$\leq (\mathbb{E}Z^2)^{1/2} \mathbb{E}(I_{\{Z>t\}})^{1/2} + t = (\mathbb{E}Z^2)^{1/2} \mathbb{P}(Z > t)^{1/2} + t.$$

With $t \leq \mathbb{E}Z$, this is a rearrangement of the claim. $\qquad \square$

Lemma 7.17. *If $X_1, \ldots, X_n$ are independent mean-zero random variables with variance σ^2 and fourth moment bounded from above by μ^4, then, for all $\mathbf{a} \in \mathbb{R}^n$,*

$$\mathbb{P}\left(\left|\sum_{\ell=1}^n a_\ell X_\ell\right| > t\|\mathbf{a}\|_2\right) \geq \frac{(\sigma^2 - t^2)^2}{\mu^4}, \qquad 0 \leq t \leq \sigma.$$

Proof. Setting $Z := \left(\sum_{\ell=1}^n a_\ell X_\ell\right)^2$, independence and the mean-zero assumption yield

$$\mathbb{E}Z = \mathbb{E}\left(\sum_{j=1}^n a_\ell X_\ell\right)^2 = \sum_{\ell=1}^n a_\ell^2 \mathbb{E}X_\ell^2 = \|\mathbf{a}\|_2^2 \sigma^2,$$

$$\mathbb{E}Z^2 = \mathbb{E}\left(\sum_{\ell=1}^n a_\ell X_\ell\right)^4 = \sum_{i,j,k,\ell \in [n]} a_i a_j a_k a_\ell \mathbb{E}(X_i X_j X_k X_\ell)$$

$$= \sum_{i,j \in [n]} a_i^2 a_j^2 \mathbb{E}(X_i^2 X_j^2), \tag{7.24}$$

because if a random variable X_i is not repeated in the product $X_i X_j X_k X_\ell$, then the independence of X_i, X_j, X_k, X_ℓ yields $\mathbb{E}(X_i X_j X_k X_\ell) = \mathbb{E}(X_i)\mathbb{E}(X_j X_k X_\ell) = 0$. Moreover, using the Cauchy–Schwarz inequality, we have, for $i, j \in [n]$,

$$\mathbb{E}(X_i^2 X_j^2) \leq \mathbb{E}(X_i^4)^{1/2} \mathbb{E}(X_j^4)^{1/2} \leq \mu^4.$$

We deduce that

$$\mathbb{E}Z^2 \leq \sum_{i,j \in [n]} a_i^2 a_j^2 \mu^4 = \|\mathbf{a}\|_2^4 \mu^4. \tag{7.25}$$

Substituting (7.24) and (7.25) into Lemma 7.16, we obtain, for $0 \leq t \leq \sigma$,

$$\mathbb{P}\left(\left|\sum_{\ell=1}^n a_\ell X_\ell\right| > t\|\mathbf{a}\|_2\right) = \mathbb{P}(Z > t^2\|\mathbf{a}\|_2^2) \geq \frac{(\sigma^2 - t^2)^2}{\mu^4},$$

which is the desired result.   □

7.3 Cramér's Theorem and Hoeffding's Inequality

We often encounter sums of independent mean-zero random variables. Deviation inequalities bound the tail of such sums.

We recall that the *moment-generating function* of a (real-valued) random variable X is defined by

$$\theta \mapsto \mathbb{E}\exp(\theta X),$$

for all $\theta \in \mathbb{R}$ whenever the expectation on the right-hand side is well defined. Its logarithm is the *cumulant-generating function*

$$C_X(\theta) = \ln \mathbb{E}\exp(\theta X).$$

With the help of these definitions, we can formulate Cramér's theorem.

Theorem 7.18. *Let $X_1, \ldots, X_M$ be a sequence of independent (real-valued) random variables with cumulant-generating functions C_{X_ℓ}, $\ell \in [M]$. Then, for $t > 0$,*

$$\mathbb{P}\Big(\sum_{\ell=1}^M X_\ell \geq t\Big) \leq \exp\Big(\inf_{\theta>0}\Big\{-\theta t + \sum_{\ell=1}^M C_{X_\ell}(\theta)\Big\}\Big).$$

Proof. For $\theta > 0$, Markov's inequality (Theorem 7.3) and independence yield

$$\mathbb{P}\Big(\sum_{\ell=1}^M X_\ell \geq t\Big) = \mathbb{P}\Big(\exp\big(\theta\sum_{\ell=1}^M X_\ell\big) \geq \exp(\theta t)\Big) \leq e^{-\theta t}\mathbb{E}[\exp\big(\theta\sum_{\ell=1}^M X_\ell\big)]$$

$$= e^{-\theta t}\mathbb{E}[\prod_{\ell=1}^M \exp(\theta X_\ell)] = e^{-\theta t}\prod_{\ell=1}^M \mathbb{E}[\exp(\theta X_\ell)]$$

$$= e^{-\theta t}\prod_{\ell=1}^M \exp(C_{X_\ell}(\theta)) = \exp\Big(-\theta t + \sum_{\ell=1}^M C_{X_\ell}(\theta)\Big).$$

Taking the infimum over $\theta > 0$ concludes the proof. $\qquad\square$

Remark 7.19. The function

$$t \mapsto \inf_{\theta>0}\Big\{-\theta t + \sum_{\ell=1}^M C_{X_\ell}(\theta)\Big\}$$

appearing in the exponential is closely connected to a convex conjugate function appearing in convex analysis; see Sect. B.3.

We will use this theorem several times later on. Let us state Hoeffding's inequality for the sum of almost surely bounded random variables as a first consequence.

Theorem 7.20. *Let $X_1, \ldots, X_M$ be a sequence of independent random variables such that $\mathbb{E}X_\ell = 0$ and $|X_\ell| \le B_\ell$ almost surely, $\ell \in [M]$. Then, for all $t > 0$,*

$$\mathbb{P}\left(\sum_{\ell=1}^{M} X_\ell \ge t\right) \le \exp\left(-\frac{t^2}{2\sum_{\ell=1}^{M} B_\ell^2}\right),$$

and consequently,

$$\mathbb{P}\left(\left|\sum_{\ell=1}^{M} X_\ell\right| \ge t\right) \le 2\exp\left(-\frac{t^2}{2\sum_{\ell=1}^{M} B_\ell^2}\right). \tag{7.26}$$

Proof. In view of Cramér's theorem, we estimate the moment-generating function of X_ℓ. Since (except possibly for an event of measure zero) $X_\ell \in [-B_\ell, B_\ell]$, we can write

$$X_\ell = t(-B_\ell) + (1-t)B_\ell,$$

where $t = \frac{B_\ell - X_\ell}{2B_\ell} \in [0,1]$. Since $f(x) = \exp(\theta x)$ is convex, we have

$$\exp(\theta X_\ell) = f(X_\ell) = f(t(-B_\ell)) + (1-t)B_\ell) \le tf(-B_\ell) + (1-t)f(B_\ell)$$

$$= \frac{B_\ell - X_\ell}{2B_\ell}e^{-\theta B_\ell} + \frac{B_\ell + X_\ell}{2B_\ell}e^{\theta B_\ell}. \tag{7.27}$$

Taking expectation and using the fact that $\mathbb{E}X_\ell = 0$, we arrive at

$$\mathbb{E}\exp(\theta X_\ell) \le \frac{1}{2}(\exp(-\theta B_\ell) + \exp(\theta B_\ell)) = \frac{1}{2}\left(\sum_{k=0}^{\infty}\frac{(-\theta B_\ell)^k}{k!} + \sum_{k=0}^{\infty}\frac{(\theta B_\ell)^k}{k!}\right)$$

$$= \sum_{k=0}^{\infty}\frac{(\theta B_\ell)^{2k}}{(2k)!} \le \sum_{k=0}^{\infty}\frac{(\theta B_\ell)^{2k}}{2^k k!} = \exp(\theta^2 B_\ell^2/2). \tag{7.28}$$

Therefore, the cumulant-generating function of X_ℓ satisfies

$$C_{X_\ell}(\theta) \le B_\ell^2 \theta^2/2.$$

It follows from Cramér's theorem (Theorem 7.18) that

$$\mathbb{P}(\sum_{\ell=1}^{M} X_\ell \ge t) \le \exp\left(\inf_{\theta > 0}\left\{-\theta t + \frac{\theta^2}{2}\sum_{\ell=1}^{M} B_\ell^2\right\}\right).$$

The optimal choice $\theta = t/(\sum_{\ell=1}^{M} B_\ell^2)$ in the above infimum yields

$$\mathbb{P}(\sum_{\ell=1}^{M} X_\ell \geq t) \leq \exp\left(-\frac{t^2}{2\sum_{\ell=1}^{M} B_\ell^2}\right).$$

Replacing X_ℓ by $-X_\ell$ gives the same bound, and an application of the union bound (7.1) then shows (7.26). □

A Rademacher variable (sometimes also called symmetric Bernoulli variable) is a random variable ϵ that takes the values $+1$ and -1 with equal probability. A Rademacher sequence ϵ is a vector of independent Rademacher variables (also called a Rademacher vector). We obtain the following version of Hoeffding's inequality for Rademacher sums.

Corollary 7.21. *Let* $\mathbf{a} \in \mathbb{R}^M$ *and* $\epsilon = (\epsilon_1, \ldots, \epsilon_M)$ *be a Rademacher sequence. Then, for* $u > 0$,

$$\mathbb{P}\left(|\sum_{\ell=1}^{M} \epsilon_\ell a_\ell| \geq \|\mathbf{a}\|_2 u\right) \leq 2\exp(-u^2/2). \tag{7.29}$$

Proof. The random variable $a_\ell \epsilon_\ell$ has mean zero and is bounded in absolute value by $|a_\ell|$. Therefore, the stated inequality follows immediately from Hoeffding's inequality 7.20. □

Remark 7.22. Note that specializing (7.28) to a Rademacher variable ϵ shows that its moment-generating function satisfies

$$\mathbb{E}\exp(\theta\epsilon) \leq \exp(\theta^2/2). \tag{7.30}$$

7.4 Subgaussian Random Variables

A random variable X is called *subgaussian* if there exist constants $\beta, \kappa > 0$ such that

$$\mathbb{P}(|X| \geq t) \leq \beta e^{-\kappa t^2} \quad \text{for all } t > 0. \tag{7.31}$$

It is called *subexponential* if there exist constants $\beta, \kappa > 0$ such that

$$\mathbb{P}(|X| \geq t) \leq \beta e^{-\kappa t} \quad \text{for all } t > 0.$$

According to Proposition 7.5, a standard Gaussian random variable is subgaussian with $\beta = 1$ and $\kappa = 1/2$. Furthermore, Rademacher and bounded random variables are subgaussian. According to Theorem 7.20, Rademacher sums are subgaussian random variables as well.

Clearly, a random variable X is subgaussian if and only if X^2 is subexponential. Setting $\alpha = (2e\kappa)^{-1/2}$ and $\gamma = 2$ in Proposition 7.13 shows that the moments of a subgaussian variable X satisfy

$$(\mathbb{E}|X|^p)^{1/p} \leq \tilde{C}\kappa^{-1/2}\beta^{1/p}p^{1/2} \quad \text{for all } p \geq 1 \tag{7.32}$$

with $\tilde{C} = e^{1/(2e)}C_{2,2}/\sqrt{2e} = e^{1/(2e)+1/6}\sqrt{\pi/(2e)} \approx 1.0282$, while the moments of a subexponential variable X satisfy (setting $\alpha = (e\kappa)^{-1}$ and $\gamma = 1$ in Proposition 7.13)

$$(\mathbb{E}|X|^p)^{1/p} \leq \hat{C}\kappa^{-1}\beta^{1/p}p \quad \text{for all } p \geq 1$$

with $\hat{C} = e^{1/(2e)}C_{2,1}e^{-1} = e^{1/(2e)+1/12}\sqrt{2\pi} \approx 3.1193$. Proposition 7.11 provides a statement in the converse direction. Let us give an equivalent characterization of subgaussian random variables.

Proposition 7.23. *Let X be a random variable.*

(a) If X is subgaussian, then there exist constants $c > 0, C \geq 1$ such that $\mathbb{E}[\exp(cX^2)] \leq C$.

(b) If $\mathbb{E}[\exp(cX^2)] \leq C$ for some constants $c, C > 0$, then X is subgaussian. More precisely, we have $\mathbb{P}(|X| \geq t) \leq Ce^{-ct^2}$.

Proof. (a) The moment estimate (7.21) with $\kappa = 1/(2e\alpha^2)$ yields

$$\mathbb{E}X^{2n} \leq \beta\kappa^{-n}n!.$$

Expanding the exponential function into its Taylor series and using Fubini's theorem show that

$$\mathbb{E}[\exp(cX^2)] = 1 + \sum_{n=1}^{\infty} \frac{c^n\mathbb{E}[X^{2n}]}{n!} \leq 1 + \beta\sum_{n=1}^{\infty}\frac{c^n\kappa^{-n}n!}{n!} = 1 + \frac{\beta c\kappa^{-1}}{1 - c\kappa^{-1}}.$$

provided $c < \kappa$.

(b) This statement follows from Markov's inequality (Theorem 7.3), since

$$\mathbb{P}(|X| \geq t) = \mathbb{P}(\exp(cX^2) \geq \exp(ct^2)) \leq \mathbb{E}[\exp(cX^2)]e^{-ct^2} \leq Ce^{-ct^2}.$$

This completes the proof.   □

Exercise 7.6 refines the statement of Proposition 7.23(a).

Let us study the Laplace transform (or moment-generating function) of a mean-zero subgaussian random variable.

Proposition 7.24. *Let X be a random variable.*

(a) If X is subgaussian with $\mathbb{E}X = 0$, then there exists a constant c (depending only on β and κ) such that

$$\mathbb{E}[\exp(\theta X)] \le \exp(c\theta^2) \quad \text{for all } \theta \in \mathbb{R}. \tag{7.33}$$

(b) Conversely, if (7.33) holds, then $\mathbb{E}X = 0$ and X is subgaussian with parameters $\beta = 2$ and $\kappa = 1/(4c)$.

Remark 7.25. Any valid constant c in (7.33) is called a *subgaussian parameter* of X. Of course, one preferably chooses the minimal possible c.

Proof. For the easier part (b), we take $\theta, t > 0$ and apply Markov's inequality (Theorem 7.3) to get

$$\mathbb{P}(X \ge t) = \mathbb{P}(\exp(\theta X) \ge \exp(\theta t)) \le \mathbb{E}[\exp(\theta X)]e^{-\theta t} \le e^{c\theta^2 - \theta t}.$$

The optimal choice $\theta = t/(2c)$ yields

$$\mathbb{P}(X \ge t) \le e^{-t^2/(4c)}.$$

Repeating the above computation with $-X$ instead of X shows that

$$\mathbb{P}(-X \ge t) \le e^{-t^2/(4c)},$$

and the union bound yields the desired estimate $\mathbb{P}(|X| \ge t) \le 2e^{-t^2/(4c)}$. In order to deduce that X has mean zero, we take the expectation in the inequality $1 + \theta X \le \exp(\theta X)$ to deduce, for $|\theta| < 1$,

$$1 + \theta\mathbb{E}(X) \le \mathbb{E}[\exp(\theta X)] \le \exp(c\theta^2) \le 1 + (c/2)\theta^2 + \mathcal{O}(\theta^4).$$

Letting $\theta \to 0$ shows that $\mathbb{E}X = 0$.

Let us now turn to the converse implication (a). We note that it is enough to consider $\theta \ge 0$, as the statement for $\theta < 0$ follows from exchanging X with $-X$. Expanding the exponential function into its Taylor series yields (together with Fubini's theorem)

$$\mathbb{E}[\exp(\theta X)] = 1 + \theta\mathbb{E}(X) + \sum_{n=2}^{\infty} \frac{\theta^n \mathbb{E}X^n}{n!} = 1 + \sum_{n=2}^{\infty} \frac{\theta^n \mathbb{E}|X|^n}{n!},$$

where we used the mean-zero assumption. First suppose that $0 \le \theta \le \theta_0$ for some θ_0 to be determined below. Then the moment estimate (7.32) and the consequence $n! \ge \sqrt{2\pi} n^n e^{-n}$ of Stirling's formula (C.13) yield

$$\mathbb{E}[\exp(\theta X)] \le 1 + \beta \sum_{n=2}^{\infty} \frac{\theta^n \tilde{C}^n \kappa^{-n/2} n^{n/2}}{n!} \le 1 + \frac{\beta}{\sqrt{2\pi}} \sum_{n=2}^{\infty} \frac{\tilde{C}^n \theta^n \kappa^{-n/2} n^{n/2}}{n^n e^{-n}}$$

$$\le 1 + \theta^2 \frac{\beta(\tilde{C}e)^2}{\sqrt{2\pi\kappa}} \sum_{n=0}^{\infty} (\tilde{C}e\theta_0 \kappa^{-1/2})^n$$

$$= 1 + \theta^2 \frac{\beta(\tilde{C}e)^2}{\sqrt{2\pi\kappa}} \frac{1}{1 - \tilde{C}e\theta_0 \kappa^{-1/2}}$$

$$= 1 + c_1 \theta^2 \le \exp(c_1 \theta^2),$$

provided $\tilde{C}e\theta_0 \kappa^{-1/2} < 1$. The latter is satisfied by setting

$$\theta_0 = (2\tilde{C}e)^{-1}\sqrt{\kappa},$$

which gives $c_1 = \sqrt{2}\beta\kappa^{-1}((\tilde{C}e)^2/\sqrt{\pi})$.

Let us now assume that $\theta > \theta_0$. We aim at proving $\mathbb{E}[\exp(\theta X - c_2\theta^2)] \le 1$. Observe that

$$\theta X - c_2 \theta^2 = -\left(\sqrt{c_2}\theta - \frac{X}{2\sqrt{c_2}}\right)^2 + \frac{X^2}{4c_2} \le \frac{X^2}{4c_2}.$$

Let $\tilde{c} > 0, \tilde{C} \ge 1$ be the constants from Proposition 7.23(a) and choose $c_2 = 1/(4\tilde{c})$. Then

$$\mathbb{E}[\exp(\theta X - c_2\theta^2)] \le \mathbb{E}[\exp(\tilde{c}X^2)] \le \tilde{C}.$$

Defining $\rho = \ln(\tilde{C})\theta_0^{-2}$ yields

$$\mathbb{E}[\exp(\theta X)] \le \tilde{C}\exp(c_2\theta^2) = \tilde{C}\exp(-\rho\theta^2)\exp((c_2 + \rho)\theta^2)$$

$$\le \tilde{C}\exp(-\rho\theta_0^2)e^{(c_2+\rho)\theta^2} = e^{(c_2+\rho)\theta^2}.$$

Setting $c = \max\{c_1, c_2 + \rho\}$ completes the proof.   □

Remark 7.26. For Rademacher and standard Gaussian random variables, the constant in (7.33) satisfies $c = 1/2$ by (7.11) and (7.30). Furthermore, for mean-zero random variables X with $|X| \le K$ almost surely, $c = K^2/2$ is a valid choice of the subgaussian parameter by (7.28).

The sum of independent mean-zero subgaussian variables is again subgaussian by the next statement.

Theorem 7.27. *Let $X_1, \ldots, X_M$ be a sequence of independent mean-zero subgaussian random variables with subgaussian parameter c in (7.33). For $\mathbf{a} \in \mathbb{R}^M$, the random variable $Z := \sum_{\ell=1}^{M} a_\ell X_\ell$ is subgaussian, i.e.,*

$$\mathbb{E}\exp(\theta Z) \le \exp(c\|\mathbf{a}\|_2^2 \theta^2), \tag{7.34}$$

and

$$\mathbb{P}\left(\left|\sum_{\ell=1}^{M} a_\ell X_\ell\right| \geq t\right) \leq 2 \exp\left(-\frac{t^2}{4c\|\mathbf{a}\|_2^2}\right) \quad \textit{for all } t > 0. \tag{7.35}$$

Proof. By independence, we have

$$\mathbb{E}\exp(\theta \sum_{\ell=1}^{M} a_\ell X_\ell) = \mathbb{E}\prod_{\ell=1}^{M} \exp(\theta a_\ell X_\ell) = \prod_{\ell=1}^{M} \mathbb{E}\exp(\theta a_\ell X_\ell) \leq \prod_{\ell=1}^{M} \exp(c\theta^2 a_\ell^2)$$

$$= \exp(c\|\mathbf{a}\|_2^2\theta^2).$$

This proves (7.34). The second inequality (7.35) follows then from Proposition 7.24(b). □

Remark 7.28. In particular, if $\boldsymbol{\epsilon} = (\epsilon_1, \dots, \epsilon_M)$ is a Rademacher sequence and $Z = \sum_{\ell=1}^{M} a_\ell \epsilon_\ell$, then

$$\mathbb{E}\exp(\theta Z) \leq \exp(\theta^2 \|\mathbf{a}\|_2^2 / 2).$$

The expected maximum of a finite number of subgaussian random variables can be estimated as follows.

Proposition 7.29. *Let $X_1, \dots, X_M$ be a sequence of (not necessarily independent) mean-zero subgaussian random variables satisfying $\mathbb{E}[\exp(\theta X_\ell)] \leq \exp(c_\ell \theta^2)$, $\ell \in [M]$. Then, with $c = \max_{\ell \in [M]} c_\ell$,*

$$\mathbb{E}\max_{\ell \in [M]} X_\ell \leq \sqrt{4c \ln(M)}, \tag{7.36}$$

$$\mathbb{E}\max_{\ell \in [M]} |X_\ell| \leq \sqrt{4c \ln(2M)}. \tag{7.37}$$

Proof. Since (7.36) is obvious for $M = 1$, we assume $M \geq 2$. Let $\beta > 0$ be a number to be chosen later. Using concavity of the logarithm in connection with Jensen's inequality, we obtain

$$\beta\mathbb{E}\max_{\ell \in [M]} X_\ell = \mathbb{E}\ln\max_{\ell \in [M]} \exp(\beta X_\ell) \leq \mathbb{E}\ln\left(\sum_{\ell=1}^{M} \exp(\beta X_\ell)\right)$$

$$\leq \ln(\sum_{\ell=1}^{M} \mathbb{E}\exp(\beta X_\ell)) \leq \ln(M \exp(c\beta^2)) = c\beta^2 + \ln(M).$$

Choosing $\beta = \sqrt{c^{-1}\ln(M)}$ yields

$$\sqrt{c^{-1}\ln(M)}\mathbb{E}\max_{\ell\in[M]}X_\ell \leq \ln(M)+\ln(M)$$

so that $\mathbb{E}\max_{\ell\in[M]} \leq \sqrt{4c\ln(M)}$.

For (7.37) we write $\mathbb{E}\max_{\ell\in[M]}|X_\ell| = \mathbb{E}\max\{X_1,\ldots,X_M,-X_1,\ldots,-X_M\}$ and apply (7.36).   □

The example of a sequence of standard Gaussian random variables shows that the estimates in the previous proposition are optimal up to possibly the constants; see Proposition 8.1(c) below.

7.5 Bernstein Inequalities

Bernstein inequality provides a useful generalization of Hoeffding's inequality (7.29) to sums of bounded or even unbounded independent random variables, which also takes into account the variance or higher moments. We start with the version below and then derive variations as consequences.

Theorem 7.30. *Let $X_1,\ldots,X_M$ be independent mean-zero random variables such that, for all integers $n \geq 2$,*

$$\mathbb{E}|X_\ell|^n \leq n!R^{n-2}\sigma_\ell^2/2 \quad \text{for all } \ell \in [M] \tag{7.38}$$

for some constants $R > 0$ and $\sigma_\ell > 0$, $\ell \in [M]$. Then, for all $t > 0$,

$$\mathbb{P}\left(\left|\sum_{\ell=1}^M X_\ell\right| \geq t\right) \leq 2\exp\left(-\frac{t^2/2}{\sigma^2 + Rt}\right), \tag{7.39}$$

where $\sigma^2 := \sum_{\ell=1}^M \sigma_\ell^2$.

Before providing the proof, we give two consequences. The first is the Bernstein inequality for bounded random variables.

Corollary 7.31. *Let $X_1,\ldots,X_M$ be independent random variables with zero mean such that $|X_\ell| \leq K$ almost surely for $\ell \in [M]$ and some constant $K > 0$. Furthermore assume $\mathbb{E}|X_\ell|^2 \leq \sigma_\ell^2$ for constants $\sigma_\ell > 0$, $\ell \in [M]$. Then, for all $t > 0$,*

$$\mathbb{P}\left(\left|\sum_{\ell=1}^M X_\ell\right| \geq t\right) \leq 2\exp\left(-\frac{t^2/2}{\sigma^2 + Kt/3}\right), \tag{7.40}$$

where $\sigma^2 := \sum_{\ell=1}^M \sigma_\ell^2$.

Proof. For $n = 2$, condition (7.38) is clearly satisfied. For $n \geq 3$, we have $n! \geq 3 \cdot 2^{n-2}$ and we obtain

$$\mathbb{E}|X_\ell|^n = \mathbb{E}[|X_\ell|^{n-2}X_\ell^2] \leq K^{n-2}\sigma_\ell^2 \leq \frac{n!K^{n-2}}{n!}\sigma_\ell^2 \leq \frac{n!K^{n-2}}{2 \cdot 3^{n-2}}\sigma_\ell^2. \tag{7.41}$$

In other words, condition (7.38) holds for all $n \geq 2$ with constant $R = K/3$. Hence, the statement follows from Theorem 7.30. □

As a second consequence, we present the Bernstein inequality for subexponential random variables.

Corollary 7.32. *Let $X_1, \ldots, X_M$ be independent mean-zero subexponential random variables, i.e., $\mathbb{P}(|X_\ell| \geq t) \leq \beta e^{-\kappa t}$ for some constants $\beta, \kappa > 0$ for all $t > 0$, $\ell \in [M]$. Then*

$$\mathbb{P}\left(\left|\sum_{\ell=1}^{M} X_\ell\right| \geq t\right) \leq 2\exp\left(-\frac{(\kappa t)^2/2}{2\beta M + \kappa t}\right). \tag{7.42}$$

Proof. Similarly to the proof of Proposition 7.13, we estimate, for $n \in \mathbb{N}, n \geq 2$,

$$\mathbb{E}|X_\ell|^n = n\int_0^\infty \mathbb{P}(|X_\ell| \geq t)t^{n-1}dt \leq \beta n\int_0^\infty e^{-\kappa t}t^{n-1}dt$$

$$= \beta n\kappa^{-n}\int_0^\infty e^{-u}u^{n-1}du = \beta n!\kappa^{-n} = n!\kappa^{-(n-2)}\frac{2\beta\kappa^{-2}}{2}.$$

Hereby, we have used that the integral in the second line equals $\Gamma(n) = (n-1)!$. Hence, condition (7.38) holds with $R = \kappa^{-1}$ and $\sigma_\ell^2 = 2\beta\kappa^{-2}$. The claim follows therefore from Theorem 7.30. □

Let us now turn to the proof of the Bernstein inequality in Theorem 7.30.

Proof (of Theorem 7.30). Motivated by Cramér's theorem, we estimate the moment-generating function of the X_ℓ. Expanding the exponential function into its Taylor series and using Fubini's theorem in order to interchange expectation and summation yield

$$\mathbb{E}[\exp(\theta X_\ell)] = 1 + \theta\mathbb{E}[X_\ell] + \sum_{n=2}^{\infty}\frac{\theta^n\mathbb{E}[X_\ell^n]}{n!} = 1 + \frac{\theta^2\sigma_\ell^2}{2}\sum_{n=2}^{\infty}\frac{\theta^{n-2}\mathbb{E}[X_\ell^n]}{n!\sigma_\ell^2/2},$$

where we additionally used that $\mathbb{E}[X_\ell] = 0$. Defining

$$F_\ell(\theta) = \sum_{n=2}^{\infty}\frac{\theta^{n-2}\mathbb{E}[X_\ell^n]}{n!\sigma_\ell^2/2},$$

we obtain

$$\mathbb{E}[\exp(\theta X_\ell)] = 1 + \theta^2 \sigma_\ell^2 F_\ell(\theta)/2 \le \exp(\theta^2 \sigma_\ell^2 F_\ell(\theta)/2).$$

Introducing $F(\theta) = \max_{\ell \in [M]} F_\ell(\theta)$ and recalling that $\sigma^2 = \sum_{\ell=1}^{M} \sigma_\ell^2$, we obtain from Cramér's theorem (Theorem 7.18)

$$\mathbb{P}\Big(\sum_{\ell=1}^{M} X_\ell \ge t\Big) \le \inf_{\theta > 0} \exp(\theta^2 \sigma^2 F(\theta)/2 - \theta t) \le \inf_{0 < R\theta < 1} \exp(\theta^2 \sigma^2 F(\theta)/2 - \theta t).$$

Since $\mathbb{E}[X_\ell^n] \le \mathbb{E}[|X_\ell|^n]$, the assumption (7.38) yields

$$F_\ell(\theta) \le \sum_{n=2}^{\infty} \frac{\theta^{n-2} \mathbb{E}[|X_\ell|^n]}{n! \sigma_\ell^2 / 2} \le \sum_{n=2}^{\infty} (R\theta)^{n-2} = \frac{1}{1 - R\theta}$$

provided $R\theta < 1$. Therefore, $F(\theta) \le (1 - R\theta)^{-1}$ and

$$\mathbb{P}\Big(\sum_{\ell=1}^{M} X_\ell \ge t\Big) \le \inf_{0 < \theta R < 1} \exp\left(\frac{\theta^2 \sigma^2}{2(1 - R\theta)} - \theta t \right).$$

Now we choose $\theta = t/(\sigma^2 + Rt)$, which clearly satisfies $R\theta < 1$. This yields

$$\mathbb{P}\Big(\sum_{\ell=1}^{M} X_\ell \ge t\Big) \le \exp\left(\frac{t^2 \sigma^2}{2(\sigma^2 + Rt)^2} \frac{1}{1 - \frac{Rt}{\sigma^2 + Rt}} - \frac{t^2}{\sigma^2 + Rt} \right)$$

$$= \exp\left(-\frac{t^2/2}{\sigma^2 + Rt} \right).$$

Exchanging X_ℓ with $-X_\ell$ yields the same estimate, and applying the union bound completes the proof. $\qquad\square$

Notes

Good sources for background on basic probability theory are, for instance, the monographs [243, 428]. The relation of tails and moments is well known (see, e.g., [322]), although the refinement with the parameter β in (7.18) seems to have appeared only recently [411]. Cramér proved the theorem named after him in [132]. We refer to [495] for more information on large deviation results in this spirit. Hoeffding's inequality (7.29) was derived in [276]. In the special case of random variables that take only values in $\{0, 1\}$ with probabilities p and $1 - p$, the so-called Chernoff bounds refine the Hoeffding inequalities; see, for instance, [116, 254].

Bernstein inequality was first proved in [43] and refined later by Bennett [40]. For further reading on scalar deviation inequalities, the reader is referred to [349, 495].

The notion of subgaussian random variables was introduced in [297]. It may be refined to *strictly subgaussian* random variables, for which the constant in (7.33) satisfies $c = \mathbb{E}|X|^2/2$. Gaussian and Bernoulli random variables, as well as random variables that are uniformly distributed on $[-1, 1]$, are strictly subgaussian; see Exercise 7.5. More information on subgaussian random variables can be found, for instance, in [78, 501].

Exercises

7.1. Show the relation (7.16) of the Beta function B to the Gamma function.

7.2. Prove Proposition 7.15.

7.3. Let $p > 1$. Generalize Lemma 7.16 by showing that any nonnegative random variable Z with finite pth moment satisfies

$$\mathbb{P}(Z > t) \geq \frac{(\mathbb{E}Z - t)^{p/(p-1)}}{(\mathbb{E}Z^p)^{1/(p-1)}}, \qquad 0 \leq t \leq \mathbb{E}Z.$$

Prove also that if $X_1, \ldots, X_M$ are independent mean-zero random variables with variance σ^2 and $2p$th absolute moment bounded above by μ^{2p}, then, for all $\mathbf{a} \in \mathbb{R}^M$,

$$\mathbb{P}\left(\left| \sum_{\ell=1}^{M} a_\ell X_\ell \right| > t\|\mathbf{a}\|_2 \right) \geq c_p \frac{(\sigma^2 - t^2)^2}{\mu^{2p/(p-1)}}, \qquad 0 \leq t \leq \sigma,$$

for some constant c_p to be determined.

7.4. Let X be a subgaussian random variable with $\mathbb{E}\exp(\theta X) \leq \exp(c\theta^2)$ for some constant $c > 0$. Show that its variance satisfies $\mathbb{E}X^2 \leq 2c$. (A subgaussian variable for which equality holds is called strictly subgaussian).

7.5. Let X be a random variable that is uniformly distributed on $[-1, 1]$. Show that $\mathbb{E}|X|^2 = 1/3$ and that

$$\mathbb{E}\exp(\theta X) \leq \exp(\theta^2/6) = \exp(\theta^2 \mathbb{E}|X|^2/2),$$

so that X is strictly subgaussian.

7.6. Let X be a subgaussian random variable with parameter $c > 0$, that is, $\mathbb{E}\exp(\theta X) \leq \exp(c\theta^2)$ for all $\theta \in \mathbb{R}$. Show that, for $t \in [0, 1/2]$,

$$\mathbb{E}\exp(tX^2/c^2) \leq \frac{1}{\sqrt{1 - 2t}}.$$

Chapter 8
Advanced Tools from Probability Theory

This chapter introduces further probabilistic tools that will be required for some of the more advanced results in the remainder of the book.

In Sect. 8.1, we compute the expectation of the ℓ_p-norm of a standard Gaussian vector for $p = 1, 2, \infty$ (required in Sect. 9.3). Section 8.2 presents simple results for Rademacher sums as well as the symmetrization method. This simple technique turns out to be powerful in various setups and will be needed in Sect. 12.5 and Chap. 14. Khintchine inequalities, treated in Sect. 8.3, estimate the moments of a Rademacher sum and allow to deduce Hoeffding-type inequalities for Rademacher sums in a different way than via moment-generating functions (required for Sect. 12.5 and in Chap. 14). Decoupling inequalities introduced in Sect. 8.6 replace one sequence of random variables in a double sum by an independent copy (required for Sect. 9.5 and Chap. 14). The scalar Bernstein inequality for bounded random variables (Corollary 7.31) will be extended in Sect. 8.5 to a powerful deviation inequality for the operator norm of sums of random matrices (required for Sects. 12.3, 12.4, and 14.1). Section 8.6 deals with Dudley's inequality, which is a crucial tool to estimate the expectation of a supremum of a subgaussian process by an integral over covering numbers of the index set of the process (required for the estimate of the restricted isometry constants in Sect. 12.5). Slepian's and Gordon's lemmas compare expectations of functions of two Gaussian random vectors in terms of the covariances of the two vectors. In particular, maxima as well as minima of maxima are important choices of such functions. These are treated in Sect. 8.7 and will be used in Sects. 9.2 and 9.4. Section 8.8 treats the concentration of measure phenomenon which states that a Lipschitz function of a Gaussian random vector highly concentrates around its mean (required in Sects. 9.2 and 9.4). The final section of this chapter deals with a deviation inequality for the supremum of an empirical process, which is sometimes called Talagrand's concentration inequality or Bousquet's inequality. It will be required in Chap. 12.

S. Foucart and H. Rauhut, *A Mathematical Introduction to Compressive Sensing*, Applied and Numerical Harmonic Analysis, DOI 10.1007/978-0-8176-4948-7_8, © Springer Science+Business Media New York 2013

8.1   Expectation of Norms of Gaussian Vectors

We estimate the expectation of the norms of standard Gaussian random vectors in $\ell_1, \ell_2,$ and ℓ_∞.

Proposition 8.1. *Let* $\mathbf{g} = (g_1, \ldots, g_n)$ *be a vector of (not necessarily independent) standard Gaussian random variables. Then*

(a) $\mathbb{E}\|\mathbf{g}\|_1 = \sqrt{\dfrac{2}{\pi}}n.$

(b) $\mathbb{E}\|\mathbf{g}\|_2^2 = n$ *and* $\sqrt{\dfrac{2}{\pi}}\sqrt{n} \leq \mathbb{E}\|\mathbf{g}\|_2 \leq \sqrt{n}.$

If the entries of $\mathbf{g}$ *are independent, then*

$$\frac{n}{\sqrt{n+1}} \leq \mathbb{E}\|\mathbf{g}\|_2 = \sqrt{2}\frac{\Gamma((n+1)/2)}{\Gamma(n/2)} \leq \sqrt{n}, \tag{8.1}$$

and consequently $\mathbb{E}\|\mathbf{g}\|_2 \sim \sqrt{n}$ *as* $n \to \infty.$
(c) It holds that

$$\mathbb{E}\max_{\ell \in [n]} g_\ell \leq \sqrt{2\ln(n)} \quad and \quad \mathbb{E}\|\mathbf{g}\|_\infty \leq \sqrt{2\ln(2n)}. \tag{8.2}$$

If the entries of $\mathbf{g}$ *are independent, then, for* $n \geq 2,$

$$\mathbb{E}\|\mathbf{g}\|_\infty \geq C\sqrt{\ln(n)} \tag{8.3}$$

with $C \geq 0.265.$

Proof. (a) By the formula for the density of a standard Gaussian random variable, we have

$$\mathbb{E}|g_\ell| = \frac{1}{\sqrt{2\pi}}\int_{-\infty}^{\infty}|u|\exp(-u^2/2)du = \sqrt{\frac{2}{\pi}}\int_0^{\infty}u\exp(-u^2/2)du = \sqrt{\frac{2}{\pi}}.$$

By linearity of the expectation, $\mathbb{E}\|\mathbf{g}\|_1 = \sum_{\ell=1}^{n}\mathbb{E}|g_\ell| = \sqrt{2/\pi}\,n.$
(b) Clearly, $\mathbb{E}\|\mathbf{g}\|_2^2 = \sum_{\ell=1}^{n}\mathbb{E}g_\ell^2 = n$ for standard Gaussian random variables $g_\ell.$ The Cauchy–Schwarz inequality for expectations (or Jensen's inequality) yields $\mathbb{E}\|\mathbf{g}\|_2 \leq \sqrt{\mathbb{E}\|\mathbf{g}\|_2^2} = \sqrt{n},$ while the Cauchy–Schwarz inequality for the inner product on $\mathbb{R}^n$ gives $\mathbb{E}\|\mathbf{g}\|_2 \geq \mathbb{E}\frac{1}{\sqrt{n}}\|\mathbf{g}\|_1 = \sqrt{2/\pi}\,\sqrt{n}.$

If the entries of $\mathbf{g}$ are independent, then $\|\mathbf{g}\|_2^2$ has the $\chi^2(n)$ distribution with probability density function $\phi_n(u)$ given by (7.15). Therefore,

$$\mathbb{E}\|\mathbf{g}\|_2 = \mathbb{E}\left(\sum_{\ell=1}^{n} g_\ell^2\right)^{1/2} = \int_0^\infty u^{1/2}\phi_n(u)du$$

$$= \frac{1}{2^{n/2}\Gamma(n/2)}\int_0^\infty u^{1/2}u^{(n/2)-1}e^{-u/2}du$$

$$= \frac{2^{n/2+1/2}}{2^{n/2}\Gamma(n/2)}\int_0^\infty t^{(n/2)-1/2}e^{-t}dt = \sqrt{2}\frac{\Gamma((n+1)/2)}{\Gamma(n/2)},$$

where we used the definition of the Gamma function in (C.9). The estimate $E_n := \mathbb{E}\|\mathbf{g}\|_2 \leq \sqrt{n}$ for Gaussian vector $\mathbf{g}$ of length n was already shown above. Furthermore,

$$E_{n+1}E_n = 2\frac{\Gamma(n/2+1)}{\Gamma(n/2)} = n,$$

by the functional equation (C.11) for the Gamma function so that $E_n = n/E_{n+1} \geq n/\sqrt{n+1}$ (compare also with Lemma C.4).

(c) The inequalities in (8.2) follow from Proposition 7.29 by noting that due to Lemma 7.6, we have $\mathbb{E}\exp(\beta g) = \exp(\beta^2/2)$ so that the subgaussian parameter $c = 1/2$ for Gaussian random variables.

If the g_ℓ are independent, then by Corollary 7.2

$$\mathbb{E}\|\mathbf{g}\|_\infty = \int_0^\infty \mathbb{P}\left(\max_{\ell\in[n]}|g_\ell| > u\right)du = \int_0^\infty \left(1 - \mathbb{P}(\max_{\ell\in[n]}|g_\ell| \leq u)\right)du$$

$$= \int_0^\infty \left(1 - \prod_{\ell=1}^{n}\mathbb{P}(|g_\ell| \leq u)\right)du \geq \int_0^\delta \left(1 - (1 - \mathbb{P}(|g| > u))^n\right)du$$

$$\geq \delta\left(1 - (1 - \mathbb{P}(|g| > \delta))^n\right).$$

Further,

$$\mathbb{P}(|g| > \delta) = \sqrt{\frac{2}{\pi}}\int_\delta^\infty e^{-t^2/2}dt \geq \sqrt{\frac{2}{\pi}}\int_\delta^{2\delta} e^{-t^2/2}dt \geq \sqrt{\frac{2}{\pi}}\delta e^{-2\delta^2}.$$

Now, we choose $\delta = \sqrt{\ln n/2}$. Then, for $n \geq 2$,

$$\mathbb{E}\|\mathbf{g}\|_\infty \geq \sqrt{\frac{\ln n}{2}}\left(1 - \left(1 - \sqrt{\frac{\ln n}{\pi}}\frac{1}{n}\right)^n\right) \geq \sqrt{\frac{\ln n}{2}}\left(1 - \exp\left(-\sqrt{\frac{\ln n}{\pi}}\right)\right)$$

$$\geq \frac{1 - \exp(-\sqrt{(\ln 2)/\pi})}{\sqrt{2}}\sqrt{\ln n},$$

which establishes the claim with $C = (1 - \exp(-\sqrt{(\ln 2)/\pi}))/\sqrt{2} \geq 0.265$. $\qquad\square$

Next we extend part (c) of the previous proposition to the maximum squared ℓ_2-norm of a sequence of standard Gaussian random vectors.

Proposition 8.2. *Let* $\mathbf{g}_1, \ldots, \mathbf{g}_M \in \mathbb{R}^n$ *be a sequence of (not necessarily independent) standard Gaussian random vectors. Then, for any* $\kappa > 0$,

$$\mathbb{E} \max_{\ell \in [M]} \|\mathbf{g}_\ell\|_2^2 \leq (2 + 2\kappa) \ln(M) + n(1 + \kappa) \ln(1 + \kappa^{-1}).$$

Consequently,

$$\mathbb{E} \max_{\ell \in [M]} \|\mathbf{g}_\ell\|_2^2 \leq (\sqrt{2 \ln(M)} + \sqrt{n})^2.$$

Proof. By concavity of the logarithm and Jensen's inequality, we have, for $\theta > 0$,

$$\mathbb{E} \max_{\ell \in [M]} \|\mathbf{g}_\ell\|_2^2 = \theta^{-1} \mathbb{E} \ln \max_{\ell \in [M]} \exp\left(\theta \|\mathbf{g}_\ell\|_2^2\right) \leq \theta^{-1} \ln \mathbb{E} \max_{\ell \in [M]} \exp\left(\theta \|\mathbf{g}_\ell\|_2^2\right)$$

$$\leq \theta^{-1} \ln\left(M \mathbb{E} \exp(\theta \|\mathbf{g}\|_2^2)\right),$$

where $\mathbf{g}$ denotes a standard Gaussian random vector in $\mathbb{R}^n$. In the last step we have used that $\max_{\ell \in [M]} \exp\left(\theta \|\mathbf{g}_\ell\|_2^2\right) \leq \sum_{\ell=1}^M \exp\left(\theta \|\mathbf{g}_\ell\|_2^2\right)$. By the independence of the components of $\mathbf{g}$ and Lemma 7.6,

$$\mathbb{E} \exp(\theta \|\mathbf{g}\|_2^2) = \mathbb{E} \exp\left(\theta \sum_{j=1}^n g_j^2\right) = \mathbb{E} \prod_{j=1}^n \exp(\theta g_j^2) = \prod_{j=1}^n \mathbb{E} \exp(\theta g_j^2)$$

$$= (1 - 2\theta)^{-n/2},$$

provided that $\theta < 1/2$. Therefore,

$$\mathbb{E} \max_{\ell \in [M]} \|\mathbf{g}_\ell\|_2^2 \leq \inf_{0 < \theta < 1/2} \theta^{-1} \left(\ln M + \frac{n}{2} \ln\left((1 - 2\theta)^{-1}\right)\right).$$

Substituting $\theta = (2 + 2\kappa)^{-1}$ yields the first claim. Using that $\ln(1 + \kappa^{-1}) \leq \kappa^{-1}$, we further get

$$\mathbb{E} \max_{\ell \in [M]} \|\mathbf{g}_\ell\|_2^2 \leq 2(1 + \kappa) \ln(M) + n(1 + \kappa^{-1}). \tag{8.4}$$

Making the optimal choice $\kappa = \sqrt{n/(2 \ln(M))}$ gives

$$\mathbb{E} \max_{\ell \in [M]} \|\mathbf{g}_\ell\|_2^2 \leq 2 \ln(M) + 2\sqrt{2n \ln(M)} + n = (\sqrt{2 \ln(M)} + \sqrt{n})^2.$$

This concludes the proof.   □

8.2 Rademacher Sums and Symmetrization

Recall that a Rademacher variable takes the values $+1$ or -1, each with probability $1/2$, and that a Rademacher sequence ϵ is a vector of independent Rademacher variables $\epsilon_\ell, \ell \in [M]$. In the sequel we will often consider Rademacher sums of the form

$$\sum_{\ell=1}^{M} \epsilon_\ell \mathbf{x}_\ell,$$

where $\mathbf{x}_\ell$ are scalars, vectors, or matrices and ϵ_ℓ are independent Rademacher variables.

Below we present the contraction principle for Rademacher sums and the symmetrization principle, which allows to replace a sum of independent random vectors by a corresponding Rademacher sum in moment estimates. Although rather simple, this tool will prove very effective later.

Let us first present the contraction principle.

Theorem 8.3. *Let* $\mathbf{x}_1, \ldots, \mathbf{x}_M$ *be vectors in a (finite-dimensional) vector space endowed with a norm* $\| \cdot \|$ *and let* $\alpha_1, \ldots, \alpha_M$ *be scalars satisfying* $|\alpha_\ell| \leq 1$. *If* $\epsilon \in \mathbb{R}^M$ *is a Rademacher sequence, then for any* $1 \leq p < \infty$,

$$\mathbb{E}\| \sum_{\ell=1}^{M} \alpha_\ell \epsilon_\ell \mathbf{x}_\ell \|^p \leq \mathbb{E}\| \sum_{\ell=1}^{M} \epsilon_\ell \mathbf{x}_\ell \|^p. \tag{8.5}$$

Proof. The function $(\alpha_1, \ldots, \alpha_M) \mapsto \mathbb{E}\| \sum_{\ell=1}^{M} \alpha_\ell \epsilon_\ell \mathbf{x}_\ell \|^p$ is convex. Therefore, on $[-1, 1]^M$ it attains its maximum at an extreme point, i.e., a point $\alpha = (\alpha_\ell)_{\ell=1}^{M}$ such that $\alpha_\ell = \pm 1$; see Theorem B.16. For such values of α_ℓ, both $\alpha_\ell \epsilon_\ell$ and ϵ_ℓ have the same distribution, so both terms in (8.5) are equal. $\square$

Symmetrization is a simple yet powerful technique to pass from a sum of arbitrary independent random variables to a Rademacher sum. A random vector $\mathbf{X}$ on $\mathbb{C}^n$ is called symmetric, if $\mathbf{X}$ and $-\mathbf{X}$ have the same distribution. Clearly, $\mathbb{E}\mathbf{X} = \mathbf{0}$ for a symmetric random vector $\mathbf{X}$. The crucial observation for symmetrization is that a symmetric random vector $\mathbf{X}$ and the random vector $\epsilon \mathbf{X}$, where ϵ is a Rademacher random variable independent of $\mathbf{X}$, have the same distribution.

Lemma 8.4. *Assume that* $\boldsymbol{\xi} = (\boldsymbol{\xi}_\ell)_{\ell=1}^{M}$ *is a sequence of independent random vectors in a finite-dimensional vector space* V *with norm* $\| \cdot \|$. *Let* $F : V \to \mathbb{R}$ *be a convex function. Then*

$$\mathbb{E}F\big(\sum_{\ell=1}^{M} (\boldsymbol{\xi}_\ell - \mathbb{E}[\boldsymbol{\xi}_\ell]) \big) \leq \mathbb{E}F\big(2 \sum_{\ell=1}^{M} \epsilon_\ell \boldsymbol{\xi}_\ell \big), \tag{8.6}$$

where $\epsilon = (\epsilon_\ell)_{\ell=1}^N$ *is a Rademacher sequence independent of* $\boldsymbol{\xi}$. *In particular, for* $1 \leq p < \infty$,

$$\left(\mathbb{E}\| \sum_{\ell=1}^M (\boldsymbol{\xi}_\ell - \mathbb{E}[\boldsymbol{\xi}_\ell]) \|^p\right)^{1/p} \leq 2\left(\mathbb{E}\| \sum_{\ell=1}^M \epsilon_\ell \boldsymbol{\xi}_\ell \|^p\right)^{1/p}. \tag{8.7}$$

Proof. Let $\boldsymbol{\xi}' = (\boldsymbol{\xi}'_1, \dots, \boldsymbol{\xi}'_M)$ denote an independent copy of the sequence of random vectors $(\boldsymbol{\xi}_1, \dots, \boldsymbol{\xi}_M)$, also independent of the Rademacher sequence ϵ to be used below. An application of Jensen's inequality yields

$$E := \mathbb{E}F\left(\sum_{\ell=1}^M (\boldsymbol{\xi}_\ell - \mathbb{E}[\boldsymbol{\xi}'_\ell])\right) \leq \mathbb{E}F\left(\sum_{\ell=1}^M (\boldsymbol{\xi}_\ell - \boldsymbol{\xi}'_\ell)\right).$$

Now observe that $(\boldsymbol{\xi}_\ell - \boldsymbol{\xi}'_\ell)_\ell$ is a sequence of independent symmetric random variables; hence, it has the same distribution as $(\epsilon_\ell(\boldsymbol{\xi}_\ell - \boldsymbol{\xi}'_\ell))_\ell$. Convexity of F gives

$$E \leq \mathbb{E}F\left(\sum_{\ell=1}^M \epsilon_\ell(\boldsymbol{\xi}_\ell - \boldsymbol{\xi}'_\ell)\right) \leq \mathbb{E}\left(\frac{1}{2}F\left(2\sum_{\ell=1}^M \epsilon_\ell\boldsymbol{\xi}_\ell\right) + \frac{1}{2}F\left(2\sum_{\ell=1}^M (-\epsilon_\ell)\boldsymbol{\xi}'_\ell\right)\right)$$

$$= \mathbb{E}F\left(2\sum_{\ell=1}^M \epsilon_\ell\boldsymbol{\xi}_\ell\right)$$

because ϵ is symmetric and $\boldsymbol{\xi}'$ has the same distribution as $\boldsymbol{\xi}$. Inequality (8.7) follows from choosing the convex function $F(\mathbf{x}) = \|\mathbf{x}\|^p$ for $p \in [1, \infty)$. $\qquad\square$

The lemma will be very useful because there are powerful techniques for estimating moments of Rademacher sums as we will see in the next section.

8.3 Khintchine Inequalities

The Khintchine inequalities provide estimates of the moments of Rademacher and related sums.

Theorem 8.5. *Let* $\mathbf{a} \in \mathbb{C}^M$ *and* $\epsilon = (\epsilon_1, \dots, \epsilon_M)$ *be a Rademacher sequence. Then, for all* $n \in \mathbb{N}$,

$$\mathbb{E}|\sum_{\ell=1}^M \epsilon_\ell a_\ell|^{2n} \leq \frac{(2n)!}{2^n n!}\|\mathbf{a}\|_2^{2n}. \tag{8.8}$$

Proof. First assume that the a_ℓ are real valued. Expanding the expectation on the left-hand side of (8.8) with the multinomial theorem (see Appendix C.4) yields

$$E := \mathbb{E}|\sum_{\ell=1}^{M} \epsilon_\ell a_\ell|^{2n}$$

$$= \sum_{\substack{j_1+\cdots+j_M=n \\ j_i \geq 0}} \frac{(2n)!}{(2j_1)!\cdots(2j_M)!}|a_1|^{2j_1}\cdots|a_M|^{2j_M}\mathbb{E}\epsilon_1^{2j_1}\cdots\mathbb{E}\epsilon_M^{2j_M}$$

$$= \sum_{\substack{j_1+\cdots+j_M=n \\ j_i \geq 0}} \frac{(2n)!}{(2j_1)!\cdots(2j_M)!}|a_1|^{2j_1}\cdots|a_M|^{2j_M}.$$

Hereby we used the independence of the ϵ_ℓ and the fact that $\mathbb{E}\epsilon_\ell^k = 0$ if k is an odd integer. For integers satisfying $j_1 + \cdots + j_M = n$ we have

$$2^n j_1! \times \cdots \times j_M! = 2^{j_1} j_1! \times \cdots \times 2^{j_M} j_M! \leq (2j_1)! \times \cdots \times (2j_M)!.$$

This implies

$$E \leq \frac{(2n)!}{2^n n!} \sum_{\substack{j_1+\cdots+j_M=n \\ j_i \geq 0}} \frac{n!}{j_1!\cdots j_n!}|a_1|^{2j_1}\cdots|a_M|^{2j_M}$$

$$= \frac{(2n)!}{2^n n!}\left(\sum_{j=1}^{M}|a_j|^2\right)^n = \frac{(2n)!}{2^n n!}\|\mathbf{a}\|_2^{2n}.$$

The complex case is derived by splitting into real and imaginary parts and applying the triangle inequality as follows:

$$\left(\mathbb{E}|\sum_{\ell=1}^{M} \epsilon_\ell(\text{Re}(a_\ell) + i\,\text{Im}(a_\ell))|^{2n}\right)^{1/2n}$$

$$= \left(\mathbb{E}[|\sum_{\ell=1}^{M} \epsilon_\ell \text{Re}(a_\ell)|^2 + |\sum_{\ell=1}^{M} \epsilon_\ell \text{Im}(a_\ell)|^2]^n\right)^{1/2n}$$

$$\leq \left((\mathbb{E}|\sum_{\ell=1}^{M} \epsilon_\ell \text{Re}(a_\ell)|^{2n})^{1/n} + (\mathbb{E}|\sum_{\ell=1}^{M} \epsilon_\ell \text{Im}(a_\ell)|^{2n})^{1/n}\right)^{1/2}$$

$$\leq \left(\left(\frac{(2n)!}{2^n n!}\right)^{1/n}(\|\text{Re}(\mathbf{a})\|_2^2 + \|\text{Im}(\mathbf{a})\|_2^2)\right)^{1/2} = \left(\frac{(2n)!}{2^n n!}\right)^{1/2n}\|\mathbf{a}\|_2.$$

This concludes the proof. $\qquad\square$

Remark 8.6. (a) The constant in the Khintchine inequality can be expressed as a
double factorial

$$\frac{(2n)!}{2^n n!} = (2n-1)!! := 1 \times 3 \times 5 \times 7 \times \cdots \times (2n-1).$$

(b) If $\mathbf{g} = (g_1, \ldots, g_M)$ is a standard Gaussian random vector, then the sum
$\sum_{\ell=1}^{M} a_\ell g_\ell$ with real a_ℓ is a Gaussian random variable with mean zero and
variance $\|\mathbf{a}\|_2^2$. By Corollary 7.7 its moments are given by

$$\mathbb{E}\Big|\sum_{\ell=1}^{M} a_\ell g_\ell\Big|^{2n} = \frac{(2n)!}{2^n n!} \|\mathbf{a}\|_2^{2n}.$$

In other words, if the Rademacher sequence is replaced by independent standard
normal variables, then (8.8) holds with equality. Therefore, the central limit
theorem shows that the constants in (8.8) are optimal. Moreover, it also follows
that $\mathbb{E}|\sum_{\ell=1}^{M} \epsilon_\ell a_\ell|^{2n} \leq \mathbb{E}|\sum_{\ell=1}^{M} g_\ell a_\ell|^{2n}$; compare also Exercise 8.2.

Based on Theorem 8.5, we can also estimate the general absolute pth moment of
a Rademacher sum.

Corollary 8.7. *Let* $\mathbf{a} \in \mathbb{C}^M$ *and* $\epsilon = (\epsilon_1, \ldots, \epsilon_M)$ *be a Rademacher sequence.*
Then, for all $p > 0$,

$$\Big(\mathbb{E}\Big|\sum_{\ell=1}^{M} \epsilon_\ell a_\ell\Big|^p\Big)^{1/p} \leq 2^{3/(4p)} e^{-1/2} \sqrt{p} \|\mathbf{a}\|_2. \tag{8.9}$$

Proof. We first assume that $p \geq 2$. Stirling's formula (C.13) for the factorial gives

$$\frac{(2n)!}{2^n n!} = \frac{\sqrt{2\pi 2n}(2n/e)^{2n} e^{R_{2n}}}{2^n \sqrt{2\pi n}(n/e)^n e^{R_n}} \leq \sqrt{2}\,(2/e)^n n^n. \tag{8.10}$$

where $1/(12n+1) \leq R_n \leq 1/(12n)$. An application of Hölder's inequality yields,
for $\theta \in [0, 1]$, and an arbitrary random variable Z,

$$\mathbb{E}|Z|^{2n+2\theta} = \mathbb{E}[|Z|^{(1-\theta)2n}|Z|^{\theta(2n+2)}] \leq (\mathbb{E}|Z|^{2n})^{1-\theta}(\mathbb{E}|Z|^{2n+2})^\theta. \tag{8.11}$$

Without loss of generality we may assume $\|\mathbf{a}\|_2 = 1$. Combining the two estimates
above yields

$$\mathbb{E}\Big|\sum_{\ell=1}^{M} \epsilon_\ell a_\ell\Big|^{2n+2\theta} \leq (\mathbb{E}\Big|\sum_{\ell=1}^{M} \epsilon_\ell a_\ell\Big|^{2n})^{1-\theta}(\mathbb{E}\Big|\sum_{\ell=1}^{M} \epsilon_\ell a_\ell\Big|^{2n+2})^\theta$$

$$\leq (\sqrt{2}(2/e)^n n^n)^{1-\theta}(\sqrt{2}(2/e)^{n+1}(n+1)^{n+1})^\theta$$

$$= \sqrt{2}(2/e)^{n+\theta} n^{(1-\theta)n} (n+1)^{\theta(n+1)}$$

$$= \sqrt{2}(2/e)^{n+\theta} (n^{1-\theta}(n+1)^{\theta})^{n+\theta} \left(\frac{n+1}{n}\right)^{\theta(1-\theta)}$$

$$\leq \sqrt{2}(2/e)^{n+\theta} (n+\theta)^{n+\theta} \left(\frac{n+1}{n}\right)^{\theta(1-\theta)}$$

$$\leq 2^{3/4}(2/e)^{n+\theta} (n+\theta)^{n+\theta}. \tag{8.12}$$

In the second line from below the inequality between the geometric and arithmetic mean was applied. The last step used that $(n+1)/n \leq 2$ and $\theta(1-\theta) \leq 1/4$. Replacing $n + \theta$ by $p/2$ completes the proof of (8.9) for $p \geq 2$.

For the case $0 < p \leq 2$ we observe that Hölder's inequality gives

$$(\mathbb{E}|\sum_{\ell=1}^{M} \epsilon_\ell a_\ell|^p)^{1/p} \leq (\mathbb{E}|\sum_{\ell=1}^{M} \epsilon_\ell a_\ell|^2)^{1/2} = 1.$$

It is an elementary exercise to show that the function $f(p) = 2^{3/(4p)} e^{-1/2} \sqrt{p}$ takes its minimum at the point $p_0 = 3(\ln 2)/2$ and $f(p_0) \approx 1.0197 > 1$. Therefore, we have (8.9) also for $p < 2$. $\qquad\square$

We obtain the following version of *Hoeffding's inequality* for complex Rademacher sums.

Corollary 8.8. *Let* a $\in \mathbb{C}^M$ *and* $\epsilon = (\epsilon_1, \ldots, \epsilon_M)$ *be a Rademacher sequence. Then, for $u > 0$,*

$$\mathbb{P}(|\sum_{\ell=1}^{M} \epsilon_\ell a_\ell| \geq \|\mathbf{a}\|_2 u) \leq 2\exp(-u^2/2). \tag{8.13}$$

Proof. We combine (8.9) with Proposition 7.11 to obtain

$$\mathbb{P}(|\sum_{\ell=1}^{M} \epsilon_\ell a_\ell| \geq \|\mathbf{a}\|_2 u) \leq 2^{3/4} \exp(-u^2/2), \quad u > 0,$$

which is even slightly better (but less appealing) than the claimed estimate. $\qquad\square$

A complex random variable which is uniformly distributed on the torus $\mathbb{T} = \{z \in \mathbb{C}, |z| = 1\}$ is called a Steinhaus variable. A sequence $\epsilon = (\epsilon_1, \ldots, \epsilon_N)$ of independent Steinhaus variables is called a *Steinhaus sequence*. There is also a version of the Khintchine inequality for Steinhaus sequences.

Theorem 8.9. *Let* $\mathbf{a} \in \mathbb{C}^M$ *and* $\epsilon = (\epsilon_1, \ldots, \epsilon_M)$ *be a Steinhaus sequence. Then, for all* $n \in \mathbb{N}$,

$$\mathbb{E}\Big|\sum_{\ell=1}^{M} \epsilon_\ell a_\ell\Big|^{2n} \leq n! \|\mathbf{a}\|_2^{2n}.$$

Proof. We expand the moments of the Steinhaus sum using the multinomial theorem

$$\mathbb{E}\Big|\sum_{\ell=1}^{M} \epsilon_\ell a_\ell\Big|^{2n} = \mathbb{E}\left[\Big(\sum_{\ell=1}^{M} \epsilon_\ell a_\ell\Big)^n \Big(\sum_{\ell=1}^{M} \overline{\epsilon_\ell a_\ell}\Big)^n\right]$$

$$= \mathbb{E}\left[\sum_{\substack{j_1+\cdots+j_M=n \\ j_\ell \geq 0}} \frac{n!}{j_1! \cdots j_M!} a_1^{j_1} \cdots a_M^{j_M} \epsilon_1^{j_1} \cdots \epsilon_M^{j_M}\right.$$

$$\times \left.\sum_{\substack{k_1+\cdots+k_M=n \\ k_\ell \geq 0}} \frac{n!}{k_1! \cdots k_M!} \overline{a_1^{k_1}} \cdots \overline{a_M^{k_M}} \overline{\epsilon_1^{k_1}} \cdots \overline{\epsilon_M^{k_M}}\right]$$

$$= \sum_{\substack{j_1+\cdots+j_M=n \\ k_1+\cdots+k_M=n \\ j_\ell, k_\ell \geq 0}} \frac{n!}{j_1! \cdots j_M!} \frac{n!}{k_1! \cdots k_M!} a_1^{j_1} \overline{a_1^{k_1}} \cdots a_M^{j_M} \overline{a_M^{k_M}} \mathbb{E}[\epsilon_1^{j_1} \overline{\epsilon_1^{k_1}} \cdots \epsilon_M^{j_M} \overline{\epsilon_M^{k_M}}].$$

Since the ϵ_j are independent and uniformly distributed on the torus we have

$$\mathbb{E}[\epsilon_1^{j_1} \overline{\epsilon_1^{k_1}} \cdots \epsilon_M^{j_M} \overline{\epsilon_M^{k_M}}] = \mathbb{E}[\epsilon_1^{j_1} \overline{\epsilon_1^{k_1}}] \times \cdots \times \mathbb{E}[\epsilon_M^{j_M} \overline{\epsilon_M^{k_M}}] = \delta_{j_1,k_1} \times \cdots \times \delta_{j_M,k_M}.$$

This yields

$$\mathbb{E}\Big|\sum_{\ell=1}^{M} \epsilon_\ell a_\ell\Big|^{2n} = \sum_{\substack{k_1+\cdots+k_M=n \\ k_\ell \geq 0}} \Big(\frac{n!}{k_1! \cdots k_M!}\Big)^2 |a_1|^{2k_1} \cdots |a_M|^{2k_M}$$

$$\leq n! \sum_{\substack{k_1+\cdots+k_M=n \\ k_\ell \geq 0}} \frac{n!}{k_1! \cdots k_M!} |a_1|^{2k_1} \cdots |a_M|^{2k_M}$$

$$= n! \Big(\sum_{\ell=1}^{M} |a_\ell|^2\Big)^{2n},$$

where the multinomial theorem was applied in the last step. $\square$

The above moment estimate leads to a Hoeffding-type inequality for Steinhaus sums.

Corollary 8.10. *Let* $\mathbf{a} \in \mathbb{C}^M$ *and* $\epsilon = (\epsilon_1, \ldots, \epsilon_M)$ *be a Steinhaus sequence. For any* $0 < \lambda < 1$,

$$\mathbb{P}(|\sum_{\ell=1}^{M} \epsilon_\ell a_\ell| \geq u\|\mathbf{a}\|_2) \leq \frac{1}{1-\lambda} e^{-\lambda u^2} \quad \text{for all } u > 0. \tag{8.14}$$

In particular, using the optimal choice $\lambda = 1 - u^{-2}$,

$$\mathbb{P}(|\sum_{\ell=1}^{M} \epsilon_\ell a_\ell| \geq u\|\mathbf{a}\|_2) \leq \exp(-u^2 + \ln(u^2) + 1) \quad \text{for all } u \geq 1. \tag{8.15}$$

Proof. Without loss of generality we assume that $\|\mathbf{a}\|_2 = 1$. Markov's inequality gives

$$\mathbb{P}(|\sum_{\ell=1}^{M} \epsilon_\ell a_\ell| \geq u) = \mathbb{P}\left(\exp(\lambda|\sum_{\ell=1}^{M} \epsilon_\ell a_\ell|^2) \geq \exp(\lambda u^2)\right)$$

$$\leq \mathbb{E}[\exp(\lambda|\sum_{\ell=1}^{M} \epsilon_\ell a_\ell|^2)] \exp(-\lambda u^2) = \exp(-\lambda u^2) \sum_{n=0}^{\infty} \frac{\lambda^n \mathbb{E}|\sum_{\ell=1}^{M} \epsilon_\ell a_\ell|^{2n}}{n!}$$

$$\leq \exp(-\lambda u^2) \sum_{n=0}^{\infty} \lambda^n = \frac{1}{1-\lambda} e^{-\lambda u^2}.$$

In the second line Fubini's theorem and in the third line Theorem 8.9 were applied.
□

8.4 Decoupling

Decoupling is a technique that reduces stochastic dependencies in certain sums of random variables, called chaos variables. A typical example is a sum of the form $\sum_{j \neq k} \epsilon_j \epsilon_k \mathbf{x}_{j,k}$ where $\mathbf{x}_{j,k}$ are some vectors and $\epsilon = (\epsilon_j)$ is a Rademacher sequence. Such a sum is called a homogeneous Rademacher chaos of order 2. The term homogeneous refers to the fact that the diagonal terms in this double sum are missing so that its expectation is zero. The following statement provides a way of "decoupling" the sum.

Theorem 8.11. *Let* $\xi = (\xi_1, \ldots, \xi_M)$ *be a sequence of independent random variables with* $\mathbb{E}\xi_j = 0$ *for all* $j \in [M]$. *Let* $\mathbf{x}_{j,k}$, $j, k \in [M]$, *be a double sequence of elements in a finite-dimensional vector space* V. *If* $F : V \to \mathbb{R}$ *is a convex function, then*

$$\mathbb{E}F\left(\sum_{\substack{j,k=1\\j\neq k}}^{M}\xi_j\xi_k\mathbf{x}_{j,k}\right) \leq \mathbb{E}F\left(4\sum_{j,k=1}^{M}\xi_j\xi_k'\mathbf{x}_{j,k}\right), \tag{8.16}$$

where $\boldsymbol{\xi}'$ denotes an independent copy of $\boldsymbol{\xi}$.

Proof. Introduce a sequence $\boldsymbol{\delta} = (\delta_j)_{j=1}^{M}$ of independent random variables δ_j taking the values 0 and 1 with probability $1/2$. Then, for $j \neq k$,

$$\mathbb{E}\delta_k(1-\delta_j) = 1/4. \tag{8.17}$$

This gives

$$E := \mathbb{E}F\left(\sum_{j\neq k}^{M}\xi_j\xi_k\mathbf{x}_{j,k}\right) = \mathbb{E}_{\boldsymbol{\xi}}F\left(4\sum_{j\neq k}^{M}\mathbb{E}_{\boldsymbol{\delta}}[\delta_j(1-\delta_k)]\xi_j\xi_k\mathbf{x}_{j,k}\right)$$

$$\leq \mathbb{E}_{\boldsymbol{\xi}}\mathbb{E}_{\boldsymbol{\delta}}F\left(4\sum_{j\neq k}^{M}\delta_j(1-\delta_k)\xi_j\xi_k\mathbf{x}_{j,k}\right),$$

where Jensen's inequality was applied in the last step. Now let

$$\sigma(\boldsymbol{\delta}) := \{j \in [M] : \delta_j = 1\}.$$

Then, by Fubini's theorem,

$$E \leq \mathbb{E}_{\boldsymbol{\delta}}\mathbb{E}_{\boldsymbol{\xi}}F\left(4\sum_{j\in\sigma(\boldsymbol{\delta})}\sum_{k\notin\sigma(\boldsymbol{\delta})}\xi_j\xi_k\mathbf{x}_{j,k}\right).$$

For fixed $\boldsymbol{\delta}$ the sequences $(\xi_j)_{j\in\sigma(\boldsymbol{\delta})}$ and $(\xi_k)_{k\notin\sigma(\boldsymbol{\delta})}$ are independent; hence, we can replace ξ_k, $k \notin \sigma(\boldsymbol{\delta})$, by an independent copy ξ_k' and obtain

$$E \leq \mathbb{E}_{\boldsymbol{\delta}}\mathbb{E}_{\boldsymbol{\xi}}\mathbb{E}_{\boldsymbol{\xi}'}F\left(4\sum_{j\in\sigma(\boldsymbol{\delta})}\sum_{k\notin\sigma(\boldsymbol{\delta})}\xi_j\xi_k'\mathbf{x}_{j,k}\right).$$

This implies the existence of a $\boldsymbol{\delta}^* \in \{0,1\}^M$, and hence, a $\sigma = \sigma(\boldsymbol{\delta}^*)$ such that

$$E \leq \mathbb{E}_{\boldsymbol{\xi}}\mathbb{E}_{\boldsymbol{\xi}'}F\left(4\sum_{j\in\sigma}\sum_{k\notin\sigma}\xi_j\xi_k'\mathbf{x}_{j,k}\right).$$

Since $\mathbb{E}\xi_j = \mathbb{E}\xi_j' = 0$, an application of Jensen's inequality yields

$$E \leq \mathbb{E}F\left(4\sum_{j\in\sigma}\left(\sum_{k\notin\sigma}\xi_j\xi_k'\mathbf{x}_{j,k} + \sum_{k\in\sigma}\xi_j\mathbb{E}[\xi_k']\mathbf{x}_{j,k}\right) + 4\sum_{j\notin\sigma}\mathbb{E}[\xi_j]\sum_{k=1}^{M}\xi_k'\mathbf{x}_{j,k}\right)$$

$$\leq \mathbb{E}F\left(4\sum_{j=1}^{M}\sum_{k=1}^{M}\xi_j\xi_k'\mathbf{x}_{j,k}\right)$$

and the proof is complete. □

The sum $\sum_{j,k}\xi_j\xi_k'\mathbf{x}_{j,k}$ on the right-hand side of (8.16) is called a decoupled chaos. It is important that the double sum on the left-hand side of (8.16) runs only over indices $j \neq k$. Moreover, since the left-hand side of (8.16) is independent of the diagonal entries $\mathbf{x}_{j,j}$, they can be chosen arbitrarily on the right-hand side. Sometimes it is convenient to choose them as $\mathbf{x}_{j,j} = 0$, but other choices may simplify computations.

An important special case of the above theorem is $F(\mathbf{x}) = \|\mathbf{x}\|^p$ with $p \geq 1$ and some (semi-)norm $\|\cdot\|$. Then (8.16) implies

$$\left(\mathbb{E}\|\sum_{j\neq k}\xi_j\xi_k\mathbf{x}_{j,k}\|^p\right)^{1/p} \leq 4\left(\mathbb{E}\|\sum_{j,k}\xi_j\xi_k'\mathbf{x}_{j,k}\|^p\right)^{1/p}.$$

The mean-zero assumption above for the random variables ξ_j can be removed after possibly adjusting constants. We will exemplify this for the following special case involving the operator norm where, additionally, the constant can be improved.

Theorem 8.12. *Let $\hat{\mathbf{H}} \in \mathbb{C}^{M\times M}$ be self-adjoint and let $\mathbf{H}$ be the matrix $\hat{\mathbf{H}}$ with the diagonal entries put to zero. Let ξ_j, $j \in [M]$, be a sequence of independent random variables. Introduce the random diagonal matrix $\mathbf{D}_\xi = \mathrm{diag}[\xi_1,\ldots,\xi_M]$. If $F : \mathbb{R}_+ \to \mathbb{R}$ is a convex nondecreasing function, then*

$$\mathbb{E}F(\|\mathbf{D}_\xi\mathbf{H}\mathbf{D}_\xi\|_{2\to2}) \leq \mathbb{E}F(2\|\mathbf{D}_\xi\hat{\mathbf{H}}\mathbf{D}_{\xi'}\|_{2\to2}), \tag{8.18}$$

where ξ' denotes an independent copy of ξ.

Proof. Let $\mathbf{H}_{jk} \in \mathbb{C}^{M\times M}$ be the matrix with entry $\hat{H}_{jk}$ in position (j,k) and zero elsewhere. Let δ_j, $j \in [M]$, be independent Bernoulli random variables taking the values 0 and 1 both with probability $1/2$. The function $\mathbf{x} \mapsto F(\|\mathbf{x}\|_{2\to2})$ is convex by Proposition B.9(b) so that Jensen's inequality and (8.17) yield

$$\mathbb{E}F(\|\mathbf{D}_\xi\mathbf{H}\mathbf{D}_\xi\|_{2\to2}) = \mathbb{E}F(\|\sum_{j<k}\xi_j\xi_k(\mathbf{H}_{jk} + \mathbf{H}_{kj})\|_{2\to2})$$

$$= \mathbb{E}_\xi F\left(2\|\mathbb{E}_\delta\sum_{j<k}[\delta_j(1-\delta_k) + \delta_k(1-\delta_j)]\xi_j\xi_k(\mathbf{H}_{jk} + \mathbf{H}_{kj})\|_{2\to2}\right)$$

$$\leq \mathbb{E}_\xi\mathbb{E}_\delta F\left(2\|\sum_{j<k}[\delta_j(1-\delta_k) + \delta_k(1-\delta_j)]\xi_j\xi_k(\mathbf{H}_{jk} + \mathbf{H}_{kj})\|_{2\to2}\right). \tag{8.19}$$

Therefore, there exists a vector δ^* with entries in $\{0, 1\}$ such that

$$\mathbb{E}F\big(\|\mathbf{D}_\xi \mathbf{H}\mathbf{D}_\xi\|_{2\to 2}\big) \le \mathbb{E}F\Big(2\|\sum_{j<k}[\delta_j^*(1-\delta_k^*)+\delta_k^*(1-\delta_j^*)]\xi_j\xi_k(\mathbf{H}_{jk}+\mathbf{H}_{kj})\|_{2\to 2}\Big).$$

Let $\sigma = \sigma(\delta^*) = \{j \in [M], \delta_j^* = 1\}$. Then

$$\mathbb{E}F\big(\|\mathbf{D}_\xi \mathbf{H}\mathbf{D}_\xi\|_{2\to 2}\big) \le \mathbb{E}F\Big(2\|\sum_{j\in\sigma, k\in\overline\sigma} \xi_j\xi_k(\mathbf{H}_{jk}+\mathbf{H}_{kj})\|_{2\to 2}\Big).$$

By rearranging the index set, we may assume that $\sigma = \{1, \dots, \text{card}(\sigma)\}$ and $\overline\sigma = \{\text{card}(\sigma)+1, \dots, M\}$. Then we can write

$$\sum_{j\in\sigma, k\in\overline\sigma} \xi_j\xi_k(\mathbf{H}_{jk}+\mathbf{H}_{kj}) = \begin{pmatrix} \mathbf{0} & \mathbf{B} \\ \mathbf{B}^* & \mathbf{0} \end{pmatrix}$$

with $\mathbf{B} \in \mathbb{C}^{\text{card}(\sigma)\times\text{card}(\overline\sigma)}$ being the restriction of $\sum_{j\in\sigma, k\in\overline\sigma} \xi_j\xi_k\mathbf{H}_{jk}$ to the indices in $\sigma \times \overline\sigma$. Using

$$\left\| \begin{pmatrix} \mathbf{0} & \mathbf{B} \\ \mathbf{B}^* & \mathbf{0} \end{pmatrix} \right\|_{2\to 2} = \|\mathbf{B}\|_{2\to 2}$$

we arrive at

$$\mathbb{E}F\big(\|\mathbf{D}_\xi \mathbf{H}\mathbf{D}_\xi\|_{2\to 2}\big) \le \mathbb{E}F\Big(2\|\sum_{j\in\sigma, k\in\overline\sigma} \xi_j\xi_k\mathbf{H}_{jk}\|_{2\to 2}\Big)$$

$$= \mathbb{E}F\Big(2\|\sum_{j\in\sigma, k\in\overline\sigma} \xi_j\xi_k'\mathbf{H}_{jk}\|_{2\to 2}\Big),$$

where ξ' is an independent copy of ξ. Since the operator norm of a submatrix is bounded by the operator norm of the full matrix (see Lemma A.9), we reinsert the missing entries to obtain

$$\mathbb{E}F\big(\|\mathbf{D}_\xi \mathbf{H}\mathbf{D}_\xi\|_{2\to 2}\big) \le \mathbb{E}F\Big(2\|\sum_{j,k} \xi_j\xi_k'\mathbf{H}_{jk}\|_{2\to 2}\Big) = \mathbb{E}F\big(2\|\mathbf{D}_\xi \hat{\mathbf{H}}\mathbf{D}_{\xi'}\|_{2\to 2}\big),$$

where we used that F is nondecreasing. This completes the argument. $\square$

We finish this section with an application to tail bounds for scalar Rademacher chaos. Let $\epsilon = (\epsilon_1, \dots, \epsilon_M)$ be a Rademacher vector. For a self-adjoint matrix $\mathbf{A} \in \mathbb{C}^{M\times M}$ with zero diagonal we consider the homogeneous Rademacher chaos

$$X := \epsilon^*\mathbf{A}\epsilon = \sum_{j\ne k} \epsilon_j\epsilon_k A_{jk}. \tag{8.20}$$

Note that by self-adjointness, X is real valued even if $\mathbf{A}$ is complex valued. This fact allows to reduce our considerations to real-valued symmetric matrices $\mathbf{A} \in \mathbb{R}^{M \times M}$ since $X = \mathrm{Re}(X) = \epsilon^* \mathrm{Re}(\mathbf{A})\epsilon$. The next result states that a homogeneous Rademacher chaos obeys a mixture of subgaussian and subexponential tail behavior, similar to Bernstein inequalities. The subgaussian part is determined by the Frobenius norm $\|\mathbf{A}\|_F^2 = \mathrm{tr}\,(\mathbf{A}^*\mathbf{A})$ (see (A.16)), while the operator norm $\|\mathbf{A}\|_{2 \to 2}$ controls the subexponential part.

Proposition 8.13. *Let* $\mathbf{A} \in \mathbb{R}^{M \times M}$ *be a symmetric matrix with zero diagonal and let* ϵ *a Rademacher vector. Then the homogeneous Rademacher chaos X defined in (8.20) satisfies, for $t > 0$,*

$$\mathbb{P}(|X| \geq t) \leq 2 \exp\left(-\min\left\{\frac{3t^2}{128\|\mathbf{A}\|_F^2}, \frac{t}{32\|\mathbf{A}\|_{2 \to 2}}\right\}\right)$$

$$= \begin{cases} 2 \exp\left(-\frac{3t^2}{128\|\mathbf{A}\|_F^2}\right) & \text{if } 0 < t \leq \frac{4\|\mathbf{A}\|_F^2}{3\|\mathbf{A}\|_{2 \to 2}}, \\ 2 \exp\left(-\frac{t}{32\|\mathbf{A}\|_{2 \to 2}}\right) & \text{if } t > \frac{4\|\mathbf{A}\|_F^2}{3\|\mathbf{A}\|_{2 \to 2}}. \end{cases}$$

Proof. The proof is based on an estimate of the moment-generating function of X. For $\theta > 0$, convexity of $x \mapsto \exp(\theta x)$ combined with the decoupling inequality (8.16) yields

$$\mathbb{E}\exp(\theta X) = \mathbb{E}\exp\left(\theta \sum_{j \neq k} \epsilon_j \epsilon_k A_{jk}\right) \leq \mathbb{E}\exp\left(4\theta \sum_{j,k} \epsilon_j \epsilon_k' A_{jk}\right)$$

$$= \mathbb{E}_\epsilon \mathbb{E}_{\epsilon'} \exp\left(4\theta \sum_k \epsilon_k' \sum_j \epsilon_j A_{jk}\right) \leq \mathbb{E}\exp\left(8\theta^2 \sum_k \left(\sum_j \epsilon_j A_{jk}\right)^2\right). \quad (8.21)$$

In the last step we have applied Theorem 7.27 conditionally on ϵ, using that $c = 1/2$ for Rademacher variables; see Remark 7.26. Observe that by symmetry of $\mathbf{A}$,

$$\sum_k \left(\sum_j \epsilon_j A_{jk}\right)^2 = \sum_k \sum_j \epsilon_j A_{jk} \sum_\ell \epsilon_\ell A_{\ell k} = \sum_{j,\ell} \epsilon_j \epsilon_\ell \sum_k A_{jk} A_{k\ell} = \epsilon^* \mathbf{A}^2 \epsilon.$$

Set $\mathbf{B} = \mathbf{A}^2$. The moment-generating function of the positive semidefinite chaos $\epsilon^* \mathbf{B} \epsilon$ can be estimated by

$$\mathbb{E}\exp(\kappa \epsilon^* \mathbf{B} \epsilon) = \mathbb{E}\exp\left(\kappa \sum_j B_{jj} + \kappa \sum_{j \neq k} \epsilon_j \epsilon_k B_{jk}\right)$$

$$\leq \exp(\kappa \mathrm{tr}\,(\mathbf{B}))\mathbb{E}\exp\left(4\kappa \sum_{j,k} \epsilon_j \epsilon_k' B_{jk}\right)$$

$$\leq \exp(\kappa \mathrm{tr}\,(\mathbf{B}))\mathbb{E}\exp\left(8\kappa^2 \sum_k \left(\sum_j \epsilon_j B_{j,k}\right)^2\right),$$

where we have again applied the decoupling inequality (8.16) together with Theorem 7.27 conditionally on ϵ. Now, positive semidefiniteness of $\mathbf{B} = \mathbf{A}^*\mathbf{A}$ allows to take the square root of $\mathbf{B}$ so that

$$\sum_k \left(\sum_j \epsilon_j B_{j,k}\right)^2 = \epsilon^*\mathbf{B}^2\epsilon = (\mathbf{B}^{1/2}\epsilon)^*\mathbf{B}(\mathbf{B}^{1/2}\epsilon) \le \|\mathbf{B}\|_{2\to2}\epsilon^*\mathbf{B}\epsilon.$$

If $8\kappa\|\mathbf{B}\|_{2\to2} < 1$, then Hölder's (or Jensen's) inequality yields

$$\mathbb{E}\exp(\kappa\epsilon^*\mathbf{B}\epsilon) \le \exp(\kappa\operatorname{tr}(\mathbf{B}))\mathbb{E}\exp(8\kappa^2\|\mathbf{B}\|_{2\to2}\epsilon^*\mathbf{B}\epsilon)$$

$$\le \exp(\kappa\operatorname{tr}(\mathbf{B}))\left(\mathbb{E}\exp(\kappa\epsilon^*\mathbf{B}\epsilon)\right)^{8\kappa\|\mathbf{B}\|_{2\to2}}.$$

After rearranging we deduce that

$$\mathbb{E}\exp(\kappa\epsilon^*\mathbf{B}\epsilon) \le \exp\left(\frac{\kappa\operatorname{tr}(\mathbf{B})}{1 - 8\kappa\|\mathbf{B}\|_{2\to2}}\right), \quad 0 < \kappa < (8\|\mathbf{B}\|_{2\to2})^{-1}. \quad (8.22)$$

Setting $\kappa = 8\theta^2$ and plugging into (8.21) yield, for $0 < \theta < (8\|\mathbf{A}\|_{2\to2})^{-1}$,

$$\mathbb{E}\exp(\theta X) \le \exp\left(\frac{8\theta^2\operatorname{tr}(\mathbf{A}^2)}{1 - 64\theta^2\|\mathbf{A}^2\|_{2\to2}}\right) = \exp\left(\frac{8\theta^2\|\mathbf{A}\|_F^2}{1 - 64\theta^2\|\mathbf{A}\|_{2\to2}^2}\right).$$

Next we use Markov's inequality to deduce, for $0 < \theta \le (16\|\mathbf{A}\|_{2\to2})^{-1}$,

$$\mathbb{P}(X \ge t) = \mathbb{P}(\exp(\theta X) \ge \exp(\theta t)) \le \exp(-\theta t)\mathbb{E}\exp(\theta X)$$

$$\le \exp\left(-\theta t + \frac{8\theta^2\|\mathbf{A}\|_F^2}{1 - 64\theta^2\|\mathbf{A}\|_{2\to2}^2}\right) \le \exp\left(-\theta t + \frac{8\theta^2\|\mathbf{A}\|_F^2}{1 - 1/4}\right)$$

$$= \exp\left(-\theta t + 32\theta^2\|\mathbf{A}\|_F^2/3\right).$$

The optimal choice $\theta = 3t/(64\|\mathbf{A}\|_F^2)$ satisfies $\theta \le (16\|\mathbf{A}\|_{2\to2})^{-1}$ provided that $t \le 4\|\mathbf{A}\|_F^2/(3\|\mathbf{A}\|_{2\to2})$. In this regime, we therefore obtain

$$\mathbb{P}(X \ge t) \le \exp\left(-\frac{3t^2}{128\|\mathbf{A}\|_F^2}\right).$$

In the other regime where $t > 4\|\mathbf{A}\|_F^2/(3\|\mathbf{A}\|_{2\to2})$ we set $\theta = (16\|\mathbf{A}\|_{2\to2})^{-1}$ so that $\theta < 3t/(64\|\mathbf{A}\|_F^2)$. Then

$$\mathbb{P}(X \ge t) \le \exp\left(-\theta t + 32\theta^2\|\mathbf{A}\|_F^2/3\right) \le \exp\left(-\theta t + \theta t/2\right) = \exp(-\theta t/2)$$

$$= \exp(-t/(32\|\mathbf{A}\|_{2\to2})).$$

Since X has the same distribution as $-X$, we get the same bounds for $\mathbb{P}(X \le -t)$, and the union bound completes the proof. $\square$

8.5 Noncommutative Bernstein Inequality

The scalar Bernstein inequalities from the previous chapter have a powerful extension to sums of random matrices. We present one version below. Another version is treated in Exercise 8.8. We denote by $\lambda_{\max}(\mathbf{X})$ the maximal eigenvalue of a self-adjoint square matrix $\mathbf{X}$. Furthermore, we introduce the function

$$h(x) := (1+x)\ln(1+x) - x, \quad x \geq 0. \tag{8.23}$$

Theorem 8.14. *Let* $\mathbf{X}_1, \ldots, \mathbf{X}_M \in \mathbb{C}^{d \times d}$ *be independent mean-zero self-adjoint random matrices. Assume that the largest eigenvalue of* $\mathbf{X}_\ell$ *satisfies*

$$\lambda_{\max}(\mathbf{X}_\ell) \leq K \quad \text{almost surely for all } \ell \in [M] \tag{8.24}$$

and set

$$\sigma^2 := \left\| \sum_{\ell=1}^M \mathbb{E}(\mathbf{X}_\ell^2) \right\|_{2 \to 2}.$$

Then, for $t > 0$,

$$\mathbb{P}\left(\lambda_{\max}\left(\sum_{\ell=1}^M \mathbf{X}_\ell \right) \geq t \right) \leq d \exp\left(-\frac{\sigma^2}{K^2} h\left(\frac{Kt}{\sigma^2} \right) \right) \tag{8.25}$$

$$\leq d \exp\left(-\frac{t^2/2}{\sigma^2 + Kt/3} \right). \tag{8.26}$$

The inequality (8.25) is also referred to as the matrix Bennett inequality. Although it is slightly stronger than the matrix Bernstein inequality (8.26), the latter is usually more convenient to use. Clearly, the difference with respect to the scalar Bernstein inequalities of the previous chapter is only the appearance of the dimensional factor d in front of the exponential. In general, this factor cannot be avoided; see also Exercise 8.6(e).

Since for a self-adjoint matrix $\|\mathbf{A}\|_{2 \to 2} = \max\{\lambda_{\max}(\mathbf{A}), \lambda_{\max}(-\mathbf{A})\}$, we obtain the next statement as a simple consequence.

Corollary 8.15. *Let* $\mathbf{X}_1, \ldots, \mathbf{X}_M \in \mathbb{C}^{d \times d}$ *be independent mean-zero self-adjoint random matrices. Assume that*

$$\|\mathbf{X}_\ell\|_{2 \to 2} \leq K \quad \text{almost surely for all } \ell \in [M], \tag{8.27}$$

and set

$$\sigma^2 := \left\| \sum_{\ell=1}^{M} \mathbb{E}(\mathbf{X}_\ell^2) \right\|_{2 \to 2}. \tag{8.28}$$

Then, for $t > 0$,

$$\mathbb{P} \left(\left\| \sum_{\ell=1}^{M} \mathbf{X}_\ell \right\|_{2 \to 2} \geq t \right) \leq 2d \exp \left(-\frac{\sigma^2}{K^2} h \left(\frac{Kt}{\sigma^2} \right) \right) \tag{8.29}$$

$$\leq 2d \exp \left(-\frac{t^2/2}{\sigma^2 + Kt/3} \right). \tag{8.30}$$

An extension to rectangular (and not necessarily self-adjoint) matrices is developed in Exercise 8.7.

The essential steps of the proof proceed in the same way as the ones of the scalar Bernstein inequality, but since we are dealing with matrices, we encounter some additional complications. We will use an extension of the Laplace transform method (or moment-generating function method) to matrices. A crucial ingredient is Lieb's concavity theorem (Theorem B.31).

We start with a simple consequence of Markov's inequality. It uses the matrix exponential $\mathbf{A} \mapsto \exp(\mathbf{A})$ defined in (A.45). We refer to Appendix A.5 for basic facts on matrix functions.

Proposition 8.16. *Let $\mathbf{Y} \in \mathbb{C}^{d \times d}$ be a self-adjoint random matrix. Then, for $t \in \mathbb{R}$,*

$$\mathbb{P}(\lambda_{\max}(\mathbf{Y}) \geq t) \leq \inf_{\theta > 0} \left\{ e^{-\theta t} \mathbb{E} \mathrm{tr} \, \exp(\theta \mathbf{Y}) \right\}. \tag{8.31}$$

Proof. For any $\theta > 0$, Markov's inequality (Theorem 7.3) yields

$$\mathbb{P}(\lambda_{\max}(\mathbf{Y}) \geq t) = \mathbb{P} \left(e^{\lambda_{\max}(\theta \mathbf{Y})} \geq e^{\theta t} \right) \leq e^{-\theta t} \mathbb{E} \left[e^{\lambda_{\max}(\theta \mathbf{Y})} \right]. \tag{8.32}$$

By the spectral mapping theorem (A.42) (or by the definition of a matrix function), and positivity of the exponential function, we have

$$e^{\lambda_{\max}(\theta \mathbf{Y})} = \lambda_{\max}(e^{\theta \mathbf{Y}}) \leq \sum_{j=1}^{d} \lambda_j(e^{\theta \mathbf{Y}}) = \mathrm{tr} \, e^{\theta \mathbf{Y}},$$

where $\lambda_j(e^{\theta \mathbf{Y}}) \geq 0$, $j \in [d]$, are the eigenvalues of $e^{\theta \mathbf{Y}}$ (possibly with repetitions). Combined with the previous estimate we reach

$$\mathbb{P}(\lambda_{\max}(\mathbf{Y}) \geq t) \leq e^{-\theta t} \mathbb{E} \mathrm{tr} \, e^{\theta \mathbf{Y}}.$$

Taking the infimum over all positive θ concludes the proof.   □

The previous proposition suggests to study the expectation of the trace exponential $\theta \mapsto \mathbb{E}\operatorname{tr} e^{\theta \mathbf{Y}}$. The next result provides a useful tool for analyzing it and is a consequence of Lieb's Theorem B.31. We will use the matrix logarithm introduced in Appendix A.5; see (A.50).

Proposition 8.17. *Let* $\mathbf{H} \in \mathbb{C}^{d \times d}$ *be a fixed self-adjoint matrix and let* $\mathbf{Y} \in \mathbb{C}^{d \times d}$ *be a self-adjoint random matrix. Then*

$$\mathbb{E}\operatorname{tr} \exp(\mathbf{H} + \mathbf{Y}) \le \operatorname{tr} \exp\left(\mathbf{H} + \ln\left(\mathbb{E}e^{\mathbf{Y}}\right)\right). \tag{8.33}$$

Proof. With the positive definite matrix $\mathbf{X} = e^{\mathbf{Y}}$ we have $\mathbf{Y} = \ln(\mathbf{X})$ by (A.50). By Lieb's Theorem B.31, the function $\mathbf{X} \mapsto \operatorname{tr} \exp(\mathbf{H} + \ln(\mathbf{X}))$ is concave on the set of positive semidefinite matrices. Jensen's inequality (7.17) therefore gives

$$\mathbb{E}\operatorname{tr} \exp(\mathbf{H} + \mathbf{Y}) = \mathbb{E}\operatorname{tr} \exp(\mathbf{H} + \ln(\mathbf{X})) \le \operatorname{tr} \exp(\mathbf{H} + \ln(\mathbb{E}\mathbf{X}))$$
$$= \operatorname{tr} \exp\left(\mathbf{H} + \ln\left(\mathbb{E}e^{\mathbf{Y}}\right)\right).$$

This concludes the proof. □

The next tool extends the previous inequality to a sequence of independent random matrices.

Proposition 8.18. *Let* $\mathbf{X}_1, \ldots, \mathbf{X}_M \in \mathbb{C}^{d \times d}$ *be independent, self-adjoint random matrices. Then, for* $\theta \in \mathbb{R}$,

$$\mathbb{E}\operatorname{tr} \exp\left(\theta \sum_{\ell=1}^{M} \mathbf{X}_\ell\right) \le \operatorname{tr} \exp\left(\sum_{\ell=1}^{M} \ln \mathbb{E}\exp(\theta \mathbf{X}_\ell)\right). \tag{8.34}$$

Proof. Without loss of generality we may assume that $\theta = 1$. We denote

$$\mathbf{Z}_\ell := \ln \mathbb{E}\exp(\mathbf{X}_\ell).$$

Since $\mathbf{X}_\ell$ are independent, we are in the position to write $\mathbb{E}_{\mathbf{X}_\ell}$ for the expectation with respect to $\mathbf{X}_\ell$ (or in other words, the expectation conditional on $\mathbf{X}_1, \ldots, \mathbf{X}_{\ell-1}, \mathbf{X}_{\ell+1}, \ldots, \mathbf{X}_M$). Using Fubini's theorem and Proposition 8.17, we arrive at

$$\mathbb{E}\operatorname{tr} \exp\left(\sum_{\ell=1}^{M} \mathbf{X}_\ell\right) = \mathbb{E}_{\mathbf{X}_1} \cdots \mathbb{E}_{\mathbf{X}_M} \operatorname{tr} \exp\left(\sum_{\ell=1}^{M-1} \mathbf{X}_\ell + \mathbf{X}_M\right)$$
$$\le \mathbb{E}_{\mathbf{X}_1} \cdots \mathbb{E}_{\mathbf{X}_{M-1}} \operatorname{tr} \exp\left(\sum_{\ell=1}^{M-1} \mathbf{X}_\ell + \ln \mathbb{E}\exp(\mathbf{X}_M)\right)$$

$$= \mathbb{E}_{\mathbf{X}_1} \cdots \mathbb{E}_{\mathbf{X}_{M-1}} \operatorname{tr} \exp \left(\sum_{\ell=1}^{M-2} \mathbf{X}_\ell + \mathbf{Z}_M + \mathbf{X}_{M-1} \right)$$

$$\leq \mathbb{E}_{\mathbf{X}_1} \cdots \mathbb{E}_{\mathbf{X}_{M-2}} \operatorname{tr} \exp \left(\sum_{\ell=1}^{M-2} \mathbf{X}_\ell + \mathbf{Z}_M + \mathbf{Z}_{M-1} \right)$$

$$\leq \cdots \leq \operatorname{tr} \exp \left(\sum_{\ell=1}^{M} \mathbf{Z}_\ell \right).$$

The application of Proposition 8.17 at step $k \in [M]$ with the matrices

$$\mathbf{H}_k = \sum_{\ell=1}^{k-1} \mathbf{X}_\ell + \sum_{\ell=k+1}^{M} \mathbf{Z}_\ell$$

is permitted since $\mathbf{H}_k$ does not depend on $\mathbf{X}_k$.   □

Before giving the next intermediate result, we recall from Appendix A.5 that we write $\mathbf{A} \preccurlyeq \mathbf{B}$ for two self-adjoint matrices $\mathbf{A}, \mathbf{B} \in \mathbb{C}^{d \times d}$ if $\mathbf{B} - \mathbf{A}$ is positive semidefinite.

Let us provide a matrix version of Cramér's theorem (Theorem 7.18).

Proposition 8.19. *Let* $\mathbf{X}_1, \ldots, \mathbf{X}_M \in \mathbb{C}^{d \times d}$ *be independent, self-adjoint random matrices. Assume that there exist a function* $g : (0, \infty) \to [0, \infty)$ *and fixed self-adjoint matrices* $\mathbf{A}_1, \ldots, \mathbf{A}_M$ *such that*

$$\mathbb{E} \exp(\theta \mathbf{X}_k) \preccurlyeq \exp(g(\theta) \mathbf{A}_k), \quad \text{for all } \theta > 0 \text{ and all } k \in [M]. \tag{8.35}$$

Then, with $\rho := \lambda_{\max} \left(\sum_{\ell=1}^{M} \mathbf{A}_\ell \right)$,

$$\mathbb{P} \left(\lambda_{\max} \left(\sum_{\ell=1}^{M} \mathbf{X}_\ell \right) \geq t \right) \leq d \inf_{\theta > 0} e^{-\theta t + g(\theta) \rho}, \quad t \in \mathbb{R}.$$

Proof. Substituting (8.34) into (8.31) yields

$$\mathbb{P} \left(\lambda_{\max} \left(\sum_{\ell=1}^{M} \mathbf{X}_\ell \right) \geq t \right) \leq \inf_{\theta > 0} \left\{ e^{-\theta t} \operatorname{tr} \exp \left(\sum_{\ell=1}^{M} \ln \mathbb{E} \exp(\theta \mathbf{X}_\ell) \right) \right\}.$$

By Proposition A.34, the matrix logarithm is matrix monotone, so that (8.35) implies

$$\ln \mathbb{E} \exp(\theta \mathbf{X}_\ell) \preccurlyeq g(\theta) \mathbf{A}_\ell \quad \text{for all } \theta > 0 \text{ and all } k \in [M].$$

Since the trace exponential is monotone (see (A.48)), a combination of the above facts yields, for each $\theta > 0$,

$$\mathbb{P}\left(\lambda_{\max}\left(\sum_{\ell=1}^{M} \mathbf{X}_\ell\right) \geq t\right) \leq e^{-\theta t} \mathrm{tr} \, \exp\left(g(\theta) \sum_{\ell=1}^{M} \mathbf{A}_\ell\right)$$

$$\leq e^{-\theta t} d \, \lambda_{\max}\left(\exp\left(g(\theta) \sum_{\ell=1}^{M} \mathbf{A}_\ell\right)\right) = d \, e^{-\theta t} \exp\left(g(\theta) \lambda_{\max}\left(\sum_{\ell=1}^{M} \mathbf{A}_\ell\right)\right).$$

The second inequality is valid because, for a positive definite $d \times d$ matrix $\mathbf{B}$, we have $\mathrm{tr} \, \mathbf{B} = \sum_{j=1}^{d} \lambda_j(\mathbf{B}) \leq d \, \lambda_{\max}(\mathbf{B})$, where $\lambda_j(\mathbf{B})$, $j \in [d]$, denote the eigenvalues of $\mathbf{B}$ (with possible repetitions). Taking the infimum over all positive θ and using the definition of ρ, we arrive at the statement of the proposition. $\qquad\square$

Before we pass to the proof of the noncommutative Bernstein inequality, we note the following deviation inequality for matrix-valued Rademacher sums, i.e., the matrix-valued analog of Hoeffding's inequality for scalar Rademacher sums in Corollaries 7.21 and 8.8.

Proposition 8.20. *Let $\epsilon = (\epsilon_1, \ldots, \epsilon_M)$ be a Rademacher sequence and $\mathbf{B}_1, \ldots,$ $\mathbf{B}_M \in \mathbb{C}^{d \times d}$ be self-adjoint matrices. Set*

$$\sigma^2 := \|\sum_{\ell=1}^{M} \mathbf{B}_\ell^2\|_{2 \to 2}.$$

Then, for $t > 0$,

$$\mathbb{P}(\|\sum_{\ell=1}^{M} \epsilon_\ell \mathbf{B}_\ell\|_{2 \to 2} \geq t) \leq 2d \exp\left(-t^2/(2\sigma^2)\right). \tag{8.36}$$

Proof. Proposition 8.19 requires to estimate $\mathbb{E} \exp(\theta \epsilon \mathbf{B})$ for a Rademacher variable ϵ and a self-adjoint matrix $\mathbf{B}$. Similarly to the scalar case in (7.28), we obtain

$$\mathbb{E} \exp(\theta \epsilon \mathbf{B}) = \frac{1}{2}\left(\exp(\theta \mathbf{B}) + \exp(-\theta \mathbf{B})\right) = \sum_{k=0}^{\infty} \frac{(\theta \mathbf{B})^{2k}}{(2k)!}$$

$$\preccurlyeq \sum_{k=0}^{\infty} \frac{(\theta \mathbf{B})^{2k}}{2^k k!} = \exp(\theta^2 \mathbf{B}^2/2)$$

because $\mathbf{B}^2$ is positive semidefinite. Therefore, (8.35) holds with $g(\theta) = \theta^2/2$ and $\mathbf{A}_\ell = \mathbf{B}_\ell^2$. The parameter ρ in Proposition 8.19 is given by

$$\rho = \| \sum_{\ell=1}^{M} \mathbf{B}_\ell^2 \|_{2\to2} = \sigma^2$$

because $\sum_{\ell=1}^{M} \mathbf{B}_\ell^2$ is positive semidefinite. Therefore,

$$\mathbb{P}(\| \sum_{\ell=1}^{M} \epsilon_\ell \mathbf{B}_\ell \|_{2\to2} \geq t) \leq \mathbb{P}(\lambda_{\max}(\sum_{\ell=1}^{M} \epsilon_\ell \mathbf{B}_\ell) \geq t) + \mathbb{P}(\lambda_{\max}(- \sum_{\ell=1}^{M} \epsilon_\ell \mathbf{B}_\ell) \geq t)$$

$$\leq 2d \inf_{\theta>0} e^{-\theta t + \theta^2 \sigma^2/2} = 2d\, e^{-t^2/(2\sigma^2)}.$$

Here, the optimal choice of θ was $\theta = t/\sigma^2$.   □

The case $d = 1$ reduces to the Hoeffding-type inequality of Corollary 8.8. The same deviation inequality also holds for matrix-valued Gaussian sums; see Exercise 8.6.

Now we are prepared to prove the noncommutative Bernstein inequality.

Proof (of Theorem 8.14). Proposition 8.19 requires to establish (8.35) for an appropriate function g and appropriate matrices $\mathbf{A}_k$. We may assume that the bound K on the maximal eigenvalue of $\mathbf{X}_\ell$, $\ell \in [M]$, satisfies $K = 1$. The general case then follows from applying the result to the rescaled matrices $\tilde{\mathbf{X}}_\ell = \mathbf{X}_\ell/K$.

We fix $\theta > 0$ and define the smooth function $f : \mathbb{R} \to \mathbb{R}$ by

$$f(x) = x^{-2}(e^{\theta x} - \theta x - 1) \text{ for } x \neq 0 \quad \text{and } f(0) = \theta^2/2.$$

Clearly, $f(x) = \theta^2 \sum_{k=2}^{\infty} \frac{(\theta x)^{k-2}}{k!}$. The derivative is given by

$$f'(x) = \theta^2 \sum_{k=3}^{\infty} \frac{\theta^{k-2}(k-2)x^{k-3}}{k!} = \frac{(\theta x - 2)e^{\theta x} + (\theta x + 2)}{x^3}.$$

We claim that $f'(x) \geq 0$ for all $x \in \mathbb{R}$, i.e., f is nondecreasing. Indeed, for $x \geq 0$ this follows from the power series expansion of f' as all coefficients are positive. For $x \in (-2/\theta, 0)$ one verifies that the absolute values $\theta^{k-2}(k-2)|x|^{k-3}/k!$, $k \geq 3$, of the terms in the power series of f' are monotonically decreasing in k and the term for $k = 3$ is positive. Since the signs of the power series are alternating, $f'(x) \geq 0$ holds also in this case. For $x \leq -2/\theta$ the nonnegativity of f' follows from the explicit formula above, where both the numerator and denominator are easily seen to be negative.

In particular, we have proven that $f(x) \leq f(1)$ whenever $x \leq 1$. All the eigenvalues of $\mathbf{X}_\ell$ are bounded by 1, so by the definition of the extension of f to matrices (A.42) and by the rule (A.43), it follows that

$$f(\mathbf{X}_\ell) \preccurlyeq f(1)\mathbf{Id}.$$

The identity $\exp(\theta x) = 1 + \theta x + x^2 f(x)$ and the fact that $f(\mathbf{X})$ commutes with $\mathbf{X}$ yield together with (A.43) that

$$\exp(\theta \mathbf{X}_\ell) = \mathbf{Id} + \theta \mathbf{X}_\ell + \mathbf{X}_\ell f(\mathbf{X}_\ell)\mathbf{X}_\ell \preccurlyeq \mathbf{Id} + \theta \mathbf{X}_\ell + f(1)\mathbf{X}_\ell^2.$$

Here we used additionally the elementary fact that $\mathbf{A} \preccurlyeq \mathbf{B}$ implies $\mathbf{HAH}^* \preccurlyeq \mathbf{HBH}^*$ for any matrix $\mathbf{H}$ of matching dimension (Lemma A.31), together with the self-adjointness of $\mathbf{X}_\ell$.

Taking expectations in the above semidefinite bound, and using $\mathbb{E}\mathbf{X}_k = \mathbf{0}$, we obtain

$$\mathbb{E}\exp(\theta \mathbf{X}_\ell) \preccurlyeq \mathbf{Id} + f(1)\mathbb{E}\mathbf{X}_\ell^2 \preccurlyeq \exp(f(1)\mathbb{E}\mathbf{X}_\ell^2) = \exp\left((e^\theta - \theta - 1)\mathbb{E}\mathbf{X}_\ell^2\right).$$

The second semidefinite bound follows from the general bound (A.46) for the matrix exponential. Setting $g(\theta) = e^\theta - \theta - 1$, it follows from Proposition 8.19 that, for $t \in \mathbb{R}$,

$$\mathbb{P}\left(\lambda_{\max}\left(\sum_{\ell=1}^M \mathbf{X}_\ell\right) \geq t\right) \leq d \inf_{\theta > 0} \left\{e^{-\theta t + g(\theta)\sigma^2}\right\}, \qquad (8.37)$$

where we have used that $\lambda_{\max}\left(\sum_{\ell=1}^M \mathbb{E}\mathbf{X}_\ell^2\right) = \sigma^2$ by positive semidefiniteness of $\sum_{\ell=1}^M \mathbb{E}\mathbf{X}_\ell^2$. Then both the Bennett-type inequality (8.25) and the Bernstein-type inequality (8.26) follow from Lemma 8.21 below.   □

Lemma 8.21. *Let* $h(x) = (1 + x)\ln(1 + x) - x$ *and* $g(\theta) = e^\theta - \theta - 1$. *Then, for* $a > 0$,

$$\inf_{\theta > 0} \{-\theta x + g(\theta)a\} = -ah(x/a), \quad x \geq 0,$$

and

$$h(x) \geq \frac{x^2/2}{1 + x/3} \quad \text{for all } x \geq 0.$$

Proof. The function $r(\theta) := g(\theta)a - \theta x$ attains its minimal value for $\theta = \ln(1 + x/a)$, and

$$r(\ln(1 + x/a)) = (x/a - \ln(1 + x/a))a - x\ln(1 + x/a) = -ah(x/a).$$

For the second statement, we first note that

$$g(\theta) = e^\theta - \theta - 1 = \sum_{k=2}^\infty \frac{\theta^k}{k!} = \frac{\theta^2}{2}\sum_{k=2}^\infty \frac{2\theta^{k-2}}{k!}.$$

By induction, it follows that $2/k! \leq (1/3)^{k-2}$ for all $k \geq 2$. Therefore, for $\theta < 3$,

$$g(\theta) \leq \frac{\theta^2}{2} \sum_{k=0}^{\infty} (\theta/3)^k = \frac{\theta^2/2}{1 - \theta/3}.$$

Making the specific choice $\theta = \dfrac{x}{1 + x/3} < 3$ shows that

$$-h(x) = \inf_{\theta > 0} (g(\theta) - \theta x) \leq \inf_{\theta \in (0,3)} \left(\frac{\theta^2/2}{1 - \theta/3} - \theta x \right)$$

$$\leq \frac{x^2/2}{(1 + x/3)^2 \left(1 - \frac{x/3}{1+x/3} \right)} - \frac{x^2}{1 + x/3} = -\frac{x^2/2}{1 + x/3}. \tag{8.38}$$

This completes the proof.   □

8.6   Dudley's Inequality

A stochastic process is a collection $X_t, t \in T$, of random variables indexed by some set T. We are interested in bounding the expectation of the supremum of a real-valued stochastic process. In order to avoid measurability issues (the supremum of an uncountable number of random variables may not be measurable), we define the so-called lattice supremum

$$\mathbb{E} \sup_{t \in T} X_t := \sup\{\mathbb{E} \sup_{t \in F} X_t, F \subset T, F \text{ finite}\}. \tag{8.39}$$

Note that for a countable index set T, where no measurability problems can arise, $\mathbb{E}(\sup_{t \in T} X_t)$ equals the right-hand side above (see Exercise 8.9), so that this definition is consistent. Also, if $t \mapsto X_t$ is almost surely continuous on T (as it will always be the case in the situations we encounter), and T is separable, then $\sup_{t \in T} X_t$ coincides with the supremum over a dense countable subset of T, so that in this case the lattice supremum coincides with $\mathbb{E}(\sup_{t \in T} X_t)$ as well.

We will always assume that the process is centered, i.e.,

$$\mathbb{E} X_t = 0 \quad \text{for all } t \in T. \tag{8.40}$$

Associated to the process $X_t, t \in T$, we define the pseudometric

$$d(s, t) := \left(\mathbb{E} |X_s - X_t|^2 \right)^{1/2}, \quad s, t \in T. \tag{8.41}$$

We refer to Definition A.2 for the notion of a pseudometric.

Definition 8.22. A centered stochastic process X_t, $t \in T$, is called subgaussian if

$$\mathbb{E} \exp(\theta(X_s - X_t)) \le \exp(\theta^2 d(s,t)^2/2), \quad s,t \in T, \theta > 0, \qquad (8.42)$$

with d being the pseudometric defined in (8.41).

Clearly, one may replace the constant $1/2$ in (8.42) by a general constant c, but for our purposes it is enough to consider $c = 1/2$.

Examples of subgaussian processes include Gaussian and Rademacher processes. A process X_t is called a centered Gaussian process if for each finite collection $t_1, \ldots, t_n \in T$, the random vector $(X_{t_1}, \ldots, X_{t_n})$ is a mean-zero Gaussian random vector. This implies, in particular, that $X_t - X_s$ is a univariate Gaussian with $\mathbb{E}(X_t - X_s) = 0$ (by (8.40)) and variance $\mathbb{E}|X_s - X_t|^2$. It follows from (7.11) (or Remark 7.26 and Theorem 7.27) that a Gaussian process is a subgaussian process in the sense of Definition 8.22. A typical example of a Gaussian process takes the form

$$X_t = \sum_{j=1}^{M} g_j x_j(t),$$

where $\mathbf{g} = (g_1, \ldots, g_M)$ is a standard Gaussian random vector and $x_j : T \to \mathbb{R}$, $j \in [M]$, are arbitrary functions.

A Rademacher process has the form

$$X_t = \sum_{j=1}^{M} \epsilon_j x_j(t), \qquad (8.43)$$

where $\epsilon = (\epsilon_1, \ldots, \epsilon_M)$ is a Rademacher sequence. Clearly, such a process satisfies (8.40). By Remark 7.26 and Theorem 7.27, it is also a subgaussian process. Observe that

$$\mathbb{E}|X_t - X_s|^2 = \mathbb{E}\left| \sum_{j=1}^{M} \epsilon_j (x_j(t) - x_j(s)) \right|^2 = \sum_{j=1}^{M} (x_j(t) - x_j(s))^2 = \|\mathbf{x}(t) - \mathbf{x}(s)\|_2^2,$$

where $\mathbf{x}(t)$ denotes the vector with components $x_j(t), j \in [M]$. Therefore, the pseudometric associated to X_t is given by

$$d(s,t) = \left(\mathbb{E}|X_t - X_s|^2 \right)^{1/2} = \|\mathbf{x}(t) - \mathbf{x}(s)\|_2. \qquad (8.44)$$

It follows from Theorem 7.27 that the increments of a subgaussian process X_t satisfy the tail estimate

$$\mathbb{P}(|X_s - X_t| \ge u d(s,t)) \le 2 \exp(-u^2/2). \qquad (8.45)$$

By Proposition 7.24(b) this inequality could as well be taken for the definition of subgaussian processes.

Dudley's inequality below relates the stochastic quantity of the lattice supremum (8.39) to the geometric concept of covering numbers. We recall from Sect. C.2 that the covering number $\mathcal{N}(T, d, \varepsilon)$ is defined as the smallest integer N such that there exists a subset F of T with $\mathrm{card}(F) = N$ and $\min_{s \in F} d(t, s) \leq \varepsilon$ for all $t \in T$. We denote the radius of T by

$$\Delta(T) = \sup_{t \in T} \sqrt{\mathbb{E}|X_t|^2}. \tag{8.46}$$

Dudley's inequality for subgaussian processes reads as follows.

Theorem 8.23. *Let X_t, $t \in T$, be a centered subgaussian process with associated pseudometric d. Then, for any $t_0 \in T$,*

$$\mathbb{E} \sup_{t \in T} X_t \leq 4\sqrt{2} \int_0^{\Delta(T)/2} \sqrt{\ln(\mathcal{N}(T, d, u))} du, \tag{8.47}$$

$$\mathbb{E} \sup_{t \in T} |X_t| \leq 4\sqrt{2} \int_0^{\Delta(T)/2} \sqrt{\ln(2\mathcal{N}(T, d, u))} du. \tag{8.48}$$

Remark 8.24. Inequality (8.48) with constant $8\sqrt{2}$ (but without the factor 2 inside the logarithm) follows also directly from (8.47) in the case of symmetric processes; see Exercise 8.10. It is known that these inequalities are sharp up to logarithmic factors if $(X_t)_{t \in T}$ is a Gaussian process, T is a subset of a finite-dimensional space, and d is induced by a norm; see also the Notes section.

Proof. We write $\Delta = \Delta(T)$ for convenience. Without loss of generality we may assume that $\Delta < \infty$ because otherwise it is straightforward to check that the right-hand sides in (8.47) and (8.48) are infinite so that there is nothing to prove.

We set $\varepsilon_n := 2^{-n}\Delta$ and $N_n := \mathcal{N}(T, d, \varepsilon_n)$. By definition of the covering numbers, we can find subsets $T_n \subset T$ of cardinality at most N_n such that for all $t \in T$ there exists $s \in T_n$ such that $d(t, s) \leq \varepsilon_n$. We write $s = \phi_n(t)$ for this particular s.

Let us first bound $\mathbb{E} \max_{t \in T_n} X_t$. To this end we observe that

$$\max_{t \in T_n} X_t = \max_{t \in T_n}(X_t - X_{\phi_{n-1}(t)} + X_{\phi_{n-1}(t)}) \leq \max_{t \in T_n}(X_t - X_{\phi_{n-1}(t)}) + \max_{t \in T_{n-1}} X_t.$$

Note that $\sqrt{\mathbb{E}(X_t - X_{\phi_{n-1}(t)})^2} \leq \varepsilon_{n-1}$ for all $t \in T_n$ by definition of the pseudometric d and ε_n. Therefore, Proposition 7.29 yields

$$\mathbb{E} \max_{t \in T_n} X_t \leq \mathbb{E} \max_{t \in T_{n-1}} X_t + \sqrt{2 \ln N_n} \varepsilon_{n-1}.$$

Moreover, $\mathbb{E}\max_{t\in T_1} X_t \le \sqrt{2\ln N_1}\Delta$. By induction and since $\mathcal{N}(T,d,\varepsilon_n) \le \mathcal{N}(T,d,u)$ for all $u \in [\varepsilon_{n+1},\varepsilon_n]$ we obtain

$$\mathbb{E}\max_{t\in T_n} X_t \le \sqrt{2\ln N_1}\Delta + \sum_{j=2}^{n} \sqrt{2\ln N_j}\varepsilon_{j-1}$$

$$= 2\sqrt{2\ln N_1}\varepsilon_1 + \sum_{j=2}^{n} 4\sqrt{2\ln N_j}(\varepsilon_j - \epsilon_{j+1})$$

$$\le 4\sqrt{2}\sum_{j=1}^{n}\int_{\varepsilon_{j+1}}^{\varepsilon_j} \sqrt{\ln\mathcal{N}(T,d,u)}du \le 4\sqrt{2}\int_0^{\Delta/2} \sqrt{\ln\mathcal{N}(T,d,u)}du.$$

By using (7.37) we similarly obtain

$$\mathbb{E}\max_{t\in T_n} |X_t| \le 4\sqrt{2}\int_0^{\Delta/2} \sqrt{\ln(2\mathcal{N}(T,d,u))}du.$$

It remains to pass from T_n to the full set T. If T is finite, then $T_n = T$ for some large enough n and we are done. If T is infinite, then by definition of the lattice supremum (8.39), we consider an arbitrary finite subset F of T. Then, for any $n \in \mathbb{N}$, by Proposition 7.29

$$\mathbb{E}\sup_{t\in F} X_t \le \mathbb{E}\sup_{t\in T_n} X_t + \mathbb{E}\sup_{t\in F}(X_t - X_{\phi_n(t)})$$

$$\le 4\sqrt{2}\int_0^{\Delta/2} \sqrt{\ln(\mathcal{N}(T,d,u))}du + \sqrt{2\ln(\text{card}(F))}\varepsilon_n.$$

Since this bound is valid for any $n \in \mathbb{N}$ and since $\varepsilon_n \to 0$ as n tends to ∞, we obtain the claimed bound for any finite subset F of T. Taking the supremum over all finite F establishes the claim for $\mathbb{E}\sup_{t\in T} X_t$. The one for $\mathbb{E}\sup_{t\in T} |X_t|$ is deduced in the same way. $\qquad\square$

The technique in the previous proof is usually referred to as *chaining*.

8.7 Slepian's and Gordon's Lemmas

Slepian's lemma and its generalization due to Gordon compare extrema of two families of Gaussian random variables. The basic idea is that the distribution of a mean-zero Gaussian vector is completely determined by its covariance structure. This suggests to compare expectations of functions of the two families by means of comparing the covariances.

Slepian's lemma reads as follows.

Lemma 8.25. *Let* $\mathbf{X}, \mathbf{Y}$ *be mean-zero Gaussian random vectors on* $\mathbb{R}^m$. *If*

$$\mathbb{E}|X_i - X_j|^2 \leq \mathbb{E}|Y_i - Y_j|^2 \quad \text{for all } i, j \in [m], \tag{8.49}$$

then

$$\mathbb{E} \max_{j \in [m]} X_j \leq \mathbb{E} \max_{j \in [m]} Y_j.$$

Remark 8.26. The L_2 distances above can be written in terms of the covariances

$$\mathbb{E}|X_i - X_j|^2 = \mathbb{E}X_i^2 - 2\mathbb{E}X_i X_j + \mathbb{E}X_j^2.$$

Under the additional assumption that $\mathbb{E}X_j^2 = \mathbb{E}Y_j^2$, condition (8.49) reads therefore $\mathbb{E}X_j X_k \geq \mathbb{E}Y_j Y_k$. In particular, comparison of the covariance structures of $\mathbf{X}$ and $\mathbf{Y}$ allows to compare the expected maxima of the two Gaussian vectors as claimed above.

Gordon's lemma stated next compares expected minima of maxima of Gaussian vectors. Slepian's lemma is the special case that $n = 1$.

Lemma 8.27. *Let* $X_{i,j}, Y_{i,j}, i \in [n], j \in [m]$, *be two finite families of mean-zero Gaussian random variables. If*

$$\mathbb{E}|X_{i,j} - X_{k,\ell}|^2 \leq \mathbb{E}|Y_{i,j} - Y_{k,\ell}|^2 \quad \text{for all } i \neq k \text{ and } j, \ell, \tag{8.50}$$

$$\mathbb{E}|X_{i,j} - X_{i,\ell}|^2 \geq \mathbb{E}|Y_{i,j} - Y_{i,\ell}|^2 \quad \text{for } i, j, \ell, \tag{8.51}$$

then

$$\mathbb{E} \min_{i \in [n]} \max_{j \in [m]} X_{i,j} \geq \mathbb{E} \min_{i \in [n]} \max_{j \in [m]} Y_{i,j}.$$

Remark 8.28. Both Slepian's and Gordon's lemmas extend to Gaussian processes indexed by possibly infinite sets. In particular, if $\mathbf{X} = (X_t)_{t \in T}, \mathbf{Y} = (Y_t)_{t \in T}$ are Gaussian processes (recall that by definition this means that any restriction $\mathbf{X}_{T_0} = (X_t)_{t \in T_0}$ to a finite subset $T_0 \subset T$ yields a Gaussian random vector) and if $\mathbb{E}|X_s - X_t|^2 \leq \mathbb{E}|Y_s - Y_t|^2$ for all $s, t \in T$, then Slepian's lemma states that

$$\mathbb{E} \sup_{t \in T} X_t \leq \mathbb{E} \sup_{t \in T} Y_t,$$

where the suprema are understood in the sense of a lattice supremum (8.39). Indeed, by the finite-dimensional version in Lemma 8.25, this relation holds for the restriction to any finite subset T_0 so that the above inequality holds.

In a similar sense, Gordon's lemma extends to doubly indexed Gaussian processes.

The proof of Gordon's lemma requires some preparation. We say that a function $F : \mathbb{R}^m \to \mathbb{R}$ is of *moderate growth* if for each $\beta > 0$,

$$\lim_{\|\mathbf{x}\|_2 \to \infty} F(\mathbf{x}) \exp(-\beta \|\mathbf{x}\|_2^2) = 0. \tag{8.52}$$

Our first technical tool is the Gaussian integration by parts formula—also known as Stein's lemma—and its generalization to higher dimensions.

Proposition 8.29. *Let* $F : \mathbb{R}^m \to \mathbb{R}$ *be a differentiable function such that* F *and all its first-order partial derivatives are of moderate growth.*

(a) Let g *be a mean-zero Gaussian random variable and* $m = 1$. *Then*

$$\mathbb{E}[gF(g)] = \mathbb{E}g^2 \mathbb{E}F'(g). \tag{8.53}$$

(b) Let $\mathbf{g} = (g_1, \ldots, g_m)$ *and* $\tilde{g}$ *(not necessarily independent of* $\mathbf{g}$*) be such that* $(\tilde{g}, \mathbf{g})$ *is a Gaussian random vector. Then*

$$\mathbb{E}\tilde{g}F(\mathbf{g}) = \sum_{j=1}^{m} \mathbb{E}(\tilde{g}g_j)\mathbb{E}\left[\frac{\partial F}{\partial x_j}(\mathbf{g})\right]. \tag{8.54}$$

Proof. (a) Setting $\tau^2 = \mathbb{E}g^2$ and using integration by parts yield

$$\mathbb{E}gF(g) = \frac{1}{\sqrt{2\pi}\tau} \int_{-\infty}^{\infty} t \exp(-t^2/(2\tau^2)) F(t) dt$$

$$= \frac{\tau^2}{\sqrt{2\pi}\tau} \int_{-\infty}^{\infty} \exp(-t^2/(2\tau^2)) F'(t) dt = \mathbb{E}g^2 \mathbb{E}F'(g).$$

The moderate growth condition ensures that all integrals are well defined and that $\exp(-t^2/(2\tau^2)) F(t) \big|_{-\infty}^{\infty} = 0$.

(b) Consider the random variables $g_j' = g_j - \tilde{g}\frac{\mathbb{E}g_j\tilde{g}}{\mathbb{E}\tilde{g}^2}$. They satisfy $\mathbb{E}g_j'\tilde{g} = 0$, and therefore, the Gaussian random vector $\mathbf{g}' = (g_1', \ldots, g_m')$ is independent of $\tilde{g}$. Using Fubini's theorem and applying (8.53) to $t \mapsto F(g_1' + t\mathbb{E}[\tilde{g}g_1]/\mathbb{E}\tilde{g}^2, \ldots, g_m' + t\mathbb{E}[\tilde{g}g_m]/\mathbb{E}\tilde{g}^2)$ conditionally on $\mathbf{g}'$ yield

$$\mathbb{E}\tilde{g}F(\mathbf{g}) = \mathbb{E}\tilde{g}F\left(g_1' + \tilde{g}\frac{\mathbb{E}\tilde{g}g_1}{\mathbb{E}\tilde{g}^2}, \ldots, g_m' + \tilde{g}\frac{\mathbb{E}\tilde{g}g_m}{\mathbb{E}\tilde{g}^2}\right)$$

$$= \mathbb{E}\tilde{g}^2 \sum_{j=1}^{m} \frac{\mathbb{E}\tilde{g}g_j}{\mathbb{E}\tilde{g}^2} \mathbb{E}\frac{\partial F}{\partial x_j}\left(g_1' + \tilde{g}\frac{\mathbb{E}\tilde{g}g_1}{\mathbb{E}\tilde{g}^2}, \ldots, g_m' + \tilde{g}\frac{\mathbb{E}\tilde{g}g_m}{\mathbb{E}\tilde{g}^2}\right)$$

$$= \sum_{j=1}^{m} \mathbb{E}[\tilde{g}g_j]\mathbb{E}\left[\frac{\partial F}{\partial x_j}(\mathbf{g})\right].$$

This completes the proof.   □

We will also require the following standard result in integration theory.

Proposition 8.30. *Let $\psi : J \times \Omega \to \mathbb{R}$ be a function, with $J \subset \mathbb{R}$ being on open interval. Let X be a random variable with values in Ω such that $t \mapsto \psi(t, X)$ is almost surely continuously differentiable in J. Assume that for each compact subinterval $I \subset J$,*

$$\mathbb{E} \sup_{t \in I} |\psi'(t, X)| < \infty. \tag{8.55}$$

Then the function $t \mapsto \phi(t) = \mathbb{E}\psi(t, X)$ is continuously differentiable and

$$\phi'(t) = \mathbb{E}\psi'(t, X). \tag{8.56}$$

Proof. Let t be in the interior of J and consider a compact subinterval $I \subset J$ containing t in its interior. For $h \in \mathbb{R} \setminus \{0\}$ such that $t + h \in I$ we consider the difference quotients

$$\phi_h(t) := \frac{\phi(t + h) - \phi(t)}{h}, \quad \psi_h(t, X) := \frac{\psi(t + h, X) - \psi(t, X)}{h}.$$

By the mean value theorem there exists $\xi \in [t, t + h]$ (or $\xi \in [t + h, t]$ if $h < 0$) such that $\psi'(\xi, X) = \psi_h(t, X)$. Therefore, $|\psi_h(t, X)| \leq \sup_{t \in I} |\psi'(t, X)|$, and by (8.55), $\psi_h(t, X)$ has an integrable majorant. By Lebesgue's dominated convergence theorem, we have

$$\lim_{h \to 0} \phi_h(t) = \mathbb{E} \lim_{h \to 0} \psi_h(t, X) = \mathbb{E}\psi'(t, X),$$

so that ϕ is continuously differentiable and (8.56) holds.   □

The crucial tool in the proof of Slepian's and Gordon's lemmas is stated next.

Proposition 8.31. *Let $F : \mathbb{R}^m \to \mathbb{R}$ be a differentiable function such that F and all its partial derivatives of first order are of moderate growth. Let $\mathbf{X} = (X_1, \ldots, X_m)$ and $\mathbf{Y} = (Y_1, \ldots, Y_m)$ be two independent mean-zero Gaussian vectors. For $t \in [0, 1]$, we define a new random vector $\mathbf{U}(t) = (U_1(t), \ldots, U_m(t))$ with components*

$$U_i(t) = \sqrt{t}X_i + \sqrt{1 - t}Y_i, \quad i \in [m]. \tag{8.57}$$

Then the function

$$\phi(t) = \mathbb{E}F(\mathbf{U}(t))$$

has derivative

$$\phi'(t) = \sum_{i=1}^{m} \mathbb{E}\left[U_i'(t)\frac{\partial F}{\partial x_i}(\mathbf{U}(t))\right]. \tag{8.58}$$

If, in addition, F is twice differentiable with all partial derivatives of second order of moderate growth, then

$$\phi'(t) = \frac{1}{2}\sum_{i,j=1}^{m}(\mathbb{E}X_iX_j - \mathbb{E}Y_iY_j)\mathbb{E}\left[\frac{\partial^2 F}{\partial x_i\partial x_j}(\mathbf{U}(t))\right]. \tag{8.59}$$

Proof. We note that

$$\frac{d}{dt}F(\mathbf{U}(t)) = \sum_{i=1}^{m}U_i'(t)\frac{\partial F}{\partial x_i}(\mathbf{U}(t)),$$

where clearly

$$U_i'(t) = \frac{d}{dt}U_i(t) = \frac{1}{2\sqrt{t}}X_i - \frac{1}{2\sqrt{1-t}}Y_i.$$

By Proposition 8.30 it therefore suffices to verify (8.55). For a compact subinterval $I = [a, b] \subset (0, 1)$, we have

$$\mathbb{E}\sup_{t\in I}|U_i'(t)\frac{\partial F}{\partial x_i}(\mathbf{U}(t))| \le \mathbb{E}\sup_{t\in I}|U_i'(t)|\sup_{t\in I}|\frac{\partial F}{\partial x_i}(\mathbf{U}(t))|$$

$$\le \sqrt{\mathbb{E}\sup_{t\in I}|U_i'(t)|^2}\sqrt{\mathbb{E}\sup_{t\in I}|\frac{\partial F}{\partial x_i}(\mathbf{U}(t))|^2},$$

where the last inequality follows from the Cauchy–Schwarz inequality. We treat both expectations above separately. The triangle inequality gives

$$\sqrt{\mathbb{E}\sup_{t\in I}|U_i'(t)|^2} \le \sqrt{\mathbb{E}\frac{1}{4a}X_i^2} + \sqrt{\mathbb{E}\frac{1}{4(1-b)}Y_i^2} < \infty.$$

For the second expectation choose $\beta > 0$. Since $\frac{\partial F}{\partial x_i}$ is of moderate growth, there exists $A > 0$ such that

$$\left|\frac{\partial F}{\partial x_i}(x)\right| \le A\exp(\beta\|x\|_2^2) \quad \text{for all } x \in \mathbb{R}^m.$$

Furthermore,

$$\|\mathbf{U}(t)\|_2 \le \sqrt{t}\|\mathbf{X}\|_2 + \sqrt{1-t}\|\mathbf{Y}\|_2 \le 2\max\{\|\mathbf{X}\|_2, \|\mathbf{Y}\|_2\},$$

and hence,

$$\sup_{t \in I} \left| \frac{\partial F}{\partial x_i}(\mathbf{U}(t)) \right| \le A \max\{\exp(4\beta \|\mathbf{X}\|_2^2), \exp(4\beta \|\mathbf{Y}\|_2^2)\}.$$

Since $\mathbf{X}$ and $\mathbf{Y}$ are mean-zero Gaussian vectors, there exist matrices $\mathbf{\Gamma}, \mathbf{\Gamma}'$ such that $\mathbf{X} = \mathbf{\Gamma}\mathbf{g}$ and $\mathbf{Y} = \mathbf{\Gamma}'\mathbf{g}'$ where $\mathbf{g}, \mathbf{g}'$ are independent standard Gaussian vectors. Therefore,

$$\mathbb{E} \sup_{t \in I} \left| \frac{\partial F}{\partial x_i}(\mathbf{U}(t)) \right| \le A \mathbb{E} \left[\exp \left(4\beta \|\mathbf{\Gamma}\|_{2 \to 2}^2 \|\mathbf{g}\|_2^2 + 4\beta \|\mathbf{\Gamma}'\|_{2 \to 2}^2 \|\mathbf{g}'\|_2^2 \right) \right]$$

$$= A \prod_{i=1}^{m} \mathbb{E} \left[\exp(4\beta \|\mathbf{\Gamma}\|_{2 \to 2}^2 g_i^2) \right] \prod_{j=1}^{m} \mathbb{E} \left[\exp(4\beta \|\mathbf{\Gamma}'\|_{2 \to 2}^2 (g_j')^2) \right]$$

$$= A(1 - 8\beta \|\mathbf{\Gamma}\|_{2 \to 2}^2)^{-m/2} (1 - 8\beta \|\mathbf{\Gamma}'\|_{2 \to 2}^2)^{-m/2} < \infty.$$

The last equality follows from Lemma 7.6 with $\theta = 0$ and a choice of $\beta > 0$ such that $8\beta \max\{\|\mathbf{\Gamma}\|_{2 \to 2}^2, \|\mathbf{\Gamma}'\|_{2 \to 2}^2\} < 1$. (Recall that $\beta > 0$ can be chosen arbitrarily and influences only the constant A). This completes the proof of (8.58).

For (8.59) we observe that $\mathbb{E}U_i'(t)U_j(t) = \frac{1}{2}(\mathbb{E}X_iX_j - \mathbb{E}Y_iY_j)$. The Gaussian integration by parts formula (8.54) yields

$$\mathbb{E} \left[U_i'(t) \frac{\partial F}{\partial x_i}(\mathbf{U}(t)) \right] = \frac{1}{2} \sum_{j=1}^{m} (\mathbb{E}X_iX_j - \mathbb{E}Y_iY_j) \mathbb{E} \frac{\partial^2 F}{\partial x_i \partial x_j}(\mathbf{U}(t)).$$

This completes the proof.   □

The next result is a generalized version of Gordon's lemma. Since we will require it also for not necessarily differentiable functions F, we work with the distributional derivative; see Sect. C.9. In particular, we say that a function F has positive distributional derivatives and write $\frac{\partial^2 F}{\partial x_i \partial x_j} \ge 0$ if, for all nonnegative twice differentiable functions g with compact support,

$$\int_{\mathbb{R}^d} F(\mathbf{x}) \frac{\partial^2 g}{\partial x_i \partial x_j}(\mathbf{x}) d\mathbf{x} \ge 0.$$

Integration by parts shows that this definition is consistent with positivity of $\frac{\partial^2 F}{\partial x_i \partial x_j}$ when F is twice differentiable.

Lemma 8.32. *Let* $F : \mathbb{R}^m \to \mathbb{R}$ *be a Lipschitz function, i.e.,* $|F(\mathbf{x}) - F(\mathbf{y})| \le L\|\mathbf{x} - \mathbf{y}\|_2$ *for all* $\mathbf{x}, \mathbf{y} \in \mathbb{R}^m$ *and some constant* $L > 0$*. Let* $\mathbf{X} = (X_1, \ldots, X_m)$ *and* $\mathbf{Y} = (Y_1, \ldots, Y_m)$ *be two mean-zero Gaussian vectors. Assume that (in the distributional sense)*

$$(\mathbb{E}|X_i - X_j|^2 - \mathbb{E}|Y_i - Y_j|^2)\frac{\partial^2 F}{\partial x_i \partial x_j} \geq 0 \quad \text{for all } i, j \in [m], \qquad (8.60)$$

and

$$F(\mathbf{x} + t\mathbf{e}) = F(\mathbf{x}) + ct \quad \text{for all } \mathbf{x} \in \mathbb{R}^m \qquad (8.61)$$

where $\mathbf{e} = (1, 1, \ldots, 1) \in \mathbb{R}^m$ and c is some constant. Then

$$\mathbb{E}F(\mathbf{X}) \leq \mathbb{E}F(\mathbf{Y}).$$

Proof. Observe that the Lipschitz assumption implies

$$|F(\mathbf{x})| \leq |F(\mathbf{0})| + L\|\mathbf{x}\|_2, \quad \mathbf{x} \in \mathbb{R}^m, \qquad (8.62)$$

so that F is of moderate growth.

We first assume that F is twice continuously differentiable such that its derivatives up to second order are of moderate growth. We note that (8.61) implies

$$\sum_{j=1}^m \frac{\partial^2 F}{\partial x_i \partial x_j}(\mathbf{x}) = 0 \quad \text{for all } i \in [m], \mathbf{x} \in \mathbb{R}^m. \qquad (8.63)$$

(In fact, (8.63) and (8.61) are equivalent.) This observation implies that $\frac{\partial^2 F}{\partial x_i^2} = -\sum_{j=1, j \neq i} \frac{\partial^2 F}{\partial x_i \partial x_j}$, and we obtain

$$\sum_{i,j=1}^m (\mathbb{E}X_i X_j - \mathbb{E}Y_i Y_j)\frac{\partial^2 F}{\partial x_i \partial x_j}$$

$$= -\sum_{i=1}^m (\mathbb{E}X_i^2 - \mathbb{E}Y_i^2) \sum_{j=1, j \neq i}^m \frac{\partial^2 F}{\partial x_i \partial x_j} + \sum_{i \neq j} (\mathbb{E}X_i X_j - \mathbb{E}Y_i Y_j)\frac{\partial^2 F}{\partial x_i \partial x_j}$$

$$= -\frac{1}{2}\sum_{i \neq j}(\mathbb{E}X_i^2 - \mathbb{E}Y_i^2 + \mathbb{E}X_j^2 - \mathbb{E}Y_j^2 - 2(\mathbb{E}X_i X_j - \mathbb{E}Y_i Y_j))\frac{\partial^2 F}{\partial x_i \partial x_j}$$

$$= -\frac{1}{2}\sum_{i,j=1}^m (\mathbb{E}|X_i - X_j|^2 - \mathbb{E}|Y_i - Y_j|^2)\frac{\partial^2 F}{\partial x_i \partial x_j} \leq 0$$

by (8.60). Therefore, the function ϕ of Proposition 8.31 has nonpositive derivative and $\mathbb{E}F(\mathbf{X}) \leq \mathbb{E}F(\mathbf{Y})$ (noting that we can assume without loss of generality that the random vectors $\mathbf{X}$ and $\mathbf{Y}$ are independent).

In the general case where F is not necessarily twice continuously differentiable, we approximate F by twice continuously differentiable functions. To this end, we choose a nonnegative twice continuously differentiable function ψ with support in $B(\mathbf{0}, 1)$, where we denote $B(\mathbf{y}, h) = \{\mathbf{x} \in \mathbb{R}^m : \|\mathbf{x} - \mathbf{y}\|_2 \leq h\}$, such that $\int_{\mathbb{R}^m} \psi(\mathbf{x}) d\mathbf{x} = 1$. Let $\psi_h = h^{-m} \psi(\mathbf{x}/h)$, $h > 0$, which also satisfies $\int_{\mathbb{R}^m} \psi_h(\mathbf{x}) d\mathbf{x} = 1$. We introduce smoothed versions F_h of the function F via convolution

$$F_h(\mathbf{x}) = F * \psi_h(\mathbf{x}) = \int_{\mathbb{R}^m} F(\mathbf{y}) \psi_h(\mathbf{x} - \mathbf{y}) d\mathbf{y}. \tag{8.64}$$

Since $\int_{\mathbb{R}^m} \psi_h(\mathbf{x}) d\mathbf{x} = 1$ and $\operatorname{supp} \psi_h \subset B(\mathbf{0}, h) = \{\mathbf{x} \in \mathbb{R}^m : \|\mathbf{x}\|_2 \leq h\}$, we have

$$|F_h(\mathbf{x}) - F(\mathbf{x})| = \left| \int_{\mathbb{R}^m} (F(\mathbf{y}) - F(\mathbf{x})) \psi_h(\mathbf{x} - \mathbf{y}) d\mathbf{y} \right|$$

$$\leq \int_{B(\mathbf{x},h)} |F(\mathbf{y}) - F(\mathbf{x})| \psi_h(\mathbf{x} - \mathbf{y}) d\mathbf{y} \leq \int_{B(\mathbf{x},h)} L\|\mathbf{y} - \mathbf{x}\|_2 \psi_h(\mathbf{x} - \mathbf{y}) d\mathbf{y}$$

$$\leq Lh,$$

where we have also used the Lipschitz assumption. In particular, F_h converges uniformly to F when $h \to 0$. Moreover, Lebesgue's dominated convergence theorem allows to interchange the integral and derivatives, so that F_h is twice continuously differentiable, and

$$\frac{\partial F_h}{\partial x_i} = F * \left(\frac{\partial \psi_h}{\partial x_i} \right), \quad \text{and} \quad \frac{\partial^2 F_h}{\partial x_i \partial x_j} = F * \left(\frac{\partial^2 \psi_h}{\partial x_i \partial x_j} \right).$$

By (8.62) and since ψ_h has compact support and is twice continuously differentiable, it is straightforward to verify from the definition of the convolution (8.64) that the partial derivatives of F_h up to second order are of moderate growth. Furthermore, for any nonnegative twice continuously differentiable function g on $\mathbb{R}^m$ with compact support, it follows from Fubini's theorem that

$$\int_{\mathbb{R}^m} F_h(\mathbf{x}) \frac{\partial^2 g}{\partial x_i \partial x_j}(\mathbf{x}) d\mathbf{x} = \int_{\mathbb{R}^m} \int_{\mathbb{R}^m} F(\mathbf{y}) \psi_h(\mathbf{y} - \mathbf{x}) d\mathbf{y} \frac{\partial^2 g}{\partial x_i \partial x_j}(\mathbf{x}) d\mathbf{x}$$

$$= \int_{\mathbb{R}^m} F(\mathbf{y}) \int_{\mathbb{R}^m} \psi_h(\mathbf{y} - \mathbf{x}) \frac{\partial^2 g}{\partial x_i \partial x_j}(\mathbf{x}) d\mathbf{x} d\mathbf{y}$$

$$= \int_{\mathbb{R}^m} F(\mathbf{y}) \frac{\partial^2}{\partial x_i \partial x_j}(\psi_h * g)(\mathbf{y}) d\mathbf{y}. \tag{8.65}$$

The last identity, i.e., the interchange of taking derivatives and convolution, is justified again by Lebesgue's dominated convergence theorem. Since both ψ_h and g are nonnegative, the function $\psi_h * g$ is nonnegative as well. It follows from (8.65) and from the assumption (8.60) on the distributional derivatives of F that (8.60) is valid also for F_h in place of F. Also the property (8.61) extends to F_h by the following calculation:

$$F_h(\mathbf{x} + t\mathbf{e}) = \int_{\mathbb{R}^m} F(\mathbf{x} + t\mathbf{e} - \mathbf{y})\psi_h(\mathbf{y})d\mathbf{y} = \int_{\mathbb{R}^m} (F(\mathbf{x} - \mathbf{y}) + ct)\psi_h(\mathbf{y})d\mathbf{y}$$

$$= F_h(\mathbf{x}) + ct \int_{\mathbb{R}^m} \psi_h(\mathbf{x})d\mathbf{x} = F_h(\mathbf{x}) + ct.$$

From the already proven statement for twice continuously differentiable functions it follows that $\mathbb{E}F_h(\mathbf{X}) \leq \mathbb{E}F_h(\mathbf{Y})$ for all $h > 0$. By uniform convergence of F_h to F we have

$$\mathbb{E}F(\mathbf{X}) = \lim_{h \to 0} \mathbb{E}F_h(\mathbf{X}) \leq \lim_{h \to 0} \mathbb{E}F_h(\mathbf{Y}) = \mathbb{E}F(\mathbf{Y}).$$

This completes the proof. □

Remark 8.33. The Lipschitz assumption in the previous lemma is not essential but simplifies the proof. The result can also be shown under other conditions on F—in particular, as used in the proof, for twice differentiable F such that F together with all its derivatives up to second order is of moderate growth.

Now we are prepared for the proof of Gordon's lemma, which in turn implies Slepian's lemma as a special case.

Proof (of Lemma 8.27). Let

$$F(\mathbf{x}) = \min_{i \in [n]} \max_{j \in [m]} x_{ij},$$

where $\mathbf{x} = (x_{ij})_{i \in [n], j \in [m]}$ is a doubly indexed vector. Then F is a Lipschitz function (with Lipschitz constant 1). We first aim at verifying (8.60). Since this condition involves only derivatives in two variables at a time, we can fix the other variables for the moment, which simplifies the notational burden. Setting $t = x_{ij}$ and $s = x_{k\ell}$ and fixing all other variables, we realize that F takes the form

$$F(\mathbf{x}) = A(t, s) := \max\{\alpha(t), \beta(s)\} \quad \text{if } i = k,$$

or

$$F(\mathbf{x}) = B(t, s) := \min\{\alpha(t), \beta(s)\} \quad \text{if } i \neq k,$$

where both α and β are functions of the form

$$g(t) = \begin{cases} a & \text{if } t < a, \\ t & \text{if } a \leq t \leq b, \\ b & \text{if } t > b. \end{cases} \tag{8.66}$$

Here $a \leq b$ are some numbers that may possibly take the values $a = -\infty$ and $b = +\infty$. We claim that the distributional derivatives of A and B are nonnegative. To prove this, for A we note that

$$A(t, s) = \frac{1}{2}(\alpha(t) + \beta(s) + |\alpha(t) - \beta(s)|).$$

Therefore, a partial weak derivative of A is given by (see Sect. C.9 for the notion of weak derivative, which coincides with the distributional derivative once it exists)

$$\frac{\partial}{\partial t} A(t, s) = \frac{1}{2}(\alpha'(t) + \alpha'(t)\operatorname{sgn}(\alpha(t) - \beta(s)))$$

$$= \begin{cases} 0 & \text{if } t \notin [a, b], \\ \frac{1}{2} + \frac{1}{2}\operatorname{sgn}(t - \beta(s)) & \text{if } t \in [a, b], \end{cases} \tag{8.67}$$

where α' is a weak derivative of α (see Exercise 8.12) and a, b are the numbers defining α (see (8.66)). The function $s \mapsto \operatorname{sgn}(t - \beta(s))$ is nonincreasing in s, and therefore, the distributional derivative $\frac{\partial^2}{\partial s \partial t} A$ is nonpositive as claimed; see also Exercise 8.12(c).

Nonnegativity of $\frac{\partial^2}{\partial s \partial t} B$ follows similarly by writing

$$B(s, t) = \min\{\alpha(t), \beta(s)\} = (\alpha(t) + \beta(s) - |\alpha(t) - \beta(s)|)/2.$$

Therefore, we showed that (in the sense of distributional derivatives)

$$\frac{\partial^2 F}{\partial x_{ij} \partial x_{k\ell}} \leq 0 \quad \text{if } i = k,$$

$$\frac{\partial^2 F}{\partial x_{ij} \partial x_{k\ell}} \geq 0 \quad \text{if } i \neq k.$$

It follows from assumptions (8.50), (8.51) that

$$(\mathbb{E}|X_{i,j} - X_{k,\ell}|^2 - \mathbb{E}|Y_{i,j} - Y_{k,\ell}|^2)\frac{\partial^2 F}{\partial x_{ij} \partial x_{k\ell}} \leq 0 \quad \text{for all } i, j, k, \ell. \tag{8.68}$$

Moreover, the function F satisfies $F(\mathbf{x} + t\mathbf{e}) = F(\mathbf{x}) + t$. The conditions of Lemma 8.32 are therefore satisfied for $-F$, and we conclude that $\mathbb{E}F(\mathbf{X}) \geq \mathbb{E}F(\mathbf{Y})$. $\qquad\square$

8.8 Concentration of Measure

Concentration of measure describes the phenomenon that Lipschitz functions on high-dimensional probability spaces concentrate well around their expectation. We present a precise statement for Gaussian measures. The proof of our first theorem uses the auxiliary tools developed in the previous section and is rather short, but only gives the nonoptimal constant 4 in the probability decay; see (8.70). With a more sophisticated technique based on the entropy of a random variable, we provide the optimal constant 2 in Theorem 8.40 below.

Theorem 8.34. *Let $f : \mathbb{R}^n \to \mathbb{R}$ be a Lipschitz function, i.e.,*

$$|f(\mathbf{x}) - f(\mathbf{y})| \leq L\|\mathbf{x} - \mathbf{y}\|_2 \quad \text{for all } \mathbf{x}, \mathbf{y} \in \mathbb{R}^n, \tag{8.69}$$

for a constant $L > 0$. Let $\mathbf{g}$ be a standard Gaussian random vector. Then for all $t > 0$

$$\mathbb{P}(f(\mathbf{g}) - \mathbb{E}[f(\mathbf{g})] \geq t) \leq \exp\left(-\frac{t^2}{4L^2}\right), \tag{8.70}$$

and consequently

$$\mathbb{P}(|f(\mathbf{g}) - \mathbb{E}[f(\mathbf{g})]| \geq t) \leq 2\exp\left(-\frac{t^2}{4L^2}\right). \tag{8.71}$$

Proof. We first assume that f is differentiable. Let $\mathbf{X}, \mathbf{Y}$ be independent copies of $\mathbf{g}$. We use the Laplace transform method which, for a parameter $\theta > 0$, requires to bound

$$\psi(\theta) := \mathbb{E}\exp(\theta(f(\mathbf{X}) - \mathbb{E}[f(\mathbf{Y})])).$$

The convexity of $t \mapsto \exp(-\theta t)$ and Jensen's inequality yield

$$\psi(\theta) \leq \mathbb{E}\exp(\theta(f(\mathbf{X}) - f(\mathbf{Y}))) = \mathbb{E}G_\theta(\mathbf{X}, \mathbf{Y}),$$

where we have set $G_\theta(\mathbf{x}, \mathbf{y}) = \exp(\theta(f(\mathbf{x}) - f(\mathbf{y})))$. The concatenated vector $\mathbf{Z} = (\mathbf{X}, \mathbf{Y})$ is a standard Gaussian vector of length $2n$. Let $\mathbf{X}'$ denote an independent copy of $\mathbf{X}$ and let $\mathbf{W} = (\mathbf{X}', \mathbf{X}')$. For $0 \leq t \leq 1$, define $\mathbf{U}(t) = \sqrt{t}\mathbf{Z} + \sqrt{1-t}\mathbf{W}$ and $\phi(t) = \mathbb{E}G_\theta(\mathbf{U}(t))$. Clearly, $\phi(0) = \mathbb{E}G_\theta(\mathbf{X}', \mathbf{X}') = \mathbb{E}\exp(\theta(f(\mathbf{X}') - f(\mathbf{X}'))) = 1$. As the next step, we use Proposition 8.31 to compute the derivative of ϕ. To this end, we note that $\mathbb{E}X_iX_j = \mathbb{E}X_i'X_j' = \delta_{ij}$ and $\mathbb{E}X_iY_j = 0$ for all i, j. Furthermore, it follows from the Lipschitz assumption (8.69) that G_θ is of moderate growth; see (8.52). Therefore, (8.59) yields

$$\phi'(t) = \frac{1}{2} \sum_{i,j \in [2n]} (\mathbb{E} Z_i Z_j - \mathbb{E} W_i W_j) \mathbb{E} \left[\frac{\partial^2 G_\theta}{\partial z_i \partial z_j} (\mathbf{U}(t)) \right]$$

$$= -\mathbb{E} \sum_{i=1}^{n} \frac{\partial^2 G_\theta}{\partial x_i \partial y_i} (\mathbf{U}(t)).$$

The partial derivatives of G_θ are given by

$$\frac{\partial^2 G_\theta}{\partial x_i \partial y_i} (\mathbf{x}, \mathbf{y}) = -\theta^2 \frac{\partial f}{\partial x_i}(\mathbf{x}) \frac{\partial f}{\partial y_i}(\mathbf{y}) \, G_\theta(\mathbf{x}, \mathbf{y}), \qquad \mathbf{x}, \mathbf{y} \in \mathbb{R}^n.$$

Since we assumed f to be differentiable it follows from the Lipschitz assumption (8.69) that

$$\| \nabla f(\mathbf{x}) \|_2^2 = \sum_{i=1}^{n} \left| \frac{\partial f}{\partial x_i}(\mathbf{x}) \right|^2 \leq L^2 \qquad \text{for all } \mathbf{x} \in \mathbb{R}^n,$$

so that the Cauchy–Schwarz inequality yields

$$\phi'(t) = \theta^2 \mathbb{E} \sum_{i=1}^{n} \frac{\partial f}{\partial x_i}(\mathbf{X}) \frac{\partial f}{\partial y_i}(\mathbf{Y}) \, G_\theta(\mathbf{U}(t))$$

$$\leq \theta^2 \mathbb{E} \| \nabla f(\mathbf{X}) \|_2 \| \nabla f(\mathbf{Y}) \|_2 G_\theta(\mathbf{U}(t)) \leq \theta^2 L^2 \mathbb{E} G_\theta(\mathbf{U}(t)) = \theta^2 L^2 \phi(t).$$

Since $\phi(t) > 0$, we may divide by it, and setting $\tau(t) := \ln \phi(t)$ shows that

$$\tau'(t) \leq \theta^2 L^2.$$

Together with $\phi(0) = 1$, this differential inequality implies by integration that

$$\tau(1) \leq \int_0^1 \theta^2 L^2 dt = \theta^2 L^2,$$

and consequently,

$$\psi(\theta) \leq \phi(1) = \exp(\tau(1)) \leq \exp(\theta^2 L^2).$$

For $t, \theta > 0$, Markov's inequality yields

$$\mathbb{P}(f(\mathbf{X}) - \mathbb{E} f(\mathbf{X}) \geq t) \leq \psi(\theta) e^{-\theta t} \leq \exp(\theta^2 L^2 - \theta t).$$

Choosing $\theta = t/(2L^2)$ (thus minimizing the last term) yields the claimed inequality (8.70).

In the general case where f is not necessarily differentiable, for each $\varepsilon > 0$, we can find a differentiable Lipschitz function g with the same Lipschitz constant L such that $|f(\mathbf{x}) - g(\mathbf{x})| \leq \epsilon$ for all $\mathbf{x} \in \mathbb{R}^n$; see Theorem C.12. It then follows that

$$\mathbb{P}(f(\mathbf{X}) - \mathbb{E}f(\mathbf{X}) \geq t) \leq \mathbb{P}(g(\mathbf{X}) - \mathbb{E}g(\mathbf{X}) \geq t - 2\varepsilon)$$

$$\leq \exp(-(t - 2\varepsilon)^2/(4L^2)).$$

Since $\varepsilon > 0$ is arbitrary, (8.70) follows also for general, not necessarily differentiable Lipschitz functions.

By replacing f with $-f$ we obtain $\mathbb{P}(\mathbb{E}f(\mathbf{X}) - f(\mathbf{X}) \geq t) \leq \exp(-t^2/(4L^2))$. Then the union bound gives (8.71).   □

In order to improve on the constant 4 in (8.70) we will use an alternative approach based on the concept of *entropy* of a random variable. We introduce the convex function

$$\phi(x) := x \ln(x), \quad x > 0,$$

which is continuously extended to $x = 0$ by $\phi(0) = 0$. For a nonnegative random variable X on some probability space $(\Omega, \Sigma, \mathbb{P})$, we then define the entropy as

$$\mathscr{E}(X) := \mathbb{E}[\phi(X)] - \phi(\mathbb{E}X) = \mathbb{E}[X \ln X] - \mathbb{E}X \ln(\mathbb{E}X). \tag{8.72}$$

If the first term is infinite, then we set $\mathscr{E}(X) = \infty$. By convexity of ϕ, it follows from Jensen's inequality that $\mathscr{E}(X) \geq 0$. The entropy is homogeneous, i.e., for a scalar $t > 0$,

$$\mathscr{E}(tX) = \mathbb{E}[tX \ln(tX)] - \mathbb{E}[tX] \ln(t\mathbb{E}X)$$

$$= t\mathbb{E}[X \ln X] + t\mathbb{E}[X \ln t] - t\mathbb{E}[X] \ln t - t\mathbb{E}[X] \ln(\mathbb{E}[X]) = t\mathscr{E}(X).$$

The basic idea of the entropy method for deducing a concentration inequality for a real-valued random variable X is to derive a bound on the entropy of $e^{\theta X}$, for $\theta > 0$, of the form

$$\mathscr{E}(e^{\theta X}) \leq g(\theta)\mathbb{E}[e^{\theta X}] \tag{8.73}$$

for some appropriate function g. Setting $F(\theta) := \mathbb{E}[e^{\theta X}]$ such an inequality is equivalent to

$$\mathscr{E}(e^{\theta X}) = \theta F'(\theta) - F(\theta) \ln F(\theta) \leq g(\theta)F(\theta). \tag{8.74}$$

Setting further $G(\theta) = \theta^{-1} \ln F(\theta)$ yields

$$G'(\theta) \leq \theta^{-2}g(\theta).$$

Noting that $G(0) = \lim_{\theta \to 0} \theta^{-1} \ln F(\theta) = F'(0)/F(0) = \mathbb{E}[X]$, this shows by integration that $G(\theta) - \mathbb{E}[X] \le \int_0^\theta t^{-2} g(t) dt$ or

$$\mathbb{E}[e^{\theta(X - \mathbb{E}[X])}] \le \exp\left(\theta \int_0^\theta t^{-2} g(t) dt\right), \quad \theta > 0. \tag{8.75}$$

Then one uses Markov's inequality to derive a tail bound for $X - \mathbb{E}[X]$. This way of reasoning is usually called the Herbst argument.

Before applying this idea in our specific situation, we provide some first results for the entropy. We start with a dual characterization.

Lemma 8.35. *Let X be a nonnegative random variable satisfying $\mathbb{E}[X] < \infty$. Then*

$$\mathscr{E}(X) = \sup\left\{\mathbb{E}[XY] : \mathbb{E}[\exp(Y)] \le 1\right\}. \tag{8.76}$$

Proof. We first deal with the case that X is strictly positive. By homogeneity of the entropy we may and do assume $\mathbb{E}X = 1$. The Fenchel inequality (B.10) yields, for Y satisfying $\mathbb{E}[\exp(Y)] \le 1$,

$$\mathbb{E}[XY] \le \mathbb{E}[\exp(Y)] + \mathbb{E}[X \ln X] - \mathbb{E}[X] \le \mathbb{E}[X \ln X] = \mathscr{E}(X).$$

This shows that the right-hand side in (8.76) is smaller than or equal to the left-hand side. For the converse direction, we choose $Y = \ln X - \ln(\mathbb{E}X)$, so that $\mathscr{E}(X) = \mathbb{E}[XY]$. This choice satisfies

$$\mathbb{E} \exp(Y) = \mathbb{E}[X] \exp\left(-\ln(\mathbb{E}X)\right) = 1.$$

Therefore, the right-hand side in (8.76) majorizes $\mathscr{E}(X)$.

The case of a general nonnegative random variable X follows from continuity of ϕ in 0 and from a simple approximation argument. $\qquad\square$

Remark 8.36. (a) Substituting $Y = \ln(Z/\mathbb{E}Z)$ for a positive random variable Z in (8.76) shows that

$$\mathscr{E}(X) = \sup\left\{\mathbb{E}[X \ln(Z)] - \mathbb{E}[X] \ln(\mathbb{E}[Z]) : Z > 0\right\}, \tag{8.77}$$

where the supremum is taken over all positive integrable random variables Z.
(b) A simple consequence of the previous lemma is the subadditivity of the entropy, i.e., for two nonnegative random variables X and Z, we have $\mathscr{E}(X + Z) \le \mathscr{E}(X) + \mathscr{E}(Z)$.

For a sequence $\mathbf{X} = (X_1, \ldots, X_n)$ we introduce, for any $i \in [n]$, the notation

$$\mathbf{X}^{(i)} = (X_1, \ldots, X_{i-1}, X_{i+1}, \ldots, X_n).$$

For a function f of $\mathbf{X}$, we write

$$\mathbb{E}_{X_i} f(\mathbf{X}) = \mathbb{E}_{X_i} [f(X_1, \ldots, X_i, \ldots, X_n)] := \mathbb{E}\left[f(\mathbf{X}) | \mathbf{X}^{(i)}\right] \qquad (8.78)$$

for the conditional expectation, which is still a function of the random variables $X_1, \ldots, X_{i-1}, X_{i+1}, \ldots, X_n$. In other words, $\mathbb{E}_{X_i} f(\mathbf{X})$ "integrates out" the dependence on X_i and is constant with respect to X_i. Then we define the conditional entropy of $f(\mathbf{X})$ as

$$\mathscr{E}_{X_i}(f(\mathbf{X})) := \mathscr{E}\left(f(\mathbf{X}) | \mathbf{X}^{(i)}\right) := \mathbb{E}_{X_i}(\phi(f(\mathbf{X}))) - \phi\left(\mathbb{E}_{X_i}(f(\mathbf{X}))\right)$$

$$= \mathbb{E}_{X_i}[f(\mathbf{X}) \ln f(\mathbf{X})] - \mathbb{E}_{X_i}[f(\mathbf{X})] \ln\left(\mathbb{E}_{X_i}[f(\mathbf{X})]\right) .$$

Clearly, $\mathscr{E}_{X_i}(f(\mathbf{X}))$ is still a random variable that depends on $\mathbf{X}^{(i)}$, i.e., the entropy is taken only with respect to X_i. The tensorization inequality for the entropy reads as follows.

Proposition 8.37. *Let* $\mathbf{X} = (X_1, \ldots, X_n)$ *be a vector of independent random variables and let* f *be a nonnegative function satisfying* $\mathbb{E}[f(\mathbf{X})] < \infty$. *Then*

$$\mathscr{E}(f(\mathbf{X})) \le \mathbb{E}\left[\sum_{i=1}^{n} \mathscr{E}_{X_i}(f(\mathbf{X}))\right] . \qquad (8.79)$$

Proof. We may assume that f is strictly positive. The general case follows from an approximation argument exploiting the continuity of ϕ at 0. We introduce the conditional expectation operator $\mathbb{E}^i$ defined by

$$\mathbb{E}^i[f(\mathbf{X})] := \mathbb{E}_{X_1, \ldots, X_{i-1}}[f(\mathbf{X})] = \mathbb{E}[f(\mathbf{X}) | X_i, \ldots, X_n],$$

which "integrates out" the dependence on the first $i-1$ random variables $X_1, \ldots, X_{i-1}$. We have $\mathbb{E}^1[f(\mathbf{X})] = f(\mathbf{X})$ and $\mathbb{E}^{n+1}[f(\mathbf{X})] = \mathbb{E}[f(\mathbf{X})]$. A decomposition by a telescoping sum gives

$$\ln(f(\mathbf{X})) - \ln(\mathbb{E}[f(\mathbf{X})]) = \sum_{i=1}^{n} (\ln(\mathbb{E}^i[f(\mathbf{X})]) - \ln(\mathbb{E}^{i+1}[f(\mathbf{X})])). \qquad (8.80)$$

The duality formula (8.77) with $Z = \mathbb{E}^i[f(\mathbf{X})] > 0$ yields

$$\mathbb{E}_{X_i}\left[f(\mathbf{X}) \left(\ln(\mathbb{E}^i[f(\mathbf{X})]) - \ln(\mathbb{E}_{X_i}[\mathbb{E}^i[f(\mathbf{X})]])\right)\right] \le \mathscr{E}_{X_i}(f(\mathbf{X})) .$$

Observe that by independence and Fubini's theorem

$$\mathbb{E}_{X_i}[\mathbb{E}^i[f(\mathbf{X})]] = \mathbb{E}_{X_i}\mathbb{E}_{X_1, \ldots, X_{i-1}}[f(\mathbf{X})] = \mathbb{E}^{i+1}[f(\mathbf{X})].$$

Multiplying by $f(\mathbf{X})$ and taking expectations on both sides of (8.80) yield

$$\mathscr{E}(f(\mathbf{X})) = \mathbb{E}[f(\mathbf{X})(\ln(f(\mathbf{X})) - \ln(\mathbb{E}[f(\mathbf{X})]))]$$

$$= \sum_{i=1}^{n} \mathbb{E}\left[\mathbb{E}_{X_i}\left[f(\mathbf{X})(\ln(\mathbb{E}^i[f(\mathbf{X})]) - \ln(\mathbb{E}_{X_i}[\mathbb{E}^i[f(\mathbf{X})]]))\right]\right]$$

$$\leq \sum_{i=1}^{n} \mathbb{E}\left[\mathscr{E}_{X_i}(f(\mathbf{X}))\right].$$

This completes the proof.   □

The key to establishing Gaussian concentration of measure is the logarithmic Sobolev inequality. We start with the one for Rademacher vectors.

Theorem 8.38. *Let* $f : \{-1, 1\}^n \to \mathbb{R}$ *be a real-valued function and* ϵ *be an* n-*dimensional Rademacher vector. Then*

$$\mathscr{E}\left(f^2(\epsilon)\right) \leq \frac{1}{2}\mathbb{E}\left[\sum_{i=1}^{n}\left(f(\epsilon) - f(\bar{\epsilon}^{(i)})\right)^2\right], \tag{8.81}$$

where $\bar{\epsilon}^{(i)} = (\epsilon_1, \ldots, \epsilon_{i-1}, -\epsilon_i, \epsilon_{i+1}, \ldots, \epsilon_n)$ *is obtained from* ϵ *by flipping the* ith *entry.*

Proof. By the tensorization inequality (Proposition 8.37) we have

$$\mathscr{E}\left(f^2(\epsilon)\right) \leq \mathbb{E}\left[\sum_{i=1}^{n}\mathscr{E}_{\epsilon_i}\left(f^2(\epsilon)\right)\right].$$

Therefore, it suffices to show that, for each $i \in [n]$,

$$\mathscr{E}_{\epsilon_i}(f(\epsilon)^2) \leq \frac{1}{2}\mathbb{E}_{\epsilon_i}\left[\left(f(\epsilon) - f(\bar{\epsilon}^{(i)})\right)^2\right]. \tag{8.82}$$

Given any realization of $(\epsilon_1, \ldots, \epsilon_{i-1}, \epsilon_{i+1}, \ldots, \epsilon_n)$, $f(\epsilon)$ (as well as $f(\bar{\epsilon}^{(i)})$) can take only two possible values denoted by $a, b \in \mathbb{R}$. Then, by definition of the entropy, the desired inequality (8.82) takes the form

$$\frac{1}{2}\left(a^2\ln(a^2) + b^2\ln(b^2)\right) - \frac{a^2 + b^2}{2}\ln\left(\frac{a^2 + b^2}{2}\right) \leq \frac{1}{2}(a - b)^2.$$

In order to establish this scalar inequality, we define, for fixed b, the function

$$H(a) = \frac{1}{2}\left(a^2 \ln(a^2) + b^2 \ln(b^2)\right) - \frac{a^2 + b^2}{2} \ln\left(\frac{a^2 + b^2}{2}\right) - \frac{1}{2}(a - b)^2.$$

The first and second derivatives of H are given by

$$H'(a) = a \ln\left(\frac{2a^2}{a^2 + b^2}\right) - (a - b),$$

$$H''(a) = \ln\left(\frac{2a^2}{a^2 + b^2}\right) + 1 - \frac{2a^2}{a^2 + b^2}.$$

We see that $H(b) = 0$ and $H'(b) = 0$. Using $\ln x \le x - 1$, we further obtain $H''(a) \le 0$ for all $a \in \mathbb{R}$ so that H is concave. It follows that $H(a) \le 0$ for all $a \in \mathbb{R}$, which finally establishes (8.82), and thereby the claim. $\square$

The right-hand side of the logarithmic Sobolev inequality (8.81) for Rademacher vectors involves a discrete version of a gradient. The Gaussian logarithmic Sobolev inequality stated next features a true gradient.

Theorem 8.39. *Let* $f : \mathbb{R}^n \to \mathbb{R}$ *be a continuously differentiable function satisfying* $\mathbb{E}\left[\phi(f^2(\mathbf{g}))\right] < \infty$ *for a standard Gaussian vector* $\mathbf{g}$ *on* $\mathbb{R}^n$. *Then*

$$\mathscr{E}\left(f^2(\mathbf{g})\right) \le 2\mathbb{E}\left[\|\nabla f(\mathbf{g})\|_2^2\right]. \tag{8.83}$$

Proof. We first prove the theorem for $n = 1$ and g being a standard normal random variable. We start by dealing with a compactly supported f. Since f' is uniformly continuous in this case, its modulus of continuity $\omega(f', \delta) := \sup_{|t-u| \le \delta} |f'(t) - f'(u)|$ satisfies $\omega(f', \delta) \to 0$ as $\delta \to 0$.

Let $\epsilon = (\epsilon_1, \ldots, \epsilon_m)$ be a Rademacher vector and set

$$S_m := \frac{1}{\sqrt{m}} \sum_{j=1}^{m} \epsilon_j.$$

The idea is to use Theorem 8.38 and the fact that, by the central limit theorem, S_m converges in distribution to a standard normal random variable. An application of (8.81) to $\tilde{f}(\epsilon) := f(S_m)$ yields

$$\mathscr{E}\left(f^2(S_m)\right) \le \frac{1}{2}\mathbb{E}\left[\sum_{i=1}^{m}\left(\tilde{f}(\epsilon) - \tilde{f}(\bar{\epsilon}^{(i)})\right)^2\right]$$

$$= \frac{1}{2}\mathbb{E}\left[\sum_{i=1}^{m}\left(f(S_m) - f\left(S_m - \frac{2\epsilon_i}{\sqrt{m}}\right)\right)^2\right].$$

For each $i \in [m]$, we notice that

$$\left| f(S_m) - f\left(S_m - \frac{2\epsilon_i}{\sqrt{m}}\right) \right| = \left| \frac{2\epsilon_i}{\sqrt{m}} f'(S_m) + \int_{S_m - 2\epsilon_i/\sqrt{m}}^{S_m} (f'(t) - f'(S_m)) dt \right|$$

$$\leq \frac{2}{\sqrt{m}} |f'(S_m)| + \frac{2}{\sqrt{m}} \omega\left(f', \frac{2}{\sqrt{m}}\right).$$

It follows that

$$\sum_{i=1}^{m} \left(f(S_m) - f\left(S_m - \frac{2\epsilon_i}{\sqrt{m}}\right)\right)^2$$

$$\leq 4 \left(f'(S_m)^2 + 2|f'(S_m)| \omega\left(f', \frac{2}{\sqrt{m}}\right) + \omega\left(f', \frac{2}{\sqrt{m}}\right)^2 \right).$$

By the boundedness of f and f', the central limit theorem (Theorem 7.8) implies that $\mathbb{E}[f'(S_m)^2] \to \mathbb{E}[f'(g)^2]$ and $\mathscr{E}(f^2(S_m)) \to \mathscr{E}(f^2(g))$ as $m \to \infty$. Altogether, we obtain the desired inequality

$$\mathscr{E}(f^2(g)) \leq 2\mathbb{E}\left[f'(g)^2\right].$$

Let us now turn to the general case where f does not necessarily have compact support. Given a small $\varepsilon > 0$, the assumption that $\mathbb{E}\left[\phi(f^2(g))\right] < \infty$ ensures the existence of $T > 0$ such that for any subset I of $\mathbb{R} \setminus [-T, T]$,

$$\frac{1}{\sqrt{2\pi}} \int_I \left| \phi(f(t)^2)\right| e^{-t^2/2} dt \leq \varepsilon \quad \text{and} \quad \frac{1}{\sqrt{2\pi}} \int_I e^{-t^2/2} dt \leq \varepsilon.$$

Let us consider a continuously differentiable function h satisfying $0 \leq h(t) \leq 1$ for all $t \in \mathbb{R}$, as well as $h(t) = 1$ for all $t \in [-T, T]$ and $h(t) = 0$ for all $t \notin [-T-1, T+1]$. We set $\hat{f} := fh$, which is a continuously differentiable function with compact support. The previous result applies to give

$$\mathscr{E}(\hat{f}^2(g)) \leq 2\mathbb{E}[\hat{f}'(g)^2].$$

By subadditivity of the entropy (see Remark 8.36(b)), we have

$$\mathscr{E}\left(f^2(g)\right) = \mathscr{E}\left(\hat{f}^2(g) + f^2(g)(1 - h^2(g))\right) \leq \mathscr{E}\left(\hat{f}^2(g)\right) + \mathscr{E}\left(f^2(g)(1 - h^2(g))\right).$$

Let us now introduce the sets $I_1 := \{t \in \mathbb{R} : |t| \geq T, f(t)^2 < e\}$ and $I_2 := \{t \in \mathbb{R} : |t| \geq T, f(t)^2 \geq e\}$. Then

$$\mathbb{E}\left[f^2(g)(1 - h^2(g))\right] = \frac{1}{\sqrt{2\pi}} \int_{\mathbb{R}\setminus[-T,T]} f^2(t)(1 - h^2(t))e^{-t^2} dt$$

$$\leq \frac{1}{\sqrt{2\pi}} \int_{I_1} f^2(t)e^{-t^2} dt + \frac{1}{\sqrt{2\pi}} \int_{I_2} f^2(t)e^{-t^2} dt$$

$$\leq \frac{e}{\sqrt{2\pi}} \int_{I_1} e^{-t^2/2} dt + \frac{1}{\sqrt{2\pi}} \int_{I_2} \phi(f^2(t))e^{-t^2} dt \leq (e+1)\varepsilon. \qquad (8.84)$$

Therefore, $\left|\phi\left(\mathbb{E}\left[f^2(g)(1 - h^2(g))\right]\right)\right| \leq |\phi((1 + e)\varepsilon)|$ if ε is sufficiently small. Introducing the sets $\tilde{I}_1 := \{t \in \mathbb{R} : |t| \geq T, f(t)^2(1 - h^2(t)) < e\}$ and $\tilde{I}_2 := \{t \in \mathbb{R} : |t| \geq T, f(t)^2(1 - h^2(t)) \geq e\}$ and denoting $\kappa = \max_{t \in [0,e]} |\phi(t)| = e^{-1}$, we have

$$\left|\mathbb{E}\left[\phi(f^2(g)(1 - h^2(g)))\right]\right|$$

$$\leq \frac{\kappa}{\sqrt{2\pi}} \int_{\tilde{I}_1} e^{-t^2/2} dt + \frac{1}{\sqrt{2\pi}} \int_{\tilde{I}_2} \phi(f^2(t)(1 - h^2(t)))e^{-t^2/2} dt$$

$$\leq \kappa\varepsilon + \frac{1}{\sqrt{2\pi}} \int_{\tilde{I}_2} \phi(f^2(t))e^{-t^2/2} dt \leq (\kappa + 1)\varepsilon.$$

By the definition of the entropy, this gives

$$\mathscr{E}\left(f^2(g)(1 - h^2(g))\right) \leq \left|\phi\left(\mathbb{E}\left[f^2(g)(1 - h^2(g))\right]\right)\right| + \left|\mathbb{E}\left[\phi(f^2(g)(1 - h^2(g)))\right]\right|$$
$$\leq |\phi((e+1)\varepsilon)| + (\kappa + 1)\varepsilon.$$

At this point, we have obtained

$$\mathscr{E}(f^2(g)) \leq 2\mathbb{E}[\hat{f}'(g)^2] + (1 + \kappa)\varepsilon + |\phi((1 + e)\varepsilon)|.$$

Now, using the triangle inequality, we have

$$\mathbb{E}[\hat{f}'(g)^2]^{1/2} = \mathbb{E}[(f'h + fh')(g)^2]^{1/2} \leq \mathbb{E}[(f'h)(g)^2]^{1/2} + \mathbb{E}[(fh')(g)^2]^{1/2}.$$

We remark that $\mathbb{E}[(f'h)(g)^2] = \mathbb{E}[f'(g)^2 h(g)^2] \leq \mathbb{E}[f'(g)^2]$ and that

$$\mathbb{E}[(fh')(g)^2] = \frac{1}{\sqrt{2\pi}} \int_{-\infty}^{\infty} f(t)^2 h'(t)^2 e^{-t^2/2} dt \leq \frac{\|h'\|_\infty^2}{\sqrt{2\pi}} \int_{I_1 \cup I_2} f(t)^2 e^{-t^2/2} dt$$

$$\leq (e + 1)\|h'\|_\infty^2 \varepsilon, \qquad (8.85)$$

where we have used (8.84). We conclude that $\mathbb{E}[(fh')(g)^2] \leq \|h'\|_\infty^2(e+1)\varepsilon$. As a result, we obtain $\mathbb{E}[\hat{f}'(g)^2]^{1/2} \leq \mathbb{E}[f'(g)^2]^{1/2} + \|h'\|_\infty ((e+1)\varepsilon)^{1/2}$. Overall, we have shown that

$$\mathscr{E}(f^2(g)) \le 2 \left(\mathbb{E}[f'(g)^2]^{1/2} + \|h'\|_\infty \left((e+1)\varepsilon\right)^{1/2} \right)^2 + |\phi((1+e)\varepsilon)| + (\kappa+1)\varepsilon.$$

Since this result is valid for all sufficiently small $\varepsilon > 0$ and since $\lim_{t \to 0} \phi(t) = \phi(0) = 0$, we conclude that

$$\mathscr{E}(f^2(g)) \le \mathbb{E}[f'(g)^2].$$

This is the desired claim for $n = 1$.

For arbitrary n, we apply the tensorization inequality (8.79) to obtain

$$\mathscr{E}(f^2(\mathbf{g})) \le \mathbb{E}\left[\sum_{i=1}^n \mathscr{E}_{g_i}(f^2(\mathbf{g})) \right] \le 2\mathbb{E}\left[\sum_{i=1}^n \left(\frac{\partial f}{\partial x_i}(\mathbf{g}) \right)^2 \right] = 2\mathbb{E}\left[\|\nabla f(\mathbf{g})\|_2^2 \right].$$

This completes the proof.   □

With the Gaussian logarithmic Sobolev inequality at hand, we are prepared to show the improved version of the concentration of measure inequality for Lipschitz functions.

Theorem 8.40. *Let* $f : \mathbb{R}^n \to \mathbb{R}$ *be a Lipschitz function with constant* L *in* (8.69). *Let* $\mathbf{g} = (g_1, \dots, g_n)$ *be a vector of independent standard Gaussian random variables. Then, for all* $t > 0$,

$$\mathbb{P}(f(\mathbf{g}) - \mathbb{E}[f(\mathbf{g})] \ge t) \le \exp\left(-\frac{t^2}{2L^2} \right), \tag{8.86}$$

and consequently

$$\mathbb{P}(|f(\mathbf{g}) - \mathbb{E}[f(\mathbf{g})]| \ge t) \le 2\exp\left(-\frac{t^2}{2L^2} \right). \tag{8.87}$$

Remark 8.41. The constant 2 in (8.86) is optimal in general. This follows from the case $n = 1$, $f(x) = x$, and the lower tail bound for a standard Gaussian random variable in Proposition 7.5.

Proof. We may assume that f is differentiable, so that the Lipschitz condition implies $\|\nabla f(\mathbf{x})\|_2 \le L$ for all $\mathbf{x} \in \mathbb{R}^n$. The general case follows similarly to the end of the proof of Theorem 8.34.

For $\theta > 0$, applying the Gaussian logarithmic Sobolev inequality (8.83) to the function $e^{\theta f/2}$ yields

$$\mathscr{E}(e^{\theta f(\mathbf{g})}) \le 2\mathbb{E}\left[\|\nabla e^{\theta f(\mathbf{g})/2}\|_2^2 \right] = \frac{\theta^2}{2}\mathbb{E}\left[e^{\theta f(\mathbf{g})} \|\nabla f(\mathbf{g})\|_2^2 \right] \le \frac{\theta^2 L^2}{2}\mathbb{E}\left[e^{\theta f(\mathbf{g})} \right], \tag{8.88}$$

where the Lipschitz assumption was applied in the last step. (Note that the hypothesis $\mathbb{E}\left[\phi(e^{2\theta f(\mathbf{g})})\right] < \infty$ required for Theorem 8.39 holds because, again by the Lipschitz assumption, we have $e^{2\theta f(\mathbf{x})} \leq e^{2\theta |f(0)|} e^{L\theta \|\mathbf{x}\|_2}$ for all $\mathbf{x} \in \mathbb{R}^n$.) The inequality (8.88) is of the form (8.73) with $g(\theta) = L^2\theta^2/2$. Therefore, the Herbst argument leads to (8.75) in the form

$$\mathbb{E}[e^{\theta(f(\mathbf{g}) - \mathbb{E}[f(\mathbf{g})])}] \leq \exp\left(\theta \int_0^\theta t^{-2} g(t) dt\right) = \exp(\theta^2 L^2/2).$$

By Markov's inequality, we obtain

$$\mathbb{P}(f(\mathbf{g}) - \mathbb{E}[f(\mathbf{g})] \geq t) \leq \inf_{\theta > 0} \exp(-\theta t) \mathbb{E}[e^{\theta(f(\mathbf{g}) - \mathbb{E}[f(\mathbf{g})])}]$$

$$\leq \inf_{\theta > 0} \exp(-\theta t + \theta^2 L^2/2) = \exp(-t^2/(2L^2)),$$

where the infimum is realized for $\theta = t/L^2$. Replacing f by $-f$ yields the same bound for $\mathbb{P}(\mathbb{E}[f(\mathbf{g})] - f(\mathbf{g}) \geq t)$, and the union bound finally implies (8.87). $\square$

We close this section with the useful special case of the Lipschitz function $\|\cdot\|_2$, which has constant $L = 1$. If $\mathbf{g} \in \mathbb{R}^n$ is a standard Gaussian vector, then it follows from Theorem 8.40 and Proposition 8.1 that

$$\mathbb{P}(\|\mathbf{g}\|_2 \geq \sqrt{n} + t) \leq \mathbb{P}(\|\mathbf{g}\|_2 \geq \mathbb{E}\|\mathbf{g}\|_2 + t) \leq e^{-t^2/2}. \tag{8.89}$$

8.9 Bernstein Inequality for Suprema of Empirical Processes

In this section, we present a deviation inequality for suprema of empirical processes above their mean, which will become very useful in Chap. 12. Let $Y_1, \ldots, Y_M$ be independent random vectors in $\mathbb{C}^n$ and let $\mathcal{F}$ be a countable collection of measurable functions from $\mathbb{C}^n$ into $\mathbb{R}$. We are interested in the random variable $Z = \sup_{\mathsf{F} \in \mathcal{F}} \sum_{\ell=1}^M \mathsf{F}(Y_\ell)$, i.e., the supremum of an empirical process. In particular, we study its deviation from its mean $\mathbb{E}Z$.

Theorem 8.42. *Let $\mathcal{F}$ be a countable set of functions $\mathsf{F} : \mathbb{C}^n \to \mathbb{R}$. Let $Y_1, \ldots, Y_M$ be independent random vectors in $\mathbb{C}^n$ such that $\mathbb{E}\mathsf{F}(Y_\ell) = 0$ and $\mathsf{F}(Y_\ell) \leq K$ almost surely for all $\ell \in [M]$ and for all $\mathsf{F} \in \mathcal{F}$ for some constant $K > 0$. Introduce*

$$Z = \sup_{\mathsf{F} \in \mathcal{F}} \sum_{\ell=1}^M \mathsf{F}(Y_\ell). \tag{8.90}$$

Let $\sigma_\ell^2 > 0$ such that $\mathbb{E}\left[F(Y_\ell)^2\right] \leq \sigma_\ell^2$ for all $F \in \mathcal{F}$ and $\ell \in [M]$. Then, for all $t > 0$,

$$\mathbb{P}(Z \geq \mathbb{E}Z + t) \leq \exp\left(-\frac{t^2/2}{\sigma^2 + 2K\mathbb{E}Z + tK/3}\right), \qquad (8.91)$$

where $\sigma^2 = \sum_{\ell=1}^M \sigma_\ell^2$.

Remark 8.43. (a) If $\mathcal{F}$ consists only of a single function, then inequality (8.91) reduces to the standard Bernstein inequality in Corollary 7.31. It is remarkable that Theorem 8.42 reproduces the same constants in this more general setting.

(b) The deviation inequality (8.91) can be extended to a concentration inequality, which is sometimes referred to as Talagrand's inequality; see the Notes section.

(c) Theorem 8.42 holds without change if Z is replaced by

$$\tilde{Z} = \sup_{F \in \mathcal{F}} \left|\sum_{\ell=1}^M F(Y_\ell)\right|.$$

Before turning to the proof of the theorem, we present the following Bernstein-type inequality for the norm of the sum of independent mean-zero random vectors. Its formulation uses the dual norm; see Definition A.3 and in particular (A.5).

Corollary 8.44. *Let $\mathbf{Y}_1, \ldots, \mathbf{Y}_M$ be independent copies of a random vector $\mathbf{Y}$ on $\mathbb{C}^n$ satisfying $\mathbb{E}\mathbf{Y} = \mathbf{0}$. Assume $\|\mathbf{Y}\| \leq K$ for some $K > 0$ and some norm $\|\cdot\|$ on $\mathbb{C}^n$. Let*

$$Z = \left\|\sum_{\ell=1}^M \mathbf{Y}_\ell\right\|$$

and

$$\sigma^2 = \sup_{\mathbf{x} \in B^*} \mathbb{E}|\langle \mathbf{x}, \mathbf{Y} \rangle|^2, \qquad (8.92)$$

where $B^* = \{\mathbf{x} \in \mathbb{C}^n, \|\mathbf{x}\|_* \leq 1\}$ denotes the unit ball in the dual norm $\|\cdot\|_*$. Then, for $t > 0$,

$$\mathbb{P}(Z \geq \mathbb{E}Z + t) \leq \exp\left(-\frac{t^2/2}{M\sigma^2 + 2K\mathbb{E}Z + tK/3}\right). \qquad (8.93)$$

Proof. Introduce the random functions $F_{\mathbf{x}}(\mathbf{Y}) := \text{Re}(\langle \mathbf{x}, \mathbf{Y} \rangle)$, $\mathbf{x} \in \widetilde{B^*}$. By the characterization (A.5) of a norm by its dual norm we have

$$Z = \sup_{\mathbf{x} \in B^*} \text{Re} \left\langle \mathbf{x}, \sum_{\ell=1}^{M} \mathbf{Y}_\ell \right\rangle = \sup_{\mathbf{x} \in B^*} \sum_{\ell=1}^{M} \text{Re}\left(\langle \mathbf{x}, \mathbf{Y}_\ell \rangle\right) = \sup_{\mathbf{x} \in B^*} \sum_{\ell=1}^{M} \mathsf{F}_{\mathbf{x}}(\mathbf{Y}_\ell).$$

Let $\widetilde{B}^*$ be a dense countable subset of B^*. Then $Z = \sup_{\mathbf{x} \in \widetilde{B}^*} \sum_{\ell=1}^{M} \mathsf{F}_{\mathbf{x}}(\mathbf{Y}_\ell)$ and

$$\sup_{\mathbf{x} \in \widetilde{B}^*} \mathbb{E}\mathsf{F}_{\mathbf{x}}(\mathbf{Y}_\ell)^2 = \sup_{\mathbf{x} \in B^*} \mathbb{E}|\langle \mathbf{x}, \mathbf{Y} \rangle|^2 = \sigma^2.$$

The random variables $\mathsf{F}_{\mathbf{x}}(\mathbf{Y}) := \text{Re}(\langle \mathbf{x}, \mathbf{Y} \rangle)$, $\mathbf{x} \in \widetilde{B}^*$, satisfy $\mathbb{E}\mathsf{F}_{\mathbf{x}}(\mathbf{Y}) = 0$ and are almost surely bounded, $|\mathsf{F}_{\mathbf{x}}(\mathbf{Y})| \leq \|\mathbf{x}\|_* \|\mathbf{Y}\| \leq K$. The conclusion follows therefore from Theorem 8.42. □

We specialize to the case of the ℓ_2-norm in the next statement.

Corollary 8.45. *Let* $\mathbf{Y}_1, \ldots, \mathbf{Y}_M$ *be independent copies of a random vector* $\mathbf{Y}$ *on* $\mathbb{C}^n$ *satisfying* $\mathbb{E}\mathbf{Y} = \mathbf{0}$. *Assume* $\|\mathbf{Y}\|_2 \leq K$ *for some* $K > 0$. *Let*

$$Z = \left\| \sum_{\ell=1}^{M} \mathbf{Y}_\ell \right\|_2, \qquad \mathbb{E}Z^2 = M\mathbb{E}\|\mathbf{Y}\|_2^2, \tag{8.94}$$

and

$$\sigma^2 = \sup_{\|\mathbf{x}\|_2 \leq 1} \mathbb{E}|\langle \mathbf{x}, \mathbf{Y} \rangle|^2.$$

Then, for $t > 0$,

$$\mathbb{P}(Z \geq \sqrt{\mathbb{E}Z^2} + t) \leq \exp\left(-\frac{t^2/2}{M\sigma^2 + 2K\sqrt{\mathbb{E}Z^2} + tK/3} \right). \tag{8.95}$$

Proof. The formula for $\mathbb{E}Z^2$ in (8.94) follows from independence, and since $\mathbb{E}\mathbf{Y}_\ell = \mathbf{0}$,

$$\mathbb{E}Z^2 = \sum_{\ell,k=1}^{M} \mathbb{E}\langle \mathbf{Y}_\ell, \mathbf{Y}_k \rangle = \sum_{\ell=1}^{M} \mathbb{E}\|\mathbf{Y}_\ell\|_2^2 = M\mathbb{E}\|\mathbf{Y}\|_2^2.$$

By Hölder's inequality, we have $\mathbb{E}Z \leq \sqrt{\mathbb{E}Z^2}$. Therefore, the claim is a consequence of Corollary 8.44. □

The so-called weak variance σ^2 in (8.92) can be estimated by

$$\sigma^2 = \sup_{\mathbf{x} \in B^*} \mathbb{E}|\langle \mathbf{x}, \mathbf{Y} \rangle|^2 \leq \mathbb{E} \sup_{\mathbf{x} \in B^*} |\langle \mathbf{x}, \mathbf{Y} \rangle|^2 = \mathbb{E}\|\mathbf{Y}\|^2. \tag{8.96}$$

Hence, the variance term σ^2 can be replaced by $\mathbb{E}\|\mathbf{Y}\|^2$ in Theorem 8.42 and Corollaries 8.44 and 8.45. Usually, however, σ^2 provides better estimates than $\mathbb{E}\|\mathbf{Y}\|^2$. In any case, noting that $\|\mathbf{Y}\| \leq K$ almost surely implies $\sigma^2 \leq \mathbb{E}\|\mathbf{Y}\|^2 \leq K^2$ yields the next consequence of Corollary 8.44.

Corollary 8.46. *Let* $\mathbf{Y}_1, \ldots, \mathbf{Y}_M$ *be independent copies of a random vector* $\mathbf{Y}$ *on* $\mathbb{C}^n$ *satisfying* $\mathbb{E}\mathbf{Y} = \mathbf{0}$. *Assume* $\|\mathbf{Y}\| \leq K$ *for some constant* $K > 0$ *and some norm* $\| \cdot \|$ *on* $\mathbb{C}^n$. *Let* $Z = \left\|\sum_{\ell=1}^M \mathbf{Y}_\ell\right\|$. *Then, for* $t > 0$,

$$\mathbb{P}(Z \geq \mathbb{E}Z + t) \leq \exp\left(-\frac{t^2/2}{MK^2 + 2K\mathbb{E}Z + Kt/3}\right). \tag{8.97}$$

We will derive the Bernstein-type inequality for suprema of empirical processes as a consequence of a more general deviation inequality for functions in independent random variables. Its formulation needs some notation.

For a sequence $\mathbf{X} = (X_1, \ldots, X_n)$ of independent random variables, we recall the notation $\mathbf{X}^{(i)} = (X_1, \ldots, X_{i-1}, X_{i+1}, \ldots, X_n)$. (Note that below the X_i are allowed to be also random vectors. Since we take functions of those, it will, however, not be important in which set the X_i take their values, so that we will simply refer to them as random variables.) Also, we recall the notation (8.78) of the conditional expectation $\mathbb{E}_{X_i} f(\mathbf{X})$ and the function h defined in (8.23), i.e.,

$$h(x) := (1 + x)\ln(1 + x) - x.$$

Then the Bernstein-type inequality for functions of independent random variables reads as follows.

Theorem 8.47. *Let* $\mathbf{X} = (X_1, \ldots, X_n)$ *be a sequence of independent random variables. Let* f, g_i, $i \in [n]$, *be measurable functions of* $\mathbf{X}$ *and* f_i, $i \in [n]$, *be measurable functions of* $\mathbf{X}^{(i)}$. *Assume that*

$$g_i(\mathbf{X}) \leq f(\mathbf{X}) - f_i(\mathbf{X}^{(i)}) \leq 1, \quad i \in [n], \tag{8.98}$$

and $\quad \mathbb{E}_{X_i}[g_i(\mathbf{X})] \geq 0, \quad i \in [n],$ \hfill (8.99)

as well as

$$\sum_{i=1}^n (f(\mathbf{X}) - f_i(\mathbf{X}^{(i)})) \leq f(\mathbf{X}). \tag{8.100}$$

Suppose further that there exist $B, \sigma > 0$ *such that*

$$g_i(\mathbf{X}) \leq B, \quad i \in [n], \quad \text{and} \quad \frac{1}{n}\sum_{i=1}^n \mathbb{E}_{X_i}[g_i(\mathbf{X})^2] \leq \sigma^2. \tag{8.101}$$

Set $v = (1 + B)\mathbb{E}[f(\mathbf{X})] + n\sigma^2$. Then, for all $\lambda > 0$,

$$\ln \mathbb{E}\left[e^{\lambda(f(\mathbf{X}) - \mathbb{E}[f(\mathbf{X})])}\right] \leq v(e^\lambda - \lambda - 1). \tag{8.102}$$

As a consequence, for $t > 0$,

$$\mathbb{P}\left(f(\mathbf{X}) \geq \mathbb{E}[f(\mathbf{X})] + t\right) \leq \exp\left(-vh\left(\frac{t}{v}\right)\right) \leq \exp\left(-\frac{t^2}{2v + 2t/3}\right). \tag{8.103}$$

Before we prove this theorem, we show how it implies the Bernstein-type inequality (8.91) for suprema of empirical processes.

Proof (of Theorem 8.42). We assume that $K = 1$. The general case is deduced via replacing F by F/K.

Suppose first that $\mathcal{F}$ is a finite set. Let $\mathbf{Y} = (Y_1, \ldots, Y_M)$. We define

$$f(\mathbf{Y}) := \sup_{\mathsf{F} \in \mathcal{F}} \sum_{\ell=1}^{M} \mathsf{F}(Y_\ell) = Z$$

and, for $i \in [M]$, we set

$$f_i(\mathbf{Y}^{(i)}) := \sup_{\mathsf{F} \in \mathcal{F}} \sum_{\ell \neq i} \mathsf{F}(Y_\ell).$$

Let F_i be the function in $\mathcal{F}$ for which the supremum is attained in the definition of f_i (recall that $\mathcal{F}$ is assumed to be finite). Note that F_i may depend on $\mathbf{Y}^{(i)}$, but not on Y_i. Then we define

$$g_i(\mathbf{Y}) := \left(\sum_{\ell=1}^{M} \mathsf{F}_i(Y_\ell)\right) - f_i(\mathbf{Y}^{(i)}) = \mathsf{F}_i(Y_i).$$

Further, F_0 denotes the function in $\mathcal{F}$ for which the supremum is attained in the definition of f. We obtain

$$g_i(\mathbf{Y}) \leq f(\mathbf{Y}) - f_i(\mathbf{Y}^{(i)}) \leq \sum_{\ell=1}^{M} \mathsf{F}_0(Y_\ell) - \sum_{\ell \neq i} \mathsf{F}_0(Y_\ell) = \mathsf{F}_0(Y_i) \leq 1.$$

This verifies condition (8.98) and the first condition in (8.101) with $B = 1$. Moreover, since F_i is independent of Y_i and $\mathbb{E}[\mathsf{F}_i(Y_i)] = 0$,

$$\mathbb{E}_{Y_i} g_i(\mathbf{Y}) = \mathbb{E}_{Y_i}\left[\sum_{\ell=1}^{M} \mathsf{F}_i(Y_\ell) - f_i(\mathbf{Y}^{(i)})\right] = \sum_{\ell \neq i} \mathsf{F}_i(Y_\ell) - f_i(\mathbf{Y}^{(i)}) = 0,$$

which shows (8.99). Moreover,

$$(M-1)f(\mathbf{Y}) = \sum_{i=1}^{M}\sum_{\ell \neq i} \mathsf{F}_0(Y_\ell) \le \sum_{i=1}^{M} f_i(\mathbf{Y}^{(i)}),$$

so that also (8.100) is satisfied. Finally,

$$\sum_{i=1}^{M} \mathbb{E}_{Y_i}[g_i(\mathbf{Y})^2] = \sum_{i=1}^{M} \mathbb{E}_{Y_i}[\mathsf{F}_i(Y_i)^2] \le \sum_{i=1}^{M} \sigma_i^2,$$

which shows that we can choose σ as desired. An application of Theorem 8.47 yields (8.91) for finite $\mathcal{F}$.

To conclude the proof for countably infinite $\mathcal{F}$, we let $G_n \subset \mathcal{F}$, $n \in \mathbb{N}$, be a sequence of finite subsets, such that $G_n \subset G_{n+1}$ and $\cup_{n \in N} G_n = \mathcal{F}$. Introduce the random variables

$$Z_n := \sup_{\mathsf{F} \in G_n} \sum_{\ell=1}^{M} \mathsf{F}(Y_\ell)$$

and, for $t > 0$, the indicator random variables $\chi_n := I_{\{Z_n - \mathbb{E}Z_n > t\}}$. By monotone convergence, $\lim_{n\to\infty} \mathbb{E}Z_n = \mathbb{E}Z$ (note that the statement of the theorem implicitly assumes that Z is integrable), and we have the pointwise limit

$$\lim_{n\to\infty} \chi_n = \chi,$$

where χ is the indicator random variable of the event $\{Z - \mathbb{E}Z > t\}$. Clearly, $\chi_n \le 1$, so that the sequence χ_n has the integrable majorant 1. It follows from Lebesgue's dominated convergence theorem that

$$\mathbb{P}(Z > \mathbb{E}Z + t) = \mathbb{P}(\sup_n(Z_n - \mathbb{E}Z_n) > t) = \mathbb{E}\left[\lim_{n\to\infty} \chi_n\right] = \lim_{n\to\infty} \mathbb{E}\chi_n$$

$$= \lim_{n\to\infty} \mathbb{P}\left(\sup_{\mathsf{F} \in G_n} \sum_{\ell=1}^{M} \mathsf{F}(Y_\ell) > \mathbb{E}\left[\sup_{\mathsf{F} \in G_n} \sum_{\ell=1}^{M} \mathsf{F}(Y_\ell)\right] + t\right)$$

$$\le \exp\left(-\frac{t^2/2}{\sigma^2 + 2\mathbb{E}Z + t/3}\right),$$

where we have used the just established estimate for finite sets of functions in the last inequality.   □

The proof of Theorem 8.47 uses again the concept of entropy of a random variable as in the proof of the concentration of measure inequality for Lipschitz functions in Sect. 8.8. We will use a variation of the Herbst argument outlined in (8.73) and the discussion afterwards. Setting $F(\theta) = \mathbb{E}[\exp(\theta f(\mathbf{X}))]$, we will develop a bound for the entropy $\mathscr{E}(e^{\theta f(\mathbf{X})}) = \theta F'(\theta) - F(\theta) \ln F(\theta)$ in terms of $F(\theta) = \mathbb{E}[e^{\theta f(\mathbf{X})}]$. This will lead to a differential inequality of the form

$$\theta F'(\theta) - F(\theta) \ln F(\theta) \leq \gamma(\theta) F(\theta) \ln F(\theta) + \rho(\theta) F(\theta)$$

for some functions γ and ρ; see (8.112) below. We will find a bound for any solution of this differential inequality using Lemma C.13. This arguments lead then to the statement of Theorem 8.47.

For our purposes we need some further properties of the entropy in addition to the ones discussed in the previous section. We start with another characterization.

Lemma 8.48. *Let X be a strictly positive and integrable random variable. Then*

$$\mathscr{E}(X) = \inf_{u>0} \mathbb{E}[\phi(X) - \phi(u) - (X - u)\phi'(u)],$$

where $\phi'(x) = \ln(x) + 1$.

Proof. Convexity of ϕ implies that, for $u > 0$,

$$\phi(\mathbb{E}X) \geq \phi(u) + \phi'(u)(\mathbb{E}X - u).$$

By the definition of the entropy this yields

$$\mathscr{E}(X) = \mathbb{E}[\phi(X)] - \phi(\mathbb{E}X) \leq \mathbb{E}[\phi(X)] - \phi(u) - \phi'(u)(\mathbb{E}X - u)$$
$$= \mathbb{E}[\phi(X) - \phi(u) - \phi'(u)(X - u)]. \tag{8.104}$$

Choosing $u = \mathbb{E}X$ yields an equality above, which proves the claim. $\qquad\square$

We also require the following consequence of the tensorization inequality of Proposition 8.37.

Corollary 8.49. *Let $\mathbf{X} = (X_1, \ldots, X_n)$ be a sequence of independent random vectors. Let f be a measurable function of $\mathbf{X}$ and f_i, $i \in [n]$, be measurable functions of $\mathbf{X}^{(i)}$ (i.e., constant in X_i). Set $F(\theta) = \mathbb{E}[\exp(\theta f(\mathbf{X}))]$ and let θ_0 be such that $F(\theta_0) < \infty$. Then, for any $\theta < \theta_0$,*

$$\theta F'(\theta) - F(\theta) \ln(F(\theta)) \leq \sum_{i=1}^{n} \mathbb{E}\left[e^{\theta f(\mathbf{X})} \psi\left(-\theta(f(\mathbf{X}) - f_i(\mathbf{X}^{(i)}))\right)\right],$$

where $\psi(x) := e^x - x - 1$.

Proof. Suppose g is a positive measurable function of $\mathbf{X}$ and g_i are positive measurable functions of $\mathbf{X}^{(i)}$, $i \in [n]$. Taking the entropy conditionally with respect to X_i in Lemma 8.48 (i.e., choosing $u = g_i(\mathbf{X}^{(i)})$ in (8.104), so that u does not depend on X_i) yields

$$\mathscr{E}_{X_i}(g(\mathbf{X})) \leq \mathbb{E}_{X_i} \left[\phi(g(\mathbf{X})) - \phi(g_i(\mathbf{X}^{(i)})) - (g(\mathbf{X}) - g_i(\mathbf{X}^{(i)}))\phi'(g_i(\mathbf{X}^{(i)})) \right]$$

$$= \mathbb{E}_{X_i} \left[g(\mathbf{X})(\ln(g(\mathbf{X})) - \ln(g_i(\mathbf{X}^{(i)}))) - (g(\mathbf{X}) - g_i(\mathbf{X}^{(i)})) \right]. \qquad (8.105)$$

We apply the above inequality to $g(\mathbf{X}) = e^{\theta f(\mathbf{X})}$ and $g_i(\mathbf{X}) = e^{\theta f_i(\mathbf{X}^{(i)})}$ to obtain

$$\mathscr{E}_{X_i}(g(\mathbf{X})) = \theta \mathbb{E}_{X_i} \left[f(\mathbf{X})e^{\theta f(\mathbf{X})} \right] - \mathbb{E}_{X_i} \left[e^{\theta f(\mathbf{X})} \right] \ln \mathbb{E}_{X_i} \left[e^{\theta f(\mathbf{X})} \right]$$

$$\leq \mathbb{E}_{X_i} \left[e^{\theta f(\mathbf{X})}(\theta f(\mathbf{X}) - \theta f_i(\mathbf{X}^{(i)})) - (e^{\theta f(\mathbf{X})} - e^{\theta f_i(\mathbf{X}^{(i)})}) \right]$$

$$= \mathbb{E}_{X_i} \left[e^{\theta f(\mathbf{X})}\psi(-\theta(f(\mathbf{X}) - f_i(\mathbf{X}^{(i)}))) \right].$$

Observe that $F'(\theta) = \mathbb{E}[f(\mathbf{X})e^{\theta f(\mathbf{X})}]$ for $\theta < \theta_0$. Therefore, an application of the tensorization inequality (8.79) shows that

$$\theta F'(\theta) - F(\theta) \ln(F(\theta)) \leq \theta \mathbb{E} \left[f(\mathbf{X})e^{\theta f(\mathbf{X})} \right] - \mathbb{E} \left[e^{\theta f(\mathbf{X})} \right] \ln \mathbb{E} \left[e^{\theta f(\mathbf{X})} \right]$$

$$= \mathscr{E}(g(\mathbf{X})) \leq \mathbb{E} \left[\sum_{i=1}^{n} \mathscr{E}_{X_i}(g(\mathbf{X})) \right]$$

$$\leq \mathbb{E} \left[\sum_{i=1}^{n} \mathbb{E}_{X_i} \left[e^{\theta f(\mathbf{X})}\psi(-\theta(f(\mathbf{X}) - f_i(\mathbf{X}^{(i)}))) \right] \right]$$

$$= \sum_{i=1}^{n} \mathbb{E} \left[e^{\theta f(\mathbf{X})}\psi(-\theta(f(\mathbf{X}) - f_i(\mathbf{X}^{(i)}))) \right].$$

This completes the proof.   □

As another auxiliary tool we need the following decoupling inequality.

Lemma 8.50. *Let* Y, Z *be random variables on a probability space* $(\Omega, \Sigma, \mathbb{P})$ *and* $\theta > 0$ *be such that* $e^{\theta Y}$ *and* $e^{\theta Z}$ *are* $\mathbb{P}$ *integrable. Then*

$$\theta \mathbb{E} \left[Ye^{\theta Z} \right] \leq \theta \mathbb{E} \left[Ze^{\theta Z} \right] - \mathbb{E} \left[e^{\theta Z} \right] \ln \mathbb{E} \left[e^{\theta Z} \right] + \mathbb{E} \left[e^{\theta Z} \right] \ln \mathbb{E} \left[e^{\theta Y} \right].$$

Proof. Let $\mathbb{Q}$ be the probability measure defined via $d\mathbb{Q} = \frac{e^{\theta Z}}{\mathbb{E}[e^{\theta Z}]} d\mathbb{P}$ with associated expectation given by

$$\mathbb{E}_{\mathbb{Q}}[X] := \frac{\mathbb{E}[Xe^{\theta Z}]}{\mathbb{E}[e^{\theta Z}]},$$

where $\mathbb{E}$ is the expectation with respect to $\mathbb{P}$. Jensen's inequality yields

$$\theta \mathbb{E}_{\mathbb{Q}}[Y - Z] = \mathbb{E}_{\mathbb{Q}}\left[\ln(e^{\theta(Y-Z)})\right] \leq \ln \mathbb{E}_{\mathbb{Q}}\left[e^{\theta(Y-Z)}\right].$$

By definition of $\mathbb{E}_{\mathbb{Q}}$ this translates into

$$\frac{\theta \mathbb{E}[(Y - Z)e^{\theta Z}]}{\mathbb{E}[e^{\theta Z}]} \leq \ln \mathbb{E}[e^{\theta Y}] - \ln \mathbb{E}[e^{\theta Z}],$$

which is equivalent to the claim.   □

The next statement is a consequence of Lemma 8.50 and Corollary 8.49.

Lemma 8.51. *Let* $\mathbf{X} = (X_1, \ldots, X_n)$ *be a sequence of independent random variables. Let f be a measurable function of $\mathbf{X}$ and f_i, $i \in [n]$, be measurable functions of $\mathbf{X}^{(i)}$. Let further g be a measurable function of $\mathbf{X}$ such that*

$$\sum_{i=1}^{n} \left(f(\mathbf{X}) - f_i(\mathbf{X}^{(i)}) \right) \leq g(\mathbf{X}). \tag{8.106}$$

Then, for all $\theta > 0$,

$$\sum_{i=1}^{n} \mathbb{E}\left[e^{\theta f(\mathbf{X})} - e^{\theta f_i(\mathbf{X}^{(i)})}\right] \leq \mathbb{E}\left[e^{\theta f(\mathbf{X})}\right] \ln \mathbb{E}\left[e^{\theta g(\mathbf{X})}\right].$$

Proof. Denote $F(\theta) = \mathbb{E}\left[e^{\theta f(\mathbf{X})}\right]$ and $G(\theta) = \mathbb{E}\left[e^{\theta g(\mathbf{X})}\right]$. We apply Corollary 8.49 to $f(\mathbf{X})$ and $\tilde{f}_i(\mathbf{X}^{(i)}) = f_i(\mathbf{X}^{(i)}) + \frac{1}{n\theta} \ln G(\theta)$, $i \in [n]$, to obtain

$$\theta F'(\theta) - F(\theta) \ln F(\theta)$$

$$\leq \sum_{i=1}^{n} \mathbb{E}\left[e^{\theta f(\mathbf{X})} \psi\left(-\theta(f(\mathbf{X}) - f_i(\mathbf{X}^{(i)}) - \frac{\ln(G(\theta))}{n\theta})\right)\right]$$

$$= \sum_{i=1}^{n} \mathbb{E}\left[G(\theta)^{1/n} e^{\theta f_i(\mathbf{X}^{(i)})} - e^{\theta f(\mathbf{X})} + e^{\theta f(\mathbf{X})}(\theta(f(\mathbf{X}) - f_i(\mathbf{X}^{(i)})))\right]$$

$$\quad - \mathbb{E}\left[e^{\theta f(\mathbf{X})}\right] \ln G(\theta)$$

$$\leq G(\theta)^{1/n} \left(\sum_{i=1}^{n} \mathbb{E}\left[e^{\theta f_i(\mathbf{X}^{(i)})}\right]\right) - nF(\theta) + \theta \mathbb{E}\left[e^{\theta f(\mathbf{X})} \sum_{i=1}^{n}(f(\mathbf{X}) - f_i(\mathbf{X}^{(i)}))\right]$$

$$\quad - F(\theta) \ln G(\theta)$$

$$\leq G(\theta)^{1/n} \left(\sum_{i=1}^{n} \mathbb{E}\left[e^{\theta f_i(\mathbf{X}^{(i)})}\right]\right) - nF(\theta) + \theta \mathbb{E}\left[e^{\theta f(\mathbf{X})} g(\mathbf{X})\right] - F(\theta) \ln G(\theta)$$

$$\leq G(\theta)^{1/n} \left(\sum_{i=1}^{n} \mathbb{E}\left[e^{\theta f_i(\mathbf{X}^{(i)})} \right] \right) - nF(\theta) + \theta \mathbb{E}\left[f(\mathbf{X}) e^{\theta f(\mathbf{X})} \right]$$

$$- \mathbb{E}\left[e^{\theta f(\mathbf{X})} \right] \ln \mathbb{E}\left[e^{\theta f(\mathbf{X})} \right] + \mathbb{E}\left[e^{\theta f(\mathbf{X})} \right] \ln \mathbb{E}\left[e^{\theta g(\mathbf{X})} \right] - F(\theta) \ln G(\theta)$$

$$= G(\theta)^{1/n} \left(\sum_{i=1}^{n} \mathbb{E}\left[e^{\theta f_i(\mathbf{X}^{(i)})} \right] \right) - nF(\theta) + \theta F'(\theta) - F(\theta) \ln F(\theta).$$

Hereby, we used the assumption (8.106) in the fourth step and Lemma 8.50 with $Y = g(\mathbf{X})$ and $Z = f(\mathbf{X})$ in the fifth step. We rewrite this as

$$nF(\theta) \leq G(\theta)^{1/n} \sum_{i=1}^{n} \mathbb{E}\left[e^{\theta f_i(\mathbf{X}^{(i)})} \right],$$

which in turn is equivalent to

$$\sum_{i=1}^{n} \mathbb{E}\left[e^{\theta f(\mathbf{X})} - e^{\theta f_i(\mathbf{X}^{(i)})} \right] \leq nF(\theta)(1 - G(\theta)^{-1/n}).$$

The inequality $e^x \geq 1 + x$ implies then that

$$n(1 - G(\theta)^{-1/n}) = n(1 - e^{-\frac{1}{n} \ln G(\theta)}) \leq \ln G(\theta),$$

so that

$$\sum_{i=1}^{n} \mathbb{E}\left[e^{\theta f(\mathbf{X})} - e^{\theta f_i(\mathbf{X}^{(i)})} \right] \leq F(\theta) \ln G(\theta).$$

This completes the proof.   □

Based on this preparation, we can now prove Theorem 8.47.

Proof (of Theorem 8.47). We define $\alpha(x) := 1 - (1 + x)e^{-x}$, $\beta(x) := \psi(-x) = e^{-x} - 1 + x$, and for $\tau > 0$ to be specified later,

$$\gamma(\theta) := \frac{\alpha(-\theta)}{\beta(-\theta) + \theta\tau}.$$

Step 1: We prove that, for $x \leq 1$, $\theta, \tau > 0$,

$$\beta(\theta x) \leq \gamma(\theta) \left(\alpha(\theta x) + \theta\tau x^2 e^{-\theta x} \right). \tag{8.107}$$

To this end we introduce the function

$$b(x) := \beta(\theta x) - \gamma(\theta) \left(\alpha(\theta x) + \theta \tau x^2 e^{-\theta x}\right).$$

Note that $\alpha(0) = \beta(0) = \alpha'(0) = \beta'(0) = 0$ so that $b(0) = b'(0) = 0$. Furthermore,

$$\beta(-\theta) + \theta \tau = e^\theta(1 - e^{-\theta} - \theta e^{-\theta} + \theta \tau e^{-\theta}) = e^\theta(\alpha(\theta) + \tau \theta e^{-\theta}),$$

which implies that $b(1) = 0$. Furthermore,

$$b'(x) = \theta \left(1 - e^{-\theta x} - \gamma(\theta)(\theta x e^{-\theta x} + 2\tau x e^{-\theta x} - \tau \theta x^2 e^{-\theta x})\right).$$

Therefore, $\lim_{x \to +\infty} b'(x) = \theta$ and $\lim_{x \to -\infty} b'(x) = +\infty$. Next, observe that we can write $b''(x) = e^{-\theta x} p(x)$ with a second-degree polynomial p with leading coefficient $-\theta^3 \gamma(\theta)\tau$. If follows that $b''(x) = 0$ has at most two solutions. If there is no solution, then b' is decreasing, which is a contradiction to $\lim_{x \to -\infty} b'(x) = +\infty$, $b'(0) = 0$ and $\lim_{x \to +\infty} b'(x) = \theta$. So let x_1, x_2 with $x_1 \le x_2$ be the (possibly equal) solutions. Then b' is decreasing in $(-\infty, x_1) \cup (x_2, \infty)$ and increasing in (x_1, x_2). Since $\lim_{x \to +\infty} b'(x) = \theta > 0$, the equation $b'(x) = 0$ can have at most two solutions, one in $(-\infty, x_1)$ and one in $[x_1, x_2)$. Recall that $b'(0) = 0$, so denote by x_3 the other solution to $b'(x) = 0$. If $x_3 \le 0$, then b is increasing in $(0, \infty)$, which is a contradiction to $b(0) = b(1) = 0$. Therefore, $x_3 > 0$, and b is increasing in $(-\infty, 0)$, decreasing in $(0, x_3)$, and increasing in (x_3, ∞). Since $b(0) = b(1) = 0$ this shows that $b(x) \le 0$ for $x \le 1$, which implies the claimed inequality (8.107).

Step 2:   Next we use (8.107) with $x = f(\mathbf{X}) - f_i(\mathbf{X}^{(i)})$ and that $\alpha(x)e^x = \beta(-x)$ to obtain

$$\beta(\theta(f(\mathbf{X}) - f(\mathbf{X}^{(i)}))) e^{\theta f(\mathbf{X})}$$

$$\le \gamma(\theta) \left(\beta(-\theta(f(\mathbf{X}) - f_i(\mathbf{X}^{(i)}))) + \theta \tau (f(\mathbf{X}) - f_i(\mathbf{X}^{(i)}))^2\right) e^{\theta f_i(\mathbf{X}^{(i)})}$$

$$= \gamma(\theta) \left(e^{\theta f(\mathbf{X})} - e^{\theta f_i(\mathbf{X}^{(i)})}\right)$$

$$+ \theta \gamma(\theta) e^{\theta f(\mathbf{X}^{(i)})} \left(\tau(f(\mathbf{X}) - f_i(\mathbf{X}^{(i)}))^2 - (f(\mathbf{X}) - f_i(\mathbf{X}^{(i)}))\right). \quad (8.108)$$

Now we choose $\tau = 1/(1 + B)$. Note that if $y \le x \le 1$ and $y \le B$, then

$$\tau x^2 - x \le \tau y^2 - y. \quad (8.109)$$

Indeed, under these conditions,

$$\tau(x^2 - y^2) = \tau(x + y)(x - y) \le \tau(1 + B)(x - y) = x - y.$$

Using the assumption $g_i(\mathbf{X}) \leq f(\mathbf{X}) - f_i(\mathbf{X}^{(i)}) \leq 1$ and $g_i(\mathbf{X}) \leq B$ in (8.108) and exploiting (8.109), we obtain

$$\beta\left(\theta(f(\mathbf{X}) - f(\mathbf{X}^{(i)}))\right) e^{\theta f(\mathbf{X})} \leq \gamma(\theta)\left(e^{\theta f(\mathbf{X})} - e^{\theta f_i(\mathbf{X}^{(i)})}\right)$$

$$+ \theta\gamma(\theta)e^{\theta f_i(\mathbf{X}^{(i)})}\left(\tau g_i^2(\mathbf{X}) - g_i(\mathbf{X})\right). \tag{8.110}$$

Since $f_i(\mathbf{X}^{(i)})$ does not depend on X_i, the assumption $\mathbb{E}_{X_i} g_i(\mathbf{X}) \geq 0$ yields

$$\mathbb{E}\left[e^{\theta f_i(\mathbf{X}^{(i)})} g_i(\mathbf{X})\right] = \mathbb{E}\left[\mathbb{E}_{X_i} e^{\theta f_i(\mathbf{X}^{(i)})} g_i(\mathbf{X})\right] = \mathbb{E}\left[e^{\theta f_i(\mathbf{X}^{(i)})} \mathbb{E}_{X_i} g_i(\mathbf{X})\right]$$

$$\geq \mathbb{E}\left[e^{\theta f_i(\mathbf{X}^{(i)})}\right] \geq 0. \tag{8.111}$$

Further note that (8.98) and (8.99) imply that

$$\mathbb{E}_{X_i}[f(\mathbf{X})] \geq \mathbb{E}_{X_i}[f_i(\mathbf{X}^{(i)}) + g_i(\mathbf{X})] \geq f_i(\mathbf{X}^{(i)}),$$

and by Jensen's inequality this yields

$$e^{\theta f_i(\mathbf{X}^{(i)})} \leq e^{\theta \mathbb{E}_{X_i} f(\mathbf{X})} \leq \mathbb{E}_{X_i}\left[e^{\theta f(\mathbf{X})}\right].$$

By taking expectations in (8.110) we therefore reach, using (8.111) in the first step together with $\theta\gamma(\theta) \geq 0$ for all $\theta \geq 0$,

$$\mathbb{E}\left[\beta\left(\theta(f(\mathbf{X}) - f_i(\mathbf{X}^{(i)}))\right) e^{\theta f(\mathbf{X})}\right]$$

$$\leq \gamma(\theta)\mathbb{E}\left[e^{\theta f(\mathbf{X})} - e^{\theta f_i(\mathbf{X}^{(i)})}\right] + \frac{\theta\gamma(\theta)}{1+B}\mathbb{E}\left[e^{\theta f_i(\mathbf{X}^{(i)})} g_i^2(\mathbf{X})\right]$$

$$= \gamma(\theta)\mathbb{E}\left[e^{\theta f(\mathbf{X})} - e^{\theta f_i(\mathbf{X}^{(i)})}\right] + \frac{\theta\gamma(\theta)}{1+B}\mathbb{E}\left[e^{\theta f_i(\mathbf{X}^{(i)})} \mathbb{E}_{X_i}\left[g_i^2(\mathbf{X})\right]\right]$$

$$\leq \gamma(\theta)\mathbb{E}\left[e^{\theta f(\mathbf{X})} - e^{\theta f_i(\mathbf{X}^{(i)})}\right] + \frac{\theta\gamma(\theta)}{1+B}\mathbb{E}\left[e^{\theta f(\mathbf{X})} \mathbb{E}_{X_i}\left[g_i^2(\mathbf{X})\right]\right].$$

Hereby, we used twice that $\mathbb{E}[\cdot] = \mathbb{E}\mathbb{E}_{X_i}[\cdot]$. Now denote $F(\theta) = \mathbb{E}[e^{\theta f(\mathbf{X})}]$. Then Corollary 8.49 together with (8.101) implies that

$$\theta F'(\theta) - F(\theta)\ln F(\theta) \leq \sum_{i=1}^{n} \mathbb{E}\left[\beta\left(\theta(f(\mathbf{X}) - f(\mathbf{X}^{(i)}))\right) e^{\theta f(\mathbf{X})}\right]$$

$$\leq \gamma(\theta)\sum_{i=1}^{n} \mathbb{E}\left[e^{\theta f(\mathbf{X})} - e^{\theta f_i(\mathbf{X}^{(i)})}\right] + \frac{\theta\gamma(\theta)}{1+B}\mathbb{E}\left[e^{\theta f(\mathbf{X})} \sum_{i=1}^{n}\mathbb{E}_{X_i}\left[g_i^2(\mathbf{X})\right]\right]$$

$$\leq \gamma(\theta)\mathbb{E}\left[e^{\theta f(\mathbf{X})}\right]\ln \mathbb{E}\left[e^{\theta f(\mathbf{X})}\right] + \frac{\theta\gamma(\theta)n\sigma^2}{1+B}\mathbb{E}\left[e^{\theta f(\mathbf{X})}\right],$$

$$= \gamma(\theta)F(\theta)\ln F(\theta) + \frac{\theta\gamma(\theta)n\sigma^2}{1+B}F(\theta), \tag{8.112}$$

where we used in the last inequality that $\sum_{i=1}^{n}(f(\mathbf{X}) - f_i(\mathbf{X}^{(i)})) \leq f(\mathbf{X})$ in combination with Lemma 8.51.

Step 3 (Modified Herbst argument): Set $G(\theta) = \mathbb{E}\left[e^{\theta(f(\mathbf{X})-\mathbb{E}[f(\mathbf{X})])}\right] = F(\theta)$ $e^{-\theta\mathbb{E}[f(\mathbf{X})]}$. Then

$$G'(\theta) = e^{-\theta\mathbb{E}[f(\mathbf{X})]}\left(F'(\theta) - \mathbb{E}[f(\mathbf{X})]F(\theta)\right),$$

$$\ln G(\theta) = \ln F(\theta) - \theta\mathbb{E}[f(\mathbf{X})],$$

$$\text{and} \quad \frac{G'(\theta)}{G(\theta)} = \frac{F'(\theta)}{F(\theta)} - \mathbb{E}[f(\mathbf{X})].$$

Therefore, (8.112) can be rewritten as

$$\theta\frac{G'(\theta)}{G(\theta)} - \ln G(\theta) \leq \gamma(\theta)\left(\ln G(\theta) + \theta\mathbb{E}[f(\mathbf{X})]\right) + \frac{n\sigma^2\theta\gamma(\theta)}{1+B}.$$

Introducing $L(\theta) = \ln G(\theta)$, the above inequality is in turn equivalent to

$$\theta L'(\theta) - (1+\gamma(\theta))L(\theta) \leq \frac{n\sigma^2 + (1+B)\mathbb{E}[f(\mathbf{X})]}{1+B}\theta\gamma(\theta) = \frac{v}{1+B}\theta\gamma(\theta).$$

Recall that we have set $\tau = 1/(1+B)$ in the definition of the function γ, so that

$$\gamma(\theta) = \frac{\alpha(-\theta)}{\beta(-\theta) + \theta/(1+B)}.$$

We claim that $L_0(\theta) := v\beta(-\theta) = v(e^{\theta} - 1 - \theta)$ is a solution to the associated differential equation

$$\theta L'(\theta) - (1+\gamma(\theta))L(\theta) = \frac{v}{1+B}\theta\gamma(\theta),$$

with the initial conditions $L_0(0) = L_0'(0) = 0$. Indeed,

$$v^{-1}\left(\theta L_0'(\theta) - (1+\gamma(\theta))L_0(\theta)\right)$$

$$= \theta(e^{\theta} - 1) - e^{\theta} + \theta + 1 - \frac{\alpha(-\theta)\beta(-\theta)}{\beta(-\theta) + \theta/(1+B)}$$

$$= \alpha(-\theta) - \frac{\alpha(-\theta)(\beta(-\theta) + \theta/(1 + B))}{\beta(-\theta) + \theta/(1 + B)} + \frac{\alpha(-\theta)\theta/(1 + B)}{\beta(-\theta) + \theta/(1 + B)}$$

$$= \frac{\theta\gamma(\theta)}{1 + B}.$$

It follows from Lemma C.13 that $L(\theta) \leq L_0(\theta)$, i.e.,

$$\ln \mathbb{E}[e^{\theta(f(\mathbf{X}) - \mathbb{E}[f(\mathbf{X})])}] \leq v(e^\theta - 1 - \theta).$$

This completes the proof of (8.102).

Step 4:   To deduce the tail inequalities in (8.103) we use Markov's inequality (Theorem 7.3) to obtain, for $\theta > 0$,

$$\mathbb{P}(f(\mathbf{X}) \geq \mathbb{E}[f(\mathbf{X})] + t) = \mathbb{P}\left(e^{\theta(f(\mathbf{X}) - \mathbb{E}[f(\mathbf{X})])} \geq e^{\theta t}\right)$$

$$\leq e^{-\theta t}\mathbb{E}[e^{\theta(f(\mathbf{X}) - \mathbb{E}[f(\mathbf{X})])}] \leq e^{-\theta t}e^{v(e^\theta - 1 - \theta)}$$

$$= e^{v(e^\theta - 1 - \theta) - \theta t}. \tag{8.113}$$

It follows from Lemma 8.21 that

$$\inf_{\theta > 0}(v(e^\theta - \theta - 1) - \theta t) = -vh(x/t),$$

where we recall that $h(t) = (1 + t)\ln(1 + t) - t$. Together with (8.113) this shows the first estimate in (8.103). The second part of Lemma 8.21 implies that $vh(t/v) \geq \frac{t^2}{2v + 2t/3}$, which yields the second inequality in (8.103). $\square$

Notes

Many results of this chapter also hold in infinite dimensional Banach spaces. Introducing random vectors in general Banach spaces, however, requires additional technicalities that we preferred to avoid here. For such details and many more results on probability in Banach spaces we refer to the monograph [322] by Ledoux and Talagrand and to the collection of articles in [296]. In particular, the relation between moments and tails and an introduction to Rademacher sums and to symmetrization are contained in [322].

The Khintchine inequalities are named after the Russian mathematician Khintchine (also spelled Khinchin) who was the first to show Theorem 8.5 in [301]. Our proof essentially follows his ideas. We have only provided estimates from above for the absolute moments of a Rademacher sum. Estimates from below have also been investigated, and the optimal constants for both lower and upper estimates for all $p > 0$ have been derived in [253]; see also [360] for simplified proofs. We have already noted that, for $p = 2n$, $n \in \mathbb{N}$, the constant $C_{2n} = (2n)!/(2^n n!)$ for the upper estimate provided in Theorem 8.5 is optimal. In case of general $p \geq 2$ (which is much harder than the even integer case), the best constant is $C_p = 2^{\frac{p-1}{2}} \Gamma(p/2)/\Gamma(3/2)$. This value is very close to the estimate in (8.9).

The proof of the Khintchine inequality for Steinhaus sums in Theorem 8.9 is slightly shorter than the one given in [380]. The technique for the proof of Corollary 8.10 for Steinhaus sums was taken from [380, 481]. An overview on (scalar) Khintchine and related inequalities can be found in [381]. An extension of the Khintchine inequalities to sums of independent random vectors that are uniformly distributed on spheres is provided in [305]. Using a similar technique as in Corollary 8.10 the following Hoeffding-type inequality has been deduced in [184] for $\mathbf{X}_1, \dots, \mathbf{X}_M \in \mathbb{R}^n$ being independent random vectors, uniformly distributed on the unit sphere $S^{n-1} = \{\mathbf{x} \in \mathbb{R}^n, \|\mathbf{x}\|_2 = 1\}$,

$$\mathbb{P}\left(\|\sum_{\ell=1}^{M} a_\ell \mathbf{X}_\ell\|_2 \geq \|\mathbf{a}\|_2 u\right) \leq \exp\left(-\frac{n}{2}(u^2 - \log(u^2) - 1)\right) \quad \text{for all } u > 1.$$

The noncommutative version of Bernstein inequality was proven by Tropp in [486] by refining an approach to the Laplace transform method for matrices due to Ahlswede and Winter [6]; see also [373, 374]. Based on the method of exchangeable pairs, a different approach to its proof, which does not require Lieb's concavity theorem (nor a similar result on matrix convexity), is presented in [339]. The more traditional approach for studying tail bounds for random matrices uses the noncommutative Khintchine inequality, which first appeared in the work of Lust-Piquard [336]; see also [337]. These inequalities work with the Schatten $2n$-norms $\|\mathbf{A}\|_{S_{2n}} = \|\sigma(\mathbf{A})\|_{2n} = (\text{tr}\,((\mathbf{A}^*\mathbf{A})^n))^{1/(2n)}$, $n \in \mathbb{N}$, where $\sigma(\mathbf{A})$ is the vector of singular values of $\mathbf{A}$, and provide bounds for matrix-valued Rademacher sums

$$\mathbb{E}\|\sum_{j=1}^{M} \epsilon_j \mathbf{B}_j\|_{S_{2n}}^{2n}$$

$$\leq \frac{(2n)!}{2^n n!} \max\left\{\left\|\left(\sum_{j=1}^{M} \mathbf{B}_j \mathbf{B}_j^*\right)^{1/2}\right\|_{S_{2n}}^{2n}, \left\|\left(\sum_{j=1}^{M} \mathbf{B}_j^* \mathbf{B}_j\right)^{1/2}\right\|_{S_{2n}}^{2n}\right\}. \quad (8.114)$$

The optimal constants for these inequalities for $p = 2n$ match the scalar case in Theorem 8.5 and were derived by Buchholz in [74, 75]; see also [411]. Rudelson showed a lemma now named after him in [430] (see also [374, 411]), which allows to derive tail bounds and moment bounds for sums of random rank-one matrices. While Rudelson's original proof used chaining methods, it was pointed out to him by Pisier that a simpler proof (contained in the published paper) can be obtained via noncommutative Khintchine inequalities. The approach to random matrices via the noncommutative Khintchine inequality has the drawback that one needs significant practice in order to apply them; see, for instance, [411, 500]. In contrast, the noncommutative Bernstein inequality of Theorem 8.14 is easy to apply and provides very good constants.

The decoupling inequality of Theorem 8.11, including its proof, is essentially taken from [66]. The variant of Theorem 8.12 for the operator norm was shown by Tropp in [480, 481]. Decoupling techniques can be extended to higher-order chaos and also to sums of the form $\sum_{j \neq k} h_{j,k}(\mathbf{X}_j, \mathbf{X}_k)$, where $\mathbf{X}_k$ are independent random vectors and $h_{j,k}$ are vector-valued functions. Moreover, decoupling inequalities do not only apply for expectations and moments. Also, a probability estimate of the form

$$\mathbb{P}(\| \sum_{j \neq k} h_{j,k}(\mathbf{X}_j, \mathbf{X}_k)\| \geq t) \leq C\mathbb{P}(\| \sum_{j \neq k} h_{j,k}(\mathbf{X}_j, \mathbf{X}_k')\| \geq t/C),$$

can be shown, where $(\mathbf{X}_k')$ is an independent copy of $(\mathbf{X}_k)$ and $C > 1$ is an appropriate constant. We refer the interested reader to [146] for further information.

The tail bounds for Rademacher chaos (Theorem 8.13) and quadratic forms in more general subgaussian random vectors have first been obtained by Hanson and Wright in [260]. The proof given here follows arguments from an unpublished work of Rauhut and Tropp. For Gaussian chaos better constants are available in [35], and yet another proof of the tail inequality appears in [466, Sect. 2.5].

The inequality of Theorem 8.23 is named after Dudley who proved his inequality in [175]. The proof in Sect. 8.6 follows the argument in [326]; see also [392]. Further proofs can be found in [22, 193, 194, 411, 466]. The nice exposition in Talagrand's book [466] leads to more powerful generic chaining inequalities, also called majorizing measure inequalities. These use the so-called γ_α functional of a metric space (T, d), which is defined as

$$\gamma_\alpha(T, d) = \inf \sup_{t \in T} \sum_{r=0}^{\infty} 2^{r/\alpha} d(t, T_r), \quad \alpha > 0,$$

where the infimum is taken over all sequences T_r, $r \in \mathbb{N}_0$, of subsets of T with cardinalities $\text{card}(T_0) = 1$, $\text{card}(T_r) \leq 2^{2^r}$, $r \geq 1$, and where $d(t, T_r) = \inf_{s \in T_r} d(t, s)$. Given a Gaussian processes X_t, $t \in T$, with associated pseudometric d defined by (8.41), Talagrand's majorizing measures theorem [459, 462, 465, 466] states that

$$C_1 \gamma_2(T, d) \leq \mathbb{E} \sup_{t \in T} X_t \leq C_2 \gamma_2(T, d),$$

for universal constants $C_1, C_2 > 0$. In particular, the lower bound is remarkable. The upper bound also holds for more general subgaussian processes. Since $\gamma_2(T, d)$ is bounded by a constant times the Dudley-type integral in (8.47) (see [466]), the above inequality implies also Dudley's inequality (with possibly a different constant). In general, $\gamma_2(T, d)$ may provide sharper bounds than Dudley's integral. However, if

T is a subset of $\mathbb{R}^N$ and d is induced by a norm, then one loses at most a factor of $\ln(N)$ when passing from the $\gamma_2(T, d)$ functional to Dudley's integral. The latter has the advantage that it is usually easier to estimate. Another type of lower bound for Gaussian processes is Sudakov's minoration; see, e.g., [322, 348].

Dudley's inequality extends to moments; see, for instance, [411]. Indeed, one also has the following inequality (using the same notation as Theorem 8.23):

$$\left(\mathbb{E} \sup_{t \in T} |X_t|^p \right)^{1/p} \leq C\sqrt{p} \int_0^{\Delta(T)/2} \sqrt{\ln(\mathcal{N}(T, d, u))} du.$$

A generalization of Dudley's inequality [306, 322, 391] holds in the framework of Orlicz spaces. A Young function is a positive convex function ψ that satisfies $\psi(0) = 0$ and $\lim_{x \to \infty} \psi(x) = \infty$. The Orlicz space L_ψ consists of all random variables X for which $\mathbb{E}\psi(|X|/c) < \infty$ for some $c > 0$. The norm

$$\|X\|_\psi = \inf\{c > 0, \mathbb{E}\psi(|X|/c) \leq 1\}$$

turns L_ψ into a Banach space [312]. Suppose that X_t, $t \in T$, is a stochastic process indexed by a (pseudo)metric d of diameter Δ such that

$$\|X_s - X_t\|_\psi \leq d(s, t).$$

Then the generalization of Dudley's inequality [322, Theorem 11.1] states that

$$\mathbb{E} \sup_{s,t} |X_s - X_t| \leq 8 \int_0^\Delta \psi^{-1}(\mathcal{N}(T, d, u)) du,$$

where ψ^{-1} is the inverse function of ψ. Taking $\psi(x) = \exp(x^2) - 1$ yields Theorem 8.23 (up to the constant). Further important special cases are $\psi(x) = \exp(x) - 1$ (exponential tail of the increments) and $\psi(x) = x^p$ (resulting in L^p-spaces of random variables).

A chaos process is of the form

$$X_\mathbf{B} = \sum_{j \neq k} \xi_j \xi_k B_{j,k}, \quad \mathbf{B} \in \mathcal{B},$$

where $\mathcal{B} \subset \mathbb{R}^{n \times n}$ is a set of square matrices and $\boldsymbol{\xi} = (\xi_1, \ldots, \xi_n)$ is a vector of independent mean-zero subgaussian random variables of unit variance. We refer to [307, 413] and the Notes section of Chap. 12 for their appearance in compressive sensing. Talagrand [466] showed the bound

$$\mathbb{E} \sup_{\mathbf{B} \in \mathcal{B}} |X_\mathbf{B}| \leq C_1 \gamma_2(\mathcal{B}, \|\cdot\|_F) + C_2 \gamma_1(\mathcal{B}, \|\cdot\|_{2 \to 2}).$$

The basis for the proof of this inequality is the tail inequality for a single chaos $X_{\mathbf{B}}$ of Theorem 8.13 and its extension to subgaussian random vectors. Due to the appearance of the γ_1 functional in the right-hand side, this bound is not sharp in many cases; see, e.g., [466] for a discussion. In fact, for a special case of relevance for compressive sensing, [307] provides an alternative bound, which we describe next. For a set of matrices $\mathcal{A} \subset \mathbb{C}^{m \times n}$ and a vector $\boldsymbol{\xi}$ of independent mean-zero subgaussian random variables of unit variance, we consider the random variable

$$Y = \sup_{\mathbf{A} \in \mathcal{A}} \left| \|\mathbf{A}\boldsymbol{\xi}\|_2^2 - \mathbb{E}\|\mathbf{A}\boldsymbol{\xi}\|_2^2 \right|.$$

If $\boldsymbol{\xi} = \boldsymbol{\epsilon}$ is a Rademacher vector, then $Y = \sup_{\mathbf{A} \in \mathcal{A}} \left| \sum_{j \neq k} \epsilon_j \epsilon_k (\mathbf{A}^* \mathbf{A})_{j,k} \right|$ so that we recover a chaos process indexed by $\mathcal{B} = \{\mathbf{A}^* \mathbf{A} : \mathbf{A} \in \mathcal{A}\}$. In [307] it is shown that (under the additional assumption $\mathcal{A} = -\mathcal{A}$)

$$\mathbb{E} \sup_{\mathbf{A} \in \mathcal{A}} \left| \|\mathbf{A}\boldsymbol{\xi}\|_2^2 - \mathbb{E}\|\mathbf{A}\boldsymbol{\xi}\|_2^2 \right| \leq C_1 \gamma_2(\mathcal{A}, \|\cdot\|_{2\to2})^2 + C_2 d_F(\mathcal{A}) \gamma_2(\mathcal{A}, \|\cdot\|_{2\to2}),$$

$$(8.115)$$

where $d_F(\mathcal{A}) = \sup_{\mathbf{A} \in \mathcal{A}} \|\mathbf{A}\|_F$ is the diameter of $\mathcal{A}$ in the Frobenius norm. Also a tail bound is provided in [307].

In slightly different form, Slepian's lemma appeared for the first time in [448]; see also Fernique's notes [193]. Other references on Slepian's lemma include [322, Corollary 3.14], [348, Theorem 3.14], and [343]. Gordon's Lemma 8.27 appeared in [233, 234]. Stein's lemma (Proposition 8.29), which was used in the proof of Slepian's and Gordon's lemma, has important applications in statistics. For instance, it is at the basis of Stein's unbiased risk estimator (SURE). We refer, e.g., to [490], for more details.

Many more details and references on the general theory of concentration of measure such as connections to isoperimetric inequalities are provided in the expositions by Barvinok [33], Ledoux [321], and Boucheron et al. [61]. The proof of Theorem 8.40 based on entropy is a variant of the one found in [61, 348], while the proof of Theorem 8.34 follows [467, Theorem 1.3.4]. The Gaussian logarithmic Sobolev inequality is originally due to Gross [247]. The argument attributed to Herbst (unpublished) to derive concentration inequalities from logarithmic Sobolev inequalities appeared in [142, Theorem A.8]. A different proof of concentration of measure for Lipschitz functions (Theorem 8.40) based on the so-called Ornstein–Uhlenbeck semigroup can be found in [321]. By using the rotation invariance of the Gaussian distribution, concentration of measure for the uniform distribution on the sphere (or on the ball) can be deduced from the Gaussian case (and vice versa); see, for instance, [33, 321, 322].

Concentration of measure inequalities are valid also for independent random variables $X_1, \ldots, X_n$ with values in $[-1, 1]$. However, one has to impose the assumption that the function $F : [-1, 1]^n \to \mathbb{R}$ is convex, in addition to being L-Lipschitz. If M is a median of $F(X_1, \ldots, X_n)$, then [321, 461]

$$\mathbb{P}(|F(X_1, \ldots, X_n) - M| \geq t) \leq 4 \exp(-t^2/(4L)^2).$$

The median can be replaced by the mean via general principles outlined in [321].

Deviation inequalities for suprema of empirical processes were already investigated in the 1980s by Massart and others; see, e.g., [11, 346]. Talagrand achieved major breakthroughs in [460,463]. In particular, he showed a concentration inequality similar to (8.91) in [463]; see also [321, Theorem 7.6]. Ledoux noticed in [320] that deviation and concentration inequalities may be deduced using entropy. The constants in the deviation and concentration inequalities were subsequently improved in [68, 69, 304, 347, 421, 422]. The proof of Theorem 8.42 follows [68]; see also [69]. Concentration below the expected supremum of an empirical process can be shown as well [69, 320, 463]. A version for not necessarily identically distributed random vectors is presented in [304], and collections $\mathcal{F}$ of unbounded functions are treated in [2, 319].

Versions of Corollary 8.46 can already be found in the monograph by Ledoux and Talagrand [322, Theorems 6.17 and 6.19], however, with nonoptimal constants. More general deviation and concentration inequalities for suprema of empirical processes and other functions of independent variables are derived, for instance, in [59, 60, 321], in particular, a version for Rademacher chaos processes is stated in [60].

Exercises

8.1. Let $\mathbf{X} = (X_1, \ldots, X_n)$ be a vector of mean-zero Gaussian random variables with variances $\sigma_\ell^2 = \mathbb{E}g_\ell^2$, $\ell \in [n]$. Show that

$$\mathbb{E}\max_{\ell\in[n]} X_\ell \leq \sqrt{2\ln(n)}\max_{\ell\in[n]}\sigma_\ell.$$

8.2. Comparison principle.
Let $\boldsymbol{\epsilon} = (\epsilon_1, \ldots, \epsilon_M)$ be a Rademacher sequence and $\mathbf{g} = (g_1, \ldots, g_N)$ be a standard Gaussian vector. Let $\mathbf{x}_1, \ldots, \mathbf{x}_M$ be vectors in a normed space.

(a) Let $\boldsymbol{\xi} = (\xi_1, \ldots, \xi_M)$ be a sequence of independent and symmetric real-valued random variables with $\mathbb{E}|\xi_\ell| < \infty$ for all $\ell \in [M]$. Show that, for $p \in [1, \infty)$,

$$\left(\min_{\ell\in[M]}\mathbb{E}|\xi_\ell|\right)\left(\mathbb{E}\Big\|\sum_{\ell=1}^{M}\epsilon_\ell\mathbf{x}_\ell\Big\|^p\right)^{1/p} \leq \left(\mathbb{E}\Big\|\sum_{\ell=1}^{M}\xi_\ell\mathbf{x}_\ell\Big\|^p\right)^{1/p}.$$

Conclude that

$$\mathbb{E}\Big\|\sum_{\ell=1}^{N}\epsilon_\ell\mathbf{x}_\ell\Big\| \leq \sqrt{\frac{\pi}{2}}\,\mathbb{E}\Big\|\sum_{\ell=1}^{N}g_\ell\mathbf{x}_\ell\Big\|.$$

(b) Show that

$$\mathbb{E}\|\sum_{\ell=1}^{M} g_\ell \mathbf{x}_\ell\| \leq \sqrt{2\log(2M)}\,\mathbb{E}\|\sum_{\ell=1}^{M} \epsilon_\ell \mathbf{x}_\ell\|.$$

Find an example which shows that the logarithmic factor above cannot be removed in general.

8.3. Let $\mathbf{a} \in \mathbb{C}^N$ and $\epsilon = (\epsilon_1,\ldots,\epsilon_N)$ be a Steinhaus sequence. Show a moment estimate of the form

$$\left(\mathbb{E}|\sum_{\ell=1}^{N} \epsilon_\ell a_\ell|^p\right)^{1/p} \leq \alpha\beta^{1/p}\sqrt{p}\|\mathbf{a}\|_2, \quad p \geq 2,$$

in two ways: (a) by using the method of Corollary 8.7 and (b) by using Proposition 7.13. Provide small values of α and β.

8.4. Hoeffdings's inequality for complex random variables.
Let $\mathbf{X} = (X_1,\ldots,X_N)$ be a vector of complex-valued mean-zero symmetric random variables, i.e., X_ℓ has the same distribution as $-X_\ell$. Assume that $|X_\ell| \leq 1$ almost surely for all $\ell \in [M]$. Show that, for $\mathbf{a} \in \mathbb{C}^M$ and $u > 0$,

$$\mathbb{P}(|\sum_{j=1}^{M} a_j X_j| \geq u\|\mathbf{a}\|_2) \leq 2\exp(-u^2/2).$$

Provide a version of this inequality when the symmetry assumption is removed.

8.5. Let $\mathbf{A} \in \mathbb{C}^{m\times N}$ be a fixed matrix.

(a) Let $\mathbf{g}$ be a standard Gaussian random vector. Show that, for $t > 0$,

$$\mathbb{P}(\|\mathbf{A}\mathbf{g}\|_2 \geq \|\mathbf{A}\|_F + t\|\mathbf{A}\|_{2\to2}) \leq e^{-t^2/2}.$$

(b) Let ϵ be a Rademacher vector. Show that, for $t > 0$,

$$\mathbb{P}(\|\mathbf{A}\epsilon\|_2 \geq c_1\|\mathbf{A}\|_F + c_2 t\|\mathbf{A}\|_{2\to2}) \leq e^{-t^2/2}.$$

Provide appropriate values of the constants $c_1, c_2 > 0$.

8.6. Deviation for matrix-valued Gaussian sums.

(a) Let g be a standard Gaussian variable and let $\mathbf{B} \in \mathbb{C}^{d\times d}$ be a self-adjoint matrix. Show that $\mathbb{E}\exp(g\theta\mathbf{B}) = \exp(\theta^2\mathbf{B}^2/2)$.
(b) Let $\mathbf{g} = (g_1,\ldots,g_M)$ be a vector of independent standard Gaussian variables and $\mathbf{B}_1,\ldots,\mathbf{B}_M \in \mathbb{C}^{d\times d}$ be self-adjoint matrices. Introduce $\sigma^2 = \|\sum_{j=1}^{M} \mathbf{B}_j^2\|_{2\to2}$. Show that

$$\mathbb{E}\exp(\theta\|\sum_{j=1}^{M} g_j \mathbf{B}_j\|_{2\to 2}) \le 2d\exp(\theta^2\sigma^2/2) \quad \text{for } \theta > 0,$$

and

$$\mathbb{P}(\|\sum_{j=1}^{M} g_j \mathbf{B}_j\|_{2\to 2} \ge t) \le 2d\exp\left(\frac{-t^2}{2\sigma^2}\right), \quad t > 0.$$

(c) For a random variable X, show that $\mathbb{E}X \le \inf_{\theta>0} \theta^{-1}\ln\mathbb{E}[\exp(\theta X)]$.

(d) Show that

$$\mathbb{E}\|\sum_{j=1}^{M} g_j \mathbf{B}_j\|_{2\to 2} \le \sqrt{2\ln(2d)}\|\sum_{j=1}^{M}\mathbf{B}_j^2\|_{2\to 2}^{1/2}, \tag{8.116}$$

and, for a Rademacher sequence $\epsilon = (\epsilon_1, \dots, \epsilon_M)$,

$$\mathbb{E}\|\sum_{j=1}^{M} \epsilon_j \mathbf{B}_j\|_{2\to 2} \le \sqrt{2\ln(2d)}\|\sum_{j=1}^{M}\mathbf{B}_j^2\|_{2\to 2}^{1/2}. \tag{8.117}$$

(e) Give an example that shows that the factor $\sqrt{\ln(2d)}$ cannot be removed from (8.116) in general.

8.7. Deviation inequalities for sums of rectangular random matrices.

(a) The self-adjoint dilation of a matrix $\mathbf{A} \in \mathbb{C}^{d_1 \times d_2}$ is defined as

$$S(\mathbf{A}) = \begin{pmatrix} \mathbf{0} & \mathbf{A} \\ \mathbf{A}^* & \mathbf{0} \end{pmatrix}.$$

Prove that $S(\mathbf{A}) \in \mathbb{C}^{(d_1+d_2)\times(d_1+d_2)}$ is self-adjoint and $\|S(\mathbf{A})\|_{2\to 2} = \|\mathbf{A}\|_{2\to 2}$.

(b) Let $\mathbf{X}_1, \dots, \mathbf{X}_M$ be a sequence of $d_1 \times d_2$ random matrices with

$$\|\mathbf{X}_\ell\|_{2\to 2} \le K \quad \text{for all } \ell \in [M],$$

and set

$$\sigma^2 := \max\{\|\sum_{\ell=1}^{M}\mathbb{E}(\mathbf{X}_\ell\mathbf{X}_\ell^*)\|_{2\to 2}, \|\sum_{\ell=1}^{M}\mathbb{E}(\mathbf{X}_\ell^*\mathbf{X}_\ell)\|_{2\to 2}\}. \tag{8.118}$$

Show that, for $t > 0$,

$$\mathbb{P}(\|\sum_{\ell=1}^{M} \mathbf{X}_\ell\|_{2\to2} \geq t) \leq 2(d_1 + d_2) \exp\left(-\frac{t^2/2}{\sigma^2 + Kt/3}\right). \qquad (8.119)$$

8.8. Noncommutative Bernstein inequality, subexponential version.
Let $\mathbf{X}_1, \ldots, \mathbf{X}_M \in \mathbb{C}^{d\times d}$ be independent mean-zero self-adjoint random matrices.
Assume that

$$\mathbb{E}[\mathbf{X}_\ell^n] \preccurlyeq n! R^{n-2} \sigma_\ell^2 \mathbf{B}_\ell^2/2, \quad \ell \in [M],$$

for some self-adjoint matrices $\mathbf{B}_\ell$ and set

$$\sigma^2 := \left\|\sum_{\ell=1}^{M} \mathbf{B}_\ell^2\right\|_{2\to2}.$$

Show that, for $t > 0$,

$$\mathbb{P}\left(\lambda_{\max}\left(\sum_{\ell=1}^{M} \mathbf{X}_\ell\right) \geq t\right) \leq d \exp\left(-\frac{t^2/2}{\sigma^2 + Rt}\right).$$

8.9. Let T be a countable index set. Show the consistency of the definition (8.39) of the lattice supremum in this case, i.e., show that

$$\mathbb{E}(\sup_{t\in T} X_t) = \sup\{\mathbb{E}(\sup_{t\in F} X_t), F \subset T, F \text{ finite}\}.$$

8.10. Let $X_t, t \in T$, be a symmetric random process, i.e., X_t has the same distribution as $-X_t$ for all $t \in T$. Show that, for an arbitrary $t_0 \in T$,

$$\mathbb{E}\sup_{t\in T} X_t \leq \mathbb{E}\sup_{t\in T} |X_t - X_{t_0}| \leq 2\mathbb{E}\sup_{t\in T} X_t = \mathbb{E}\sup_{s,t\in T} |X_s - X_t|.$$

8.11. Derive the following generalization of Dudley's inequality: Let $X_t, t \in T$, be a subgaussian process with associated pseudometric d, i.e.,

$$\mathbb{P}(|X_s - X_t| \geq ud(s,t)) \leq 2e^{-cu^2}.$$

Then, for some arbitrary $t_0 \in T$ and $p \geq 1$,

$$\left(\mathbb{E}\sup_{t\in T} |X_t - X_{t_0}|^p\right)^{1/p} \leq C\sqrt{p}\int_0^\infty \sqrt{\log(\mathcal{N}(T,d,u))}du,$$

for some appropriate constant $C > 0$ depending only on c.

8.12. Weak and distributional derivatives.

Recall the notion of weak and distributional derivatives from Sect. C.9.

(a) Show that the function $f(t) = |t|$ has weak derivative

$$f'(t) = \text{sgn}(t) = \begin{cases} -1 & \text{if } t < 0, \\ 0 & \text{if } t = 0, \\ 1 & \text{if } t > 0. \end{cases}$$

(b) Let g be a function of the form (8.66). Show that a weak derivative is given by

$$g'(t) = \chi_{[a,b]}(t) = \begin{cases} 1 & \text{if } t \in [a, b], \\ 0 & \text{if } t \notin [a, b]. \end{cases}$$

(c) Let f be nondecreasing and differentiable except possibly at finitely many points. Show that f has a nonnegative distributional derivative.

(d) Assume that f has a nonnegative weak derivative. Show that f is nondecreasing.

Chapter 9
Sparse Recovery with Random Matrices

It was shown in Chap. 6 that recovery of s-sparse vectors via various algorithms including ℓ_1-minimization is guaranteed if the restricted isometry constants of the measurement matrix satisfy $\delta_{\kappa s} \leq \delta_*$ for an appropriate small integer κ and some $\delta_* \in (0, 1)$. The derived condition for ℓ_1-minimization is, for instance, $\delta_{2s} < 0.6246$. In Chap. 5, we have seen explicit $m \times m^2$ matrices that satisfy such a condition once $m \geq Cs^2$. This bound relies on the estimate $\delta_s \leq (s-1)\mu$ in terms of the coherence μ; see also the discussion at the end of Sect. 6.1. But at this point of the theoretical development, $m \times N$ matrices with small δ_s are not known to exist when m is significantly smaller than Cs^2. The purpose of this chapter is to show their existence when $m \geq C_\delta s \ln(N/s)$ using probabilistic arguments. We consider subgaussian random matrices whose entries are drawn independently according to a subgaussian distribution. This includes Gaussian and Rademacher variables. For such matrices, the restricted isometry property holds with high probability in the stated parameter regime. We refer to Theorem 9.12 for an exact statement.

We also discuss a nonuniform setting where we analyze the probability that a fixed s-sparse vector $\mathbf{x}$ is recovered from $\mathbf{y} = \mathbf{A}\mathbf{x}$ via ℓ_1-minimization using a random draw of a subgaussian matrix $\mathbf{A}$. In this setting, the proof has the advantage of being simple and providing good constants in the estimate for the required number m of measurements (although the term $\ln(N/s)$ is replaced by $\ln N$ in the first instance). Then we restrict our considerations to Gaussian matrices. Using the Slepian and Gordon lemmas as well as concentration of measure, we derive in the nonuniform setting "roughly" (that is, for large dimensions) the sufficient condition

$$m > 2s \ln(N/s).$$

Returning to the uniform setting, we further obtain bounds for the conditioning of Gaussian random matrices and, as a consequence, explicit bounds for the restricted isometry constants. In the Gaussian case, we can also show the null space property directly without invoking the restricted isometry property.

S. Foucart and H. Rauhut, *A Mathematical Introduction to Compressive Sensing*, Applied and Numerical Harmonic Analysis, DOI 10.1007/978-0-8176-4948-7_9, © Springer Science+Business Media New York 2013

Finally, we make a small detour to the Johnson–Lindenstrauss lemma, which states that a finite set of points in a high-dimensional Euclidean space can be mapped to a lower-dimensional space via a linear map without significantly perturbing their mutual distances. This linear map can be chosen as the realization of a subgaussian random matrix. This fact follows immediately from a concentration inequality that is crucial in the proof of the restricted isometry property for subgaussian matrices; see (9.6). For this reason, a matrix satisfying the concentration inequality is sometimes even referred to as a Johnson–Lindenstrauss mapping, and in this sense, the Johnson–Lindenstrauss lemma implies the restricted isometry property. We will also show the converse statement that a matrix satisfying the restricted isometry property provides a Johnson–Lindenstrauss mapping when the column signs are randomized.

9.1 Restricted Isometry Property for Subgaussian Matrices

We consider a matrix $\mathbf{A} \in \mathbb{R}^{m \times N}$ having random variables as their entries. Such $\mathbf{A}$ is called a *random matrix* or random matrix ensemble.

Definition 9.1. Let $\mathbf{A}$ be an $m \times N$ random matrix.

(a) If the entries of $\mathbf{A}$ are independent Rademacher variables (i.e., taking values ± 1 with equal probability), then $\mathbf{A}$ is called a *Bernoulli random matrix*.

(b) If the entries of $\mathbf{A}$ are independent standard Gaussian random variables, then $\mathbf{A}$ is called a *Gaussian random matrix*.

(c) If the entries of $\mathbf{A}$ are independent mean-zero subgaussian random variables with variance 1 and same subgaussian parameters β, κ in (7.31), i.e.,

$$\mathbb{P}(|A_{j,k}| \geq t) \leq \beta e^{-\kappa t^2} \quad \text{for all } t > 0, \quad j \in [m], k \in [N], \qquad (9.1)$$

then $\mathbf{A}$ is called a *subgaussian random matrix*.

Clearly, Gaussian and Bernoulli random matrices are subgaussian. Also note that the entries of a subgaussian matrix do not necessarily have to be identically distributed. Equivalently to (9.1), we may require that

$$\mathbb{E}[\exp(\theta A_{j,k})] \leq \exp(c\theta^2) \quad \text{for all } \theta \in \mathbb{R}, \quad j \in [m], k \in [N], \qquad (9.2)$$

for some constant c that is independent of j, k and N; see Proposition 7.24. Note that the smallest possible value of the constant is $c = 1/2$ due to the normalization $\mathbb{E}A_{j,k}^2 = 1$ and Exercise 7.4.

We start by stating the main result on the restricted isometry property for subgaussian random matrices.

Theorem 9.2. *Let* $\mathbf{A}$ *be an* $m \times N$ *subgaussian random matrix. Then there exists a constant* $C > 0$ *(depending only on the subgaussian parameters* β, κ*) such that the restricted isometry constant of* $\frac{1}{\sqrt{m}}\mathbf{A}$ *satisfies* $\delta_s \leq \delta$ *with probability at least* $1 - \varepsilon$ *provided*

$$m \geq C\delta^{-2}\big(s\ln(eN/s) + \ln(2\varepsilon^{-1})\big). \tag{9.3}$$

Setting $\varepsilon = 2\exp(-\delta^2 m/(2C))$ yields the condition

$$m \geq 2C\delta^{-2}s\ln(eN/s),$$

which guarantees that $\delta_s \leq \delta$ with probability at least $1 - 2\exp\big(-\delta^2 m/(2C)\big)$. This is the statement often found in the literature.

The normalization $\frac{1}{\sqrt{m}}\mathbf{A}$ is natural because $\mathbb{E}\|\frac{1}{\sqrt{m}}\mathbf{A}\mathbf{x}\|_2^2 = \|\mathbf{x}\|_2^2$ for a fixed vector $\mathbf{x}$ and a subgaussian random matrix $\mathbf{A}$ (where by convention all entries have variance 1). Expressed differently, the squared ℓ_2-norm of the columns of $\mathbf{A}$ is one in expectation. Therefore, the restricted isometry constant δ_s measures the deviation of $\|\frac{1}{\sqrt{m}}\mathbf{A}\mathbf{x}\|_2^2$ from its mean, uniformly over all s-sparse vectors $\mathbf{x}$.

The following is an important special case of Theorem 9.2.

Corollary 9.3. *Let* $\mathbf{A}$ *be an* $m \times N$ *Gaussian or Bernoulli random matrix. Then there exists a universal constant* $C > 0$ *such that the restricted isometry constant of* $\frac{1}{\sqrt{m}}\mathbf{A}$ *satisfies* $\delta_s \leq \delta$ *with probability at least* $1 - \varepsilon$ *provided*

$$m \geq C\delta^{-2}\big(s\ln(eN/s) + \ln(2\varepsilon^{-1})\big). \tag{9.4}$$

For Gaussian random matrices, we will provide a slightly improved estimate with explicit constants in Sect. 9.3.

Subgaussian matrices belong to a larger class of random matrices introduced below. Theorem 9.2 will follow from its generalization to this larger class. We start with some definitions.

Definition 9.4. Let $\mathbf{Y}$ be a random vector on $\mathbb{R}^N$.

(a) If $\mathbb{E}|\langle\mathbf{Y}, \mathbf{x}\rangle|^2 = \|\mathbf{x}\|_2^2$ for all $\mathbf{x} \in \mathbb{R}^N$, then $\mathbf{Y}$ is called *isotropic*.
(b) If, for all $\mathbf{x} \in \mathbb{R}^N$ with $\|\mathbf{x}\|_2 = 1$, the random variable $\langle\mathbf{Y}, \mathbf{x}\rangle$ is subgaussian with subgaussian parameter c being independent of $\mathbf{x}$, that is,

$$\mathbb{E}[\exp(\theta\langle\mathbf{Y}, \mathbf{x}\rangle)] \leq \exp(c\theta^2), \quad \text{for all } \theta \in \mathbb{R}, \quad \|\mathbf{x}\|_2 = 1, \tag{9.5}$$

then $\mathbf{Y}$ is called a subgaussian random vector.

Remark 9.5. Isotropic subgaussian random vectors do not necessarily have independent entries. The constant c in (9.5) should ideally be independent of N. Lemma 9.7 provides an example where this is the case. Note that if $\mathbf{Y}$ is a randomly selected row or column of an orthogonal matrix, then $\mathbf{Y}$ is isotropic and subgaussian, but c depends on N.

We consider random matrices $\mathbf{A} \in \mathbb{R}^{m \times N}$ of the form

$$\mathbf{A} = \begin{pmatrix} \mathbf{Y}_1^\top \\ \vdots \\ \mathbf{Y}_m^\top \end{pmatrix}$$

with independent subgaussian and isotropic rows $\mathbf{Y}_1, \ldots, \mathbf{Y}_m \in \mathbb{R}^N$. The following result establishes the restricted isometry property for such matrices.

Theorem 9.6. *Let $\mathbf{A}$ be an $m \times N$ random matrix with independent, isotropic, and subgaussian rows with the same subgaussian parameter c in (9.5). If*

$$m \geq C\delta^{-2}\big(s\ln(eN/s) + \ln(2\varepsilon^{-1})\big),$$

then the restricted isometry constant of $\frac{1}{\sqrt{m}}\mathbf{A}$ satisfies $\delta_s \leq \delta$ with probability at least $1 - \varepsilon$.

The proof of this theorem is given below. Theorem 9.2 follows then from a combination with the following lemma.

Lemma 9.7. *Let $\mathbf{Y} \in \mathbb{R}^N$ be a random vector with independent, mean-zero, and subgaussian entries with variance 1 and the same subgaussian parameter c in (7.33). Then $\mathbf{Y}$ is an isotropic and subgaussian random vector with the subgaussian parameter in (9.5) equal to c.*

Proof. Let $\mathbf{x} \in \mathbb{R}^N$ with $\|\mathbf{x}\|_2 = 1$. Since the Y_ℓ are independent, mean-zero, and of variance 1, we have

$$\mathbb{E}|\langle \mathbf{Y}, \mathbf{x}\rangle|^2 = \sum_{\ell,\ell'=1}^{N} x_\ell x_{\ell'} \mathbb{E}Y_\ell Y_{\ell'} = \sum_{\ell=1}^{N} x_\ell^2 = \|\mathbf{x}\|_2^2.$$

Therefore, $\mathbf{Y}$ is isotropic. Furthermore, according to Theorem 7.27, the random variable $Z = \langle \mathbf{Y}, \mathbf{x}\rangle = \sum_{\ell=1}^{N} x_\ell Y_\ell$ is subgaussian with parameter c. Hence, $\mathbf{Y}$ is a subgaussian random vector with parameters independent of N. □

Concentration Inequality

The proof of Theorem 9.6 heavily relies on the following concentration inequality for random matrices. The latter in turn is a consequence of Bernstein inequality for subexponential random variables, which arise when forming the ℓ_2-norm by summing up squares of subgaussian random variables.

Lemma 9.8. *Let* $\mathbf{A}$ *be an* $m \times N$ *random matrix with independent, isotropic, and subgaussian rows with the same subgaussian parameter* c *in* (9.5). *Then, for all* $\mathbf{x} \in \mathbb{R}^N$ *and every* $t \in (0,1)$,

$$\mathbb{P}\left(\left| m^{-1} \|\mathbf{Ax}\|_2^2 - \|\mathbf{x}\|_2^2 \right| \geq t \|\mathbf{x}\|_2^2 \right) \leq 2 \exp(-\tilde{c}t^2 m), \tag{9.6}$$

where $\tilde{c}$ *depends only on* c.

Proof. Let $\mathbf{x} \in \mathbb{R}^N$. Without loss of generality we may assume that $\|\mathbf{x}\|_2 = 1$. Denote the rows of $\mathbf{A}$ by $\mathbf{Y}_1, \ldots, \mathbf{Y}_m \in \mathbb{R}^N$ and consider the random variables

$$Z_\ell = |\langle \mathbf{Y}_\ell, \mathbf{x} \rangle|^2 - \|\mathbf{x}\|_2^2, \quad \ell \in [m].$$

Since $\mathbf{Y}_\ell$ is isotropic we have $\mathbb{E}Z_\ell = 0$. Further, Z_ℓ is subexponential because $\langle \mathbf{Y}_\ell, \mathbf{x} \rangle$ is subgaussian, that is, $\mathbb{P}(|Z_\ell| \geq r) \leq \beta \exp(-\kappa r)$ for all $r > 0$ and some parameters β, κ depending only on c. Observe now that

$$m^{-1} \|\mathbf{Ax}\|_2^2 - \|\mathbf{x}\|_2^2 = \frac{1}{m} \sum_{\ell=1}^m \left(|\langle \mathbf{Y}_\ell, \mathbf{x} \rangle|^2 - \|\mathbf{x}\|_2^2 \right) = \frac{1}{m} \sum_{\ell=1}^m Z_\ell.$$

By independence of the $\mathbf{Y}_\ell$, also the Z_ℓ are independent. Therefore, it follows from Bernstein inequality for subexponential random variables (Corollary 7.32) that

$$\mathbb{P}\left(\left| m^{-1} \sum_{\ell=1}^m Z_\ell \right| \geq t \right) = \mathbb{P}\left(\left| \sum_{\ell=1}^m Z_\ell \right| \geq tm \right) \leq 2 \exp\left(-\frac{\kappa^2 m^2 t^2 / 2}{2\beta m + \kappa m t} \right)$$

$$\leq 2 \exp\left(-\frac{\kappa^2}{4\beta + 2\kappa} m t^2 \right),$$

where we used that $t \in (0,1)$ in the last step. Hence, the claim follows with $\tilde{c} = \kappa^2 / (4\beta + 2\kappa)$. □

We note that the normalized random matrix $\tilde{\mathbf{A}} = \frac{1}{\sqrt{m}} \mathbf{A}$, with $\mathbf{A}$ satisfying the assumptions of the previous lemma, obeys, for $\mathbf{x} \in \mathbb{R}^N$ and all $t \in (0,1)$,

$$\mathbb{P}\left(\left| \|\tilde{\mathbf{A}}\mathbf{x}\|_2^2 - \|\mathbf{x}\|_2^2 \right| \geq t \|\mathbf{x}\|_2^2 \right) \leq 2 \exp(-\tilde{c}t^2 m). \tag{9.7}$$

This is the starting point of the proof of the restricted isometry property.

Proof of the RIP

Now we show that a random matrix satisfying the concentration inequality (9.7) also satisfies the sth order restricted isometry property, provided that its number of rows scales at least like s times a logarithmic factor. We first show that a single column submatrix of a random matrix is well conditioned under an appropriate condition on its size.

Theorem 9.9. *Suppose that an $m \times N$ random matrix $\mathbf{A}$ is drawn according to a probability distribution for which the concentration inequality (9.7) holds, that is, for $t \in (0, 1)$,*

$$\mathbb{P}\left(\left|\|\mathbf{A}\mathbf{x}\|_2^2 - \|\mathbf{x}\|_2^2\right| \geq t\|\mathbf{x}\|_2^2\right) \leq 2\exp\left(-\tilde{c}t^2\,m\right) \quad \text{for all } \mathbf{x} \in \mathbb{R}^N. \tag{9.8}$$

For $S \subset [N]$ with $\mathrm{card}(S) = s$ and $\delta, \varepsilon \in (0, 1)$, if

$$m \geq C\delta^{-2}(7s + 2\ln(2\varepsilon^{-1})), \tag{9.9}$$

where $C = 2/(3\tilde{c})$, then with probability at least $1 - \varepsilon$

$$\|\mathbf{A}_S^*\mathbf{A}_S - \mathbf{Id}\|_{2\to 2} < \delta.$$

Proof. According to Proposition C.3, for $\rho \in (0, 1/2)$, there exists a finite subset U of the unit ball $B_S = \{\mathbf{x} \in \mathbb{R}^N, \mathrm{supp}\,\mathbf{x} \subset S, \|\mathbf{x}\|_2 \leq 1\}$ which satisfies

$$\mathrm{card}(U) \leq \left(1 + \frac{2}{\rho}\right)^s \quad \text{and} \quad \min_{\mathbf{u} \in U} \|\mathbf{z} - \mathbf{u}\|_2 \leq \rho \quad \text{for all } \mathbf{z} \in B_S.$$

The concentration inequality (9.8) gives, for $t \in (0, 1)$ depending on δ and ρ to be determined later,

$$\mathbb{P}\left(\left|\|\mathbf{A}\mathbf{u}\|_2^2 - \|\mathbf{u}\|_2^2\right| \geq t\|\mathbf{u}\|_2^2 \quad \text{for some } \mathbf{u} \in U\right)$$
$$\leq \sum_{\mathbf{u} \in U} \mathbb{P}\left(\left|\|\mathbf{A}\mathbf{u}\|_2^2 - \|\mathbf{u}\|_2^2\right| \geq t\|\mathbf{u}\|_2^2\right) \leq 2\,\mathrm{card}(U)\exp\left(-\tilde{c}t^2\,m\right)$$
$$\leq 2\left(1 + \frac{2}{\rho}\right)^s \exp\left(-\tilde{c}t^2\,m\right).$$

Let us assume now that the realization of the random matrix $\mathbf{A}$ yields

$$\left|\|\mathbf{A}\mathbf{u}\|_2^2 - \|\mathbf{u}\|_2^2\right| < t\|\mathbf{u}\|_2^2 \quad \text{for all } \mathbf{u} \in U. \tag{9.10}$$

By the above, this occurs with probability exceeding

$$1 - 2\left(1 + \frac{2}{\rho}\right)^s \exp\left(-\tilde{c}t^2\,m\right). \tag{9.11}$$

For a proper choice of ρ and t, we are going to prove that (9.10) implies $\left|\|\mathbf{A}\mathbf{x}\|_2^2 - \|\mathbf{x}\|_2^2\right| \leq \delta$ for all $\mathbf{x} \in B_S$, i.e., $\|\mathbf{A}_S^*\mathbf{A}_S - \mathbf{Id}\|_{2\to 2} \leq \delta$. With $\mathbf{B} = \mathbf{A}_S^*\mathbf{A}_S - \mathbf{Id}$, (9.10) means that $|\langle \mathbf{B}\mathbf{u}, \mathbf{u}\rangle| < t$ for all $\mathbf{u} \in U$. Now consider a vector $\mathbf{x} \in B_S$ and choose a vector $\mathbf{u} \in U$ satisfying $\|\mathbf{x} - \mathbf{u}\|_2 \leq \rho < 1/2$. We obtain

$$|\langle \mathbf{Bx}, \mathbf{x} \rangle| = |\langle \mathbf{Bu}, \mathbf{u} \rangle + \langle \mathbf{B}(\mathbf{x} + \mathbf{u}), \mathbf{x} - \mathbf{u} \rangle| \leq |\langle \mathbf{Bu}, \mathbf{u} \rangle| + |\langle \mathbf{B}(\mathbf{x} + \mathbf{u}), \mathbf{x} - \mathbf{u} \rangle|$$
$$< t + \|\mathbf{B}\|_{2 \to 2} \|\mathbf{x} + \mathbf{u}\|_2 \|\mathbf{x} - \mathbf{u}\|_2 \leq t + 2\|\mathbf{B}\|_{2 \to 2}\, \rho.$$

Taking the maximum over all $\mathbf{x} \in B_S$, we deduce that

$$\|\mathbf{B}\|_{2 \to 2} < t + 2\|\mathbf{B}\|_{2 \to 2}\, \rho, \qquad \text{i.e.,} \qquad \|\mathbf{B}\|_{2 \to 2} \leq \frac{t}{1 - 2\rho}.$$

We therefore choose $t := (1 - 2\rho)\delta$, so that $\|\mathbf{B}\|_{2 \to 2} < \delta$. By (9.11), we conclude that

$$\mathbb{P}\left(\|\mathbf{A}_S^*\mathbf{A}_S - \mathbf{Id}\|_{2 \to 2} \geq \delta\right) \leq 2\left(1 + \frac{2}{\rho}\right)^s \exp\left(-\tilde{c}(1 - 2\rho)^2\delta^2 m\right). \qquad (9.12)$$

It follows that $\|\mathbf{A}_S^*\mathbf{A}_S - \mathbf{Id}\|_{2 \to 2} \leq \delta$ with probability at least $1 - \varepsilon$ provided

$$m \geq \frac{1}{\tilde{c}(1 - 2\rho)^2}\delta^{-2}\left(\ln(1 + 2/\rho)s + \ln(2\varepsilon^{-1})\right). \qquad (9.13)$$

We now choose $\rho = 2/(e^{7/2} - 1) \approx 0.0623$ so that $1/(1 - 2\rho)^2 \leq 4/3$ and $\ln(1 + 2/\rho)/(1 - 2\rho)^2 \leq 14/3$. Thus, (9.13) is fulfilled when

$$m \geq \frac{2}{3\tilde{c}}\delta^{-2}\left(7s + 2\ln(2\varepsilon^{-1})\right). \qquad (9.14)$$

This concludes the proof. $\square$

Remark 9.10. (a) The attentive reader may have noticed that the above proof remains valid if one passes from coordinate subspaces indexed by S to restrictions of $\mathbf{A}$ to arbitrary s-dimensional subspaces of $\mathbb{R}^N$.

(b) The statement (and proof) does not depend on the columns of $\mathbf{A}$ outside S. Therefore, it follows that for an $m \times s$ subgaussian random matrix $\mathbf{B}$, we have

$$\|\frac{1}{m}\mathbf{B}^*\mathbf{B} - \mathbf{Id}\|_{2 \to 2} < \delta$$

with probability at least $1 - \varepsilon$ provided that (9.9) holds, or equivalently,

$$\mathbb{P}(\|m^{-1}\mathbf{B}^*\mathbf{B} - \mathbf{Id}\|_{2 \to 2} \geq \delta) \leq 2\exp\left(-\frac{3\tilde{c}}{4}\delta^2 m + \frac{7}{2}s\right). \qquad (9.15)$$

Let us now pass to the restricted isometry property.

Theorem 9.11. *Suppose that an $m \times N$ random matrix $\mathbf{A}$ is drawn according to a probability distribution for which the concentration inequality (9.7) holds, that is, for $t \in (0, 1)$,*

$$\mathbb{P}\left(\left|\|\mathbf{A}\mathbf{x}\|_2^2 - \|\mathbf{x}\|_2^2\right| \geq t\|\mathbf{x}\|_2^2\right) \leq 2\exp\left(-\tilde{c}t^2 m\right) \quad \text{for all } \mathbf{x} \in \mathbb{R}^N.$$

If, for $\delta, \varepsilon \in (0, 1)$,

$$m \geq C\delta^{-2}\left[s\left(9 + 2\ln(N/s)\right) + 2\ln(2\varepsilon^{-1})\right],$$

where $C = 2/(3\tilde{c})$, then with probability at least $1 - \varepsilon$, the restricted isometry constant δ_s of $\mathbf{A}$ satisfies $\delta_s < \delta$.

Proof. We recall from (6.2) that $\delta_s = \sup_{S \subset [N], \mathrm{card}(S)=s} \|\mathbf{A}_S^*\mathbf{A}_S - \mathbf{Id}\|_{2\to 2}$. With the same notation as in the proof of Theorem 9.9, using (9.12) and taking the union bound over all $\binom{N}{s}$ subsets $S \subset [N]$ of cardinality s yields

$$\mathbb{P}(\delta_s \geq \delta) \leq \sum_{S \subset [N], \mathrm{card}(S)=s} \mathbb{P}\left(\|\mathbf{A}_S^*\mathbf{A}_S - \mathbf{Id}\|_{2\to 2} \geq \delta\right)$$

$$\leq 2\binom{N}{s}\left(1 + \frac{2}{\rho}\right)^s \exp\left(-\tilde{c}\delta^2(1 - 2\rho)^2 m\right)$$

$$\leq 2\left(\frac{eN}{s}\right)^s \left(1 + \frac{2}{\rho}\right)^s \exp\left(-\tilde{c}\delta^2(1 - 2\rho)^2 m\right),$$

where we have additionally applied Lemma C.5 in the last step. Making the choice $\rho = 2/(e^{7/2} - 1)$ as before yields that $\delta_s < \delta$ with probability at least $1 - \varepsilon$ if

$$m \geq \frac{1}{\tilde{c}\delta^2}\left(\frac{4}{3}s\ln(eN/s) + \frac{14}{3}s + \frac{4}{3}\ln(2\varepsilon^{-1})\right),$$

which is a reformulation of the desired condition.   □

By possibly adjusting constants, the above theorem combined with Lemma 9.8 implies Theorem 9.6, and together with Lemma 9.7, this proves Theorem 9.2 and Corollary 9.3.

We now gather the results of this section to conclude with the main theorem about sparse reconstruction via ℓ_1-minimization from random measurements.

Theorem 9.12. *Let $\mathbf{A}$ be an $m \times N$ subgaussian random matrix. There exist constants $C_1, C_2 > 0$ only depending on the subgaussian parameters β, κ such that if, for $\varepsilon \in (0, 1)$,*

$$m \geq C_1 s\ln(eN/s) + C_2\ln(2\varepsilon^{-1}),$$

then with probability at least $1 - \varepsilon$ *every s-sparse vector* $\mathbf{x}$ *is recovered from* $\mathbf{y} = \mathbf{A}\mathbf{x}$ *via* ℓ_1*-minimization.*

Proof. The statement follows from a combination of Theorems 9.2 and 6.9 (or alternatively, Theorem 6.12) by additionally noting that exact sparse recovery is independent of the normalization of the matrix. $\square$

This result extends to stable and robust recovery.

Theorem 9.13. *Let* $\mathbf{A}$ *be an* $m \times N$ *subgaussian random matrix. There exist constants* $C_1, C_2 > 0$ *depending only on the subgaussian parameters* β, κ *and universal constants* $D_1, D_2 > 0$ *such that if, for* $\varepsilon \in (0, 1)$,

$$m \geq C_1 s \ln(eN/s) + C_2 \ln(2\varepsilon^{-1}),$$

then the following statement holds with probability at least $1 - \varepsilon$ *uniformly for every s-sparse vector* $\mathbf{x} \in \mathbb{C}^N$: *given* $\mathbf{y} = \mathbf{A}\mathbf{x} + \mathbf{e}$ *with* $\|\mathbf{e}\|_2 \leq \sqrt{m}\,\eta$ *for some* $\eta \geq 0$, *a solution* $\mathbf{x}^\sharp$ *of*

$$\underset{\mathbf{z} \in \mathbb{C}^N}{\text{minimize}} \; \|\mathbf{z}\|_1 \quad \text{subject to} \; \|\mathbf{A}\mathbf{z} - \mathbf{y}\|_2 \leq \sqrt{m}\,\eta \qquad (9.16)$$

satisfies

$$\|\mathbf{x} - \mathbf{x}^\sharp\|_2 \leq D_1 \frac{\sigma_s(\mathbf{x})_1}{\sqrt{s}} + D_2\,\eta,$$

$$\|\mathbf{x} - \mathbf{x}^\sharp\|_1 \leq D_1 \sigma_s(\mathbf{x})_1 + D_2 \sqrt{s}\,\eta.$$

Proof. The optimization problem (9.16) is equivalent to

$$\underset{\mathbf{z} \in \mathbb{C}^N}{\text{minimize}} \; \|\mathbf{z}\|_1 \quad \text{subject to} \quad \|\frac{1}{\sqrt{m}}\mathbf{A}\mathbf{z} - \frac{1}{\sqrt{m}}\mathbf{y}\|_2 \leq \eta$$

involving the rescaled matrix $\mathbf{A}/\sqrt{m}$. A combination of Theorems 9.2 and 6.12 yields the result. $\square$

Remark 9.14. Setting $\varepsilon = 2\exp(-m/(2C_2))$ shows stable and robust recovery of all s-sparse vectors via ℓ_1-minimization with probability at least $1 - 2\exp(-m/(2C_2))$ using an $m \times N$ subgaussian random matrix for which

$$m \geq 2C_1 s \ln(eN/s).$$

We will see in Chap. 10 that this condition on the required number of measurement cannot be improved.

The condition $\|\mathbf{e}\|_2 \leq \sqrt{m}\,\eta$ in Theorem 9.13 is natural for a vector $\mathbf{e} \in \mathbb{C}^m$. For instance, it is implied by the entrywise bound $|e_j| \leq \eta$ for all $j \in [m]$.

We obtain the same type of uniform recovery results for other algorithms guaranteed to succeed under restricted isometry conditions; see Chap. 6. This includes iterative hard thresholding, hard thresholding pursuit, orthogonal matching pursuit, and compressive sampling matching pursuit. Moreover, such recovery guarantees hold as well for general random matrices satisfying the concentration inequality (9.8), such as random matrices with independent isotropic subgaussian rows.

Universality

Often sparsity does not occur with respect to the canonical basis but rather with respect to some other orthonormal basis. This means that the vector of interest can be written as $\mathbf{z} = \mathbf{U}\mathbf{x}$ with an $N \times N$ orthogonal matrix $\mathbf{U}$ and an s-sparse vector $\mathbf{x} \in \mathbb{C}^N$. Taking measurements of $\mathbf{z}$ with a random matrix $\mathbf{A}$ can be written as

$$\mathbf{y} = \mathbf{A}\mathbf{z} = \mathbf{A}\mathbf{U}\mathbf{x}.$$

In order to recover $\mathbf{z}$, it suffices to first recover the sparse vector $\mathbf{x}$ and then to form $\mathbf{z} = \mathbf{U}\mathbf{x}$. Therefore, this more general problem reduces to the standard compressive sensing problem with measurement matrix $\mathbf{A}' = \mathbf{A}\mathbf{U}$. Hence, we consider this model with a random $m \times N$ matrix $\mathbf{A}$ and a fixed (deterministic) orthogonal matrix $\mathbf{U} \in \mathbb{R}^{N \times N}$ as a new measurement matrix of interest in this context. It turns out that the previous analysis can easily be adapted to this more general situation.

Theorem 9.15. *Let $\mathbf{U} \in \mathbb{R}^{N \times N}$ be a (fixed) orthogonal matrix. Suppose that an $m \times N$ random matrix $\mathbf{A}$ is drawn according to a probability distribution for which the concentration inequality*

$$\mathbb{P}\left(\left|\|\mathbf{A}\mathbf{x}\|_2^2 - \|\mathbf{x}\|_2^2\right| \geq t\|\mathbf{x}\|_2^2\right) \leq 2\exp\left(-\tilde{c}t^2 m\right) \tag{9.17}$$

holds for all $t \in (0,1)$ and $\mathbf{x} \in \mathbb{R}^N$. Then, given $\delta, \varepsilon \in (0,1)$, the restricted isometry constant δ_s of $\mathbf{A}\mathbf{U}$ satisfies $\delta_s < \delta$ with probability at least $1 - \varepsilon$ provided

$$m \geq C\delta^{-2}\left[s\left(9 + 2\ln(N/s)\right) + 2\ln(2\varepsilon^{-1})\right]$$

with $C = 2/(3\tilde{c})$.

Proof. The crucial point of the proof is that the concentration inequality (9.17) holds also with $\mathbf{A}$ replaced by $\mathbf{A}\mathbf{U}$. Indeed, let $\mathbf{x} \in \mathbb{R}^N$ and set $\mathbf{x}' = \mathbf{U}\mathbf{x}$. The fact that $\mathbf{U}$ is orthogonal yields

$$\mathbb{P}\left(\left|\|\mathbf{A}\mathbf{U}\mathbf{x}\|_2^2 - \|\mathbf{x}\|_2^2\right| \geq t\|\mathbf{x}\|_2^2\right) = \mathbb{P}\left(\left|\|\mathbf{A}\mathbf{x}'\|_2^2 - \|\mathbf{x}'\|_2^2\right| \geq t\|\mathbf{x}'\|_2^2\right)$$

$$\leq 2\exp\left(-\tilde{c}t^2 m\right).$$

Therefore, the statement follows from Theorem 9.11. $\qquad\square$

In particular, since the orthogonal matrix $\mathbf{U}$ is arbitrary in the above theorem, sparse recovery with subgaussian matrices is universal with respect to the orthonormal basis in which signals are sparse. It even means that at the encoding stage when measurements $\mathbf{y} = \mathbf{A}\mathbf{U}\mathbf{x}$ are taken, the orthogonal matrix $\mathbf{U}$ does not need to be known. It is only used at the decoding stage when a recovery algorithm is applied.

We emphasize, however, that universality does not mean that a single (fixed) measurement matrix $\mathbf{A}$ is able to deal with sparsity in any orthonormal basis. It is actually straightforward to see that this is impossible because once $\mathbf{A}$ is fixed, one can construct an orthogonal $\mathbf{U}$ for which sparse recovery fails. The theorem only states that for a fixed orthogonal $\mathbf{U}$, a random choice of $\mathbf{A}$ will work well with high probability.

9.2 Nonuniform Recovery

In this section, we consider the probability that a fixed sparse vector $\mathbf{x}$ is recovered via ℓ_1-minimization from $\mathbf{y} = \mathbf{A}\mathbf{x}$ using a random draw of a subgaussian matrix $\mathbf{A}$. We first discuss differences between uniform and nonuniform recovery. Then we give a first simple estimate for subgaussian matrices with good constants followed by an improved version for the special case of Gaussian matrices.

Uniform Versus Nonuniform Recovery

One may pursue different strategies in order to obtain rigorous recovery results. We distinguish between uniform and nonuniform recovery guarantees. A uniform recovery guarantee means that, with high probability on the choice of a random matrix, all sparse signals can be recovered using the same matrix. The bounds for the restricted isometry property that we have just derived indeed imply *uniform recovery* for subgaussian random matrices. A nonuniform recovery result only states that a fixed sparse signal can be recovered with high probability using a random draw of the matrix. In particular, such weaker nonuniform results allow in principle the small exceptional set of matrices for which recovery is not necessarily guaranteed to depend on the signal, in contrast to a uniform statement. Clearly, uniform recovery implies nonuniform recovery, but the converse is not true. In mathematical terms, a uniform recovery guarantee provides a lower probability estimate of the form

$$\mathbb{P}(\forall s\text{-sparse } \mathbf{x}, \text{ recovery of } \mathbf{x} \text{ is successful using } \mathbf{A}) \geq 1 - \varepsilon,$$

while nonuniform recovery gives a statement of the form

$$\forall s\text{-sparse } \mathbf{x} : \mathbb{P}(\text{recovery of } \mathbf{x} \text{ is successful using } \mathbf{A}) \geq 1 - \varepsilon,$$

where in both cases the probability is over the random draw of $\mathbf{A}$. Due to the appearance of the quantifier $\forall \mathbf{x}$ at different places, the two types of statements are clearly different.

For subgaussian random matrices, a nonuniform analysis enables to provide explicit and good constants in estimates for the required number of measurements—although the asymptotic analysis is essentially the same in both types of recovery guarantees. The advantage of the nonuniform approach will become more apparent later in Chap. 12, where we will see that such type of results will be easier to prove for structured random matrices and will provide better estimates both in terms of constants and asymptotic behavior.

Subgaussian Random Matrices

Our first nonuniform recovery results for ℓ_1-minimization concerns subgaussian random matrices.

Theorem 9.16. *Let* $\mathbf{x} \in \mathbb{C}^N$ *be an s-sparse vector. Let* $\mathbf{A} \in \mathbb{R}^{m \times N}$ *be a randomly drawn subgaussian matrix with subgaussian parameter c in (9.2). If, for some* $\varepsilon \in (0, 1)$,

$$m \geq \frac{4c}{1-\delta} s \ln(2N/\varepsilon), \quad with \; \delta = \sqrt{\frac{C}{4c}\left(\frac{7}{\ln(2N/\varepsilon)} + \frac{2}{s}\right)} \qquad (9.18)$$

(assuming N and s are large enough so that $\delta < 1$), then with probability at least $1 - \varepsilon$ the vector $\mathbf{x}$ is the unique minimizer of $\|\mathbf{z}\|_1$ subject to $\mathbf{Az} = \mathbf{Ax}$.

The constant $C = 2/(3\tilde{c})$ depends only on the subgaussian parameter c through $\tilde{c}$ in (9.6).

Remark 9.17. The parameter δ in (9.18) tends to zero as N and s get large so that, roughly speaking, sparse recovery is successful provided $m > 4cs \ln(2N/\varepsilon)$. In the Gaussian and Bernoulli case where $c = 1/2$, we roughly obtain the sufficient condition

$$m > 2s \ln(2N/\varepsilon). \qquad (9.19)$$

Later we will replace the factor $\ln(2N/\varepsilon)$ by $\ln(N/s)$ in the Gaussian case.

Proof (of Theorem 9.16). Set $S := \operatorname{supp} \mathbf{x}$ and note that $\operatorname{card}(S) \leq s$. By Corollary 4.28, it is sufficient to show that $\mathbf{A}_S$ is injective and that

$$|\langle (\mathbf{A}_S)^\dagger \mathbf{a}_\ell, \operatorname{sgn}(\mathbf{x}_S) \rangle| = |\langle \mathbf{a}_\ell, (\mathbf{A}_S^\dagger)^* \operatorname{sgn}(\mathbf{x}_S) \rangle| < 1 \quad \text{for all } \ell \in \overline{S}.$$

Therefore, the probability of failure of recovery is bounded, for any $\alpha > 0$, by

$$P := \mathbb{P}\left(\exists \ell \in \overline{S} : |\langle \mathbf{a}_\ell, (\mathbf{A}_S^\dagger)^* \mathrm{sgn}(\mathbf{x}_S)\rangle| \geq 1\right)$$

$$\leq \mathbb{P}\left(\exists \ell \in \overline{S} : |\langle \mathbf{a}_\ell, (\mathbf{A}_S^\dagger)^* \mathrm{sgn}(\mathbf{x}_S)\rangle| \geq 1 \,\Big|\, \|(\mathbf{A}_S^\dagger)^* \mathrm{sgn}(\mathbf{x}_S)\|_2 < \alpha\right) \quad (9.20)$$

$$+ \mathbb{P}(\|(\mathbf{A}_S^\dagger)^* \mathrm{sgn}(\mathbf{x}_S)\|_2 \geq \alpha). \quad (9.21)$$

The term in (9.20) above is estimated using Corollary 8.8. (In the real case, we may alternatively use Theorem 7.27.) Hereby, we additionally use the independence of all the entries of $\mathbf{A}$ so that, in particular, $\mathbf{a}_\ell$ and $\mathbf{A}_S$ are independent for $\ell \notin S$. Conditioning on the event that $\|(\mathbf{A}_S^\dagger)^* \mathrm{sgn}(\mathbf{x}_S)\|_2 < \alpha$, we obtain

$$\mathbb{P}\left(|\langle \mathbf{a}_\ell, (\mathbf{A}_S^\dagger)^* \mathrm{sgn}(\mathbf{x}_S)\rangle| \geq 1\right) = \mathbb{P}\left(\left|\sum_{j=1}^{m}(\mathbf{a}_\ell)_j [(\mathbf{A}_S^\dagger)^* \mathrm{sgn}(\mathbf{x}_S)]_j\right| \geq 1\right)$$

$$\leq 2\exp\left(-\frac{1}{4c\alpha^2}\right).$$

The term in (9.20) can be estimated by $2(N-s)\exp(-1/(4c\alpha^2))$ due to the union bound, which in turn is no larger than $(N-s)\varepsilon/N \leq (N-1)\varepsilon/N$ when

$$\alpha^{-2} = 4c\ln(2N/\varepsilon).$$

For the term in (9.21), we observe that

$$\|(\mathbf{A}_S^\dagger)^* \mathrm{sgn}(\mathbf{x}_S)\|_2^2 \leq \sigma_{\min}^{-2}(\mathbf{A}_S)\|\mathrm{sgn}(\mathbf{x}_S)\|_2^2 \leq \sigma_{\min}^{-2}(\mathbf{A}_S)\, s,$$

where $\sigma_{\min}$ denotes the smallest singular value; see (A.26). Therefore,

$$\mathbb{P}(\|(\mathbf{A}_S^\dagger)^* \mathrm{sgn}(\mathbf{x}_S)\|_2 \geq \alpha) \leq \mathbb{P}\left(\sigma_{\min}^2(\mathbf{A}_S/\sqrt{m}) \leq \frac{s}{\alpha^2 m}\right)$$

$$\leq \mathbb{P}(\sigma_{\min}^2(\mathbf{A}_S/\sqrt{m}) \leq 1-\delta),$$

where we have used that $\delta \in (0,1)$ is such that

$$m \geq \frac{4c}{1-\delta}s\ln(2N/\varepsilon) = \frac{s}{\alpha^2(1-\delta)}.$$

By Theorem 9.9 and Remark 9.10(b), the matrix $\mathbf{B} = \mathbf{A}_S/\sqrt{m}$ satisfies

$$\mathbb{P}(\sigma_{\min}^2(\mathbf{B}) \leq 1-\delta) \leq \mathbb{P}(\|\mathbf{B}^*\mathbf{B} - \mathbf{Id}\|_{2\to 2} \geq \delta) \leq 2\exp\left(-\frac{\delta^2 m}{2C} + \frac{7}{2}s\right).$$

The constant $C = 2/(3\tilde{c})$ depends only on the subgaussian parameter c through $\tilde{c}$ in (9.6). The term in (9.21) is now bounded by ε/N provided

$$m \geq C\delta^{-2}\left(7s + 2\ln(2N/\varepsilon)\right). \tag{9.22}$$

We see that (9.22) is implied by (9.18) with the given choice of δ and the desired probability of failure satisfies $P \leq (N-1)\varepsilon/N + \varepsilon/N = \varepsilon$. $\square$

Gaussian Random Matrices

Now we improve on the logarithmic factor in (9.18) and make recovery also robust under measurement error. For ease of presentation, we restrict to recovery of real vectors but note that extensions to the complex case are possible.

Theorem 9.18. *Let $\mathbf{x} \in \mathbb{R}^N$ be an s-sparse vector. Let $\mathbf{A} \in \mathbb{R}^{m \times N}$ be a randomly drawn Gaussian matrix. If, for some $\varepsilon \in (0, 1)$,*

$$\frac{m^2}{m+1} \geq 2s\left(\sqrt{\ln(eN/s)} + \sqrt{\frac{\ln(\varepsilon^{-1})}{s}}\right)^2, \tag{9.23}$$

then with probability at least $1 - \varepsilon$ the vector $\mathbf{x}$ is the unique minimizer of $\|\mathbf{z}\|_1$ subject to $\mathbf{Az} = \mathbf{Ax}$.

Remark 9.19. The proof enables one to even deduce a slightly more precise (but more complicated) condition; see Remark 9.25. Roughly speaking, for large N and mildly large s, condition (9.23) requires

$$m > 2s\ln(eN/s). \tag{9.24}$$

Theorem 9.18 extends to robust recovery.

Theorem 9.20. *Let $\mathbf{x} \in \mathbb{R}^N$ be an s-sparse vector. Let $\mathbf{A} \in \mathbb{R}^{m \times N}$ be a randomly drawn Gaussian matrix. Assume that noisy measurements $\mathbf{y} = \mathbf{Ax} + \mathbf{e}$ are taken with $\|\mathbf{e}\|_2 \leq \eta$. If, for $\varepsilon \in (0, 1)$ and $\tau > 0$,*

$$\frac{m^2}{m+1} \geq 2s\left(\sqrt{\ln(eN/s)} + \sqrt{\frac{\ln(\varepsilon^{-1})}{s}} + \frac{\tau}{\sqrt{s}}\right)^2,$$

then with probability at least $1 - \varepsilon$, every minimizer $\mathbf{x}^\sharp$ of $\|\mathbf{z}\|_1$ subject to $\|\mathbf{Az} - \mathbf{y}\|_2 \leq \eta$ satisfies

$$\|\mathbf{x} - \mathbf{x}^\sharp\|_2 \leq \frac{2\eta}{\tau}.$$

We develop the proof of these theorems in several steps. The basic ingredients are the recovery conditions of Theorems 4.35 and 4.36 based on the tangent cone $T(\mathbf{x})$ of the ℓ_1-norm defined in (4.34).

We start our analysis with a general concentration of measure result for Gaussian random matrices. We recall from Proposition 8.1(b) that a standard Gaussian random vector $\mathbf{g} \in \mathbb{R}^m$ satisfies

$$E_m := \mathbb{E}\|\mathbf{g}\|_2 = \sqrt{2}\frac{\Gamma((m+1)/2)}{\Gamma(m/2)} \tag{9.25}$$

and $m/\sqrt{m+1} \leq E_m \leq \sqrt{m}$. For a set $T \subset \mathbb{R}^N$, we introduce its *Gaussian width* by

$$\ell(T) := \mathbb{E} \sup_{\mathbf{x} \in T} \langle \mathbf{g}, \mathbf{x} \rangle, \tag{9.26}$$

where $\mathbf{g} \in \mathbb{R}^N$ is a standard Gaussian random vector. The following result is known as Gordon's *escape through the mesh* theorem.

Theorem 9.21. *Let* $\mathbf{A} \in \mathbb{R}^{m \times N}$ *be a Gaussian random matrix and* T *be a subset of the unit sphere* $S^{N-1} = \{\mathbf{x} \in \mathbb{R}^N, \|\mathbf{x}\|_2 = 1\}$. *Then, for* $t > 0$,

$$\mathbb{P}\left(\inf_{\mathbf{x} \in T} \|\mathbf{A}\mathbf{x}\|_2 \leq E_m - \ell(T) - t\right) \leq e^{-t^2/2}.$$

Proof. We first aim to estimate the expectation $\mathbb{E} \inf_{\mathbf{x} \in T} \|\mathbf{A}\mathbf{x}\|_2$ via Gordon's lemma (Lemma 8.27). For $\mathbf{x} \in T$ and $\mathbf{y} \in S^{m-1}$, we define the Gaussian process

$$X_{\mathbf{x},\mathbf{y}} := \langle \mathbf{A}\mathbf{x}, \mathbf{y} \rangle = \sum_{\ell=1}^{m}\sum_{j=1}^{N} A_{\ell j} x_j y_\ell. \tag{9.27}$$

Then $\inf_{\mathbf{x} \in T} \|\mathbf{A}\mathbf{x}\|_2 = \inf_{\mathbf{x} \in T} \max_{\mathbf{y} \in S^{m-1}} X_{\mathbf{x},\mathbf{y}}$. The key idea is to compare $X_{\mathbf{x},\mathbf{y}}$ to another Gaussian process $Y_{\mathbf{x},\mathbf{y}}$, namely, to

$$Y_{\mathbf{x},\mathbf{y}} := \langle \mathbf{g}, \mathbf{x} \rangle + \langle \mathbf{h}, \mathbf{y} \rangle,$$

where $\mathbf{g} \in \mathbb{R}^N$ and $\mathbf{h} \in \mathbb{R}^m$ are independent standard Gaussian random vectors. Let $\mathbf{x}, \mathbf{x}' \in S^{N-1}$ and $\mathbf{y}, \mathbf{y}' \in S^{m-1}$. One the one hand, since the $A_{\ell j}$ are independent and of variance 1, we have

$$\mathbb{E}|X_{\mathbf{x},\mathbf{y}} - X_{\mathbf{x}',\mathbf{y}'}|^2 = \mathbb{E}\left|\sum_{\ell=1}^{m}\sum_{j=1}^{N} A_{\ell j}(x_j y_\ell - x_j' y_\ell')\right|^2 = \sum_{\ell=1}^{m}\sum_{j=1}^{N}(x_j y_\ell - x_j' y_\ell')^2$$

$$= \sum_{\ell=1}^{m}\sum_{j=1}^{N}\left(x_j^2 y_\ell^2 + (x_j')^2(y_\ell')^2 - 2x_j x_j' y_\ell y_\ell'\right)$$

$$= \|\mathbf{x}\|_2^2\|\mathbf{y}\|_2^2 + \|\mathbf{x}'\|_2^2\|\mathbf{y}'\|_2^2 - 2\langle\mathbf{x},\mathbf{x}'\rangle\langle\mathbf{y},\mathbf{y}'\rangle$$
$$= 2 - 2\langle\mathbf{x},\mathbf{x}'\rangle\langle\mathbf{y},\mathbf{y}'\rangle.$$

On the other hand, using independence and the isotropicity of the standard multivariate Gaussian distribution (see Lemma 9.7), we have

$$
\begin{aligned}
\mathbb{E}|Y_{\mathbf{x},\mathbf{y}} - Y_{\mathbf{x}',\mathbf{y}'}|^2 &= \mathbb{E}\,|\langle\mathbf{g},\mathbf{x}-\mathbf{x}'\rangle + \langle\mathbf{h},\mathbf{y}-\mathbf{y}'\rangle|^2 \\
&= \mathbb{E}|\langle\mathbf{g},\mathbf{x}-\mathbf{x}'\rangle|^2 + \mathbb{E}|\langle\mathbf{h},\mathbf{y}-\mathbf{y}'\rangle|^2 \\
&= \|\mathbf{x}-\mathbf{x}'\|_2^2 + \|\mathbf{y}-\mathbf{y}'\|_2^2 \\
&= \|\mathbf{x}\|_2^2 + \|\mathbf{x}'\|_2^2 - 2\langle\mathbf{x},\mathbf{x}'\rangle + \|\mathbf{y}\|_2^2 + \|\mathbf{y}'\|_2^2 - 2\langle\mathbf{y},\mathbf{y}'\rangle \\
&= 4 - 2\langle\mathbf{x},\mathbf{x}'\rangle - 2\langle\mathbf{y},\mathbf{y}'\rangle.
\end{aligned}
$$

We obtain

$$
\begin{aligned}
\mathbb{E}|Y_{\mathbf{x},\mathbf{y}} - Y_{\mathbf{x}',\mathbf{y}'}|^2 - \mathbb{E}|X_{\mathbf{x},\mathbf{y}} - X_{\mathbf{x}',\mathbf{y}'}|^2 &= 2(1 - \langle\mathbf{x},\mathbf{x}'\rangle - \langle\mathbf{y},\mathbf{y}'\rangle + \langle\mathbf{x},\mathbf{x}'\rangle\langle\mathbf{y},\mathbf{y}'\rangle) \\
&= 2(1 - \langle\mathbf{x},\mathbf{x}'\rangle)(1 - \langle\mathbf{y},\mathbf{y}'\rangle) \geq 0,
\end{aligned}
$$

where the last inequality follows from the Cauchy–Schwarz inequality and the fact that all vectors are unit norm. Equality holds if and only if $\langle\mathbf{x},\mathbf{x}'\rangle = 1$ or $\langle\mathbf{y},\mathbf{y}'\rangle = 1$. Therefore, we have shown that

$$\mathbb{E}|X_{\mathbf{x},\mathbf{y}} - X_{\mathbf{x}',\mathbf{y}'}|^2 \leq \mathbb{E}|Y_{\mathbf{x},\mathbf{y}} - Y_{\mathbf{x}',\mathbf{y}'}|^2, \tag{9.28}$$

$$\mathbb{E}|X_{\mathbf{x},\mathbf{y}} - X_{\mathbf{x},\mathbf{y}'}|^2 = \mathbb{E}|Y_{\mathbf{x},\mathbf{y}} - Y_{\mathbf{x},\mathbf{y}'}|^2. \tag{9.29}$$

It follows from Gordon's lemma (Lemma 8.27) and Remark 8.28 that

$$
\begin{aligned}
\mathbb{E}\inf_{\mathbf{x}\in T}\|\mathbf{A}\mathbf{x}\|_2 = \mathbb{E}\inf_{\mathbf{x}\in T}\max_{\mathbf{y}\in S^{m-1}} X_{\mathbf{x},\mathbf{y}} &\geq \mathbb{E}\inf_{\mathbf{x}\in T}\max_{\mathbf{y}\in S^{m-1}} Y_{\mathbf{x},\mathbf{y}} \\
&= \mathbb{E}\inf_{\mathbf{x}\in T}\max_{\mathbf{y}\in S^{m-1}} \{\langle\mathbf{g},\mathbf{x}\rangle + \langle\mathbf{h},\mathbf{y}\rangle\} = \mathbb{E}\inf_{\mathbf{x}\in T}\{\langle\mathbf{g},\mathbf{x}\rangle + \|\mathbf{h}\|_2\} \\
&= \mathbb{E}\|\mathbf{h}\|_2 - \mathbb{E}\sup_{\mathbf{x}\in T}\langle\mathbf{g},\mathbf{x}\rangle = E_m - \ell(T),
\end{aligned}
$$

where we have once applied the symmetry of a standard Gaussian vector.

We notice that $F(\mathbf{A}) := \inf_{\mathbf{x}\in T}\|\mathbf{A}\mathbf{x}\|_2$ defines a Lipschitz function with respect to the Frobenius norm (which corresponds to the ℓ_2-norm by identifying $\mathbb{R}^{m\times N}$ with $\mathbb{R}^{mN}$). Indeed, for two matrices $\mathbf{A},\mathbf{B}\in\mathbb{R}^{m\times N}$,

$$\inf_{\mathbf{x}\in T}\|\mathbf{A}\mathbf{x}\|_2 \leq \inf_{\mathbf{x}\in T}(\|\mathbf{B}\mathbf{x}\|_2 + \|(\mathbf{A}-\mathbf{B})\mathbf{x}\|_2) \leq \inf_{\mathbf{x}\in T}(\|\mathbf{B}\mathbf{x}\|_2 + \|\mathbf{A}-\mathbf{B}\|_{2\to2})$$

$$\leq \inf_{\mathbf{x}\in T}\|\mathbf{B}\mathbf{x}\|_2 + \|\mathbf{A}-\mathbf{B}\|_F.$$

Hereby, we have used that $T \subset S^{N-1}$ and that the operator norm is bounded by the Frobenius norm; see (A.17). Interchanging the role of $\mathbf{A}$ and $\mathbf{B}$, we conclude that $|F(\mathbf{A}) - F(\mathbf{B})| \leq \|\mathbf{A} - \mathbf{B}\|_F$. It follows from concentration of measure (Theorem 8.40) that

$$\mathbb{P}(\inf_{\mathbf{x} \in T} \|\mathbf{A}\mathbf{x}\|_2 \leq \mathbb{E} \inf_{\mathbf{x} \in T} \|\mathbf{A}\mathbf{x}\|_2 - t) \leq e^{-t^2/2}.$$

Substituting the estimation on the expectation $\mathbb{E} \inf_{\mathbf{x} \in T} \|\mathbf{A}\mathbf{x}\|_2$ derived above concludes the proof. $\qquad \square$

According to Theorem 4.35, to prove that a random vector $\mathbf{x} \in \mathbb{R}^N$ is recovered from $\mathbf{A}\mathbf{x}$ via ℓ_1-minimization, we need to establish that $\ker \mathbf{A} \cap T(\mathbf{x}) = \{\mathbf{0}\}$, where

$$T(\mathbf{x}) = \text{cone}\{\mathbf{z} - \mathbf{x} : \mathbf{z} \in \mathbb{R}^N, \|\mathbf{z}\|_1 \leq \|\mathbf{x}\|_1\}.$$

Equivalently, we need to establish that

$$\inf_{\mathbf{z} \in T(\mathbf{x}) \cap S^{N-1}} \|\mathbf{A}\mathbf{z}\|_2 > 0.$$

To this end, Theorem 9.21 is useful if we find an upper bound for the Gaussian width of $T := T(\mathbf{x}) \cap S^{N-1}$. The next result gives such a bound involving the normal cone of the ℓ_1-norm at $\mathbf{x}$, i.e., the polar cone $\mathcal{N}(\mathbf{x}) = T(\mathbf{x})^\circ$ of $T(\mathbf{x})$ (see (B.3)), namely,

$$\mathcal{N}(\mathbf{x}) = \{\mathbf{z} \in \mathbb{R}^N : \langle \mathbf{z}, \mathbf{w} - \mathbf{x} \rangle \leq 0 \text{ for all } \mathbf{w} \text{ such that } \|\mathbf{w}\|_1 \leq \|\mathbf{x}\|_1\}. \quad (9.30)$$

Proposition 9.22. *Let* $\mathbf{g} \in \mathbb{R}^N$ *be a standard Gaussian random vector. Then*

$$\ell(T(\mathbf{x}) \cap S^{N-1}) \leq \mathbb{E} \min_{\mathbf{z} \in \mathcal{N}(\mathbf{x})} \|\mathbf{g} - \mathbf{z}\|_2. \quad (9.31)$$

Proof. It follows from (B.40) that

$$\ell(T(\mathbf{x}) \cap S^{N-1}) = \mathbb{E} \max_{\mathbf{z} \in T(\mathbf{x}), \|\mathbf{z}\|_2 = 1} \langle \mathbf{g}, \mathbf{z} \rangle \leq \mathbb{E} \max_{\mathbf{z} \in T(\mathbf{x}), \|\mathbf{z}\|_2 \leq 1} \langle \mathbf{g}, \mathbf{z} \rangle$$

$$\leq \mathbb{E} \min_{\mathbf{z} \in T(\mathbf{x})^\circ} \|\mathbf{g} - \mathbf{z}\|_2.$$

By definition of the normal cone, this establishes the claim. $\qquad \square$

The previous result leads us to computing the normal cone of the ℓ_1-norm at a sparse vector.

Lemma 9.23. *Let* $\mathbf{x} \in \mathbb{R}^N$ *with* $\text{supp}(\mathbf{x}) = S \subset [N]$. *Then*

$$\mathcal{N}(\mathbf{x}) = \bigcup_{t \geq 0} \{\mathbf{z} \in \mathbb{R}^N, z_\ell = t \, \text{sgn}(x_\ell) \text{ for } \ell \in S, |z_\ell| \leq t \text{ for } \ell \in \overline{S}\}. \quad (9.32)$$

Proof. If $\mathbf{z}$ is contained in the right-hand side of (9.32), then, for $\mathbf{w}$ such that $\|\mathbf{w}\|_1 \leq \|\mathbf{x}\|_1$,

$$\langle \mathbf{z}, \mathbf{w} - \mathbf{x} \rangle = \langle \mathbf{z}, \mathbf{w} \rangle - \langle \mathbf{z}, \mathbf{x} \rangle \leq \|\mathbf{z}\|_\infty \|\mathbf{w}\|_1 - \|\mathbf{z}\|_\infty \|\mathbf{x}\|_1 = \|\mathbf{z}\|_\infty (\|\mathbf{w}\|_1 - \|\mathbf{x}\|_1)$$
$$\leq 0,$$

hence, $\mathbf{z} \in \mathcal{N}(\mathbf{x})$.

Conversely, assume that $\mathbf{z} \in \mathcal{N}(\mathbf{x})$, i.e., that $\langle \mathbf{z}, \mathbf{w} \rangle \leq \langle \mathbf{z}, \mathbf{x} \rangle$ for all $\mathbf{w}$ with $\|\mathbf{w}\|_1 \leq \|\mathbf{x}\|_1$. Consider a vector $\mathbf{w}$ with $\|\mathbf{w}\|_1 = \|\mathbf{x}\|_1$ such that $\mathrm{sgn}(w_j) = \mathrm{sgn}(z_j)$ whenever $|z_j| = \|\mathbf{z}\|_\infty$ and $w_j = 0$ otherwise. Then

$$\|\mathbf{z}\|_\infty \|\mathbf{w}\|_1 = \langle \mathbf{z}, \mathbf{w} \rangle \leq \langle \mathbf{z}, \mathbf{x} \rangle \leq \|\mathbf{z}\|_\infty \|\mathbf{x}\|_1 = \|\mathbf{z}\|_\infty \|\mathbf{w}\|_1.$$

In particular, $\langle \mathbf{z}, \mathbf{x} \rangle = \|\mathbf{z}\|_\infty \|\mathbf{x}\|_1$, which means that $z_\ell = \mathrm{sgn}(x_\ell)\|\mathbf{z}\|_\infty$ for all $\ell \in S$. Since obviously also $|z_\ell| \leq \|\mathbf{z}\|_\infty$ for all $\ell \in \overline{S}$, we choose $t = \|\mathbf{z}\|_\infty$ to see that $\mathbf{z}$ is contained in right-hand side of (9.32). $\qquad\square$

Now we are in a position to estimate the desired Gaussian width.

Proposition 9.24. *If $\mathbf{x} \in \mathbb{R}^N$ is an s-sparse vector, then*

$$\left(\ell(T(\mathbf{x}) \cap S^{N-1}) \right)^2 \leq 2s \ln(eN/s). \tag{9.33}$$

Proof. It follows from Proposition 9.22 and Hölder's inequality that

$$(\ell(T(\mathbf{x}) \cap S^{N-1}))^2 \leq \left(\mathbb{E} \min_{\mathbf{z} \in \mathcal{N}(\mathbf{x})} \|\mathbf{g} - \mathbf{z}\|_2 \right)^2 \leq \mathbb{E} \min_{\mathbf{z} \in \mathcal{N}(\mathbf{x})} \|\mathbf{g} - \mathbf{z}\|_2^2. \tag{9.34}$$

Let $S = \mathrm{supp}(\mathbf{x})$ so that $\mathrm{card}(S) \leq s$. According to (9.32), we have

$$\min_{\mathbf{z} \in \mathcal{N}(\mathbf{x})} \|\mathbf{g} - \mathbf{z}\|_2^2 = \min_{\substack{t \geq 0 \\ |z_\ell| \leq t, \ell \in \overline{S}}} \sum_{\ell \in S} (g_\ell - t\,\mathrm{sgn}(\mathbf{x}_\ell))^2 + \sum_{\ell \in \overline{S}} (g_\ell - z_\ell)^2. \tag{9.35}$$

A straightforward computation shows that (see also Exercise 15.2)

$$\min_{|z_\ell| \leq t} (g_\ell - z_\ell)^2 = S_t(g_\ell)^2,$$

where S_t is the soft thresholding operator (B.18) given by

$$S_t(u) = \begin{cases} u - t & \text{if } u \geq t, \\ 0 & \text{if } |u| \leq t, \\ u + t & \text{if } u \leq -t. \end{cases}$$

Hence, for a fixed $t > 0$ independent of $\mathbf{g}$,

$$\min_{\mathbf{z}\in\mathcal{N}(\mathbf{x})} \|\mathbf{g} - \mathbf{z}\|_2^2 \leq \mathbb{E}\left[\sum_{\ell\in S}(g_\ell - t\mathrm{sgn}(x_\ell))^2\right] + \mathbb{E}\left[\sum_{\ell\in\overline{S}} S_t(g_\ell)^2\right]$$

$$\leq s\mathbb{E}(g+t)^2 + \sum_{\ell\in\overline{S}}\mathbb{E}S_t(g_\ell)^2 = s(1+t^2) + (N-s)\mathbb{E}S_t(g)^2,$$

where g is a (univariate) standard Gaussian random variable. It remains to estimate $\mathbb{E}S_t(g)^2$. Applying symmetry of g and S_t as well as a change of variables, we get

$$\mathbb{E}S_t(g)^2 = \frac{2}{\sqrt{2\pi}}\int_0^\infty S_t(u)^2 e^{-u^2/2}du = \sqrt{\frac{2}{\pi}}\int_t^\infty (u-t)^2 e^{-u^2/2}du$$

$$= \sqrt{\frac{2}{\pi}}\int_0^\infty v^2 e^{-(v+t)^2/2}dv = e^{-t^2/2}\sqrt{\frac{2}{\pi}}\int_0^\infty v^2 e^{-v^2/2}e^{-vt}dv \quad (9.36)$$

$$\leq e^{-t^2/2}\sqrt{\frac{2}{\pi}}\int_0^\infty v^2 e^{-v^2/2}dv = e^{-t^2/2}\mathbb{E}[g^2] = e^{-t^2/2}.$$

Combining the above arguments gives

$$(\ell(T(\mathbf{x})\cap S^{N-1}))^2 \leq \min_{t\geq 0}\left\{s(1+t^2) + (N-s)e^{-t^2/2}\right\}$$

$$\leq \min_{t\geq 0}\left\{s(1+t^2) + Ne^{-t^2/2}\right\}.$$

The choice $t = \sqrt{2\ln(N/s)}$ shows that

$$(\ell(T(\mathbf{x})\cap S^{N-1}))^2 \leq s(1 + 2\ln(N/s)) + s = 2s\ln(eN/s).$$

This completes the proof. □

Let us finally complete the proofs of the nonuniform recovery results for Gaussian random matrices.

Proof (of Theorem 9.18). Let us set $t = \sqrt{2\ln(\varepsilon^{-1})}$. By Proposition 9.24 and since $E_m \geq m/\sqrt{m+1}$ (see Proposition 8.1(b)), Condition (9.23) ensures that

$$E_m - \ell(T(\mathbf{x})\cap S^{N-1}) - t \geq 0.$$

It follows from Theorem 9.21 that

$$\mathbb{P}\left(\min_{\mathbf{z}\in T(\mathbf{x})\cap S^{N-1}}\|\mathbf{A}\mathbf{z}\|_2 > 0\right)$$

$$\geq \mathbb{P}\left(\min_{\mathbf{z}\in T(\mathbf{x})\cap S^{N-1}}\|\mathbf{A}\mathbf{z}\|_2 > E_m - \ell(T(\mathbf{x})\cap S^{N-1}) - t\right)$$

$$\geq 1 - e^{-t^2/2} = 1 - \varepsilon. \quad (9.37)$$

This implies that $T(\mathbf{x}) \cap \ker \mathbf{A} = \{\mathbf{0}\}$ with probability at least $1 - \varepsilon$. An application of Theorem 4.35 concludes the proof. □

Proof (of Theorem 9.20). With the same notation as in the previous proof, the assumptions of Theorem 9.20 imply

$$E_m - \ell(T(\mathbf{x}) \cap S^{N-1}) - \tau - t \geq 0.$$

As in (9.37), we conclude that

$$\mathbb{P} \left(\min_{\mathbf{z} \in T(\mathbf{x}) \cap S^{N-1}} \|\mathbf{A}\mathbf{z}\|_2 \geq \tau \right) \geq 1 - \varepsilon.$$

The claim follows then from Theorem 4.36. □

Remark 9.25. We can also achieve a slightly better (but more complicated) bound for the Gaussian width in Proposition 9.24. Instead of the bound $\mathbb{E}S_t(g)^2 \leq e^{-t^2/2}$ derived in the proof of Proposition 9.24, we use the elementary inequality $v e^{-v^2/2} \leq e^{-1/2}$ in (9.36) to obtain

$$\mathbb{E}S_t(g)^2 = e^{-t^2/2} \sqrt{\frac{2}{\pi}} \int_0^\infty v^2 e^{-v^2/2} e^{-vt} dv \leq \sqrt{\frac{2}{\pi}} e^{-1/2} e^{-t^2/2} \int_0^\infty v e^{-vt} dv$$

$$= \sqrt{\frac{2}{\pi e}} t^{-2} e^{-t^2/2}. \tag{9.38}$$

Following the arguments in the proof of Proposition 9.24, the choice $t = \sqrt{2\ln(N/s) - 1} = \sqrt{2\ln(N/(\sqrt{e}s))}$, which is valid if $N > \sqrt{e}s$, yields

$$\left(\ell(T(\mathbf{x}) \cap S^{N-1}) \right)^2 \leq 2s \ln(N/s) + \sqrt{\frac{1}{2\pi}} \frac{s}{\ln(N/(\sqrt{e}s))}.$$

Therefore, if $N > \sqrt{e}s$, then the recovery condition (9.23) can be refined to

$$\frac{m^2}{m+1} \geq 2s \left(\sqrt{\ln(N/s) + \frac{1}{\sqrt{8\pi} \ln(N/(\sqrt{e}s))}} + \sqrt{\frac{\ln(\varepsilon^{-1})}{s}} \right)^2. \tag{9.39}$$

Roughly speaking, for large N, mildly large s, and large ratio N/s, we obtain the "asymptotic" recovery condition

$$m > 2s \ln(N/s). \tag{9.40}$$

This is the general rule of thumb for compressive sensing and reflects well empirical tests for sparse recovery using Gaussian matrices but also different random matrices. However, our proof of this result is restricted to the Gaussian case.

9.3 Restricted Isometry Property for Gaussian Matrices

We return to the uniform setting specialized to Gaussian matrices, for which we can provide explicit constants in the estimate for the required number of measurements. We first treat the restricted isometry property. In the next section, we provide a direct proof for the null space property of Gaussian matrices resulting in very reasonable constants.

The approach for the restricted isometry property of Gaussian random matrices in this section (as well as the one for the null space property in the next section) is based on concentration of measure (Theorem 8.40) and on the Slepian–Gordon lemmas, therefore it does not generalize to subgaussian matrices.

We start with estimates for the extremal singular values of a Gaussian random matrix.

Theorem 9.26. *Let* $\mathbf{A}$ *be an* $m \times s$ *Gaussian matrix with* $m > s$ *and let* $\sigma_{\min}$ *and* $\sigma_{\max}$ *be the smallest and largest singular value of the renormalized matrix* $\frac{1}{\sqrt{m}}\mathbf{A}$. *Then, for* $t > 0$,

$$\mathbb{P}(\sigma_{\max} \geq 1 + \sqrt{s/m} + t) \leq e^{-mt^2/2}, \tag{9.41}$$

$$\mathbb{P}(\sigma_{\min} \leq 1 - \sqrt{s/m} - t) \leq e^{-mt^2/2}. \tag{9.42}$$

Proof. By Proposition A.16, the extremal singular values are 1-Lipschitz functions with respect to the Frobenius norm (which corresponds to the ℓ_2-norm by identifying $\mathbb{R}^{m \times s}$ with $\mathbb{R}^{ms}$). Therefore, it follows from concentration of measure (Theorem 8.40) that in particular the largest singular value of the non-normalized matrix $\mathbf{A}$ satisfies

$$\mathbb{P}(\sigma_{\max}(\mathbf{A}) \geq \mathbb{E}[\sigma_{\max}(\mathbf{A})] + r) \leq e^{-r^2/2}. \tag{9.43}$$

Let us estimate the expectation above using Slepian's lemma (Lemma 8.25). Observe that

$$\sigma_{\max}(\mathbf{A}) = \sup_{\mathbf{x} \in S^{s-1}} \sup_{\mathbf{y} \in S^{m-1}} \langle \mathbf{A}\mathbf{x}, \mathbf{y} \rangle. \tag{9.44}$$

As in the proof of Theorem 9.21, we introduce two Gaussian processes by

$$X_{\mathbf{x},\mathbf{y}} := \langle \mathbf{A}\mathbf{x}, \mathbf{y} \rangle,$$

$$Y_{\mathbf{x},\mathbf{y}} := \langle \mathbf{g}, \mathbf{x} \rangle + \langle \mathbf{h}, \mathbf{y} \rangle, \tag{9.45}$$

where $\mathbf{g} \in \mathbb{R}^s, \mathbf{h} \in \mathbb{R}^m$ are two independent standard Gaussian random vectors. By (9.28), for $\mathbf{x}, \mathbf{x}' \in S^{s-1}$ and $\mathbf{y}, \mathbf{y}' \in S^{m-1}$, we have

$$\mathbb{E}|X_{\mathbf{x},\mathbf{y}} - X_{\mathbf{x}',\mathbf{y}'}|^2 \leq \mathbb{E}|Y_{\mathbf{x},\mathbf{y}} - Y_{\mathbf{x}',\mathbf{y}'}|^2. \tag{9.46}$$

Slepian's lemma (Lemma 8.25) together with Remark 8.28 implies that

$$\mathbb{E}\,\sigma_{\max}(\mathbf{A}) = \mathbb{E} \sup_{\mathbf{x}\in S^{s-1}, \mathbf{y}\in S^{m-1}} X_{\mathbf{x},\mathbf{y}} \le \mathbb{E} \sup_{\mathbf{x}\in S^{s-1}, \mathbf{y}\in S^{m-1}} Y_{\mathbf{x},\mathbf{y}}$$

$$= \mathbb{E} \sup_{\mathbf{x}\in S^{s-1}} \langle \mathbf{g}, \mathbf{x}\rangle + \mathbb{E} \sup_{\mathbf{y}\in S^{m-1}} \langle \mathbf{h}, \mathbf{y}\rangle = \mathbb{E}\|\mathbf{g}\|_2 + \mathbb{E}\|\mathbf{h}\|_2 \le \sqrt{s} + \sqrt{m}.$$

The last inequality follows from Proposition 8.1(b). Plugging this estimate into (9.43) shows that

$$\mathbb{P}(\sigma_{\max}(\mathbf{A}) \ge \sqrt{m} + \sqrt{s} + r) \le e^{-r^2/2}.$$

Rescaling by $\frac{1}{\sqrt{m}}$ shows the estimate for the largest singular value of $\frac{1}{\sqrt{m}}\mathbf{A}$.

The smallest singular value $\sigma_{\min}(\mathbf{A}) = \inf_{\mathbf{x}\in S^{s-1}} \|\mathbf{A}\mathbf{x}\|_2$ can be estimated with the help of Theorem 9.21 (which relies on concentration of measure for Lipschitz functions as well). The required Gaussian width of $T = S^{s-1}$ is given, for a standard Gaussian vector $\mathbf{g}$ in $\mathbb{R}^s$, by

$$\ell(S^{s-1}) = \mathbb{E} \sup_{\mathbf{x}\in S^{s-1}} \langle \mathbf{g}, \mathbf{x}\rangle = \mathbb{E}\|\mathbf{g}\|_2 = E_s.$$

By Proposition 8.1(b) and Lemma C.4, we further obtain

$$E_m - \ell(S^{s-1}) = \sqrt{2}\frac{\Gamma((m+1)/2)}{\Gamma(m/2)} - \sqrt{2}\frac{\Gamma((s+1)/2)}{\Gamma(s/2)} \ge \sqrt{m} - \sqrt{s}.$$

Applying Theorem 9.21 and rescaling concludes the proof.   □

With this tool at hand, we can easily show the restricted isometry property for Gaussian matrices.

Theorem 9.27. *Let $\mathbf{A}$ be an $m \times N$ Gaussian matrix with $m < N$. For $\eta, \varepsilon \in (0,1)$, assume that*

$$m \ge 2\eta^{-2}\left(s\ln(eN/s) + \ln(2\varepsilon^{-1})\right). \qquad (9.47)$$

Then with probability at least $1 - \varepsilon$ the restricted isometry constant δ_s of $\frac{1}{\sqrt{m}}\mathbf{A}$ satisfies

$$\delta_s \le 2\left(1 + \frac{1}{\sqrt{2\ln(eN/s)}}\right)\eta + \left(1 + \frac{1}{\sqrt{2\ln(eN/s)}}\right)^2 \eta^2. \qquad (9.48)$$

Remark 9.28. Note that (9.48) implies the simpler inequality $\delta_s \le C\eta$ with $C = 2(1 + \sqrt{1/2}) + (1 + \sqrt{1/2})^2 \approx 6.3284$. In other words, the condition

$$m \ge \tilde{C}\delta^{-2}\left(s\ln(eN/s) + \ln(2\varepsilon^{-1})\right)$$

with $\tilde{C} = 2C^2 \approx 80.098$ implies $\delta_s \leq \delta$. In most situations, i.e., when $s \ll N$, the statement of the theorem provides better constants. For instance, if $2\ln(eN/(2s)) \geq 8$, that is, $N/s \geq 2e^3 \approx 40.171$ and $\eta = 1/5$, then $\delta_{2s} \leq 0.6147 < 0.6248$ (compare with Theorem 6.12 concerning ℓ_1-minimization) provided

$$m \geq 50\left(2s\ln(eN/(2s)) + \ln(2\varepsilon^{-1})\right).$$

Further, in the limit $N/s \to \infty$, we obtain $\delta_s \leq 2\eta + \eta^2$. Then the choice $\eta = 0.27$ yields $\delta_{2s} \leq 0.6129$ under the condition $m \geq 27.434\left(2s\ln(eN/(2s)) + \ln(2\varepsilon^{-1})\right)$ in this asymptotic regime.

Proof (of Theorem 9.27). We proceed similarly to the proof of Theorem 9.11. Let $S \subset [N]$ be of cardinality s. The submatrix $\mathbf{A}_S$ is an $m \times s$ Gaussian matrix, and the eigenvalues of $\frac{1}{m}\mathbf{A}_S^*\mathbf{A}_S - \mathbf{Id}$ are contained in $[\sigma_{\min}^2 - 1, \sigma_{\max}^2 - 1]$ where $\sigma_{\min}$ and $\sigma_{\max}$ are the extremal singular values of $\tilde{\mathbf{A}}_S = \frac{1}{\sqrt{m}}\mathbf{A}_S$. Theorem 9.26 implies that with probability at least $1 - 2\exp(-m\eta^2/2)$

$$\|\tilde{\mathbf{A}}_S^*\tilde{\mathbf{A}}_S - \mathbf{Id}\|_{2\to 2} \leq \max\left\{(1 + \sqrt{s/m} + \eta)^2 - 1, 1 - (1 - (\sqrt{s/m} + \eta))^2\right\}$$

$$= 2(\sqrt{s/m} + \eta) + (\sqrt{s/m} + \eta)^2.$$

Taking the union bound over all $\binom{N}{s}$ subsets of $[N]$ of cardinality s, and in view of $\delta_s = \max_{S\subset[N],\mathrm{card}(S)=s}\|\tilde{\mathbf{A}}_S^*\tilde{\mathbf{A}}_S - \mathbf{Id}\|_{2\to 2}$, we obtain

$$\mathbb{P}\left(\delta_s > 2(\sqrt{s/m} + \eta) + (\sqrt{s/m} + \eta)^2\right) \leq 2\binom{N}{s}e^{-m\eta^2/2}$$

$$\leq 2\left(\frac{eN}{s}\right)^s e^{-m\eta^2/2}.$$

In the second inequality, we have applied Lemma C.5. The last term is dominated by ε due to condition (9.47), which also implies $\sqrt{s/m} \leq \frac{\eta}{\sqrt{2\ln(eN/s)}}$. The conclusion of the theorem follows. $\qquad\square$

9.4 Null Space Property for Gaussian Matrices

Our next theorem states stable uniform recovery with Gaussian random matrices via ℓ_1-minimization and provides good constants. It is established by showing the stable null space property of Definition 4.11 directly rather than by relying on the restricted isometry property.

Theorem 9.29. *Let* $\mathbf{A} \in \mathbb{R}^{m \times N}$ *be a random draw of a Gaussian matrix. Assume that*

$$\frac{m^2}{m+1} \geq 2s \ln(eN/s) \left(1 + \rho^{-1} + D(s/N) + \sqrt{\frac{\ln(\varepsilon^{-1})}{s \ln(eN/s)}}\right)^2, \qquad (9.49)$$

where D *is a function that satisfies* $D(\alpha) \leq 0.92$ *for all* $\alpha \in (0, 1]$ *and* $\lim_{\alpha \to 0} D(\alpha) = 0$. *Then, with probability at least* $1 - \varepsilon$, *every vector* $\mathbf{x} \in \mathbb{R}^N$ *is approximated by a minimizer* $\mathbf{x}^{\sharp}$ *of* $\|\mathbf{z}\|_1$ *subject to* $\mathbf{A}\mathbf{z} = \mathbf{A}\mathbf{x}$ *in the sense that*

$$\|\mathbf{x} - \mathbf{x}^{\sharp}\|_1 \leq \frac{2(1 + \rho)}{1 - \rho} \sigma_s(\mathbf{x})_1.$$

Remark 9.30. (a) The function D in the above theorem is given by

$$D(\alpha) = \frac{1}{\sqrt{2 \ln(e\alpha^{-1})}} + \frac{1}{(8\pi e^3)^{1/4} \ln(e\alpha^{-1})}. \qquad (9.50)$$

Note that the theorem is only interesting for $N \geq 2s$ because otherwise it is impossible to fulfill the null space property. In this range, D satisfies the slightly better upper bound

$$D(\alpha) \leq D(1/2) = \frac{1}{\sqrt{2 \ln(2e)}} + \frac{1}{(8\pi e^3)^{1/4} \ln(2e)} \approx 0.668.$$

(b) Roughly speaking, for large N, mildly large s and small ratio s/N (which is the situation of most interest in compressive sensing), Condition (9.49) turns into

$$m > 2(1 + \rho^{-1})^2 s \ln(eN/s).$$

(c) The proof proceeds by establishing the null space property of order s with constant ρ. Letting $\rho = 1$ yields exact recovery of all s-sparse vectors roughly under the condition

$$m > 8s \ln(eN/s). \qquad (9.51)$$

The proof strategy is similar to Sect. 9.2. In particular, we use Gordon's escape through the mesh theorem (Theorem 9.21). For $\rho \in (0, 1]$, we introduce the set

$$T_{\rho,s} := \left\{\mathbf{w} \in \mathbb{R}^N : \|\mathbf{w}_S\|_1 \geq \rho \|\mathbf{w}_{\overline{S}}\|_1 \text{ for some } S \subset [N], \mathrm{card}(S) = s\right\}.$$

If we show that

$$\min\{\|\mathbf{A}\mathbf{w}\|_2 : \mathbf{w} \in T_{\rho,s} \cap S^{N-1}\} > 0, \qquad (9.52)$$

then we can conclude that

$$\|\mathbf{v}_S\|_1 < \rho\|\mathbf{v}_{\overline{S}}\|_1 \quad \text{for all } \mathbf{v} \in \ker \mathbf{A} \setminus \{\mathbf{0}\} \text{ and all } S \subset [N] \text{ with } \operatorname{card}(S) = s,$$

so that the stable null space property holds. If $\rho = 1$, this implies exact recovery of all s-sparse vectors by Theorem 4.4. If $\rho < 1$, we even have stability of the reconstruction by Theorem 4.12. Following Theorem 9.21, we are led to study the Gaussian widths of the set $T_{\rho,s} \cap S^{N-1}$. As a first step, we relate this problem to the following convex cone

$$K_{\rho,s} := \left\{ \mathbf{u} \in \mathbb{R}^N : u_\ell \geq 0 \text{ for all } \ell \in [N], \sum_{\ell=1}^{s} u_\ell \geq \rho \sum_{\ell=s+1}^{N} u_\ell \right\}. \qquad (9.53)$$

Our next result is similar to Proposition 9.22. We recall that the nonincreasing rearrangement $\mathbf{g}^*$ of a vector $\mathbf{g}$ has entries $g_j^* = |g_{\ell_j}|$ with a permutation $j \mapsto \ell_j$ of $[N]$ such that $g_1^* \geq g_2^* \geq \cdots \geq g_N^* \geq 0$; see Definition 2.4. Also, recall the definition (B.2) of the dual cone

$$K_{\rho,s}^* = \{\mathbf{z} \in \mathbb{R}^N : \langle \mathbf{z}, \mathbf{u} \rangle \geq 0 \text{ for all } \mathbf{u} \in K_{\rho,s}\}.$$

Proposition 9.31. *Let* $\mathbf{g} \in \mathbb{R}^N$ *be a standard Gaussian vector and* $\mathbf{g}^*$ *its nonincreasing rearrangement. Then*

$$\ell(T_{\rho,s} \cap S^{N-1}) \leq \mathbb{E} \min_{\mathbf{z} \in K_{\rho,s}^*} \|\mathbf{g}^* + \mathbf{z}\|_2.$$

Proof. Consider the maximization problem $\max_{\mathbf{w} \in T_{\rho,s} \cap S^{N-1}} \langle \mathbf{g}, \mathbf{w} \rangle$ appearing in the definition of the Gaussian widths (9.26) (since $T_{\rho,s} \cap S^{N-1}$ is compact the maximum is attained). By invariance of $T_{\rho,s}$ under permutation of the indices and under entrywise sign changes, we have

$$\max_{\mathbf{w} \in T_{\rho,s} \cap S^{N-1}} \langle \mathbf{g}, \mathbf{w} \rangle = \max_{\mathbf{w} \in T_{\rho,s} \cap S^{N-1}} \langle \mathbf{g}^*, \mathbf{w} \rangle = \max_{\mathbf{u} \in K_{\rho,s} \cap S^{N-1}} \langle \mathbf{g}^*, \mathbf{u} \rangle$$

$$\leq \min_{\mathbf{z} \in K_{\rho,s}^*} \|\mathbf{g}^* + \mathbf{z}\|_2,$$

where the inequality uses (B.39). The claim follows by taking expectations. □

The previous results motivate us to investigate the dual cone $K_{\rho,s}^*$.

Lemma 9.32. *The dual cone of* $K_{\rho,s}$ *defined in (9.53) satisfies*

$$K_{\rho,s}^* \supset Q_{\rho,s} := \bigcup_{t \geq 0} \{\mathbf{z} \in \mathbb{R}^N : z_\ell = t, \ell \in [s], z_\ell \geq -\rho t, \ell = s+1, \ldots, N\}.$$

Proof. Take a vector $\mathbf{z} \in Q_{\rho,s}$. Then, for any $\mathbf{u} \in K_{\rho,s}$,

$$\langle \mathbf{z}, \mathbf{u} \rangle = \sum_{\ell=1}^{s} z_\ell u_\ell + \sum_{\ell=s+1}^{N} z_\ell u_\ell \geq t \sum_{\ell=1}^{s} u_\ell - t\rho \sum_{\ell=s+1}^{N} u_\ell \geq 0. \qquad (9.54)$$

Therefore, $\mathbf{z} \in K_{\rho,s}^*$.   □

We are now in a position to estimate the Gaussian widths of $T_{\rho,s} \cap S^{N-1}$.

Proposition 9.33. *The Gaussian width of $T_{\rho,s} \cap S^{N-1}$ satisfies*

$$\ell(T_{\rho,s} \cap S^{N-1}) \leq \sqrt{2s \ln(eN/s)} \left(1 + \rho^{-1} + D(s/N) \right),$$

where D is the function in (9.50).

Proof. By Proposition 9.31, it suffices to estimate $E := \mathbb{E} \min_{\mathbf{z} \in K_{\rho,s}^*} \|\mathbf{g}^* + \mathbf{z}\|_2$. Replacing $K_{\rho,s}^*$ by its subset $Q_{\rho,s}$ from Lemma 9.32 yields

$$E \leq \mathbb{E} \min_{\mathbf{z} \in Q_{\rho,s}} \|\mathbf{g}^* + \mathbf{z}\|_2$$

$$\leq \mathbb{E} \min_{\substack{t \geq 0 \\ z_\ell \geq -\rho t, \ell = s+1,\ldots,N}} \sqrt{\sum_{\ell=1}^{s} (g_\ell^* + t)^2 + \sum_{\ell=s+1}^{N} (g_\ell^* + z_\ell)^2}.$$

Consider a fixed $t \geq 0$. Then

$$E \leq \mathbb{E} \left[\sum_{\ell=1}^{s} (g_\ell^* + t)^2 \right]^{1/2} + \mathbb{E} \left[\min_{z_\ell \geq -\rho t} \sum_{\ell=s+1}^{N} (g_\ell^* + z_\ell)^2 \right]^{1/2}$$

$$\leq \mathbb{E} \sqrt{\sum_{\ell=1}^{s} (g_\ell^*)^2} + t\sqrt{s} + \mathbb{E} \left[\sum_{\ell=s+1}^{N} S_{\rho t}(g_\ell^*)^2 \right]^{1/2}, \qquad (9.55)$$

where $S_{\rho t}$ is the soft thresholding operator (B.18). It follows from Hölder's inequality, from the fact that the sum below is over the $N - s$ smallest elements of the sequence $(S_{\rho t}(g_\ell))_\ell$, and from (9.38) that the third term in (9.55) can be estimated by

$$\mathbb{E} \left[\sum_{\ell=s+1}^{N} S_{\rho t}(g_\ell^*)^2 \right]^{1/2} \leq \left[\mathbb{E} \sum_{\ell=s+1}^{N} S_{\rho t}(g_\ell)^2 \right]^{1/2} \leq \sqrt{(N-s) \sqrt{\frac{2}{\pi e}} \frac{e^{-(\rho t)^2/2}}{(\rho t)^2}}.$$

It remains to estimate the first summand in (9.55). By Hölder's inequality, Proposition 8.2, and Lemma C.5,

$$\mathbb{E}\sqrt{\sum_{\ell=1}^{s}(g_\ell^*)^2} = \mathbb{E}\max_{S\subset[N],\text{card}(S)=s}\|\mathbf{g}_S\|_2 \le \sqrt{\mathbb{E}\max_{S\subset[N],\text{card}(S)=s}\|\mathbf{g}_S\|_2^2}$$

$$\le \sqrt{2\ln\binom{N}{s}} + \sqrt{s} \le \sqrt{2s\ln(eN/s)} + \sqrt{s}$$

$$= \sqrt{2s\ln(eN/s)}\left(1 + \frac{1}{\sqrt{2\ln(eN/s)}}\right).$$

Altogether we have derived that

$$E \le \sqrt{2s\ln(eN/s)}\left(1 + \frac{1}{\sqrt{2\ln(eN/s)}}\right) + t\sqrt{s} + \sqrt{(N-s)\sqrt{\frac{2}{\pi e}}\frac{e^{-(\rho t)^2/2}}{(\rho t)^2}}.$$

We choose $t = \rho^{-1}\sqrt{2\ln(eN/s)}$ to obtain

$$E \le \sqrt{2s\ln(eN/s)}\left(1 + \rho^{-1} + \frac{1}{\sqrt{2\ln(eN/s)}}\right)$$

$$+ \sqrt{(N-s)\sqrt{\frac{2}{\pi e}}\frac{s}{2eN\ln(eN/s)}}$$

$$\le \sqrt{2s\ln(eN/s)}\left(1 + \rho^{-1} + \frac{1}{\sqrt{2\ln(eN/s)}} + \frac{1}{(8\pi e^3)^{1/4}\ln(eN/s)}\right).$$

This completes the proof by the definition (9.50) of the function D. $\square$

In view of Theorem 4.12, the uniform recovery result of Theorem 9.29 is now an immediate consequence of the following statement.

Corollary 9.34. *Let* $\mathbf{A} \in \mathbb{R}^{m\times N}$ *be a random draw of a Gaussian matrix. Let* $s < N, \rho \in (0,1], \varepsilon \in (0,1)$ *such that*

$$\frac{m^2}{m+1} \ge 2s\ln(eN/s)\left(1 + \rho^{-1} + D(s/N) + \sqrt{\frac{\ln(\varepsilon^{-1})}{s\ln(eN/s)}}\right)^2.$$

Then with probability at least $1 - \varepsilon$ *the matrix* $\mathbf{A}$ *satisfies the stable null space property of order* s *with constant* ρ.

Proof. Taking into account the preceding results, the proof is a variation of the one of Theorem 9.18; see also Exercise 9.7. $\square$

9.5   Relation to Johnson–Lindenstrauss Embeddings

The Johnson–Lindenstrauss lemma is not a statement connected with sparsity per se, but it is closely related to the concentration inequality (9.6) for subgaussian matrices leading to the restricted isometry property. Assume that we are given a finite set $\{x_1, \ldots, x_M\} \subset \mathbb{R}^N$ of points. If N is large, then it is usually computationally expensive to process these points. Therefore, it is of interest to project them into a lower-dimensional space while preserving essential geometrical properties such as mutual distances. The Johnson–Lindenstrauss lemma states that such lower-dimensional embeddings exist. For simplicity, we state our results for the real case, but note that it has immediate extensions to $\mathbb{C}^N$ (for instance, simply by identifying $\mathbb{C}^N$ with $\mathbb{R}^{2N}$).

Lemma 9.35. *Let* $x_1, \ldots, x_M \in \mathbb{R}^N$ *be an arbitrary set of points and* $\eta > 0$. *If* $m > C\eta^{-2} \ln(M)$, *then there exists a matrix* $B \in \mathbb{R}^{m \times N}$ *such that*

$$(1 - \eta)\|x_j - x_\ell\|_2^2 \leq \|B(x_j - x_\ell)\|_2^2 \leq (1 + \eta)\|x_j - x_\ell\|_2^2$$

for all $j, \ell \in [M]$. *The constant* $C > 0$ *is universal.*

Proof. Considering the set

$$E = \{x_j - x_\ell : 1 \leq j < \ell \leq M\}$$

of cardinality $\text{card}(E) \leq M(M - 1)/2$, it is enough to show the existence of B such that

$$(1 - \eta)\|x\|_2^2 \leq \|Bx\|_2^2 \leq (1 + \eta)\|x\|_2^2 \quad \text{for all } x \in E. \tag{9.56}$$

We take $B = \frac{1}{\sqrt{m}} A \in \mathbb{R}^{m \times N}$, where A is a random draw of a subgaussian matrix. Then (9.6) implies that, for any fixed $x \in E$ and an appropriate constant $\tilde{c}$,

$$\mathbb{P}\left(\|Bx\|_2^2 - \|x\|_2^2 \geq \eta\|x\|_2^2 \right) \leq 2\exp(-\tilde{c}m\eta^2).$$

By the union bound, (9.56) holds simultaneously for all $x \in E$ with probability at least

$$1 - M^2 e^{-\tilde{c}m\eta^2}.$$

We take $m \geq \tilde{c}^{-1}\eta^{-2} \ln(M^2/\varepsilon)$ so that $M^2 \exp(-\tilde{c}m\eta^2) \leq \varepsilon$. Then inequality (9.56) holds with probability at least $1 - \varepsilon$, and existence of a map with the desired property is established when $\varepsilon < 1$. Considering the limit $\varepsilon \to 1$, this gives the claim with $C = 2\tilde{c}^{-1}$. $\qquad\square$

This proof shows that the concentration inequality (9.6) is closely related to the Johnson–Lindenstrauss lemma. As (9.6) implies the restricted isometry property by Theorem 9.11, one may even say that in this sense the Johnson–Lindenstrauss lemma implies the restricted isometry property. We show next that in some sense also the converse holds: Given a matrix $\mathbf{A}$ satisfying the restricted isometry property, randomization of the column signs of $\mathbf{A}$ provides a Johnson–Lindenstrauss embedding via (9.56). In what follows, for a vector $\mathbf{u}$, the symbol $\mathbf{D_u} = \mathrm{diag}(\mathbf{u})$ stands for the diagonal matrix with $\mathbf{u}$ on the diagonal.

Theorem 9.36. *Let $E \subset \mathbb{R}^N$ be a finite point set of cardinality $\mathrm{card}(E) = M$. For $\eta, \varepsilon \in (0,1)$, let $\mathbf{A} \in \mathbb{R}^{m \times N}$ with restricted isometry constant satisfying $\delta_{2s} \leq \eta/4$ for some $s \geq 16 \ln(4M/\varepsilon)$ and let $\epsilon = (\epsilon_1, \ldots, \epsilon_N)$ be a Rademacher sequence. Then, with probability exceeding $1 - \varepsilon$,*

$$(1 - \eta)\|\mathbf{x}\|_2^2 \leq \|\mathbf{AD}_\epsilon \mathbf{x}\|_2^2 \leq (1 + \eta)\|\mathbf{x}\|_2^2 \quad \textit{for all } \mathbf{x} \in E.$$

Remark 9.37. (a) Without randomization of the column signs, the theorem is false. Indeed, there is no assumption on the point set E. Therefore, if we choose the points of E from the kernel of the matrix $\mathbf{A}$ (which is not assumed random here), there is no chance for the lower bound to hold. Randomization of the column signs ensures that the probability of E intersecting the kernel of $\mathbf{AD}_\epsilon$ is very small.

(b) There is no direct condition on the embedding dimension m in the previous theorem, but the requirement $\delta_{2s} \leq \eta/4$ for $\mathbf{A} \in \mathbb{R}^{m \times N}$ imposes an indirect condition. For "good" matrices, one expects that this requires $m \geq C\eta^{-2}s\ln^\alpha(N)$, say, so that the condition on s in the previous result turns into $m \geq C\eta^{-2}\ln^\alpha(N)\ln(M/\varepsilon)$. In comparison to the original Johnson–Lindenstrauss lemma (Lemma 9.35), we only observe an additional factor of $\ln^\alpha(N)$.

(c) The theorem allows one to derive Johnson–Lindenstrauss embeddings for other types of random matrices not being subgaussian. In Chap. 12, we will indeed deal with different types of matrices $\mathbf{A}$ satisfying the restricted isometry property, for instance, random partial Fourier matrices, so that $\mathbf{AD}_\epsilon$ will provide a Johnson–Lindenstrauss embedding. It is currently not known how to show the Johnson–Lindenstrauss embedding directly for such type of matrices.

Proof. Without loss of generality, we may assume that all $\mathbf{x} \in E$ are normalized as $\|\mathbf{x}\|_2 = 1$. Consider a fixed $\mathbf{x} \in E$. Similarly to the proof of Theorem 6.9, we partition $[N]$ into blocks of size s according to the nonincreasing rearrangement of $\mathbf{x}$. More precisely, $S_1 \subset [N]$ is an index set of s largest absolute entries of the vector $\mathbf{x}$, $S_2 \subset [N] \setminus S_1$ is an index set of s largest absolute entries of $\mathbf{x}$ in $[N] \setminus S_1$, and so on. As usual, the expression $\mathbf{x}_S$ (and similar ones below) can mean both the restriction of the vector $\mathbf{x}$ to the indices in S and the vector whose entries are set to zero outside of S. We write

$$\|\mathbf{AD}_\epsilon \mathbf{x}\|_2^2 = \|\mathbf{AD}_\epsilon \sum_{j\geq 1} \mathbf{x}_{S_j}\|_2^2$$

$$= \sum_{j\geq 1} \|\mathbf{AD}_\epsilon \mathbf{x}_{S_j}\|_2^2 + 2\langle \mathbf{AD}_\epsilon \mathbf{x}_{S_1}, \mathbf{AD}_\epsilon \mathbf{x}_{\overline{S_1}}\rangle + \sum_{\substack{j,\ell \geq 2 \\ j\neq \ell}} \langle \mathbf{AD}_\epsilon \mathbf{x}_{S_j}, \mathbf{AD}_\epsilon \mathbf{x}_{S_\ell}\rangle. \quad (9.57)$$

Since $\mathbf{A}$ possesses the restricted isometry property with $\delta_s \leq \eta/4$ and since $\|\mathbf{D}_\epsilon \mathbf{x}_{S_j}\|_2 = \|\mathbf{x}_{S_j}\|_2$, the first term in (9.57) satisfies

$$(1 - \eta/4)\|\mathbf{x}\|_2^2 = (1 - \eta/4) \sum_{j\geq 1} \|\mathbf{x}_{S_j}\|_2^2 \leq \sum_{j\geq 1} \|\mathbf{AD}_\epsilon \mathbf{x}_{S_j}\|_2^2 \leq (1 + \eta/4)\|\mathbf{x}\|_2^2.$$

To estimate the second term in (9.57), we consider the random variable

$$X := \langle \mathbf{AD}_\epsilon \mathbf{x}_{S_1}, \mathbf{AD}_\epsilon \mathbf{x}_{\overline{S_1}}\rangle = \langle \mathbf{v}, \epsilon_{\overline{S_1}}\rangle = \sum_{\ell \notin S_1} \epsilon_\ell v_\ell,$$

where $\mathbf{v} \in \mathbb{R}^{\overline{S_1}}$ is given by

$$\mathbf{v} = \mathbf{D}_{\mathbf{x}_{\overline{S_1}}} \mathbf{A}_{\overline{S_1}}^* \mathbf{A}_{S_1} \mathbf{D}_{\mathbf{x}_{S_1}} \epsilon_{S_1}.$$

Hereby, we have exploited the fact that $D_\epsilon \mathbf{x} = D_\mathbf{x} \epsilon$. Observing that $\mathbf{v}$ and $\epsilon_{\overline{S_1}}$ are stochastically independent, we aim at applying Hoeffding's inequality (Corollary 7.21). To this end we estimate the 2-norm of the vector $\mathbf{v}$ as

$$\|\mathbf{v}\|_2 = \sup_{\|\mathbf{z}\|_2 \leq 1} \langle \mathbf{z}, \mathbf{v}\rangle = \sup_{\|\mathbf{z}\|_2 \leq 1} \sum_{j\geq 2} \langle \mathbf{z}_{S_j}, \mathbf{D}_{\mathbf{x}_{S_j}} \mathbf{A}_{S_j}^* \mathbf{A}_{S_1} \mathbf{D}_{\epsilon_{S_1}} \mathbf{x}_{S_1}\rangle$$

$$\leq \sup_{\|\mathbf{z}\|_2 \leq 1} \sum_{j\geq 2} \|\mathbf{z}_{S_j}\|_2 \|\mathbf{D}_{\mathbf{x}_{S_j}} \mathbf{A}_{S_j}^* \mathbf{A}_{S_1} \mathbf{D}_{\epsilon_{S_1}}\|_{2\to 2} \|\mathbf{x}_{S_1}\|_2$$

$$\leq \sup_{\|\mathbf{z}\|_2 \leq 1} \sum_{j\geq 2} \|\mathbf{A}_{S_j}^* \mathbf{A}_{S_1}\|_{2\to 2} \|\mathbf{z}_{S_j}\|_2 \|\mathbf{x}_{S_j}\|_\infty \|\mathbf{x}_{S_1}\|_2,$$

where we have used that $\|\mathbf{D}_\mathbf{u}\|_{2\to 2} = \|\mathbf{u}\|_\infty$ and $\|\epsilon\|_\infty = 1$. In view of the construction of $S_1, S_2, \ldots$, it follows from Lemma 6.10 that $\|\mathbf{x}_{S_j}\|_\infty \leq s^{-1/2}\|\mathbf{x}_{S_{j-1}}\|_2$. Moreover, $\|\mathbf{x}_{S_1}\|_2 \leq \|\mathbf{x}\|_2 \leq 1$ and $\|\mathbf{A}_{S_j}^* \mathbf{A}_{S_1}\|_{2\to 2} \leq \delta_{2s}$ for $j \geq 2$ by Proposition 6.3. We continue our estimation with

$$\|\mathbf{v}\|_2 \leq \frac{\delta_{2s}}{\sqrt{s}} \sup_{\|\mathbf{z}\|_2 \leq 1} \sum_{j\geq 2} \|\mathbf{z}_{S_j}\|_2 \|\mathbf{x}_{S_{j-1}}\|_2$$

$$\leq \frac{\delta_{2s}}{\sqrt{s}} \sup_{\|\mathbf{z}\|_2 \leq 1} \sum_{j\geq 2} \frac{1}{2} \left(\|\mathbf{z}_{S_j}\|_2^2 + \|\mathbf{x}_{S_{j-1}}\|_2^2\right) \leq \frac{\delta_{2s}}{\sqrt{s}},$$

where we have used the inequality $2ab \leq a^2 + b^2$ and the fact that $\sum_{j \geq 2} \|\mathbf{x}_{S_j}\|_2^2 \leq \|\mathbf{x}\|_2^2 = 1$. Applying Hoeffding's inequality (7.29) conditionally on ϵ_{S_1} yields, for any $t > 0$,

$$\mathbb{P}(|X| \geq t) \leq 2 \exp\left(-\frac{t^2 s}{2\delta_{2s}^2}\right) \leq 2 \exp\left(-\frac{8 \, st^2}{\eta^2}\right). \tag{9.58}$$

Next we consider the third term in (9.57), which can be written as

$$Y := \sum_{\substack{j,\ell \geq 2 \\ j \neq \ell}} \langle \mathbf{AD}_\epsilon \mathbf{x}_{S_j}, \mathbf{AD}_\epsilon \mathbf{x}_{S_\ell} \rangle = \sum_{j,\ell = s+1}^{N} \epsilon_j \epsilon_\ell B_{j,\ell} = \boldsymbol{\epsilon}^* \mathbf{B} \boldsymbol{\epsilon},$$

where $\mathbf{B} \in \mathbb{R}^{N \times N}$ is a symmetric matrix with zero diagonal given entrywise by

$$B_{j,\ell} = \begin{cases} x_j \mathbf{a}_j^* \mathbf{a}_\ell x_\ell & \text{if } j, \ell \in \overline{S_1} \text{ and } j, \ell \text{ belong to different blocks } S_k, \\ 0 & \text{otherwise.} \end{cases}$$

Here, the $\mathbf{a}_j$, $j \in [N]$, denote the columns of $\mathbf{A}$ as usual. We are thus led to estimate the tail of a Rademacher chaos, which by Proposition 8.13 requires a bound on the spectral and on the Frobenius norm of $\mathbf{B}$. By symmetry of $\mathbf{B}$, the spectral norm can be estimated by

$$\|\mathbf{B}\|_{2 \to 2} = \sup_{\|\mathbf{z}\|_2 \leq 1} \langle \mathbf{Bz}, \mathbf{z} \rangle = \sup_{\|\mathbf{z}\|_2 \leq 1} \sum_{\substack{j,\ell \geq 2 \\ j \neq \ell}} \langle \mathbf{z}_{S_j}, \mathbf{D}_{\mathbf{x}_{S_j}} \mathbf{A}_{S_j}^* \mathbf{A}_{S_\ell} \mathbf{D}_{\mathbf{x}_{S_\ell}} \mathbf{z}_{S_\ell} \rangle$$

$$\leq \sup_{\|\mathbf{z}\|_2 \leq 1} \sum_{\substack{j,\ell \geq 2 \\ j \neq \ell}} \|\mathbf{z}_{S_j}\|_2 \|\mathbf{z}_{S_\ell}\|_2 \|\mathbf{x}_{S_j}\|_\infty \|\mathbf{x}_{S_\ell}\|_\infty \|\mathbf{A}_{S_j}^* \mathbf{A}_{S_\ell}\|_{2 \to 2}$$

$$\leq \delta_{2s} \sup_{\|\mathbf{z}\|_2 \leq 1} \sum_{\substack{j,\ell \geq 2 \\ j \neq \ell}} \|\mathbf{z}_{S_j}\|_2 \|\mathbf{z}_{S_\ell}\|_2 s^{-1/2} \|\mathbf{x}_{S_{j-1}}\|_2 s^{-1/2} \|\mathbf{x}_{S_{\ell-1}}\|_2$$

$$\leq \frac{\delta_{2s}}{4s} \sup_{\|\mathbf{z}\|_2 \leq 1} \sum_{\substack{j,\ell \geq 2 \\ j \neq \ell}} \left(\|\mathbf{x}_{S_{j-1}}\|_2^2 + \|\mathbf{z}_{S_j}\|_2^2 \right) \left(\|\mathbf{x}_{S_{\ell-1}}\|_2^2 + \|\mathbf{z}_{S_\ell}\|_2^2 \right) \leq \frac{\delta_{2s}}{s}.$$

The Frobenius norm obeys the bound

$$\|\mathbf{B}\|_F^2 = \sum_{\substack{j,k \geq 2 \\ j \neq k}} \sum_{i \in S_j} \sum_{\ell \in S_k} (x_i \mathbf{a}_i^* \mathbf{a}_\ell x_\ell)^2 = \sum_{\substack{j,k \geq 2 \\ j \neq k}} \sum_{i \in S_j} x_i^2 \|\mathbf{D}_{\mathbf{x}_{S_k}} \mathbf{A}_{S_k}^* \mathbf{a}_i\|_2^2$$

$$\leq \sum_{\substack{j,k \geq 2 \\ j \neq k}} \sum_{i \in S_j} x_i^2 \|\mathbf{x}_{S_k}\|_\infty^2 \|\mathbf{A}_{S_k}^* \mathbf{a}_i\|_2^2 \leq \delta_{2s}^2 \sum_{\substack{j,k \geq 2 \\ j \neq k}} \|\mathbf{x}_{S_j}\|_2^2 s^{-1} \|\mathbf{x}_{S_k}\|_2^2 \leq \frac{\delta_{2s}^2}{s}.$$

Hereby, we have made use of $\|A_{S_k}^* a_i\|_2 = \|A_{S_k}^* a_i\|_{2\to 2} \le \delta_{s+1} \le \delta_{2s}$; see Proposition 6.5. It follows from Proposition 8.13 that the tail of the third term in (9.57) can be estimated, for any $r > 0$, by

$$\mathbb{P}(|Y| \ge r) \le 2\exp\left(-\min\left\{\frac{3\,r^2}{128\,\|B\|_F^2}, \frac{r}{32\,\|B\|_{2\to 2}}\right\}\right)$$

$$\le 2\exp\left(-\min\left\{\frac{3\,sr^2}{128\,\delta_{2s}^2}, \frac{sr}{32\,\delta_{2s}}\right\}\right).$$

$$\le 2\exp\left(-s\min\left\{\frac{3\,r^2}{8\,\eta^2}, \frac{r}{8\,\eta}\right\}\right). \tag{9.59}$$

Now, with the choice $t = \eta/8$ in (9.58) and $r = \eta/2$ in (9.59) and taking into account (9.57), we arrive at

$$(1-\eta)\|x\|_2^2 \le \|AD_\epsilon x\|_2^2 \le (1+\eta)\|x\|_2^2 \tag{9.60}$$

for our fixed $x \in E$ with probability at least

$$1 - 2\exp(-s/8) - 2\exp(-s\min\{3/32, 1/16\}) \ge 1 - 4\exp(-s/16).$$

Taking the union bound over all $x \in E$ shows that (9.60) holds for all $x \in E$ simultaneously with probability at least

$$1 - 4M\exp(-s/16) \ge 1 - \varepsilon$$

under the condition $s \ge 16\ln(4M/\varepsilon)$. This concludes the proof. $\square$

Notes

Even though the papers [299] by Kashin and [219] by Garnaev and Gluskin did not prove the RIP, they already contained in substance the arguments needed to derive it for Gaussian and Bernoulli matrices. Section 9.1 follows the general idea of the paper [351] by Mendelson et al., and independently developed in [31] by Baraniuk et al. There, however, the restricted isometry property is considered without squares on the ℓ_2-norms. As a result, the proof given here is slightly different. Similar techniques were also used in extensions [414], including the D-RIP [86] covered in Exercise 9.10 and the corresponding notion of the restricted isometry property in low-rank matrix recovery [89, 418]; see Exercise 9.12. Candès and Tao were the first to show the restricted isometry property for Gaussian matrices in [97]. They essentially followed the approach given in Sect. 9.3. They relied on the condition number estimate for Gaussian random matrices of Theorem 9.26. The proof method of the latter, based on Slepian's and Gordon's lemmas as well as on concentration of measure, follows [141].

The nonuniform recovery result of Theorem 9.16 has been shown in [21]; see also [91] for a very similar approach. The accurate estimate of the required number of samples in the Gaussian case (Theorem 9.18) appeared in a slightly different form in [108], where far-reaching extensions to other situations such as low-rank matrix recovery are also treated. The estimate of the null space property for Gaussian random matrices (Theorem 9.29) has not appeared elsewhere in this form. Similar ideas, however, were used in [433, 454].

The escape through the mesh theorem (Theorem 9.21) is essentially due to Gordon [235], where it appeared with slightly worse constants. It was used first in compressed sensing by Rudelson and Vershynin in [433]; see also [108, 454].

The Johnson–Lindenstrauss lemma appeared for the first time in [295]. A different proof was given in [136]. Theorem 9.36 on the relation of the restricted isometry property to the Johnson–Lindenstrauss lemma was shown by Krahmer and Ward in [309].

Random matrices were initially introduced in the context of statistics by Wishart and in mathematical physics by Wigner. There is a large body of literature on the asymptotic analysis of the spectrum of random matrices when the matrix dimension tends to infinity. A well-known result in this context states that the empirical distribution of Wigner random matrices (Hermitian random matrices with independent entries up to symmetries) converges to the famous semicircle law. We refer to the monographs [13, 25] for further information on asymptotic random matrix theory.

The methods employed in this chapter belong to the area of nonasymptotic random matrix theory [435, 501], which considers spectral properties of random matrices in fixed (but usually large) dimension. Rudelson and Vershynin [434] exploited methods developed in compressive sensing (among other techniques) and established a result on the smallest singular value of square Bernoulli random matrices, which was an open conjecture for a long time. By distinguishing the action of the matrix on compressible and on incompressible vectors, they were able to achieve their breakthrough. The compressible vectors are handled in the same way as in the proof of the restricted isometry property for rectangular random matrices in Sect. 9.1.

Sparse recovery with Gaussian matrices via polytope geometry. Donoho and Tanner [154, 165–167] approach the analysis of sparse recovery via ℓ_1-minimization using Gaussian random matrices through the geometric characterization of Corollary 4.39. They consider an asymptotic scenario where the dimension N tends to infinity and where $m = m_N$ and $s = s_N$ satisfy

$$\lim_{N\to\infty} \frac{m_N}{N} = \delta \quad \text{and} \quad \lim_{N\to\infty} \frac{s_N}{m_N} = \rho$$

for some $\delta, \rho \in [0, 1]$. They show that there exist thresholds that separate regions in the plane $[0, 1]^2$ of parameters (δ, ρ), where recovery succeeds and recovery

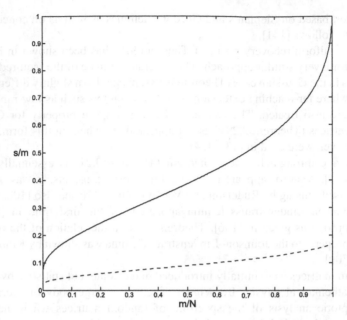

Fig. 9.1 Strong threshold $\rho_S = \rho_S(\delta)$ (*dashed curve*), weak threshold $\rho_W = \rho_W(\delta)$ (*solid curve*), $\delta = m/N$, $\rho = s/m$

fails with probability tending to 1 as $N \to \infty$. In other words, a phase transition phenomenon occurs for high dimensions N. They distinguish a strong threshold $\rho_S = \rho_S(\delta)$ and a weak threshold $\rho_W = \rho_W(\delta)$.

In our terminology, the strong threshold corresponds to uniform recovery via ℓ_1-minimization. In the limit as $N \to \infty$, if $s \leq \rho m$ with $\rho < \rho_S(\delta)$ and $\delta = m/N$, then recovery of all s-sparse vectors is successful with high probability. Moreover, if $s \geq \rho m$ with $\rho > \rho_S(\delta)$, then recovery of all s-sparse vectors fails with high probability.

The weak threshold corresponds to nonuniform recovery. (The formulation in [154, 165–167] is slightly different than our notion of nonuniform recovery, but for Gaussian random matrices, both notions are equivalent.) In the limit as $N \to \infty$, if $s \leq \rho m$ with $\rho < \rho_W(\delta)$ and $\delta = m/N$, then a fixed s-sparse vector is recovered from $\mathbf{y} = \mathbf{A}\mathbf{x}$ via ℓ_1-minimization with high probability using a draw of an $m \times N$ Gaussian random matrix $\mathbf{A}$. Conversely, if $s \geq \rho m$ with $\rho > \rho_W(\delta)$, then ℓ_1-minimization fails to recover a given s-sparse vector from $\mathbf{y} = \mathbf{A}\mathbf{x}$ with high probability.

Unfortunately, no closed forms for the functions ρ_W and ρ_S are available. Nevertheless, Donoho and Tanner [154, 167] provide complicated implicit expressions and compute these functions numerically; see Fig. 9.1. Moreover, they derive the asymptotic behavior of $\rho_W(\delta)$ and $\rho_S(\delta)$ when $\delta \to 0$, that is, in the relevant scenario when m is significantly smaller than N. These are

$$\rho_S(\delta) \sim \frac{1}{2e\ln((\sqrt{\pi}\delta)^{-1})} \qquad \text{and} \qquad \rho_W(\delta) \sim \frac{1}{2\ln(\delta^{-1})} \qquad \delta \to 0.$$

As a consequence, we roughly obtain the following statements for large N:

- *Uniform recovery.* If

$$m > 2es\ln(N/(\sqrt{\pi}m)),$$

then with high probability on the draw of a Gaussian random matrix, every s-sparse vector $\mathbf{x}$ is recovered from $\mathbf{y} = \mathbf{A}\mathbf{x}$ via ℓ_1-minimization. Conversely, if $m < 2es\ln(N/(\sqrt{\pi}m))$, then recovery of all s-sparse vectors fails with high probability.

- *Nonuniform recovery.* If

$$m > 2s\ln(N/m),$$

then a fixed s-sparse vector $\mathbf{x}$ is recovered from $\mathbf{y} = \mathbf{A}\mathbf{x}$ via ℓ_1-minimization with high probability on the draw of a Gaussian random matrix $\mathbf{A}$. Conversely, if $m < 2s\ln(N/m)$, then with high probability, ℓ_1-minimization fails to recover a fixed s-sparse vector using a random draw of a Gaussian matrix.

The involved analysis of Donoho and Tanner builds on the characterization of sparse recovery in Corollary 4.39. Stated in slightly different notation, s-sparse recovery is equivalent to s-neighborliness of the projected polytope $\mathbf{A}B_1^N$: Every set of s vertices of $\mathbf{A}B_1^N$ (not containing antipodal points) spans an $(s-1)$-face of $\mathbf{A}B_1^N$; see also Exercise 4.16. This property is investigated directly using works by Affentranger and Schneider [3] and by Vershik and Sporyshev [498] on random polytopes. Additionally, Donoho and Tanner provide thresholds for the case where it is known a priori that the sparse vector has only nonnegative entries [165,167]. This information can be used as an additional constraint in the ℓ_1-minimization problem and in this case one has to analyze the projected simplex $\mathbf{A}S^N$, where S^N is the standard simplex, that is, the convex hull of the canonical unit vectors and the zero vector.

It is currently not clear whether this approach can be extended to further random matrices such as Bernoulli matrices, although the same weak threshold is observed empirically for a variety of random matrices [168]. For illustration, an empirical phase diagram is shown in Fig. 9.2. It is not clear either whether the polytope approach can cover stability and robustness of reconstruction.

The fact that this analysis provides precise statements about the failure of recovery via ℓ_1-minimization allows one to deduce that the constant 2 in our nonuniform recovery analysis for Gaussian random matrices in Sect. 9.2 is optimal; see (9.23). Moreover, the constant 8 appearing in our analysis of the null space property in Theorem 9.29 (see also (9.51)) is not optimal, but at least it is not too far from the optimal value $2e$. In contrast to the polytope approach, however, Theorem 9.29 also covers the stability of reconstruction.

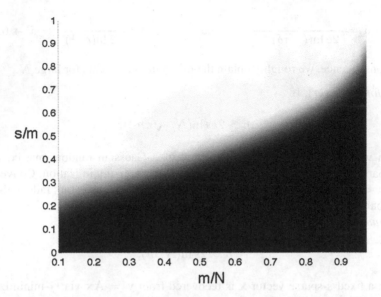

Fig. 9.2 Empirically observed weak threshold. *Black* corresponds to 100 % empirical success probability, *white* to 0 % empirical success probability (Image Courtesy by Jared Tanner, University of Oxford)

A precise phase transition analysis of the restricted isometry constants of Gaussian random matrices has been performed in [24, 53].

Message passing algorithms [162] in connection with Gaussian random matrices also allow for a precise asymptotic analysis.

Exercises

9.1. Coherence of a Bernoulli random matrix.
Let $\mathbf{A} = [\mathbf{a}_1 | \mathbf{a}_2 | \cdots | \mathbf{a}_N]$ be an $m \times N$ Bernoulli matrix. Let μ be the coherence of $m^{-1/2} \mathbf{A}$, i.e., $\mu = m^{-1} \max_{j \neq k} |\langle \mathbf{a}_j, \mathbf{a}_k \rangle|$. Show that

$$\mu \leq 2\sqrt{\frac{\ln(N/\varepsilon)}{m}}$$

with probability at least $1 - \varepsilon^2$.

9.2. Concentration inequality for Gaussian matrices.
Let $\mathbf{A}$ be an $m \times N$ Gaussian random matrix. Show that, for $\mathbf{x} \in \mathbb{R}^N$ and $t \in (0, 1)$,

$$\mathbb{P}\left(|m^{-1} \|\mathbf{A}\mathbf{x}\|_2^2 - \|\mathbf{x}\|_2^2| \geq t \|\mathbf{x}\|_2^2 \right) \leq 2 \exp\left(-m(t^2/4 - t^3/6)\right). \qquad (9.61)$$

Show that this concentration inequality holds as well for a Bernoulli random matrix.

9.3. Smallest singular value of a subgaussian matrix.
Let $\mathbf{B}$ be an $m \times s$ subgaussian random matrix, and let $\sigma_{\min}$ be the smallest singular value of $\frac{1}{\sqrt{m}}\mathbf{B}$. Show that, for $t \in (0, 1)$,

$$\mathbb{P}\left(\sigma_{\min} \leq 1 - c_1\sqrt{\frac{s}{m}} - t\right) \leq 2\exp(-c_2 mt^2).$$

Provide values for the constants $c_1, c_2 > 0$, possibly in terms of $\tilde{c}$ in (9.6).

9.4. Largest singular value via Dudley's inequality.
Let $\mathbf{A}$ be an $m \times s$ (unnormalized) subgaussian random matrix. Use Dudley's inequality (Theorem 8.23) to show that

$$\mathbb{E}\|\mathbf{A}\|_{2 \to 2} \leq C(\sqrt{m} + \sqrt{s}).$$

9.5. Extremal singular values of complex Gaussian matrices.
For $s < m$, let $\mathbf{A}$ be an $m \times s$ random matrix with entries $A_{jk} = g_{jk} + ih_{jk}$ where the (g_{jk}) and (h_{jk}) are independent mean-zero Gaussian random variables of variance 1. Show that the largest singular value $\sigma_{\max}$ and the smallest singular value $\sigma_{\min}$ of the renormalized matrix $\frac{1}{\sqrt{2m}}\mathbf{A}$ satisfy

$$\mathbb{P}(\sigma_{\max} \geq 1 + \sqrt{s/m} + t) \leq e^{-mt^2},$$
$$\mathbb{P}(\sigma_{\min} \leq 1 - \sqrt{s/m} - t) \leq e^{-mt^2}.$$

9.6. Let $\mathbf{A}$ be an $m \times N$ Gaussian random matrix. For $\delta \in (0, 1)$, prove that with high probability the modified restricted isometry property stating that

$$(1 - \delta)\sqrt{m}\|\mathbf{x}\|_2 \leq \|\mathbf{A}\mathbf{x}\|_1 \leq (1 + \delta)\sqrt{m}\|\mathbf{x}\|_2 \qquad \text{for all } s\text{-sparse } \mathbf{x} \in \mathbb{R}^N$$

is fulfilled provided that

$$m \geq c(\delta)\, s\, \ln(eN/s).$$

9.7. Verify Corollary 9.34 in detail.

9.8. Let $\mathbf{A} \in \mathbb{R}^{m \times N}$ be a random matrix satisfying the concentration inequality (9.7). Given $\delta \in (0, 1)$, prove that with probability at least $1 - N^{-c_1}$, the matrix $\mathbf{A}$ satisfies the *homogeneous restricted isometry property*

$$\left(1 - \sqrt{\frac{r}{s}}\delta\right)\|\mathbf{x}\|_2^2 \leq \|\mathbf{A}\mathbf{x}\|_2^2 \leq \left(1 + \sqrt{\frac{r}{s}}\delta\right)\|\mathbf{x}\|_2^2$$

for all r-sparse $\mathbf{x} \in \mathbb{C}^N$ and all $r \leq s$ provided that

$$m \geq c_2\delta^{-2}s\ln(N).$$

9.9. Let $\mathbf{A} \in \mathbb{R}^{m \times N}$ be a random matrix whose columns are independent and uniformly distributed on the sphere S^{m-1}. Show that its restricted isometry constant satisfies $\delta_s < \delta$ with probability at least $1 - \varepsilon$ provided

$$m \geq C\delta^{-2} \left(s \ln(eN/s) + \ln(2\varepsilon^{-1})\right),$$

where $C > 0$ is an appropriate universal constant.

9.10. D-RIP
Let $\mathbf{D} \in \mathbb{R}^{N \times M}$ (the dictionary) with $M \geq N$ and let $\mathbf{A} \in \mathbb{R}^{m \times N}$ (the measurement matrix). The restricted isometry constants δ_s adapted to $\mathbf{D}$ are defined as the smallest constants such that

$$(1 - \delta_s)\|\mathbf{z}\|_2^2 \leq \|\mathbf{Az}\|_2^2 \leq (1 + \delta_s)\|\mathbf{z}\|_2^2$$

for all $\mathbf{z} \in \mathbb{R}^N$ of the form $\mathbf{z} = \mathbf{Dx}$ for some s-sparse $\mathbf{x} \in \mathbb{R}^M$.

If $\mathbf{A}$ is an $m \times N$ subgaussian random matrix, show that the restricted isometry constants adapted to $\mathbf{D}$ of $m^{-1/2}\mathbf{A}$ satisfy $\delta_s \leq \delta$ with probability at least $1 - \varepsilon$ provided that

$$m \geq C\delta^{-2} \left(s \ln(eM/s) + \ln(2\varepsilon^{-1})\right).$$

9.11. Recovery with random Gaussian noise on the measurements.

(a) Let $\mathbf{e} \in \mathbb{R}^m$ be a random vector with independent mean-zero Gaussian entries of variance σ^2. Show that $\|\mathbf{e}\|_2 \leq 2\sigma\sqrt{m}$ with probability at least $1 - e^{-m/2}$.
(b) Let $\mathbf{A} \in \mathbb{R}^{m \times N}$ be a subgaussian random matrix and $\mathbf{e}$ be a random vector with independent mean-zero Gaussian entries of variance σ^2. For an s-sparse vector $\mathbf{x} \in \mathbb{R}^N$, set $\mathbf{y} = \mathbf{Ax} + \mathbf{e}$ and assume that

$$m \geq Cs \ln(eN/s).$$

Let $\mathbf{x}^{\sharp} \in \mathbb{R}^N$ be a solution of

$$\underset{\mathbf{z} \in \mathbb{R}^N}{\text{minimize}} \|\mathbf{z}\|_1 \quad \text{subject to } \|\mathbf{Az} - \mathbf{y}\|_2 \leq 2\sigma\sqrt{m}.$$

Show that

$$\|\mathbf{x} - \mathbf{x}^{\sharp}\|_2 \leq C'\sigma \tag{9.62}$$

with probability at least $1 - \exp(-cm)$. Here, $C, C', c > 0$ are suitable absolute constants.

(c) Let $\mathbf{A} \in \mathbb{R}^{m \times N}$ be a matrix with ℓ_2-normalized columns, i.e., $\|\mathbf{a}_j\|_2 = 1$ for all $j \in [N]$. Let $\mathbf{e} \in \mathbb{R}^m$ be a random vector with independent mean-zero Gaussian entries of variance σ^2. Show that

$$\|\mathbf{A}^*\mathbf{e}\|_\infty \leq 2\sigma\sqrt{\ln(2N)}$$

with probability at least $1 - 1/(2N)$.

(d) Let $\mathbf{A} \in \mathbb{R}^{m \times N}$ be a subgaussian random matrix and $\mathbf{e}$ be a random vector with independent mean-zero Gaussian entries of variance σ^2. For an s-sparse vector $\mathbf{x} \in \mathbb{R}^N$, set $\mathbf{y} = \mathbf{A}\mathbf{x} + \mathbf{e}$ and assume that

$$m \geq Cs\ln(eN/s).$$

Let $\mathbf{x}^\sharp \in \mathbb{R}^N$ be a minimizer of the Dantzig selector

$$\underset{\mathbf{z} \in \mathbb{R}^N}{\text{minimize}}\,\|\mathbf{z}\|_1 \quad \text{subject to } \|\mathbf{A}^*(\mathbf{A}\mathbf{z} - \mathbf{y})\|_\infty \leq 2\sigma\sqrt{m\ln(2N)}.$$

Show that

$$\|\mathbf{x} - \mathbf{x}^\sharp\|_2 \leq C'\sigma\sqrt{\frac{s\ln(2N)}{m}} \tag{9.63}$$

with probability at least $1 - N^{-1}$. Here, $C, C' > 0$ are suitable absolute constants. Hint: Exercise 6.18.

(e) Extend (d) to subgaussian random matrices $\mathbf{A} \in \mathbb{R}^{m \times N}$ and mean-zero subgaussian random vectors with independent entries of variance σ^2.

The subtle difference between the estimates (9.62) and (9.63) consists in the fact that once $m \geq Cs\ln(2N)$, then the right-hand side of (9.63) decreases with growing m, while (9.62) remains constant. For this reason, the Dantzig selector is often preferred over quadratically constrained ℓ_1-minimization in a statistical context.

9.12. Rank-RIP for subgaussian measurement maps.

For a measurement map $\mathcal{A} : \mathbb{R}^{n_1 \times n_2} \to \mathbb{R}^m$, the rank restricted isometry constant δ_s is defined as the smallest number such that

$$(1 - \delta_s)\|\mathbf{X}\|_F^2 \leq \|\mathcal{A}(\mathbf{X})\|_2^2 \leq (1 + \delta_s)\|\mathbf{X}\|_F^2 \quad \text{for all } \mathbf{X} \text{ of rank at most } s;$$

see also Exercise 6.25. A measurement map $\mathcal{A}$ is called subgaussian if all the entries $\mathcal{A}_{jk\ell}$ in the representation

$$\mathcal{A}(\mathbf{X})_j = \sum_{k,\ell} \mathcal{A}_{jk\ell} X_{k\ell}$$

are independent mean-zero subgaussian random variables with variance 1 and the same subgaussian parameter c. Show that the restricted isometry constants of $1/\sqrt{m}\mathcal{A}$ satisfy $\delta_s \leq \delta$ with probability at least $1 - \varepsilon$ provided that

$$m \geq C_\delta\left(s(n_1 + n_2) + \ln(2\varepsilon^{-1})\right),$$

and explain why this bound is optimal. As a first step show that the covering numbers of the set $D_s = \{\mathbf{X} \in \mathbb{R}^{n_1 \times n_2} : \|\mathbf{X}\|_F \leq 1, \text{rank}(\mathbf{X}) \leq s\}$ satisfy

$$\mathcal{N}(D_s, \|\cdot\|_F, \rho) \leq (1 + 6/\rho)^{(n_1+n_2+1)s}.$$

Hint: Use the (reduced) singular value decomposition $\mathbf{X} = \mathbf{UDV}^*$, where $\mathbf{U} \in \mathbb{R}^{n_1 \times s}$ and $\mathbf{V} \in \mathbb{R}^{n_2 \times s}$ have orthonormal columns and $\mathbf{D} \in \mathbb{R}^{s \times s}$ is diagonal. Cover the sets of the three components $\mathbf{U}, \mathbf{V}, \mathbf{D}$ separately with respect to suitable norms.

Chapter 10
Gelfand Widths of ℓ_1-Balls

In this chapter, we make a detour to the geometry of ℓ_1^N in order to underline the optimality of random sensing in terms of the number of measurements. In particular, we show that the minimal number of measurements required for stable recovery via any method is $m \geq Cs\ln(eN/s)$, which matches the bound for random measurements and ℓ_1-minimization stated in Theorem 9.12. In Sect. 10.1, we introduce several notions of widths and show that Gelfand widths are closely related to the worst-case reconstruction error of compressive sensing methods over classes of vectors. In Sect. 10.2, we establish upper and lower bounds for the Gelfand widths of ℓ_1-balls. In fact, the methods developed so far turn out to be appropriate tools to tackle this venerable problem originating from pure mathematics. We give further instances of methods from compressive sensing being used successfully in Banach space geometry in Sect. 10.3, where we establish lower and upper bounds of certain Kolmogorov widths as well as Kashin's decomposition theorem. For convenience, we only consider real vector spaces in this chapter, but straightforward extensions to the complex case are possible.

10.1 Definitions and Relation to Compressive Sensing

We introduce in this section several notions of widths. We start with the classical notion of Gelfand widths.

Definition 10.1. The Gelfand m-width of a subset K of a normed space X is defined as

$$d^m(K, X) := \inf \left\{ \sup_{\mathbf{x} \in K \cap L^m} \|\mathbf{x}\|, \ L^m \text{ subspace of } X \text{ with } \operatorname{codim}(L^m) \leq m \right\}.$$

Since a subspace L^m of X is of codimension at most m if and only if there exists linear functionals $\lambda_1, \ldots, \lambda_m : X \to \mathbb{R}$ in the dual space X^* such that

S. Foucart and H. Rauhut, *A Mathematical Introduction to Compressive Sensing*, Applied and Numerical Harmonic Analysis, DOI 10.1007/978-0-8176-4948-7_10, © Springer Science+Business Media New York 2013

$$L^m = \{\mathbf{x} \in X : \lambda_i(\mathbf{x}) = 0 \text{ for all } i \in [m]\} = \ker \mathbf{A},$$

where $\mathbf{A} : X \to \mathbb{R}^m, \mathbf{x} \mapsto [\lambda_1(\mathbf{x}), \ldots, \lambda_m(\mathbf{x})]^\top$, we also have the representation

$$d^m(K, X) = \inf \left\{ \sup_{\mathbf{x} \in K \cap \ker \mathbf{A}} \|\mathbf{x}\|, \ \mathbf{A} : X \to \mathbb{R}^m \text{ linear} \right\}.$$

We readily observe that the sequence $(d^m(K, X))_{m \geq 0}$ is nonincreasing. Its first term is $d^0(K, X) = \sup_{\mathbf{x} \in K} \|\mathbf{x}\|$. If $N := \dim(X)$ is finite, then $d^m(K, X) = \inf_{\mathbf{A} \in \mathbb{R}^{m \times N}} \sup_{\mathbf{x} \in K \cap \ker \mathbf{A}} \|\mathbf{x}\|$ and if K contains the zero vector, then we have $d^m(K, X) = 0$ for all $m \geq N$. If otherwise $\dim(X)$ is infinite, then we have $\lim_{m \to \infty} d^m$ $(K, X) = 0$ as soon as the set K is compact—see Exercise 10.2.

We now highlight the pivotal role of Gelfand widths in compressive sensing. To do so, we show that they are comparable to quantities that measure the worst-case reconstruction errors of optimal measurement/reconstruction schemes. We call the first of these quantities the (nonadaptive) compressive widths.

Definition 10.2. The compressive m-width of a subset K of a normed space X is defined as

$$E^m(K, X) := \inf \left\{ \sup_{\mathbf{x} \in K} \|\mathbf{x} - \Delta(\mathbf{A}\mathbf{x})\|, \ \mathbf{A} : X \to \mathbb{R}^m \text{ linear}, \Delta : \mathbb{R}^m \to X \right\}.$$

We emphasize that the reconstruction map $\Delta : \mathbb{R}^m \to X$ used in this definition is kept arbitrary. No assumptions are made on tractability and even NP-hard algorithms including ℓ_0-minimization are allowed. The measurement scheme associated to the linear map $\mathbf{A}$ is *nonadaptive*, in the sense that the m linear functionals $\lambda_1, \ldots, \lambda_m \in X^*$ given by $\mathbf{A}\mathbf{x} = [\lambda_1(\mathbf{x}), \ldots, \lambda_m(\mathbf{x})]^\top$ are chosen once and for all (which is the situation usually encountered in compressive sensing). In contrast, we may also consider the *adaptive* setting, where the choice of a measurement depends on the result of previous measurements according to a specific rule. In this way, the measurement scheme is represented by the *adaptive* map $F : X \to \mathbb{R}^m$ defined by

$$F(\mathbf{x}) = \begin{bmatrix} \lambda_1(\mathbf{x}) \\ \lambda_{2;\lambda_1(\mathbf{x})}(\mathbf{x}) \\ \vdots \\ \lambda_{m;\lambda_1(\mathbf{x}),\ldots,\lambda_{m-1}(\mathbf{x})}(\mathbf{x}) \end{bmatrix}, \tag{10.1}$$

where the functionals $\lambda_1, \lambda_{2;\lambda_1(\mathbf{x})}, \ldots, \lambda_{m;\lambda_1(\mathbf{x}),\ldots,\lambda_{m-1}(\mathbf{x})}$ are all linear and $\lambda_{\ell;\lambda_1(\mathbf{x}),\ldots,\lambda_{\ell-1}(\mathbf{x})}$ is allowed to depend on the previous evaluations $\lambda_1(\mathbf{x}), \ldots,$ $\lambda_{\ell-1;\lambda_1(\mathbf{x}),\ldots,\lambda_{\ell-2}(\mathbf{x})}$. This leads to the introduction of the adaptive compressive widths.

Definition 10.3. The adaptive compressive m-width of a subset K of a normed space X is defined as

$$E_{\mathrm{ada}}^m(K, X)$$

$$:= \inf \left\{ \sup_{\mathbf{x} \in K} \|\mathbf{x} - \Delta(F(\mathbf{x}))\|, F : X \to \mathbb{R}^m \text{ adaptive}, \Delta : \mathbb{R}^m \to X \right\}. \quad (10.2)$$

The intuitive expectation that adaptivity improves the performance of the measurement/reconstruction scheme is false, at least when considering worst cases over K. The following theorem indeed shows that, under some mild conditions, the nonadaptive and the adaptive compressive sensing widths are equivalent and that they are both comparable to the Gelfand widths.

Theorem 10.4. *If K is a subset of a normed space X, then*

$$E_{\mathrm{ada}}^m(K, X) \le E^m(K, X).$$

If the subset K satisfies $-K = K$, then

$$d^m(K, X) \le E_{\mathrm{ada}}^m(K, X).$$

If the set K further satisfies $K + K \subset a\,K$ for some positive constant a, then

$$E^m(K, X) \le a\,d^m(K, X).$$

Proof. The first inequality is straightforward, because any linear measurement map $\mathbf{A} : X \to \mathbb{R}^m$ can be considered adaptive.

Let us now assume that the set K satisfies $-K = K$. We consider an adaptive map $F : X \to \mathbb{R}^m$ of the form (10.1) and a reconstruction map $\Delta : \mathbb{R}^m \to X$. We define the linear map $\mathbf{A} : X \to \mathbb{R}^m$ by $\mathbf{A}(\mathbf{x}) = [\lambda_1(\mathbf{x}), \lambda_{2;0}(\mathbf{x}), \dots, \lambda_{m;0,\dots,0}(\mathbf{x})]^\top$ and we set $L^m := \ker \mathbf{A}$. Since this is a subspace of X satisfying $\mathrm{codim}(L^m) \le m$, the definition of the Gelfand widths implies

$$d^m(K, X) \le \sup_{\mathbf{v} \in K \cap \ker \mathbf{A}} \|\mathbf{v}\|. \quad (10.3)$$

We notice that, for $\mathbf{v} \in \ker \mathbf{A}$, we have $\lambda_1(\mathbf{v}) = 0$. Then $\lambda_{2;\lambda_1(\mathbf{v})}(\mathbf{v}) = \lambda_{2;0}(\mathbf{v})$, and so on until $\lambda_{m;\lambda_1(\mathbf{v}),\dots,\lambda_{m-1}(\mathbf{v})}(\mathbf{v}) = \lambda_{m;0,\dots,0}(\mathbf{v}) = 0$, so that $F(\mathbf{v}) = \mathbf{0}$. Thus, for any $\mathbf{v} \in K \cap \ker \mathbf{A}$, we observe that

$$\|\mathbf{v} - \Delta(\mathbf{0})\| = \|\mathbf{v} - \Delta(F(\mathbf{v}))\| \le \sup_{\mathbf{x} \in K} \|\mathbf{x} - \Delta(F(\mathbf{x}))\|,$$

and likewise, since $-\mathbf{v} \in K \cap \ker \mathbf{A}$, that

$$\| - \mathbf{v} - \Delta(\mathbf{0})\| = \| - \mathbf{v} - \Delta(F(-\mathbf{v}))\| \leq \sup_{\mathbf{x} \in K} \|\mathbf{x} - \Delta(F(\mathbf{x}))\|.$$

We derive that, for any $\mathbf{v} \in K \cap \ker \mathbf{A}$,

$$\|\mathbf{v}\| = \left\| \frac{1}{2}(\mathbf{v} - \Delta(\mathbf{0})) - \frac{1}{2}(-\mathbf{v} - \Delta(\mathbf{0})) \right\| \leq \frac{1}{2}\|\mathbf{v} - \Delta(\mathbf{0})\| + \frac{1}{2}\| - \mathbf{v} - \Delta(\mathbf{0})\|$$

$$\leq \sup_{\mathbf{x} \in K} \|\mathbf{x} - \Delta(F(\mathbf{x}))\|. \tag{10.4}$$

According to (10.3) and (10.4), we have

$$d^m(K, X) \leq \sup_{\mathbf{x} \in K} \|\mathbf{x} - \Delta(F(\mathbf{x}))\|.$$

The inequality $d^m(K, X) \leq E_{\mathrm{ada}}^m(K, X)$ follows by taking the infimum over all possible F and Δ.

Let us finally also assume that $K + K \subset a\,K$ for some positive constant a. We consider a subspace L^m of the space X with $\mathrm{codim}(L^m) \leq m$. We choose a linear map $\mathbf{A} : X \to \mathbb{R}^m$ such that $\ker \mathbf{A} = L^m$, and we define a map $\Delta : \mathbb{R}^m \to X$ in such a way that

$$\Delta(\mathbf{y}) \in K \cap \mathbf{A}^{-1}(\mathbf{y}) \qquad \text{for all } \mathbf{y} \in \mathbf{A}(K).$$

With this choice, we have

$$E^m(K, X) \leq \sup_{\mathbf{x} \in K} \|\mathbf{x} - \Delta(\mathbf{A}\mathbf{x})\| \leq \sup_{\mathbf{x} \in K} \left[\sup_{\mathbf{z} \in K \cap \mathbf{A}^{-1}(\mathbf{A}\mathbf{x})} \|\mathbf{x} - \mathbf{z}\| \right].$$

For $\mathbf{x} \in K$ and $\mathbf{z} \in K \cap \mathbf{A}^{-1}(\mathbf{A}\mathbf{x})$, we observe that the vector $\mathbf{x} - \mathbf{z}$ belongs to $K + (-K) \subset a\,K$ and to $\ker \mathbf{A} = L^m$ as well. Therefore, we obtain

$$E^m(K, X) \leq \sup_{\mathbf{u} \in aK \cap L^m} \|\mathbf{u}\| = a \sup_{\mathbf{v} \in K \cap L^m} \|\mathbf{v}\|.$$

Taking the infimum over L^m, we conclude that $E^m(K, X) \leq a\,d^m(K, X)$. $\qquad \square$

In the context of compressive sensing, it is natural to consider sets K of compressible vectors. As indicated by Proposition 2.3, the unit balls B_q^N in ℓ_q^N with small q are good models of compressible vectors. However, in the case $q < 1$, the ℓ_q-quasinorm only satisfies a quasi-triangle, which poses additional difficulties in estimating the corresponding Gelfand widths. In order to avoid such complications, we only treat the case $q = 1$ here, which nevertheless allows us to draw the same conclusions concerning the minimal number of required measurements for sparse recovery. Moreover, the proof methods can be extended to the quasinorm case $q < 1$; see Exercise 10.10.

In the next section, we give matching upper and lower bounds for the Gelfand width $d^m(B_1^N, \ell_p^N)$ of ℓ_1-balls in ℓ_p^N when $1 < p \leq 2$; see Theorems 10.9 and 10.10. They provide the following result.

Theorem 10.5. *For $1 < p \leq 2$ and $m < N$, there exist constants $c_1, c_2 > 0$ depending only on p such that*

$$c_1 \min\left\{1, \frac{\ln(eN/m)}{m}\right\}^{1-1/p} \leq d^m(B_1^N, \ell_p^N) \leq c_2 \min\left\{1, \frac{\ln(eN/m)}{m}\right\}^{1-1/p}. \tag{10.5}$$

We immediately obtain corresponding estimates for the compressive widths, where we recall that $A \asymp B$ means that there exist absolute constants c_1, c_2 such that $c_1 A \leq B \leq c_2 A$.

Corollary 10.6. *For $1 < p \leq 2$ and $m < N$, the adaptive and nonadaptive compressive widths satisfy*

$$E_{\mathrm{ada}}^m(B_1^N, \ell_p^N) \asymp E^m(B_1^N, \ell_p^N) \asymp \min\left\{1, \frac{\ln(eN/m)}{m}\right\}^{1-1/p}.$$

Proof. Since $-B_1^N = B_1^N$ and $B_1^N + B_1^N \subset 2B_1^N$, Theorem 10.4 implies

$$d^m(B_1^N, \ell_p^N) \leq E_{\mathrm{ada}}^m(B_1^N, \ell_p^N) \leq E^m(B_1^N, \ell_p^N) \leq 2\, d^m(B_1^N, \ell_p^N).$$

Theorem 10.5 therefore concludes the proof.   □

The lower estimate is of particular significance for compressive sensing. Indeed, under the condition

$$m \geq c\, s \ln\left(\frac{eN}{s}\right), \tag{10.6}$$

we have seen that there are matrices $\mathbf{A} \in \mathbb{R}^{m \times N}$ with small restricted isometry constants and reconstruction maps providing the stability estimate

$$\|\mathbf{x} - \Delta(\mathbf{A}\mathbf{x})\|_p \leq \frac{C}{s^{1-1/p}} \sigma_s(\mathbf{x})_1 \qquad \text{for all } \mathbf{x} \in \mathbb{R}^N.$$

Such reconstruction maps include, for instance, basis pursuit, iterative hard thresholding, or orthogonal matching pursuit; see Chap. 6. Conversely, we can now show that the existence of Δ and $\mathbf{A}$—or Δ and an adaptive F—providing such a stability estimate forces the number of measurements to be bounded from below as in (10.6).

Proposition 10.7. *Let $1 < p \leq 2$. Suppose that there exist a matrix $\mathbf{A} \in \mathbb{R}^{m \times N}$ and a map $\Delta : \mathbb{R}^m \to \mathbb{R}^N$ such that, for all $\mathbf{x} \in \mathbb{R}^N$,*

$$\|\mathbf{x} - \Delta(\mathbf{A}\mathbf{x})\|_p \leq \frac{C}{s^{1-1/p}} \sigma_s(\mathbf{x})_1. \tag{10.7}$$

Then, for some constant $c_1, c_2 > 0$ depending only on C,

$$m \geq c_1 \, s \ln\left(\frac{eN}{s}\right),$$

provided $s > c_2$.

The same statement holds true for an adaptive map $F : \mathbb{R}^N \to \mathbb{R}^m$ in place of a linear map $\mathbf{A}$.

Proof. It is enough to prove the statement for an adaptive map $F : \mathbb{R}^N \to \mathbb{R}^m$. We notice that (10.7) implies

$$E_{\text{ada}}^m(B_1^N, \ell_p^N) \leq \frac{C}{s^{1-1/p}} \sup_{\mathbf{x} \in B_1^N} \sigma_s(\mathbf{x})_1 \leq \frac{C}{s^{1-1/p}}.$$

But, in view of Corollary 10.6, there is a constant $c > 0$ such that

$$c \min\left\{1, \frac{\ln(eN/m)}{m}\right\}^{1-1/p} \leq E_{\text{ada}}^m(B_1^N, \ell_p^N).$$

Thus, for some constant $c' > 0$,

$$c' \min\left\{1, \frac{\ln(eN/m)}{m}\right\} \leq \frac{1}{s}.$$

We derive either $s \leq 1/c'$ or $m \geq c' s \ln(eN/m)$. The hypothesis $s > c_2 := 1/c'$ allows us to discard the first alternative. Calling upon Lemma C.6, the second alternative gives $m \geq c_1 s \ln(eN/s)$ with $c_1 = c'e/(1 + e)$. This is the desired result. $\qquad\square$

The restrictions $s > c_2$ and $p > 1$ will be removed in the nonadaptive setting by Theorem 11.7 in the next chapter. Accepting that this theorem is true for now, we can state the following result on the minimal number of measurements needed to enforce the restricted isometry property.

Corollary 10.8. *If the $2s$th restricted isometry constant of $\mathbf{A} \in \mathbb{R}^{m \times N}$ satisfies $\delta_{2s} < 0.6246$, say, then necessarily*

$$m \geq c \, s \ln\left(\frac{eN}{s}\right)$$

for some constant $c > 0$ depending only on δ_{2s}.

Proof. If $\delta_{2s} < 0.6246$ and if Δ is the ℓ_1-minimization reconstruction map, we know from Theorem 6.12 that (10.7) with $p = 2$ holds for some constant C depending only on δ_{2s}. The conclusion follows from Proposition 10.7 if $s > c_2$. The case $s \leq c_2$ (of minor importance) is handled with Theorem 11.7. $\qquad\square$

10.2 Estimate for the Gelfand Widths of ℓ_1-Balls

In this section, we establish the two-sided estimate of Theorem 10.5 for the Gelfand widths of the unit ℓ_1-balls in ℓ_p^N when $1 \leq p \leq 2$. We separate the lower and upper estimates.

Upper Bound

With the results that we have established in previous chapters, it is quite simple to bound the Gelfand widths from above. For instance, recovery theorems such as Theorems 6.12, 6.21, and 6.28, applied to (subgaussian random) matrices with the restricted isometry property, imply that

$$E^m(B_1^N, \ell_p^N) \leq \frac{C}{s^{1-1/p}} \sup_{\mathbf{x} \in B_1^N} \sigma_s(\mathbf{x})_1 \leq \frac{C}{s^{1-1/p}}$$

when m is of the order of $s \ln(eN/s)$ (see Theorem 9.2) or, equivalently (see Lemma C.6), if s is of the order of $m/\ln(eN/m)$. Then, using Theorem 10.4, we get

$$d^m(B_1^N, \ell_p^N) \leq E^m(B_1^N, \ell_p^N) \leq C' \left\{ \frac{\ln(eN/m)}{m} \right\}^{1-1/p}.$$

A more rigorous and self-contained argument (not relying on any recovery theorems) is given below. It is strongly inspired by the proof techniques developed in Chap. 6.

Theorem 10.9. *There is a constant $C > 0$ such that, for $1 < p \leq 2$ and $m < N$,*

$$d^m(B_1^N, \ell_p^N) \leq C \min\left\{1, \frac{\ln(eN/m)}{m}\right\}^{1-1/p}.$$

Proof. Using the inequality $\|\mathbf{x}\|_p \leq \|\mathbf{x}\|_1$, $\mathbf{x} \in \mathbb{R}^N$, in the definition of the Gelfand width immediately gives

$$d^m(B_1^N, \ell_p^N) \leq d^0(B_1^N, \ell_1^N) = 1.$$

As a result, if $m \leq c \ln(eN/m)$ then

$$d^m(B_1^N, \ell_p^N) \leq \min\left\{1, \frac{c\ln(eN/m)}{m}\right\}^{1-1/p}. \qquad (10.8)$$

On the other hand, if $m > c\ln(eN/m)$ with $c := 144(1 + e^{-1})$, we define $s \geq 1$ to be the largest integer smaller than $m/(c\ln(eN/m))$, so that

$$\frac{m}{2c\ln(eN/m)} \leq s < \frac{m}{c\ln(eN/m)}.$$

Note that $m > cs\ln(eN/m)$ yields $m > c's\ln(eN/s)$ with $c' = 144$; see Lemma C.6. Then Theorem 9.27 concerning the restricted isometry property of Gaussian random matrices with $\eta = 1/6$ and $\varepsilon = 2\exp(-m/144)$ guarantees the existence of a measurement matrix $\mathbf{A} \in \mathbb{R}^{m \times N}$ with restricted isometry constant

$$\delta_s(\mathbf{A}) \leq \delta := 4\eta + 4\eta^2 = 7/9,$$

since $m \geq 72(s\ln(eN/s) - m/144)$, i.e., $m \geq 144s\ln(eN/s)$. (Instead of Theorem 9.27, we could alternatively use the easier Theorem 9.2 or 9.11 on the restricted isometry property of subgaussian random matrices, which however does not specify the constants.)

Partitioning the index set $[N]$ as the disjoint union $S_0 \cup S_1 \cup S_2 \cup \ldots$ of index sets of size s in such a way that $|x_i| \geq |x_j|$ whenever $i \in S_{k-1}, j \in S_k$, and $k \geq 1$, we recall from Lemma 6.10 that $\|\mathbf{x}_{S_k}\|_2 \leq \|\mathbf{x}_{S_{k-1}}\|_1/\sqrt{s}$ for all $k \geq 1$. Therefore, for $\mathbf{x} \in L^m := \ker \mathbf{A}$, we have

$$\|\mathbf{x}\|_p \leq \sum_{k\geq 0} \|\mathbf{x}_{S_k}\|_p \leq \sum_{k\geq 0} s^{1/p-1/2} \|\mathbf{x}_{S_k}\|_2 \leq \sum_{k\geq 0} \frac{s^{1/p-1/2}}{\sqrt{1-\delta}} \|\mathbf{A}(\mathbf{x}_{S_k})\|_2$$

$$= \frac{s^{1/p-1/2}}{\sqrt{1-\delta}}\left[\|\mathbf{A}(-\sum_{k\geq 1}\mathbf{x}_{S_k})\|_2 + \sum_{k\geq 1}\|\mathbf{A}(\mathbf{x}_{S_k})\|_2\right]$$

$$\leq \frac{s^{1/p-1/2}}{\sqrt{1-\delta}}\left[2\sum_{k\geq 1}\|\mathbf{A}(\mathbf{x}_{S_k})\|_2\right] \leq 2\sqrt{\frac{1+\delta}{1-\delta}}\, s^{1/p-1/2}\sum_{k\geq 1}\|\mathbf{x}_{S_k}\|_2$$

$$\leq 2\sqrt{\frac{1+\delta}{1-\delta}}\, s^{1/p-1/2}\sum_{k\geq 1}\|\mathbf{x}_{S_{k-1}}\|_1/\sqrt{s} = 2\sqrt{\frac{1+\delta}{1-\delta}}\,\frac{1}{s^{1-1/p}}\sum_{k\geq 1}\|\mathbf{x}_{S_{k-1}}\|_1$$

$$\leq 2\sqrt{\frac{1+\delta}{1-\delta}}\left(\frac{2c\ln(eN/m)}{m}\right)^{1-1/p}\|\mathbf{x}\|_1. \qquad (10.9)$$

Using $\delta = 7/9$ and $2^{1-1/p} \leq 2$, it follows that, for all $\mathbf{x} \in B_1^N \cap L^m$,

$$\|\mathbf{x}\|_p \leq 8\sqrt{2}\left\{\frac{c\ln(eN/m)}{m}\right\}^{1-1/p}.$$

This shows that, if $m > c \ln(eN/m)$, then

$$d^m(B_1^N, \ell_p^N) \leq 8\sqrt{2} \min\left\{1, \frac{c \ln(eN/m)}{m}\right\}^{1-1/p}. \tag{10.10}$$

Combining (10.8) and (10.10), we conclude

$$d^m(B_1^N, \ell_p^N) \leq C \min\left\{1, \frac{\ln(eN/m)}{m}\right\}^{1-1/p}$$

with $C = 8\sqrt{2}c = 1152\sqrt{2}(1 + e^{-1})$, which is the desired upper bound. □

Lower Bound

We now establish the lower bound for the Gelfand widths of ℓ_1-balls in ℓ_p^N for $1 < p \leq \infty$. This bound matches the previous upper bound up to a multiplicative constant. We point out that a lower bound where the minimum in (10.5) does not appear would be invalid, since the width $d^m(B_1^N, \ell_p^N)$ is bounded above by one; hence, it cannot exceed $c \ln(eN/m)/m$ for small m and large N.

Theorem 10.10. *There is a constant $c > 0$ such that, for $1 < p \leq \infty$ and $m < N$,*

$$d^m(B_1^N, \ell_p^N) \geq c \min\left\{1, \frac{\ln(eN/m)}{m}\right\}^{1-1/p}.$$

Somewhat unexpectedly, the proof of this proposition relies on the following estimate of the necessary number of measurements required for the exact recovery of s-sparse vectors via ℓ_1-minimization. Of course, this important result is of independent interest.

Theorem 10.11. *Given a matrix $\mathbf{A} \in \mathbb{R}^{m \times N}$, if every $2s$-sparse vector $\mathbf{x} \in \mathbb{R}^N$ is a minimizer of $\|\mathbf{z}\|_1$ subject to $\mathbf{Az} = \mathbf{Ax}$, then*

$$m \geq c_1 s \ln\left(\frac{N}{c_2 s}\right),$$

where $c_1 = 1/\ln 9$ and $c_2 = 4$.

This is based on the key combinatorial lemma that follows.

Lemma 10.12. *Given integers $s < N$, there exist*

$$n \geq \left(\frac{N}{4s}\right)^{s/2} \tag{10.11}$$

subsets $S_1, \ldots, S_n$ of $[N]$ such that each S_j has cardinality s and

$$\operatorname{card}(S_i \cap S_j) < \frac{s}{2} \qquad \text{whenever } i \neq j. \tag{10.12}$$

Proof. We may assume that $s \leq N/4$, for otherwise it suffices to take $n = 1$ subset of $[N]$. Let $\mathcal{B}$ denote the family of subsets of $[N]$ having cardinality s. We draw an element $S_1 \in \mathcal{B}$, and we collect in a family $\mathcal{A}_1$ all the sets $S \in \mathcal{B}$ such that $\operatorname{card}(S_1 \cap S) \geq s/2$. We have

$$\operatorname{card}(\mathcal{A}_1) = \sum_{k=\lceil s/2 \rceil}^{s} \binom{s}{k} \binom{N-s}{s-k} \leq 2^s \max_{\lceil s/2 \rceil \leq k \leq s} \binom{N-s}{s-k} = 2^s \binom{N-s}{\lfloor s/2 \rfloor},$$

where the last equality holds because $\lfloor s/2 \rfloor \leq (N-s)/2$ when $s \leq N/2$. We observe that any set $S \in \mathcal{B} \setminus \mathcal{A}_1$ satisfies $\operatorname{card}(S_1 \cap S) < s/2$. Next, we draw an element $S_2 \in \mathcal{B} \setminus \mathcal{A}_1$, provided that the latter is nonempty. As before, we collect in a family $\mathcal{A}_2$ all the sets $S \in \mathcal{B} \setminus \mathcal{A}_1$ such that $\operatorname{card}(S_2 \cap S) \geq s/2$, we remark that

$$\operatorname{card}(\mathcal{A}_2) \leq 2^s \binom{N-s}{\lfloor s/2 \rfloor},$$

and we observe that any set $S \in \mathcal{B} \setminus (\mathcal{A}_1 \cup \mathcal{A}_2)$ satisfies $\operatorname{card}(S_1 \cap S) < s/2$ and $\operatorname{card}(S_2 \cap S) < s/2$. We repeat the procedure of selecting sets $S_1, \ldots, S_n$ until $\mathcal{B} \setminus (\mathcal{A}_1 \cup \cdots \cup \mathcal{A}_n)$ is empty. In this way, (10.12) is automatically fulfilled. Moreover,

$$n \geq \frac{\operatorname{card}(\mathcal{B})}{\max_{1 \leq i \leq n} \operatorname{card}(\mathcal{A}_i)} \geq \frac{\binom{N}{s}}{2^s \binom{N-s}{\lfloor s/2 \rfloor}}$$

$$= \frac{1}{2^s} \frac{N(N-1) \cdots (N-s+1)}{(N-s)(N-s-1) \cdots (N-s-\lfloor s/2 \rfloor + 1)} \frac{1}{s(s-1) \cdots (\lfloor s/2 \rfloor + 1)}$$

$$\geq \frac{1}{2^s} \frac{N(N-1) \cdots (N - \lceil s/2 \rceil + 1)}{s(s-1) \cdots (s - \lceil s/2 \rceil + 1)} \geq \frac{1}{2^s} \left(\frac{N}{s}\right)^{\lceil s/2 \rceil} \geq \left(\frac{N}{4s}\right)^{s/2}.$$

This shows that (10.11) is fulfilled, too, and concludes the proof. $\square$

With this lemma at hand, we can turn to the proof of the theorem.

Proof (of Theorem 10.11). Let us consider the quotient space

$$X := \ell_1^N / \ker \mathbf{A} = \{ [\mathbf{x}] := \mathbf{x} + \ker \mathbf{A}, \mathbf{x} \in \mathbb{R}^N \},$$

which is normed with

$$\| [\mathbf{x}] \| := \inf_{\mathbf{v} \in \ker \mathbf{A}} \| \mathbf{x} - \mathbf{v} \|_1, \qquad \mathbf{x} \in \mathbb{R}^N.$$

Given a $2s$-sparse vector $\mathbf{x} \in \mathbb{R}^N$, we notice that every vector $\mathbf{z} = \mathbf{x} - \mathbf{v}$ with $\mathbf{v} \in \ker \mathbf{A}$ satisfies $\mathbf{Az} = \mathbf{Ax}$. Thus, our assumption gives $\|[\mathbf{x}]\| = \|\mathbf{x}\|_1$. Let $S_1, \dots, S_n$ be the sets introduced in Lemma 10.12, and let us define s-sparse vectors $\mathbf{x}^1, \dots, \mathbf{x}^n \in \mathbb{R}^N$ with unit ℓ_1-norms by

$$x_k^i = \begin{cases} 1/s & \text{if } k \in S_i, \\ 0 & \text{if } k \notin S_i. \end{cases} \tag{10.13}$$

For $1 \le i \ne j \le n$, we have $\|[\mathbf{x}^i] - [\mathbf{x}^j]\| = \|[\mathbf{x}^i - \mathbf{x}^j]\| = \|\mathbf{x}^i - \mathbf{x}^j\|_1$, since the vector $\mathbf{x}^i - \mathbf{x}^j$ is $2s$-sparse. We also have $\|\mathbf{x}^i - \mathbf{x}^j\|_1 > 1$, since $|x_k^i - x_k^j|$ equals $1/s$ if $k \in S_i \Delta S_j = (S_i \cup S_j) \setminus (S_i \cap S_j)$ and vanishes otherwise and since $\operatorname{card}(S_i \Delta S_j) > s$. We conclude that

$$\|[\mathbf{x}^i] - [\mathbf{x}^j]\| > 1 \qquad \text{for all } 1 \le i \ne j \le n.$$

This shows that $\{[\mathbf{x}^1], \dots, [\mathbf{x}^n]\}$ is a 1-separating subset of the unit sphere of X, which has dimension $r := \operatorname{rank}(\mathbf{A}) \le m$. According to Proposition C.3, this implies that $n \le 3^r \le 3^m$. In view of (10.11), we obtain

$$\left(\frac{N}{4s}\right)^{s/2} \le 3^m.$$

Taking the logarithm on both sides gives the desired result.   □

We are now ready to prove the main result of this section.

Proof (of Theorem 10.10). With $c' := 2/(1 + 4\ln 9)$, we are going to show that

$$d^m(B_1^N, \ell_p^N) \ge \frac{\mu^{1-1/p}}{2^{2-1/p}}, \qquad \text{where } \mu := \min\left\{1, \frac{c' \ln(eN/m)}{m}\right\}.$$

The result will then follow with $c = \min\{1, c'\}^{1-1/p}/2^{2-1/p} \ge \min\{1, c'\}/4$. By way of contradiction, we assume that $d^m(B_1^N, \ell_p^N) < \mu^{1-1/p}/2^{2-1/p}$. This implies the existence of a subspace L^m of $\mathbb{R}^N$ with $\operatorname{codim}(L^m) \le m$ such that, for all $\mathbf{v} \in L^m \setminus \{\mathbf{0}\}$,

$$\|\mathbf{v}\|_p < \frac{\mu^{1-1/p}}{2^{2-1/p}}\|\mathbf{v}\|_1.$$

Let us consider a matrix $\mathbf{A} \in \mathbb{R}^{m \times N}$ such that $\ker \mathbf{A} = L^m$. Let us also define an integer $s \ge 1$ by $s := \lfloor 1/\mu \rfloor$, so that

$$\frac{1}{2\mu} < s \le \frac{1}{\mu}.$$

In the way, we have, for all $\mathbf{v} \in \ker \mathbf{A} \setminus \{0\}$,

$$\|\mathbf{v}\|_p < \frac{1}{2}\left(\frac{1}{2s}\right)^{1-1/p}\|\mathbf{v}\|_1.$$

The inequality $\|\mathbf{v}\|_1 \leq N^{1-1/p}\|\mathbf{v}\|_p$ ensures that $1 < (N/2s)^{1-1/p}/2$, hence that $2s < N$. Then, for $S \subset [N]$ with $\mathrm{card}(S) \leq 2s$ and for $\mathbf{v} \in \ker \mathbf{A} \setminus \{0\}$, we have

$$\|\mathbf{v}_S\|_1 \leq (2s)^{1-1/p}\|\mathbf{v}_S\|_p \leq (2s)^{1-1/p}\|\mathbf{v}\|_p < \frac{1}{2}\|\mathbf{v}\|_1.$$

This is the null space property (4.2) of order $2s$. Thus, according to Theorem 4.5, every $2s$-sparse vector $\mathbf{x} \in \mathbb{R}^N$ is uniquely recovered from $\mathbf{y} = \mathbf{Ax}$ by ℓ_1-minimization. Theorem 10.11 now implies that

$$m \geq c_1 s \ln\left(\frac{N}{c_2 s}\right), \qquad c_1 = \frac{1}{\ln 9}, \qquad c_2 = 4.$$

Theorem 2.13 also implies that $m \geq 2(2s) = c_2 s$. It follows that

$$m \geq c_1 s \ln\left(\frac{N}{m}\right) = c_1 s \ln\left(\frac{eN}{m}\right) - c_1 s > \frac{c_1}{2\mu}\ln\left(\frac{eN}{m}\right) - \frac{c_1}{4}m.$$

After rearrangement, we deduce

$$m > \frac{2c_1}{4 + c_1}\frac{\ln(eN/m)}{\min\left\{1, c'\ln(eN/m)/m\right\}} \geq \frac{2c_1}{4 + c_1}\frac{\ln(eN/m)}{c'\ln(eN/m)/m} = m.$$

This is the desired contradiction. $\square$

10.3   Applications to the Geometry of Banach Spaces

Let us now highlight two applications of the previous results and their proofs in the geometry of Banach spaces. By relating the Gelfand widths to their duals, the Kolmogorov widths, we first obtain lower and upper bounds for the latter. Next, we show that $\mathbb{R}^{2m}$ can be split into two orthogonal subspaces on which the ℓ_1-norm and the ℓ_2-norm are essentially equivalent. This is called a Kashin splitting.

Kolmogorov Widths

Let us start with the definition.

Definition 10.13. The Kolmogorov m-width of a subset K of a normed space X is defined as

$$d_m(K, X) := \inf \left\{ \sup_{\mathbf{x} \in K} \inf_{\mathbf{z} \in X_m} \|\mathbf{x} - \mathbf{z}\|, \ X_m \text{ subspace of } X \text{ with } \dim(X_m) \le m \right\}.$$

The Kolmogorov widths of ℓ_p-balls in ℓ_q are closely related to certain Gelfand widths as shown by the following duality result.

Theorem 10.14. *For $1 \le p, q \le \infty$, let p^*, q^* be such that $1/p^* + 1/p = 1$ and $1/q^* + 1/q = 1$. Then*

$$d_m(B_p^N, \ell_q^N) = d^m(B_{q^*}^N, \ell_{p^*}^N).$$

The proof uses a classical observation about best approximation. Below $\| \cdot \|_*$ denotes the dual of some norm $\| \cdot \|$; see Definition A.3.

Lemma 10.15. *Let Y be a finite-dimensional subspace of a normed space X. Given $\mathbf{x} \in X \setminus Y$ and $\mathbf{y}^* \in Y$, the following properties are equivalent:*

(a) $\mathbf{y}^*$ *is a best approximation to* $\mathbf{x}$ *from* Y.
(b) $\|\mathbf{x} - \mathbf{y}^*\| = \lambda(\mathbf{x})$ *for some linear functional λ vanishing on Y and satisfying* $\|\lambda\|_* \le 1$.

Proof. Let us first assume that (b) holds. To derive (a), we observe that $\lambda(\mathbf{y}) = 0$ for all $\mathbf{y} \in Y$, so that

$$\|\mathbf{x} - \mathbf{y}^*\| = \lambda(\mathbf{x}) = \lambda(\mathbf{x} - \mathbf{y}) \le \|\lambda\| \|\mathbf{x} - \mathbf{y}\| \le \|\mathbf{x} - \mathbf{y}\| \quad \text{for all } \mathbf{y} \in Y.$$

Conversely, let us assume that (a) holds. We define a linear functional $\tilde{\lambda}$ on the space $[Y \oplus \operatorname{span}(\mathbf{x})]$ by

$$\tilde{\lambda}(\mathbf{y} + t\mathbf{x}) = t \|\mathbf{x} - \mathbf{y}^*\| \quad \text{for all } \mathbf{y} \in Y \text{ and } t \in \mathbb{R}.$$

Setting $t = 0$, it is seen that $\tilde{\lambda}$ vanishes on Y. Besides, for $\mathbf{y} \in Y$ and $t \ne 0$, we have

$$|\tilde{\lambda}(\mathbf{y} + t\mathbf{x})| = |t| \|\mathbf{x} - \mathbf{y}^*\| \le |t| \|\mathbf{x} - (-\mathbf{y}/t)\| = \|\mathbf{y} + t\mathbf{x}\|.$$

This inequality—which remains valid for $t = 0$—implies $\|\tilde{\lambda}\| \le 1$. The linear functional λ required in (b) is the Hahn–Banach extension of the linear functional $\tilde{\lambda}$ to the whole space X (see also Appendix C.7). $\qquad \square$

Proof (of Theorem 10.14). Given a subspace X_m of ℓ_q^N with $\dim(X_m) \le m$ and a vector $\mathbf{x} \in B_p^N$, Lemma 10.15 shows that

$$\inf_{z \in X_m} \|x - z\|_q \le \sup_{u \in B_{q^*}^N \cap X_m^\perp} \langle u, x \rangle.$$

Moreover, for all $u \in B_{q^*}^N \cap X_m^\perp$ and any $z \in X_m$, we have

$$\langle u, x \rangle = \langle u, x - z \rangle \le \|u\|_{q^*} \|x - z\|_q.$$

We deduce the equality

$$\inf_{z \in X_m} \|x - z\|_q = \sup_{u \in B_{q^*}^N \cap X_m^\perp} \langle u, x \rangle.$$

It follows that

$$\sup_{x \in B_p^N} \inf_{z \in X_m} \|x - z\|_q = \sup_{x \in B_p^N} \sup_{u \in B_{q^*}^N \cap X_m^\perp} \langle u, x \rangle = \sup_{u \in B_{q^*}^N \cap X_m^\perp} \sup_{x \in B_p^N} \langle u, x \rangle$$

$$= \sup_{u \in B_{q^*}^N \cap X_m^\perp} \|u\|_{p^*}.$$

Taking the infimum over all subspaces X_m with $\dim(X_m) \le m$ and noticing the one-to-one correspondence between the subspaces $X_m^\perp$ and the subspaces L^m with $\mathrm{codim}(L^m) \le m$, we conclude

$$d_m(B_p^N, \ell_q^N) = d^m(B_{q^*}^N, \ell_{p^*}^N).$$

This is the desired identity.   □

Our estimate on the Gelfand widths in Theorem 10.5 immediately implies now the following estimate of the Kolmogorov widths of ℓ_p^N-balls in ℓ_∞^N for $p \in [2, \infty)$.

Theorem 10.16. *For $2 \le p < \infty$ and $m < N$, there exist constants $c_1, c_2 > 0$ depending only on p such that the Kolmogorov widths d_m satisfy*

$$c_1 \min \left\{ 1, \frac{\ln(eN/m)}{m} \right\}^{1/p} \le d_m(B_p^N, \ell_\infty^N) \le c_2 \min \left\{ 1, \frac{\ln(eN/m)}{m} \right\}^{1/p}.$$

Kashin's Decomposition Theorem

Specifying the upper estimate of the Gelfand width of the unit ℓ_1-ball in ℓ_2^N to the case $N = 2m$, we obtain $d^m(B_1^{2m}, \ell_2^{2m}) \le C/\sqrt{m}$, which says that there is a subspace E of $\mathbb{R}^{2m}$ of dimension m such that

$$\|x\|_2 \le \frac{C}{\sqrt{m}} \|x\|_1 \qquad \text{for all } x \in E.$$

Together with $\|\mathbf{x}\|_1 \leq \sqrt{2m}\,\|\mathbf{x}\|_2$, which is valid for any $\mathbf{x} \in \mathbb{R}^{2m}$, this means that the norms $\|\cdot\|_1/\sqrt{m}$ and $\|\cdot\|_2$ are comparable on E. In other words, as a subspace of ℓ_1^{2m}, the m-dimensional space E is almost Euclidean. Kashin's decomposition theorem states even more, namely, that one can find an m-dimensional space E such that both E and its orthogonal complement $E^\perp$, as subspaces of ℓ_1^{2m}, are almost Euclidean.

Theorem 10.17. *There exist universal constants $\alpha, \beta > 0$ such that, for any $m \geq 1$, the space $\mathbb{R}^{2m}$ can be split into an orthogonal sum of two m-dimensional subspaces E and $E^\perp$ such that*

$$\alpha \sqrt{m}\,\|\mathbf{x}\|_2 \leq \|\mathbf{x}\|_1 \leq \beta \sqrt{m}\,\|\mathbf{x}\|_2 \tag{10.14}$$

for all $\mathbf{x} \in E$ and for all $\mathbf{x} \in E^\perp$.

Proof. The second inequality in (10.14) holds with $\beta := \sqrt{2}$ regardless of the subspace E of $\mathbb{R}^{2m}$ considered, so we focus on the first inequality. Let $\mathbf{G}$ be a scaled $m \times m$ Gaussian random matrix whose entries are independent Gaussian random variables with mean zero and variance $1/m$. We define two full-rank $m \times (2m)$ matrices by

$$\mathbf{A} := \begin{bmatrix} \mathbf{Id} \mid \mathbf{G} \end{bmatrix}, \qquad \mathbf{B} := \begin{bmatrix} \mathbf{G}^* \mid -\mathbf{Id} \end{bmatrix},$$

and we consider the m-dimensional space $E := \ker \mathbf{A}$. In view of $\mathbf{B}\mathbf{A}^* = \mathbf{0}$, we have $E^\perp = \operatorname{ran} \mathbf{A}^* \subset \ker \mathbf{B}$, and $E^\perp = \ker \mathbf{B}$ follows from consideration of dimensions. We are going to show that, given any $t \in (0, 1)$ and any $\mathbf{x} \in \mathbb{R}^{2m}$, the matrices $\mathbf{M} = \mathbf{A}$ and $\mathbf{M} = \mathbf{B}$ satisfy the concentration inequality

$$\mathbb{P}\big(\big|\|\mathbf{M}\mathbf{x}\|_2^2 - \|\mathbf{x}\|_2^2\big| \geq t\|\mathbf{x}\|_2^2\big) \leq 2\exp\big(-\tilde{c}t^2 m\big) \tag{10.15}$$

for some constant $\tilde{c} > 0$. Fixing $0 < \delta < 1$, say $\delta := \sqrt{2/3}$, Theorem 9.11 with $\varepsilon = 2\exp(-\tilde{c}m/4)$ implies that $\delta_s(\mathbf{A}) \leq \delta$ and $\delta_s(\mathbf{B}) \leq \delta$ with probability at least $1 - 4\exp(-\tilde{c}m/4)$ provided

$$m \geq \frac{2}{3\tilde{c}\delta^2}[s(9+2\ln(2m/s))+\tilde{c}m/2], \quad \text{i.e.,} \quad \tilde{c}m \geq 2s(9+2\ln(2m/s)). \tag{10.16}$$

The above probability is positive if also $m > 8\ln(2)/\tilde{c}$ and we further require $m > 1/(2\gamma)$ for a constant γ small enough to have $4\gamma(9 + 2\ln(2/\gamma)) \leq \tilde{c}$. In this way, the integer $s := \lfloor 2\gamma m \rfloor \geq 1$ satisfies $\gamma m \leq s \leq 2\gamma m$, and (10.16) is therefore fulfilled. Let now $\mathbf{x} \in E \cup E^\perp$, i.e., $\mathbf{x} \in \ker \mathbf{M}$ for $\mathbf{M} = \mathbf{A}$ or $\mathbf{M} = \mathbf{B}$. Reproducing the argument in the proof of Proposition 10.9 (see (10.9)) starting with the partition $[N] = S_0 \cup S_1 \cup S_2 \cup \cdots$, we arrive at

$$\|\mathbf{x}\|_2 \leq 2\sqrt{\frac{1+\delta}{1-\delta}}\frac{\|\mathbf{x}\|_1}{\sqrt{s}} \leq \frac{2(\sqrt{2}+\sqrt{3})}{\sqrt{\gamma m}}\|\mathbf{x}\|_1.$$

This is the desired inequality with $\sqrt{\gamma}/(2(\sqrt{2}+\sqrt{3}))$ taking the role of α when $m >$ $m_* := \max\{8\ln(2)/\tilde{c}, 1/(2\gamma)\}$. When $m \le m_*$, the desired inequality simply follows from $\|\mathbf{x}_1\| \ge \|\mathbf{x}\|_2 \ge \sqrt{m}\|\mathbf{x}\|_2/\sqrt{m_*}$. The result is therefore acquired with $\alpha := \min\{\sqrt{\gamma}/(2(\sqrt{2}+\sqrt{3})), 1/\sqrt{m_*}\}$. It remains to establish the concentration inequality (10.15). In the case $\mathbf{M} = \mathbf{A}$—the case $\mathbf{M} = \mathbf{B}$ being similar—we notice that, with $\mathbf{x} = [\mathbf{u}, \mathbf{v}]^\top$,

$$\left|\|\mathbf{A}\mathbf{x}\|_2^2 - \|\mathbf{x}\|_2^2\right| = \left|\|\mathbf{u} + \mathbf{G}\mathbf{v}\|_2^2 - \|\mathbf{u}\|_2^2 - \|\mathbf{v}\|_2^2\right| = \left|2\langle\mathbf{u}, \mathbf{G}\mathbf{v}\rangle + \|\mathbf{G}\mathbf{v}\|_2^2 - \|\mathbf{v}\|_2^2\right|$$

$$\le 2\left|\langle\mathbf{u}, \mathbf{G}\mathbf{v}\rangle\right| + \left|\|\mathbf{G}\mathbf{v}\|_2^2 - \|\mathbf{v}\|_2^2\right|.$$

Thus, if $\left|\|\mathbf{A}\mathbf{x}\|_2^2 - \|\mathbf{x}\|_2^2\right| \ge t\|\mathbf{x}\|_2^2$, at least one of the following two alternatives holds:

$$2\left|\langle\mathbf{u}, \mathbf{G}\mathbf{v}\rangle\right| \ge \frac{t}{2}(\|\mathbf{u}\|_2^2 + \|\mathbf{v}\|_2^2), \quad \text{or} \quad \left|\|\mathbf{G}\mathbf{v}\|_2^2 - \|\mathbf{v}\|_2^2\right| \ge \frac{t}{2}(\|\mathbf{u}\|_2^2 + \|\mathbf{v}\|_2^2).$$

The first inequality implies $\left|\langle\mathbf{u}, \mathbf{G}\mathbf{v}\rangle\right| \ge \frac{t}{2}\|\mathbf{u}\|_2\|\mathbf{v}\|_2$ and the second one $\left|\|\mathbf{G}\mathbf{v}\|_2^2 - \|\mathbf{v}\|_2^2\right| \ge \frac{t}{2}\|\mathbf{v}\|_2^2$. In terms of probability, we have

$$\mathbb{P}\left(\left|\|\mathbf{A}\mathbf{x}\|_2^2 - \|\mathbf{x}\|_2^2\right| \ge t\|\mathbf{x}\|_2^2\right)$$

$$\le \mathbb{P}\left(\left|\langle\mathbf{u}, \mathbf{G}\mathbf{v}\rangle\right| \ge t\|\mathbf{u}\|_2\|\mathbf{v}\|_2/2\right) + \mathbb{P}\left(\left|\|\mathbf{G}\mathbf{v}\|_2^2 - \|\mathbf{v}\|_2^2\right| \ge t\|\mathbf{v}\|_2^2/2\right).$$

For the first probability, we observe that $\langle\mathbf{u}, \mathbf{G}\mathbf{v}\rangle = \sum_{i=1}^m u_i \sum_{j=1}^m G_{i,j} v_j$ has the same distribution as a mean zero Gaussian random variable of variance $\|\mathbf{u}\|_2^2\|\mathbf{v}\|_2^2/m$ so that the standard tail estimate of Proposition 7.5 gives

$$\mathbb{P}\left(\left|\langle\mathbf{u}, \mathbf{G}\mathbf{v}\rangle\right| \ge t\|\mathbf{u}\|_2\|\mathbf{v}\|_2/2\right) = \mathbb{P}\left(|g| \ge t\sqrt{m}/2\right) \le \exp\left(-t^2 m/8\right).$$

For the second probability, we recall from Exercise 9.2 (see also Lemma 9.8, where the constant is not specified) that

$$\mathbb{P}\left(\left|\|\mathbf{G}\mathbf{v}\|_2^2 - \|\mathbf{v}\|_2^2\right| \ge t\|\mathbf{v}\|_2^2/2\right) \le 2\exp\left(-(t^2/16 - t^3/48)m\right)$$

$$\le 2\exp\left(-t^2 m/24\right).$$

As a consequence of the previous estimates, we obtain

$$\mathbb{P}\left(\left|\|\mathbf{A}\mathbf{x}\|_2^2 - \|\mathbf{x}\|_2^2\right| \ge t\|\mathbf{x}\|_2^2\right) \le \exp\left(-t^2 m/8\right) + \min\left\{1, 2\exp\left(-t^2 m/24\right)\right\}.$$

To complete the proof, it remains to notice that the latter is smaller than $2\exp$ $(-\tilde{c}t^2 m)$ for the properly chosen constant $\tilde{c} = \ln(4/3)/\ln(2^{12}) \approx 0.0346$. □

Notes

The definition of Gelfand widths sometimes appears with $\text{codim}(L^m) = m$ instead of $\text{codim}(L^m) \leq m$; see, for instance, Pinkus' book [387]. This is of course equivalent to the definition we have used.

We have coined here the terms nonadaptive and adaptive compressive widths for the quantities $E^m(C, X)$ and $E^m_{\text{ada}}(C, X)$, but note that this is not a standard terminology in the literature. In the compressive sensing literature, the nonadaptive compressive width appeared (without name), along with the corresponding part of Theorem 10.4 in [123]; see also [152, 213, 370]. The other part of Theorem 10.4 is an instance of general results from information-based complexity showing that "adaptivity does not help"; see [371].

As outlined before Theorem 10.9, the upper bound for the Gelfand width of the ℓ_1-ball in ℓ_p^N, $1 < p \leq 2$, can be deduced from the existence of a matrix $\mathbf{A} \in \mathbb{R}^{m \times N}$ with the property that $\|\mathbf{x} - \Delta_1(\mathbf{A}\mathbf{x})\|_p \leq C\|\mathbf{x}\|_1/s^{1-1/p}$ for all $\mathbf{x} \in \mathbb{R}^N$ where s is of the order of $m/\ln(eN/m)$. This property is implied by the property that $\|\mathbf{x} - \Delta_1(\mathbf{A}\mathbf{x})\|_p \leq C\sigma_s(\mathbf{x})_1/s^{1-1/p}$ for all $\mathbf{x} \in \mathbb{R}^N$. The two properties are in fact equivalent (up to a change of the constant C), as observed in [300]; see also Exercise 11.5.

The lower estimate for Gelfand widths of ℓ_1-balls given in Proposition 10.10 was obtained by Garnaev and Gluskin in [219]. Their original proof, which is reproduced in Exercise 10.9, dealt with the dual Kolmogorov width. The proof relying only on compressive sensing techniques presented here was proposed in [213], where the more general case of ℓ_p-balls, $0 < p \leq 1$, was treated in a similar way. Exercise 10.10 asks to work out this extension. The key combinatorial lemma, namely, Lemma 10.12, follows [213, 350], but it had also been used in other areas before; see, e.g., [77, 237, 368].

For $1 < q < p \leq \infty$, the order of the Gelfand widths of ℓ_q-balls in ℓ_p^N is known; see [333, pp. 481–482] and the references therein for the dual statement about Kolmogorov widths. Precisely, for $1 \leq m < N$, we have

- If $1 < q < p \leq 2$,

$$d^m(B_q^N, \ell_p^N) \asymp \min\left\{1, \frac{N^{1-1/q}}{m^{1/2}}\right\}^{\frac{1/q-1/p}{1/q-1/2}}.$$

- If $1 < q \leq 2 < p \leq \infty$,

$$d^m(B_q^N, \ell_p^N) \asymp \max\left\{\frac{1}{N^{1/q-1/p}}, \left(1 - \frac{m}{N}\right)^{1/2} \min\left(1, \frac{N^{1-1/q}}{m^{1/2}}\right)\right\}.$$

- If $2 \leq q < p \leq \infty$,

$$d^m(B_q^N, \ell_p^N) \asymp \max\left\{\frac{1}{N^{1/q-1/p}}, \left(1 - \frac{m}{N}\right)^{\frac{1/q-1/p}{1-2/p}}\right\}.$$

Theorem 10.17 was first established by Kashin in [299]. Szarek then gave a shorter proof in [457]. The argument presented here is close to a proof given by Schechtman in [442], which implicitly contains a few ideas now familiar in compressive sensing.

Exercises

10.1. Determine the Gelfand widths $d^m(B_1^N, \ell_2^N)$, $1 \le m < N$, of the unit ℓ_1-ball in the Euclidean space ℓ_2^N when $N = 2$ and $N = 3$.

10.2. For a compact subset K of an infinite-dimensional normed space K, prove that $\lim_{m\to\infty} d^m(C, X) = 0$.

10.3. Let $L_2(\mathbb{T})$ denote the space of square-integrable 2π-periodic functions on $\mathbb{R}$ and let $C^1(\mathbb{T})$ be the space of continuously differentiable 2π-periodic functions on $\mathbb{R}$. Consider the subset K of $L_2(\mathbb{T})$ defined by

$$K := \{g \in C^1(\mathbb{T}) : \|g'\|_2 \le 1\}.$$

Prove that

$$d_0(K, L_2(\mathbb{T})) = \infty, \qquad d_{2n-1}(K, L_2(\mathbb{T})) = d_{2n}(K, L_2(\mathbb{T})) = \frac{1}{n} \quad \text{for } n \ge 1.$$

Evaluate first the quantity

$$\sup_{f\in L_2(\mathbb{T})} \inf_{g\in\mathcal{T}_{n-1}} \|f - g\|_2,$$

where

$$\mathcal{T}_{n-1} := \text{span}\{1, \sin(x), \cos(x), \ldots, \sin((n-1)x), \cos((n-1)x)\}$$

is the space of trigonometric polynomials of degree at most $n - 1$.

10.4. Prove that

$$d^m(B_{X_n}, X) = 1, \quad X_n \text{ an } n\text{-dimensional subspace of } X, \quad m < n. \quad (10.17)$$

Prove also that

$$d_m(B_{X_n}, X) = 1, \quad X_n \text{ an } n\text{-dimensional subspace of } X, \quad m < n. \quad (10.18)$$

For (10.18), use the so-called *theorem of deviation of subspaces* (which you should derive from the Borsuk–Ulam theorem; see Exercise 2.9):

If U and V are two finite-dimensional subspaces of a normed space X with $\dim(V) > \dim(U)$, then there exists a nonzero vector $\mathbf{v} \in V$ to which $\mathbf{0}$ is a best approximation from U, i.e.,

$$\|\mathbf{v}\| \leq \|\mathbf{v} - \mathbf{u}\| \qquad \text{for all} \quad \mathbf{u} \in U.$$

10.5. Let K be a subset of a normed space X with $\mathbf{0} \in K$. Prove that

$$d^m(K, X) \leq 2E_{\text{ada}}^m(K, X).$$

10.6. Let $B_{1,+}^N$ be the subset of the unit ball B_1^N consisting of all nonnegative vectors, i.e.,

$$B_{1,+}^N = \{\mathbf{x} \in B_1^N : x_j \geq 0 \text{ for all } j \in [N]\}.$$

Prove that

$$d^m(B_1^N, \ell_2^N) \leq 2E^m(B_{1,+}^N, \ell_2^N),$$

and deduce that

$$E^m(B_{1,+}^N, \ell_2^N) \asymp \min\left\{1, \frac{\ln(eN/m)}{m}\right\}^{1/2}.$$

10.7. For $1 \leq p < q \leq \infty$ and $m < N$, prove that

$$d^m(B_p^N, \ell_q^N) \geq \frac{1}{(m+1)^{1/p - 1/q}}.$$

10.8. For $\mathbf{A} \in \mathbb{R}^{m \times N}$ and $s \geq 2$, show that if every s-sparse vector $\mathbf{x} \in \mathbb{R}^N$ is a minimizer of $\|\mathbf{z}\|_1$ subject to $\mathbf{A}\mathbf{z} = \mathbf{A}\mathbf{x}$, then $m \geq cs\ln(eN/s)$ for some constant $c > 0$, but that this does not hold for $s = 1$.

10.9. Original proof of the lower bound
This problem aims at establishing the lower bound of Proposition 10.10 by way of the Kolmogorov width $d_m(B_p^N, \ell_\infty^N)$, $1 \leq p \leq \infty$.

(a) Given a subset C of the normed space X, for $\varepsilon > 2d_m(C, X)$ and $t > 0$, prove that the maximal number of points in $C \cap tB_X$ with mutual distance in X exceeding ε satisfies

$$\mathcal{P}(C \cap tB_X, X, \varepsilon) \leq \left(1 + 2\frac{t + d_m(C, X)}{\varepsilon - 2d_m(C, X)}\right)^m.$$

(b) For $1 \leq k \leq N$ and $0 < \varepsilon < k^{-1/p}$, prove that

$$\mathcal{P}(B_p^N \cap k^{-1/p} B_\infty^N, \ell_\infty^N, \varepsilon) \geq 2^k \binom{N}{k}.$$

(c) Conclude that, for $1 \leq m < N$,

$$d_m(B_p^N, \ell_\infty^N) \geq \frac{1}{3} \min \left\{ 1, \frac{\ln(3N/m)}{6m} \right\}^{1/p}.$$

10.10. Gelfand widths of ℓ_p-balls for $p < 1$
This problem aims at extending the upper and lower bounds in Theorem 10.5 to the
Gelfand widths $d^m(B_p^N, \ell_q^N)$ with $0 < p \leq 1$ and $p < q \leq 2$.

(a) Let $\mathbf{A} \in \mathbb{R}^{m \times N}$ and $0 < p \leq 1$. Show that if every $2s$-sparse vector $\mathbf{x} \in \mathbb{R}^N$ is
the unique minimizer of $\|\mathbf{z}\|_p$ subject to $\mathbf{Az} = \mathbf{Ax}$, then $m \geq c_1 p s \ln(c_2 N/s)$
for appropriate constants $c_1, c_2 > 0$.
(b) Let $0 < p \leq 1$ and $p < q \leq 2$. Show that there exists constants $c_{p,q}$ and $C_{p,q}$
only depending on p and q such that

$$c_{p,q} \min \left\{ 1, \frac{\ln(eN/m)}{m} \right\}^{1/p - 1/q} \leq d^m(B_p^N, \ell_q^N)$$

$$\leq C_{p,q} \min \left\{ 1, \frac{\ln(eN/m)}{m} \right\}^{1/p - 1/q}.$$

10.11. Observe that Kashin's decomposition theorem also applies to ℓ_p^{2m} with $1 <
p \leq 2$ instead of ℓ_1^{2m}, i.e., observe that there are orthogonal subspaces E and $E^\perp$ of
dimension m such that

$$\alpha \, m^{1/p - 1/2} \|\mathbf{x}\|_2 \leq \|\mathbf{x}\|_p \leq \beta \, m^{1/p - 1/2} \|\mathbf{x}\|_2$$

for all $\mathbf{x} \in E$ and all $\mathbf{x} \in E^\perp$, where $\alpha, \beta > 0$ are absolute constants.

Chapter 11
Instance Optimality and Quotient Property

This chapter examines further fundamental limits of sparse recovery and properties of ℓ_1-minimization. In Sect. 11.1, we introduce the concept of ℓ_p-instance optimality for a pair of measurement matrix and reconstruction map, which requires that the reconstruction error in ℓ_p can be compared with the best s-term approximation error in ℓ_p. The minimal number of measurements to achieve ℓ_1-instance optimality is determined, complementing some results from Chap. 10. It is also revealed that ℓ_2-instance optimality is not a valid concept for the range of parameters typical to compressive sensing. In retrospect, this explains why the term $\sigma_s(\mathbf{x})_1/\sqrt{s}$ instead of $\sigma_s(\mathbf{x})_2$ appears in (uniform) ℓ_2-estimates for the reconstruction error (see, for instance, (4.16) or Theorem 6.12). Nonetheless, ℓ_1-minimization allows for a weaker form of ℓ_2-instance optimality when the measurement matrix is a subgaussian random matrix, as established in Sect. 11.4. The tools needed for the analysis of this *nonuniform instance optimality* are developed in Sects. 11.2 and 11.3, which at the same time explore another important topic. In fact, when the measurements are corrupted, a quadratically constrained ℓ_1-minimization may be used (see, for instance, Theorem 6.12). However, this requires a good estimate of the ℓ_2-norm of the measurement error. Under certain assumptions, we prove somewhat unexpectedly that equality-constrained ℓ_1-minimization also leads to very similar reconstruction estimates, even though knowledge of the measurement error is not required for this approach. The key to such results is the concept of quotient property introduced in Sect. 11.2. Stability and robustness estimates for equality-constrained ℓ_1-minimization are provided under the quotient property. It is then shown in Sect. 11.3 that different versions of the quotient property hold with high probability for Gaussian matrices and for subgaussian matrices.

S. Foucart and H. Rauhut, *A Mathematical Introduction to Compressive Sensing*,
Applied and Numerical Harmonic Analysis, DOI 10.1007/978-0-8176-4948-7_11,
© Springer Science+Business Media New York 2013

11.1   Uniform Instance Optimality

When a measurement–reconstruction scheme is assessed for s-sparse recovery, it is natural to compare the reconstruction error for a vector $\mathbf{x} \in \mathbb{C}^N$ to the error of best s-term approximation

$$\sigma_s(\mathbf{x})_p = \inf \left\{ \|\mathbf{x} - \mathbf{z}\|_p, \; \mathbf{z} \in \mathbb{C}^N \text{ is } s\text{-sparse} \right\}.$$

This motivates the introduction of the *instance optimality* concept.

Definition 11.1. Given $p \geq 1$, a pair of measurement matrix $\mathbf{A} \in \mathbb{C}^{m \times N}$ and reconstruction map $\Delta : \mathbb{C}^m \to \mathbb{C}^N$ is called ℓ_p-*instance optimal* of order s with constant $C > 0$ if

$$\|\mathbf{x} - \Delta(\mathbf{A}\mathbf{x})\|_p \leq C \, \sigma_s(\mathbf{x})_p \qquad \text{for all } \mathbf{x} \in \mathbb{C}^N.$$

In Theorems 6.12, 6.21, and 6.25, we have seen examples of ℓ_1-instance optimal pairs, i.e., a matrix $\mathbf{A}$ with small restricted isometry constants δ_{2s}, δ_{6s}, or δ_{26s}, together with a reconstruction map Δ corresponding to basis pursuit, iterative hard thresholding, or orthogonal matching pursuit, respectively. In fact, more general statements have been established where the reconstruction error was measured in ℓ_q for $q \geq 1$. With the following terminology, the previous pairs $(\mathbf{A}, \Delta)$ are mixed (ℓ_q, ℓ_1)-instance optimal.

Definition 11.2. Given $q \geq p \geq 1$, a pair of measurement matrix $\mathbf{A} \in \mathbb{C}^{m \times N}$ and reconstruction map $\Delta : \mathbb{C}^m \to \mathbb{C}^N$ is called mixed (ℓ_q, ℓ_p)-*instance optimal* of order s with constant $C > 0$ if

$$\|\mathbf{x} - \Delta(\mathbf{A}\mathbf{x})\|_q \leq \frac{C}{s^{1/p-1/q}} \, \sigma_s(\mathbf{x})_p \qquad \text{for all } \mathbf{x} \in \mathbb{C}^N.$$

Remark 11.3. The term $s^{1/p-1/q}$ in this definition is not only motivated by the results mentioned above. Indeed, since we are mainly interested in the reconstruction of compressible vectors, we want to compare the reconstruction error $\|\mathbf{x} - \Delta(\mathbf{A}\mathbf{x})\|_q$ to the error of best approximation $\sigma_s(\mathbf{x})_q$ for vectors $\mathbf{x} \in \mathbb{C}^N$ belonging to balls B_r^N or $B_{r,\infty}^N$ with $r < 1$. By considering the nonincreasing rearrangements of such vectors, we observe that

$$\sup_{\mathbf{x} \in B_{r,\infty}^N} \sigma_s(\mathbf{x})_q \asymp \frac{1}{s^{1/r-1/q}} \asymp \frac{1}{s^{1/p-1/q}} \sup_{\mathbf{x} \in B_{r,\infty}^N} \sigma_s(\mathbf{x})_p. \tag{11.1}$$

This justifies that we should compare $\|\mathbf{x} - \Delta(\mathbf{A}\mathbf{x})\|_q$ to $\sigma_s(\mathbf{x})_p / s^{1/p-1/q}$.

Our goal is to determine conditions on the number of measurements under which instance optimality can be achieved for some pair of measurement matrix and reconstruction map. We start with a useful characterization for the existence of instance optimal pairs.

Theorem 11.4. *Let $q \geq p \geq 1$ and a measurement matrix $\mathbf{A} \in \mathbb{C}^{m \times N}$ be given. If there exists a reconstruction map Δ making the pair $(\mathbf{A}, \Delta)$ mixed (ℓ_q, ℓ_p)-instance optimal of order s with constant C, then*

$$\|\mathbf{v}\|_q \leq \frac{C}{s^{1/p-1/q}} \sigma_{2s}(\mathbf{v})_p \qquad \text{for all } \mathbf{v} \in \ker \mathbf{A}. \tag{11.2}$$

Conversely, if (11.2) holds, then there exists a reconstruction map Δ making the pair $(\mathbf{A}, \Delta)$ mixed (ℓ_q, ℓ_p)-instance optimal of order s with constant $2C$.

Proof. Let us first assume that $(\mathbf{A}, \Delta)$ is a mixed (ℓ_q, ℓ_p)-instance optimal pair of order s with constant C. Given $\mathbf{v} \in \ker \mathbf{A}$, let S be an index set of s largest absolute entries of $\mathbf{v}$. The instance optimality implies $-\mathbf{v}_S = \Delta(\mathbf{A}(-\mathbf{v}_S))$. Since $\mathbf{A}(-\mathbf{v}_S) = \mathbf{A}(\mathbf{v}_{\overline{S}})$, we have $-\mathbf{v}_S = \Delta(\mathbf{A}(\mathbf{v}_{\overline{S}}))$. We now derive (11.2) from

$$\|\mathbf{v}\|_q = \|\mathbf{v}_{\overline{S}} + \mathbf{v}_S\|_q = \|\mathbf{v}_{\overline{S}} - \Delta(\mathbf{A}(\mathbf{v}_{\overline{S}}))\|_q$$

$$\leq \frac{C}{s^{1/p-1/q}} \sigma_s(\mathbf{v}_{\overline{S}})_p = \frac{C}{s^{1/p-1/q}} \sigma_{2s}(\mathbf{v})_p.$$

Conversely, let us assume that (11.2) holds for some measurement matrix $\mathbf{A}$. We define a reconstruction map by

$$\Delta(\mathbf{y}) := \operatorname{argmin}\{\sigma_s(\mathbf{z})_p \text{ subject to } \mathbf{A}\mathbf{z} = \mathbf{y}\}.$$

For $\mathbf{x} \in \mathbb{C}^N$, applying (11.2) to $\mathbf{v} := \mathbf{x} - \Delta(\mathbf{A}\mathbf{x}) \in \ker \mathbf{A}$ yields

$$\|\mathbf{x} - \Delta(\mathbf{A}\mathbf{x})\|_q \leq \frac{C}{s^{1/p-1/q}} \sigma_{2s}(\mathbf{x} - \Delta(\mathbf{A}\mathbf{x}))_p$$

$$\leq \frac{C}{s^{1/p-1/q}} \left(\sigma_s(\mathbf{x})_p + \sigma_s(\Delta(\mathbf{A}\mathbf{x}))_p\right) \leq \frac{2C}{s^{1/p-1/q}} \sigma_s(\mathbf{x})_p,$$

where we have used the triangle inequality $\sigma_{2s}(\mathbf{u} + \mathbf{v})_p \leq \sigma_s(\mathbf{u})_p + \sigma_s(\mathbf{v})_p$ and the definition of $\Delta(\mathbf{A}\mathbf{x})$. This proves that $(\mathbf{A}, \Delta)$ is a mixed (ℓ_q, ℓ_p)-instance optimal pair of order s with constant $2C$.   □

We stress that in the case $q = p = 1$ the condition (11.2) reduces to

$$\|\mathbf{v}\|_1 \leq C \sigma_{2s}(\mathbf{v})_1 \qquad \text{for all } \mathbf{v} \in \ker \mathbf{A}.$$

This is reminiscent of the null space property of order $2s$ for recovery via ℓ_1-minimization as formulated in (4.3), namely,

$$\|\mathbf{v}\|_1 < 2 \sigma_{2s}(\mathbf{v})_1 \qquad \text{for all } \mathbf{v} \in \ker \mathbf{A}.$$

The link between arbitrary instance optimal pairs $(\mathbf{A}, \Delta)$ and the pair $(\mathbf{A}, \Delta_1)$, where Δ_1 denotes the ℓ_1-minimization reconstruction map, will be further investigated in Exercise 11.5.

Theorem 11.4 enables us to prove that ℓ_2-instance optimality is not a proper concept in compressive sensing, since ℓ_2-instance optimal pairs—even of order $s = 1$—can only exist if the number m of measurements is comparable to the dimension N. Note that this assertion will be moderated in Theorems 11.23 and 11.25, where we switch from a uniform setting to a nonuniform setting.

Theorem 11.5. *If a pair of measurement matrix and reconstruction map is ℓ_2-instance optimal of order $s \geq 1$ with constant C, then*

$$m \geq c\, N \tag{11.3}$$

for some constant c depending only on C.

Proof. According to Theorem 11.4, the measurement matrix $\mathbf{A}$ in the instance optimal pair satisfies

$$\|\mathbf{v}\|_2 \leq C\,\sigma_s(\mathbf{v})_2 \qquad \text{for all } \mathbf{v} \in \ker \mathbf{A}.$$

In particular, specifying this condition to $s = 1$ yields $\|\mathbf{v}\|_2^2 \leq C^2(\|\mathbf{v}\|_2^2 - |v_j|^2)$ for all $\mathbf{v} \in \ker \mathbf{A}$ and all $j \in [N]$, i.e., $C^2|v_j|^2 \leq (C^2 - 1)\|\mathbf{v}\|_2^2$. If $(\mathbf{e}_1, \ldots, \mathbf{e}_N)$ denotes the canonical basis of $\mathbb{C}^N$, this means that $|\langle \mathbf{v}, \mathbf{e}_j \rangle| \leq C'\|\mathbf{v}\|_2$ for all $\mathbf{v} \in \ker \mathbf{A}$ and all $j \in [N]$, where $C' := \sqrt{(C^2 - 1)/C^2}$. Thus, if $\mathbf{P}$ represents the orthogonal projector onto $\ker \mathbf{A}$, we have

$$N - m \leq \dim(\ker \mathbf{A}) = \operatorname{tr}(\mathbf{P}) = \sum_{j=1}^{N} \langle \mathbf{P}\mathbf{e}_j, \mathbf{e}_j \rangle \leq \sum_{j=1}^{N} C'\|\mathbf{P}\mathbf{e}_j\|_2 \leq N\,C'.$$

This immediately implies the desired result with $c = 1 - \sqrt{(C^2 - 1)/C^2}$. $\square$

We now turn our attention to ℓ_1-instance optimality and (ℓ_q, ℓ_1)-instance optimality for $q \geq 1$. As already recalled, we have established in Chap. 6 that several reconstruction algorithms give rise to mixed (ℓ_q, ℓ_1)-instance optimal pairs $(\mathbf{A}, \Delta)$, provided the measurement matrix $\mathbf{A}$ has small restricted isometry constants. Moreover, Theorem 9.6 guarantees that this occurs with high probability for subgaussian random matrices $\mathbf{A}$ provided $m \geq c\,s \ln(eN/s)$ for some constant $c > 0$. Theorems 11.6 and 11.7 below show that a smaller number m of measurements is impossible. Thereby, we remove the earlier restrictions of Theorem 10.7 that $q > 1$ and that s is larger than some constant. In the case $q = 1$, Gelfand width estimates can no longer be used for the proof, but the tools developed in Chap. 10 are still appropriate to deal with this more delicate case.

Theorem 11.6. *If a pair of measurement matrix $\mathbf{A} \in \mathbb{C}^{m \times N}$ and reconstruction map $\Delta : \mathbb{C}^m \to \mathbb{C}^N$ is ℓ_1-instance optimal of order s with constant C, then*

$$m \geq c\,s\,\ln(eN/s) \tag{11.4}$$

for some constant c depending only on C.

Proof. By Lemma 10.12, there exist $n \geq (N/4s)^{s/2}$ index sets $S_1, \ldots, S_n$ of size s satisfying $\mathrm{card}(S_i \cap S_j) < s/2$ for all $1 \leq i \neq j \leq n$. We consider the s-sparse vectors $\mathbf{x}^1, \ldots, \mathbf{x}^n$ already defined in (10.13) by

$$
x_k^i = \begin{cases} 1/s & \text{if } k \in S_i, \\ 0 & \text{if } k \notin S_i. \end{cases}
$$

We notice that $\|\mathbf{x}^i\|_1 = 1$ and that $\|\mathbf{x}^i - \mathbf{x}^j\|_1 > 1$ for all $1 \leq i \neq j \leq n$. Setting $\rho := 1/(2(C+1))$, we claim that $\{\mathbf{A}(\mathbf{x}^i + \rho B_1^N), i \in [n]\}$ is a disjoint collection of subsets of $\mathbf{A}(\mathbb{C}^N)$, which has dimension $d \leq m$. Indeed, if there existed indices $i \neq j$ and vectors $\mathbf{z}, \mathbf{z}' \in \rho B_1^N$ such that $\mathbf{A}(\mathbf{x}^i + \mathbf{z}) = \mathbf{A}(\mathbf{x}^j + \mathbf{z}')$, then a contradiction would follow from

$$
\|\mathbf{x}^i - \mathbf{x}^j\|_1 = \|(\mathbf{x}^i + \mathbf{z} - \Delta(\mathbf{A}(\mathbf{x}^i + \mathbf{z}))) - (\mathbf{x}^j + \mathbf{z}' - \Delta(\mathbf{A}(\mathbf{x}^j + \mathbf{z}'))) - \mathbf{z} + \mathbf{z}'\|_1
$$

$$
\leq \|\mathbf{x}^i + \mathbf{z} - \Delta(\mathbf{A}(\mathbf{x}^i + \mathbf{z}))\|_1 + \|\mathbf{x}^j + \mathbf{z}' - \Delta(\mathbf{A}(\mathbf{x}^j + \mathbf{z}'))\|_1 + \|\mathbf{z}\|_1 + \|\mathbf{z}'\|_1
$$

$$
\leq C\,\sigma_s(\mathbf{x}^i + \mathbf{z})_1 + C\,\sigma_s(\mathbf{x}^j + \mathbf{z}')_1 + \|\mathbf{z}\|_1 + \|\mathbf{z}'\|_1
$$

$$
\leq C\,\|\mathbf{z}\|_1 + C\,\|\mathbf{z}'\|_1 + \|\mathbf{z}\|_1 + \|\mathbf{z}'\|_1 \leq 2\,(C+1)\,\rho = 1.
$$

Next, we observe that the collection $\{\mathbf{A}(\mathbf{x}^i + \rho B_1^N), i \in [n]\}$ is contained in $(1 + \rho)\mathbf{A}(B_1^N)$. Therefore, considering the volume of this collection (in the space $\mathbf{A}(\mathbb{C}^N)$), we deduce

$$
\sum_{i \in [n]} \mathrm{vol}\left(\mathbf{A}(\mathbf{x}^i + \rho B_1^N)\right) \leq \mathrm{vol}\left((1 + \rho)\mathbf{A}(B_1^N)\right).
$$

Using homogeneity and translation invariance of the volume and noting the d-dimensional complex case is equivalent to the $2d$-dimensional real case, we derive

$$
n\,\rho^{2d}\,\mathrm{vol}\left(\mathbf{A}(B_1^N)\right) \leq (1 + \rho)^{2d}\,\mathrm{vol}\left(\mathbf{A}(B_1^N)\right).
$$

This yields

$$
\left(\frac{N}{4s}\right)^{s/2} \leq n \leq \left(1 + \frac{1}{\rho}\right)^{2d} = (2C + 3)^{2d} \leq (2C + 3)^{2m}. \tag{11.5}
$$

Taking the logarithms in (11.5) on the one hand and on the other hand remarking that the pair $(\mathbf{A}, \Delta)$ ensures exact recovery of s-sparse vectors, we obtain

$$
\frac{m}{s} \geq \frac{\ln(N/4s)}{4\ln(2C+3)}, \qquad \frac{m}{s} \geq 2.
$$

Combining these two inequalities leads to

$$\left(4\ln(2C+3)+2\right)\frac{m}{s} \geq \ln(N/4s)+\ln(e^4) = \ln(e^4 N/4s) \geq \ln(eN/s).$$

This is the desired result where $c = 1/(4(\ln(2C+3)+2))$.   □

With the help of Theorem 11.6, we can prove that the requirement (11.4) on the number of measurements is also imposed by mixed (ℓ_q, ℓ_1)-instance optimality when $q > 1$. This is formally stated in the following theorem.

Theorem 11.7. *Given* $q > 1$, *if a pair of measurement matrix and reconstruction map is mixed* (ℓ_q, ℓ_1)-*instance optimal of order* s *with constant* C, *then*

$$m \geq c\, s\, \ln(eN/s)$$

for some constant c *depending only on* C.

The proof is a simple consequence of Theorem 11.6 and of the following lemma, which roughly says that mixed (ℓ_q, ℓ_1)-instance optimality is preserved when decreasing q.

Lemma 11.8. *Given* $q \geq q' \geq p \geq 1$, *if a pair* $(\mathbf{A}, \Delta)$ *is mixed* (ℓ_q, ℓ_p)-*instance optimal of order* s *with constant* C, *then there is a reconstruction map* Δ' *making the pair* $(\mathbf{A}, \Delta')$ *mixed* $(\ell_{q'}, \ell_p)$-*instance optimal of order* s *with constant* C' *depending only on* C.

Proof. Let us consider a vector $\mathbf{v} \in \ker \mathbf{A}$. Since the pair $(\mathbf{A}, \Delta)$ is mixed (ℓ_q, ℓ_p)-instance optimal of order s with constant C, Theorem 11.4 yields

$$\|\mathbf{v}\|_q \leq \frac{C}{s^{1/p-1/q}}\sigma_{2s}(\mathbf{v})_p.$$

Let S denote an index set of $3s$ largest entries of $\mathbf{v}$ in modulus. We have

$$\|\mathbf{v}_S\|_{q'} \leq (3s)^{1/q'-1/q}\|\mathbf{v}_S\|_q \leq (3s)^{1/q'-1/q}\|\mathbf{v}\|_q$$

$$\leq (3s)^{1/q'-1/q}\frac{C}{s^{1/p-1/q}}\sigma_{2s}(\mathbf{v})_p = \frac{3^{1/q'-1/q}\,C}{s^{1/p-1/q'}}\sigma_{2s}(\mathbf{v})_p$$

$$\leq \frac{3\,C}{s^{1/p-1/q'}}\sigma_{2s}(\mathbf{v})_p.$$

Moreover, we derive from Proposition 2.3 that

$$\|\mathbf{v}_{\overline{S}}\|_{q'} \leq \frac{1}{s^{1/p-1/q'}}\sigma_{2s}(\mathbf{v})_p.$$

Thus, we obtain

$$\|\mathbf{v}\|_{q'} \leq \|\mathbf{v}_S\|_{q'} + \|\mathbf{v}_{\overline{S}}\|_{q'} \leq \frac{3\,C+1}{s^{1/p-1/q'}}\sigma_{2s}(\mathbf{v})_p.$$

In view of the converse part of Theorem 11.4, the desired result holds with $C' = 2(3\,C + 1)$. □

In parallel with Lemma 11.8, it can be proved that mixed (ℓ_q, ℓ_p)-instance optimality is also preserved when decreasing p instead of q; see Exercise 11.2. This exercise also complements Theorem 11.7 by determining a lower bound on the number of measurements to achieve mixed (ℓ_q, ℓ_p)-instance optimality with $p > 1$, hence showing that it is not achievable in the regime $m \asymp s\ln(eN/m)$.

11.2 Robustness and Quotient Property

In addition to stability—or instance optimality in the terminology of the previous section—the robustness of recovery algorithms under measurement error is very important. We have seen in Chaps. 4 and 6 that quadratically constrained basis pursuit, that is, the reconstruction map given by

$$\Delta = \Delta_{1,\eta,\mathbf{A}}(\mathbf{y}) := \operatorname{argmin}\{\|\mathbf{z}\|_1 \text{ subject to } \|\mathbf{A}\mathbf{z} - \mathbf{y}\|_2 \leq \eta\} \qquad (11.6)$$

applied to the inaccurate measurements $\mathbf{y} = \mathbf{A}\mathbf{x} + \mathbf{e}$ with $\|\mathbf{e}\|_2 \leq \eta$ provides a stability estimate of the form

$$\|\mathbf{x} - \Delta(\mathbf{A}\mathbf{x} + \mathbf{e})\|_2 \leq \frac{C}{\sqrt{s}}\sigma_s(\mathbf{x})_1 + D\eta \qquad (11.7)$$

under suitable hypotheses on $\mathbf{A}$—for instance, if its restricted isometry constants satisfy $\delta_{2s} < 0.62$; see Theorem 6.12. The drawback of this result is that running quadratically constrained basis pursuit requires a proper guess of the level η of measurement error, but such an estimate may not a priori be available.

In contrast, Theorems 6.21, 6.25, and 6.28 showed that other algorithms such as iterative hard thresholding, orthogonal matching pursuit, and compressive sampling matching pursuit do provide robustness estimates of the type

$$\|\mathbf{x} - \Delta(\mathbf{A}\mathbf{x} + \mathbf{e})\|_2 \leq \frac{C}{\sqrt{s}}\sigma_s(\mathbf{x})_1 + D\|\mathbf{e}\|_2, \qquad (11.8)$$

without the need to estimate $\|\mathbf{e}\|_2$. However, these algorithms require as part of their input an estimate of the sparsity level s, which is a priori not available either.

Ideally, we would like to have a tractable reconstruction map Δ (together with a measurement matrix $\mathbf{A}$) that provides reconstruction estimates of the form (11.7) without the need to specify neither the level η of measurement error nor the sparsity s. In this context, we will investigate the use of equality-constrained ℓ_1-minimization

$$\Delta_1(\mathbf{y}) = \Delta_{1,\mathbf{A}}(\mathbf{y}) := \operatorname{argmin}\{\|\mathbf{z}\|_1 \text{ subject to } \mathbf{Az} = \mathbf{y}\}, \tag{11.9}$$

with subgaussian random matrices $\mathbf{A}$ when flawed measurements $\mathbf{y} = \mathbf{Ax} + \mathbf{e}$ are given. Obviously, Δ_1 neither requires an estimate of $\|\mathbf{e}\|_2$ (or some other characteristics of $\mathbf{e}$) nor of the sparsity level s.

In order to put our results into context, we first recall the error estimates obtained for quadratically constrained basis pursuit $\Delta_{1,\eta}$. Let $\mathbf{A}$ be the realization of a renormalized $m \times N$ subgaussian matrix with $m \geq cs\ln(eN/m)$ and let $1 \leq p \leq 2$. Then, with high probability, the robustness estimate takes the form

$$\|\mathbf{x} - \Delta_{1,\eta,\mathbf{A}}(\mathbf{Ax} + \mathbf{e})\|_p \leq \frac{C}{s^{1-1/p}}\sigma_s(\mathbf{x})_1 + Ds^{1/p-1/2}\eta, \tag{11.10}$$

valid for all $\mathbf{x} \in \mathbb{C}^N$ and $\mathbf{e} \in \mathbb{C}^m$ with $\|\mathbf{e}\|_2 \leq \eta$. Thus, setting

$$s_* := s_*(m, N) := \frac{m}{\ln(eN/m)},$$

we derive that, for any $1 \leq p \leq 2$, if $s \leq s_*/c$, then

$$\|\mathbf{x} - \Delta_{1,\eta,\mathbf{A}}(\mathbf{Ax} + \mathbf{e})\|_p \leq \frac{C}{s^{1-1/p}}\sigma_s(\mathbf{x})_1 + D's_*^{1/p-1/2}\eta$$

holds for all $\mathbf{x} \in \mathbb{C}^N$ and $\mathbf{e} \in \mathbb{C}^m$ with $\|\mathbf{e}\|_2 \leq \eta$.

The purpose of this section and the next one is to show that such robustness results can also be achieved for the equality-constrained ℓ_1-minimization $\Delta_1 = \Delta_{1,\mathbf{A}}$. (Note that we will omit to write the subscript $\mathbf{A}$ since it is always clear from the context, for instance, when writing the pair $(\mathbf{A}, \Delta_1)$ for the pair $(\mathbf{A}, \Delta_{1,\mathbf{A}})$.) The measurement process involves Gaussian and subgaussian matrices. These matrices, introduced in Definition 9.1, are required to have entries with variance 1. Here, we use renormalized measurement matrices $\tilde{\mathbf{A}}$ with entries of variance $1/m$, so that $\tilde{\mathbf{A}} = m^{-1/2}\mathbf{A}$, where $\mathbf{A}$ is an unnormalized subgaussian matrix. The first main result pertains to the Gaussian case.

Theorem 11.9. *Let $1 \leq p \leq 2$ and let $\tilde{\mathbf{A}} = \frac{1}{\sqrt{m}}\mathbf{A}$ where $\mathbf{A}$ is an $m \times N$ Gaussian random matrix. If*

$$N \geq c_2 m \qquad and \qquad s \leq c_3 s_* = \frac{c_3 m}{\ln(eN/m)},$$

then with probability at least $1 - 3\exp(-c_1 m)$ *the* ℓ_p-*error estimates*

$$\|\mathbf{x} - \Delta_1(\tilde{\mathbf{A}}\mathbf{x} + \mathbf{e})\|_p \leq \frac{C}{s^{1-1/p}}\sigma_s(\mathbf{x})_1 + Ds_*^{1/p-1/2}\|\mathbf{e}\|_2 \tag{11.11}$$

are valid for all $\mathbf{x} \in \mathbb{C}^N$ *and* $\mathbf{e} \in \mathbb{C}^m$. *The constants* $c_1, c_2, c_3, C, D > 0$ *are universal.*

The above assumption that $N \geq c_2 m$ is not a severe restriction in compressive sensing, where the number of measurements is typically much smaller than the ambient dimension.

The second main result concerns the more general subgaussian matrices. In this case, the ℓ_2-norm on the measurement error has to be slightly adjusted to

$$\|\mathbf{e}\|^{\left(\sqrt{\ln(eN/m)}\right)} := \max\left\{\|\mathbf{e}\|_2, \sqrt{\ln(eN/m)}\|\mathbf{e}\|_\infty\right\}.$$

In fact, for Bernoulli matrices, the statement of the previous theorem does not hold without this modification of the ℓ_2-norm as argued in the next section; see p. 350. We note that in practice the modified norm $\|\mathbf{e}\|^{\left(\sqrt{\ln(eN/m)}\right)}$ does not usually differ much from the ℓ_2-norm. In fact, the two norms are only different for approximately sparse vectors $\mathbf{e}$. For instance, if $\mathbf{e}$ is a random noise vector, then this occurs with very small probability.

Theorem 11.10. *Let* $1 \leq p \leq 2$ *and let* $\tilde{\mathbf{A}} = \frac{1}{\sqrt{m}}\mathbf{A}$ *where* $\mathbf{A}$ *is an* $m \times N$ *subgaussian matrix with symmetric entries and with subgaussian parameter* c *in* (9.2). *There exist constants* $c_1, c_2, c_3, C, D > 0$ *depending only on* c *such that if*

$$N \geq c_2 m \qquad and \qquad s \leq c_3 s_* = \frac{c_3 m}{\ln(eN/m)},$$

then with probability at least $1 - 5\exp(-c_1 m)$ *the* ℓ_p-*error estimates*

$$\|\mathbf{x} - \Delta_1(\tilde{\mathbf{A}}\mathbf{x} + \mathbf{e})\|_p \leq \frac{C}{s^{1-1/p}}\sigma_s(\mathbf{x})_1 + Ds_*^{1/p-1/2}\|\mathbf{e}\|^{\left(\sqrt{\ln(eN/m)}\right)} \tag{11.12}$$

are valid for all $\mathbf{x} \in \mathbb{C}^N$ *and* $\mathbf{e} \in \mathbb{C}^m$.

The symmetry assumption on the entries of $\mathbf{A}$ is not a severe restriction and is satisfied in particular for Bernoulli matrices.

The fundamental tool for establishing Theorems 11.9 and 11.10 is a new property of the measurement matrix called the *quotient property*. In this section, we show that the estimates (11.11) and (11.12) are implied by the quotient property, and in the next section we establish the quotient property for random matrices.

Definition 11.11. Given $q \geq 1$, a measurement matrix $\mathbf{A} \in \mathbb{C}^{m \times N}$ is said to possess the ℓ_q-*quotient property* with constant d relative to a norm $\|\cdot\|$ on $\mathbb{C}^m$ if, for all $\mathbf{e} \in \mathbb{C}^m$, there exists $\mathbf{u} \in \mathbb{C}^N$ with

$$\mathbf{Au} = \mathbf{e} \quad \text{and} \quad \|\mathbf{u}\|_q \le d\, s_*^{1/q-1/2} \|\mathbf{e}\|,$$

where $s_* = m/\ln(eN/m)$.

We point out that the quotient property is a natural assumption to make, since it is implied by the robustness estimate

$$\|\mathbf{x} - \Delta_1(\mathbf{Ax} + \mathbf{e})\|_q \le \frac{C}{s^{1-1/q}} \sigma_s(\mathbf{x})_1 + D s_*^{1/q-1/2} \|\mathbf{e}\|. \tag{11.13}$$

Indeed, setting $\mathbf{x} = \mathbf{0}$ in (11.13) gives $\|\Delta_1(\mathbf{e})\|_q \le D s_*^{1/q-1/2} \|\mathbf{e}\|$. This implies—if $q = 1$, then it is equivalent to—the ℓ_q-quotient property by taking $\mathbf{u} = \Delta_1(\mathbf{e})$. The ℓ_1-quotient property asserts that the image under $\mathbf{A}$ of the ℓ_1-ball of radius $d\sqrt{s_*}$ covers the unit ball relative to $\|\cdot\|$. The terminology *quotient property* is explained by a reformulation involving the quotient norm of the set $[\mathbf{e}] = \mathbf{u} + \ker \mathbf{A}$ of preimages of a vector $\mathbf{e} = \mathbf{Au} \in \mathbb{C}^m$, namely,

$$\|[\mathbf{e}]\|_{\ell_q/\ker \mathbf{A}} := \inf \left\{ \|\mathbf{u} + \mathbf{v}\|_q, \mathbf{v} \in \ker \mathbf{A} \right\} = \inf \left\{ \|\mathbf{z}\|_q, \mathbf{Az} = \mathbf{e} \right\}.$$

Thus, the ℓ_q-quotient property is equivalent to

$$\|[\mathbf{e}]\|_{\ell_q/\ker \mathbf{A}} \le d s_*^{1/q-1/2} \|\mathbf{e}\| \quad \text{for all } \mathbf{e} \in \mathbb{C}^m.$$

Another reformulation, used in Sect. 11.3 to establish the quotient property for random matrices, involves the dual norm of the norm $\|\cdot\|$, but is not needed at this point. The rest of this section is of deterministic nature, and Theorems 11.9 and 11.10 will become simple consequences of Theorem 11.12 below as soon as we verify that its two hypotheses hold with high probability for random matrices. Note that the first hypothesis—the robust null space property—is already acquired. Incidentally, we point out that the robust null space property is also a natural assumption to make, since it is necessary for the desired estimate (11.13), as we can see by setting $\mathbf{x} = \mathbf{v} \in \mathbb{C}^N$ and $\mathbf{e} = -\mathbf{Av} \in \mathbb{C}^m$—see also Remark 4.24.

Theorem 11.12. *If a matrix* $\mathbf{A} \in \mathbb{C}^{m \times N}$ *satisfies*

- *The* ℓ_2-*robust null space property of order* $c s_*$ *with* $s_* = m/\ln(eN/m)$ *and constants* $0 < \rho < 1$ *and* $\tau > 0$ *relative to a norm* $\|\cdot\|$,
- *The* ℓ_1-*quotient property with constant* d *relative to the norm* $\|\cdot\|$,

then, for any $s \le c s_*$ *and for all* $\mathbf{x} \in \mathbb{C}^N$ *and* $\mathbf{e} \in \mathbb{C}^m$,

$$\|\mathbf{x} - \Delta_1(\mathbf{Ax} + \mathbf{e})\|_q \le \frac{C}{s^{1-1/q}} \sigma_s(\mathbf{x})_1 + D s_*^{1/q-1/2} \|\mathbf{e}\|, \qquad 1 \le q \le 2.$$

The constants C *and* D *depend only on* ρ, τ, c, *and* d.

Remark 11.13. The specific value of $s_* = m/\ln(eN/m)$ is not very important for the statement of this theorem. If s_* is kept as a free parameter in the definition of the quotient property, then the theorem remains true for an arbitrary value of s_*. Treating s_* as a parameter, we realize that the first requirement—the ℓ_2-robust null space property of order cs_*—becomes easier to satisfy for smaller values of s_*, while the second requirement—the ℓ_1-quotient property—becomes easier to satisfy for larger values of s_*. Therefore, it is a priori not clear whether there exists a value of s_* for which both requirements are satisfied simultaneously. We will see that this is indeed the case for subgaussian random matrices, for certain norms $\|\cdot\|$, and for our choice $s_* = m/\ln(eN/m)$.

The next two lemmas account for Theorem 11.12. The first lemma asserts that the mixed instance optimality and the simultaneous quotient property—to be defined below—together yield the desired robustness estimates. The second lemma asserts that the robust null space property and the ℓ_1-quotient property together yield the simultaneous quotient property.

Definition 11.14. Given $q \geq 1$, a matrix $\mathbf{A} \in \mathbb{C}^{m \times N}$ is said to have the *simultaneous (ℓ_q, ℓ_1)-quotient property* with constants d and d' relative to a norm $\|\cdot\|$ on $\mathbb{C}^m$ if, for all $\mathbf{e} \in \mathbb{C}^m$, there exists $\mathbf{u} \in \mathbb{C}^N$ with

$$\mathbf{A}\mathbf{u} = \mathbf{e} \quad \text{and} \quad \begin{cases} \|\mathbf{u}\|_q \leq d\, s_*^{1/q-1/2}\|\mathbf{e}\|, \\ \|\mathbf{u}\|_1 \leq \quad d'\, s_*^{1/2}\|\mathbf{e}\|. \end{cases}$$

The two lemmas mentioned above read as follows.

Lemma 11.15. *Given $q \geq 1$, if a measurement matrix $\mathbf{A} \in \mathbb{C}^{m \times N}$ and a reconstruction map Δ are such that*

- *$(\mathbf{A}, \Delta)$ is mixed (ℓ_q, ℓ_1)-instance optimal of order cs_* with constant C,*
- *$\mathbf{A}$ has the simultaneous (ℓ_q, ℓ_1)-quotient property with constants d and d' relative to a norm $\|\cdot\|$,*

then, for all $\mathbf{x} \in \mathbb{C}^N$, $\mathbf{e} \in \mathbb{C}^m$ and $s \leq cs_*$,

$$\|\mathbf{x} - \Delta(\mathbf{A}\mathbf{x} + \mathbf{e})\|_q \leq \frac{C}{s^{1-1/q}}\sigma_s(\mathbf{x})_1 + Ds_*^{1/q-1/2}\|\mathbf{e}\|, \qquad D := \frac{Cd'}{c^{1-1/q}} + d.$$

Proof. For $\mathbf{x} \in \mathbb{C}^N$ and $\mathbf{e} \in \mathbb{C}^m$, the simultaneous (ℓ_q, ℓ_1)-quotient property ensures the existence of $\mathbf{u} \in \mathbb{C}^N$ satisfying

$$\mathbf{A}\mathbf{u} = \mathbf{e} \quad \text{and} \quad \begin{cases} \|\mathbf{u}\|_q \leq d\, s_*^{1/q-1/2}\|\mathbf{e}\|, \\ \|\mathbf{u}\|_1 \leq \quad d'\, s_*^{1/2}\|\mathbf{e}\|. \end{cases} \tag{11.14}$$

Using the instance optimality, we then derive

$$\|\mathbf{x} - \Delta(\mathbf{A}\mathbf{x} + \mathbf{e})\|_q = \|\mathbf{x} - \Delta(\mathbf{A}(\mathbf{x} + \mathbf{u}))\|_q \le \|\mathbf{x} + \mathbf{u} - \Delta(\mathbf{A}(\mathbf{x} + \mathbf{u}))\|_q + \|\mathbf{u}\|_q$$

$$\le \frac{C}{s^{1-1/q}} \sigma_s(\mathbf{x} + \mathbf{u})_1 + \|\mathbf{u}\|_q$$

$$\le \frac{C}{s^{1-1/q}} (\sigma_s(\mathbf{x})_1 + \|\mathbf{u}\|_1) + \|\mathbf{u}\|_q.$$

Substituting the inequalities of (11.14) into the latter yields the result for $s = cs_*$. The result for $s \le cs_*$ follows by monotonicity of $\sigma_s(\mathbf{x})_1 / s^{1-1/q}$. □

Lemma 11.16. *Given $q \ge 1$ and a norm $\| \cdot \|$ on $\mathbb{C}^m$, if a measurement matrix $\mathbf{A} \in \mathbb{C}^{m \times N}$ satisfies*

- *The ℓ_q-robust null space property of order $c s_*$ with constants $\rho > 0$ and $\tau > 0$ relative to $s_*^{1/q-1/2} \| \cdot \|$,*
- *The ℓ_1-quotient property with constant d relative to $\| \cdot \|$,*

then the matrix $\mathbf{A}$ also satisfies the simultaneous (ℓ_q, ℓ_1)-quotient property relative to $\| \cdot \|$ with constants $D := (1 + \rho)d/c^{1-1/q} + \tau$ and $D' := d$.

Proof. Let us consider a vector $\mathbf{e} \in \mathbb{C}^m$. By the ℓ_1-quotient property, there exists $\mathbf{u} \in \mathbb{C}^N$ such that $\mathbf{A}\mathbf{u} = \mathbf{e}$ and $\|\mathbf{u}\|_1 \le ds_*^{1/2}\|\mathbf{e}\|$. Next, we establish the estimate $\|\mathbf{u}\|_q \le Ds_*^{1/q-1/2}\|\mathbf{e}\|$ for some constant D. For an index set S of cs_* largest absolute entries of $\mathbf{u}$, we first use Proposition 2.3 to derive

$$\|\mathbf{u}_{\overline{S}}\|_q \le \frac{1}{(c\,s_*)^{1-1/q}} \|\mathbf{u}\|_1.$$

The ℓ_q-robust null space property of order $c s_*$ yields

$$\|\mathbf{u}_S\|_q \le \frac{\rho}{(c\,s_*)^{1-1/q}} \|\mathbf{u}_{\overline{S}}\|_1 + \tau s_*^{1/q-1/2}\|\mathbf{A}\mathbf{u}\| \le \frac{\rho}{(c\,s_*)^{1-1/q}} \|\mathbf{u}\|_1 + \tau s_*^{1/q-1/2}\|\mathbf{e}\|.$$

It follows that

$$\|\mathbf{u}\|_q = \|\mathbf{u}_{\overline{S}} + \mathbf{u}_S\|_q \le \|\mathbf{u}_{\overline{S}}\|_q + \|\mathbf{u}_S\|_q \le \frac{1+\rho}{(c\,s_*)^{1-1/q}} \|\mathbf{u}\|_1 + \tau s_*^{1/q-1/2}\|\mathbf{e}\|.$$

The estimate $\|\mathbf{u}\|_1 \le ds_*^{1/2}\|\mathbf{e}\|$ yields the desired result. □

Now that Lemmas 11.15 and 11.16 have been established, Theorem 11.12 can be derived with the help of results from Chap. 4.

Proof (of Theorem 11.12). We assume that $\mathbf{A} \in \mathbb{C}^{m \times N}$ satisfies the ℓ_2-robust null space property of order $c s_*$ with constant $0 < \rho < 1$ and $\tau > 0$ relative to $\| \cdot \|$, as well as the ℓ_1-quotient property with constant d relative to $\| \cdot \|$. Then, for any

$1 \leq q \leq 2$, Definition 4.21 and the considerations afterwards ensure that $\mathbf{A}$ satisfies the ℓ_q-robust null space property of order cs_* with constant $0 < \rho < 1$ and $\tau c^{1/q-1/2} > 0$ relative to $s_*^{1/q-1/2} \| \cdot \|$. Lemma 11.16 now implies that $\mathbf{A}$ satisfies the simultaneous (ℓ_q, ℓ_1)-quotient property with constants $D = (1+\rho)d/c^{1-1/q} + \tau c^{1/q-1/2} \leq (1+\rho)d/\min\{1,c\}^{1/2} + \tau \max\{1,c\}^{1/2}$ and $D' = d$. Next, for any $1 \leq q \leq 2$, Theorem 4.25 ensures that the pair $(\mathbf{A}, \Delta_1)$ is mixed (ℓ_q, ℓ_1)-instance optimal of any order $s \leq cs_*$ with constant $C = (1+\rho)^2/(1+\rho)$. Lemma 11.15 finally yields the desired estimate with constants depending on ρ, τ, c, and d.   □

11.3   Quotient Property for Random Matrices

In this section, we prove the ℓ_1-quotient property for certain random matrices. First, we focus on Gaussian matrices, where the ℓ_1-quotient property holds relative to the ℓ_2-norm. Second, we analyze general subgaussian random matrices, where the ℓ_1-quotient property holds relative to a slight alteration of the ℓ_2-norm. The basis of both arguments is a convenient reformulation of the quotient property involving the dual norm of a norm $\| \cdot \|$, i.e., $\|\mathbf{e}\|_* = \sup_{\|\mathbf{y}\|=1} |\langle \mathbf{y}, \mathbf{e}\rangle|$ for $\mathbf{e} \in \mathbb{C}^m$; see Definition A.3.

Lemma 11.17. *For $q \geq 1$, a matrix $\mathbf{A} \in \mathbb{C}^{m \times N}$ has the ℓ_q-quotient property with constant d relative to a norm $\| \cdot \|$ if and only if*

$$\|\mathbf{e}\|_* \leq d\, s_*^{1/q-1/2} \|\mathbf{A}^*\mathbf{e}\|_{q^*} \qquad \text{for all } \mathbf{e} \in \mathbb{C}^m, \tag{11.15}$$

where $s_* = \dfrac{m}{\ln(eN/m)}$ and where $q^* = \dfrac{q}{q-1}$ is the conjugate exponent of q.

Proof. Let us assume that $\mathbf{A}$ has the ℓ_q-quotient property. For $\mathbf{e} \in \mathbb{C}^m$, we have $\|\mathbf{e}\|_* = \langle \mathbf{y}, \mathbf{e}\rangle$ for some $\mathbf{y} \in \mathbb{C}^m$ with $\|\mathbf{y}\| = 1$. The vector $\mathbf{y}$ can be written as $\mathbf{y} = \mathbf{A}\mathbf{u}$ for some $\mathbf{u} \in \mathbb{C}^N$ with $\|\mathbf{u}\|_q \leq ds_*^{1/q-1/2}$. We deduce (11.15) from

$$\|\mathbf{e}\|_* = \langle \mathbf{A}\mathbf{u}, \mathbf{e}\rangle = \langle \mathbf{u}, \mathbf{A}^*\mathbf{e}\rangle \leq \|\mathbf{u}\|_q \|\mathbf{A}^*\mathbf{e}\|_{q^*} \leq ds_*^{1/q-1/2} \|\mathbf{A}^*\mathbf{e}\|_{q^*}.$$

Conversely, let us assume that (11.15) holds. We consider the case $q > 1$ first. If $\mathbf{e} = \mathbf{0}$, then the statement is trivial. Thus, we choose $\mathbf{e} \in \mathbb{C}^m \setminus \{\mathbf{0}\}$ and let $\mathbf{u} \in \mathbb{C}^N \setminus \{\mathbf{0}\}$ be a minimizer of $\|\mathbf{z}\|_q$ subject to $\mathbf{A}\mathbf{z} = \mathbf{e}$. Our goal is to show that $\|\mathbf{u}\|_q \leq ds_*^{1/q-1/2} \|\mathbf{e}\|$. Let us fix a vector $\mathbf{v} \in \ker \mathbf{A}$. Given $\tau = te^{i\theta}$ with $t > 0$ small enough to have $\mathbf{u} + \tau\mathbf{v} \neq \mathbf{0}$, we consider the vector $\mathbf{w}^\tau \in \mathbb{C}^N$ whose entries are given by

$$w_j^\tau := \frac{\operatorname{sgn}(u_j + \tau v_j)\,|u_j + \tau v_j|^{q-1}}{\|\mathbf{u} + \tau\mathbf{v}\|_q^{q-1}}, \qquad j \in [N].$$

We notice that $\langle \mathbf{w}^\tau, \mathbf{u} + \tau\mathbf{v} \rangle = \|\mathbf{u} + \tau\mathbf{v}\|_q$ with $\|\mathbf{w}^\tau\|_{q^*} = 1$. The vector $\mathbf{w} := \lim_{\tau \to 0} \mathbf{w}^\tau$ is well defined and independent of $\mathbf{v}$, thanks to the assumption $q > 1$. It satisfies $\langle \mathbf{w}, \mathbf{u} \rangle = \|\mathbf{u}\|_q$ with $\|\mathbf{w}\|_{q^*} = 1$. The definition of $\mathbf{u}$ yields

$$\mathrm{Re}\langle \mathbf{w}^\tau, \mathbf{u} \rangle \leq \|\mathbf{u}\|_q \leq \|\mathbf{u} + \tau\mathbf{v}\|_q = \mathrm{Re}\langle \mathbf{w}^\tau, \mathbf{u} + \tau\mathbf{v} \rangle,$$

so that $\mathrm{Re}\langle \mathbf{w}^\tau, e^{i\theta}\mathbf{v} \rangle \geq 0$. Taking the limit as t tends to zero, we obtain $\mathrm{Re}\langle \mathbf{w}, e^{i\theta}\mathbf{v} \rangle \geq 0$ independently of θ; hence, $\langle \mathbf{w}, \mathbf{v} \rangle = 0$. Since this is true for all $\mathbf{v} \in \ker \mathbf{A}$, we have $\mathbf{w} \in (\ker \mathbf{A})^\perp = \mathrm{ran}\,\mathbf{A}^*$. Therefore, we can write $\mathbf{w} = \mathbf{A}^*\mathbf{y}$ for some $\mathbf{y} \in \mathbb{C}^m$. According to (11.15) and $\|\mathbf{w}\|_{q^*} = 1$, we have $\|\mathbf{y}\|_* \leq ds_*^{1/q-1/2}$. It follows that

$$\|\mathbf{u}\|_q = \langle \mathbf{w}, \mathbf{u} \rangle = \langle \mathbf{A}^*\mathbf{y}, \mathbf{u} \rangle = \langle \mathbf{y}, \mathbf{A}\mathbf{u} \rangle = \langle \mathbf{y}, \mathbf{e} \rangle \leq \|\mathbf{y}\|_*\|\mathbf{e}\| \leq ds_*^{1/q-1/2}\|\mathbf{e}\|.$$

This establishes the ℓ_q-quotient property in the case $q > 1$. We use an approximation argument for the case $q = 1$. To this end, let us consider a sequence of numbers $q_n > 1$ converging to 1. For each n, in view of $\|\mathbf{A}^*\mathbf{e}\|_\infty \leq \|\mathbf{A}^*\mathbf{e}\|_{q_n^*}$, the property (11.15) for $q = 1$ implies a corresponding property for $q = q_n$ when d is changed to ds_*^{1/q_n^*}. Given $\mathbf{e} \in \mathbb{C}^m$, the preceding argument yields a sequence of vectors $\mathbf{u}^n \in \mathbb{C}^N$ with $\mathbf{A}\mathbf{u}^n = \mathbf{e}$ and $\|\mathbf{u}^n\|_{q_n} \leq ds_*^{1/q_n^*}s_*^{1/q_n-1/2}\|\mathbf{e}\| = ds_*^{1/2}\|\mathbf{e}\|$. Since the sequence $(\mathbf{u}^n)$ is bounded in the ℓ_∞-norm, it has a convergent subsequence. Denoting by $\mathbf{u} \in \mathbb{C}^N$ its limit, we obtain $\mathbf{A}\mathbf{u} = \mathbf{e}$ and $\|\mathbf{u}\|_1 = \lim_{n\to\infty} \|\mathbf{u}^n\|_{q_n} \leq ds_*^{1/2}\|\mathbf{e}\|$. This settles the case $q = 1$. $\square$

Remark 11.18. In the case of a real matrix $\mathbf{A}$, we can also consider a real version of the quotient property, i.e., for all $\mathbf{e} \in \mathbb{R}^m$, there exists $\mathbf{u} \in \mathbb{R}^N$ with

$$\mathbf{A}\mathbf{u} = \mathbf{e} \quad \text{and} \quad \|\mathbf{u}\|_q \leq d\,s_*^{1/q-1/2}\|\mathbf{e}\|.$$

The real and complex versions are in fact equivalent, up to a possible change of the constant d. A real version of Lemma 11.17 also holds. Exercise 11.6 asks for a detailed verification of these statements. When we establish the quotient property for random matrices, we actually prove the real analog of (11.15), that is, $\|\mathbf{e}\|_* \leq ds_*^{1/q-1/2}\|\mathbf{A}^*\mathbf{e}\|_{q^*}$ for all $\mathbf{e} \in \mathbb{R}^m$.

Gaussian Random Matrices

We are now in the position to prove the ℓ_1-quotient property for Gaussian random matrices and then to deduce Theorem 11.9 concerning robustness of equality-constrained ℓ_1-minimization. We point out that the numerical constants in the following theorems have not been optimized, they have simply been chosen for convenience. Moreover, these constants rely on estimates for random matrices from

Chap. 9 such as Theorem 9.26 that are particular to the Gaussian case. In the proofs below, one may as well use the easier theorems for subgaussian matrices from Sect. 9.1, which, however, do not specify constants.

Theorem 11.19. *For $N \geq 2m$, if $\mathbf{A}$ is a draw of an $m \times N$ Gaussian random matrix, then the matrix $\tilde{\mathbf{A}} = \frac{1}{\sqrt{m}}\mathbf{A}$ possesses the ℓ_1-quotient property with constant $D = 34$ relative to the ℓ_2-norm with probability at least $1 - \exp(-m/100)$.*

Proof. According to Lemma 11.17 and Remark 11.18, we need to prove that

$$\mathbb{P}\big(\|\mathbf{e}\|_2 \leq D\sqrt{s_*}\|\tilde{\mathbf{A}}^*\mathbf{e}\|_\infty \text{ for all } \mathbf{e} \in \mathbb{R}^m\big) \geq 1 - \exp(-m/100). \quad (11.16)$$

To this end, we separate two cases: $2m \leq N < Cm$ and $N \geq Cm$, where $C = 165^6$ for reasons that will become apparent later. In the first case, by considering the renormalized matrix $\mathbf{B} := \sqrt{m/N}\tilde{\mathbf{A}}^* = \mathbf{A}^*/\sqrt{N} \in \mathbb{R}^{N \times m}$, we notice that the existence of $\mathbf{e} \in \mathbb{R}^m$ such that $\|\mathbf{e}\|_2 > D\sqrt{s_*}\|\tilde{\mathbf{A}}^*\mathbf{e}\|_\infty$ implies

$$\|\mathbf{e}\|_2 > D\sqrt{\frac{s_*N}{m}}\|\mathbf{Be}\|_\infty \geq D\sqrt{\frac{s_*}{m}}\|\mathbf{Be}\|_2 \geq \frac{D}{\sqrt{\ln(eN/m)}}\sigma_{\min}(\mathbf{B})\|\mathbf{e}\|_2.$$

In view of $N < Cm$, we derive

$$\sigma_{\min}(\mathbf{B}) < \frac{\sqrt{\ln(eC)}}{D} =: 1 - \sqrt{\frac{m}{N}} - t.$$

If $D \geq 6\sqrt{\ln(eC)}$ (which holds for $D = 34$ and $C = 165^6$), then the number t satisfies

$$t = 1 - \sqrt{\frac{m}{N}} - \frac{\sqrt{\ln(eC)}}{D} \geq 1 - \sqrt{\frac{1}{2}} - \frac{1}{6} \geq \frac{1}{10}.$$

Calling upon Theorem 9.26, we obtain

$$\mathbb{P}\big(\|\mathbf{e}\|_2 > 34\sqrt{s_*}\|\tilde{\mathbf{A}}^*\mathbf{e}\|_\infty \text{ for some } \mathbf{e} \in \mathbb{R}^m\big) \leq \mathbb{P}\bigg(\sigma_{\min}(\mathbf{B}) < 1 - \sqrt{\frac{m}{N}} - t\bigg)$$

$$\leq \exp\bigg(-\frac{t^2N}{2}\bigg) \leq \exp\bigg(-\frac{N}{200}\bigg) \leq \exp\bigg(-\frac{m}{100}\bigg).$$

This establishes (11.16) in the case $2m \leq N < Cm$.

The case $N \geq Cm$ is more delicate. Here, with $D = 8$, we will prove the stronger statement

$$\mathbb{P}\big(\|\mathbf{e}\|_2 > D\sqrt{s_*}\|\tilde{\mathbf{A}}^*\mathbf{e}\| \text{ for some } \mathbf{e} \in \mathbb{R}^m\big) \leq \exp\big(-m/3\big). \quad (11.17)$$

The norm $\|\cdot\|$ appearing in this statement is defined by

$$\|\mathbf{z}\| := \frac{1}{2h} \sum_{\ell=1}^{2h} \|\mathbf{z}_{T_\ell}\|_\infty, \qquad \mathbf{z} \in \mathbb{R}^N, \tag{11.18}$$

for some natural number $h \le N/2$ to be chosen below and some fixed partition $T_1, \ldots, T_{2h}$ of $[N]$ with $\mathrm{card}(T_\ell) \in \{\lfloor N/h \rfloor, \lfloor N/h \rfloor + 1\}$. The straightforward inequality

$$\|\mathbf{z}\| \le \|\mathbf{z}\|_\infty$$

explains why (11.17) implies (11.16). Another key property of the norm defined in (11.18) is that, for any $\mathbf{z} \in \mathbb{R}^N$, there exists a subset L of $[2h]$ of size h such that

$$\|\mathbf{z}_{T_\ell}\|_\infty \le 2\|\mathbf{z}\| \quad \text{for all } \ell \in L.$$

Indeed, the inequality

$$\|\mathbf{z}\| \ge \frac{1}{2h} \sum_{\ell : \|\mathbf{z}_{T_\ell}\|_\infty > 2\|\mathbf{z}\|} \|\mathbf{z}_{T_\ell}\|_\infty \ge \frac{1}{h} \, \mathrm{card}(\{\ell : \|\mathbf{z}_{T_\ell}\|_\infty > 2\|\mathbf{z}\|\}) \, \|\mathbf{z}\|$$

implies $\mathrm{card}(\{\ell : \|\mathbf{z}_{T_\ell}\|_\infty > 2\|\mathbf{z}\|\}) \le h$, i.e., $\mathrm{card}(\{\ell : \|\mathbf{z}_{T_\ell}\|_\infty \le 2\|\mathbf{z}\|\}) \ge h$. Therefore, for a fixed $\mathbf{e} \in \mathbb{R}^m$ and $d := D/2$, we have

$$\mathbb{P}\big(\|\mathbf{e}\|_2 > d\sqrt{s_*}\|\tilde{\mathbf{A}}^*\mathbf{e}\|\big)$$

$$\le \mathbb{P}\left(\|(\tilde{\mathbf{A}}^*\mathbf{e})_{T_\ell}\|_\infty < \frac{2\|\mathbf{e}\|_2}{d\sqrt{s_*}} \text{ for all } \ell \text{ in some } L \subset [2h], \mathrm{card}(L) = h \right)$$

$$\le \sum_{L \subset [2h], \mathrm{card}(L) = h} \mathbb{P}\left(\max_{j \in T_\ell} \left|(\tilde{\mathbf{A}}^*\mathbf{e})_j\right| < \frac{2\|\mathbf{e}\|_2}{d\sqrt{s_*}} \text{ for all } \ell \in L \right)$$

$$= \sum_{L \subset [2h], \mathrm{card}(L) = h} \mathbb{P}\left(\left|(\tilde{\mathbf{A}}^*\mathbf{e})_j\right| < \frac{2\|\mathbf{e}\|_2}{d\sqrt{s_*}} \text{ for all } j \in \cup_{\ell \in L} T_\ell \right)$$

$$= \sum_{L \subset [2h], \mathrm{card}(L) = h} \prod_{j \in \cup_{\ell \in L} T_\ell} \mathbb{P}\left(\left|(\tilde{\mathbf{A}}^*\mathbf{e})_j\right| < \frac{2\|\mathbf{e}\|_2}{d\sqrt{s_*}} \right).$$

In the last step, we have used the independence of the $(\tilde{\mathbf{A}}^*\mathbf{e})_j$, which follows from the independence of the columns of $\mathbf{A}$. For each $j \in \cup_{\ell \in L} T_\ell$, we notice that $(\tilde{\mathbf{A}}^*\mathbf{e})_j = \sum_{i=1}^m A_{i,j} e_i / \sqrt{m}$ is a mean-zero Gaussian random variable with variance $\|\mathbf{e}\|_2^2 / m$. Therefore, denoting by g a standard normal random variable, we obtain

$$\mathbb{P}(\|\mathbf{e}\|_2 > d\sqrt{s_*}\|\tilde{\mathbf{A}}^*\mathbf{e}\|) \leq \sum_{L\subset[2h],\text{card}(L)=h} \prod_{j\in\cup_{\ell\in L}T_\ell} \mathbb{P}\left(|g| < \frac{2\sqrt{m/s_*}}{d}\right)$$

$$= \sum_{L\subset[2h],\text{card}(L)=h} \left(1 - \mathbb{P}\left(|g| \geq \frac{2\sqrt{m/s_*}}{d}\right)\right)^{\text{card}(\cup_{\ell\in L}T_\ell)}$$

$$\leq \binom{2h}{h}\left(1 - \mathbb{P}\left(|g| \geq \frac{2\sqrt{m/s_*}}{d}\right)\right)^{N/2}. \tag{11.19}$$

At this point, we bound the tail of a standard normal variable from below as

$$\mathbb{P}\left(|g| \geq \frac{2\sqrt{m/s_*}}{d}\right) = \sqrt{\frac{2}{\pi}} \int_{2\sqrt{m/s_*}/d}^{\infty} \exp(-t^2/2)dt$$

$$\geq \sqrt{\frac{2}{\pi}} \int_{2\sqrt{m/s_*}/d}^{4\sqrt{m/s_*}/d} \exp(-t^2/2)dt \geq \sqrt{\frac{2}{\pi}} \frac{2\sqrt{m/s_*}}{d} \exp\left(-\frac{8m/s_*}{d^2}\right)$$

$$\geq \frac{\sqrt{8/\pi}}{d} \exp\left(-\frac{8}{d^2}\ln\left(\frac{eN}{m}\right)\right) = \frac{\sqrt{8/\pi}}{d}\left(\frac{m}{eN}\right)^{8/d^2}. \tag{11.20}$$

In the third step, we have used the inequality $\int_a^b f(t)dt \geq (b-a)f(b)$ for a decreasing function f. Substituting (11.20) into (11.19) and using the inequalities $\binom{n}{k} \leq \left(\frac{en}{k}\right)^k$ (see Lemma C.5) as well as $1 - x \leq \exp(-x)$, we derive

$$\mathbb{P}(\|\mathbf{e}\|_2 > d\sqrt{s_*}\|\tilde{\mathbf{A}}^*\mathbf{e}\|) \leq (2e)^h \exp\left(-\frac{\sqrt{8/\pi}}{d}\left(\frac{m}{eN}\right)^{8/d^2}\right)^{N/2}$$

$$= \exp\left(\ln(2e)h - \frac{\sqrt{2/\pi}}{de^{8/d^2}}m^{8/d^2}N^{1-8/d^2}\right). \tag{11.21}$$

We now use covering arguments to deduce a probability estimate valid for all $\mathbf{e} \in \mathbb{R}^m$ simultaneously. According to Proposition C.3, with $0 < \delta < 1$ to be chosen later, we can find a δ-covering $\{\mathbf{f}_1,\ldots,\mathbf{f}_n\}$ of the unit sphere of ℓ_2^m with cardinality $n \leq (1+2/\delta)^m$. Let us suppose that there exists $\mathbf{e} \in \mathbb{R}^m$ with $\|\mathbf{e}\|_2 > D\sqrt{s_*}\|\tilde{\mathbf{A}}^*\mathbf{e}\|$. Without loss of generality, we may assume that $\|\mathbf{e}\|_2 = 1$; hence, $\|\mathbf{e} - \mathbf{f}_i\|_2 \leq \delta$ for some $i \in [n]$. It follows that

$$D\sqrt{s_*}\|\tilde{\mathbf{A}}^*\mathbf{f}_i\| \leq D\sqrt{s_*}\|\tilde{\mathbf{A}}^*\mathbf{e}\| + D\sqrt{s_*}\|\tilde{\mathbf{A}}^*(\mathbf{e}-\mathbf{f}_i)\|$$

$$< 1 + D\frac{\sqrt{s_*}}{2h}\sum_{\ell=1}^{2h}\|(\tilde{\mathbf{A}}^*(\mathbf{e}-\mathbf{f}_i))_{T_\ell}\|_\infty$$

$$\leq 1 + D\frac{\sqrt{s_*}}{2h}\sum_{\ell=1}^{2h}\|(\tilde{\mathbf{A}}^*(\mathbf{e}-\mathbf{f}_i))_{T_\ell}\|_2$$

$$\leq 1 + D\sqrt{\frac{s_*}{2h}}\|\tilde{\mathbf{A}}^*(\mathbf{e}-\mathbf{f}_i)\|_2. \qquad (11.22)$$

Applying Theorem 9.26 to the renormalized matrix $\mathbf{B} = \mathbf{A}^*/\sqrt{N} \in \mathbb{R}^{N\times m}$, we obtain

$$\mathbb{P}\left(\sigma_{\max}(\mathbf{B}) > 1 + 2\sqrt{\frac{m}{N}}\right) \leq \exp\left(-\frac{m}{2}\right). \qquad (11.23)$$

Thus, in the likely case $\sigma_{\max}(\mathbf{B}) \leq 1+2\sqrt{m/N}$, whence $\sigma_{\max}(\mathbf{B}) \leq \sqrt{2}$ provided $N \geq Cm$ with $C \geq 12 + 8\sqrt{2}$ (which is satisfied for $C = 165^6$), we have

$$\|\tilde{\mathbf{A}}^*(\mathbf{e}-\mathbf{f}_i)\|_2 \leq \sigma_{\max}(\tilde{\mathbf{A}}^*)\|\mathbf{e}-\mathbf{f}_i\|_2 = \sqrt{\frac{N}{m}}\,\sigma_{\max}(\mathbf{B})\|\mathbf{e}-\mathbf{f}_i\|_2 \leq \sqrt{\frac{2N}{m}}\delta.$$

In turn, we deduce

$$d\sqrt{s_*}\|\tilde{\mathbf{A}}^*\mathbf{f}_i\| = \frac{1}{2}\left(D\sqrt{s_*}\|\tilde{\mathbf{A}}^*\mathbf{f}_i\|\right) \leq \frac{1}{2}\left(1 + D\sqrt{\frac{s_*N}{hm}}\,\delta\right).$$

The choice $\delta := D^{-1}\sqrt{h/N}$ together with the fact that $s_* \leq m$ yields

$$d\sqrt{s_*}\|\tilde{\mathbf{A}}^*\mathbf{f}_i\| \leq \|\mathbf{f}_i\|_2.$$

Summarizing the previous considerations yields

$$\mathbb{P}\left(\|\mathbf{e}\|_2 > D\sqrt{s_*}\|\tilde{\mathbf{A}}^*\mathbf{e}\| \text{ for some } \mathbf{e} \in \mathbb{R}^m\right)$$

$$= \mathbb{P}\left(\|\mathbf{e}\|_2 > D\sqrt{s_*}\|\tilde{\mathbf{A}}^*\mathbf{e}\| \text{ for some } \mathbf{e} \in \mathbb{R}^m \text{ and } \sigma_{\max}(B) > 1 + 2\sqrt{\frac{m}{N}}\right)$$

$$+ \mathbb{P}\left(\|\mathbf{e}\|_2 > D\sqrt{s_*}\|\tilde{\mathbf{A}}^*\mathbf{e}\| \text{ for some } \mathbf{e} \in \mathbb{R}^m \text{ and } \sigma_{\max}(B) \leq 1 + 2\sqrt{\frac{m}{N}}\right)$$

$$\leq \mathbb{P}\left(\sigma_{\max}(\mathbf{B}) > 1 + 2\sqrt{\frac{m}{N}}\right) + \mathbb{P}(\|\mathbf{f}_i\|_2 > d\sqrt{s_*}\|\tilde{\mathbf{A}}^*\mathbf{f}_i\| \text{ for some } i \in [n]).$$

By (11.23), the first term on the right-hand side is bounded by $\exp(-m/2)$. Moreover, a union bound over the n elements of the covering, the inequality $n \leq (1 + 2/\delta)^m \leq \exp(2m/\delta)$, and the probability estimate (11.21) applied to a fixed $\mathbf{f}_i \in \mathbb{R}^m$ show that the second term is bounded by

$$\exp\left(2D\sqrt{\frac{N}{h}}\,m + \ln(2e)h - \frac{\sqrt{2/\pi}}{de^{8/d^2}}\,m^{8/d^2}N^{1-8/d^2}\right). \qquad (11.24)$$

Now we choose $d = 4$ (corresponding to $D = 8$) and

$$h = \lceil m^{2/3}N^{1/3}\rceil$$

(so that $h \le N/2$ when $N \ge Cm$ with $C = 165^6$) to bound the second term by

$$\exp\left(16\,m^{2/3}N^{1/3} + 2\ln(2e)\,m^{2/3}N^{1/3} - \frac{1}{\sqrt{8\pi e}}m^{1/2}N^{1/2}\right)$$

$$= \exp\left(-\left[\frac{1}{\sqrt{8\pi e}} - \frac{2\ln(2e^9)}{(N/m)^{1/6}}\right]m^{1/2}N^{1/2}\right)$$

$$\le \exp\left(-\left[\frac{1}{\sqrt{8\pi e}} - \frac{2\ln(2e^9)}{C^{1/6}}\right]m^{1/2}N^{1/2}\right) \le \exp\left(-\frac{m^{1/2}N^{1/2}}{300}\right).$$

The choice $C = 165^6$ accounts for the last inequality. Putting the two bounds together, we obtain

$$\mathbb{P}\big(\|\mathbf{e}\|_2 > 8\sqrt{s_*}\|\tilde{\mathbf{A}}^*\mathbf{e}\| \text{ for some } \mathbf{e} \in \mathbb{R}^m\big)$$

$$\le \exp\left(-\frac{m}{2}\right) + \exp\left(-\frac{m^{1/2}N^{1/2}}{300}\right) \le \exp\left(-\frac{m}{3}\right).$$

This establishes (11.17) for the case $N \ge Cm$ and concludes the proof. $\square$

Remark 11.20. The introduction of the auxiliary norm in (11.18) may seem strange at first sight, as one would be inclined to work with the ℓ_∞-norm directly. In fact, this would correspond (up to constants) to choosing $h = 1$ in the proof. The reader is invited to verify that setting $h = 1$ in (11.24) does not lead to the desired conclusion.

We now prove the main robustness estimate for Gaussian random matrices.

Proof (of Theorem 11.9). Under the assumption $N \ge c_2 m$ with $c_2 := 2$, Theorem 11.19 guarantees that the matrix $\tilde{\mathbf{A}}$ has the ℓ_1-quotient property relative to the ℓ_2-norm with probability at least $1 - \exp(-m/100)$. Moreover, the assumption $s \le c_3 s_*$ with $c_3 := 1/1400$ reads $m \ge 1400s\ln(eN/m)$. Lemma C.6, in view of $1400/\ln(1400e) \ge 160$ and $\ln(eN/s) \ge \ln(eN/(2s))$, implies the inequality $m \ge 80(2s)\ln(eN/(2s))$. This is equivalent to

$$\frac{5m}{4} \ge 80(2s)\ln\left(\frac{eN}{2s}\right) + \frac{m}{4}, \qquad \text{i.e.,} \quad m \ge \frac{2}{\eta^2}(2s)\ln\left(\frac{eN}{2s}\right) + \frac{2}{\eta^2}\ln\left(\frac{2}{\varepsilon}\right),$$

where $\eta := 1/\sqrt{32}$ and $\varepsilon := 2\exp(-m/320)$. Theorem 9.27 implies that, with probability at least $1 - 2\exp(-m/320)$, the restricted isometry constant of the matrix $\tilde{A}$ satisfies

$$\delta_{2s} \leq 2\left(1 + \frac{1}{\sqrt{2\ln(eN/(2s))}}\right)\eta + \left(1 + \frac{1}{\sqrt{2\ln(eN/(2s))}}\right)^2 \eta^2$$

$$\leq 2\left(1 + \frac{1}{\sqrt{2\ln(1400e)}}\right)\frac{1}{\sqrt{32}} + \left(1 + \frac{1}{\sqrt{2\ln(1400e)}}\right)^2 \frac{1}{32} \approx 0.489.$$

Then Theorem 6.13 ensures that the matrix $\tilde{A}$ has the ℓ_2-robust null space property of order s. Thus, with probability at least

$$1 - \exp(-m/100) - 2\exp(m/320) \geq 1 - 3\exp(-c_1 m), \qquad c_1 := 1/320,$$

the matrix $\tilde{A}$ satisfies both the ℓ_1-quotient property relative to the ℓ_2-norm and the ℓ_2-robust null space property of order $s \leq c_3 s_*$. The conclusion now follows from Theorem 11.12. $\qquad\qquad\square$

Subgaussian Random Matrices

Let us now consider more general subgaussian random matrices. For the special case of renormalized Bernoulli matrices, the ℓ_1-quotient property relative to the ℓ_2-norm, namely,

for all $e \in \mathbb{C}^m$, there exists $u \in \mathbb{C}^N$ with $\tilde{A}u = e$ and $\|u\|_1 \leq d\sqrt{s_*}\|e\|_2$,

cannot be true. Indeed, such a matrix $\tilde{A}$ has entries $\tilde{A}_{i,j} = \pm 1/\sqrt{m}$, so that the ℓ_1-quotient property applied to a unit vector $e_i = [0,\ldots,0,1,0,\ldots,0]^\top \in \mathbb{C}^m$ would give rise to a vector $u \in \mathbb{C}^N$ for which

$$1 = (\tilde{A}u)_i = \sum_{j=1}^{N} \tilde{A}_{i,j}u_j \leq \frac{\|u\|_1}{\sqrt{m}} \leq \frac{d\sqrt{s_*}\|e\|_2}{\sqrt{m}} = \frac{d}{\sqrt{\ln(eN/m)}},$$

a contradiction for large enough N.

In order to nevertheless obtain robustness estimates, the strategy consists in eliminating the troublesome vectors e_i by clipping the ℓ_2-ball around them. This motivates the introduction of the following norm defined, for $\alpha \geq 1$, by

$$\|y\|^{(\alpha)} := \max\{\|y\|_2, \alpha\|y\|_\infty\}, \qquad y \in \mathbb{C}^m. \qquad (11.25)$$

Then the ℓ_1-quotient property relative to this norm applied to the vectors $\mathbf{e}_i = [0, \ldots, 0, 1, 0, \ldots, 0]^\top \in \mathbb{C}^m$ yields

$$1 \leq \frac{d\sqrt{s_*}\|\mathbf{e}\|^{(\alpha)}}{\sqrt{m}} = \frac{d\alpha}{\sqrt{\ln(eN/m)}}.$$

This dictates the choice $\alpha = \sqrt{\ln(eN/m)} \geq 1$ for Bernoulli matrices.

Here is the precise statement about the ℓ_1-quotient property for certain random matrices including Bernoulli matrices as a special case.

Theorem 11.21. *Let $\mathbf{A}$ be an $m \times N$ matrix whose entries are independent symmetric random variables with variance 1 and fourth moment bounded by some $\mu^4 \geq 1$, for which the concentration inequality*

$$\mathbb{P}\big(\big|N^{-1}\|\mathbf{A}^*\mathbf{y}\|_2^2 - \|\mathbf{y}\|_2^2\big| > t\|\mathbf{y}\|_2^2\big) \leq 2\exp(-\tilde{c}t^2 N) \tag{11.26}$$

holds for all $\mathbf{y} \in \mathbb{R}^m$ and $t \in (0, 1)$. Then there exist constants $C, D > 0$ depending only on μ and $\tilde{c}$ such that, if $N \geq C\, m$, then with probability at least $1 - 3\exp(-m)$ the matrix $\tilde{\mathbf{A}} := \frac{1}{\sqrt{m}}\mathbf{A}$ has the ℓ_1-quotient property with constant D relative to the norm $\|\cdot\|^{(\alpha)}$, $\alpha := \sqrt{\ln(eN/m)}$.

The arguments follow the same lines as the Gaussian case. In particular, estimates from below for tail probabilities involving the dual norm of $\|\cdot\|^{(\alpha)}$ are needed. We start by comparing this dual norm to a more tractable norm.

Lemma 11.22. *For an integer $k \geq 1$, the dual norm of $\|\cdot\|^{(\sqrt{k})}$ is comparable with the norm $|\cdot|_k$ defined by*

$$|\mathbf{y}|_k := \max\left\{\sum_{\ell=1}^{k} \|\mathbf{y}_{B_\ell}\|_2, \ B_1, \ldots, B_k \text{ form a partition of } [m]\right\}, \tag{11.27}$$

in the sense that

$$\sqrt{\frac{1}{k}}\,|\mathbf{y}|_k \leq \|\mathbf{y}\|_*^{(\sqrt{k})} \leq \sqrt{\frac{2}{k}}\,|\mathbf{y}|_k, \qquad \mathbf{y} \in \mathbb{C}^m. \tag{11.28}$$

Proof. We define a norm on $\mathbb{C}^m \times \mathbb{C}^m$ by

$$\|(\mathbf{u}, \mathbf{v})\| := \max\big\{\|\mathbf{u}\|_2, \sqrt{k}\|\mathbf{v}\|_\infty\big\}.$$

This makes the linear map $T : (\mathbb{C}^m, \|\cdot\|^{(\sqrt{k})}) \to (\mathbb{C}^m \times \mathbb{C}^m, \|(\cdot, \cdot)\|)$, $\mathbf{z} \mapsto (\mathbf{z}, \mathbf{z})$ an isometry from $\mathbb{C}^m$ onto $X := T(\mathbb{C}^m)$. Let us now fix a vector $\mathbf{y} \in \mathbb{C}^m$. We have

$$\|\mathbf{y}\|_*^{(\sqrt{k})} = \max_{\|\mathbf{u}\|^{(\sqrt{k})}=1} |\langle \mathbf{u}, \mathbf{y}\rangle| = \max_{\|(\mathbf{u},\mathbf{u})\|=1} |\langle T^{-1}((\mathbf{u}, \mathbf{u})), \mathbf{y}\rangle| = \|\lambda\|_{X^*},$$

where we have defined the linear functional λ on X by $\lambda(\mathbf{x}) := \langle T^{-1}(\mathbf{x}), \mathbf{y} \rangle$. The *Hahn–Banach extension* theorem (see Appendix C.7) ensures the existence of a linear functional $\tilde{\lambda}$ defined on $\mathbb{C}^m \times \mathbb{C}^m$ such that $\tilde{\lambda}(\mathbf{x}) = \lambda(\mathbf{x})$ for all $\mathbf{x} \in X$ and $\|\tilde{\lambda}\|_* = \|\lambda\|_{X^*}$, where $\| \cdot \|_*$ denotes the dual norm of $\|(\cdot, \cdot)\|$. There exists $(\mathbf{y}', \mathbf{y}'') \in \mathbb{C}^m \times \mathbb{C}^m$ such that this functional can be written as $\tilde{\lambda}(\mathbf{u}, \mathbf{v}) = \langle (\mathbf{u}, \mathbf{v}), (\mathbf{y}', \mathbf{y}'') \rangle = \langle \mathbf{u}, \mathbf{y}' \rangle + \langle \mathbf{v}, \mathbf{y}'' \rangle$ for all $(\mathbf{u}, \mathbf{v}) \in \mathbb{C}^m \times \mathbb{C}^m$. The identity $\tilde{\lambda}(T(\mathbf{z})) = \lambda(T(\mathbf{z}))$, i.e., $\langle \mathbf{z}, \mathbf{y}' + \mathbf{y}'' \rangle = \langle \mathbf{z}, \mathbf{y} \rangle$, valid for all $\mathbf{z} \in \mathbb{C}^m$, yields $\mathbf{y}' + \mathbf{y}'' = \mathbf{y}$. Moreover, the dual norm is given by $\|\tilde{\lambda}\|_* = \|\mathbf{y}'\|_2 + \|\mathbf{y}''\|_1 / \sqrt{k}$, which yields $\|\mathbf{y}\|_*^{(\sqrt{k})} = \|\mathbf{y}'\|_2 + \|\mathbf{y}''\|_1 / \sqrt{k}$. Now, choosing optimal partitions $B_1', \ldots, B_k'$ and $B_1'', \ldots, B_k''$ of $[m]$ for the vectors $\mathbf{y}'$ and $\mathbf{y}''$ (see (11.27)), we observe that

$$|\mathbf{y}'|_k = \sum_{\ell=1}^{k} \|\mathbf{y}'_{B_\ell'}\|_2 \leq \sqrt{k} \sqrt{\sum_{\ell=1}^{k} \|\mathbf{y}'_{B_\ell'}\|_2^2} = \sqrt{k} \|\mathbf{y}'\|_2,$$

$$|\mathbf{y}''|_k = \sum_{\ell=1}^{k} \|\mathbf{y}''_{B_\ell''}\|_2 \leq \sum_{\ell=1}^{k} \|\mathbf{y}''_{B_\ell''}\|_1 = \|\mathbf{y}''\|_1.$$

It follows that

$$|\mathbf{y}|_k = |\mathbf{y}' + \mathbf{y}''|_k \leq |\mathbf{y}'|_k + |\mathbf{y}''|_k \leq \sqrt{k} \left(\|\mathbf{y}'\|_2 + \|\mathbf{y}''\|_1 / \sqrt{k} \right) = \sqrt{k} \|\mathbf{y}\|_*^{(\sqrt{k})}.$$

This proves the leftmost inequality in (11.28).

For the rightmost inequality, given a fixed vector $\mathbf{y} \in \mathbb{C}^m$, we consider a vector $\mathbf{u} \in \mathbb{C}^m$ with $\|\mathbf{u}\|^{(\sqrt{k})} = 1$ such that $\|\mathbf{y}\|_*^{(\sqrt{k})} = \langle \mathbf{u}, \mathbf{y} \rangle$. The definition of $\|\mathbf{u}\|^{(\sqrt{k})}$ implies that $\|\mathbf{u}\|_2 \leq 1$ and that $\|\mathbf{u}\|_\infty \leq 1 / \sqrt{k}$. Let $m_0 := 0$ and iteratively define $m_\ell \geq m_{\ell-1} + 1$ with $m_\ell \leq m$ for $\ell = 1, 2, \ldots$ to be the smallest integer such that

$$\sum_{i=m_{\ell-1}+1}^{m_\ell} |u_i|^2 > 1/k$$

as long as this is possible. Then

$$\sum_{i=m_{\ell-1}+1}^{m_\ell - 1} |u_i|^2 \leq \frac{1}{k}, \quad \text{and} \quad \sum_{i=m_{\ell-1}+1}^{m_\ell} |u_i|^2 \leq \frac{2}{k}.$$

We notice that the last m_h defined in this way has index $h < k$. Indeed, if m_k was defined, we would obtain a contradiction from

$$\|\mathbf{u}\|_2^2 \geq \sum_{\ell=1}^{k} \sum_{i=m_{\ell-1}+1}^{m_\ell} |u_i|^2 > \sum_{\ell=1}^{k} \frac{1}{k} = 1.$$

We also notice that, because m_{h+1} is undefined, we have

$$\sum_{i=m_h+1}^{m} |u_i|^2 \leq \frac{1}{k}.$$

We now set $B_\ell = \{m_{\ell-1}+1, \ldots, m_\ell\}$ for $1 \leq \ell \leq h$, $B_{h+1} := \{m_h+1, \ldots, m\}$, and $B_\ell = \emptyset$ for $h+2 \leq \ell \leq k$. In view of $\|\mathbf{u}_{B_\ell}\|_2 \leq \sqrt{2/k}$ for all $1 \leq \ell \leq k$, we derive

$$\|\mathbf{y}\|_*^{(\sqrt{k})} = \langle \mathbf{u}, \mathbf{y} \rangle = \sum_{\ell=1}^{k} \langle \mathbf{u}_{B_\ell}, \mathbf{y}_{B_\ell} \rangle \leq \sum_{\ell=1}^{k} \|\mathbf{u}_{B_\ell}\|_2 \|\mathbf{y}_{B_\ell}\|_2 \leq \sqrt{\frac{2}{k}} \sum_{\ell=1}^{k} \|\mathbf{y}_{B_\ell}\|_2$$

$$\leq \sqrt{\frac{2}{k}} |\mathbf{y}|_k.$$

This proves the rightmost inequality in (11.28).   □

We are now ready to carry on with the proof of Theorem 11.21.

Proof (of Theorem 11.21). Let us suppose that $N \geq C m$, where the constant $C \geq 1$ has to meet three requirements specified below. We set

$$\beta := \sqrt{\frac{\ln(eN/m)}{\ln(eC)}} \geq 1.$$

Since $\beta \leq \alpha$, hence $\|\mathbf{e}\|^{(\beta)} \leq \|\mathbf{e}\|^{(\alpha)}$, the ℓ_1-quotient property relative to $\|\cdot\|^{(\beta)}$ implies the one relative to $\|\cdot\|^{(\alpha)}$, so we concentrate on the ℓ_1-quotient property relative to the norm $\|\cdot\|^{(\beta)}$. According to Lemma 11.17 and Remark 11.18, we need to prove that

$$\mathbb{P}\big(\|\mathbf{e}\|_*^{(\beta)} \leq D\sqrt{s_*}\|\tilde{\mathbf{A}}^*\mathbf{e}\|_\infty \text{ for all } \mathbf{e} \in \mathbb{R}^m\big) \geq 1 - 3\exp\big(-m\big).$$

As in the proof of the Gaussian case, we show the stronger statement

$$\mathbb{P}\big(\|\mathbf{e}\|_*^{(\beta)} \leq D\sqrt{s_*}\|\tilde{\mathbf{A}}^*\mathbf{e}\| \text{ for all } \mathbf{e} \in \mathbb{R}^m\big) \geq 1 - 3\exp\big(-m\big) \qquad (11.29)$$

with $D := 16\sqrt{\ln(eC)}$. The norm $\|\cdot\| \leq \|\cdot\|_\infty$ is the norm defined in (11.18) with some natural number $h \leq N/2$ to be determined below. To this end, we assume that there exists $\mathbf{e} \in \mathbb{R}^m$ such that

$$\|\mathbf{e}\|_*^{(\beta)} > D\sqrt{s_*}\|\tilde{\mathbf{A}}^*\mathbf{e}\|.$$

Introducing the integer $k := \lfloor \beta^2 \rfloor \geq 1$, for which

$$k \leq \beta^2 < 2k,$$

we have $\|\mathbf{y}\|^{(\sqrt{k})} \leq \|\mathbf{y}\|^{(\beta)}$ for all $\mathbf{y} \in \mathbb{R}^m$, and in turn the dual norms satisfy $\|\mathbf{y}\|_*^{(\sqrt{k})} \geq \|\mathbf{y}\|_*^{(\beta)}$ for all $\mathbf{y} \in \mathbb{R}^m$. Assuming without loss of generality that $\|\mathbf{e}\|_*^{(\sqrt{k})} = 1$, we obtain $D\sqrt{s_*}\|\tilde{\mathbf{A}}^*\mathbf{e}\| < 1$. Moreover, for $0 < \delta < 1$ to be chosen later, Lemma C.3 ensures the existence of a δ-covering $\{\mathbf{f}_1, \ldots, \mathbf{f}_n\}$ with respect to $\|\cdot\|_*^{(\sqrt{k})}$ of the unit sphere of $(\mathbb{R}^m, \|\cdot\|_*^{(\sqrt{k})})$ with cardinality $n \leq (1 + 2/\delta)^m$. Selecting an integer $i \in [n]$ such that $\|\mathbf{e} - \mathbf{f}_i\|_*^{(\sqrt{k})} \leq \delta$, it follows as in (11.22) that

$$D\sqrt{s_*}\|\tilde{\mathbf{A}}^*\mathbf{f}_i\| \leq 1 + D\frac{\sqrt{s_*}}{\sqrt{2h}}\|\tilde{\mathbf{A}}^*(\mathbf{e} - \mathbf{f}_i)\|_2$$

$$\leq 1 + D\sqrt{\frac{s_* N}{2hm}}\sigma_{\max}(\mathbf{B})\|\mathbf{e} - \mathbf{f}_i\|_2, \qquad (11.30)$$

where $\mathbf{B} \in \mathbb{R}^{N \times m}$ is the renormalized matrix $\mathbf{B} := \sqrt{m/N}\,\tilde{\mathbf{A}}^* = \mathbf{A}^*/\sqrt{N}$. We observe that if $B_1, \ldots, B_k$ is an optimal partition for $|\mathbf{e} - \mathbf{f}_i|_k$, then

$$\|\mathbf{e} - \mathbf{f}_i\|_2 \leq \sum_{\ell=1}^{k} \|(\mathbf{e} - \mathbf{f}_i)_{B_\ell}\|_2 = |\mathbf{e} - \mathbf{f}_i|_k \leq \sqrt{k}\|\mathbf{e} - \mathbf{f}_i\|_*^{(\sqrt{k})} \leq \sqrt{k}\,\delta, \quad (11.31)$$

where we have used Lemma 11.22. Thus, under the assumption that $\sigma_{\max}(\mathbf{B}) \leq \sqrt{2}$, (11.30) and (11.31) yield

$$D\sqrt{s_*}\|\tilde{\mathbf{A}}^*\mathbf{f}_i\| < 1 + D\sqrt{\frac{ks_*}{m}}\sqrt{\frac{N}{h}}\delta \leq 1 + \frac{D}{\sqrt{\ln(eC)}}\sqrt{\frac{N}{h}}\delta.$$

Now we choose

$$\delta := \frac{\sqrt{\ln(eC)}}{D}\sqrt{\frac{h}{N}} \qquad (11.32)$$

and exploit the fact that $1 = \|\mathbf{f}_i\|_*^{(\sqrt{k})} \leq 2|\mathbf{f}_i|_k/\sqrt{k}$ to obtain

$$D\sqrt{s_*}\|\tilde{\mathbf{A}}^*\mathbf{f}_i\| < 2 \leq 4|\mathbf{f}_i|_k/\sqrt{k}.$$

Summarizing the previous considerations and setting $d := D/4$ gives

$$\mathbb{P}\big(\|\mathbf{e}\|_*^{(\beta)} > D\sqrt{s_*}\|\tilde{\mathbf{A}}^*\mathbf{e}\|_\infty \text{ for some } \mathbf{e} \in \mathbb{R}^m\big)$$

$$\leq \mathbb{P}\big(\sigma_{\max}(B) > \sqrt{2}\big) + \mathbb{P}\big(|\mathbf{f}_i|_k > d\sqrt{ks_*}\|\tilde{\mathbf{A}}^*\mathbf{f}_i\| \text{ for some } i \in [n]\big). \quad (11.33)$$

For the first term in (11.33), we call upon Theorem 9.9 to obtain

$$\mathbb{P}\big(\sigma_{\max}(B) > \sqrt{2}\big) = \mathbb{P}\big(\sigma^2_{\max}(B) > 2\big) \leq \mathbb{P}\big(\|\mathbf{B}^*\mathbf{B} - \mathbf{Id}\|_{2\to 2} > 1\big)$$

$$\leq 2\exp\left(-\frac{\tilde{c}N}{2}\right)$$

provided

$$N \geq \frac{2}{3\tilde{c}}\left(7m + \tilde{c}N\right), \qquad \text{i.e.,} \qquad N \geq \frac{14}{\tilde{c}}m.$$

The first requirement imposed on C is therefore $C \geq 14/\tilde{c}$. In this case, we have the bound

$$\mathbb{P}\big(\sigma_{\max}(B) > \sqrt{2}\big) \leq 2\exp\big(-7m\big). \qquad (11.34)$$

For the second term in (11.33), we begin by bounding the probability $\mathbb{P}(|e|_k > d\sqrt{ks_*}\|\tilde{\mathbf{A}}^*\mathbf{e}\|)$ for a fixed vector $\mathbf{e} \in \mathbb{R}^m$. As in the proof of Theorem 11.19, using the existence of a subset L of $[2h]$ of size h such that $\|\mathbf{z}_{T_\ell}\|_\infty \leq 2\|\mathbf{z}\|$ for any $\mathbf{z} \in \mathbb{R}^N$ and any $\ell \in L$, we observe that

$$\mathbb{P}\big(|e|_k > d\sqrt{ks_*}\,\|\tilde{\mathbf{A}}^*\mathbf{e}\|\big)$$

$$\leq \mathbb{P}\left(\|(\tilde{\mathbf{A}}^*\mathbf{e})_{T_\ell}\|_\infty < \frac{2|e|_k}{d\sqrt{ks_*}} \text{ for all } \ell \text{ in some } L \subset [2h], \operatorname{card}(L) = h\right)$$

$$\leq \sum_{L\subset[2h],\operatorname{card}(L)=h} \mathbb{P}\left(\max_{j\in T_\ell}|(\tilde{\mathbf{A}}^*\mathbf{e})_j| < \frac{2|e|_k}{d\sqrt{ks_*}} \text{ for all } \ell \in L\right)$$

$$= \sum_{L\subset[2h],\operatorname{card}(L)=h} \prod_{j\in\cup_{\ell\in L}T_\ell} \mathbb{P}\left(|(\tilde{\mathbf{A}}^*\mathbf{e})_j| < \frac{2|e|_k}{d\sqrt{ks_*}}\right)$$

$$= \sum_{L\subset[2h],\operatorname{card}(L)=h} \prod_{j\in\cup_{\ell\in L}T_\ell} \left(1 - 2\mathbb{P}\left((\tilde{\mathbf{A}}^*\mathbf{e})_j \geq \frac{2|e|_k}{d\sqrt{ks_*}}\right)\right), \quad (11.35)$$

where the independence of the random variables $A_{i,j}$ was used in the third step and their symmetry in the last step. Let now $B_1,\dots,B_k$ denote a partition of $[m]$ such that $|e|_k = \sum_{\ell=1}^k \|\mathbf{e}_{B_\ell}\|_2$. For each $j \in \cup_{\ell\in L}T_\ell$, we have

$$\mathbb{P}\left((\tilde{\mathbf{A}}^*\mathbf{e})_j \geq \frac{2|e|_k}{d\sqrt{ks_*}}\right) = \mathbb{P}\left(\sum_{\ell=1}^k \sum_{i\in B_\ell} \frac{A_{i,j}}{\sqrt{m}}e_i \geq \sum_{\ell=1}^k \frac{2\|\mathbf{e}_{B_\ell}\|_2}{d\sqrt{ks_*}}\right)$$

$$\geq \mathbb{P}\left(\sum_{i\in B_\ell} A_{i,j}e_i \geq \frac{2}{d}\sqrt{\frac{m}{ks_*}}\|\mathbf{e}_{B_\ell}\|_2 \text{ for all } \ell \in [k]\right)$$

$$= \prod_{\ell \in [k]} \mathbb{P}\Big(\sum_{i \in B_\ell} A_{i,j} e_i \geq \frac{2}{d} \sqrt{\frac{m}{ks_*}} \|e_{B_\ell}\|_2 \Big)$$

$$= \prod_{\ell \in [k]} \frac{1}{2} \mathbb{P}\Big(\Big| \sum_{i \in B_\ell} A_{i,j} e_i \Big| \geq \frac{2}{d} \sqrt{\frac{m}{ks_*}} \|e_{B_\ell}\|_2 \Big),$$

where we have again used the independence and symmetry of the random variables $A_{i,j}$ in the last two steps. For each $\ell \in [k]$, we use Lemma 7.17 to obtain

$$\mathbb{P}\Big(\Big| \sum_{i \in B_\ell} A_{i,j} e_i \Big| \geq \frac{2}{d} \sqrt{\frac{m}{ks_*}} \|e_{B_\ell}\|_2 \Big) \geq \frac{1}{\mu^4} \Big(1 - \frac{4m}{d^2 ks_*} \Big)^2 \geq \frac{1}{\mu^4} \Big(1 - \frac{8m}{d^2 \beta^2 s_*} \Big)^2$$

$$= \frac{1}{\mu^4} \Big(1 - \frac{8\ln(eC)}{d^2} \Big)^2 = \frac{1}{4\mu^4},$$

where we have chosen the value $d = D/4 = 4\sqrt{\ln(eC)}$ and have used the definitions of s_* and β. Under a second requirement on C, namely, $C \geq 64\mu^8/e$, it follows that

$$\mathbb{P}\Big((\tilde{\mathbf{A}}^* e)_j \geq \frac{2|e|_k}{d\sqrt{ks_*}} \Big) \geq \Big(\frac{1}{8\mu^4} \Big)^k \geq \Big(\frac{1}{8\mu^4} \Big)^{\beta^2} = \exp\big(-\beta^2 \ln(8\mu^4) \big)$$

$$= \exp\Big(-\ln\Big(\frac{eN}{m} \Big) \frac{\ln(8\mu^4)}{\ln(eC)} \Big) \geq \Big(\frac{m}{eN} \Big)^{1/2}. \qquad (11.36)$$

Substituting (11.36) into (11.35) and using $1 - x \leq \exp(-x)$, we obtain

$$\mathbb{P}\big(|e|_k > d\sqrt{ks_*} \|\tilde{\mathbf{A}}^* e\| \big) \leq \sum_{L \subset [2h], \text{card}(L)=h} \exp\Big(-2\Big(\frac{m}{eN} \Big)^{1/2} \Big)^{\text{card}(\cup_{\ell \in L} T_\ell)}$$

$$\leq \binom{2h}{h} \exp\Big(-2\Big(\frac{m}{eN} \Big)^{1/2} \Big)^{N/2} \leq \exp\Big(\ln(2e)h - \frac{1}{e^{1/2}} m^{1/2} N^{1/2} \Big),$$

where we have also used the fact that $\binom{2h}{h} \leq (2e)^h$ by Lemma C.5. Thus, in view of $n \leq (1 + 2/\delta)^m \leq \exp(2m/\delta)$, we derive

$$\mathbb{P}\big(|\mathbf{f}_i|_k > d\sqrt{ks_*} \|\tilde{\mathbf{A}}^* \mathbf{f}_i\| \text{ for some } i \in [n] \big) \leq n \, \mathbb{P}\big(|e|_k > d\sqrt{ks_*} \|\tilde{\mathbf{A}}^* e\| \big)$$

$$\leq \exp\Big(\frac{2}{\delta} m + \ln(2e)h - \frac{1}{e^{1/2}} m^{1/2} N^{1/2} \Big).$$

We now choose $h := \lceil m^{2/3} N^{2/3} \rceil$ (so that $h \leq N/2$ when $N \geq Cm$ with $C \geq 64\mu^8/e$). We then have $h \leq 2\, m^{2/3} N^{2/3}$ and the choice (11.32) of δ together with $D = 16\sqrt{\ln(eC)}$ implies $2/\delta = 32(N/h)^{1/2} \leq 32(N/m)^{1/3}$. It follows that

$$\mathbb{P}\big(|\mathbf{f}_i|_k > d\sqrt{ks_*}\|\tilde{\mathbf{A}}^*\mathbf{f}_i\| \text{ for some } i \in [n]\big)$$

$$\leq \exp\left(32 m^{2/3} N^{1/3} + 2\ln(2e)m^{2/3}N^{1/3} - \frac{1}{e^{1/2}}m^{1/2}N^{1/2}\right)$$

$$\leq \exp\left(-\left[\frac{1}{e^{1/2}} - \frac{2\ln(2e^{17})}{(N/m)^{1/6}}\right]m^{1/2}N^{1/2}\right)$$

$$\leq \exp\left(-\left[\frac{1}{e^{1/2}} - \frac{2\ln(2e^{17})}{C^{1/6}}\right]m^{1/2}N^{1/2}\right).$$

A third requirement on C, namely $C^{1/6} \geq 4e^{1/2}\ln(2e^{17})$, implies that

$$\mathbb{P}\big(|\mathbf{f}_i|_k > d\sqrt{ks_*}\|\tilde{\mathbf{A}}^*\mathbf{f}_i\| \text{ for some } i \in [n]\big)$$

$$\leq \exp\left(-\frac{m^{1/2}N^{1/2}}{2e^{1/2}}\right) \leq \exp\left(-\frac{4}{e}m\right). \tag{11.37}$$

In the last step, we have used that $N \geq Cm$ together with the second requirement $C \geq 64\mu^8/e$. Finally, substituting (11.37) and (11.34) into (11.33), we conclude that

$$\mathbb{P}\big(\|\mathbf{e}\|_*^{(\beta)} > D\sqrt{s_*}\|\tilde{\mathbf{A}}^*\mathbf{e}\|_\infty \text{ for some } \mathbf{e} \in \mathbb{R}^m\big)$$

$$\leq 2\exp\big(-7m\big) + \exp\big(-4m/e\big) \leq 3\exp\big(-m\big).$$

We have proved the desired estimate (11.29). $\qquad\square$

We now prove the main robustness estimate for subgaussian random matrices.

Proof (of Theorem 11.10). According to the definition of subgaussian random matrices and to the bound of moments in terms of tail probabilities, i.e., Definition 9.1 and Proposition 7.13, the symmetric entries of the subgaussian matrix $\mathbf{A}$ have fourth moments bounded by some $\mu^4 \geq 1$. Moreover, according to Lemma 9.8, the concentration inequality (11.26) is satisfied. Thus, by choosing c_2 properly, Theorem 11.21 guarantees that, with probability at least $1 - 3\exp(-m)$, the matrix $\tilde{\mathbf{A}} = \frac{1}{\sqrt{m}}\mathbf{A}$ has the ℓ_1-quotient property relative to the norm $\|\cdot\|^{(\alpha)}$, $\alpha := \sqrt{\ln(eN/m)}$. Furthermore, according to Theorem 9.11, there is a constant $\tilde{c} > 0$ such that $\delta_{2s}(\tilde{\mathbf{A}}) < 1/3$ with probability at least $1 - 2\exp(-\tilde{c}m/15)$ provided

$$m \geq \frac{6}{\tilde{c}}\left[s\left(18 + 4\ln\left(\frac{N}{2s}\right)\right) + \frac{2\tilde{c}}{15}m\right], \quad \text{i.e.,} \quad m \geq \frac{60}{\tilde{c}}s\left(9 + 2\ln\left(\frac{N}{2s}\right)\right).$$

Since $9 + 2\ln(N/2s) \leq 9\ln(eN/s)$, this is implied by $m \geq (540/\tilde{c})s\ln(eN/s)$. Using Lemma C.6, we observe that this condition is in turn implied by the condition $s \leq c_3 s_*$—which is equivalent to $m \geq (1/c_3)s\ln(eN/m)$—provided c_3 is chosen small enough to have $c_3\ln(e/c_3) \leq \tilde{c}/540$. Theorem 6.13 now ensures that the matrix $\tilde{\mathbf{A}}$ satisfies the ℓ_2-robust null space property of order s relative to $\|\cdot\|_2$. Since $\|\cdot\|_2 \leq \|\cdot\|^{(\alpha)}$, it also satisfies the ℓ_2-robust null space property of order s relative to $\|\cdot\|^{(\alpha)}$. Thus, with probability at least

$$1 - 3\exp(-m) - 2\exp(-\tilde{c}m/15) \geq 1 - 5\exp(-c_1 m), \quad c_1 := \min\{1, \tilde{c}/15\},$$

the matrix $\tilde{\mathbf{A}}$ satisfies both the ℓ_1-quotient property and the ℓ_2-robust null space property of order $s \leq s_*/c_3$ relative to the norm $\|\cdot\|^{(\alpha)}$. The conclusion now follows from Theorem 11.12. $\square$

11.4 Nonuniform Instance Optimality

In Theorem 11.5, we have established that the uniform ℓ_2-instance optimality— the property that $\|\mathbf{x} - \Delta(\mathbf{A}\mathbf{x})\|_2 \leq C\sigma_s(\mathbf{x})_2$ for all $\mathbf{x} \in \mathbb{C}^N$—is only possible in the case $m \geq cN$, which is essentially irrelevant for compressive sensing. In this section, we change the point of view by fixing $\mathbf{x} \in \mathbb{C}^N$ at the start. We are going to prove for the ℓ_1-minimization map that the *nonuniform ℓ_2-instance optimality*—the property that $\|\mathbf{x} - \Delta_1(\mathbf{A}\mathbf{x})\|_2 \leq C\sigma_s(\mathbf{x})_2$ for this fixed $\mathbf{x} \in \mathbb{C}^N$— occurs with high probability on the draw of an $m \times N$ random matrix $\mathbf{A}$, provided $m \geq cs\ln(eN/s)$.

We notice that such estimates hold for other algorithms as well such as iterative hard thresholding, hard thresholding pursuit, orthogonal matching pursuit, and compressive sampling matching pursuit. In fact, the proof for these algorithms is rather simple: under suitable restricted isometry conditions specified in Theorems 6.21, 6.25, and 6.28, we have $\|\mathbf{x} - \Delta(\mathbf{A}\mathbf{x})\|_2 \leq C\|\mathbf{A}\mathbf{x}_{\overline{S}}\|_2$, where S denotes an index set of s largest absolute entries of $\mathbf{x}$. Then the desired estimate follows from the concentration inequality (9.7) for subgaussian random matrices, which implies that $\|\mathbf{A}\mathbf{x}_{\overline{S}}\|_2 \leq 2\|\mathbf{x}_{\overline{S}}\|_2 = 2\sigma_s(\mathbf{x})_2$ with high probability. However, these algorithms (except perhaps orthogonal matching pursuit) require s as an input. Advantageously, the ℓ_1-minimization does not.

For ℓ_1-minimization, the key to proving the nonuniform ℓ_2-instance optimality lies in the stable and robust estimates of Theorems 11.9 and 11.10. We begin with the easier case of Gaussian matrices. Again, the result also incorporates measurement error even though equality-constrained ℓ_1-minimization Δ_1 is used rather than the seemingly more natural quadratically constrained ℓ_1-minimization $\Delta_{1,\eta}$.

Theorem 11.23. *Let* $\mathbf{x} \in \mathbb{C}^N$ *be a fixed vector and let* $\tilde{\mathbf{A}} = \frac{1}{\sqrt{m}}\mathbf{A}$ *where* $\mathbf{A}$ *is a random draw of an* $m \times N$ *Gaussian matrix. If*

$$N \geq c_2 m, \qquad s \leq c_3 s_* = \frac{c_3 m}{\ln(eN/m)},$$

then with probability at least $1 - 5\exp(-c_1 m)$ *the* ℓ_2*-error estimates*

$$\|\mathbf{x} - \Delta_1(\tilde{\mathbf{A}}\mathbf{x} + \mathbf{e})\|_2 \leq C\sigma_s(\mathbf{x})_2 + D\|\mathbf{e}\|_2 \tag{11.38}$$

are valid for all $\mathbf{e} \in \mathbb{C}^m$. *The constants* $c_1, c_2, c_3, C, D > 0$ *are universal.*

Proof. Let S denote a set of s largest absolute entries of $\mathbf{x}$. We have

$$\|\mathbf{x} - \Delta_1(\tilde{\mathbf{A}}\mathbf{x} + \mathbf{e})\|_2 \leq \|\mathbf{x}_{\overline{S}}\|_2 + \|\mathbf{x}_S - \Delta_1(\tilde{\mathbf{A}}\mathbf{x} + \mathbf{e})\|_2$$

$$= \sigma_s(\mathbf{x})_2 + \|\mathbf{x}_S - \Delta_1(\tilde{\mathbf{A}}\mathbf{x}_S + \mathbf{e}')\|_2, \tag{11.39}$$

where $\mathbf{e}' := \tilde{\mathbf{A}}\mathbf{x}_{\overline{S}} + \mathbf{e}$. Taking the conditions $N \geq c_2 m$ and $s \leq c_3 s_*$ into account, Theorem 11.9 applied to $\mathbf{x}_S \in \mathbb{C}^N$ and $\mathbf{e}' \in \mathbb{C}^m$ yields

$$\|\mathbf{x}_S - \Delta_1(\tilde{\mathbf{A}}\mathbf{x}_S + \mathbf{e}')\|_2 \leq D\|\mathbf{e}'\|_2 \leq D\|\tilde{\mathbf{A}}\mathbf{x}_{\overline{S}}\|_2 + D\|\mathbf{e}\|_2 \tag{11.40}$$

with probability at least $1 - 3\exp(-c_1' m)$ for some constant $c_1' > 0$. Next, the concentration inequality for Gaussian random matrices (see Exercise 9.2 or Lemma 9.8, which, however, does not specify constants) ensures that

$$\|\tilde{\mathbf{A}}\mathbf{x}_{\overline{S}}\|_2 \leq 2\|\mathbf{x}_{\overline{S}}\|_2 = 2\sigma_s(\mathbf{x})_2 \tag{11.41}$$

with probability at least $1 - 2\exp(-m/12)$. We finally derive (11.38) by combining the inequalities (11.39), (11.40), and (11.41). The desired probability is at least $1 - 3\exp(-c_1' m) - 2\exp(-m/12) \geq 1 - 5\exp(-c_1 m)$, $c_1 := \min\{c_1', 1/12\}$. $\quad\square$

In the same spirit, a nonuniform mixed (ℓ_q, ℓ_p)-instance optimality result for Gaussian matrices can be proved for any $1 \leq p \leq q \leq 2$. It is worth recalling that, in the regime $m \asymp s\ln(eN/m)$, a matching result cannot be obtained in the uniform setting; see Exercise 11.2.

Theorem 11.24. *Let* $1 \leq p \leq q \leq 2$, *let* $\mathbf{x} \in \mathbb{C}^N$ *be a fixed vector, and let* $\tilde{\mathbf{A}} = \frac{1}{\sqrt{m}}\mathbf{A}$ *where* $\mathbf{A}$ *is a random draw of an* $m \times N$ *Gaussian matrix. If*

$$N \geq c_2 m, \qquad s \leq c_3 s_* = \frac{c_3 m}{\ln(eN/m)},$$

then with probability at least $1 - 5\exp(-c_1 m)$ *the error estimates*

$$\|\mathbf{x} - \Delta_1(\tilde{\mathbf{A}}\mathbf{x} + \mathbf{e})\|_q \leq \frac{C}{s^{1/p-1/q}}\sigma_s(\mathbf{x})_p + Ds_*^{1/q-1/2}\|\mathbf{e}\|_2$$

are valid for all $\mathbf{e} \in \mathbb{C}^m$. *The constants* $c_1, c_2, c_3, C, D > 0$ *are universal.*

Proof. If $c_1', c_2', c_3', C', D' > 0$ are the constants of Theorem 11.9, we define $c_3 := c_3'/3$. Then, for $s \leq c_3 s_*$, we consider an index set S of s largest absolute entries of $\mathbf{x}$, and an index set T of $t := \lceil c_3 s_* \rceil \geq s$ next largest absolute entries of $\mathbf{x}$. We have

$$\|\mathbf{x} - \Delta_1(\tilde{\mathbf{A}}\mathbf{x} + \mathbf{e})\|_q \leq \|\mathbf{x}_{\overline{S \cup T}}\|_q + \|\mathbf{x}_{S \cup T} - \Delta_1(\tilde{\mathbf{A}}\mathbf{x} + \mathbf{e})\|_q$$

$$\leq \frac{1}{t^{1/p - 1/q}}\|\mathbf{x}_{\overline{S}}\|_p + \|\mathbf{x}_{S \cup T} - \Delta_1(\tilde{\mathbf{A}}\mathbf{x}_{S \cup T} + \mathbf{e}')\|_q, \qquad (11.42)$$

where we have used Proposition 2.3 and have set $\mathbf{e}' := \tilde{\mathbf{A}}\mathbf{x}_{\overline{S \cup T}} + \mathbf{e}$ in the last inequality. Taking $c_2 = c_2'$ and noticing that $s + t \leq c_3' s_*$, Theorem 11.9 applied to $\mathbf{x}_{S \cup T} \in \mathbb{C}^N$ and $\mathbf{e}' \in \mathbb{C}^m$ yields

$$\|\mathbf{x}_{S \cup T} - \Delta_1(\tilde{\mathbf{A}}\mathbf{x}_{S \cup T} + \mathbf{e}')\|_q \leq D s_*^{1/q - 1/2}\|\mathbf{e}'\|_2$$

$$\leq D s_*^{1/q - 1/2}\|\tilde{\mathbf{A}}\mathbf{x}_{\overline{S \cup T}}\|_2 + D s_*^{1/q - 1/2}\|\mathbf{e}\|_2$$

$$\leq D \frac{t^{1/q - 1/2}}{c_3^{1/q - 1/2}}\|\tilde{\mathbf{A}}\mathbf{x}_{\overline{S \cup T}}\|_2 + D s_*^{1/q - 1/2}\|\mathbf{e}\|_2 \qquad (11.43)$$

with probability at least $1 - 3\exp(-c_1'm)$. The concentration inequality for Gaussian matrices (see Exercise 9.2 or Lemma 9.8), in conjunction with Proposition 2.3, gives

$$\|\tilde{\mathbf{A}}\mathbf{x}_{\overline{S \cup T}}\|_2 \leq 2\|\mathbf{x}_{\overline{S \cup T}}\|_2 \leq \frac{2}{t^{1/p - 1/2}}\|\mathbf{x}_{\overline{S}}\|_p \qquad (11.44)$$

with probability at least $1 - 2\exp(-m/12)$. Combining (11.42)–(11.44), we deduce

$$\|\mathbf{x} - \Delta_1(\tilde{\mathbf{A}}\mathbf{x} + \mathbf{e})\|_q \leq \frac{1 + 2Dc_3^{1/2 - 1/q}}{t^{1/p - 1/2}}\|\mathbf{x}_{\overline{S}}\|_p + D s_*^{1/q - 1/2}\|\mathbf{e}\|_2$$

$$\leq \frac{1 + 2Dc_3^{-1/2}}{s^{1/p - 1/2}}\sigma_s(\mathbf{x})_p + D s_*^{1/q - 1/2}\|\mathbf{e}\|_2.$$

The desired probability is $1 - 3\exp(-c_1'm) - 2\exp(-m/12) \geq 1 - 5\exp(-c_1 m)$ with $c_1 := \min\{c_1', 1/12\}$. $\square$

The previous results extend to subgaussian matrices. We do not isolate the ℓ_2-instance optimality here, as we state the nonuniform mixed instance optimality directly. As in Sect. 11.3, the ℓ_2-norm on the measurement error is replaced by $\|\mathbf{e}\|^{(\sqrt{\ln(eN/m)})} = \max\{\|\mathbf{e}\|_2, \sqrt{\ln(eN/m)}\|\mathbf{e}\|_\infty\}$. The condition on N in (11.45) below will be satisfied in reasonable practical scenarios.

Theorem 11.25. *Let* $1 \le p \le q \le 2$, *let* $\mathbf{x} \in \mathbb{C}^N$ *be a fixed vector, and let* $\tilde{\mathbf{A}} = \frac{1}{\sqrt{m}} \mathbf{A}$ *where* $\mathbf{A}$ *is a random draw of an* $m \times N$ *subgaussian matrix with symmetric entries and with subgaussian parameter* c *in* (9.2). *There exist constants* $c_1, c_2, c_3, c_4, C, D > 0$ *depending only on* c *such that if*

$$c_2 m \le N \le \frac{m}{e} \exp(c_3 \sqrt{m}), \qquad s \le c_4 s_* = \frac{c_4 m}{\ln(eN/m)}, \qquad (11.45)$$

then with probability at least $1 - 9\exp(-c_1 \sqrt{m})$ *the error estimates*

$$\|\mathbf{x} - \Delta_1(\tilde{\mathbf{A}}\mathbf{x} + \mathbf{e})\|_q \le \frac{C}{s^{1/p-1/q}} \sigma_s(\mathbf{x})_p + D s_*^{1/q-1/2} \|\mathbf{e}\| \left(\sqrt{\ln(eN/m)}\right)$$

are valid for all $\mathbf{e} \in \mathbb{C}^m$.

Proof. The argument is similar to the one used in the proof of Theorem 11.24, with the addition of step (11.48) below. Let $c_1', c_2', c_3', C', D' > 0$ be the constants of Theorem 11.10, and define $c_4 := c_3'/3$. Then, for $s \le c_4 s_*$, we consider an index set S of s largest absolute entries of $\mathbf{x}$, and an index set T of $t := \lceil c_4 s_* \rceil \ge s$ next largest absolute entries of $\mathbf{x}$. We have

$$\|\mathbf{x} - \Delta_1(\tilde{\mathbf{A}}\mathbf{x} + \mathbf{e})\|_q \le \|\mathbf{x}_{\overline{SUT}}\|_q + \|\mathbf{x}_{SUT} - \Delta_1(\tilde{\mathbf{A}}\mathbf{x} + \mathbf{e})\|_q$$

$$\le \frac{1}{t^{1/p-1/q}} \|\mathbf{x}_{\overline{S}}\|_p + \|\mathbf{x}_{SUT} - \Delta_1(\tilde{\mathbf{A}}\mathbf{x}_{SUT} + \mathbf{e}')\|_q, \qquad (11.46)$$

where we have applied Proposition 2.3 and have set $\mathbf{e}' := \tilde{\mathbf{A}}\mathbf{x}_{\overline{SUT}} + \mathbf{e}$. Taking $c_2 = c_2'$ and noticing that $s + t \le c_3' s_*$, Theorem 11.10 applied to $\mathbf{x}_{SUT} \in \mathbb{C}^N$ and $\mathbf{e}' \in \mathbb{C}^m$ yields with $D = D'$

$$\|\mathbf{x}_{SUT} - \Delta_1(\tilde{\mathbf{A}}\mathbf{x}_{SUT} + \mathbf{e}')\|_q \le D s_*^{1/q-1/2} \|\mathbf{e}'\| \left(\sqrt{\ln(eN/m)}\right)$$

$$\le D s_*^{1/q-1/2} \|\tilde{\mathbf{A}}\mathbf{x}_{\overline{SUT}}\| \left(\sqrt{\ln(eN/m)}\right) + D s_*^{1/q-1/2} \|\mathbf{e}\| \left(\sqrt{\ln(eN/m)}\right)$$

$$\le D \frac{t^{1/q-1/2}}{c_4^{1/q-1/2}} \|\tilde{\mathbf{A}}\mathbf{x}_{\overline{SUT}}\| \left(\sqrt{\ln(eN/m)}\right) + D s_*^{1/q-1/2} \|\mathbf{e}\| \left(\sqrt{\ln(eN/m)}\right) \qquad (11.47)$$

with probability at least $1 - 5\exp(-c_1'm)$. By the concentration inequality of Lemma 9.8, we have

$$\|\tilde{\mathbf{A}}\mathbf{x}_{\overline{SUT}}\|_2 \le 2\|\mathbf{x}_{\overline{SUT}}\|_2$$

with probability at least $1 - 2\exp(-\tilde{c}m)$ for a constant $\tilde{c}$ depending only on the subgaussian parameter c. Moreover, for each $i \in [m]$, Theorem 7.27 guarantees that the inequality

$$|(\tilde{\mathbf{A}}\mathbf{x}_{\overline{SUT}})_i| \le \frac{2}{\sqrt{\ln(eN/m)}} \|\mathbf{x}_{\overline{SUT}}\|_2$$

holds with probability at least $1 - 2\exp\left(-c^{-1}m/\ln(eN/m)\right)$. By the union bound, it follows that

$$\|\tilde{\mathbf{A}}\mathbf{x}_{\overline{SUT}}\|_\infty \leq \frac{2}{\sqrt{\ln(eN/m)}}\|\mathbf{x}_{\overline{SUT}}\|_2 \qquad (11.48)$$

with probability at least $1 - 2m\exp\left(-c^{-1}m/\ln(eN/m)\right)$. We note that this probability is at least $1 - 2\exp(-\sqrt{m})$ when $N \leq m\exp(c_3\sqrt{m})/e$ with $c_3 := 1/(2c)$, since

$$m\exp\left(\frac{-m}{c\ln(eN/m)}\right) \leq m\exp\left(\frac{-m}{cc_3\sqrt{m}}\right) = \exp\left(\ln(m)-2\sqrt{m}\right) \leq \exp\left(-\sqrt{m}\right).$$

We have obtained

$$\|\tilde{\mathbf{A}}\mathbf{x}_{\overline{SUT}}\|^{\left(\sqrt{\ln(eN/m)}\right)} = \max\{\|\tilde{\mathbf{A}}\mathbf{x}_{\overline{SUT}}\|_2, \sqrt{\ln(eN/m)}\|\tilde{\mathbf{A}}\mathbf{x}_{\overline{SUT}}\|_\infty\}$$

$$\leq 2\|\mathbf{x}_{\overline{SUT}}\|_2 \leq \frac{2}{t^{1/p-1/2}}\|\mathbf{x}_{\overline{S}}\|_p,$$

with probability at least $1 - 2\exp(-\tilde{c}m) - 2\exp(-\sqrt{m}) \geq 1 - 4\exp(-c'\sqrt{m})$, $c' := \min\{\tilde{c}, 1\}$. Combining this with (11.46) and (11.47), we deduce

$$\|\mathbf{x}-\Delta_1(\tilde{\mathbf{A}}\mathbf{x} + \mathbf{e})\|_q \leq \frac{1+2Dc_4^{1/2-1/q}}{t^{1/p-1/2}}\|\mathbf{x}_{\overline{S}}\|_p + Ds_*^{1/q-1/2}\|\mathbf{e}\|^{\left(\sqrt{\ln(eN/m)}\right)}$$

$$\leq \frac{1+2Dc_4^{-1/2}}{s^{1/p-1/2}}\sigma_s(\mathbf{x})_p + Ds_*^{1/q-1/2}\|\mathbf{e}\|^{\left(\sqrt{\ln(eN/m)}\right)}.$$

The desired probability is at least

$$1 - 5\exp(-c_1'm) - 4\exp(-c'\sqrt{m}) \geq 1 - 9\exp(-c_1\sqrt{m})$$

with $c_1 := \min\{c_1', c'\}$. $\square$

Notes

The notions of instance optimality and mixed instance optimality were introduced by Cohen et al. in [123]. Theorems 11.4 and 11.5 are taken from this article. The other major theorem of Sect. 11.1, namely, Theorem 11.6 on the minimal number of measurements for ℓ_1-instance optimality, is taken from [213].

The ℓ_1-quotient property was introduced in the context of compressive sensing by Wojtaszczyk in [509]. The content of Sect. 11.2 essentially follows the ideas of this article, except that the restricted isometry property is replaced by the weaker notion of robust null space property and that error estimates in ℓ_q-norm for all $1 \leq q \leq 2$ are given. The ℓ_1-quotient property for Gaussian random matrices was proved in [509], too, save for the extra requirement that $N \geq cm \ln^\xi(m)$ for some $\xi > 0$. This issue was resolved here with the use of the norm defined in (11.18). As a matter of fact, the ℓ_1-quotient property for Gaussian matrices had been established earlier in a different context by Gluskin in [228], where a certain optimality of the probability estimate was also proved.

Gaussian matrices are not the only random matrices to possess the ℓ_1-quotient property relative to the ℓ_2-norm and in turn to provide the reconstruction estimates of Theorem 11.9. It was established in [210] that *Weibull random matrices* also do. These are (rescaled versions) of matrices whose entries are independent symmetric random variables that satisfy, for some $r \geq 1$ and some $c > 0$,

$$\mathbb{P}(|A_{i,j}| \geq t) = \exp(-ct^r), \qquad t \geq 0.$$

For matrices satisfying the restricted isometry property, Wojtaszczyk also showed in [510] that the estimates of Theorem 11.9 for Gaussian random matrices can be obtained with a modified version of ℓ_1-minimization in which one artificially adds columns to the matrix $\mathbf{A}$.

The ℓ_1-quotient property relative to the norm $\max\{\|\cdot\|_2, \alpha\|\cdot\|_\infty\}$ was introduced in the context of compressive sensing by DeVore, Petrova, and Wojtaszczyk in [148], where it was established for Bernoulli random matrices. The ℓ_1-quotient property had in fact been shown earlier in a different context by Litvak et al. in [329]. We followed the proof of [329], because of a slight flaw in the proof in [148], namely, that the vectors in the δ-covering depend on the random matrix; hence, the concentration inequality cannot be applied directly to them. The key Lemma 11.22 was proved by Montgomery-Smith in [356].

The results given in Sect. 11.4 on the nonuniform ℓ_2-instance optimality appeared (under the terminology of instance optimality in probability) in [148, 509].

Exercises

11.1. Verify in details the observation made in (11.1).

11.2. For $q \geq p \geq p' \geq 1$, prove that if a pair $(\mathbf{A}, \Delta)$ is mixed (ℓ_q, ℓ_p)-instance optimal of order s with constant C, then it is also mixed $(\ell_q, \ell_{p'})$-instance optimal of order $\lceil s/2 \rceil$ with constant C' depending only on C.

Combine this observation with Theorem 11.7 to derive that mixed (ℓ_q, ℓ_p)-instance optimal pairs $(\mathbf{A}, \Delta)$ of order s, where $\mathbf{A} \in \mathbb{C}^{m \times N}$ and $\Delta : \mathbb{C}^m \to \mathbb{C}^N$, can only exist if $m \geq cs \ln(eN/s)$. For $q > p > 1$, improve this bound using the estimate for the Gelfand width $d^m(B_p^N, \ell_q^N)$ given on page 327.

11.3. Prove that if the coherence of a matrix $\mathbf{A} \in \mathbb{C}^{m \times N}$ with ℓ_2-normalized columns satisfies $\mu(\mathbf{A}) < 1/4$, then the operator norm $\|\mathbf{A}\|_{2 \to 2}$ cannot be bounded by an absolute constant $C > 0$ unless $m \geq cN$ for some constant $c > 0$ depending only on C.

11.4. Let a measurement matrix $\mathbf{A} \in \mathbb{C}^{m \times N}$ be given and let $0 < p < 1$. Prove that if there is a reconstruction map Δ such that $\|\mathbf{x} - \Delta(\mathbf{Ax})\|_p \leq C\sigma_{2s}(\mathbf{x})_p$ for all $\mathbf{x} \in \mathbb{C}^N$, then $\|\mathbf{v}\|_p \leq C\sigma_{2s}(\mathbf{v})_p$ for all $\mathbf{v} \in \ker \mathbf{A}$. Prove conversely that if $\|\mathbf{v}\|_p \leq C\sigma_{2s}(\mathbf{v})_p$ for all $\mathbf{v} \in \ker \mathbf{A}$, then there is a reconstruction map Δ such that $\|\mathbf{x} - \Delta(\mathbf{Ax})\|_p \leq 2^{1/p}C\sigma_{2s}(\mathbf{x})_p$ for all $\mathbf{x} \in \mathbb{C}^N$.

11.5. Let a measurement matrix $\mathbf{A} \in \mathbb{C}^{m \times N}$ be given. Suppose that, for some integer $s \geq 1$ and some constant $C \geq 1$, there exists a reconstruction map $\Delta : \mathbb{C}^m \to \mathbb{C}^N$ such that

$$\|\mathbf{x} - \Delta(\mathbf{Ax})\|_2 \leq \frac{C}{\sqrt{s}}\|\mathbf{x}\|_1 \qquad \text{for all } \mathbf{x} \in \mathbb{C}^N.$$

Prove that the pair $(\mathbf{A}, \Delta_1)$ is mixed (ℓ_2, ℓ_1)-instance optimal of order t with constant $(2+\rho)/(1-\rho)$ provided $\rho := 2C\sqrt{t/s} < 1$. Deduce that the existence of a pair $(\mathbf{A}, \Delta)$ which is mixed (ℓ_2, ℓ_1)-instance optimal of order $\lceil 9C^2t \rceil$ with constant C implies that the pair $(\mathbf{A}, \Delta_1)$ is mixed (ℓ_2, ℓ_1)-instance optimal of order t with constant 8.

11.6. Let $\mathbf{A} \in \mathbb{R}^{m \times N}$ and let $\| \cdot \|$ be a norm on $\mathbb{C}^m$ invariant under complex conjugation, i.e., satisfying $\|\overline{\mathbf{y}}\| = \|\mathbf{y}\|$ for all $\mathbf{y} \in \mathbb{C}^m$. For $q \geq 1$, prove that the real and complex versions of the ℓ_q-quotient property, namely,

$$\forall \mathbf{e} \in \mathbb{R}^m, \exists \mathbf{u} \in \mathbb{R}^N : \mathbf{Au} = \mathbf{e}, \|\mathbf{u}\|_q \leq d\, s_*^{1/q-1/2}\|\mathbf{e}\|, \tag{11.49}$$

$$\forall \mathbf{e} \in \mathbb{C}^m, \exists \mathbf{u} \in \mathbb{C}^N : \mathbf{Au} = \mathbf{e}, \|\mathbf{u}\|_q \leq d\, s_*^{1/q-1/2}\|\mathbf{e}\|, \tag{11.50}$$

are equivalent, in the sense that (11.50) implies (11.49) with the same constant d and that (11.49) implies (11.50) with the constant d replaced by $2d$.

11.7. Prove Lemma 11.17 for the case $q = 1$ without using limiting arguments.

11.8. Prove that the dual norm of the norm $\| \cdot \|^{(\alpha)}$ introduced in (11.25) can be expressed as

$$\|\mathbf{y}\|_*^{(\alpha)} = \inf \left\{ \|\mathbf{y}'\|_2 + \frac{1}{\alpha}\|\mathbf{y}''\|_1, \mathbf{y}' + \mathbf{y}'' = \mathbf{y} \right\}, \qquad \mathbf{y} \in \mathbb{C}^m.$$

11.9. Let $q \geq 1$ and let $\| \cdot \|$ be a norm on $\mathbb{C}^m$. Given a matrix $\mathbf{A} \in \mathbb{C}^{m \times N}$, suppose that there exist $D > 0$ and $0 < \rho < 1$ such that, for each $\mathbf{e} \in \mathbb{C}^m$, one can find $\mathbf{u} \in \mathbb{C}^N$ with $\|\mathbf{Au} - \mathbf{e}\| \leq \rho\|\mathbf{e}\|$ and $\|\mathbf{u}\|_q \leq Ds_*^{1/q-1/2}\|\mathbf{e}\|$. Prove that the matrix $\mathbf{A}$ satisfies the ℓ_q-quotient property with constant $D/(1 - \rho)$ relative to the norm $\| \cdot \|$.

11.10. Let $q \geq 1$ and let $\| \cdot \|$ be a norm on $\mathbb{C}^m$. Suppose that a pair of measurement matrix $\mathbf{A} \in \mathbb{C}^{m \times N}$ and reconstruction map $\Delta : \mathbb{C}^m \to \mathbb{C}^N$ is mixed (ℓ_q, ℓ_1)-instance optimal of order $s \leq c s_*$ and that $\mathbf{A}$ has the simultaneous (ℓ_q, ℓ_1)-quotient property relative to $\| \cdot \|$; see Definition 11.14. Prove that there is a constant $D > 0$ such that

$$\| \mathbf{x} - \Delta(\mathbf{A}\mathbf{x}) \|_q \leq \| \mathbf{x}_{\overline{S}} \|_q + D s_*^{1/q - 1/2} \| \mathbf{A}\mathbf{x}_{\overline{S}} \|$$

for any $\mathbf{x} \in \mathbb{C}^N$ and any index set $S \subset [N]$ of size s.

Chapter 12
Random Sampling in Bounded Orthonormal Systems

We have seen in the previous chapters that subgaussian random matrices provide optimal measurement matrices for compressive sensing. While this is a very important insight for the theory, the use of such "completely random" matrices, where all entries are independent, is limited for practical purposes. Indeed, subgaussian random matrices do not possess any structure. However, structure is important for several reasons:

- Applications may impose certain structure on the measurement matrix due to physical or other constraints. We refer the reader to Section 1.2 for some examples.
- Structure of the measurement matrix often allows to have fast matrix–vector multiplication algorithms for both the matrix itself and its adjoint by exploiting, for instance, the fast Fourier transform (FFT). This is crucial for speedups in recovery algorithms (including ℓ_1-minimization), and only in this situation can large-scale problems be treated with compressive sensing techniques.
- For large unstructured matrices, difficulties in storing all the matrix entries arise, while a structured matrix is usually generated by a number of parameters much smaller than the number of matrix entries, so that it is much easier to store.

From this point of view, it is important to investigate whether certain structured random matrices may provide recovery guarantees similar to the ones for subgaussian random matrices. By a structured random matrix, we mean a structured matrix that is generated by a random choice of parameters.

The important setup at the core and the origin of the field that we will study exclusively below occurs when randomly sampling functions whose expansion in a bounded orthonormal system (see the precise definition below) is sparse or compressible. Special cases consist in randomly sampling sparse trigonometric polynomials and in taking random samples from the Fourier transform of a sparse vector. In the latter case, the associated measurement matrix is a random partial Fourier matrix, and it has a fast matrix–vector multiplication routine using the FFT. The analysis of the resulting random measurement matrices becomes more involved than the analysis for subgaussian random matrices because the entries are

S. Foucart and H. Rauhut, *A Mathematical Introduction to Compressive Sensing*,
Applied and Numerical Harmonic Analysis, DOI 10.1007/978-0-8176-4948-7_12,
© Springer Science+Business Media New York 2013

not independent anymore. In this context, nonuniform recovery results are simpler to derive than uniform recovery results based on the restricted isometry property. We will proceed by increasing difficulty of the proofs.

Other types of structured random matrices, including partial random circulant matrices, are discussed briefly in the Notes section.

12.1 Bounded Orthonormal Systems

An important class of structured random matrices is connected with random sampling in certain finite-dimensional function spaces. We require an orthonormal basis whose elements are uniformly bounded in the L_∞-norm. The most prominent example consists of the trigonometric system. In a discrete setup, the resulting matrix is a random partial Fourier matrix, which was the first structured random matrix investigated in compressive sensing.

Let $\mathcal{D} \subset \mathbb{R}^d$ be endowed with a probability measure ν. Further, let $\Phi = \{\phi_1, \ldots, \phi_N\}$ be an orthonormal system of complex-valued functions on $\mathcal{D}$, that is, for $j, k \in [N]$,

$$\int_{\mathcal{D}} \phi_j(\mathbf{t})\overline{\phi_k(\mathbf{t})}d\nu(\mathbf{t}) = \delta_{j,k} = \begin{cases} 0 & \text{if } j \neq k, \\ 1 & \text{if } j = k. \end{cases} \tag{12.1}$$

Definition 12.1. We call $\Phi = \{\phi_1, \ldots, \phi_N\}$ a *bounded orthonormal system* (BOS) with constant K if it satisfies (12.1) and if

$$\|\phi_j\|_\infty := \sup_{\mathbf{t} \in \mathcal{D}} |\phi_j(\mathbf{t})| \leq K \quad \text{for all } j \in [N]. \tag{12.2}$$

The smallest value that the constant K can take is $K = 1$. Indeed,

$$1 = \int_{\mathcal{D}} |\phi_j(\mathbf{t})|^2 d\nu(\mathbf{t}) \leq \sup_{\mathbf{t} \in \mathcal{D}} |\phi_j(\mathbf{t})|^2 \int_{\mathcal{D}} d\nu(\mathbf{t}) \leq K^2.$$

In the extreme case $K = 1$, we necessarily have $|\phi_j(\mathbf{t})| = 1$ for ν-almost all $\mathbf{t} \in \mathcal{D}$ as revealed by the same chain of inequalities.

Note that *some* bound K can be found for most reasonable sets of functions $\{\phi_j, j \in [N]\}$. The crucial point of the boundedness condition (12.2) is that K should ideally be independent of N. Intuitively, such a condition excludes, for instance, that the functions ϕ_j are very localized in small regions of $\mathcal{D}$.

We consider functions of the form

$$f(\mathbf{t}) = \sum_{k=1}^N x_k \phi_k(\mathbf{t}), \quad \mathbf{t} \in \mathcal{D}. \tag{12.3}$$

Let $\mathbf{t}_1, \ldots, \mathbf{t}_m \in \mathcal{D}$ be some sampling points and suppose we are given the sample values

$$y_\ell = f(\mathbf{t}_\ell) = \sum_{k=1}^{N} x_k \phi_k(\mathbf{t}_\ell), \quad \ell \in [m].$$

Introducing the *sampling matrix* $\mathbf{A} \in \mathbb{C}^{m \times N}$ with entries

$$A_{\ell,k} = \phi_k(\mathbf{t}_\ell), \qquad \ell \in [m], k \in [N], \tag{12.4}$$

the vector $\mathbf{y} = [y_1, \ldots, y_m]^\top$ of sample values (measurements) can be written in the form

$$\mathbf{y} = \mathbf{A}\mathbf{x}, \tag{12.5}$$

where $\mathbf{x} = [x_1, \ldots, x_N]^\top$ is the vector of coefficients in (12.3).

Our task is to reconstruct the function f, or equivalently its vector $\mathbf{x}$ of coefficients, from the vector of samples $\mathbf{y}$. We wish to perform this task with as few samples as possible. Without further assumptions, this is impossible if $m < N$. As common in this book, we therefore introduce sparsity. A function f of the form (12.3) is called s-sparse with respect to $(\phi_1, \ldots, \phi_N)$ if its coefficient vector $\mathbf{x}$ is s-sparse. Sparse functions appear in a variety of applications; see Sect. 1.2 for some motivating examples. The problem of recovering an s-sparse function from m sample values reduces to the compressive sensing problem with the measurement matrix given by the sampling matrix $\mathbf{A}$ in (12.4).

Since the deterministic construction of suitable matrices for compressive sensing is to date an open problem, we now introduce randomness. We assume that the sampling points $\mathbf{t}_1, \ldots, \mathbf{t}_m$ are selected independently at random according to the probability measure ν. This means that $\mathbb{P}(\mathbf{t}_\ell \in B) = \nu(B)$, $\ell \in [m]$, for a measurable subset $B \subset \mathcal{D}$. We then call the associated matrix (12.4) the *random sampling matrix* associated to a BOS with constant $K \geq 1$. Note that this matrix has stochastically independent rows, but the entries within each row are not independent. Indeed, for fixed ℓ, the entries $A_{\ell,k}$, $k \in [N]$, all depend on the single random sampling point $\mathbf{t}_\ell$.

Before continuing with the general theory, we present some important examples of BOSs.

1. *Trigonometric polynomials.* Let $\mathcal{D} = [0, 1]$ and set, for $k \in \mathbb{Z}$,

$$\phi_k(t) = e^{2\pi i k t}.$$

The probability measure ν is the Lebesgue measure on $[0, 1]$. Then, for all $j, k \in \mathbb{Z}$,

$$\int_0^1 \phi_k(t)\overline{\phi_j(t)}dt = \delta_{j,k}. \tag{12.6}$$

The constant in (12.2) is $K = 1$. For a subset $\Gamma \subset \mathbb{Z}$ of size N, we then consider the trigonometric polynomials of the form

$$f(t) = \sum_{k \in \Gamma} x_k \phi_k(t) = \sum_{k \in \Gamma} x_k e^{2\pi i k t}.$$

The common choice $\Gamma = \{-q, -q + 1, \ldots, q - 1, q\}$ results in trigonometric polynomials of degree at most q (then $N = 2q + 1$). We emphasize, however, that an arbitrary choice of $\Gamma \subset \mathbb{Z}$ of size $\operatorname{card}(\Gamma) = N$ is possible. Introducing sparsity on the coefficient vector $\mathbf{x} \in \mathbb{C}^N$ then leads to the notion of s-sparse trigonometric polynomials.

The sampling points $t_1, \ldots, t_m$ are chosen independently and uniformly at random from $[0, 1]$. The entries of the associated structured random matrix $\mathbf{A}$ are given by

$$A_{\ell,k} = e^{2\pi i k t_\ell}, \quad \ell \in [m], \quad k \in \Gamma. \tag{12.7}$$

The matrix $\mathbf{A}$ is a Fourier type matrix, sometimes also called *nonequispaced Fourier matrix*.

This example extends to multivariate trigonometric polynomials on $[0, 1]^d$, $d \in \mathbb{N}$. Indeed, the monomials $\phi_{\mathbf{k}}(\mathbf{t}) = e^{2\pi i \langle \mathbf{k}, \mathbf{t} \rangle}$, $\mathbf{k} \in \mathbb{Z}^d$, $\mathbf{t} \in [0, 1]^d$, form an orthonormal system on $[0, 1]^d$. For readers familiar with abstract harmonic analysis we mention that this example can be further generalized to characters of a compact commutative group. The corresponding measure is the Haar measure of the group.

We recall that Figs. 1.2 and 1.3 in Chap. 1 show an example of exact recovery of a 5-sparse trigonometric polynomial in dimension 64 from 16 Fourier samples using ℓ_1-minimization.

2. *Real trigonometric polynomials.* Instead of the complex exponentials above, we may also take the real functions

$$\phi_{2k}(t) = \sqrt{2} \cos(2\pi k t), \quad k \in \mathbb{N}, \quad \phi_0(t) = 1,$$

$$\phi_{2k-1}(t) = \sqrt{2} \sin(2\pi k t), \quad k \in \mathbb{N}. \tag{12.8}$$

They also form an orthonormal system on $\mathcal{D} = [0, 1]$ with respect to the Lebesgue measure and the constant in (12.2) is $K = \sqrt{2}$. The sampling points $t_1, \ldots, t_m$ are chosen again according to the uniform distribution on $[0, 1]$.

Figure 12.1 presents a phase transition plot for recovery of sparse cosine expansions (i.e., only the functions ϕ_{2k} are considered) from random samples via iteratively reweighted least squares, an algorithm to be studied in Sect. 15.3. Black means 100 % empirical success probability, while white means 0 % empirical success probability. Here, the vector length is chosen as $N = 300$. (For more information on phase transition plots in general we refer to the Notes section of Chap. 9.)

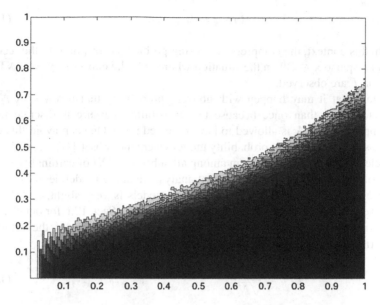

Fig. 12.1 Empirical recovery rates for random sampling associated to the cosine system and reconstruction via the iteratively reweighted least squares algorithm; horizontal axis m/N, vertical axis s/m

3. *Discrete orthonormal systems.* Let $\mathbf{U} \in \mathbb{C}^{N \times N}$ be a unitary matrix. The normalized columns $\sqrt{N}\mathbf{u}_k \in \mathbb{C}^N$, $k \in [N]$, form an orthonormal system with respect to the discrete uniform measure on $[N]$ given by $\nu(B) = \operatorname{card}(B)/N$ for $B \subset [N]$, i.e.,

$$\frac{1}{N} \sum_{t=1}^{N} \sqrt{N}\mathbf{u}_k(t) \overline{\sqrt{N}\mathbf{u}_\ell(t)} = \langle \mathbf{u}_k, \mathbf{u}_\ell \rangle = \delta_{k,\ell}, \quad k, \ell \in [N].$$

Here, $\mathbf{u}_k(t) := U_{t,k}$ denotes the tth entry of the kth column of $\mathbf{U}$. The boundedness condition (12.2) requires that the normalized entries of $\mathbf{U}$ are bounded, i.e.,

$$\sqrt{N} \max_{k,t \in [N]} |U_{t,k}| = \max_{k,t \in [N]} |\sqrt{N}\mathbf{u}_k(t)| \le K. \tag{12.9}$$

Choosing the points $t_1, \ldots, t_m$ independently and uniformly at random from $[N]$ generates the random matrix $\mathbf{A}$ by selecting its rows independently and uniformly at random from the rows of $\sqrt{N}\mathbf{U}$, that is,

$$\mathbf{A} = \sqrt{N}\mathbf{R}_T\mathbf{U},$$

where $T = \{t_1, \ldots, t_m\}$ and $\mathbf{R}_T : \mathbb{C}^N \to \mathbb{C}^m$ denote the random subsampling operator

$$(\mathbf{R}_T \mathbf{z})_\ell = z_{t_\ell}, \quad \ell \in [m]. \tag{12.10}$$

In this context, the compressive sensing problem corresponds to the recovery of an s-sparse $\mathbf{x} \in \mathbb{C}^N$ in the situation where only the entries of $\tilde{\mathbf{y}} = \sqrt{N}\mathbf{U}\mathbf{x} \in \mathbb{C}^N$ on T are observed.

Note that it may happen with nonzero probability that a row of $\sqrt{N}\mathbf{U}$ is selected more than once because the probability measure is discrete in this example. Hence, $\mathbf{A}$ is allowed to have repeated rows. One can avoid this effect by passing to a different probability model where the subset $\{t_1, \ldots, t_m\} \subset [N]$ is selected uniformly at random among all subsets of $[N]$ of cardinality m. This probability model requires a different analysis than the model described above. However, the difference between the two models is very slight, and the final recovery results are essentially the same. We refer to Sect. 12.6 for details.

4. *Partial discrete Fourier transform.* An important example of the setup just described is the discrete Fourier matrix $\mathbf{F} \in \mathbb{C}^{N \times N}$ with entries

$$F_{\ell,k} = \frac{1}{\sqrt{N}} e^{2\pi i (\ell-1)(k-1)/N}, \quad \ell, k \in [N]. \tag{12.11}$$

The Fourier matrix $\mathbf{F}$ is unitary; see Exercise 12.1. The constant in the boundedness condition (12.9) is clearly $K = 1$. The vector $\hat{\mathbf{x}} = \mathbf{F}\mathbf{x}$ is called the discrete Fourier transform of $\mathbf{x}$. The considerations of the previous example applied to this situation lead to the problem of reconstructing a sparse vector $\mathbf{x}$ from m independent and uniformly distributed random entries of its discrete Fourier transform $\hat{\mathbf{x}}$. The resulting matrix $\mathbf{A}$ is called *random partial Fourier matrix*. This matrix can also be viewed as a special case of the nonequispaced Fourier matrix in (12.7) with the points t_ℓ being chosen from the grid $\mathbb{Z}_N/N = \{0, 1/N, 2/N, \ldots, (N-1)/N\}$ instead of the whole interval $[0, 1]$. Note that the discrete Fourier matrix in (12.11) can also be extended to higher dimensions, i.e., to grids $(\mathbb{Z}_N/N)^d$ for $d \in \mathbb{N}$.

A crucial point for applications is that computations can be performed quickly using the FFT. It evaluates the Fourier transform of a vector $\mathbf{x} \in \mathbb{C}^N$ in complexity $\mathcal{O}(N \ln N)$; see Appendix C.1. Then a partial Fourier matrix $\mathbf{A} = \mathbf{R}_T \mathbf{F}$ also has a fast matrix–vector multiplication: Simply compute $\mathbf{F}\mathbf{x}$ via the FFT and then omit all entries outside T. Similarly, multiplication by the adjoint to form $\mathbf{A}^* \mathbf{y}$ can be performed fast: extend the vector $\mathbf{y}$ with zeros outside T and apply $\mathbf{F}^*$, which can also be done via the FFT.

5. *Hadamard transform.* The Hadamard transform $\mathbf{H} = \mathbf{H}_n \in \mathbb{R}^{2^n \times 2^n}$ can be viewed as a Fourier transform on $\mathbb{Z}_2^n = \{0, 1\}^n$. Writing out indices $j, \ell \in [2^n]$ into a binary expansion,

$$j = \sum_{k=1}^{n} j_k 2^{k-1} + 1 \quad \text{and} \quad \ell = \sum_{k=1}^{n} \ell_k 2^{k-1} + 1$$

with $j_k, \ell_k \in \{0, 1\}$, an entry $H_{j,\ell}$ of the Hadamard matrix $\mathbf{H}_n$ is given by

$$H_{j,\ell} = \frac{1}{2^{n/2}}(-1)^{\sum_{k=1}^{n} j_k \ell_k}.$$

The Hadamard matrix is self-adjoint and orthogonal, that is, $\mathbf{H}_n = \mathbf{H}_n^* = \mathbf{H}_n^{-1}$. The constant in (12.2) or (12.9) is once more $K = 1$. The Hadamard transform also comes with a fast matrix–vector multiplication algorithm, which operates in complexity $\mathcal{O}(N \ln N)$, where $N = 2^n$. The algorithm recursively uses the identity

$$\mathbf{H}_n = \frac{1}{\sqrt{2}} \begin{pmatrix} \mathbf{H}_{n-1} & \mathbf{H}_{n-1} \\ \mathbf{H}_{n-1} & -\mathbf{H}_{n-1} \end{pmatrix}, \qquad \mathbf{H}_0 = 1,$$

which can be taken as an alternative definition for the Hadamard matrix. A slightly different description of the Hadamard transform will be discussed in Sect. 12.2.

6. *Incoherent bases.* Let $\mathbf{V} = [\mathbf{v}_1 | \ldots | \mathbf{v}_N]$, $\mathbf{W} = [\mathbf{w}_1 | \ldots | \mathbf{w}_N] \in \mathbb{C}^{N \times N}$ be two unitary matrices. Their columns form two orthonormal bases of $\mathbb{C}^N$. Assume that a vector $\mathbf{z} \in \mathbb{C}^N$ is sparse with respect to $\mathbf{V}$ rather than with respect to the canonical basis, that is, $\mathbf{z} = \mathbf{V}\mathbf{x}$ for a sparse vector $\mathbf{x}$. Further, assume that $\mathbf{z}$ is sampled with respect to $\mathbf{W}$, i.e., we obtain measurements

$$y_k = \langle \mathbf{z}, \mathbf{w}_{t_k} \rangle, \quad k \in [m],$$

with $T := \{t_1, \ldots, t_m\} \subset [N]$. In matrix vector form, this can be written as

$$\mathbf{y} = \mathbf{R}_T \mathbf{W}^* \mathbf{z} = \mathbf{R}_T \mathbf{W}^* \mathbf{V}\mathbf{x},$$

where $\mathbf{R}_T$ is again the sampling operator (12.10). Defining the unitary matrix $\mathbf{U} := \mathbf{W}^* \mathbf{V} \in \mathbb{C}^{N \times N}$, we return to the situation of Example 3. Condition (12.9) now reads

$$\sqrt{N} \max_{\ell, k \in [N]} |\langle \mathbf{v}_\ell, \mathbf{w}_k \rangle| \leq K. \tag{12.12}$$

The bases $(\mathbf{v}_\ell)$ and $(\mathbf{w}_\ell)$ are called incoherent if K can be chosen small. Examples 4 and 5 fall into this setting by choosing one of the bases as the canonical basis, say $\mathbf{W} = \mathbf{Id} \in \mathbb{C}^{N \times N}$. The Fourier basis and the canonical basis are actually maximally incoherent, since $K = 1$ in this case.

Further examples, namely, Haar wavelets coupled with noiselets and Legendre polynomials, will be mentioned in the Notes section.

In the remainder of this chapter, we develop rigorous recovery results for the described setup, that is, for the sampling matrix in (12.4) formed with randomly chosen sampling points. We will proceed by increasing difficulty of the proofs. Let us already summarize the main findings at this point.

- *Nonuniform recovery.* A fixed s-sparse vector $\mathbf{x}$ can be reconstructed via ℓ_1-minimization with high probability using a random draw of the sampling matrix $\mathbf{A} \in \mathbb{C}^{m \times N}$ in (12.4) associated to a BOS with constant $K \geq 1$ in (12.2) provided that

$$m \geq CK^2 s \ln(N).$$

  We first show two easier versions of this results (Theorems 12.11 and 12.18) in which the nonzero coefficients of $\mathbf{x}$ are additionally assumed to have random signs. We remove this randomness assumption in Theorem 12.20. We further provide a weak stability and robustness estimate for the reconstruction in Theorem 12.22.

- *Uniform recovery.* Theorem 12.31 states that the restricted isometry constants of the rescaled random sampling matrix $\frac{1}{\sqrt{m}}\mathbf{A}$ satisfy $\delta_s \leq \delta$ with high probability provided

$$m \geq CK^2 \delta^{-2} s \ln^4(N).$$

  This implies uniform, stable, and robust s-sparse recovery via ℓ_1-minimization and the other recovery algorithms discussed in Chap. 6 with high probability when $m \geq C'K^2 s \ln^4(N)$.

Before turning to the proofs of these results in Sects. 12.3–12.5, we investigate lower bounds on the required number of samples and discuss connections with uncertainty principle in the next section.

12.2   Uncertainty Principles and Lower Bounds

In this section, we concentrate essentially on the Fourier system of Example 4 and on the Hadamard matrix of Example 5 in order to illustrate some basic facts and bounds that arise in random sampling of BOSs. In particular, we provide lower bounds on the minimal number of measurements (see (12.29)) which are slightly stronger than the general bounds obtained in Chap. 10 using Gelfand widths.

We recall that $\mathbf{F} \in \mathbb{C}^{N \times N}$ is the Fourier transform matrix with entries

$$F_{\ell,k} = \frac{1}{\sqrt{N}} e^{2\pi i(\ell-1)(k-1)/N}, \quad \ell, k \in [N].$$

With the stated normalization, the matrix $\mathbf{F}$ is unitary. For a vector $\mathbf{x} \in \mathbb{C}^N$, its discrete Fourier transform is denoted

$$\hat{\mathbf{x}} = \mathbf{F}\mathbf{x}.$$

Uncertainty principles state that a vector cannot be simultaneously localized both in time and frequency. In other words, it is impossible that both $\mathbf{x}$ and $\hat{\mathbf{x}}$ are concentrated in a small portion of $[N]$. Various versions of the uncertainty principle make the notion of localization precise. We present a general discrete version for incoherent bases (see Example 6 above). Let $\mathbf{V} = [\mathbf{v}_1 | \cdots | \mathbf{v}_N]$ and $\mathbf{W} = [\mathbf{w}_1 | \cdots | \mathbf{w}_N] \in \mathbb{C}^{N \times N}$ be two unitary matrices that are mutually incoherent, that is,

$$\sqrt{N} \max_{\ell,k \in [N]} |\langle \mathbf{v}_\ell, \mathbf{w}_k \rangle| \leq K \qquad (12.13)$$

for some $K \geq 1$. Taking the Fourier/identity matrix pair $(\mathbf{V} = \mathbf{F}, \mathbf{W} = \mathbf{Id})$, we obtain the optimal constant $K = 1$.

Theorem 12.2. *Let $\mathbf{V}$ and $\mathbf{W} \in \mathbb{C}^{N \times N}$ be two mutually incoherent unitary matrices with parameter K in (12.13). Let $\mathbf{y} \in \mathbb{C}^N \setminus \{0\}$ and $\mathbf{x}, \mathbf{z} \in \mathbb{C}^N$ be the representation coefficients in $\mathbf{y} = \mathbf{Vx} = \mathbf{Wz}$. Then*

$$\|\mathbf{x}\|_0 + \|\mathbf{z}\|_0 \geq \frac{2\sqrt{N}}{K}. \qquad (12.14)$$

Proof. Since $\mathbf{V}$ is unitary, left multiplication of the identity $\mathbf{Vx} = \mathbf{Wz}$ by $\mathbf{V}^*$ yields $\mathbf{x} = \mathbf{V}^*\mathbf{Wz}$. An entry of $\mathbf{x}$ satisfies

$$|x_k| = |(\mathbf{V}^*\mathbf{Wz})_k| = |\sum_\ell (\mathbf{V}^*\mathbf{W})_{k,\ell} z_\ell| \leq \sum_\ell |\langle \mathbf{w}_\ell, \mathbf{v}_k \rangle| |z_\ell|$$

$$\leq \max_\ell |\langle \mathbf{w}_\ell, \mathbf{v}_k \rangle| \, \|\mathbf{z}\|_1 \leq \frac{K}{\sqrt{N}} \|\mathbf{z}\|_1.$$

Summation over $k \in \mathrm{supp}(\mathbf{x})$ yields

$$\|\mathbf{x}\|_1 \leq \|\mathbf{x}\|_0 \frac{K}{\sqrt{N}} \|\mathbf{z}\|_1.$$

Multiplying the identity $\mathbf{Vx} = \mathbf{Wz}$ from the left by $\mathbf{W}^*$ similarly gives

$$\|\mathbf{z}\|_1 \leq \|\mathbf{z}\|_0 \frac{K}{\sqrt{N}} \|\mathbf{x}\|_1.$$

Multiplication of both inequalities and division by $\|\mathbf{x}\|_1 \|\mathbf{z}\|_1$ imply the inequality $1 \leq \|\mathbf{z}\|_0 \|\mathbf{x}\|_0 K^2 / N$ or expressed differently

$$\sqrt{\|\mathbf{z}\|_0 \|\mathbf{x}\|_0} \geq \frac{\sqrt{N}}{K}.$$

Using the fact that the arithmetic mean dominates the geometric mean, we obtain

$$\frac{\|\mathbf{z}\|_0 + \|\mathbf{x}\|_0}{2} \geq \sqrt{\|\mathbf{z}\|_0 \|\mathbf{x}\|_0} \geq \frac{\sqrt{N}}{K}.$$

This completes the proof.   □

Specializing to the Fourier/identity matrix pair, for which $K = 1$, we arrive at the following result.

Corollary 12.3. *Let* $\mathbf{x} \in \mathbb{C}^N \setminus \{0\}$. *Then*

$$\|\mathbf{x}\|_0 + \|\hat{\mathbf{x}}\|_0 \geq 2\sqrt{N}, \tag{12.15}$$

where $\hat{\mathbf{x}} = \mathbf{F}\mathbf{x}$ *is the discrete Fourier transform of* $\mathbf{x}$.

This uncertainty principle has consequences for signal separation (Exercise 12.3), and it implies a weak result concerning recovery from undersampled measurements (Exercise 12.4). However, our interest in the above statements is rather due to the fact that they have companion results suggesting the use of random sets of samples. Indeed, the bound (12.15) cannot be improved in general, as the next proposition shows that it is sharp for so-called *delta trains*.

Proposition 12.4. *Let* $N = n^2$ *be a square. Set* $\mathbf{x} \in \mathbb{C}^N$ *to be the vector with entries*

$$x_\ell = \begin{cases} 1 & \text{if } \ell = 1 \bmod n, \\ 0 & \text{otherwise.} \end{cases} \tag{12.16}$$

Then $\hat{\mathbf{x}} = \mathbf{x}$ *and* $\|\hat{\mathbf{x}}\|_0 = \|\mathbf{x}\|_0 = \sqrt{N}$.

Proof. By definition of the Fourier transform, we have, for $j \in [n^2]$,

$$\hat{x}_j = \frac{1}{n} \sum_{\ell=1}^{n^2} x_\ell e^{2\pi i (\ell-1)(j-1)/n^2} = \frac{1}{n} \sum_{k=1}^{n} e^{2\pi i (k-1)(j-1)/n} = \begin{cases} 1 & \text{if } j = 1 \bmod n, \\ 0 & \text{otherwise.} \end{cases}$$

This shows that $\hat{\mathbf{x}} = \mathbf{x}$ and that $\|\hat{\mathbf{x}}\|_0 = \|\mathbf{x}\|_0 = n = \sqrt{N}$.   □

Delta trains illustrate why arbitrary sampling sets $T \subset [N]$ are not suitable for sparse recovery from Fourier measurements. Suppose that $N = n^2$ is a square, and let $\mathbf{x}$ be defined as in (12.16). We consider the set of sampling points $T := [n^2] \setminus \{1, n+1, 2n+1, \ldots, (n-1)n+1\}$. Then by the previous proposition, the restriction of $\hat{\mathbf{x}}$ to T is the zero vector, that is,

$$\mathbf{y} = \mathbf{R}_T \mathbf{F} \mathbf{x} = 0.$$

Any reasonable algorithm will output $\mathbf{x}^\sharp = \mathbf{0}$ from $\mathbf{y} = \mathbf{0}$. In other words, this sampling scheme cannot distinguish $\mathbf{x}$ from the zero vector. Observe that $s = \|\mathbf{x}\|_0 = n$, but the number of samples satisfies

$$m = \operatorname{card}(T) = n^2 - n.$$

In conclusion, for this choice of sampling set not even $m = s^2 - s$ samples are sufficient. This example somewhat suggests to consider random choices of sampling sets T. Indeed, the described sampling set is very structured, and this is essentially the reason why it allows for counterexamples. Good sampling sets instead possess only very little additive structure, and the simplest way to construct an unstructured set of numbers is to choose the set at random.

Next, we investigate a general lower bound on the number m of samples for s-sparse recovery in dimension N. We have seen in Chap. 10 that for a general stable sparse recovery problem, we have the lower bound

$$m \geq Cs \ln(eN/s).$$

We will construct an example that shows that the term $\ln(eN/s)$ has to be replaced by $\ln N$ in the context of random sampling in BOSs. To this end, we use the Hadamard transform $\mathbf{H}$ introduced in Example 5.

The Hadamard transform is related to Fourier analysis on the additive group $\mathbb{Z}_2^n = (\{0,1\}^n, +)$, where addition is understood modulo 2. We give here a slightly different description of the Hadamard matrix than in the previous section. The constant function $\chi_0(t) = 1$ on $\mathbb{Z}_2$ and the function

$$\chi_1(t) = \begin{cases} 1 & \text{if } t = 0, \\ -1 & \text{if } t = 1, \end{cases}$$

are the characters on $\mathbb{Z}_2$, meaning that $\chi_j(t+r) = \chi_j(t)\chi_j(r)$ for all $j, t, r \in \{0,1\}$. We also observe that $\chi_{j+k}(t) = \chi_j(t)\chi_k(t)$ for all $j, k \in \{0,1\}$. One easily checks that the characters are orthonormal with respect to the normalized counting measure on $\mathbb{Z}_2$, that is,

$$\langle \chi_j, \chi_k \rangle := \frac{1}{2} \sum_{t \in \{0,1\}} \chi_j(t)\chi_k(t) = \delta_{j,k}.$$

For $\mathbf{j}, \mathbf{t} \in \mathbb{Z}_2^n$, we define a character on $\mathbb{Z}_2^n$ as the tensor product

$$\chi_{\mathbf{j}}(\mathbf{t}) = \prod_{\ell=1}^{n} \chi_{j_\ell}(t_\ell).$$

Using the corresponding properties on $\mathbb{Z}_2$, we see that

$$\chi_{\mathbf{j}}(\mathbf{t} + \mathbf{r}) = \chi_{\mathbf{j}}(\mathbf{t})\chi_{\mathbf{j}}(\mathbf{r}) \quad \text{and} \quad \chi_{\mathbf{j}+\mathbf{k}}(\mathbf{t}) = \chi_{\mathbf{j}}(\mathbf{t})\chi_{\mathbf{k}}(\mathbf{t}). \tag{12.17}$$

It follows from the orthonormality of χ_0 and χ_1 that the functions $\chi_{\mathbf{j}}$ are orthonormal with respect to the counting measure on $\mathbb{Z}_2^n$, that is,

$$\langle \chi_{\mathbf{j}}, \chi_{\mathbf{k}} \rangle = 2^{-n} \sum_{\mathbf{t} \in \mathbb{Z}_2^n} \chi_{\mathbf{j}}(\mathbf{t})\chi_{\mathbf{k}}(\mathbf{t}) = \delta_{\mathbf{j},\mathbf{k}}. \tag{12.18}$$

The constant in the uniform bound (12.2) of these functions is $K = 1$. The (unnormalized) Hadamard transform (Fourier transform on $\mathbb{Z}_2^n$) of a vector $\mathbf{x}$ indexed by $\mathbb{Z}_2^n$ is then defined entrywise as

$$z_{\mathbf{j}} = (\mathbf{H}\mathbf{x})_{\mathbf{j}} = \sum_{\mathbf{t} \in \mathbb{Z}_2^n} x_{\mathbf{t}}\chi_{\mathbf{j}}(\mathbf{t}).$$

Key to our lower estimate is the fact that an arbitrary subset of $\mathbb{Z}_2^n$ contains (the translate of) a large subgroup of $\mathbb{Z}_2^n$.

Lemma 12.5. *For any subset Λ of $\mathbb{Z}_2^n$, if $N := \operatorname{card}(\mathbb{Z}_2^n) = 2^n$ and if $\kappa := \operatorname{card}(\Lambda)/N$ satisfies $\log_2(\kappa^{-1}) \geq 10\, N^{-3/4}$, then there exist an element $\mathbf{b} \in \mathbb{Z}_2^n$ and a subgroup Γ of $\mathbb{Z}_2^n$ such that*

$$\mathbf{b} + \Gamma \subset \Lambda \quad \text{and} \quad \operatorname{card}(\Gamma) \geq \frac{n}{8\log_2(\kappa^{-1})}. \tag{12.19}$$

Proof. We iteratively construct elements $\gamma_0, \gamma_1, \ldots, \gamma_p \in \mathbb{Z}_2^n$ and subsets $\Lambda_0, \Lambda_1, \ldots, \Lambda_p$ of $\mathbb{Z}_2^n$ as follows: we set $\gamma_0 = \mathbf{0}$ and $\Lambda_0 := \Lambda$, and, for $j \geq 1$, with $G(\gamma_0, \ldots, \gamma_{j-1})$ denoting the group generated by $\gamma_0, \ldots, \gamma_{j-1}$, we define

$$\gamma_j := \operatorname{argmax}\{\operatorname{card}((\gamma + \Lambda_{j-1}) \cap \Lambda_{j-1}) : \gamma \notin G(\gamma_0, \ldots, \gamma_{j-1})\}, \tag{12.20}$$

$$\Lambda_j := (\gamma_j + \Lambda_{j-1}) \cap \Lambda_{j-1}. \tag{12.21}$$

The condition $\gamma \notin G(\gamma_0, \ldots, \gamma_{j-1})$ guarantees that $G(\gamma_0, \ldots, \gamma_j)$ is twice as large as $G(\gamma_0, \ldots, \gamma_{j-1})$, so that $\operatorname{card}(G(\gamma_0, \ldots, \gamma_j)) = 2^j$ follows by induction. Therefore, the construction of $\gamma_1, \ldots, \gamma_p$ via (12.20) is possible as long as $2^{p-1} < N$ and in particular for p chosen as in (12.23) below. Let us now show that property (12.21) implies, for $j \geq 1$,

$$\Lambda_j + G(\gamma_0, \ldots, \gamma_j) \subset \Lambda_{j-1} + G(\gamma_0, \ldots, \gamma_{j-1}). \tag{12.22}$$

Indeed, for $\mathbf{g} \in \Lambda_j + G(\gamma_0, \ldots, \gamma_j)$, we write $\mathbf{g} = \lambda_j + \sum_{\ell=0}^{j} \delta_\ell \gamma_\ell$ for some $\lambda_j \in \Lambda_j$ and some $\delta_0, \ldots, \delta_j \in \{0, 1\}$. In view of $\Lambda_j = (\gamma_j + \Lambda_{j-1}) \cap \Lambda_{j-1}$, we can always write $\lambda_j = \lambda_{j-1} + \delta_j \gamma_j$ for some $\lambda_{j-1} \in \Lambda_{j-1}$—if $\delta_j = 0$, we use $\lambda_j \in \Lambda_{j-1}$ and if $\delta_j = 1$, we use $\lambda_j \in \gamma_j + \Lambda_{j-1}$. It follows that

$$\mathbf{g} = \lambda_{j-1} + \delta_j \gamma_j + \sum_{\ell=0}^{j} \delta_\ell \gamma_\ell = \lambda_{j-1} + \sum_{\ell=0}^{j-1} \delta_\ell \gamma_\ell \in \Lambda_{j-1} + G(\gamma_0, \ldots, \gamma_{j-1}).$$

This establishes (12.22). We derive that $\Lambda_p + G(\gamma_0, \ldots, \gamma_p) \subset \Lambda_0 + G(\gamma_0) = \Lambda$ by induction. Thus, choosing $\Gamma = G(\gamma_0, \ldots, \gamma_p)$ and picking any $\mathbf{b} \in \Lambda_p$, we have $\mathbf{b} + \Gamma \subset \Lambda$. It remains to prove that the size of Γ is large and that an element $\mathbf{b} \in \Lambda_p$ does exist. By considering $p \geq 0$ such that

$$2^{p-1} < \frac{n}{8 \log_2(\kappa^{-1})} \leq 2^p, \tag{12.23}$$

we immediately obtain the second part of (12.19). To show that $\mathrm{card}(\Lambda_p) > 0$, we use property (12.21). For $j \geq 1$, the observation that the maximum is larger than the average leads to

$$\mathrm{card}(\Lambda_j) \geq \frac{1}{N - 2^{j-1}} \sum_{\gamma \in \mathbb{Z}_2^n \backslash G(\gamma_0, \ldots, \gamma_{j-1})} \mathrm{card}((\gamma + \Lambda_{j-1}) \cap \Lambda_{j-1})$$

$$= \frac{1}{N - 2^{j-1}} \Bigg[\sum_{\gamma \in \mathbb{Z}_2^n} \mathrm{card}((\gamma + \Lambda_{j-1}) \cap \Lambda_{j-1})$$

$$- \sum_{\gamma \in G(\gamma_0, \ldots, \gamma_{j-1})} \mathrm{card}((\gamma + \Lambda_{j-1}) \cap \Lambda_{j-1}) \Bigg].$$

On the one hand, we have

$$\sum_{\gamma \in G(\gamma_0, \ldots, \gamma_{j-1})} \mathrm{card}((\gamma + \Lambda_{j-1}) \cap \Lambda_{j-1}) \leq \sum_{\gamma \in G(\gamma_0, \ldots, \gamma_{j-1})} \mathrm{card}(\Lambda_{j-1}) = 2^{j-1} \mathrm{card}(\Lambda_{j-1}).$$

On the other hand, with $\mathbb{1}_A$ denoting the characteristic function of a set A, we have

$$\sum_{\gamma \in \mathbb{Z}_2^n} \mathrm{card}((\gamma + \Lambda_{j-1}) \cap \Lambda_{j-1}) = \sum_{\gamma \in \mathbb{Z}_2^n} \sum_{\mathbf{h} \in \Lambda_{j-1}} \mathbb{1}_{\gamma + \Lambda_{j-1}}(\mathbf{h})$$

$$= \sum_{\mathbf{h} \in \Lambda_{j-1}} \sum_{\gamma \in \mathbb{Z}_2^n} \mathbb{1}_{\mathbf{h} - \Lambda_{j-1}}(\gamma) = \sum_{\mathbf{h} \in \Lambda_{j-1}} \mathrm{card}(\Lambda_{j-1}) = \mathrm{card}(\Lambda_{j-1})^2.$$

As a result, we obtain

$$\mathrm{card}(\Lambda_j) \geq \frac{\mathrm{card}(\Lambda_{j-1})}{N - 2^{j-1}} \Big[\mathrm{card}(\Lambda_{j-1}) - 2^{j-1} \Big].$$

By induction, this implies the estimate

$$\text{card}(\Lambda_j) \geq \kappa^{2^j} N \left(1 - \frac{2^{j-1}}{N} \sum_{\ell=0}^{j-1} \kappa^{-2^\ell} \right). \tag{12.24}$$

Indeed, this holds for $j = 0$, and if it holds for $j - 1$, then

$\text{card}(\Lambda_j)$

$$\geq \kappa^{2^{j-1}} N \left(1 - \frac{2^{j-2}}{N} \sum_{\ell=0}^{j-2} \kappa^{-2^\ell} \right) \frac{\kappa^{2^{j-1}} N}{N - 2^{j-1}} \left(1 - \frac{2^{j-2}}{N} \sum_{\ell=0}^{j-2} \kappa^{-2^\ell} - \frac{2^{j-1}}{\kappa^{2^{j-1}} N} \right)$$

$$\geq \kappa^{2^j} N \left(1 - \frac{2^{j-1}}{N} \sum_{\ell=0}^{j-2} \frac{1}{2\kappa^{2^\ell}} \right) \left(1 - \frac{2^{j-1}}{N} \left(\sum_{\ell=0}^{j-2} \frac{1}{2\kappa^{2^\ell}} + \frac{1}{\kappa^{2^{j-1}}} \right) \right)$$

$$\geq \kappa^{2^j} N \left(1 - \frac{2^{j-1}}{N} \sum_{\ell=0}^{j-1} \frac{1}{\kappa^{2^\ell}} \right).$$

This finishes the inductive justification of (12.24). Since $\sum_{\ell=0}^{p-1} \kappa^{-2^\ell} \leq p\kappa^{-2^{p-1}}$, we derive in particular

$$\text{card}(\Lambda_p) \geq \kappa^{2^p} N \left(1 - \frac{2^{p-1}}{N} p \kappa^{-2^{p-1}} \right) = \kappa^{2^{p-1}} \left(\kappa^{2^{p-1}} N - 2^{p-1} p \right).$$

Using the leftmost inequality in (12.23), as well as $p \leq n$ and the assumption $\log_2(\kappa^{-1}) \geq 10 N^{-3/4}$, we obtain

$$\text{card}(\Lambda_p) \geq \kappa^{2^{p-1}} \left(\kappa^{n/(8 \log_2(\kappa^{-1}))} 2^n - \frac{n^2}{8 \log_2(\kappa^{-1})} \right)$$

$$\geq \kappa^{2^{p-1}} \left(2^{n(1-1/8)} - \frac{n^2 \, 2^{3n/4}}{80} \right) = \kappa^{2^{p-1}} 2^{3n/4} \left(2^{n/8} - \frac{n^2}{80} \right) > 0.$$

The proof is now complete.   □

Remark 12.6. The condition $\log_2(\kappa^{-1}) \geq 10 N^{-3/4}$ in the previous lemma can be replaced by any condition of the type $\log_2(\kappa^{-1}) \geq c_\beta N^{-\beta}, 0 < \beta < 1$. This only requires an adjustment of the constants.

The next result is analogous to Proposition 12.4.

Proposition 12.7. *Given a subgroup G of $\mathbb{Z}_2^n$, the set*

$$G^\perp := \{ \boldsymbol{\lambda} \in \mathbb{Z}_2^n : \sum_{\mathbf{g} \in G} \chi_{\boldsymbol{\lambda}}(\mathbf{g}) \neq 0 \} \tag{12.25}$$

forms another subgroup of $\mathbb{Z}_2^n$. Furthermore, the unnormalized Hadamard transform of the vector $\mathbf{z} \in \mathbb{C}^{\mathbb{Z}_2^n}$ with entries

$$z_{\mathbf{j}} = \begin{cases} 1 & \text{if } \mathbf{j} \in G, \\ 0 & \text{otherwise,} \end{cases}$$

is given by

$$\hat{z}_{\mathbf{k}} = \begin{cases} \operatorname{card}(G) & \text{if } \mathbf{k} \in G^{\perp}, \\ 0 & \text{otherwise.} \end{cases}$$

In particular, $\|\mathbf{z}\|_0 \cdot \|\hat{\mathbf{z}}\|_0 = \operatorname{card}(G) \cdot \operatorname{card}(G^{\perp}) = 2^n$. Finally, $(G^{\perp})^{\perp} = G$.

Proof. First, we observe that $\mathbf{0} \in G^{\perp}$ because $\chi_0 = 1$ is the constant function and that $G^{\perp}$ is trivially closed under inversion because any element of $\mathbb{Z}_2^n$ is its own inverse. Then, using the fact that G is a group, we obtain, for $\mathbf{h} \in G$ and $\lambda \in G^{\perp}$,

$$\sum_{\mathbf{g} \in G} \chi_{\lambda}(\mathbf{g}) = \sum_{\mathbf{g} \in G} \chi_{\lambda}(\mathbf{h} + \mathbf{g}) = \chi_{\lambda}(\mathbf{h}) \sum_{\mathbf{g} \in G} \chi_{\lambda}(\mathbf{g}).$$

In view of $\sum_{\mathbf{g} \in G} \chi_{\lambda}(\mathbf{g}) \neq 0$, we deduce

$$\chi_{\lambda}(\mathbf{h}) = 1 \qquad \text{for all } \mathbf{h} \in G \text{ and } \lambda \in G^{\perp}. \tag{12.26}$$

In particular, given $\lambda, \rho \in G^{\perp}$, we derive

$$\sum_{\mathbf{g} \in G} \chi_{\lambda + \rho}(\mathbf{g}) = \sum_{\mathbf{g} \in G} \chi_{\lambda}(\mathbf{g}) \chi_{\rho}(\mathbf{g}) = \sum_{\mathbf{g} \in G} 1 = \operatorname{card}(G) \neq 0,$$

which shows that $\lambda + \rho \in G^{\perp}$. We have now established that $G^{\perp}$ is a group. The special case $\rho = \mathbf{0}$ of the previous identity reads

$$\sum_{\mathbf{g} \in G} \chi_{\lambda}(\mathbf{g}) = \operatorname{card}(G) \qquad \text{for all } \lambda \in G^{\perp}. \tag{12.27}$$

The definition of the unnormalized Hadamard transform then yields

$$\hat{z}_{\mathbf{k}} = \sum_{\mathbf{g} \in G} \chi_{\mathbf{k}}(\mathbf{g}) = \begin{cases} \operatorname{card}(G) & \text{if } \mathbf{k} \in G^{\perp}, \\ 0 & \text{otherwise.} \end{cases}$$

Summing (12.27) over all $\lambda \in G^{\perp}$ and using the orthogonality relation (12.18), we obtain

$$\mathrm{card}(G) \cdot \mathrm{card}(G^\perp) = \sum_{\lambda \in G^\perp} \sum_{g \in G} \chi_\lambda(g) = \sum_{\lambda \in \mathbb{Z}_2^n} \sum_{g \in G} \chi_\lambda(g)$$

$$= \sum_{g \in G} \sum_{\lambda \in \mathbb{Z}_2^n} \chi_g(\lambda) = 2^n \sum_{g \in G} \langle \chi_g, \chi_0 \rangle = 2^n.$$

For the final statement, observe that $\chi_\lambda(h) = \chi_h(\lambda)$ for all λ, h. The relation (12.26) yields $\sum_{\lambda \in G^\perp} \chi_h(\lambda) = \mathrm{card}(G^\perp) \neq 0$ for $h \in G$ so that $G \subset (G^\perp)^\perp$. Moreover, $\mathrm{card}(G) \cdot \mathrm{card}(G^\perp) = 2^n = \mathrm{card}(G^\perp) \cdot \mathrm{card}((G^\perp)^\perp)$ as just shown. It follows that $(G^\perp)^\perp = G$. This completes the proof.   □

Now we are in the position to provide a lower bound on the number m of measurements for recovery of s-sparse vectors in $\mathbb{C}^N$, $N = 2^n$, from samples of the Hadamard transform. The bound applies to an arbitrary (not necessarily random) set of m samples.

Theorem 12.8. *Let T be an arbitrary subset of $\mathbb{Z}_2^n$ of size m. If m satisfies $cN^{1/4} \leq m \leq N/2$ with $N = 2^n$ and $c = 10\ln(2) \approx 6.93$, then there exists a nonzero vector $\mathbf{x} \in \mathbb{C}^N$ whose Hadamard transform vanishes on T and whose sparsity obeys*

$$\|\mathbf{x}\|_0 \leq \frac{16\,m}{\log_2(N)}. \tag{12.28}$$

Proof. We consider the set $\Lambda := \mathbb{Z}_2^n \setminus T$. With $\kappa := \mathrm{card}(\Lambda)/N = 1 - \frac{m}{N}$, the concavity of the logarithm together with the assumption on m yields

$$\log_2(\kappa^{-1}) = -\log_2\left(1 - \frac{m}{N}\right) \begin{cases} \geq \dfrac{m}{\ln(2)\,N} \geq 10\,N^{-3/4}, \\[2mm] \leq \dfrac{2\,m}{N}. \end{cases}$$

Thus, Lemma 12.5 guarantees the existence of an element $\mathbf{b} \in \mathbb{Z}_2^n$ and a subgroup Γ of $\mathbb{Z}_2^n$ such that $\mathbf{b} + \Gamma \subset \Lambda$ and $\mathrm{card}(\Gamma) \geq n/(8\log_2(\kappa^{-1}))$. The vector $\mathbf{z} \in \mathbb{C}^{\mathbb{Z}_2^n}$ introduced in Proposition 12.7 with $G := \Gamma^\perp$ satisfies

$$\|\mathbf{z}\|_0 = \mathrm{card}(\Gamma^\perp) = \frac{N}{\mathrm{card}(\Gamma)} \leq \frac{8\log_2(\kappa^{-1})\,N}{n} \leq \frac{16\,m}{n},$$

and consequently so does the vector $\mathbf{x} \in \mathbb{C}^{\mathbb{Z}_2^n}$ defined by $x_{\mathbf{k}} = \chi_{\mathbf{b}}(\mathbf{k}) z_{\mathbf{k}}$. It remains to verify that the Hadamard transform of $\mathbf{x}$ vanishes on T. For this purpose, we notice that, for any $\mathbf{j} \in \mathbb{Z}_2^n$,

$$\hat{x}_{\mathbf{j}} = \sum_{\mathbf{t} \in \mathbb{Z}_2^n} x_{\mathbf{t}} \chi_{\mathbf{j}}(\mathbf{t}) = \sum_{\mathbf{t} \in \mathbb{Z}_2^n} z_{\mathbf{t}} \chi_{\mathbf{j}+\mathbf{b}}(\mathbf{t}) = \hat{z}_{\mathbf{j}+\mathbf{b}}.$$

Hence, according to Proposition 12.7, we have $\hat{x}_j = 0$ if $\mathbf{j} + \mathbf{b} \notin G^{\perp}$, i.e., $\mathbf{j} \notin \mathbf{b} + \Gamma$ because $(G^{\perp})^{\perp} = G$ again by Proposition 12.7. Since this occurs when $j \in T$, the proof is now complete.   □

The result below shows that, for random sampling in general BOSs, a factor $\ln(N)$ must appear in the number of required measurements because it would otherwise contradict the special case of the Hadamard system. This is in contrast to other measurement matrices, where the logarithmic factor can be lowered to $\ln(N/s)$; see Chaps. 9 and 10.

Corollary 12.9. *Let T be an arbitrary subset of $\mathbb{Z}_2^n$ of size $m \in [cN^{1/4}, N/2]$ with $c = 10\ln(2)$. The existence of a method to recover every s-sparse vector from the samples indexed by T of its Hadamard transform requires*

$$m \geq C s \ln(N), \qquad C = \frac{1}{8\ln(2)} \approx 0.1803. \tag{12.29}$$

If only $m \leq N/2$, that is, without assuming a lower bound on m, the existence of a stable method to recover every s-sparse vector from the samples indexed by T of its Hadamard transform requires

$$m \geq C s \ln(N)$$

for some constant C depending on the stability requirement.

Remark 12.10. Recall that a stable recovery method (associated to the sampling matrix $\mathbf{A} \in \mathbb{C}^{m \times N}$) is a map $\Delta : \mathbb{C}^m \to \mathbb{C}^N$ such that, for all $\mathbf{x} \in \mathbb{C}^N$,

$$\|\mathbf{x} - \Delta(\mathbf{A}\mathbf{x})\|_1 \leq \hat{C}\sigma_s(\mathbf{x})_1 \quad \text{for some constant } \hat{C} > 0.$$

Proof. Suppose that $cN^{1/4} \leq m \leq N/2$ and that there exists a method to recover every s-sparse vector from the samples indexed by T of its Hadamard transform. Let us decompose the nonzero vector $\mathbf{x} \in \mathbb{C}^{\mathbb{Z}_2^n}$ of Theorem 12.8 as $\mathbf{x} = \mathbf{u} - \mathbf{v}$ for two distinct vectors $\mathbf{u}, \mathbf{v} \in \mathbb{C}^{\mathbb{Z}_2^n}$ of sparsity at most $(\|\mathbf{x}\|_0 + 1)/2$. Since the Hadamard transforms of $\mathbf{u}$ and $\mathbf{v}$ are identical on T, we must have $(\|\mathbf{x}\|_0 + 1)/2 > s$, hence

$$2s \leq \|\mathbf{x}\|_0 \leq \frac{16\,m}{\log_2(N)},$$

and (12.29) follows. In the case $m \leq cN^{1/4}$, we know from Theorem 11.6 that if there exists a stable method to recover every s-sparse vector from the samples indexed by T of its Hadamard transform, then there is a constant c' such that $m \geq c' s \ln(eN/s)$. Moreover, Theorem 2.13 implies that $m \geq 2s$ so that

$$m \geq c' s \ln(2eN/m) \geq c' s \ln(2eN^{3/4}/c) \geq C s \ln(N)$$

for some appropriate constant $C > 0$. This concludes the proof.   □

12.3 Nonuniform Recovery: Random Sign Patterns

We start with nonuniform recovery guarantees for random sampling in bounded orthonormal systems. In order to simplify the argument, we assume in this section that the signs of the nonzero coefficients of the vector to be recovered are random. Recall that the recovery condition in Theorem 4.26 depends only on the signs of $\mathbf{x}$ on its support, so that the magnitudes of the entries of $\mathbf{x}$ do not play any role. This is why we impose randomness only on the signs of the entries. In this way, the vector $\mathbf{x}$ certainly becomes random as well. But in contrast to Chap. 14, where we focus on recovery of random signals using deterministic matrices $\mathbf{A}$, the support of $\mathbf{x}$ is kept arbitrary here. With a deterministic support, the randomness in $\mathbf{x}$ can be considered mild. In any case, we will remove the assumption on the randomness of the signs in the next section at the cost of a more complicated approach.

Recall that we consider the random sampling matrix $\mathbf{A}$ in (12.4) associated to a BOS with constant $K \geq 1$ introduced in (12.2). The sampling points $\mathbf{t}_1, \ldots, \mathbf{t}_m$ are chosen independently at random according to the probability measure ν.

Theorem 12.11. *Let $\mathbf{x} \in \mathbb{C}^N$ be a vector supported on a set S of size s such that $\mathrm{sgn}(\mathbf{x}_S)$ forms a Rademacher or Steinhaus sequence. Let $\mathbf{A} \in \mathbb{C}^{m \times N}$ be the random sampling matrix associated to a BOS with constant $K \geq 1$. If*

$$m \geq CK^2 s \ln^2(6N/\varepsilon), \tag{12.30}$$

then with probability at least $1 - \varepsilon$, the vector $\mathbf{x}$ is the unique minimizer of $\|\mathbf{z}\|_1$ subject to $\mathbf{Az} = \mathbf{Ax}$. The constant C is no larger than 35.

We remark that the probability in this result is both with respect to $\mathbf{A}$ and with respect to $\mathbf{x}$. Theorem 12.12 and Corollary 12.14 below analyze the random matrix $\mathbf{A}$, while Proposition 12.15 exploits the randomness of $\mathrm{sgn}(\mathbf{x}_S)$.

The above result will be slightly improved in Theorem 12.18 below by replacing the exponent 2 by 1 at the logarithmic factor in (12.30).

The proof of Theorem 12.11 requires some preparatory results provided next. As a crucial tool we use the recovery condition for individual vectors of Corollary 4.28. This leads us to investigating the conditioning of the submatrix $\mathbf{A}_S$ associated to the support S of the vector to be recovered. The proof of the result below is based on the noncommutative Bernstein inequality of Theorem 8.14.

Theorem 12.12. *Let $\mathbf{A} \in \mathbb{C}^{m \times N}$ be the random sampling matrix associated to a BOS with constant $K \geq 1$. Let $S \subset [N]$ be of size s. Then, for $\delta \in (0, 1)$, the normalized matrix $\tilde{\mathbf{A}} = \frac{1}{\sqrt{m}}\mathbf{A}$ satisfies*

$$\|\tilde{\mathbf{A}}_S^* \tilde{\mathbf{A}}_S - \mathrm{Id}\|_{2 \to 2} \leq \delta$$

with probability at least

$$1 - 2s \exp\left(-\frac{3m\delta^2}{8K^2 s}\right). \tag{12.31}$$

Remark 12.13. Expressed differently, $\|\mathbf{A}_S^*\mathbf{A}_S - \mathbf{Id}\|_{2\to 2} \leq \delta$ with probability at least $1 - \varepsilon$ provided $m \geq (8/3)K^2\delta^{-2}s\ln(2s/\varepsilon)$.

Proof. Denote by $\mathbf{Y}_\ell = (\overline{\phi_j(\mathbf{t}_\ell)})_{j\in S} \in \mathbb{C}^s$ a column of $\mathbf{A}_S^*$. By independence of the $\mathbf{t}_\ell$, these are independent random vectors. Their ℓ_2-norm is bounded by

$$\|\mathbf{Y}_\ell\|_2 = \sqrt{\sum_{j\in S}|\phi_j(\mathbf{t}_\ell)|^2} \leq K\sqrt{s}. \tag{12.32}$$

Furthermore, for $j, k \in S$,

$$\mathbb{E}\left(\mathbf{Y}_\ell\mathbf{Y}_\ell^*\right)_{j,k} = \mathbb{E}\left[\overline{\phi_j(\mathbf{t}_\ell)}\phi_k(\mathbf{t}_\ell)\right] = \int_{\mathcal{D}} \overline{\phi_j(\mathbf{t})}\phi_k(\mathbf{t})d\nu(\mathbf{t}) = \delta_{j,k},$$

or, in other words, $\mathbb{E}\mathbf{Y}_\ell\mathbf{Y}_\ell^* = \mathbf{Id}$. Observe that

$$\tilde{\mathbf{A}}_S^*\tilde{\mathbf{A}}_S - \mathbf{Id} = \frac{1}{m}\sum_{\ell=1}^m (\mathbf{Y}_\ell\mathbf{Y}_\ell^* - \mathbb{E}\mathbf{Y}_\ell\mathbf{Y}_\ell^*).$$

The matrices $\mathbf{X}_\ell = \mathbf{Y}_\ell\mathbf{Y}_\ell^* - \mathbb{E}\mathbf{Y}_\ell\mathbf{Y}_\ell^* \in \mathbb{C}^{s\times s}$ have mean zero. Moreover,

$$\|\mathbf{X}_\ell\|_{2\to 2} = \max_{\|\mathbf{x}\|_2=1}|\langle\mathbf{Y}_\ell\mathbf{Y}_\ell^*\mathbf{x}, \mathbf{x}\rangle - \|\mathbf{x}\|_2^2| = |\|\mathbf{Y}_\ell\|_2^2 - 1| \leq K^2s,$$

and since $(\mathbf{Y}_\ell\mathbf{Y}_\ell^*)^2 = \mathbf{Y}_\ell(\mathbf{Y}_\ell^*\mathbf{Y}_\ell)\mathbf{Y}_\ell^* = \|\mathbf{Y}_\ell\|_2^2\mathbf{Y}_\ell\mathbf{Y}_\ell^*$, we have

$$\mathbb{E}\mathbf{X}_\ell^2 = \mathbb{E}\left(\mathbf{Y}_\ell\mathbf{Y}_\ell^*\mathbf{Y}_\ell\mathbf{Y}_\ell^* - 2\mathbf{Y}_\ell\mathbf{Y}_\ell^* + \mathbf{Id}\right) = \mathbb{E}\left((\|\mathbf{Y}_\ell\|_2^2 - 2)\mathbf{Y}_\ell\mathbf{Y}_\ell^*\right) + \mathbf{Id}$$
$$\preccurlyeq (K^2s - 2)\mathbb{E}[\mathbf{Y}_\ell\mathbf{Y}_\ell^*] + \mathbf{Id} \preccurlyeq K^2s\,\mathbf{Id}. \tag{12.33}$$

The variance parameter in (8.28) can therefore be estimated by

$$\sigma^2 := \left\|\sum_{\ell=1}^m \mathbb{E}(\mathbf{X}_\ell^2)\right\|_{2\to 2} \leq mK^2s\|\mathbf{Id}\|_{2\to 2} = K^2sm.$$

The noncommutative Bernstein inequality (8.30) yields, for $\delta \in (0, 1)$,

$$\mathbb{P}\left(\left\|\tilde{\mathbf{A}}_S^*\tilde{\mathbf{A}}_S - \mathbf{Id}\right\|_{2\to 2} > \delta\right) = \mathbb{P}\left(\|\sum_{\ell=1}^m \mathbf{X}_\ell\|_{2\to 2} > \delta m\right)$$
$$\leq 2s\exp\left(-\frac{\delta^2m^2/2}{K^2sm + K^2s\delta m/3}\right) \leq 2s\exp\left(-\frac{3}{8}\frac{\delta^2m}{K^2s}\right).$$

The proof is completed. $\square$

A bound for the coherence of $\mathbf{A}$ can be deduced from Theorem 12.12, as shown below. Coherence estimates can alternatively be shown using simpler techniques which do not require bounds on condition numbers; see Exercise 12.6. Note that below we do not require normalization of the columns (in contrast to Chap. 5).

Corollary 12.14. *Let* $\mathbf{A} \in \mathbb{C}^{m \times N}$ *be the random sampling matrix associated to a BOS with constant* $K \geq 1$. *Then the coherence* μ *of* $\tilde{\mathbf{A}} = \frac{1}{\sqrt{m}} \mathbf{A}$ *satisfies*

$$\mu \leq \sqrt{\frac{16K^2 \ln(2N^2/\varepsilon)}{3m}}$$

with probability at least $1 - \varepsilon$.

Proof. We denote the columns of $\tilde{\mathbf{A}}$ by $\tilde{\mathbf{a}}_j$, $j \in [N]$. Let $S = \{j, k\}$ with $j \neq k$ be a two element set. Then the matrix $\tilde{\mathbf{A}}_S^* \tilde{\mathbf{A}}_S - \mathbf{Id}$ contains $\langle \tilde{\mathbf{a}}_j, \tilde{\mathbf{a}}_k \rangle$ as a matrix entry. Since the absolute value of any entry of a matrix is bounded by the operator norm (Lemma A.9), we have

$$|\langle \tilde{\mathbf{a}}_j, \tilde{\mathbf{a}}_k \rangle| \leq \|\tilde{\mathbf{A}}_S^* \tilde{\mathbf{A}}_S - \mathbf{Id}\|_{2 \to 2}.$$

By Theorem 12.12, applied with $s = 2$, the probability that the operator norm on the right is not bounded by δ is at most

$$4 \exp\left(-\frac{3m\delta^2}{16K^2}\right).$$

Taking the union bound over all $N(N-1)/2 \leq N^2/2$ two element sets $S \subset [N]$ shows that

$$\mathbb{P}(\mu \geq \delta) \leq 2N^2 \exp\left(-\frac{3m\delta^2}{16K^2}\right).$$

Requiring that the right-hand side is at most ε yields the conclusion. $\square$

Proposition 12.15. *Let* $\mathbf{x} \in \mathbb{C}^N$ *be a vector supported on a set* S *such that* $\mathrm{sgn}(\mathbf{x}_S)$ *forms a Rademacher or Steinhaus sequence. If* $\mathbf{A} \in \mathbb{C}^{m \times N}$ *is such that* $\mathbf{A}_S$ *is injective and*

$$\|\mathbf{A}_S^\dagger \mathbf{a}_\ell\|_2 \leq \alpha \quad \text{for all } \ell \in \overline{S} \text{ and some } \alpha > 0, \tag{12.34}$$

then, with probability at least

$$1 - 2N \exp(-\alpha^{-2}/2),$$

the vector $\mathbf{x}$ *is the unique minimizer of* $\|\mathbf{z}\|_1$ *subject to* $\mathbf{A}\mathbf{z} = \mathbf{A}\mathbf{x}$.

Note that the statement is nontrivial only if $\alpha < \left(2\ln(2N)\right)^{-1/2}$.

Proof. In the Rademacher case, the union bound and Hoeffding's inequality (see Corollary 7.21 for the real case and Corollary 8.8 for the general complex case) yield

$$\mathbb{P}(\max_{\ell \in \overline{S}} |\langle \mathbf{A}_S^\dagger \mathbf{a}_\ell, \mathrm{sgn}(\mathbf{x}_S)\rangle| \geq 1) \leq \sum_{\ell \in \overline{S}} \mathbb{P}\left(|\langle \mathbf{A}_S^\dagger \mathbf{a}_\ell, \mathrm{sgn}(\mathbf{x}_S)\rangle| \geq \|\mathbf{A}_S^\dagger \mathbf{a}_\ell\|_2 \alpha^{-1}\right)$$

$$\leq 2N \exp(-\alpha^{-2}/2).$$

In the Steinhaus case we even obtain a better estimate from Corollary 8.10. An application of Corollary 4.28 finishes the proof. $\qquad\square$

Remark 12.16. In Chap. 14 we will choose the matrix $\mathbf{A}$ deterministically and the support set S at random. Proposition 12.15 remains applicable in this situation.

Next, we provide a condition ensuring that $\|\mathbf{A}_S^\dagger \mathbf{a}_\ell\|_2$ is small. The first condition requires that $\mathbf{A}_S$ is well conditioned and that the coherence of $\mathbf{A}$ is small. We note again that in contrast with the definition (5.1) of the coherence, we do not require the columns of $\mathbf{A}$ to be normalized here, even though this will be the case in many examples.

Proposition 12.17. *Let $\mathbf{A} \in \mathbb{C}^{m \times N}$ with coherence μ and let $S \subset [N]$ be of size s. If $\|\mathbf{A}_S^* \mathbf{A}_S - \mathbf{Id}\|_{2\to2} \leq \delta$ for some $\delta \in (0, 1)$, then*

$$\|\mathbf{A}_S^\dagger \mathbf{a}_\ell\|_2 \leq \frac{\sqrt{s}\mu}{1 - \delta} \qquad \text{for all } \ell \in \overline{S}.$$

Proof. Since $\|\mathbf{A}_S^* \mathbf{A}_S - \mathbf{Id}\|_{2\to2} \leq \delta < 1$, the matrix $\mathbf{A}_S$ is injective, and by Lemma A.12,

$$\|(\mathbf{A}_S^* \mathbf{A}_S)^{-1}\|_{2\to2} \leq \frac{1}{1 - \delta}.$$

By definition of the operator norm

$$\|\mathbf{A}_S^\dagger \mathbf{a}_\ell\|_2 = \|(\mathbf{A}_S^* \mathbf{A}_S)^{-1} \mathbf{A}_S^* \mathbf{a}_\ell\|_2 \leq \|(\mathbf{A}_S^* \mathbf{A}_S)^{-1}\|_{2\to2} \|\mathbf{A}_S^* \mathbf{a}_\ell\|_2$$

$$\leq (1 - \delta)^{-1} \|\mathbf{A}_S^* \mathbf{a}_\ell\|_2.$$

For $\ell \in \overline{S}$, we can further estimate

$$\|\mathbf{A}_S^* \mathbf{a}_\ell\|_2 = \sqrt{\sum_{j \in S} |\langle \mathbf{a}_\ell, \mathbf{a}_j\rangle|^2} \leq \sqrt{s}\mu.$$

Combining the two estimates completes the proof. $\qquad\square$

In Exercise 12.5, an alternative way of bounding the term $\|\mathbf{A}_S^\dagger \mathbf{a}_\ell\|_2$ is provided. Both bounds require that only one column submatrix of $\mathbf{A}$, or at least only a small number of them, is well conditioned in contrast with the restricted isometry property which requires that all column submatrices of a certain size to be well conditioned simultaneously. It is significantly simpler to prove well conditionedness for a single column submatrix of a structured random matrix.

Now we are in the position to prove the nonuniform recovery result stated in Theorem 12.11.

Proof (of Theorem 12.11). Set $\alpha = \sqrt{s}u/(1-\delta)$ for some $\delta, u \in (0,1)$ to be chosen later. Let μ be the coherence of $\tilde{\mathbf{A}} = \frac{1}{\sqrt{m}}\mathbf{A}$. By Propositions 12.15 and 12.17, the probability that recovery by basis pursuit fails is bounded by

$$P = 2Ne^{-\alpha^{-2}/2} + \mathbb{P}\left(\|\tilde{\mathbf{A}}_S^\dagger \tilde{\mathbf{a}}_\ell\|_2 \geq \alpha \quad \text{for some } \ell \in [N] \setminus S\right)$$

$$\leq 2Ne^{-\alpha^{-2}/2} + \mathbb{P}(\|\tilde{\mathbf{A}}_S^* \tilde{\mathbf{A}}_S - \mathbf{Id}\|_{2\to 2} > \delta) + \mathbb{P}(\mu > u). \tag{12.35}$$

Remark 12.13 yields $\mathbb{P}(\|\tilde{\mathbf{A}}_S^* \tilde{\mathbf{A}}_S - \mathbf{Id}\|_{2\to 2} > \delta) \leq \varepsilon/3$ under the condition

$$m \geq \frac{8K^2}{3\delta^2} s \ln(6s/\varepsilon). \tag{12.36}$$

Corollary 12.14 asserts that $\mathbb{P}(\mu > u) \leq \varepsilon/3$ provided

$$m \geq \frac{16K^2}{3u^2} \ln(6N^2/\varepsilon),$$

which (since $\ln(6N^2/\varepsilon) \leq 2\ln(6N/\varepsilon)$) is implied by

$$m \geq \frac{32K^2}{3u^2} \ln(6N/\varepsilon). \tag{12.37}$$

Set $u = 2\delta/\sqrt{s}$. Then (12.37) implies (12.36), and $\alpha = 2\delta/(1-\delta)$. Next we set $\delta^{-2} = 13\ln(6N/\varepsilon)$. Then the first term in (12.35) is bounded by

$$2N\exp(-\alpha^{-2}/2) = 2N\exp\left(-\frac{(1-\delta)^2}{8\delta^2}\right)$$

$$= 2N\exp\left(-(1 - (13\ln(6N/\varepsilon))^{-1/2})^2 \cdot 13\ln(6N/\varepsilon)/8\right)$$

$$\leq 2N\exp(-C\ln(6N/\varepsilon)) \leq \varepsilon/3,$$

where $C = 13(1 - (13\ln(6))^{-1/2})^2/8 \approx 1.02 \geq 1$. Plugging the value of δ into the definition of u, that is, $u = (cs\ln(6N/\varepsilon))^{-1/2}$ with $c = 13/4$, and then into (12.37), we find that recovery by basis pursuit fails with probability at most ε provided

$$m \geq CK^2 s \ln^2(6N/\varepsilon)$$

with $C = 32 \cdot 13/(3 \cdot 4) < 35$. This completes the proof.   □

The next statement improves on the exponent 2 at the logarithmic term in (12.30). Unlike the previous result, its proof does not use the coherence, but rather a sophisticated way of bounding the term $\|\tilde{\mathbf{A}}_S^* \tilde{\mathbf{a}}_j\|_2$ using Corollary 8.45.

Theorem 12.18. *Let* $\mathbf{x} \in \mathbb{C}^N$ *be an* s*-sparse vector with support* S *such that its sign sequence* $\mathrm{sgn}(\mathbf{x}_S)$ *forms a Rademacher or Steinhaus sequence. Let* $\mathbf{A} \in \mathbb{C}^{m \times N}$ *be the random sampling matrix* (12.4) *associated to a BOS with constant* $K \geq 1$. *Assume that* $m \geq cK^2 \ln^2(6N/\varepsilon)$ *for an appropriate constant* $c > 0$ *and that*

$$m \geq 18K^2 s \ln(6N/\varepsilon). \tag{12.38}$$

Then with probability at least $1 - \varepsilon$ *the vector* $\mathbf{x}$ *is the unique minimizer of* $\|\mathbf{z}\|_1$ *subject to* $\mathbf{Az} = \mathbf{Ax}$.

We start with a technical lemma.

Lemma 12.19. *With the notation of Theorem 12.18, let* $\tilde{\mathbf{A}} = \frac{1}{\sqrt{m}}\mathbf{A}$. *Then, for* $t > 0$,

$$\mathbb{P}\left(\max_{j \in \overline{S}} \|\tilde{\mathbf{A}}_S^* \tilde{\mathbf{a}}_j\|_2 \geq \sqrt{\frac{K^2 s}{m}} + t \right) \leq N \exp\left(-\frac{mt^2}{K^2\sqrt{s}} \frac{1}{\frac{2}{\sqrt{s}} + 4\sqrt{\frac{K^2 s}{m}} + \frac{2t}{3}} \right).$$

Proof. Fix $j \in \overline{S}$. We introduce the vectors $\mathbf{Y}_\ell = (\phi_k(\mathbf{t}_\ell))_{k \in S} \in \mathbb{C}^S$ and $\mathbf{Z}_\ell = \left(\phi_j(\mathbf{t}_\ell)\overline{\phi_k(\mathbf{t}_\ell)} \right)_{k \in S} = \phi_j(\mathbf{t}_\ell)\mathbf{Y}_\ell \in \mathbb{C}^S$. Then

$$\|\tilde{\mathbf{A}}_S^* \tilde{\mathbf{a}}_j\|_2 = \frac{1}{m} \left\| \sum_{\ell=1}^{m} \mathbf{Z}_\ell \right\|_2.$$

Our aim is to apply the vector-valued Bernstein inequality of Corollary 8.45. The $\mathbf{Z}_\ell$ are independent copies of a random vector $\mathbf{Z}$ that satisfies $\mathbb{E}\mathbf{Z} = \mathbf{0}$ by orthonormality of the ϕ_k and by the fact that $j \notin S$. The norm of the vector $\mathbf{Z}$ can be bounded by

$$\|\mathbf{Z}\|_2 = |\phi_j(\mathbf{t}_1)| \|\mathbf{Y}_1\|_2 \leq \sqrt{s}K^2$$

due to the boundedness condition (12.2) and since $\mathrm{card}(S) \leq s$. Furthermore,

$$\mathbb{E}\|\mathbf{Z}\|_2^2 = \mathbb{E}[|\phi_j(\mathbf{t}_1)|^2 \|\mathbf{Y}_1\|_2^2] \leq K^2 \mathbb{E}\|\mathbf{Y}_1\|_2^2 = K^2 \sum_{k \in S} \mathbb{E}|\phi_k(\mathbf{t}_1)|^2 \leq K^2 s.$$

To estimate the weak variance, we observe that for $\mathbf{z} \in \mathbb{C}^S$ with $\|\mathbf{z}\|_2 \leq 1$,

$$\mathbb{E}|\langle \mathbf{z}, \mathbf{Z} \rangle|^2 = \mathbb{E}[|\phi_j(\mathbf{t}_1)|^2 |\langle \mathbf{z}, \mathbf{Y}_1 \rangle|^2] \leq K^2 \mathbb{E}[\mathbf{z}^* \mathbf{Y}_1 \mathbf{Y}_1^* \mathbf{z}] = K^2 \|\mathbf{z}\|_2^2 \leq K^2$$

again by the orthonormality condition (12.1). Hence,

$$\sigma^2 = \sup_{\|\mathbf{z}\|_2 \leq 1} \mathbb{E}|\langle \mathbf{z}, \mathbf{Z} \rangle|^2 \leq K^2.$$

The Bernstein inequality (8.95) yields, for $t > 0$,

$$\mathbb{P}(\|\sum_{\ell=1}^m \mathbf{Z}_\ell\|_2 \geq \sqrt{msK^2} + t) \leq \exp\left(-\frac{t^2/2}{mK^2 + 2\sqrt{s}K^2\sqrt{msK^2} + t\sqrt{s}K^2/3}\right).$$

Rescaling by $1/m$ and taking the union bound over all $j \in \overline{S}$ yield the claimed probability estimate. $\square$

Proof (of Theorem 12.18). For reasons that will become apparent below, if $s < 36 \ln(6N/\varepsilon)$, then we enlarge S by some arbitrary elements from $[N] \setminus S$ to obtain S_+ such that $S \subset S_+$ and $\mathrm{card}(S_+) = \lceil 36 \ln(6N/\varepsilon) \rceil$. Moreover, we extend $\mathrm{sgn}(\mathbf{x}_S)$ such that $\mathrm{sgn}(\mathbf{x}_{S_+})$ forms a random Rademacher or Steinhaus sequence. If $s \geq 36 \ln(6N/\epsilon)$, then if $\|\mathbf{x}\|_0 = s$, we simply take $S_+ = S$ and if $\|\mathbf{x}\|_0 < s$, we again enlarge S to a set $S_+ \supset S$ with $\mathrm{card}(S_+) = s$ and $\mathrm{sgn}(\mathbf{x}_S)$ to a Rademacher or Steinhaus sequence $\mathrm{sgn}(\mathbf{x}_{S_+})$, so that altogether

$$\mathrm{card}(S_+) = s_+ := \max\{s, \lceil 36 \ln(6N/\varepsilon) \rceil\}.$$

By Remark 4.27(a), once we verify the condition (a) of Theorem 4.26 for S_+ and $\mathrm{sgn}(\mathbf{x}_{S_+})$ (for instance, via Proposition 12.15), $\mathbf{x}$ is recovered exactly via ℓ_1-minimization.

Proposition 12.15 requires us to bound the term

$$\|\tilde{\mathbf{A}}_{S_+}^\dagger \tilde{\mathbf{a}}_j\|_2 \leq \|(\tilde{\mathbf{A}}_{S_+}^* \tilde{\mathbf{A}}_{S_+})^{-1}\|_{2 \to 2} \|\tilde{\mathbf{A}}_{S_+}^* \tilde{\mathbf{a}}_j\|_2, \quad j \notin S_+,$$

where $\tilde{\mathbf{A}} = \frac{1}{\sqrt{m}} \mathbf{A}$ and $\tilde{\mathbf{a}}_1, \ldots, \tilde{\mathbf{a}}_N$ denote its columns. Note that the injectivity of $\tilde{\mathbf{A}}_{S_+}$ (as required by Proposition 12.15) will be ensured once $\tilde{\mathbf{A}}_{S_+}^* \tilde{\mathbf{A}}_{S_+}$ is merely invertible. The operator norm satisfies $\|(\tilde{\mathbf{A}}_{S_+}^* \tilde{\mathbf{A}}_{S_+})^{-1}\|_{2 \to 2} \leq (1 - \delta)^{-1}$ provided $\|\tilde{\mathbf{A}}_{S_+}^* \tilde{\mathbf{A}}_{S_+} - \mathrm{Id}\|_{2 \to 2} \leq \delta$, the latter being treated by Theorem 12.12. For the remaining term, we set $t = \beta K \sqrt{s_+/m}$ in Lemma 12.19 to obtain

$$\mathbb{P}\left(\max_{j\in\overline{S_+}}\|\tilde{\mathbf{A}}^*_{S_+}\tilde{\mathbf{a}}_j\|_2 \geq (1+\beta)K\sqrt{\frac{s_+}{m}}\right)$$

$$\leq N\exp\left(-\frac{\beta^2\sqrt{s_+}}{2/\sqrt{s_+}+(4+2\beta/3)K\sqrt{s_+/m}}\right). \tag{12.39}$$

Setting $v = (1+\beta)K\sqrt{\frac{s_+}{m}}$, if $\|\tilde{\mathbf{A}}^*_{S_+}\tilde{\mathbf{A}}_{S_+} - \mathbf{Id}\|_{2\to2} \leq \delta$ and $\max_{j\in\overline{S_+}}\|\tilde{\mathbf{A}}^*_{S_+}\tilde{\mathbf{a}}_j\|_2$
$\leq v$, then $\max_{j\in\overline{S_+}}\|\tilde{\mathbf{A}}^\dagger_{S_+}\tilde{\mathbf{a}}_j\|_2 \leq v/(1-\delta)$. Therefore, by Proposition 12.15 and
Theorem 12.12, the probability that basis pursuit fails to recover $\mathbf{x}$ is bounded by

$$\mathbb{P}(\max_{j\in\overline{S_+}}|\langle\tilde{\mathbf{A}}^\dagger_{S_+}\tilde{\mathbf{a}}_j, \mathrm{sgn}(\mathbf{x}_{S_+})\rangle| \geq 1)$$

$$\leq \mathbb{P}\left(\max_{j\in\overline{S_+}}|\langle\tilde{\mathbf{A}}^\dagger_{S_+}\tilde{\mathbf{a}}_j, \mathrm{sgn}(\mathbf{x}_{S_+})\rangle| \geq 1 \,\middle|\, \|\tilde{\mathbf{A}}^*_{S_+}\tilde{\mathbf{A}}_{S_+} - \mathbf{Id}\|_{2\to2} \leq \delta\right.$$

$$\left.\text{and } \max_{j\in\overline{S_+}}\|\tilde{\mathbf{A}}^*_{S_+}\tilde{\mathbf{a}}_j\|_2 \leq v\right)$$

$$+ \mathbb{P}(\|\tilde{\mathbf{A}}^*_{S_+}\tilde{\mathbf{A}}_{S_+} - \mathbf{Id}\|_{2\to2} \geq \delta) + \mathbb{P}(\max_{j\in\overline{S_+}}\|\tilde{\mathbf{A}}^*_{S_+}\tilde{\mathbf{a}}_j\|_2 \geq v)$$

$$\leq 2N\exp\left(-\frac{(1-\delta)^2}{2v^2}\right) + 2s_+\exp\left(-\frac{3m\delta^2}{8K^2s_+}\right) \tag{12.40}$$

$$+ N\exp\left(-\frac{\beta^2\sqrt{s_+}}{2/\sqrt{s_+}+(4+2\beta/3)K\sqrt{s_+/m}}\right). \tag{12.41}$$

Let us choose $\delta = \beta = 1/2$. Then the second term in (12.40) is bounded by $\varepsilon/3$
provided

$$m \geq \frac{32}{3}K^2s_+\ln(6s/\varepsilon). \tag{12.42}$$

The first term in (12.40) does not exceed $\varepsilon/3$ provided

$$v^{-2} \geq 2(1-\delta)^{-2}\ln(6N/\varepsilon),$$

which, by definition of v, is equivalent to

$$m \geq \frac{2(1+\beta)^2}{(1-\delta)^2}K^2s_+\ln(6N/\varepsilon) = 18K^2s_+\ln(6N/\varepsilon). \tag{12.43}$$

Suppose that this condition holds. Since $s_+ \geq 36 \ln(6N/\varepsilon)$, the term in (12.41) is bounded by

$$N \exp\left(-\frac{\sqrt{s_+}/4}{2/\sqrt{s_+} + \frac{13}{3}K\sqrt{s_+/m}}\right) \leq N \exp\left(-\frac{\sqrt{s_+}/4}{\frac{2}{6\sqrt{\ln(6N/\varepsilon)}} + \frac{13}{3\sqrt{18\ln(6N/\varepsilon)}}}\right)$$

$$= N \exp\left(-\frac{\sqrt{s_+\ln(6N/\varepsilon)}}{4(1/3 + 13/(3\sqrt{18}))}\right) \leq N \exp\left(-\frac{27\ln(6N/\varepsilon)}{6 + 13\sqrt{2}}\right)$$

$$\leq \varepsilon/6,$$

because $27/(6 + 13\sqrt{2}) \approx 1.1072 > 1$.

Since (12.42) is implied by (12.43), we have shown that the probability of recovery failure is at most $\varepsilon/3 + \varepsilon/3 + \varepsilon/6 < \varepsilon$ provided that

$$m \geq 18K^2 s_+ \ln(6N/\varepsilon).$$

This gives the desired result with $c = 18 \cdot 37 = 666$.   □

12.4   Nonuniform Recovery: Deterministic Sign Patterns

As already announced, we remove in this section the assumption that the sign pattern of the nonzero coefficients is random. This means that the coefficient vector is completely arbitrary (but fixed). Only the sampling matrix is randomly chosen. The main result of this section reads as follows.

Theorem 12.20. *Let* $\mathbf{x} \in \mathbb{C}^N$ *be* s-*sparse and let* $\mathbf{A} \in \mathbb{C}^{m \times N}$ *be the random sampling matrix associated to a BOS with constant* $K \geq 1$. *If*

$$m \geq CK^2 s \ln(N) \ln(\varepsilon^{-1}), \tag{12.44}$$

where $C > 0$ *is a universal constant, then* $\mathbf{x}$ *is the unique minimizer of* $\|\mathbf{z}\|_1$ *subject to* $\mathbf{Az} = \mathbf{Ax}$ *with probability at least* $1 - \varepsilon$.

Remark 12.21. The proof reveals the more precise condition

$$m \geq CK^2 s \left[\ln(4N)\ln(8\varepsilon^{-1}) + \ln(4)\ln(16e^8 s\varepsilon^{-4})\right].$$

with $C = 163.48$.

The previous result can be made stable under sparsity defect and robust under noise. Note that the error bound implied by the restricted isometry property shown in the next section is stronger, but requires slightly more samples.

Theorem 12.22. *Let* $\mathbf{x} \in \mathbb{C}^N$ *and let* $\mathbf{A} \in \mathbb{C}^{m \times N}$ *to be the random sampling matrix associated to a BOS with constant* $K \geq 1$. *For* $\mathbf{y} = \mathbf{A}\mathbf{x} + \mathbf{e}$ *with* $\|\mathbf{e}\|_2 \leq \eta\sqrt{m}$ *for some* $\eta \geq 0$, *let* $\mathbf{x}^\sharp$ *be a solution to*

$$\underset{\mathbf{z} \in \mathbb{C}^N}{\text{minimize}} \|\mathbf{z}\|_1 \quad \text{subject to } \|\mathbf{A}\mathbf{z} - \mathbf{y}\|_2 \leq \eta\sqrt{m}. \tag{12.45}$$

If

$$m \geq CK^2 s \ln(N) \ln(\varepsilon^{-1}), \tag{12.46}$$

then with probability at least $1 - \varepsilon$, *the reconstruction error satisfies*

$$\|\mathbf{x} - \mathbf{x}^\sharp\|_2 \leq C_1 \sigma_s(\mathbf{x})_1 + C_2\sqrt{s}\eta.$$

The constants $C, C_1, C_2 > 0$ *are universal.*

Remark 12.23. The assumption $\|\mathbf{e}\|_2 \leq \eta\sqrt{m}$ on the noise is natural. Indeed, if $f(\mathbf{t}) = \sum_{j \in [N]} x_j \phi_j(\mathbf{t})$ is the function associated with $\mathbf{x}$, then it is satisfied under the pointwise error estimate $|f(\mathbf{t}_\ell) - y_\ell| \leq \eta$ for all $\ell \in [m]$.

   In contrast to the approach of the previous section, the proof of these results relies on the recovery condition via an inexact dual of Theorem 4.32 and its extension to stable recovery (Theorem 4.33). As before, we introduce the rescaled matrix $\tilde{\mathbf{A}} = \frac{1}{\sqrt{m}}\mathbf{A}$, where $\mathbf{A}$ is the sampling matrix in (12.4). The term $\|(\tilde{\mathbf{A}}_S^*\tilde{\mathbf{A}}_S)^{-1}\|_{2 \to 2}$ in (4.25) will be treated with Theorem 12.12 by noticing that $\|\tilde{\mathbf{A}}_S^*\tilde{\mathbf{A}}_S - \mathrm{Id}\|_{2 \to 2} \leq \delta$ implies $\|(\tilde{\mathbf{A}}_S^*\tilde{\mathbf{A}}_S)^{-1}\|_{2 \to 2} \leq (1 - \delta)^{-1}$ (Lemma A.12). The other terms in Theorem 4.32 will be bounded based on the following lemmas together with some estimates from the previous section. All the following results refer to the rescaled sampling matrix $\tilde{\mathbf{A}}$ as just introduced.

Lemma 12.24. *Let* $\mathbf{v} \in \mathbb{C}^N$ *with* $\mathrm{supp}(\mathbf{v}) = S$, $\mathrm{card}(S) = s$. *Then, for* $t > 0$,

$$\mathbb{P}(\|\tilde{\mathbf{A}}_{\overline{S}}^*\tilde{\mathbf{A}}\mathbf{v}\|_\infty \geq t\|\mathbf{v}\|_2) \leq 4N\exp\left(-\frac{m}{4K^2}\frac{t^2}{1 + \sqrt{s/18}\,t}\right). \tag{12.47}$$

Proof. Note that

$$\|\tilde{\mathbf{A}}_{\overline{S}}^*\tilde{\mathbf{A}}\mathbf{v}\|_\infty = \max_{k \in \overline{S}} |\langle \mathbf{e}_k, \tilde{\mathbf{A}}^*\tilde{\mathbf{A}}\mathbf{v}\rangle|,$$

where $\mathbf{e}_k$ denotes the kth canonical vector. Without loss of generality, we may assume that $\|\mathbf{v}\|_2 = 1$. Denote

$$\mathbf{Y}_\ell = (\overline{\phi_j(\mathbf{t}_\ell)})_{j \in [N]} \in \mathbb{C}^N. \tag{12.48}$$

We fix $k \in \overline{S}$ and write

$$\langle \mathbf{e}_k, \tilde{\mathbf{A}}^* \tilde{\mathbf{A}} \mathbf{v} \rangle = \frac{1}{m} \sum_{\ell=1}^{m} \langle \mathbf{e}_k, \mathbf{Y}_\ell \mathbf{Y}_\ell^* \mathbf{v} \rangle = \frac{1}{m} \sum_{\ell=1}^{m} Z_\ell$$

with $Z_\ell = \langle \mathbf{e}_k, \mathbf{Y}_\ell \mathbf{Y}_\ell^* \mathbf{v} \rangle$. We aim to apply the Bernstein inequality of Corollary 7.31. To this end, we note that the Z_ℓ are independent and satisfy $\mathbb{E} Z_\ell = \langle \mathbf{e}_k, \mathbb{E}[\mathbf{Y}_\ell \mathbf{Y}_\ell^*] \mathbf{v} \rangle = \langle \mathbf{e}_k, \mathbf{v} \rangle = 0$ since the orthonormality relation (12.1) yields $\mathbb{E}[\mathbf{Y}_\ell \mathbf{Y}_\ell^*] = \mathrm{Id}$ and since $k \notin S = \mathrm{supp}\, \mathbf{v}$. Next it follows from the Cauchy–Schwarz inequality that

$$|Z_\ell| = |\langle \mathbf{e}_k, \mathbf{Y}_\ell \mathbf{Y}_\ell^* \mathbf{v} \rangle| = |\langle \mathbf{e}_k, \mathbf{Y}_\ell \rangle \langle \mathbf{Y}_\ell, \mathbf{v} \rangle| = |\langle \mathbf{e}_k, \mathbf{Y}_\ell \rangle| |\langle (\mathbf{Y}_\ell)_S, \mathbf{v} \rangle|$$
$$\leq |\phi_k(\mathbf{t}_\ell)| \|(\mathbf{Y}_\ell)_S\|_2 \|\mathbf{v}\|_2 \leq K^2 \sqrt{s}.$$

Hereby, we used the facts that $|\phi_k(\mathbf{t}_\ell)| \leq K$ by the boundedness condition (12.2) and that $\|(\mathbf{Y}_\ell)_S\|_2 \leq K\sqrt{s}$ as in (12.32). The variance of Z_ℓ can be estimated as

$$\mathbb{E}|Z_\ell|^2 = \mathbb{E}\left[\langle \mathbf{e}_k, \mathbf{Y}_\ell \mathbf{Y}_\ell^* \mathbf{v} \rangle \langle \mathbf{Y}_\ell \mathbf{Y}_\ell^* \mathbf{v}, \mathbf{e}_k \rangle\right] = \mathbb{E}\left[|\langle \mathbf{e}_k, \mathbf{Y}_\ell \rangle|^2 \mathbf{v}^* \mathbf{Y}_\ell \mathbf{Y}_\ell^* \mathbf{v}\right]$$
$$\leq K^2 \mathbf{v}^* \mathbb{E}[\mathbf{Y}_\ell \mathbf{Y}_\ell^*] \mathbf{v} = K^2 \|\mathbf{v}\|_2^2 = K^2.$$

Clearly, $\mathrm{Re}(Z_\ell)$ and $\mathrm{Im}(Z_\ell)$ satisfy the same bounds as Z_ℓ itself. The union bound, the fact that $|z|^2 = \mathrm{Re}(z)^2 + \mathrm{Im}(z)^2$ for any complex number z, and Bernstein's inequality (7.40) yield, for $t > 0$,

$$\mathbb{P}(|\langle \mathbf{e}_k, \tilde{\mathbf{A}}^* \tilde{\mathbf{A}} \mathbf{v} \rangle| \geq t)$$
$$\leq \mathbb{P}\left(\left|\frac{1}{m} \sum_{\ell=1}^{m} \mathrm{Re}(Z_\ell)\right| \geq t/\sqrt{2}\right) + \mathbb{P}\left(\left|\frac{1}{m} \sum_{\ell=1}^{m} \mathrm{Im}(Z_\ell)\right| \geq t/\sqrt{2}\right)$$
$$\leq 4 \exp\left(-\frac{(mt)^2/4}{mK^2 + K^2\sqrt{s}tm/(3\sqrt{2})}\right) = 4 \exp\left(-\frac{m}{4K^2}\frac{t^2}{1 + \sqrt{s/18}\,t}\right).$$

Taking the union bound over all $k \in \overline{S}$ completes the proof.   □

Note that in the real-valued case (that is, the functions ϕ_j as well as the vector $\mathbf{v}$ are real valued), the constant 4 in the probability estimate (12.47) above can be replaced by 2 in both instances.

Lemma 12.25. *Let $S \subset [N]$ with $\mathrm{card}(S) = s$ and $\mathbf{v} \in \mathbb{C}^S$ with $\|\mathbf{v}\|_2 = 1$. Then, for $t > 0$,*

$$\mathbb{P}\left(\|(\tilde{\mathbf{A}}_S^* \tilde{\mathbf{A}}_S - \mathrm{Id})\mathbf{v}\|_2 \geq \sqrt{\frac{K^2 s}{m}} + t\right) \leq \exp\left(-\frac{mt^2}{2K^2 s}\frac{1}{1 + 2\sqrt{\frac{K^2 s}{m}} + t/3}\right).$$

Remark 12.26. Note that Theorem 12.12 concerning the spectral norm of $\tilde{\mathbf{A}}_S^* \tilde{\mathbf{A}}_S - \mathrm{Id}$ implies a similar statement. We require the above results because the probability estimate does not involve a factor s in front of the exponential term in contrast to (12.31).

Proof. Similar to the previous proof we introduce vectors $\mathbf{Y}_\ell = (\overline{\phi_j(t_\ell)})_{j \in S} \in \mathbb{C}^S$. Note that

$$(\tilde{\mathbf{A}}_S^* \tilde{\mathbf{A}}_S - \mathrm{Id})\mathbf{v} = \frac{1}{m} \sum_{\ell=1}^m (\mathbf{Y}_\ell \mathbf{Y}_\ell^* - \mathrm{Id})\mathbf{v} = \frac{1}{m} \sum_{\ell=1}^m \mathbf{Z}_\ell$$

with the vectors $\mathbf{Z}_\ell = (\mathbf{Y}_\ell \mathbf{Y}_\ell^* - \mathrm{Id})\mathbf{v} \in \mathbb{C}^S$. Our aim is to apply the vector-valued Bernstein inequality of Corollary 8.45. Observe to this end that the $\mathbf{Z}_\ell$ are independent copies of a single random vector $\mathbf{Z}$ because the $\mathbf{Y}_\ell$ are independent copies of a random vector $\mathbf{Y} = (\overline{\phi_j(t)})_{j \in S}$, and they satisfy $\mathbb{E}\mathbf{Z}_\ell = \mathbb{E}\mathbf{Z} = \mathbb{E}(\mathbf{Y}\mathbf{Y}^* - \mathrm{Id})\mathbf{v} = 0$. Furthermore,

$$\mathbb{E}\|\mathbf{Z}_\ell\|_2^2 = \mathbb{E}\|(\mathbf{Y}\mathbf{Y}^* - \mathrm{Id})\mathbf{v}\|_2^2 = \mathbb{E}\left[|\langle \mathbf{Y}, \mathbf{v}\rangle|^2 \|\mathbf{Y}\|_2^2\right] - 2\mathbb{E}|\langle \mathbf{Y}, \mathbf{v}\rangle|^2 + 1.$$

Recall from (12.32) that the boundedness condition (12.2) implies that $\|\mathbf{Y}\|_2 \leq \sqrt{s}K$. The Cauchy–Schwarz inequality gives therefore

$$|\langle \mathbf{Y}, \mathbf{v}\rangle| \leq \sqrt{s}K.$$

Furthermore,

$$\mathbb{E}|\langle \mathbf{Y}, \mathbf{v}\rangle|^2 = \sum_{j,k \in S} v_j \overline{v_k} \mathbb{E}[\phi_k(t)\overline{\phi_j(t)}] = \|\mathbf{v}\|_2^2 = 1$$

by the orthogonality condition (12.1). Hence,

$$\mathbb{E}\|\mathbf{Z}\|_2^2 = \mathbb{E}\left[|\langle \mathbf{Y}, \mathbf{v}\rangle|^2 \|\mathbf{Y}\|_2^2\right] - 2\mathbb{E}|\langle \mathbf{Y}, \mathbf{v}\rangle|^2 + 1 \leq (sK^2 - 2)\mathbb{E}|\langle \mathbf{Y}, \mathbf{v}\rangle|^2 + 1$$
$$= sK^2 - 1 \leq sK^2.$$

For the uniform bound, observe that

$$\|\mathbf{Z}\|_2^2 = \|(\mathbf{Y}\mathbf{Y}^* - \mathrm{Id})\mathbf{v}\|_2^2 = |\langle \mathbf{Y}, \mathbf{v}\rangle|^2 \|\mathbf{Y}\|_2^2 - 2|\langle \mathbf{Y}, \mathbf{v}\rangle|^2 + 1$$
$$= |\langle \mathbf{Y}, \mathbf{v}\rangle|^2 (\|\mathbf{Y}\|_2^2 - 2) + 1 \leq sK^2(sK^2 - 2) + 1 \leq s^2 K^4,$$

so that $\|\mathbf{Z}\|_2 \leq sK^2$ for all realizations of $\mathbf{Z}$. Furthermore, we simply bound the weak variance by the strong variance,

$$\sigma^2 = \sup_{\|\mathbf{z}\|_2 \leq 1} \mathbb{E}|\langle \mathbf{z}, \mathbf{Z}\rangle|^2 \leq \mathbb{E}\|\mathbf{Z}\|_2^2 \leq sK^2.$$

Then the ℓ_2-valued Bernstein inequality (8.95) yields

$$\mathbb{P}(\|\sum_{\ell=1}^{m} \mathbf{Z}_\ell\|_2 \geq \sqrt{msK^2} + t) \leq \exp\left(-\frac{t^2/2}{msK^2 + 2sK^2\sqrt{msK^2} + tsK^2/3}\right),$$

so that, with t replaced by mt, we obtain

$$\mathbb{P}\left(\|(\tilde{\mathbf{A}}_S^*\tilde{\mathbf{A}}_S - \mathbf{Id})\mathbf{v}\|_2 \geq \sqrt{\frac{K^2s}{m}} + t\right) \leq \exp\left(-\frac{mt^2}{2K^2s}\frac{1}{1 + 2\sqrt{\frac{K^2s}{m}} + t/3}\right).$$

This completes the proof. $\square$

Next we provide a variant of Lemma 12.19, which is more convenient here.

Lemma 12.27. *For $S \subset [N]$ with* $\mathrm{card}(S) = s$ *and* $0 < t \leq 2\sqrt{s}$, *we have*

$$\mathbb{P}\left(\max_{j \in \overline{S}} \|\tilde{\mathbf{A}}_S^*\tilde{\mathbf{a}}_j\|_2 \geq t\right) \leq 2(s+1)N \exp\left(-\frac{3}{10}\frac{mt^2}{K^2s}\right). \tag{12.49}$$

Proof. Fix $j \in \overline{S}$. As in the proof of Lemma 12.19, we introduce the vectors $\mathbf{Y}_\ell = (\overline{\phi_k(\mathbf{t}_\ell)})_{k \in S} \in \mathbb{C}^S$ and $\mathbf{Z}_\ell = \left(\phi_j(\mathbf{t}_\ell)\overline{\phi_k(\mathbf{t}_\ell)}\right)_{k \in S} = \phi_j(\mathbf{t}_\ell)\mathbf{Y}_\ell \in \mathbb{C}^S$. Then

$$\|\tilde{\mathbf{A}}_S^*\tilde{\mathbf{a}}_j\|_2 = \frac{1}{m}\left\|\sum_{\ell=1}^{m} \mathbf{Z}_\ell\right\|_2.$$

Our goal is to apply the noncommutative Bernstein inequality of Theorem 8.14 and its extension in Exercise 8.7 by treating the $\mathbf{Z}_\ell$ as matrices and noting that the operator norm of $\mathbf{Z}_\ell$ equals its ℓ_2-norm. The $\mathbf{Z}_\ell$ are independent and satisfy $\mathbb{E}\mathbf{Z}_\ell = \mathbf{0}$ by orthonormality of the ϕ_k and by the fact that $j \notin S$. They can be bounded by

$$\|\mathbf{Z}_\ell\|_{2 \to 2} = \|\mathbf{Z}_\ell\|_2 = \|\mathbf{Y}_\ell\|_2 |\phi_j(\mathbf{t}_\ell)| \leq \sqrt{s}K^2,$$

by the boundedness condition (12.2) and since $\mathrm{card}(S) = s$. Furthermore,

$$\mathbb{E}[\mathbf{Z}_\ell^*\mathbf{Z}_\ell] = \mathbb{E}\|\mathbf{Z}_\ell\|_2^2 = \mathbb{E}[|\phi_j(\mathbf{t}_\ell)|^2\|\mathbf{Y}_\ell\|_2^2] \leq K^2\mathbb{E}\|\mathbf{Y}_\ell\|_2^2 = K^2\sum_{k \in S}\mathbb{E}|\phi_k(\mathbf{t}_\ell)|^2$$

$$= K^2s.$$

Moreover,

$$\mathbb{E}[\mathbf{Z}_\ell\mathbf{Z}_\ell^*] = \mathbb{E}[|\phi_j(\mathbf{t}_\ell)|^2\mathbf{Y}_\ell\mathbf{Y}_\ell^*] \preccurlyeq K^2\mathbb{E}[\mathbf{Y}_\ell\mathbf{Y}_\ell^*] = K^2\mathbf{Id}.$$

Therefore, the variance parameter σ^2 in (8.118) satisfies

$$\sigma^2 = \max \left\{ \| \sum_{\ell=1}^{m} \mathbb{E}[\mathbf{Z}_\ell \mathbf{Z}_\ell^*] \|_{2 \to 2}, \| \sum_{\ell=1}^{m} \mathbb{E}[\mathbf{Z}_\ell^* \mathbf{Z}_\ell] \|_{2 \to 2} \right\} \le K^2 m s.$$

The version of the noncommutative Bernstein inequality for rectangular random matrices (8.119) yields

$$\mathbb{P}\left(\| \sum_{\ell=1}^{m} \mathbf{Z}_\ell \|_2 \ge u \right) \le 2(s+1) \exp \left(-\frac{u^2/2}{K^2 m s + u\sqrt{s}K^2/3} \right).$$

Setting $u = mt$, taking the union bound over $j \in \bar{S}$, and using that $0 < t \le 2\sqrt{s}$ yields

$$\mathbb{P}\left(\max_{j \in \bar{S}} \| \tilde{\mathbf{A}}_S^* \tilde{\mathbf{a}}_j \|_2 \ge t \right) \le 2N(s+1) \exp \left(-\frac{mt^2/2}{K^2 s + t\sqrt{s}K^2/3} \right)$$

$$\le 2N(s+1) \exp \left(-\frac{3}{10} \frac{mt^2}{K^2 s} \right).$$

This completes the proof.   □

Before turning to the proof of Theorem 12.20, we provide a slightly weaker result which we strengthen afterwards.

Proposition 12.28. *Let* $\mathbf{x} \in \mathbb{C}^N$ *be* s-*sparse and let* $\mathbf{A} \in \mathbb{C}^{m \times N}$ *be the random sampling matrix associated to a BOS with constant* $K \ge 1$. *If*

$$m \ge cK^2 s \left[2\ln(4N)\ln(12\varepsilon^{-1}) + \ln(s)\ln(12e\varepsilon^{-1}\ln(s)) \right],$$

with $c = 8e^2(1 + (1/\sqrt{8} + 1/6)/e) \approx 70.43$, *then* $\mathbf{x}$ *is the unique minimizer of* $\|\mathbf{z}\|_1$ *subject to* $\mathbf{A}\mathbf{z} = \mathbf{A}\mathbf{x}$.

Remark 12.29. If $\ln(s)\ln(\ln s) \le c\ln(N)$, then the above result already implies Theorem 12.20.

Proof. The proof relies on the so-called *golfing scheme* and an application of the recovery result in Theorem 4.32 based on an inexact dual vector. We partition the m-independent samples into L disjoint blocks of sizes $m_1, \ldots, m_L$ to be specified later; in particular, $m = \sum_{j=1}^{L} m_j$. These blocks correspond to row submatrices of $\mathbf{A}$, which we denote by $\mathbf{A}^{(1)} \in \mathbb{C}^{m_1 \times N}, \ldots, \mathbf{A}^{(L)} \in \mathbb{C}^{m_L \times N}$. It will be crucial below that these submatrices are stochastically independent. As usual, we also introduce the rescaled matrix $\tilde{\mathbf{A}} = \frac{1}{\sqrt{m}}\mathbf{A}$ and the support S of $\mathbf{x}$.

We set $\mathbf{u}^{(0)} = \mathbf{0} \in \mathbb{C}^N$ and define recursively

$$\mathbf{u}^{(n)} = \frac{1}{m_n}(\mathbf{A}^{(n)})^*\mathbf{A}_S^{(n)}(\mathrm{sgn}(\mathbf{x}_S) - \mathbf{u}_S^{(n-1)}) + \mathbf{u}^{(n-1)}, \qquad (12.50)$$

for $n \in [L]$. The vector $\mathbf{u} := \mathbf{u}^{(L)}$ will serve as a candidate for the inexact dual of Theorem 4.32. By construction of $\mathbf{u}$, there exists indeed a vector $\mathbf{h} \in \mathbb{C}^m$ such that $\mathbf{u} = \mathbf{A}^*\mathbf{h}$ and by rescaling $\mathbf{u} = \tilde{\mathbf{A}}^*\tilde{\mathbf{h}}$ for some $\tilde{\mathbf{h}} \in \mathbb{C}^m$. For ease of notation, we introduce $\mathbf{w}^{(n)} = \mathrm{sgn}(\mathbf{x}_S) - \mathbf{u}_S^{(n)}$. Observe that

$$\mathbf{w}^{(n)} = \left(\mathbf{Id} - \frac{1}{m_n}(\mathbf{A}_S^{(n)})^*\mathbf{A}_S^{(n)}\right)\mathbf{w}^{(n-1)} \qquad (12.51)$$

and

$$\mathbf{u} = \sum_{n=1}^{L}\frac{1}{m_n}(\mathbf{A}^{(n)})^*\mathbf{A}_S^{(n)}\mathbf{w}^{(n-1)}. \qquad (12.52)$$

We now verify the conditions of Theorem 4.32. To this end, we use the lemmas proven above. First, we require the inequalities

$$\|\mathbf{w}^{(n)}\|_2 \le \left(\sqrt{\frac{K^2 s}{m_n}} + r_n\right)\|\mathbf{w}^{(n-1)}\|_2, \quad n \in [L], \qquad (12.53)$$

$$\left\|\frac{1}{m_n}(\mathbf{A}_{\overline{S}}^{(n)})^*\mathbf{A}_S^{(n)}\mathbf{w}^{(n-1)}\right\|_\infty \le t_n\|\mathbf{w}^{(n-1)}\|_2, \quad n \in [L], \qquad (12.54)$$

where the parameters r_n, t_n will be specified below. The probability $p_1(n)$ that (12.53) fails to hold can be bounded using Lemma 12.25 as

$$p_1(n) \le \exp\left(-\frac{m_n r_n^2}{2K^2 s}\frac{1}{1 + 2\sqrt{\frac{K^2 s}{m_n}} + r_n/3}\right).$$

Due to Lemma 12.24, the probability $p_2(n)$ that (12.54) fails is bounded by

$$p_2(n) \le 4N\exp\left(-\frac{m_n}{4K^2}\frac{t_n^2}{1 + \sqrt{s/18}\,t_n}\right). \qquad (12.55)$$

Let $r_n' := \sqrt{K^2 s/m_n} + r_n$. Then the definition of $\mathbf{w}^{(n)}$ yields

$$\|\mathrm{sgn}(\mathbf{x}_S) - \mathbf{u}_S\|_2 = \|\mathbf{w}^{(L)}\|_2 \le \|\mathrm{sgn}(\mathbf{x}_S)\|_2\prod_{n=1}^{L}r_n' \le \sqrt{s}\prod_{n=1}^{L}r_n'.$$

Furthermore, (12.52) yields

$$\|\mathbf{u}_{\overline{S}}\|_\infty \leq \sum_{n=1}^{L} \|\frac{1}{m_n}(\mathbf{A}_{\overline{S}}^{(n)})^* \mathbf{A}_S^{(n)} \mathbf{w}^{(n-1)}\|_\infty \leq \sum_{n=1}^{L} t_n \|\mathbf{w}^{(n-1)}\|_2$$

$$\leq \sqrt{s} \sum_{n=1}^{L} t_n \prod_{j=1}^{n-1} r_j',$$

with the understanding that $\prod_{j=1}^{n-1} r_j' = 1$ if $n = 1$. Next we need to set the parameters $L, m_1, \ldots, m_L, r_1, \ldots, r_L, t_1, \ldots, t_L$ such that $\|\mathrm{sgn}(\mathbf{x}_S) - \mathbf{u}_S\|_2 \leq \gamma$ and $\|\mathbf{u}_{\overline{S}}\|_\infty \leq \theta$ for some appropriate values of θ and γ in Theorem 4.32. We assume now that the parameters satisfy

$$L = \lceil \ln(s)/2 \rceil + 2,$$
$$m_1 = m_2 \geq cK^2 s \ln(4N) \ln(2\varepsilon^{-1}), \quad m_n \geq cK^2 s \ln(2L\varepsilon^{-1}), \quad n = 3, \ldots, L,$$
$$r_1 = r_2 = \frac{1}{2e\sqrt{\ln(4N)}}, \quad r_n = \frac{1}{2e}, \quad n = 3, \ldots, L,$$
$$t_1 = t_2 = \frac{1}{e\sqrt{s}}, \quad t_n = \frac{\ln(4N)}{e\sqrt{s}}, \quad n = 3, \ldots, L,$$

where $c = 8e^2(1 + (1/\sqrt{8} + 1/6)/e) \approx 70.43$ and $\varepsilon \in (0, 1/6)$. Then $r_1' = r_2' \leq 1/(e\sqrt{\ln(4N)})$ and $r_n' \leq e^{-1}, n = 3, \ldots, L$. Thus,

$$\|\mathrm{sgn}(\mathbf{x}_S) - \mathbf{u}_S\|_2 \leq \sqrt{s} \prod_{n=1}^{L} r_n' \leq \sqrt{s} e^{-\ln(s)/2-2} = e^{-2},$$

and

$$\|\mathbf{u}_{\overline{S}}\|_\infty \leq e^{-1} \left(1 + \frac{1}{e\sqrt{\ln(4N)}} + \sum_{n=2}^{L-1} e^{-n} \right) \leq \frac{e^{-1}}{1 - e^{-1}} = \frac{1}{e - 1}.$$

The probabilities $p_1(n)$ can be estimated as

$$p_1(1), p_1(2) \leq \exp\left(-\frac{m_1 r_1^2}{2K^2 s} \frac{1}{1 + 2\sqrt{K^2 s/m_1} + r_1/3} \right)$$

$$\leq \exp\left(-\frac{c\ln(4N)\ln(2\varepsilon^{-1})}{8e^2 \ln(4N)} \frac{1}{1 + 2(c\ln(4N)\ln(2\varepsilon^{-1}))^{-1/2} + 1/(6e\sqrt{\ln(4N)})} \right)$$

$$\leq \varepsilon/2, \tag{12.56}$$

where the definition of c was used in the last step. A similar estimation gives

$$p_1(n) \leq \varepsilon/(2L), \quad n = 3, \ldots, L.$$

This yields $\sum_{n=1}^{L} p_1(n) \leq 2\varepsilon$. Next, the probabilities $p_2(n)$ can be estimated as

$$p_2(1), p_2(2) \leq 4N \exp\left(-\frac{cK^2 s \ln(4N) \ln(2\varepsilon^{-1})}{4K\left(e^2 s(1 + \sqrt{s/18}/(e\sqrt{s}))\right)}\right)$$

$$= 4N \exp\left(-\frac{c}{4e^2(1 + 1/(e\sqrt{18}))} \ln(4N) \ln(2\varepsilon^{-1})\right)$$

$$\leq 4N \exp\left(-\ln(4N) - \ln(2\varepsilon^{-1})\right) = \varepsilon/2,$$

where we have used that $2ab \geq a + b$ for $a, b \geq 1$. A similar estimate gives

$$p_2(n) \leq \varepsilon/(2L) \quad \text{for } n \geq 3,$$

so that again $\sum_{n=1}^{L} p_2(n) \leq 2\varepsilon$.

If the overall number m is larger than the sum of the right-hand sides of the assumed lower bounds on the m_n's plus L (the latter accounting for the fact that the m_n are integers), then our choice of the m_n's is possible. The latter is the case if

$$m \geq 2cK^2 s \ln(4N) \ln(2\varepsilon^{-1}) + cK^2 \lceil \ln(s)/2 \rceil s \ln(2\lceil \ln(s)/2 \rceil \varepsilon^{-1}) + L,$$

which is implied by

$$m \geq cK^2 s \left[2 \ln(4N) \ln(2\varepsilon^{-1}) + \ln(s) \ln(2e\varepsilon^{-1} \ln(s))\right]. \tag{12.57}$$

By Theorem 12.12, we have $\|\tilde{\mathbf{A}}_S^* \tilde{\mathbf{A}}_S - \mathbf{Id}\|_{2\to2} \leq 1/2$ with probability at least $1 - 2s \exp\left(-\frac{3m}{32K^2 s}\right)$. Hence, the first part of condition (4.25) of Theorem 4.32, $\|(\tilde{\mathbf{A}}_S^* \tilde{\mathbf{A}}_S)^{-1}\|_{2\to2} \leq \alpha = 2$, holds with probability at least $1 - \varepsilon$ provided

$$m \geq \frac{32}{3} K^2 s \ln(2s\varepsilon^{-1}). \tag{12.58}$$

In the notation of Theorem 4.32, we have so far chosen parameters $\alpha = 2$, $\gamma = e^{-2}$, and $\theta = (e - 1)^{-1}$. The condition $\theta + \alpha\beta\gamma < 1$ together with the second part of (4.25) translates into

$$\max_{\ell \in \overline{S}} \|\tilde{\mathbf{A}}_S^* \tilde{\mathbf{a}}_\ell\|_2 \leq \beta$$

with $\beta < \gamma^{-1}\alpha^{-1}(1 - \theta) = e^2(e - 2)/(2(e - 1)) \approx 1.544$. Let us choose $\beta = 3/2$, say. Lemma 12.27 together with $(s + 1) \leq N$ implies that

$$\mathbb{P}\left(\max_{\ell \in \overline{S}} \|\tilde{\mathbf{A}}_S^* \tilde{\mathbf{a}}_\ell\|_2 \geq \beta\right) \leq 2N^2 \exp\left(-\frac{3}{10}\frac{m\beta^2}{K^2 s}\right).$$

This term is bounded by ε provided $m \geq (10/3)\beta^{-2}K^2 s \ln(2N^2/\varepsilon)$, which is implied by

$$m \geq CK^2 s \ln(2N/\varepsilon) \tag{12.59}$$

with $C = 20\beta^{-2}/3 \approx 2.96$.

Altogether we have shown that the conditions of Theorem 4.32 hold with probability at least $1 - 6\varepsilon$ provided (12.57), (12.58), (12.59) are fulfilled. Replacing ε by $\varepsilon/6$, and noting that (12.57) is stronger than (12.58) and (12.59), concludes the proof. $\qquad\square$

Remark 12.30. The proof method is called golfing scheme because the vector $\mathbf{u}^{(n)}$ attains the desired inexact dual vector after a finite number of iterations, just like in golf where the ball (ideally) approaches the hole with every stroke.

Now we modify the previous proof by a clever trick to obtain the main result of this section.

Proof (of Theorem 12.20). We use the basic structure of the previous proof. The strengthening is based on the idea that we can sample slightly more row blocks $\mathbf{A}^{(n)}$ of the matrix $\mathbf{A}$ than in the previous proof. Then we use only those for which (12.53) and (12.54) are satisfied. The probability that these inequalities hold only for a fraction of the samples is much higher than the probability that they hold simultaneously for all sampled blocks. The fact that we choose slightly more blocks will not deteriorate the overall number m of samples—in contrast, it actually decreases m because the size m_n of each block can be made smaller.

To be more precise, we choose a number $L' > L$ of row submatrices to be determined below. As in the previous proof, we set $\mathbf{u}^{(0)} = \mathbf{0}$ and define recursively $\mathbf{u}^{(1)}$ and $\mathbf{u}^{(2)}$ (for $n = 1, 2$ we do not allow replacements) via (12.50). Next we continue with the recursive definition of $\mathbf{u}^{(n)}$, but always check whether the associated vector $\mathbf{w}^{(n)} = \text{sgn}(\mathbf{x}_S) - \mathbf{u}_S^{(n)}$ satisfies (12.53) and (12.54). If these conditions are not satisfied, we discard this particular n in the sense that we replace $\mathbf{A}^{(n)}$ by $\mathbf{A}^{(n+1)}$ (and also all $\mathbf{A}^{(\ell)}$ by $\mathbf{A}^{(\ell+1)}$ for $\ell > n$). Then we redefine $\mathbf{u}^{(n)}$ and $\mathbf{w}^{(n)}$ using the modified $\mathbf{A}^{(n)}$. We continue in this way by always discarding an n when (12.53) and (12.54) are not satisfied, until we arrive at $n = L$ (below we estimate the probability that this actually happens). Since the $\mathbf{A}^{(n)}$ are independent, the events that (12.53) and (12.54) hold for a given $n \in [L']$ are independent.

In comparison with the previous proof, we use a different definition of m_n for $n \geq 3$, namely, we require

$$m_n \geq cK^2 s \ln(2\rho^{-1}),$$

for some $\rho \in (0, 1)$ to be defined below. The remaining quantities L, m_1, m_2, r_n, t_n are defined in the same way as before. Again the probabilities $p_1(1), p_1(2)$, $p_2(1), p_2(2)$ are bounded by $\varepsilon/2$. We need to determine the probability that (12.53) and (12.54) hold for at least $L - 2$ choices of $n \in \{3, 4, \ldots, L'\}$. The modified definition of m_n and similar estimates as in the previous proof give $p_1(n) \leq \rho/2$ and $p_2(n) \leq \rho/2, n \geq 3$, so that the event B_n that both (12.53) and (12.54) hold for a given $n \geq 3$ occurs with probability at least $1 - \rho$. The event that B_n occurs for at least $L - 2$ choices of n has probability larger than the event that

$$\sum_{n=3}^{L'} X_n \geq L - 2,$$

where the X_n are independent random variables that take the value 1 with probability $1 - \rho$ and the value 0 with probability ρ. Clearly, $\mathbb{E}X_n = 1 - \rho$ and $|X_n - \mathbb{E}X_n| \leq 1$ for all n. Setting $J := L' - 2$, Hoeffding's inequality (Theorem 7.20) shows that

$$\mathbb{P}\left(\sum_{n=3}^{L'} X_n < (1 - \rho)J - \sqrt{J}t\right) = \mathbb{P}\left(\sum_{n=3}^{L'} (X_n - \mathbb{E}X_n) < -\sqrt{J}t\right) \leq e^{-t^2/2}.$$

With $t = ((1 - \rho)J + 2 - L)/\sqrt{J}$, this gives

$$\mathbb{P}\left(\sum_{n=3}^{L'} X_n < L - 2\right) \leq \exp\left(-\frac{((1 - \rho)J + 2 - L)^2}{2J}\right).$$

The choice

$$J = \left\lceil \frac{2}{1 - \rho}(L - 2) + \frac{2}{(1 - \rho)^2} \ln(\tilde{\varepsilon}^{-1}) \right\rceil \tag{12.60}$$

implies that the event B_n occurs at least $L - 2$ times with probability at least $1 - \tilde{\varepsilon}$. As in the previous proof, our choice for the m_n's is possible if

$$m \geq 2cK^2 s \ln(4N) \ln(2\varepsilon^{-1}) + JcK^2 s \ln(2\rho^{-1}) + L'.$$

Choosing $\rho = 1/2$, this condition is implied by

$$m \geq 2cK^2 s \ln(4N) \ln(2\varepsilon^{-1}) + \ln(4)cK^2 s(2\ln(s) + 8\ln(\tilde{\varepsilon}^{-1}) + 8). \tag{12.61}$$

Note that with $\tilde{\varepsilon} = \varepsilon$ this condition is stronger than (12.58) and (12.59). Altogether we showed that ℓ_1-minimization recovers the vector $\mathbf{x}$ with probability at least $1 - 4\varepsilon$. Replacing ε by $\varepsilon/4$ and observing that (12.61) is implied by

$$m \geq CK^2 s \ln(N) \ln(\varepsilon^{-1})$$

with an appropriate constant C concludes the proof. □

Let us finally establish stable recovery.

Proof (of Theorem 12.22). The proof is based on the inexact dual condition of Theorem 4.33. We use the golfing scheme of the previous proof, and in particular, we make the same choices of the parameters $L, L', J, m_n, r_n, r'_n, t_n$ and ρ as there. We impose the additional constraint that

$$\frac{m}{m_n} \leq C'(r'_n)^2 \ln(4N), \quad n = 1, \ldots, L', \tag{12.62}$$

for an appropriate constant $C' > 0$. This is possible by the condition on m and by definition of the r_n. Moreover, we again choose $\tilde{\varepsilon} = \varepsilon$ in the definition of J in (12.60).

Let $S \subset [N]$ with $\operatorname{card}(S) = s$ be an index set of s largest coefficients of **x**. Condition (4.27) and the first two conditions of (4.28) of Theorem 4.33 with $\tilde{\mathbf{A}} = \frac{1}{\sqrt{m}}\mathbf{A}$ in place of **A** hold with probability at least $1 - \varepsilon$ with appropriate values of the constants $\delta, \beta, \gamma, \theta$. This follows from the arguments of the previous proofs.

It remains to verify the condition $\|\mathbf{h}\|_2 \leq \tau\sqrt{s}$ in (4.28) for the vector $\tilde{\mathbf{h}} \in \mathbb{C}^m$ used to construct $\mathbf{u} = \tilde{\mathbf{A}}^*\tilde{\mathbf{h}}$ in the previous proof. For notational simplicity, we assume that the first L values of n are taken for the construction of the inexact dual, that is, **u** is given by (12.50). Using the rescaled matrices $\tilde{\mathbf{A}}^{(n)} = \frac{1}{\sqrt{m}}\mathbf{A}^{(n)}$ gives

$$\mathbf{u} = \sum_{n=1}^{L} \frac{1}{m_n}(\mathbf{A}^{(n)})^* \mathbf{A}_S^{(n)} \mathbf{w}^{(n-1)} = \sum_{n=1}^{L} \frac{m}{m_n}(\tilde{\mathbf{A}}^{(n)})^* \tilde{\mathbf{A}}_S^{(n)} \mathbf{w}^{(n-1)}.$$

Hence, $\mathbf{u} = \tilde{\mathbf{A}}^*\mathbf{h}$ with $\tilde{\mathbf{h}} = [(\mathbf{h}^{(1)})^*, \ldots, (\mathbf{h}^{(L)})^*, 0, \ldots, 0]^*$ and

$$\mathbf{h}^{(n)} = \frac{m}{m_n}\tilde{\mathbf{A}}_S^{(n)}\mathbf{w}^{(n-1)} \in \mathbb{C}^{m_n}, \quad n = 1, \ldots, L.$$

This yields

$$\|\tilde{\mathbf{h}}\|_2^2 = \sum_{n=1}^{L}\|\mathbf{h}^{(n)}\|_2^2 = \sum_{n=1}^{L}\frac{m}{m_n}\left\|\sqrt{\frac{m}{m_n}}\tilde{\mathbf{A}}_S^{(n)}\mathbf{w}^{(n-1)}\right\|_2^2$$

$$= \sum_{n=1}^{L}\frac{m}{m_n}\left\|\sqrt{\frac{1}{m_n}}\mathbf{A}_S^{(n)}\mathbf{w}^{(n-1)}\right\|_2^2.$$

In view of the relation (12.51) for the vectors $\mathbf{w}^n$, we obtain, for $n \geq 1$,

$$\left\| \sqrt{\frac{1}{m_n}} \mathbf{A}_S^{(n)} \mathbf{w}^{(n-1)} \right\|_2^2 = \left\langle \frac{1}{m_n} (\mathbf{A}_S^{(n)})^* \mathbf{A}_S^{(n)} \mathbf{w}^{(n-1)}, \mathbf{w}^{(n-1)} \right\rangle$$

$$= \left\langle \left(\frac{1}{m_n} (\mathbf{A}_S^{(n)})^* \mathbf{A}_S^{(n)} - \mathbf{Id} \right) \mathbf{w}^{(n-1)}, \mathbf{w}^{(n-1)} \right\rangle + \| \mathbf{w}^{(n-1)} \|_2^2$$

$$= \left\langle -\mathbf{w}^{(n)}, \mathbf{w}^{(n-1)} \right\rangle + \| \mathbf{w}^{(n-1)} \|_2^2 \leq \| \mathbf{w}^{(n)} \|_2 \| \mathbf{w}^{(n-1)} \|_2 + \| \mathbf{w}^{(n-1)} \|_2^2.$$

Recall from (12.53) that $\| \mathbf{w}^{(n)} \|_2 \leq r_n' \| \mathbf{w}^{(n-1)} \|_2 \leq \| \mathbf{w}^{(n-1)} \|_2$ for all n (except for an event of probability at most ε). This gives

$$\left\| \sqrt{\frac{1}{m_n}} \mathbf{A}_S^{(n)} \mathbf{w}^{(n-1)} \right\|_2^2 \leq 2 \| \mathbf{w}^{(n-1)} \|_2^2 \leq 2 \| \mathbf{w}^{(0)} \|_2^2 \prod_{j=1}^{n-1} (r_j')^2$$

$$= 2 \| \mathrm{sgn}(\mathbf{x}_S) \|_2^2 \prod_{j=1}^{n-1} (r_j')^2 = 2s \prod_{j=1}^{n-1} (r_j')^2.$$

The definition of the constants r_n and the additional constraint (12.62) therefore yield

$$\| \tilde{\mathbf{h}} \|_2^2 \leq 2s \sum_{n=1}^{L} \frac{m}{m_n} \prod_{j=1}^{n-1} (r_j')^2 \leq C's \ln(4N) \sum_{n=1}^{L} (r_n')^2 \prod_{j=1}^{n-1} (r_j')^2$$

$$\leq C'e^{-2}s \sum_{n=1}^{L} \prod_{j=2}^{n-1} (r_j')^2 \leq C''s,$$

where we used that $\prod_{j=2}^{n} (r_j')^2 \leq e^{-2(n-1)}$ for $n \geq 2$ and the convention that $\prod_{j=2}^{1} (r_j')^2 = 1$. Therefore, all conditions of Theorem 4.33 are satisfied for $\mathbf{x}$ and $\tilde{\mathbf{A}}$ with probability at least $1 - \varepsilon$. Noting that the optimization problem

$$\underset{\mathbf{z} \in \mathbb{C}^N}{\text{minimize}} \ \| \mathbf{z} \|_1 \quad \text{subject to} \quad \left\| \tilde{\mathbf{A}} \mathbf{z} - \frac{1}{\sqrt{m}} \mathbf{y} \right\|_2 \leq \eta$$

is equivalent to (12.45) completes the proof. $\qquad\qquad \square$

12.5   Restricted Isometry Property

In this section we derive an estimate for the restricted isometry constants of the (renormalized) random sampling matrix $\mathbf{A}$ in (12.4) associated to a BOS. This leads to stable and uniform sparse recovery via ℓ_1-minimization and other algorithms. The main result reads as follows.

Theorem 12.31. *Let* $\mathbf{A} \in \mathbb{C}^{m \times N}$ *be the random sampling matrix associated to a BOS with constant* $K \geq 1$. *If, for* $\delta \in (0, 1)$,

$$m \geq CK^2\delta^{-2}s\ln^4(N), \tag{12.63}$$

then with probability at least $1 - N^{-\ln^3(N)}$ *the restricted isometry constant* δ_s *of* $\frac{1}{\sqrt{m}}\mathbf{A}$ *satisfies* $\delta_s \leq \delta$. *The constant* $C > 0$ *is universal.*

The above theorem follows from the more precise result stated next by choosing $\varepsilon = N^{-\ln^3(N)}$.

Theorem 12.32. *Let* $\mathbf{A} \in \mathbb{C}^{m \times N}$ *be the random sampling matrix associated to a BOS with constant* $K \geq 1$. *For* $\varepsilon, \eta_1, \eta_2 \in (0, 1)$, *if*

$$\frac{m}{\ln(9m)} \geq C_1 \eta_1^{-2} K^2 s \ln^2(4s) \ln(8N), \tag{12.64}$$

$$m \geq C_2 \eta_2^{-2} K^2 s \ln(\varepsilon^{-1}), \tag{12.65}$$

then with probability at least $1 - \varepsilon$ *the restricted isometry constant* δ_s *of* $\frac{1}{\sqrt{m}}\mathbf{A}$ *satisfies* $\delta_s \leq \eta_1 + \eta_1^2 + \eta_2$. *The constants may be chosen as* $C_2 = 32/3 \approx 10.66$ *and* $C_1 = c_0 C^2 \approx 5576$ *where* $c_0 = 174.24$ *and* $C = 4\sqrt{2}$ *is the constant from Dudley's inequality in Theorem 8.23.*

Remark 12.33. (a) The constant C_1 in the previous result is large and certainly not optimal. However, an improvement will probably be cumbersome and will not provide more insight.

(b) Instead of (12.64), one usually prefers a condition that features only m on the left-hand side. The conditions in (12.64) and (12.65) are in fact implied by

$$m \geq CK^2\delta^{-2}s\max\{\ln^2(s)\ln(K^2\delta^{-2}s\ln(N))\ln(N), \ln(\varepsilon^{-1})\} \tag{12.66}$$

for some appropriate constant $C > 0$, which is slightly better than (12.63). Indeed, the monotonicity of $x \mapsto x/\ln(x)$ on $[0, \infty)$ shows that (12.66) gives

$$\frac{m}{\ln(9m)} \geq CK^2\delta^{-2}\frac{s\ln^2(s)\ln(K^2\delta^{-2}s\ln(N))\ln(N)}{\ln(9C) + \ln(K^2\delta^{-2}s\ln(N)) + 2\ln(\ln(s)) + \ln(\ln(N))}$$
$$\geq C'K^2\delta^{-2}s\ln^2(s)\ln(N).$$

If $s \geq K^2\delta^{-2}\ln(N)$ so that s is not tiny, then $\ln(K^2\delta^{-2}s\ln(N)) \leq 2\ln(s)$. Therefore, in this case the simpler condition

$$m \geq C''K^2\delta^{-2}s\ln^3(s)\ln(N)$$

for some appropriate constant $C'' > 0$ implies $\delta_s \leq \delta$ with probability at least $1 - N^{-\ln^3(s)}$.

Using the theory of Chap. 6, we obtain the following result concerning recovery of sparse expansions with respect to a BOS from random samples.

Corollary 12.34. *Suppose that*

$$m \geq CK^2 s \ln^4(N)$$

for a universal constant $C > 0$. Then

(a) *With probability at least $1 - N^{-\ln^3(N)}$, every s-sparse vector $\mathbf{x} \in \mathbb{C}^N$ is exactly recovered from the samples $\mathbf{y} = \mathbf{Ax} = \left(\sum_{j=1}^{N} x_j \phi_j(t_\ell) \right)_{\ell=1}^{m}$ by basis pursuit.*

(b) *More generally, with probability at least $1 - N^{-\ln^3(N)}$, every $\mathbf{x} \in \mathbb{C}^N$ is approximately recovered from the inaccurate samples $\mathbf{y} = \mathbf{Ax} + \mathbf{e}$, $\|\mathbf{e}\|_2 \leq \eta\sqrt{m}$, as a solution $\mathbf{x}^\sharp$ of*

$$\underset{\mathbf{z} \in \mathbb{C}^N}{\text{minimize}} \; \|\mathbf{z}\|_1 \quad \text{subject to} \; \|\mathbf{Az} - \mathbf{y}\|_2 \leq \eta\sqrt{m} \qquad (12.67)$$

in the sense that

$$\|\mathbf{x} - \mathbf{x}^\sharp\|_p \leq \frac{C_1}{s^{1-1/p}} \sigma_s(\mathbf{x})_1 + C_2 s^{1/p-1/2}\eta, \quad 1 \leq p \leq 2.$$

The constants $C, C_1, C_2 > 0$ are universal.

Proof. It suffices to combine Theorem 12.32 with Theorem 6.12 for the normalized matrix $\tilde{\mathbf{A}} = \frac{1}{\sqrt{m}}\mathbf{A}$. □

Remark 12.35. (a) The assumption $\|\mathbf{e}\|_2 \leq \eta\sqrt{m}$ on the measurement error is satisfied if each sample is taken with accuracy η, that is, $|y_\ell - f(t_\ell)| \leq \eta$; see Remark 12.23.

(b) Of course, the above result applies verbatim to the other algorithms for which recovery under restricted isometry conditions has been shown in Chap. 6. This includes iterative hard thresholding, hard thresholding pursuit, orthogonal matching pursuit, and compressive sampling matching pursuit (CoSaMP).

Compared to the nonuniform recovery conditions of Theorems 12.18, 12.20, and 12.22, we pay some extra logarithmic factors, but we gain uniform recovery and we improve on the stability estimate. Compared to the condition of Theorem 9.11 concerning the restricted isometry property of subgaussian random matrices which involves only the factor $\ln(eN/s)$, we also introduce additional logarithmic factors.

In the remainder of this section, we develop the proof of Theorem 12.32. We first note that—unlike in the case of Gaussian (or subgaussian) random matrices—the strategy of taking the probabilistic bound (12.31) for the condition number of a single column submatrix and then applying the union bound over all collections of s-element subsets of the N columns of $\mathbf{A}$ only leads to a poor estimate of the number of samples m ensuring sparse recovery; see Exercise 12.7. Indeed, the bound (12.95) for m scales quadratically in s, while the desired estimate (12.63) obeys a linear scaling up to logarithmic factors. Below we pursue a different strategy that uses Dudley's inequality (Theorem 8.23) as a main tool.

Proof (of Theorem 12.32, first part). We use the characterization of the restricted isometry constants in (6.2), namely,

$$\delta_s = \max_{S \subset [N], \mathrm{card}(S)=s} \|\tilde{\mathbf{A}}_S^* \tilde{\mathbf{A}}_S - \mathbf{Id}\|_{2 \to 2},$$

where again $\tilde{\mathbf{A}} = \frac{1}{\sqrt{m}} \mathbf{A}$. Let us introduce the set

$$D_{s,N} := \{ \mathbf{z} \in \mathbb{C}^N : \|\mathbf{z}\|_2 \leq 1, \|\mathbf{z}\|_0 \leq s \} = \bigcup_{S \subset [N], \mathrm{card}(S)=s} B_S, \qquad (12.68)$$

where B_S denotes the unit sphere in $\mathbb{C}^S$ with respect to the ℓ_2-norm. The quantity

$$\|\mathbf{B}\|_s := \sup_{\mathbf{z} \in D_{s,N}} |\langle \mathbf{Bz}, \mathbf{z} \rangle|$$

defines a seminorm on $\mathbb{C}^{N \times N}$ and $\delta_s = \|\tilde{\mathbf{A}}^* \tilde{\mathbf{A}} - \mathbf{Id}\|_s$.

Let $\mathbf{X}_\ell = \left(\overline{\phi_j(\mathbf{t}_\ell)} \right)_{j=1}^N \in \mathbb{C}^N$ be a random column vector associated to the sampling point $\mathbf{t}_\ell$, $\ell \in [m]$. Note that $\mathbf{X}_\ell^*$ is a row of $\mathbf{A}$, so that $\mathbf{A}^* \mathbf{A} = \sum_{\ell=1}^m \mathbf{X}_\ell \mathbf{X}_\ell^*$. The orthogonality relation (12.1) implies $\mathbb{E} \mathbf{X}_\ell \mathbf{X}_\ell^* = \mathbf{Id}$. We can express the restricted isometry constant of $\tilde{\mathbf{A}}$ as

$$\delta_s = \|\frac{1}{m} \sum_{\ell=1}^m \mathbf{X}_\ell \mathbf{X}_\ell^* - \mathbf{Id}\|_s = \frac{1}{m} \|\sum_{\ell=1}^m (\mathbf{X}_\ell \mathbf{X}_\ell^* - \mathbb{E} \mathbf{X}_\ell \mathbf{X}_\ell^*)\|_s. \qquad (12.69)$$

Let us first consider the expectation of δ_s. Using symmetrization (Lemma 8.4), we estimate

$$\mathbb{E} \|\sum_{\ell=1}^m (\mathbf{X}_\ell \mathbf{X}_\ell^* - \mathbb{E} \mathbf{X}_\ell \mathbf{X}_\ell^*)\|_s \leq 2 \mathbb{E} \|\sum_{\ell=1}^m \epsilon_\ell \mathbf{X}_\ell \mathbf{X}_\ell^*\|_s, \qquad (12.70)$$

where $\epsilon = (\epsilon_1, \ldots, \epsilon_m)$ is a Rademacher sequence independent from the random sampling points $\mathbf{t}_\ell$, $\ell \in [m]$. The following lemma, which heavily relies on Dudley's inequality, is key to the estimate of the expectation above.

Lemma 12.36. *Let $\mathbf{x}_1, \ldots, \mathbf{x}_m$ be vectors in $\mathbb{C}^N$ with $\|\mathbf{x}_\ell\|_\infty \leq K$ for all $\ell \in [m]$. Then, for $s \leq m$,*

$$\mathbb{E}\|\sum_{\ell=1}^{m}\epsilon_\ell\mathbf{x}_\ell\mathbf{x}_\ell^*\|_s \leq C_1 K\sqrt{s}\ln(4s)\sqrt{\ln(8N)\ln(9m)}\sqrt{\|\sum_{\ell=1}^{m}\mathbf{x}_\ell\mathbf{x}_\ell^*\|_s}, \quad (12.71)$$

where $C_1 = \sqrt{2}C_0 C = 8C_0 \approx 26.4$. Here, $C = 4\sqrt{2}$ is the constant in Dudley's inequality and $C_0 = 3.3$.

Proof. Observe that

$$E := \mathbb{E}\|\sum_{\ell=1}^{m}\epsilon_\ell\mathbf{x}_\ell\mathbf{x}_\ell^*\|_s = \mathbb{E}\sup_{\mathbf{u}\in D_{s,N}}\left|\sum_{\ell=1}^{m}\epsilon_\ell|\langle\mathbf{x}_\ell,\mathbf{u}\rangle|^2\right|.$$

This is the supremum of the Rademacher process $X_\mathbf{u} = \sum_{\ell=1}^{m}\epsilon_\ell|\langle\mathbf{x}_\ell,\mathbf{u}\rangle|^2$, which has associated pseudometric

$$d(\mathbf{u},\mathbf{v}) = \left(\mathbb{E}|X_\mathbf{u} - X_\mathbf{v}|^2\right)^{1/2} = \sqrt{\sum_{\ell=1}^{m}\left(|\langle\mathbf{x}_\ell,\mathbf{u}\rangle|^2 - |\langle\mathbf{x}_\ell,\mathbf{v}\rangle|^2\right)^2},$$

see also (8.44). Then, for $\mathbf{u},\mathbf{v} \in D_{s,N}$, the triangle inequality gives

$$d(\mathbf{u},\mathbf{v}) = \left(\sum_{\ell=1}^{m}(|\langle\mathbf{x}_\ell,\mathbf{u}\rangle| - |\langle\mathbf{x}_\ell,\mathbf{v}\rangle|)^2(|\langle\mathbf{x}_\ell,\mathbf{u}\rangle| + |\langle\mathbf{x}_\ell,\mathbf{v}\rangle|)^2\right)^{1/2}$$

$$\leq \max_{\ell\in[m]}||\langle\mathbf{x}_\ell,\mathbf{u}\rangle| - |\langle\mathbf{x}_\ell,\mathbf{v}\rangle||\sup_{\mathbf{u}',\mathbf{v}'\in D_{s,N}}\sqrt{\sum_{\ell=1}^{m}(|\langle\mathbf{x}_\ell,\mathbf{u}'\rangle| + |\langle\mathbf{x}_\ell,\mathbf{v}'\rangle|)^2}$$

$$\leq 2R\max_{\ell\in[m]}|\langle\mathbf{x}_\ell,\mathbf{u}-\mathbf{v}\rangle|,$$

where

$$R = \sup_{\mathbf{u}'\in D_{s,N}}\sqrt{\sum_{\ell=1}^{m}|\langle\mathbf{x}_\ell,\mathbf{u}'\rangle|^2} = \sqrt{\|\sum_{\ell=1}^{m}\mathbf{x}_\ell\mathbf{x}_\ell^*\|_s}. \quad (12.72)$$

We further introduce the auxiliary seminorm

$$\|\mathbf{u}\|_X := \max_{\ell\in[m]}|\langle\mathbf{x}_\ell,\mathbf{u}\rangle|, \quad \mathbf{u}\in\mathbb{C}^N. \quad (12.73)$$

We derived that the rescaled process $X_\mathbf{u}/(2R)$ satisfies

$$\left(\mathbb{E}|X_\mathbf{u}/(2R) - X_\mathbf{v}/(2R)|^2\right)^{1/2} \leq \|\mathbf{u}-\mathbf{v}\|_X.$$

It follows from Dudley's inequality (8.48) that

$$E \leq 2CR \int_0^{\Delta(D_{s,N}, \|\cdot\|_X)/2} \sqrt{\ln(2\mathcal{N}(D_{s,N}, \|\cdot\|_X, t))} dt, \tag{12.74}$$

with $C = 4\sqrt{2}$ and where $\Delta(D_{s,N}, \|\cdot\|_X) = \sup_{\mathbf{u} \in D_{s,N}} \|\mathbf{u}\|_X$. By the Cauchy–Schwarz inequality, for $\mathbf{u} \in D_{s,N}$,

$$\|\mathbf{u}\|_X = \max_{\ell \in [m]} |\langle \mathbf{x}_\ell, \mathbf{u} \rangle| \leq \|\mathbf{u}\|_1 \max_{\ell \in [m]} \|\mathbf{x}_\ell\|_\infty \leq K\sqrt{s}\|\mathbf{u}\|_2 \leq K\sqrt{s}. \tag{12.75}$$

Therefore, the radius satisfies

$$\Delta(D_{s,N}, \|\cdot\|_X) \leq K\sqrt{s}.$$

Our next task is to estimate the covering numbers $\mathcal{N}(D_{s,N}, \|\cdot\|_X, t)$. We do this in two different ways. One estimate is good for small values of t and the other one for large values of t. For small values of t we use a volumetric argument, that is, Proposition C.3. Note that $\|\mathbf{x}\|_X \leq K\sqrt{s}\|\mathbf{x}\|_2$ for $\mathbf{x} \in D_{s,N}$ by (12.75). Using subadditivity (C.4) of the covering numbers, we obtain

$$\mathcal{N}(D_{s,N}, \|\cdot\|_X, t) \leq \sum_{S \subset [N], \mathrm{card}(S)=s} \mathcal{N}(B_S, K\sqrt{s}\|\cdot\|_2, t)$$

$$= \sum_{S \subset [N], \mathrm{card}(S)=s} \mathcal{N}\left(B_S, \|\cdot\|_2, \frac{t}{K\sqrt{s}}\right) \leq \binom{N}{s}\left(1 + \frac{2K\sqrt{s}}{t}\right)^{2s}$$

$$\leq \left(\frac{eN}{s}\right)^s \left(1 + \frac{2K\sqrt{s}}{t}\right)^{2s}.$$

Hereby, we have also used the covering number estimate of Lemma C.3 (noting that we treat the s-dimensional complex unit ball as real $2s$-dimensional unit ball by isometry) and the bound of the binomial coefficient in Lemma C.5. We arrive at

$$\sqrt{\ln(2\mathcal{N}(D_{s,N}, \|\cdot\|_X, t))} \leq \sqrt{2s}\sqrt{\ln(2eN/s) + \ln(1 + 2K\sqrt{s}/t)}$$

$$\leq \sqrt{2s}\left(\sqrt{\ln(2eN/s)} + \sqrt{\ln(1 + 2K\sqrt{s}/t)}\right), \quad t > 0. \tag{12.76}$$

For large values of t, we introduce the norm

$$\|\mathbf{z}\|_1^* := \sum_{j=1}^N (|\operatorname{Re}(z_j)| + |\operatorname{Im}(z_j)|), \qquad \mathbf{z} \in \mathbb{C}^N,$$

which is the usual ℓ_1-norm after identification of $\mathbb{C}^N$ with $\mathbb{R}^{2N}$. Then by the Cauchy–Schwarz inequality, we have the embedding

$$D_{s,N} \subset \sqrt{2s}B^N_{\|\cdot\|^*_1} = \{\mathbf{x} \in \mathbb{C}^N, \|\mathbf{x}\|^*_1 \le \sqrt{2s}\}.$$

Lemma 12.37 below implies that, for $0 < t < 2K\sqrt{s}$,

$$\sqrt{\ln(2\mathcal{N}(D_{s,N}, \|\cdot\|_X, t))} \le 6K\sqrt{2s}\sqrt{\ln(9m)\ln(8N)}t^{-1}. \qquad (12.77)$$

Next we combine inequalities (12.76) and (12.77) to estimate the Dudley integral in (12.74). Recalling that $\Delta(D_{s,N}, \|\cdot\|_X) \le K\sqrt{s}$, we obtain, for an arbitrary $\kappa \in (0, \Delta(D_{s,N}, \|\cdot\|_X)/2)$,

$$I := \int_0^{\Delta(D_{s,N},\|\cdot\|_X)/2} \sqrt{\ln(2\mathcal{N}(D_{s,N}, \|\cdot\|_X, t))}dt$$

$$\le \sqrt{2s}\int_0^\kappa \left(\sqrt{\ln(2eN/s)} + \sqrt{\ln\left(1 + 2K\sqrt{s}/t\right)}\right)dt$$

$$+ 6K\sqrt{2s\ln(9m)\ln(8N)}\int_\kappa^{K\sqrt{s}/2} t^{-1}dt$$

$$\le \sqrt{2s}\left(\kappa\sqrt{\ln(2eN/s)} + \kappa\sqrt{\ln(e(1 + 2K\sqrt{s}/\kappa))}\right.$$

$$\left.+ 6K\sqrt{\ln(9m)\ln(8N)}\ln(K\sqrt{s}/(2\kappa))\right).$$

Hereby, we have applied Lemma C.9. The choice $\kappa = K/3$ yields

$$I \le \sqrt{2s}K\left(\frac{1}{3}\sqrt{\ln(2eN/s)} + \frac{1}{3}\sqrt{\ln(e(1 + 6\sqrt{s}))}\right.$$

$$\left.+ 6\sqrt{\ln(9m)\ln(8N)}\frac{1}{2}\ln(9s/4)\right)$$

$$\le \sqrt{2s}KC_0\sqrt{\ln(9m)\ln(8N)}\ln(4s),$$

where (tacitly assuming $N \ge 4, m \ge 3$—otherwise the statement is not interesting)

$$C_0 := \frac{1}{3\sqrt{\ln(27)}\ln(4)} + \frac{1}{3}\sqrt{1 + \frac{\ln(7e/2)}{\ln(2)}}\frac{1}{\sqrt{\ln(27)}\ln(40)\ln(4)} + 3 < 3.3.$$

Combining the above estimates with (12.74) and the definition (12.72) of R completes the proof of Lemma 12.36 with $C_1 = \sqrt{2}C_0C = 8C_0 < 26.4$. $\square$

The above proof relied on the following covering number bound.

Lemma 12.37. *Let U be a subset of $B_{\|\cdot\|_1^*}^N$ and $\|\cdot\|_X$ be the seminorm in (12.73) defined for vectors $\mathbf{x}_\ell$ with $\|\mathbf{x}_\ell\|_\infty \leq K$, $\ell \in [m]$. Then, for $0 < t < \sqrt{2}K$,*

$$\sqrt{\ln(2\mathcal{N}(U, \|\cdot\|_X, t))} \leq 6K\sqrt{\ln(9m)\ln(8N)}t^{-1}.$$

Proof. The idea is to approximate a fixed $\mathbf{x} \in U$ by a finite set of very sparse vectors. In order to find a vector $\mathbf{z}$ from this finite set that is close to $\mathbf{x}$, we use the so-called *empirical method of Maurey*. To this end, we observe that $B_{\|\cdot\|_1^*}$ is the convex hull of $V := \{\pm\mathbf{e}_j, \pm i\mathbf{e}_j : j \in [N]\}$, where $\mathbf{e}_j$ denotes the jth canonical unit vector in $\mathbb{C}^N$. Hence, we can write $\mathbf{x}$ as the convex combination $\mathbf{x} = \sum_{j=1}^{4N} \lambda_j \mathbf{v}_j$ with $\lambda_j \geq 0$ and $\sum_{j=1}^{4N} \lambda_j = 1$ and where the $\mathbf{v}_j$ list the $4N$ elements of V. We define a random vector $\mathbf{Z}$ that takes the value $\mathbf{v}_j \in V$ with probability λ_j. Since $\sum_j \lambda_j = 1$, this is a valid probability distribution. Note that

$$\mathbb{E}\mathbf{Z} = \sum_{j=1}^{4N} \lambda_j \mathbf{v}_j = \mathbf{x}.$$

Let $\mathbf{Z}_1, \ldots, \mathbf{Z}_M$ be independent copies of $\mathbf{Z}$, where M is a number to be determined later. We attempt to approximate $\mathbf{x}$ with the M-sparse vector

$$\mathbf{z} = \frac{1}{M} \sum_{k=1}^{M} \mathbf{Z}_k.$$

We estimate the expected distance of $\mathbf{z}$ to $\mathbf{x}$ in $\|\cdot\|_X$ by first using symmetrization (Lemma 8.4) to obtain

$$\mathbb{E}\|\mathbf{z} - \mathbf{x}\|_X = \mathbb{E}\|\frac{1}{M}\sum_{k=1}^{M}(\mathbf{Z}_k - \mathbb{E}\mathbf{Z}_k)\|_X \leq \frac{2}{M}\mathbb{E}\|\sum_{k=1}^{M}\epsilon_k\mathbf{Z}_k\|_X$$

$$= \frac{2}{M}\mathbb{E}\max_{\ell\in[m]}\left|\sum_{k=1}^{M}\epsilon_k\langle\mathbf{x}_\ell,\mathbf{Z}_k\rangle\right|,$$

where ϵ is a Rademacher sequence independent of $(\mathbf{Z}_1, \ldots, \mathbf{Z}_M)$. Now we fix a realization of $(\mathbf{Z}_1, \ldots, \mathbf{Z}_M)$ and consider for the moment only expectation and probability with respect to ϵ, that is, conditional on $(\mathbf{Z}_1, \ldots, \mathbf{Z}_M)$. Since $\|\mathbf{x}_\ell\|_\infty \leq K$ and $\mathbf{Z}_k$ has exactly one nonzero component of magnitude 1, we have $|\langle\mathbf{x}_\ell, \mathbf{Z}_k\rangle| \leq \|\mathbf{x}_\ell\|_\infty\|\mathbf{Z}_k\|_1 \leq K$. It follows that

$$\|(\langle\mathbf{x}_\ell, \mathbf{Z}_k\rangle)_{k=1}^M\|_2 \leq \sqrt{M}K, \quad \ell \in [m].$$

It follows from Theorem 8.8 that the random variable $Y_\ell := \sum_{k=1}^{M} \epsilon_k \langle \mathbf{x}_\ell, \mathbf{Z}_k \rangle$ satisfies (conditional on the $\mathbf{Z}_k$),

$$\mathbb{P}_\epsilon(|Y_\ell| \geq \sqrt{M}Kt) \leq 2e^{-t^2/2}, \quad t > 0.$$

Therefore, by the union bound

$$\mathbb{P}_\epsilon(\max_{\ell \in [m]} |Y_\ell| \geq \sqrt{M}Kt) \leq 2m\, e^{-t^2/2}.$$

Proposition 7.14 yields then

$$\mathbb{E}_\epsilon \max_{\ell \in [m]} \left| \sum_{k=1}^{M} \epsilon_k \langle \mathbf{x}_\ell, \mathbf{Z}_k \rangle \right| \leq C\sqrt{M}K \sqrt{\ln(8m)}.$$

with $C = 3/2$. (In the case of a real BOS one can reduce our considerations to the real space $\mathbb{R}^N$. Then we may use Proposition 7.29 to obtain the slightly better estimate $\mathbb{E}_\epsilon \max_{\ell \in [m]} \left| \sum_{k=1}^{M} \epsilon_k \langle \mathbf{x}_\ell, \mathbf{Z}_k \rangle \right| \leq \sqrt{2MK^2 \ln(2m)}$.) By Fubini's theorem, we finally obtain

$$\mathbb{E}\|\mathbf{z} - \mathbf{x}\|_X \leq \frac{2}{M}\mathbb{E}_\mathbf{Z}\mathbb{E}_\epsilon \max_{\ell \in [m]} \left| \sum_{k=1}^{M} \epsilon_k \langle \mathbf{x}_\ell, \mathbf{Z}_k \rangle \right| \leq \frac{3K}{\sqrt{M}} \sqrt{\ln(8m)}.$$

This implies that there exists a vector of the form

$$\mathbf{z} = \frac{1}{M} \sum_{k=1}^{M} \mathbf{z}_k, \tag{12.78}$$

where each $\mathbf{z}_k$ is one of the vectors in $\{\pm\mathbf{e}_j, \pm i\mathbf{e}_j : j \in [N]\}$, such that

$$\|\mathbf{z} - \mathbf{x}\|_X \leq \frac{3K}{\sqrt{M}} \sqrt{\ln(8m)}. \tag{12.79}$$

In particular,

$$\|\mathbf{z} - \mathbf{x}\|_X \leq t/2 \tag{12.80}$$

provided

$$\frac{3K}{\sqrt{M}} \sqrt{\ln(8m)} \leq t/2. \tag{12.81}$$

For each $\mathbf{x} \in U$, we can therefore find a vector $\mathbf{z}$ of the form (12.78) such that $\|\mathbf{x} - \mathbf{z}\|_X \leq t/2$. Each $\mathbf{z}_k$ can take $4N$ values, so that $\mathbf{z}$ can take at most $(4N)^M$ values. The definition of the covering numbers requires that each point of the covering belongs to U as well, but we only know that the points $\mathbf{z}$ are contained in $B_{\|\cdot\|_1^*}^N$. We can correct for this by replacing each point $\mathbf{z}$ by a point $\mathbf{z}' \in U$ with $\|\mathbf{z} - \mathbf{z}'\|_X \leq t/2$ provided such a point exists. If such a point $\mathbf{z}'$ does not exist, then we simply discard $\mathbf{z}$ as it will not be needed for the covering of U. Then for every $\mathbf{x} \in U$ we can find a point $\mathbf{z}' \in U$ from the new covering such that $\|\mathbf{x} - \mathbf{z}'\|_X \leq \|\mathbf{x} - \mathbf{z}\|_X + \|\mathbf{z} - \mathbf{z}'\|_X \leq t$. Again the number of points $\mathbf{z}'$ of the covering is bounded by $(4N)^M$.

The choice

$$M = \left\lfloor \frac{36K^2}{t^2} \ln(9m) \right\rfloor$$

satisfies (12.81). Indeed, then

$$M \geq \frac{36K^2}{t^2} \ln(9m) - 1 \geq \frac{36K^2}{t^2} \ln(8m) + \frac{36K^2 \ln(9/8)}{t^2} - 1$$

$$\geq \frac{36K^2}{t^2} \ln(8m) + 18 \ln(9/8) - 1 \geq \frac{36K^2}{t^2} \ln(8m)$$

since $t \leq \sqrt{2}K$ and $18 \ln(9/8) \approx 2.12 > 1$. We deduce that the covering numbers can be estimated by

$$\sqrt{\ln(2N(U, \|\cdot\|_X, t))} \leq \sqrt{\ln(2(4N)^M)} \leq \sqrt{\left\lfloor \frac{36K^2}{t^2} \ln(9m) \right\rfloor \ln(8N)}$$

$$\leq 6K\sqrt{\ln(9m)\ln(8N)}t^{-1}.$$

This completes the proof of Lemma 12.37. □

Proof (of Theorem 12.32, Second Part). We proceed in two steps.

Estimate of Expectation. Recall from (12.69) that

$$E := \mathbb{E}\delta_s = m^{-1}\mathbb{E}\left\| \sum_{\ell=1}^{m} (\mathbf{X}_\ell \mathbf{X}_\ell^* - \mathrm{Id}) \right\|_s$$

Set $G(K, s, m, N) = K\sqrt{s}\ln(4s)\sqrt{\ln(8N)\ln(9m)}$. Then Fubini's theorem, (12.70) and Lemma 12.36 imply that

$$E \leq \frac{2}{m}\mathbb{E}_X\mathbb{E}_\epsilon \left\| \sum_{\ell=1}^{m} \epsilon_\ell \mathbf{X}_\ell \mathbf{X}_\ell^* \right\|_s \leq \frac{2C_1 G(K, s, m, N)}{\sqrt{m}}\mathbb{E}_X\sqrt{\left\| m^{-1}\sum_{\ell=1}^{m} \mathbf{X}_\ell \mathbf{X}_\ell^* \right\|_s}.$$

Inserting the identity **Id**, applying the triangle inequality, and using the Cauchy–Schwarz inequality for expectations we obtain

$$E \leq 2C_1 \frac{G(K,s,m,N)}{\sqrt{m}} \mathbb{E} \sqrt{m^{-1} \|\sum_{\ell=1}^{m} (\mathbf{X}_\ell \mathbf{X}_\ell^* - \mathbf{Id})\|_s + 1}$$

$$\leq 2C_1 \frac{G(K,s,m,N)}{\sqrt{m}} \sqrt{E+1}.$$

Setting $D := 2C_1 G(K,s,m,N)/\sqrt{m}$, this reads $E \leq D\sqrt{E+1}$. Squaring this inequality and completing the squares yields $(E - D^2/2)^2 \leq D^2 + D^4/4$, which gives

$$E \leq \sqrt{D^2 + D^4/4} + D^2/2 \leq D + D^2. \tag{12.82}$$

If

$$D = \frac{2C_1 K \sqrt{2s} \ln(4s) \sqrt{\ln(9m) \ln(8N)}}{\sqrt{m}} \leq \eta_1 \tag{12.83}$$

for some $\eta_1 \in (0,1)$, then

$$E = \mathbb{E}\delta_s \leq \eta_1 + \eta_1^2.$$

Probability Estimate. It remains to show that δ_s does not deviate much from its expectation with high probability. To this end, we use the deviation inequality of Theorem 8.42. By definition of the norm $\|\cdot\|_s$, we can write

$$m\delta_s = \|\sum_{\ell=1}^{m} (\mathbf{X}_\ell \mathbf{X}_\ell^* - \mathbf{Id})\|_s = \sup_{S \subset [N], \text{card}(S)=s} \|\sum_{\ell=1}^{m} ((\mathbf{X}_\ell)_S (\mathbf{X}_\ell)_S^* - \mathbf{Id}_S)\|_{2 \to 2}$$

$$= \sup_{(\mathbf{z},\mathbf{w}) \in Q_{s,N}} \text{Re} \left\langle \sum_{\ell=1}^{m} (\mathbf{X}_\ell \mathbf{X}_\ell^* - \mathbf{Id})\mathbf{z}, \mathbf{w} \right\rangle$$

$$= \sup_{(\mathbf{z},\mathbf{w}) \in Q_{s,N}^*} \sum_{\ell=1}^{m} \text{Re} \left\langle (\mathbf{X}_\ell \mathbf{X}_\ell^* - \mathbf{Id})\mathbf{z}, \mathbf{w} \right\rangle,$$

where $(\mathbf{X}_\ell)_S$ denotes the vector $\mathbf{X}_\ell$ restricted to the entries in S, and $Q_{s,N} = \bigcup_{S \subset [N], \text{card}(S) \leq s} Q_{S,N}$ with

$$Q_{S,N} = \{(\mathbf{z},\mathbf{w}) \,:\, \mathbf{z},\mathbf{w} \in \mathbb{C}^N, \|\mathbf{z}\|_2 = \|\mathbf{w}\|_2 = 1, \text{supp}\,\mathbf{z}, \text{supp}\,\mathbf{w} \subset S\}.$$

Further, $Q_{s,N}^*$ denotes a dense countable subset of $Q_{s,N}$. Introducing $f_{\mathbf{z},\mathbf{w}}(\mathbf{X}) = \text{Re}\langle(\mathbf{X}\mathbf{X}^* - \text{Id})\mathbf{z}, \mathbf{w}\rangle$, we therefore have

$$m\delta_s = \sup_{(\mathbf{z},\mathbf{w})\in Q_{s,N}^*} \sum_{\ell=1}^m f_{\mathbf{z},\mathbf{w}}(\mathbf{X}_\ell).$$

To check the boundedness of $f_{\mathbf{z},\mathbf{w}}$ for $(\mathbf{z}, \mathbf{w}) \in Q_{S,N}$ with $\text{card}(S) = s$, we notice that

$$|f_{\mathbf{z},\mathbf{w}}(\mathbf{X})| \le |\langle(\mathbf{X}\mathbf{X}^* - \text{Id})\mathbf{z}, \mathbf{w}\rangle| \le \|\mathbf{z}\|_2 \|\mathbf{w}\|_2 \|\mathbf{X}_S\mathbf{X}_S^* - \text{Id}_S\|_{2\to 2}$$

$$\le \|\mathbf{X}_S\mathbf{X}_S^* - \text{Id}_S\|_{1\to 1} = \max_{j\in S} \sum_{k\in S} |\phi_j(\mathbf{t})\overline{\phi_k(\mathbf{t})} - \delta_{j,k}|$$

$$\le sK^2$$

by the boundedness condition (12.2). Hereby, we used that the operator norm on ℓ_2 is bounded by the one on ℓ_1 for self-adjoint matrices (Lemma A.8) as well as the explicit expression (A.10) for $\|\cdot\|_{1\to 1}$. For the variance term σ_ℓ^2, we estimate

$$\mathbb{E}|f_{\mathbf{z},\mathbf{w}}(\mathbf{X}_\ell)|^2 \le \mathbb{E}|\langle(\mathbf{X}\mathbf{X}^* - \text{Id})\mathbf{z}, \mathbf{w}\rangle|^2$$

$$= \mathbb{E}\mathbf{w}^*(\mathbf{X}_S\mathbf{X}_S^* - \text{Id})\mathbf{z}((\mathbf{X}_S\mathbf{X}_S^* - \text{Id})\mathbf{z})^*\mathbf{w}$$

$$\le \|\mathbf{w}\|_2^2 \mathbb{E}\|(\mathbf{X}_S\mathbf{X}_S^* - \text{Id})\mathbf{z}((\mathbf{X}_S\mathbf{X}_S^* - \text{Id})\mathbf{z})^*\|_{2\to 2}$$

$$= \mathbb{E}\|(\mathbf{X}_S\mathbf{X}_S^* - \text{Id})\mathbf{z}\|_2^2 = \mathbb{E}\|\mathbf{X}_S\|_2^2|\langle\mathbf{X}, \mathbf{z}\rangle|^2 - 2\mathbb{E}|\langle\mathbf{X}, \mathbf{z}\rangle|^2 + 1,$$

exploiting that $\|\mathbf{u}\mathbf{u}^*\|_{2\to 2} = \|\mathbf{u}\|_2^2$; see (A.14). Observe that

$$\|\mathbf{X}_S\|_2^2 = \sum_{j\in S} |\phi_j(\mathbf{t})|^2 \le sK^2$$

by the boundedness condition (12.2). Furthermore,

$$\mathbb{E}|\langle\mathbf{X}, \mathbf{z}\rangle|^2 = \sum_{j,k\in S} z_j\overline{z_k}\mathbb{E}[\overline{\phi_k(\mathbf{t})}\phi_j(\mathbf{t})] = \|\mathbf{z}\|_2^2 = 1$$

by the orthogonality condition (12.1). Hence,

$$\mathbb{E}|f_{\mathbf{z},\mathbf{w}}(\mathbf{X}_\ell)|^2 \le \mathbb{E}\|\mathbf{X}_S\|_2^2|\langle\mathbf{X}, \mathbf{z}\rangle|^2 - 2\mathbb{E}|\langle\mathbf{X}, \mathbf{z}\rangle|^2 + 1 \le (sK^2 - 2)\mathbb{E}|\langle\mathbf{X}, \mathbf{z}\rangle|^2 + 1$$

$$= sK^2 - 1 < sK^2.$$

Now we are prepared to apply Theorem 8.42. Under Condition (12.83), this gives

$$\mathbb{P}(\delta_s \geq \eta_1 + \eta_1^2 + \eta_2) \leq \mathbb{P}(\delta_s \geq \mathbb{E}\delta_s + \eta_2)$$

$$= \mathbb{P}(\|\sum_{\ell=1}^{m}(\mathbf{X}_\ell\mathbf{X}_\ell^* - \mathbf{Id})\|_s \geq \mathbb{E}\|\sum_{\ell=1}^{m}(\mathbf{X}_\ell\mathbf{X}_\ell^* - \mathbf{Id})\|_s + \eta_2 m)$$

$$\leq \exp\left(-\frac{(\eta_2 m)^2}{2msK^2 + 4(\eta_1 + \eta_1^2)msK^2 + 2\eta_2 msK^2/3}\right)$$

$$= \exp\left(-\frac{m\eta_2^2}{K^2 s}\frac{1}{2 + 4(\eta_1 + \eta_1^2) + 2\eta_2/3}\right) \leq \exp\left(-c(\eta_1)\frac{m\eta_2^2}{K^2 s}\right),$$

with $c(\eta_1) = (2 + 4(\eta_1 + \eta_1^2) + 2/3)^{-1} \leq (2 + 8 + 2/3)^{-1} = 3/32$. The left-hand term is less than ε provided

$$m \geq \tilde{C}\eta_2^{-2}K^2 s \ln(\varepsilon^{-1})$$

with $\tilde{C} = 32/3 \approx 10.66$.

Taking also (12.83) into account, we proved that $\delta_s \leq \eta_1 + \eta_1^2 + \eta_2$ with probability at least $1 - \varepsilon$ provided that m satisfies the two conditions

$$\frac{m}{\ln(9m)} \geq \overline{C}\eta_1^{-2}K^2 s \ln^2(4s)\ln(8N),$$

$$m \geq \tilde{C}\eta_2^{-2}K^2 s \ln(\varepsilon^{-1})$$

with $\overline{C} = 8C_1^2 = 16C_0^2 C^2 = 16 \cdot 32C_0^2 \approx 5576$. Here, $C = 4\sqrt{2}$ is the constant from Dudley's inequality (Theorem 8.23). This finally completes the proof of Theorem 12.32.   □

12.6   Discrete Bounded Orthonormal Systems

The three previous sections developed bounds for sparse recovery of randomly sampled functions that have a sparse expansion in terms of a general BOS. Several important examples mentioned in Sect. 12.1 are discrete, i.e., the functions ϕ_k are the columns (or rows) of a unitary matrix $\mathbf{U} \in \mathbb{C}^{N \times N}$, $\mathbf{U}^*\mathbf{U} = \mathbf{U}\mathbf{U}^* = \mathbf{Id}$, with bounded entries,

$$\sqrt{N}\max_{k,t\in[N]}|U_{t,k}| \leq K, \tag{12.84}$$

see (12.9). Among the mentioned examples are the Fourier matrix $\mathbf{F}$ and the matrix $\mathbf{U} = \mathbf{W}^*\mathbf{V}$ resulting from two incoherent orthonormal bases $\mathbf{V}, \mathbf{W}$.

Random sampling of entries corresponds to selecting the rows of the measurement matrix $\mathbf{A}$ uniformly at random from the rows of $\mathbf{U}$. As already mentioned, the probability model of taking the samples independently and uniformly at random has the slight disadvantage that some rows may be selected more than once with nonzero probability. In order to avoid this drawback, we discuss another probability model. Let $\mathbf{u}_j \in \mathbb{C}^N$, $j \in [N]$, be the columns of $\mathbf{U}^*$, i.e., $\mathbf{u}_j^*$ are the rows of $\mathbf{U}$.

Selecting subsets uniformly at random. We choose the set $\Omega \subset [N]$ indexing the rows uniformly at random among all subsets of $[N]$ of size m. This means that each subset is selected with equal probability. Since the number $\binom{N}{m}$ of such subsets is finite, this is a valid probability model. The matrix $\mathbf{A}$ consists then of the rows $\mathbf{u}_j^*$, $j \in \Omega$. It has exactly m rows in this probability model.

The matrix $\mathbf{A}$ is called a *random partial unitary matrix*. If $\mathbf{U} = \mathbf{F} \in \mathbb{C}^N$ is the Fourier matrix, then we call $\mathbf{A}$ a random partial Fourier matrix.

The difficulty in analyzing the above probability model stems from the fact that the selections of $\mathbf{u}_j^*$, $j \in [N]$, as rows of $\mathbf{A}$, are not independent. We resolve this problem by simply relating results for this probability model to the previously derived results for the probability model where the selections of rows (as sampling points) are done independently at random. We only state the analog of the uniform recovery result in Corollary 12.34(a). Analogs of other statements in the previous sections can be derived as well.

Corollary 12.38. *Let $\mathbf{U} \in \mathbb{C}^{N \times N}$ be a unitary matrix with entries bounded by $K/\sqrt{N}$ as in (12.84). Let $\mathbf{A} \in \mathbb{C}^{m \times N}$ be the submatrix of $\mathbf{U}$ obtained by selecting a subset of rows uniformly at random. If*

$$ m \geq CK^2 s \ln^4(N), \tag{12.85} $$

then with probability at least $1 - N^{-\ln^3(N)}$ every s-sparse vector $\mathbf{x} \in \mathbb{C}^N$ is the unique minimizer of $\|\mathbf{z}\|_1$ subject to $\mathbf{Az} = \mathbf{Ax}$.

Proof. Let $T' = \{t_1', \ldots, t_m'\}$, where the $t_\ell' \in [N]$ are selected independently and uniformly at random from $[N]$. The size of T' is also random, since some of the t_ℓ may coincide. Furthermore, for $k \leq m$, let T_k be a subset of $[N]$ chosen uniformly at random among all subsets of cardinality k. For any subset $T \subset [N]$, let $F(T)$ denote the event that ℓ_1-minimization fails to recover every s-sparse $\mathbf{x}$ from the samples of $\mathbf{Ux}$ on T, that is, from $\mathbf{y} = \mathbf{R}_T \mathbf{Ux}$. It follows from Theorem 4.5 together with Remark 4.6 that $F(\hat{T}) \subset F(T)$ whenever $T \subset \hat{T}$. In other words, adding samples decreases the probability of failure. In particular, $\mathbb{P}(F(T_m)) \leq \mathbb{P}(F(T_k))$ for all $k \leq m$. Moreover, conditional on the event that $\mathrm{card}(T') = k$ for $k \leq m$, T' has the same distribution as T_k. We obtain

$$\mathbb{P}(F(T')) = \sum_{k=1}^{m} \mathbb{P}(F(T') | \operatorname{card}(T') = k) \mathbb{P}(\operatorname{card}(T') = k)$$

$$= \sum_{k=1}^{m} \mathbb{P}(F(T_k)) \mathbb{P}(\operatorname{card}(T') = k) \geq \mathbb{P}(F(T_m)) \sum_{k=1}^{m} \mathbb{P}(\operatorname{card}(T') = k)$$

$$= \mathbb{P}(F(T_m)).$$

So the failure probability in the model of selecting row subsets uniformly at random is bounded by the failure probability for the model of selecting rows uniformly and independently at random. An application of Corollary 12.34 yields the claim.   □

Another discrete probability model of interest uses Bernoulli selectors; see Exercise 12.10.

12.7   Relation to the Λ_1-Problem

In this section, we consider a discrete BOS as described in Example 3. Let $\mathbf{V} \in \mathbb{C}^{N \times N}$ be a unitary matrix and set K as in (12.9), namely,

$$K = \sqrt{N} \max_{k,t \in [N]} |V_{t,k}|.$$

We compare the ℓ_1-norm and ℓ_2-norm of expansions in terms of subsets of this discrete BOS. To be more concrete, one may think of the Fourier matrix $\mathbf{V} = \mathbf{F}$ with entries $F_{j,k} = e^{2\pi i jk/N}/\sqrt{N}$ and constant $K = 1$.

Let $\Lambda \subset [N]$ and let $\mathbf{v}_k$, $k \in [N]$, denote the columns of $\mathbf{V}^*$, i.e., the $\mathbf{v}_k^*$ are the rows of $\mathbf{V}$. In the Fourier case $(\mathbf{v}_k)_j = e^{2\pi i jk/N}/\sqrt{N}$. The Cauchy–Schwarz inequality implies that for all coefficient sequences $(b_k)_{k \in \Lambda} \in \mathbb{C}^\Lambda$,

$$\frac{1}{\sqrt{N}} \| \sum_{k \in \Lambda} b_k \mathbf{v}_k \|_1 \leq \| \sum_{k \in \Lambda} b_k \mathbf{v}_k \|_2.$$

A converse of the above inequality is given by the estimate $\| \cdot \|_2 \leq \| \cdot \|_1$. The Λ_1-problem consists of finding a large subset $\Lambda \subset [N]$ such that the much better estimate

$$\| \sum_{k \in \Lambda} b_k \mathbf{v}_k \|_2 \leq \frac{D(N)}{\sqrt{N}} \| \sum_{k \in \Lambda} b_k \mathbf{v}_k \|_1 \qquad (12.86)$$

holds for all $(b_k)_{k \in \Lambda} \in \mathbb{C}^\Lambda$ with a small constant $D(N)$, say, $D(N) = C \log^\alpha(N)$. Such a set Λ will be called a Λ_1-set. In this case, the ℓ_2-norm and the ℓ_1-norm (scaled by the factor $N^{-1/2}$) of orthogonal expansions on Λ are almost equivalent.

Any singleton $\Lambda = \{\ell\}$, $\ell \in [N]$, is a Λ_1-set because by orthonormality and uniform boundedness

$$1 = \|\mathbf{v}_\ell\|_2^2 = \sum_{j=1}^N |(\mathbf{v}_\ell)_j|^2 \le \frac{K}{\sqrt{N}} \sum_{j=1}^N |(\mathbf{v}_\ell)_j| = \frac{K}{\sqrt{N}} \|\mathbf{v}_\ell\|_1,$$

that is, $\|\mathbf{v}_\ell\|_1 \ge K^{-1}\sqrt{N}$, so that, for any $b_\ell \in \mathbb{C}$,

$$\|b_\ell \mathbf{v}_\ell\|_2 = |b_\ell| \le \frac{K}{\sqrt{N}} \|b_\ell \mathbf{v}_\ell\|_1$$

and (12.86) holds with $D(N) = K$ for $\Lambda = \{\ell\}$. However, singleton sets are of limited interest, and we would like to find large sets Λ with $\mathrm{card}(\Lambda) \ge cN$, say.

As it turns out, the Λ_1-problem is strongly related to the ℓ_2-robust null space property (Definition 4.21) and thus to the restricted isometry property.

Proposition 12.39. *Let $\mathbf{V} \in \mathbb{C}^{N \times N}$ be a unitary matrix with rows $\mathbf{v}_\ell \in \mathbb{C}^N$, and let $\Omega \subset [N]$. Assume that the matrix $\mathbf{A} = \mathbf{R}_\Omega \mathbf{V}$, that is, the restriction of $\mathbf{V}$ to the rows indexed by Ω, satisfies the ℓ_2-robust null space property of order s with constants ρ and $\tau > 0$. Then the complement $\overline{\Omega} = [N] \setminus \Omega$ is a Λ_1-set in the sense that*

$$\Big\| \sum_{\ell \in \overline{\Omega}} b_\ell \mathbf{v}_\ell \Big\|_2 \le \frac{1+\rho}{\sqrt{s}} \Big\| \sum_{\ell \in \overline{\Omega}} b_\ell \mathbf{v}_\ell \Big\|_1$$

for all $(b_\ell)_{\ell \in \overline{\Omega}} \in \mathbb{C}^{\overline{\Omega}}$.

Proof. Inequality (4.21) specialized to $p = q = 2$ and $\mathbf{z} = \mathbf{u} \in \ker \mathbf{A}$ and $\mathbf{x} = \mathbf{0}$ implies

$$\|\mathbf{u}\|_2 \le \frac{1+\rho}{\sqrt{s}} \|\mathbf{u}\|_1 \quad \text{for all } \mathbf{u} \in \ker \mathbf{A}. \tag{12.87}$$

Since $\mathbf{A}$ is the row submatrix of a unitary matrix, its kernel is spanned by the rows left out in $\mathbf{A}$, that is, by the ones indexed by $\overline{\Omega}$. Therefore, any $\mathbf{u} \in \ker \mathbf{A}$ takes the form

$$\mathbf{u} = \sum_{\ell \in \overline{\Omega}} b_\ell \mathbf{v}_\ell.$$

Combining these facts concludes the proof. $\qquad\square$

Since the restricted isometry property implies the ℓ_2-robust null space property (Theorem 6.13), we can combine the above proposition with the estimate of the restricted isometry constants of random sampling matrices associated to BOSs to arrive at the following theorem on the Λ_1-problem.

Theorem 12.40. *Let* $\{\mathbf{v}_1, \dots \mathbf{v}_N\}$ *be an orthonormal basis of* $\mathbb{C}^N$ *with uniformly bounded entries, i.e.,* $|(\mathbf{v}_j)_k| \leq K/\sqrt{N}$. *For* $c \in (0, 1)$, *there exists a set* $\Lambda \subset [N]$ *with* $\mathrm{card}(\Lambda) \geq cN$ *such that*

$$\| \sum_{\ell \in \Lambda} b_\ell \mathbf{v}_\ell \|_2 \leq \frac{CK \ln^2(N)}{\sqrt{N}} \| \sum_{\ell \in \Lambda} b_\ell \mathbf{v}_\ell \|_1 \qquad (12.88)$$

holds for all $(b_\ell)_{\ell \in \Lambda} \in \mathbb{C}^\Lambda$. *The constant* C *depends only on* c, *more precisely* $C = C'(1 - c)^{-1/2}$ *for some universal constant* C'.

A slightly better estimate in terms of the logarithmic factors is available; see the Notes section below. However, as a consequence of Lemma 12.5, the term $\ln^2(N)$ cannot be improved to something better than $\sqrt{\ln N}$ in general; see Exercise 12.11.

Proof. Let $m = \lfloor (1 - c)N \rfloor$. Then Theorem 12.31 implies the existence of a set $\Omega \subset [N]$ such that the restricted isometry constant of the matrix $\mathbf{A} = \frac{1}{\sqrt{m}}\mathbf{R}_\Omega \mathbf{V}$ satisfies $\delta_{2s} \leq \delta^* := 0.5$ for the choice

$$s = \lceil C_0 \frac{m}{K^2 \ln^4(N)} \rceil,$$

where C_0 is a universal constant. It follows now from Theorem 6.13 that $\mathbf{A}$ satisfies the ℓ_2-robust null space property with constants ρ, τ depending only on δ^*. The kernel of $\mathbf{A}$ does not depend on the scaling of $\mathbf{A}$, so that (12.87) holds for $\mathbf{R}_\Omega \mathbf{V}$ and the result of Proposition 12.39 is valid for $\Lambda = \overline{\Omega}$ which has cardinality $\mathrm{card}(\Lambda) \geq cN$. We conclude that

$$\| \sum_{\ell \in \Lambda} b_\ell \mathbf{v}_\ell \|_2 \leq \frac{1 + \rho}{\sqrt{s}} \| \sum_{\ell \in \Lambda} b_\ell \mathbf{v}_\ell \|_1.$$

Taking into account our choices of s and m, we arrive at

$$\| \sum_{\ell \in \Lambda} b_\ell \mathbf{v}_\ell \|_2 \leq \frac{1 + \rho}{\sqrt{C_0(1 - c)}} \frac{K \ln^2(N)}{\sqrt{N}} \| \sum_{\ell \in \Lambda} b_\ell \mathbf{v}_\ell \|_1.$$

This completes the proof.   □

Notes

Background on Fourier analysis (Examples 1, 4, 5) can be found, for instance, in [198, 236, 390, 452, 505]. The complex exponentials of Examples 1 can be generalized to characters of commutative groups; see, for instance, [199, 437]. The sampling matrix (12.7) arising from continuously sampling trigonometric expansions has an (approximate) fast matrix–vector multiplication called the nonequispaced FFT [400]. Like the FFT (see Appendix C.1), it has complexity $\mathcal{O}(N \log N)$.

The uncertainty principle for the discrete Fourier transform in Corollary 12.3 was shown by Donoho and Stark in [164], where they also realized that the uncertainty principle is not always a negative statement, but it can also be used to derive positive conclusions about signal separation and recovery; see also [158]. Later in [181], Elad and Bruckstein derived the discrete uncertainty principle for general pairs of bases (Theorem 12.2). Kuppinger et al. extended this further to an uncertainty principle for pairs of possibly redundant systems in [314]. An overview on uncertainty principles in general, including the classical uncertainty principles of Heisenberg and the one of Hardy, is provided in [200].

Lemma 12.5 concerning the existence of translates of large subgroups in arbitrary subsets of $\mathbb{Z}_2^n$, then leading to the lower bound (12.29) for the number of samples in undersampled Hadamard transforms featuring the factor $\ln(N)$, goes back to the work of Bourgain and Talagrand on the Λ_1-problem [464], but was published much later in [248].

The nonuniform recovery result Theorem 12.11 with random sign pattern seems to have first appeared in [411], while the improvement of Theorem 12.18 was shown by Candès and Romberg in [93]. The idea of using random signs in order to derive recovery bounds for ℓ_1-minimization appeared first in [481]. The nonuniform recovery result of Theorem 12.20, in which the randomness in the signs of the coefficient vectors is removed, was shown by Candès and Plan in [88]. The key technique in their proof, that is, the golfing scheme, was developed by Gross in [245] in the context of matrix completion and more general low-rank matrix recovery problems; see also [417]. Instead of the deviation result for sums of random vectors in ℓ_2 (Corollary 8.45) and the noncommutative Bernstein inequality (8.26), which were used in Sect. 12.4 to derive Lemmas 12.24, 12.25, 12.27, Candès and Plan used a slightly different version of the vector Bernstein inequality due to Gross [245, Theorem 11], which also allows to remove the factor $(s+1)$ in (12.49). (This factor, however, is not important as it only enters in the term $\ln(2N(s+1)) \leq \ln(2N^2) \leq 2\ln(2N)$.) Moreover, Candès and Plan showed stronger robustness estimates than the one of Theorem 12.22, in which the factor $\sqrt{s}$ can essentially be replaced by $\ln(s)^{3/2}$ while still keeping the bound (12.46) on the number of required samples (in contrast to the bound on the restricted isometry constants which involves more logarithmic factors). To do so, they introduced weak restricted isometry constants and estimated these. This requires additional steps compared to the proof of the restricted isometry property in Sect. 12.5; see [88] for details.

The special case of partial random Fourier matrices (Example 4 in Sect. 12.1) was treated already in the first contribution of Candès et al. to compressive sensing [94]. They provided a nonuniform recovery result for deterministic sign patterns (in the noiseless case), where the number m of samples scales as

$$m \geq Cs \ln(N/\varepsilon) \tag{12.89}$$

in order to achieve recovery via ℓ_1-minimization with probability at least $1 - \varepsilon$. This estimate was extended to random sampling of sparse trigonometric polynomials (as described in Example 1 in Sect. 12.1) in [408]. It is remarkable that this bound is still slightly better with regard to the dependence in ε than the result for general BOS, Theorem 12.20, where one encounters the term $\ln(N)\ln(\varepsilon^{-1})$ in contrast to $\ln(N/\varepsilon) = \ln(N) + \ln(\varepsilon^{-1})$ above. (For instance, with $\varepsilon = N^{-\gamma}$ the first term results in $\gamma \ln^2(N)$, while the second only yields $(\gamma + 1)\ln(N)$.) It is presently not clear how to arrive at a bound of the form (12.89) for general systems. The rather long proof of the sufficient condition (12.89) in [94, 408] heavily uses the algebraic structure of the Fourier system and proceeds via involved combinatorial estimates. It does not seem possible to extend this approach to general BOSs. We refer to [283, 412] for further nonuniform recovery results whose proofs also proceed via similar combinatorial arguments.

The restricted isometry property for partial random Fourier matrices (Example 4) was first analyzed by Candès and Tao in [97], where they obtained the bound $m \geq C_\delta s \ln^5(N)\ln(\varepsilon^{-1})$ for the number of required samples to achieve the restricted isometry property of order s with probability at least $1 - \varepsilon$. This estimate was then improved by Rudelson and Vershynin in [433] to the condition $m \geq C_\delta s \ln^2(s)\ln(s\ln(N))\ln(N)\ln(\varepsilon^{-1})$. A further slight improvement to $m \geq C_{\delta,\varepsilon} s \ln^3(s)\ln(N)$ appeared in [115]. (The proofs in the papers [97, 115, 433] actually apply to more general discrete orthonormal systems as described in Example 3.) In [409, 411] the analysis was generalized to possibly continuous BOSs, and the probability estimate was improved to the one stated in Theorem 12.31 by using Bernstein's inequality for suprema of empirical processes (Theorem 8.42). We followed Rudelson and Vershynin's approach in Sect. 12.5 to estimate the expected restricted isometry constants. With similar techniques it is also possible to directly establish the null space property for random sampling matrices arising from BOSs. We refer to [103] for details on this and for many other facts relating compressive sensing, random matrices, and Banach space geometry.

The Λ_1-problem was investigated in [464] by Bourgain and Talagrand, who treated the case of general (not necessarily discrete) BOSs $\{\phi_j, j \in [N]\}$, where orthonormality is with respect to a probability measure ν. The main result in [464] states the existence of a subset $\Lambda \subset [N]$ with $\text{card}(\Lambda) \geq cN$ such that

$$\left\|\sum_{\ell \in \Lambda} b_\ell \phi_\ell\right\|_{L_2(\nu)} \leq CK\sqrt{\ln(N)\ln(\ln N)} \left\|\sum_{\ell \in \Lambda} b_\ell \phi_\ell\right\|_{L_1(\nu)}.$$

(Note that the factor $1/\sqrt{N}$ has to be introduced in the discrete setting of Sect. 12.7 because the usual ℓ_1 and ℓ_2-norms are not taken with respect to a probability measure, in contrast to the spaces $L^2(\nu)$ and $L^1(\nu)$ above.) It follows from Lemma 12.5 that a factor of $\ln N$ has to be present in this estimate; see Exercise 12.11. It is conjectured, however, that the term $\ln \ln N$ can be removed, but this conjecture remains open until today. Taking this fact into account together with the relation of the restricted isometry property with the Λ_1-problem (see the proof of Theorem 12.40), it seems a very hard problem to remove all but one logarithmic factors in the estimate (12.63) for the restricted isometry property, as this would imply a positive solution to this conjecture (at least in the discrete case). Further results on the Λ_1-problem are contained in the paper [248], which also deals with Kashin-type decompositions for BOSs. The Λ_p-problem, for $p > 2$, was solved by Bourgain in [62]; see also [63] for more information on this topic.

Signal Separation

Similar mathematical techniques as the ones developed in this chapter have been used in the problem of separating a signal decomposed in two components (see p. 18 for a general description of signal separation problems). One component is supposed to be sparse and the other one sparse in the Fourier domain [92, 164, 481]. Assuming that the support set is random in at least one of the components, then one can show that separation is possible via ℓ_1-minimization provided that the sparsity s in both components does not exceed $N/\sqrt{\ln N}$, where N is the signal length [92]. The proof methods are similar to the ones used for nonuniform recovery guarantees.

Fast Johnson–Lindenstrauss Mappings

Theorem 9.36 combined with Theorem 12.32, which establishes the restricted isometry property for sampling matrices such as the random partial Fourier matrix, provides a Johnson–Lindenstrauss embedding for the mapping $\mathbf{A}\mathbf{D}_\epsilon$, where ϵ is a Rademacher vector; see Exercise 12.12. In contrast to a subgaussian random matrix, $\mathbf{A}\mathbf{D}_\epsilon$ comes with a fast matrix multiplication routine when $\mathbf{A}$ is, for instance, the partial random Fourier matrix [7, 9, 309]. Hinrichs and Vybiral investigated a similar scenario when $\mathbf{A}$ is a partial random circulant matrix [274, 502]; see also [309, 413].

Sublinear Fourier Algorithms

Even before the area of compressive sensing began to develop, algorithms that compute sparse Fourier transforms in sublinear runtime in the signal length N were known [223, 519]. Such algorithms are based on random sampling in the time domain. In contrast to the setup of this chapter, however, the samples are not independent, which allows for enough algebraic structure to enable fast computation. Although these algorithms were initially designed for speed, one can separate the sampling and the reconstruction process so that they apply also in compressive sensing setups. A very appealing construction making use of the Chinese remainder theorem was presented by Iwen in [287, 288]. He provides a deterministic version of the algorithm, which uses $m \geq Cs^2 \ln^4(N)$ samples and has runtime $\mathcal{O}(s^2 \ln^4(N))$, and a randomized variant, which requires $m \geq Cs \ln^4 N$ samples and runs in time $\mathcal{O}(s \ln^4(N))$. A numerical evaluation of sublinear Fourier algorithms is reported in [289, 445]. Another sublinear Fourier algorithm is developed in [261]. In Chap. 13, we will present a sublinear sparse recovery algorithm in the different context of lossless expanders.

Further Examples of Bounded Orthonormal Systems

We discuss three other examples for which the theory of this chapter is applicable. Since complete proofs go beyond the scope of this book, we only mention the basic facts and refer to further literature for the details.

Haar wavelets and noiselets. This example is a special case of Example 6, which is potentially useful for image processing applications. It is convenient to start with a continuous description of Haar wavelets and noiselets [126] and then pass to the discrete setup via sampling. The Haar scaling function on $\mathbb{R}$ is defined as the characteristic function of the interval $[0, 1)$,

$$\phi(x) = \chi_{[0,1)}(x) = \begin{cases} 1 & \text{if } x \in [0, 1), \\ 0 & \text{otherwise.} \end{cases} \tag{12.90}$$

The Haar wavelet is then defined as

$$\psi(x) = \phi(2x) - \phi(2x - 1) = \begin{cases} 1 & \text{if } x \in [0, 1/2), \\ -1 & \text{if } x \in [1/2, 1), \\ 0 & \text{otherwise.} \end{cases} \tag{12.91}$$

Furthermore, denote

$$\psi_{j,k}(x) = 2^{j/2}\psi(2^j x - k), \quad \phi_k(x) = \phi(x - k), \quad x \in \mathbb{R}, j \in \mathbb{Z}, k \in \mathbb{Z}. \tag{12.92}$$

It is straightforward to verify [508] that, for $n \in \mathbb{N}$, the Haar-wavelet system

$$\Psi_n := \{\phi_k, k \in \mathbb{Z}\} \cup \{\psi_{j,k}, k = 0, \ldots, 2^j - 1, j = 0, \ldots, n - 1\} \tag{12.93}$$

forms an orthonormal basis of

$$V_n = \{f \in L^2[0, 1] : f \text{ is constant on } [k2^{-n}, (k + 1)2^{-n}), k = 0, \ldots, 2^n - 1\}.$$

Haar (and more advanced) wavelets are suitable to represent piecewise smooth functions. In particular, natural images are usually sparse (or compressible) with respect to their two-dimensional wavelet representation. Therefore, Haar wavelets are very useful for image processing tasks.

Now let $N = 2^n$ for some $n \in \mathbb{N}$. Since the functions $\psi_{j,k}$ are constant on intervals of the form $[2^{-n}k, 2^{-n}(k+1))$, we conclude that the vectors $\tilde{\phi}, \tilde{\psi}^{(j,k)} \in \mathbb{C}^N, j = 0, \ldots, n-1, k = 0, \ldots, 2^j - 1$, with entries

$$\tilde{\phi}_t = 2^{-n/2}\phi(t/N), \quad t = 0, \ldots, N - 1$$

$$\tilde{\psi}_t^{(j,k)} = 2^{-n/2}\psi_{j,k}(t/N), \quad t = 0, \ldots, N - 1$$

form an orthonormal basis of $\mathbb{C}^N$. We collect these vectors as the columns of a unitary matrix $\Psi \in \mathbb{C}^{N \times N}$.

Next, we introduce the noiselet system on $[0,1]$. Let $g_1 = \phi = \chi_{[0,1)}$ be the Haar scaling function and define recursively, for $r \geq 1$, the complex-valued functions

$$g_{2r}(x) = (1-i)g_r(2x) + (1+i)g_r(2x - 1),$$

$$g_{2r+1}(x) = (1+i)g_r(2x) + (1-i)g_r(2x - 1).$$

It is shown in [126] that the functions $\{2^{-n/2}g_r, r = 2^n, \ldots, 2^{n+1} - 1\}$ form an orthonormal basis of V_n. The key property for us consists in the fact that they are maximally incoherent with respect to the Haar basis. Indeed, Lemma 10 in [126] states that

$$\left| \int_0^1 g_r(x)\psi_{j,k}(x)dx \right| = 1 \quad \text{provided } r \geq 2^j - 1, \quad 0 \leq k \leq 2^j - 1. \quad (12.94)$$

For the discrete noiselet basis on $\mathbb{C}^N$, $N = 2^n$, we take the vectors

$$\tilde{g}_t^{(r)} = 2^{-n}g_{N+r}(t/N), \quad r = 0, \ldots, N - 1, \quad t = 0, \ldots, N - 1.$$

Again, since the functions $g_{N+r}, r = 0, \ldots, N - 1$, are constant on intervals of the form $[2^{-n}k, 2^{-n}(k+1))$, it follows that the vectors $\tilde{g}^{(r)}, r = 0, \ldots, N - 1$ form an orthonormal basis of $\mathbb{C}^N$. We collect these as columns into a unitary matrix $\mathbf{G} \in \mathbb{C}^{N \times N}$. Due to (12.94), the unitary matrix $\mathbf{U} = \mathbf{G}^*\Psi \in \mathbb{C}^{N \times N}$ satisfies (12.9) with $K = 1$. In other words, the incoherence condition (12.12) for the Haar basis and the noiselet basis holds with the optimal constant $K = 1$.

Due to the their recursive definition, both the Haar-wavelet transform and the noiselet transform, that is, the application of Ψ and $\mathbf{G}$ and their adjoints, come with a fast algorithm that computes a matrix–vector multiplication in $\mathcal{O}(N \log(N))$ time.

As a simple signal model, images or other types of signals are sparse in the Haar-wavelet basis. The setup of this chapter corresponds to randomly sampling such functions with respect to noiselets. For more information on wavelets, we refer to [122, 137, 341, 508].

Legendre polynomials and other orthogonal polynomial systems. The Legendre polynomials $L_j, j = 0, 1, 2, \ldots$ form a system of orthonormal polynomials, where

L_j is a polynomial of precise degree j and orthonormality is with respect to the normalized Lebesgue measure $dx/2$ on $[-1, 1]$, that is,

$$\frac{1}{2} \int_{-1}^{1} L_j(x)L_k(x)dx = \delta_{j,k}.$$

We refer to [17, 117, 458] for details on orthogonal polynomials and in particular on Legendre polynomials. The supremum norm of Legendre polynomials is given by

$$\|L_j\|_\infty = \sup_{t\in[-1,1]} |L_j(t)| = \sqrt{2j+1},$$

so considering the polynomials $L_j, j = 0, \ldots, N-1$ yields the constant $K = \sqrt{2N-1}$ in (12.2). Unfortunately, such a K grows with N. Plugging this value of K, for instance, in the estimate (12.63) for the sufficient number of samples ensuring $\delta_s \leq \delta$, yields

$$m \geq C\delta^{-2}Ns\ln^3(s)\ln(N).$$

This estimate is useless for compressive sensing because the number of measurements is required to be larger than the signal length N.

Of course, the possibility of better estimates is not refuted, and in fact the problem can be circumvented by a simple observation; see [416]. The crucial point is that L_2-normalized Legendre polynomials L_j only grow with j near the endpoints ± 1 of the interval $[-1, 1]$. Define the function

$$v(t) = (\pi/2)^{1/2}(1-t^2)^{1/4}.$$

Then Theorem 7.3.3 in [458] implies that, for all $j \geq 0$,

$$\sup_{t\in[-1,1]} v(t)|L_j(t)| \leq \sqrt{3}.$$

We define the auxiliary orthonormal function system $Q_j = vL_j$, where orthonormality is now with respect to the Chebyshev probability measure (arcsine distribution)

$$d\nu(t) = \pi^{-1}(1-t^2)^{-1/2}dt.$$

Indeed,

$$\int_{-1}^{1} Q_j(t)Q_k(t)d\nu(t) = \frac{1}{2\pi} \int_{-1}^{1} L_j(t)L_k(t)v(t)^2(1-t^2)^{-1}dt$$

$$= \frac{1}{2} \int_{-1}^{1} L_j(t)L_k(t)dt = \delta_{j,k}.$$

Therefore, the system $\{Q_j\}_{j=0}^{N-1}$ forms a BOS with constant $K = \sqrt{3}$ with respect to the Chebyshev measure. The results derived in this chapter are hence valid for the random sampling matrix $\mathbf{B} \in \mathbb{R}^{m \times N}$ having entries

$$B_{\ell,j} = Q_j(t_\ell),$$

where the t_ℓ are sampled independently according to the Chebyshev measure ν. (This causes more sample points to lie near the endpoints $[-1, 1]$ compared to sampling from the uniform measure.) For instance, the restricted isometry constant δ_s of $\frac{1}{\sqrt{m}}\mathbf{B}$ satisfies $\delta_s \leq \delta$ with high probability provided $m \geq C\delta^{-2}s\ln^4(N)$. Multiplying with the function v can be interpreted as preconditioning of the Legendre sampling matrix. Defining $\mathbf{A} \in \mathbb{R}^{m \in N}$ and $\mathbf{D} \in \mathbb{R}^{m \times m}$ via

$$A_{\ell,j} = L_j(t_\ell), \quad \mathbf{D} = \mathrm{diag}[v(t_1), \ldots, v(t_m)],$$

we realize that $\mathbf{B} = \mathbf{DA}$. Since $\mathbf{D}$ is invertible with probability 1, the matrices $\mathbf{A}$ and $\mathbf{B}$ have the same null space almost surely. Now if $\frac{1}{\sqrt{m}}\mathbf{B}$ satisfies the restricted isometry property, say $\delta_{2s} < 0.62$, then by Theorem 6.13, it satisfies the ℓ_2-robust null space property and, in particular, the stable null space property. The latter depends only on the null space of $\mathbf{B}$, which coincides with the one of $\mathbf{A}$, so that also $\mathbf{A}$ satisfies the stable null space property. By Theorem 4.12, this in turn ensures stable sparse recovery via ℓ_1-minimization using the matrix $\mathbf{A}$. Altogether, choosing m independent random sampling points according to the Chebyshev measure ν with $m \geq C's\ln^4(N)$, the sampling matrix $\mathbf{A}$ satisfies the stable null space property of order s with high probability, and we have stable s-sparse recovery via ℓ_1-minimization.

Alternatively, given Legendre-type measurements $\mathbf{y} = \mathbf{A}x$, we may multiply by the scaled diagonal matrix to obtain transformed measurements $\mathbf{y}' = \frac{1}{\sqrt{m}}\mathbf{D}\mathbf{y} = \frac{1}{\sqrt{m}}\mathbf{B}x$ and work directly with the preconditioned matrix $\frac{1}{\sqrt{m}}\mathbf{B} = \frac{1}{\sqrt{m}}\mathbf{DA}$ and the modified $\mathbf{y}'$ in any recovery algorithm. In this way, iterative hard thresholding, hard thresholding pursuit, orthogonal matching pursuit, and CoSaMP can be used in the setup of random sampling of sparse Legendre polynomial expansions.

It is important to note that the Legendre transform matrix $\mathbf{A}$ has fast matrix–vector multiplication algorithms (see [265, 292, 398, 399, 491]) which may speed up recovery algorithms.

Extensions to other orthogonal polynomial expansions on $[-1, 1]$ are possible, where orthogonality is with respect to a weight function that satisfies a mild continuity condition. This includes, for instance, all Jacobi polynomials $P_k^{(\alpha, \beta)}$ with $\alpha, \beta \geq -1/2$ [458]. It is quite interesting that for this whole family of orthogonal polynomials, random sampling is with respect to the Chebyshev measure ν. We refer to [416] for details.

The idea of preconditioning can be applied also in a context where sparsity is with respect to wavelets and random Fourier coefficients are measured. This situation occurs in MRI applications. We refer to [310] for details.

Spherical harmonics. Extensions of the previous example to the system of spherical harmonics (an orthonormal system for $L_2(S^2)$, where S^2 is the 2-sphere in $\mathbb{R}^3$; see [17]) are given in [80]. So far, the preconditioning trick above only yields the restricted isometry property provided $m \geq C s N^{1/6} \ln^4(N)$, after an earlier bound in [415] where $N^{1/4}$ appeared instead of $N^{1/6}$. The result of [80] was established in a more general context by developing involved weighted L_∞ bounds for eigenfunctions of the Laplace operator on certain manifolds including the 2-sphere and thereby improving on estimates for associated Legendre polynomials in [311]. The key ingredient consists of identifying the right sampling measure. We refer to [80] for details.

An application of sparse spherical harmonic expansions for the inpainting problem of the cosmic microwave background is contained in [1]. Fast matrix–vector multiplication algorithms involving spherical harmonics are provided, for instance, in [265].

Further Types of Structured Random Matrices

As mentioned in the beginning of this chapter, it is important for practical purposes that measurement matrices possess some structure. While this chapter covered only random sampling matrices, there are further important types of structured random matrices. Their detailed analysis is beyond the scope of this book, but we summarize some basic facts below.

Partial random circulant matrices. For a vector $\mathbf{b} = [b_0, b_1, \ldots, b_{N-1}]^\top \in \mathbb{C}^N$, the associated circulant matrix $\mathbf{\Phi} = \mathbf{\Phi}(\mathbf{b}) \in \mathbb{C}^{N \times N}$ is defined entrywise by

$$\Phi_{k,j} = b_{j-k \bmod N}, \quad k, j \in [N].$$

The application of $\mathbf{\Phi}$ to a vector is the discrete circular convolution,

$$(\mathbf{\Phi} \mathbf{x})_j = (\mathbf{x} * \tilde{\mathbf{b}})_j = \sum_{\ell=1}^{N} x_\ell \tilde{b}_{j-\ell \bmod N},$$

where $\tilde{b}_j = b_{N-j}$. Let $\Theta \subset [N]$ be an arbitrary (deterministic) subset of cardinality m. We define the partial circulant matrix $\mathbf{\Phi}^\Theta = \mathbf{\Phi}^\Theta(\mathbf{b}) = \mathbf{R}_\Theta \mathbf{\Phi}(\mathbf{b}) \in \mathbb{C}^{m \times N}$ as the submatrix of $\mathbf{\Phi}$ consisting of the rows indexed by Θ. The application of a partial circulant matrix corresponds to convolution with $\mathbf{b}$ followed by subsampling on Θ. It is important from a computational viewpoint that circulant matrices can be diagonalized using the discrete Fourier transform; see, e.g., [231]. Therefore, there is a fast matrix–vector multiplication algorithm for partial circulant matrices of complexity $\mathcal{O}(N \log(N))$ that uses the FFT.

Choosing the generator $\mathbf{b} = \epsilon$ to be a Rademacher sequence makes the matrix $\mathbf{\Phi}^\Theta = \mathbf{\Phi}^\Theta(\epsilon)$ a structured random matrix, which is called partial random circulant matrix. It is interesting to study recovery guarantees for ℓ_1-minimization and the restricted isometry property of the resulting matrix.

Of particular relevance is the case $N = mL$ with $L \in \mathbb{N}$ and $\Theta = \{L, 2L, \ldots, mL\}$. Then the application of $\mathbf{\Phi}^\Theta(\mathbf{b})$ corresponds to convolution with the sequence $\mathbf{b}$ followed by a downsampling by a factor of L. This setting was studied numerically in [489] by Tropp et al. (using orthogonal matching pursuit). Also of interest is the case $\Theta = [m]$ which was studied in [26, 263].

Nonuniform recovery guarantees in the spirit of Theorem 12.11 for partial random circulant matrices in connection with ℓ_1-minimization were derived in [410, 411]. A sufficient condition on the number of samples is $m \geq Cs \log^2(N/\varepsilon)$ for recovery with probability at least $1 - \varepsilon$. After initial nonoptimal bounds in [26, 263, 413], the so far best estimate on the restricted isometry constants of $\mathbf{\Phi}^\Theta(\epsilon)$ obtained in [307] states that $\delta_s \leq \delta$ with high probability provided

$$m \geq C\delta^{-2}s\ln^2(s)\ln^2(N).$$

The proof uses the estimate (8.115), and the analysis of the corresponding covering numbers is based on some of the results developed in Sect. 12.5.

Time–frequency structured random matrices. Recall that the translation and modulation (frequency shift) operators on $\mathbb{C}^m$ are defined by

$$(\mathbf{T}^k\mathbf{g})_j = h_{j\ominus k} \quad \text{and} \quad (\mathbf{M}^\ell\mathbf{g})_j = e^{2\pi i \ell j/m}g_j,$$

where $\ominus$ is subtraction modulo m. The operators $\pi(\lambda) = \mathbf{M}^\ell\mathbf{T}^k$, $\lambda = (k, \ell)$, are called time–frequency shifts, and the system $\{\pi(\lambda) : \lambda \in [m]\times[m]\}$ of all time–frequency shifts forms a basis of the matrix space $\mathbb{C}^{m\times m}$ [308, 317]. Given a vector $\mathbf{g} \in \mathbb{C}^n$, the system of all possible time–frequency shifts of $\mathbf{g}$, that is,

$$\{\pi(\lambda)\mathbf{g}, \lambda \in [m] \times [m]\}$$

is called a full Gabor system with window $\mathbf{g}$; see [244]. The matrix $\mathbf{A} = \mathbf{A_g} \in \mathbb{C}^{m\times m^2}$ whose columns list the vectors $\pi(\lambda)\mathbf{g}$, $\lambda \in [m]\times[m]$, of the Gabor system is referred to as a Gabor synthesis matrix [119, 317, 412]. Note that $\mathbf{A_g}$ allows for fast matrix–vector multiplication algorithms based on the FFT; see, for instance, [191, 192]. Note that the matrix with coherence $\mu = 1/\sqrt{m}$ (in case $m \geq 5$ is prime) constructed in the proof of Proposition 5.13 is a Gabor synthesis matrix with window $g_j = \frac{1}{\sqrt{m}}e^{2\pi i j^3/m}$.

Let us now choose the vector $\mathbf{g}$ at random, that is,

$$\mathbf{g} = \frac{1}{\sqrt{m}}\epsilon,$$

where $\epsilon \in \mathbb{C}^m$ is a Steinhaus sequence. Then the matrix $\mathbf{A} = \mathbf{A_g}$ becomes a structured random matrix, and we are interested in its performance for compressive sensing. A nonuniform recovery result for a fixed s-sparse vector (with deterministic sign pattern) was shown in [412]. Exact recovery via ℓ_1-minimization occurs with high probability provided

$$m \geq cs \ln(m).$$

(Note that in this setup $N = m^2$, so that $\ln(N) = 2\ln(m)$.) After an initial nonoptimal estimate in [385], it was proved in [307] that the restricted isometry constants of $\mathbf{A_g}$ satisfy $\delta_s \leq \delta$ with high probability provided

$$m \geq c\delta^{-2}s\ln^2(s)\ln^2(m).$$

Sparse recovery with time–frequency structured random matrices has potential applications for the channel identification problem [384] in wireless communications and sonar [352, 453] as well as in radar [268]. Note that the results in [384] and [268] were derived based on coherence estimates and an analysis for random signals [481], similar to the one outlined in Chap. 14.

More background on time–frequency analysis can be found in Gröchenig's excellent book [244].

Random demodulator. In some engineering applications, hardware limitations do not allow random sampling in time, especially when the sampling rate is very high. In order to overcome this technological problem, one may instead multiply with random sign flips at a very high rate, integrate the signal over some time period, and then sample equidistantly at a relatively low sampling rate [488]. The advantage is that all these components can be realized in hardware relatively easily. In particular, performing a sign flip at a very high rate is much simpler to realize than sampling at this high rate with high accuracy. In mathematical terms, the sampling matrix modeling this sensing scenario can be described as follows. Let $\mathbf{F} \in \mathbb{C}^{N \times N}$ be the N-dimensional discrete Fourier matrix, which arises here because sparsity in the Fourier domain is assumed. Furthermore, let $\mathbf{D}_\epsilon \in \mathbb{R}^{N \times N}$ be a random diagonal matrix having a Rademacher sequence ϵ on its diagonal, and let finally $\mathbf{H} \in \mathbb{R}^{m \times N}$ model the integration process, where we assume for simplicity that m divides N. The jth row of $\mathbf{H}$ has N/m ones starting in column $(j-1)N/m + 1$ and is zero elsewhere. An example for $m = 3$ and $N = 12$ is

$$\mathbf{H} = \begin{pmatrix} 1\,1\,1\,1\,0\,0\,0\,0\,0\,0\,0\,0 \\ 0\,0\,0\,0\,1\,1\,1\,1\,0\,0\,0\,0 \\ 0\,0\,0\,0\,0\,0\,0\,0\,1\,1\,1\,1 \end{pmatrix}.$$

The measurement matrix $\mathbf{A} \in \mathbb{C}^{m \times N}$ is then the structured random matrix

$$\mathbf{A} = \mathbf{H}\mathbf{D}_\epsilon\mathbf{F},$$

where the randomness comes from the Rademacher vector ϵ on the diagonal of $\mathbf{D}_\epsilon$. It has been shown in [488] that the restricted isometry constants of a suitably rescaled version of $\mathbf{A}$ satisfy $\delta_s \leq \delta$ with high probability provided

$$m \geq C_\delta s \ln^6(N).$$

Therefore, the sampling mechanism described above can efficiently reconstruct signals that are s-sparse in the Fourier domain from m measurements using various algorithms including ℓ_1-minimization. The proof of the restricted isometry property uses parts of the analysis developed in Sect. 12.5. We refer to [488] for details.

A construction of a structured $m \times N$ random matrix with restricted isometry constants satisfying $\delta_s \leq \delta$ in the optimal parameter regime $m \geq C\delta^{-2}s\ln(N)$ under the additional condition $s \leq \sqrt{N}$ was provided in [10]. This matrix is the repeated product of discrete Fourier transforms and random diagonal matrices with ± 1 entries on the diagonal, followed by deterministic subsampling on m coordinates. We refer to [10] for details and to [8] for a similar earlier construction in the context of Johnson–Lindenstrauss embeddings.

Exercises

12.1. Show that the Fourier matrix defined in (12.11) is unitary.

12.2. Let N be an even integer and let $\mathbf{F} \in \mathbb{C}^{N \times N}$ be the Fourier matrix (12.11). Let $\mathbf{A} \in \mathbb{C}^{N/2 \times N}$ be the matrix $\mathbf{F}$ where every second row has been removed, that is, $A_{\ell,k} = F_{2\ell,k}$. Provide an example of a nonzero 2-sparse vector $\mathbf{x} \in \mathbb{C}^N$ such that $\mathbf{Ax} = 0$.

12.3. Let $\mathbf{F}$ be the Fourier matrix (12.11). Let $\mathbf{x}, \mathbf{z} \in \mathbb{C}^N$ with $\|\mathbf{x}\|_0 + \|\mathbf{z}\|_0 < \sqrt{N}$. Setting $\mathbf{y} = \mathbf{x} + \mathbf{Fz} \in \mathbb{C}^N$, show that $(\mathbf{x}, \mathbf{z})$ is the unique solution to $\mathbf{y} = \mathbf{x}' + \mathbf{Fz}'$ among all $\mathbf{x}', \mathbf{z}' \in \mathbb{C}^N$ with $\|\mathbf{x}'\|_0 + \|\mathbf{z}'\|_0 < \sqrt{N}$. In particular, the signal $\mathbf{y}$ can be separated uniquely into the components $\mathbf{x}$ and $\mathbf{Fz}$ under such sparsity assumption.

12.4. Let $T \subset [N]$ be an arbitrary subset of cardinality m. Show that every s-sparse $\mathbf{x} \in \mathbb{C}^N$ with $s < \sqrt{N}/2$ can be recovered from the samples of its Fourier transform on T, i.e., from $\mathbf{y} = \mathbf{R}_T \mathbf{Fx}$ provided $m \geq N - \sqrt{N}$.

12.5. Let $\mathbf{A} \in \mathbb{C}^{m \times N}$ and $S \subset [N]$. Assume that, for $\ell \in [N] \setminus S$,

$$\|\mathbf{A}^*_{S \cup \{\ell\}} \mathbf{A}_{S \cup \{\ell\}} - \mathbf{Id}\|_{2 \to 2} \leq \delta.$$

Show that $\|\mathbf{A}^\dagger_S \mathbf{a}_\ell\|_2 \leq \dfrac{\delta}{1 - \delta}$.

12.6. Let $\Gamma \subset \mathbb{Z}$ with $\mathrm{card}(\Gamma) = N$. Consider the nonequispaced random Fourier matrix $\mathbf{A} \in \mathbb{C}^{m \times N}$ from Example 1 in Sect. 12.1. Improve Corollary 12.14 for this case using Corollary 8.10 (with $\lambda = 4/5$) by showing that the coherence μ of the normalized matrix $\tilde{\mathbf{A}} = \frac{1}{\sqrt{m}}\mathbf{A}$ satisfies

$$\mu \leq \sqrt{\frac{5\ln(5N^2/(2\varepsilon))}{4m}}$$

with probability at least $1 - \varepsilon$.

12.7. Let $\mathbf{A} \in \mathbb{C}^{m \times N}$ be the random sampling matrix in (12.4) associated to a BOS with constant $K \geq 1$. Use the probabilistic estimate (12.31) and the union bound to show that the restricted isometry constant δ_s of $\frac{1}{\sqrt{m}}\mathbf{A}$ satisfies $\delta_s \leq \delta$ with probability at least $1 - \varepsilon$ provided

$$m \geq \frac{8K^2}{3\delta^2}s^2\left(\ln(eN/s) + \ln(2s/\varepsilon)/s\right). \tag{12.95}$$

(In other words, the union bound is not enough to provide good estimates of δ_s, since it does not provide linear scaling of m in s.)

12.8. Let $\tilde{\mathbf{A}} = \frac{1}{\sqrt{m}}\mathbf{A} \in \mathbb{C}^{m \times N}$, where $\mathbf{A}$ is the random sampling matrix in (12.4) associated to a BOS with constant $K \geq 1$. Let $\mathbf{x} \in \mathbb{C}^N$ be an s-sparse vector. Show that, for $t > 0$,

$$\mathbb{P}(|\|\tilde{\mathbf{A}}\mathbf{x}\|_2^2 - \|\mathbf{x}\|_2^2| \geq t\|\mathbf{x}\|_2^2) \leq 2\exp\left(-\frac{m}{K^2 s}\frac{t^2/2}{1+t/3}\right).$$

12.9. Maurey method.
Let $V \subset \mathbb{R}^N$ be a finite set of cardinality Q and $\|\cdot\|$ be a norm on $\mathbb{R}^N$. Assume that for every integer $M \geq 1$ and for every $(\mathbf{v}_1, \ldots, \mathbf{v}_M) \in V^M$, we have $\mathbb{E}\|\sum_{j=1}^M \epsilon_j\mathbf{v}_j\| \leq A\sqrt{L}$ for a Rademacher vector $\epsilon = (\epsilon_1, \ldots, \epsilon_M)$ and some $A > 0$. Show that the covering numbers of the convex hull of V satisfy, for $t > 0$,

$$\sqrt{\ln \mathcal{N}(\text{conv}(V), \|\cdot\|, t)} \leq CAt^{-1}\sqrt{\ln(Q)},$$

where $C > 0$ is an appropriate universal constant.

12.10. Bernoulli selectors.
Let $\mathbf{U} \in \mathbb{C}^{N \times N}$ be a unitary matrix with constant K in (12.84). Let $\delta_j, j \in [N]$, be independent Bernoulli selectors, that is, random variables that take the value 1 with probability m/N and 0 with probability $1 - m/N$. Define the random sampling set $T = \{j \in [N], \delta_j = 1\}$, and let $\mathbf{A}$ be the random submatrix of $\mathbf{U}$ defined by $\mathbf{A} = \mathbf{R}_T\mathbf{U}$.

(a) In this context, the cardinality of T is random. Show that $\mathbb{E}\,\text{card}(T) = m$ and derive an upper bound for $\mathbb{P}(|m - \text{card}(T)| \geq t)$ when $t > 0$.

(b) Let $S \subset [N]$ with $\text{card}(S) = s$. Setting $\tilde{\mathbf{A}} = \sqrt{N/m}\,\mathbf{A} = \sqrt{N/m}\mathbf{R}_T\mathbf{U}$, verify that $\tilde{\mathbf{A}}^*\tilde{\mathbf{A}} = \frac{N}{m}\sum_{j=1}^N \delta_j\mathbf{X}_j\mathbf{X}_j^*$, where $(X_j)_\ell = \overline{U_{\ell,j}}$. Use the matrix Bernstein inequality to derive an upper bound for $\mathbb{P}(\|\tilde{\mathbf{A}}_S^*\tilde{\mathbf{A}}_S - \mathbf{Id}\|_{2\to 2} \geq t)$ when $t > 0$.

12.11. Lower bound for the Λ_1-problem.

Let $\mathbf{H} \in \mathbb{C}^{N\times N}$, $N = 2^n$, be the normalized Hadamard matrix described in Example 5 and Sect. 12.2. Denote by $\mathbf{v}_1,\dots,\mathbf{v}_N \in \mathbb{C}^N$ the columns of $\mathbf{H}$. Let $\Lambda \subset [N]$ be an arbitrary subset with $\text{card}(\Lambda) = cN$ for some $c \in (0,1)$. Show that there exists a vector $\mathbf{a} \in \mathbb{C}^\Lambda \setminus \{0\}$ such that

$$\|\sum_{j\in\Lambda} a_j\mathbf{v}_j\|_2 \geq c'\sqrt{\frac{\ln(N)}{N}}\|\sum_{j\in\Lambda} a_j\mathbf{v}_j\|_1,$$

where c' is a constant that only depends on c.

Consequently, the factor $\ln(N)^2$ in (12.88) cannot be improved to a term better than $\sqrt{\ln(N)}$ in general.

12.12. Fast Johnson–Lindenstrauss mappings.

Let $\mathbf{x}_1,\dots,\mathbf{x}_M \in \mathbb{C}^N$ be an arbitrary set of points. Let $\mathbf{A}$ be the $m \times N$ random sampling matrix (12.4) associated to a BOS with constant $K \geq 1$ and $\mathbf{D}_\epsilon \in \mathbb{R}^{N\times N}$ a diagonal matrix with a Rademacher vector ϵ on the diagonal where ϵ and $\mathbf{A}$ are independent. Show that if $m \geq C\eta^{-2}\ln(M)\ln^4(N)$ then with high probability the matrix $\mathbf{\Phi} = \frac{1}{\sqrt{m}}\mathbf{A}\mathbf{D}_\epsilon \in \mathbb{C}^{m\times N}$ provides a Johnson–Lindenstrauss embedding in the sense that

$$(1-\eta)\|\mathbf{x}_j - \mathbf{x}_k\|_2^2 \leq \|\mathbf{\Phi}\mathbf{x}_j - \mathbf{\Phi}\mathbf{x}_k\|_2^2 \leq (1+\eta)\|\mathbf{x}_j - \mathbf{x}_k\|_2^2 \quad \text{for all } j,k \in [M].$$

Chapter 13
Lossless Expanders in Compressive Sensing

In this chapter, we introduce another type of matrices that can be used when reconstructing sparse vectors from a limited number of measurements. They are adjacency matrices of certain bipartite graphs called lossless expanders. These objects are defined in Sect. 13.1, where some of their useful properties are established. In Sect. 13.2, we resort to probabilistic arguments to show that lossless expanders do indeed exist. Then, in Sect. 13.3, we prove that using their adjacency matrices as measurement matrices allows for a stable and robust reconstruction of sparse vectors via ℓ_1-minimization. One of the nice features of this approach is that the robust null space property can be proved directly in the ℓ_1-setting without resorting to auxiliary tools such as restricted isometry properties. Section 13.4 shows the stability and robustness of a thresholding-based algorithm. Finally, Sect. 13.5 presents a simple sublinear-time algorithm.

13.1 Definitions and Basic Properties

Throughout this chapter, we consider *bipartite* graphs, i.e., graphs $G = (L, R, E)$ where each edge $e := \overline{ji} \in E$ connects a left vertex $j \in L$ with a right vertex $i \in R$. Removing vertices if necessary, we assume that every vertex is attached to an edge. The sets L and R are identified with $[N]$ and $[m]$, respectively, where $N := \text{card}(L)$ and $m := \text{card}(R)$. The *degree* of a left vertex is the number of right vertices it connects with. A bipartite graph is called *left regular* with degree d if all left vertices have the same degree d (see Fig. 13.1). For such *left d-regular* bipartite graphs, given a set $J \subset [N]$ of left vertices, the cardinality of the set

$$E(J) := \{\overline{ji} \in E \text{ with } j \in J\}$$

of all edges emanating from J is exactly

$$\text{card}(E(J)) = d \, \text{card}(J).$$

S. Foucart and H. Rauhut, *A Mathematical Introduction to Compressive Sensing*, Applied and Numerical Harmonic Analysis, DOI 10.1007/978-0-8176-4948-7_13, © Springer Science+Business Media New York 2013

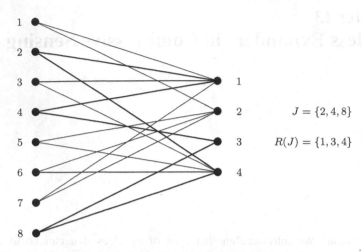

Fig. 13.1 A left regular bipartite graph with left degree two

The set

$$R(J) = \{i \in R : \text{ there is a } j \in J \text{ with } \overline{ji} \in E\}$$

of right vertices connected to J satisfies

$$\text{card}(R(J)) \leq d\, \text{card}(J).$$

Equality occurs if and only if no two edges emanating from J share a common right vertex. In the typical situation where the number N of left vertices is much larger than the number m of right vertices, such an equality cannot be met for large sets J. However, we shall see that an almost-equality can be met for small sets J. This almost-equality constitutes the expansion property, and left regular bipartite graphs with this property are called lossless expanders. The precise definition is given below. We stress the difference between this concept and the better-known concept of expanders which involves classical (nonbipartite) graphs—see Notes section.

Definition 13.1. A left regular bipartite graph with left degree d is called an (s, d, θ)-*lossless expander* if it satisfies the expansion property

$$\text{card}(R(J)) \geq (1 - \theta)\, d\, \text{card}(J) \tag{13.1}$$

for all sets J of left vertices such that $\text{card}(J) \leq s$. The smallest $\theta \geq 0$ for which the expansion property holds is called the sth *restricted expansion constant* and is denoted by θ_s.

It is readily seen that the restricted expansion constants satisfy

$$0 = \theta_1 \leq \theta_2 \leq \cdots \leq \theta_s \leq \theta_{s+1} \leq \cdots \leq \theta_N.$$

It is also possible to compare the constants θ_t of higher order in terms of the constants θ_s of lower order, similarly to Proposition 6.6 for restricted isometry constants.

Proposition 13.2. *Given a left d-regular bipartite graph, for integers $k, s \geq 1$,*

$$\theta_{ks} \leq (k-1)\theta_{2s} + \theta_s.$$

Proof. Let T be a set of left vertices satisfying $t := \mathrm{card}(T) \leq ks$. We partition T as $T = S_1 \cup \cdots \cup S_k$, where each S_ℓ satisfies $s_\ell := \mathrm{card}(S_\ell) \leq s$. We have

$$\mathrm{card}(R(T)) = \mathrm{card}\left(\bigcup_{1 \leq \ell \leq k} R(S_\ell) \right)$$

$$\geq \sum_{1 \leq \ell \leq k} \mathrm{card}(R(S_\ell)) - \sum_{1 \leq \ell_1 < \ell_2 \leq k} \mathrm{card}(R(S_{\ell_1}) \cap R(S_{\ell_2})).$$

In view of $\mathrm{card}(R(S_\ell)) \geq (1 - \theta_s)ds_\ell$ and of

$$\mathrm{card}(R(S_{\ell_1}) \cap R(S_{\ell_2})) = \mathrm{card}(R(S_{\ell_1})) + \mathrm{card}(R(S_{\ell_2})) - \mathrm{card}\,R(S_{\ell_1} \cup S_{\ell_2})$$

$$\leq ds_{\ell_1} + ds_{\ell_2} - (1 - \theta_{2s})d(s_{\ell_1} + s_{\ell_2}) = \theta_{2s}d(s_{\ell_1} + s_{\ell_2}),$$

we then obtain

$$\mathrm{card}(R(T)) \geq \sum_{1 \leq \ell \leq k} (1 - \theta_s)ds_\ell - \sum_{1 \leq \ell_1 < \ell_2 \leq k} \theta_{2s}d(s_{\ell_1} + s_{\ell_2})$$

$$= (1 - \theta_s)dt - \frac{\theta_{2s}d}{2}\left(\sum_{1 \leq \ell_1, \ell_2 \leq k} (s_{\ell_1} + s_{\ell_2}) - \sum_{1 \leq \ell_1 \leq k} (s_{\ell_1} + s_{\ell_1}) \right)$$

$$= (1 - \theta_s)dt - \frac{\theta_{2s}d}{2}\left(\sum_{1 \leq \ell_1 \leq k} (ks_{\ell_1} + t) - 2t \right)$$

$$= (1 - \theta_s)dt - \frac{\theta_{2s}d}{2}(2kt - 2t) = (1 - \theta_s - (k-1)\theta_{2s})dt.$$

This shows that $\theta_{ks} \leq \theta_s + (k-1)\theta_{2s}$, as announced. $\square$

We now formulate two lemmas and a corollary to be used in Sects. 13.3 and 13.4. They all formalize the intuition that collisions at right vertices are rare in a lossless expander.

Lemma 13.3. *Given a left d-regular bipartite graph, if disjoint sets J and K of left vertices satisfy* $\operatorname{card}(J) + \operatorname{card}(K) \leq s$, *then the set*

$$E(K; J) := \{\overline{ji} \in E(K) \text{ with } i \in R(J)\}$$

is small in the sense that

$$\operatorname{card}(E(K; J)) \leq \theta_s \, d \, s.$$

Proof. We partition the set E_0 of edges emanating from $J \cup K$ into three distinct subsets:

- The set E_1 of edges emanating from J
- The set E_2 of edges emanating from K and whose right vertices are not connected to any left vertex in J
- The set E_3 of edges emanating from K and whose right vertices are also connected to left vertices in J

We need to bound the cardinality of the set $E(K; J) = E_3$. In view of $\operatorname{card}(E_0) = d \operatorname{card}(J \cup K) = d(\operatorname{card}(J) + \operatorname{card}(K))$ and $\operatorname{card}(E_1) = d \operatorname{card}(J)$, we have

$$\operatorname{card}(E_3) = \operatorname{card}(E_0) - \operatorname{card}(E_1) - \operatorname{card}(E_2) = d \operatorname{card}(K) - \operatorname{card}(E_2). \quad (13.2)$$

We now observe that each right vertex $i \in R(K) \setminus R(J)$ gives rise to at least one edge emanating from K whose right vertex is not connected to any left vertex in J, so that

$$\operatorname{card}(E_2) \geq \operatorname{card}(R(K) \setminus R(J)) = \operatorname{card}(R(J \cup K)) - \operatorname{card}(R(J)).$$

We now take

$$\operatorname{card}(R(J)) \leq d \operatorname{card}(J)$$
$$\operatorname{card}(R(J \cup K)) \geq (1 - \theta) d \operatorname{card}(J \cup K) = (1 - \theta) \, d \, (\operatorname{card}(J) + \operatorname{card}(K))$$

into account to derive the inequality

$$\operatorname{card}(E_2) \geq (1 - \theta) d \operatorname{card}(K) - \theta d \operatorname{card}(J). \quad (13.3)$$

Substituting (13.3) into (13.2), we conclude that

$$\operatorname{card}(E_3) \leq \theta \, d \, (\operatorname{card}(K) + \operatorname{card}(J)),$$

which is the desired result.   □

Lemma 13.4. *Let S be a set of s left vertices in a left d-regular bipartite graph. For each $i \in R(S)$, if $\ell(i) \in S$ denotes a fixed left vertex connected to i, then*

$$E'(S) := \{\overline{ji} \in E(S) : j \neq \ell(i)\}$$

is small in the sense that

$$\operatorname{card}(E'(S)) \leq \theta_s \, d \, s.$$

Proof. The set $E(S)$ of edges emanating from S is partitioned as $E(S) = E'(S) \cup E''(S)$, where $E''(S) := \{\overline{\ell(i) \, i}, i \in R(S)\}$. Since $\operatorname{card}(E(S)) = d \, s$ and $\operatorname{card}(E''(S)) = \operatorname{card}(R(S)) \geq (1 - \theta_s) \, d \, s$, we conclude that $\operatorname{card}(E'(S)) = \operatorname{card}(E(S)) - \operatorname{card}(E''(S)) \leq \theta_s \, d \, s$. $\square$

Corollary 13.5. *Given a left d-regular bipartite graph, if S is a set of s left vertices, then the set*

$$R_1(S) := \{i \in R(S) : \text{there is a unique } j \in S \text{ with } \overline{ji} \in E\}$$

of right vertices connected to exactly one left vertex in S is large in the sense that

$$\operatorname{card}(R_1(S)) \geq (1 - 2\theta_s) \, d \, s.$$

Proof. Fixing a left vertex $\ell(i) \in S$ for each $i \in R(S)$ as in Lemma 13.4, any $i \in R_{\geq 2}(S) := R(S) \setminus R_1(S)$ gives rise to at least one edge in $E'(S)$. Thus, $\operatorname{card}(R_{\geq 2}(S)) \leq \operatorname{card}(E'(S)) \leq \theta_s \, ds$, and we consequently have $\operatorname{card}(R_1(S)) = \operatorname{card}(R(S)) - \operatorname{card}(R_{\geq 2}(S)) \geq (1 - 2\theta_s) \, d \, s$. $\square$

13.2 Existence of Lossless Expanders

In this section, we prove that lossless expanders with parameters relevant to compressive sensing do exist. As a matter of fact, we prove that most left regular bipartite graphs are lossless expanders, i.e., that random left regular bipartite graphs are, with high probability, lossless expanders.

Theorem 13.6. *For $0 < \varepsilon < 1/2$, the proportion of (s, d, θ)-lossless expanders among all left d-regular bipartite graphs with N left vertices and m right vertices exceeds $1 - \varepsilon$ provided that*

$$d = \left\lceil \frac{1}{\theta} \ln \left(\frac{eN}{\varepsilon s} \right) \right\rceil \quad \text{and} \quad m \geq c_\theta \, s \ln \left(\frac{eN}{\varepsilon s} \right),$$

where c_θ is a constant depending only on θ.

Proof. Since each of the left vertices $j \in [N]$ connects to a set $R(j) \subset [m]$ of d right vertices, the total number of left d-regular bipartite graphs is

$$\binom{m}{d}^N.$$

Among these graphs, a graph fails to be an (s, d, θ)-lossless expander if there is a set $J \subset [N]$ with $2 \leq j := \mathrm{card}(J) \leq s$ such that $\mathrm{card}(R(J)) < (1 - \theta)dj$, i.e.,

$$R(J) \subset I \quad \text{for some set } I \subset [m] \text{ with } \mathrm{card}(I) = r_j := \lceil (1 - \theta)dj \rceil - 1.$$

For fixed sets I and J, the number of left d-regular bipartite graphs satisfying the latter is

$$\binom{r_j}{d}^j \binom{m}{d}^{N-j}.$$

Taking the union over all possible sets I and J, we see that the number of left d-regular bipartite graphs that are not (s, d, θ)-lossless expanders is at most

$$\sum_{j=2}^{s} \binom{N}{j} \binom{m}{r_j} \binom{r_j}{d}^j \binom{m}{d}^{N-j}.$$

Therefore, the proportion of graphs that are not (s, d, θ)-lossless expanders among the left d-regular bipartite graphs is at most

$$p := \sum_{j=2}^{s} p_j, \qquad \text{where} \quad p_j := \binom{N}{j} \binom{m}{r_j} \left(\frac{\binom{r_j}{d}}{\binom{m}{d}} \right)^j.$$

Using the simple inequalities of Lemma C.5, namely,

$$\left(\frac{n}{k} \right)^k \leq \binom{n}{k} \leq \left(\frac{en}{k} \right)^k,$$

we obtain, for each $2 \leq j \leq s$,

$$p_j \leq \left(\frac{eN}{j} \right)^j \left(\frac{em}{r_j} \right)^{r_j} \left(\frac{\left(\frac{er_j}{d} \right)^d}{\left(\frac{m}{d} \right)^d} \right)^j = \left(\frac{eN}{j} \right)^j e^{r_j + dj} \left(\frac{r_j}{m} \right)^{dj - r_j}.$$

With the choice $c_\theta = 2e^{2/\theta}/\theta$, we now observe that

$$m \geq e^{2/\theta} \frac{2}{\theta} \ln \left(\frac{eN}{\varepsilon s} \right) \qquad s \geq e^{2/\theta} d s \qquad \text{and} \qquad r_j \leq (1 - \theta) d j \leq d j.$$

Taking $j \leq s$ into account, we derive

$$p_j \leq \left(\frac{eN}{j}\right)^j e^{(2-\theta)dj} \left(\frac{dj}{e^{2/\theta}ds}\right)^{dj-r_j} \leq \left(\frac{eN}{j}\right)^j e^{(2-\theta)dj} \left(\frac{j}{e^{2/\theta}s}\right)^{\theta dj}$$

$$= \left(\frac{eN}{j} e^{-\theta d} \left(\frac{j}{s}\right)^{\theta d}\right)^j \leq \left(\frac{eN}{j} \frac{\varepsilon s}{eN} \left(\frac{j}{s}\right)^{\theta d}\right)^j = \left(\varepsilon \left(\frac{j}{s}\right)^{\theta d-1}\right)^j \leq \varepsilon^j.$$

It follows that

$$p = \sum_{j=2}^{s} p_j \leq \sum_{j=2}^{s} \varepsilon^j \leq \sum_{j=2}^{\infty} \varepsilon^j = \frac{\varepsilon^2}{1-\varepsilon} < \varepsilon,$$

which is the desired result. □

To obtain a result where the targeted probability does not enter the number of measurements, one can simply make a specific choice for ε, e.g., $\varepsilon = s/(eN)$.

Corollary 13.7. *A bipartite graph with N left vertices and m right vertices drawn at random among all left d-regular graphs where $d := \lceil 2\ln(eN/s)/\theta\rceil$ satisfies $\theta_s \leq \theta$ with probability at least*

$$1 - \frac{s}{eN}$$

provided that, for some constant c'_θ depending only on θ,

$$m \geq c'_\theta \, s \, \ln\left(\frac{eN}{s}\right).$$

The constants in Theorem 13.6 and Corollary 13.7 are $c_\theta = 2e^{2/\theta}/\theta$ and $c'_\theta = 4e^{2/\theta}/\theta$. They become extremely large when θ is small. We are now going to improve this dependence on θ. To this end, we change the probability model for the selection of right vertices connected to a left vertex $j \in [N]$. Precisely, for each $j \in [N]$, we choose d random elements of $[m]$ independently, allowing repetition. The resulting graph may not be left d-regular, i.e., we may have $\text{card}(R(j)) < d$ for some $j \in [N]$. However, once the expansion property (13.1) is established, it suffices to artificially add edges to ensure that $\text{card}(R(j)) = d$ for all $j \in [N]$, hence to create an (s, d, θ)-lossless expander.

Theorem 13.8. *There exist (s, d, θ)-lossless expanders with N left vertices and m right vertices provided that*

$$d = \left\lceil \frac{2}{\theta} \ln\left(\frac{eN}{s}\right)\right\rceil \qquad \text{and} \qquad m \geq \frac{3e^2}{\theta^2} s \ln\left(\frac{eN}{s}\right).$$

Proof. Under the probability model just described, we want to prove that

$$p := \mathbb{P}\big(\operatorname{card}(R(J)) < (1 - \theta)d \operatorname{card}(J) \text{ for some } J \subset [N] \text{ with } \operatorname{card}(J) \le s\big)$$

is smaller than 1. For a fixed $J \subset [N]$ with $j = \operatorname{card}(R(J)) \le s$, we define $p_j := \mathbb{P}\big(\operatorname{card}(R(J)) < (1 - \theta)dj\big)$, so that the union bound yields

$$p \le \sum_{j=1}^{s} \binom{N}{j} p_j \le \sum_{j=1}^{s} \left(\frac{eN}{j}\right)^{j} p_j.$$

We think of the right vertices connected to J as a sequence $(i_1, i_2, \ldots, i_{dj})$ of independent random elements of $[m]$ produced one at a time. In view of possible repetitions within $(i_1, i_2, \ldots, i_{dj})$, a count of the cardinality of $R(J)$ gives dj minus the cardinality of the set I of i_ℓ that have appeared earlier. Since the probability that a fixed i_ℓ has appeared earlier is

$$\mathbb{P}\big(i_\ell \in \{i_1, \ldots, i_{\ell-1}\}\big) \le \frac{\ell - 1}{m} \le \frac{dj}{m},$$

we obtain, setting $t_j := \lfloor \theta dj \rfloor + 1$,

$$p_j = \mathbb{P}(\operatorname{card}(I) > \theta dj) = \mathbb{P}(\operatorname{card}(I) \ge t_j)$$

$$\le \binom{dj}{t_j}\left(\frac{dj}{m}\right)^{t_j} \le \left(\frac{edj}{t_j}\right)^{t_j}\left(\frac{dj}{m}\right)^{t_j} \le \left(\frac{edj}{\theta dj}\right)^{t_j}\left(\frac{dj}{m}\right)^{t_j} = \left(\frac{edj}{\theta m}\right)^{t_j}.$$

The conditions on d and m imply that

$$d < \frac{2}{\theta} \ln\left(\frac{eN}{s}\right) + 1 \le \frac{3}{\theta} \ln\left(\frac{eN}{s}\right) \qquad \text{and} \qquad m \ge \frac{e^2}{\theta} ds.$$

It now follows that

$$p_j \le \left(\frac{j}{es}\right)^{t_j} \le \left(\frac{j}{es}\right)^{\theta dj} = \left(e^{-\theta d}\left(\frac{j}{s}\right)^{\theta d}\right)^{j} \le \left(\frac{s^2}{e^2 N^2}\left(\frac{j}{s}\right)^{\theta d}\right)^{j},$$

where the last step used $d \ge 2\ln(eN/s)/\theta$. Finally, we deduce that

$$p \le \sum_{j=1}^{s}\left(\frac{eN}{j}\frac{s^2}{e^2 N^2}\left(\frac{j}{s}\right)^{\theta d}\right)^{j} \le \sum_{j=1}^{s}\left(\frac{s}{eN}\left(\frac{j}{s}\right)^{\theta d-1}\right)^{j} \le \sum_{j=1}^{\infty}\left(\frac{1}{e}\right)^{j} = \frac{1}{e - 1}.$$

It remains to remark that $1/(e - 1) < 1$ to conclude the proof. $\square$

The dependence on θ being set aside, Theorem 13.8 is optimal in the sense that the existence of a lossless expander forces the number m of right vertices to satisfy $m \ge c\, s \ln(eN/s)$ for some $c > 0$, as we shall see in Corollary 13.14.

13.3   Analysis of Basis Pursuit

In this section, we prove that lossless expanders provide suitable measurement matrices for basis pursuit. These matrices are the adjacency matrices of the bipartite graph, defined as follows.

Definition 13.9. The *adjacency matrix* of a bipartite graph $G = ([N], [m], E)$ is the $m \times N$ matrix $\mathbf{A}$ with entries

$$A_{i,j} = \begin{cases} 1 & \text{if } \overline{ji} \in E, \\ 0 & \text{if } \overline{ji} \notin E. \end{cases}$$

It is completely equivalent, and sometimes more appropriate, to think of an (s, d, θ)-lossless expander as a matrix $\mathbf{A}$ populated with zeros and ones, with d ones per column, and such that there are at least $(1 - \theta)dk$ nonzero rows in any submatrix of $\mathbf{A}$ composed of $k \leq s$ columns. Because of their zero–one structure, such matrices present some advantages over subgaussian random matrices; notably they require less storage space and they allow for faster computations. They also allow for stable and robust sparse recovery, as established below. As usual, perfect recovery is obtained in the particular case where the vector $\mathbf{x}$ is exactly s-sparse and the measurement error η equals zero.

Theorem 13.10. *Suppose that $\mathbf{A} \in \{0,1\}^{m \times N}$ is the adjacency matrix of a left d-regular bipartite graph satisfying*

$$\theta_{2s} < \frac{1}{6}.$$

For $\mathbf{x} \in \mathbb{C}^N$ and $\mathbf{e} \in \mathbb{C}^m$ with $\|\mathbf{e}\|_1 \leq \eta$, if $\mathbf{y} = \mathbf{A}\mathbf{x} + \mathbf{e}$, then a solution $\mathbf{x}^\sharp$ of

$$\underset{\mathbf{z} \in \mathbb{C}^N}{\text{minimize}} \ \|\mathbf{z}\|_1 \quad \text{subject to } \|\mathbf{A}\mathbf{z} - \mathbf{y}\|_1 \leq \eta$$

approximates the vector $\mathbf{x}$ with ℓ_1-error

$$\|\mathbf{x} - \mathbf{x}^\sharp\|_1 \leq \frac{2(1 - 2\theta)}{(1 - 6\theta)} \sigma_s(\mathbf{x})_1 + \frac{4}{(1 - 6\theta)d} \eta.$$

According to Theorem 4.19, this is a corollary of the following result.

Theorem 13.11. *The adjacency matrix $\mathbf{A} \in \{0,1\}^{m \times N}$ of a left d-regular bipartite graph satisfies the ℓ_1-robust null space property of order s provided $\theta_{2s} < 1/6$, precisely*

$$\|\mathbf{v}_S\|_1 \leq \frac{2\theta_{2s}}{1 - 4\theta_{2s}} \|\mathbf{v}_{\overline{S}}\|_1 + \frac{1}{(1 - 4\theta_{2s})d} \|\mathbf{A}\mathbf{v}\|_1 \tag{13.4}$$

for all $\mathbf{v} \in \mathbb{C}^N$ and all $S \subset [N]$ with $\mathrm{card}(S) = s$.

We isolate the following two lemmas for the proof of Theorem 13.11.

Lemma 13.12. *Let $\mathbf{A} \in \{0, 1\}^{m \times N}$ be the adjacency matrix of a left d-regular bipartite graph. If S and T are two disjoint subsets of $[N]$ and if $\mathbf{x} \in \mathbb{C}^N$, then*

$$\|(\mathbf{A}\mathbf{x}_S)_{R(T)}\|_1 \leq \theta_{s+t}\, d\,(s + t)\, \|\mathbf{x}_S\|_\infty,$$

where $s = \mathrm{card}(S)$ and $t = \mathrm{card}(T)$.

Proof. We estimate the term $\|(\mathbf{A}\mathbf{x}_S)_{R(T)}\|_1$ as

$$\|(\mathbf{A}\mathbf{x}_S)_{R(T)}\|_1 = \sum_{i \in R(T)} |(\mathbf{A}\mathbf{x}_S)_i| = \sum_{i=1}^m \mathbf{1}_{\{i \in R(T)\}} \left| \sum_{j \in S} A_{i,j} x_j \right|$$

$$\leq \sum_{i=1}^m \mathbf{1}_{\{i \in R(T)\}} \sum_{j \in S} \mathbf{1}_{\{\overline{ji} \in E\}} |x_j|$$

$$= \sum_{j \in S} \sum_{i=1}^m \mathbf{1}_{\{i \in R(T) \text{ and } \overline{ji} \in E\}} |x_j| = \sum_{\overline{ji} \in E(S;T)} |x_j|$$

$$\leq \mathrm{card}(E(S; T)) \|\mathbf{x}_S\|_\infty.$$

The conclusion follows from the bound on $\mathrm{card}(E(S; T))$ of Lemma 13.3. □

Lemma 13.13. *Let $\mathbf{A} \in \{0, 1\}^{m \times N}$ be the adjacency matrix of a left d-regular bipartite graph. Given an s-sparse vector $\mathbf{w} \in \mathbb{C}^N$, let $\mathbf{w}' \in \mathbb{C}^m$ be defined by $w'_i := w_{\ell(i)}$, $i \in [m]$, where*

$$\ell(i) := \underset{j \in [N]}{\mathrm{argmax}} \{|w_j|, \overline{ji} \in E\}.$$

Then

$$\|\mathbf{A}\mathbf{w} - \mathbf{w}'\|_1 \leq \theta_s\, d\, \|\mathbf{w}\|_1.$$

Proof. We may and do assume that the left vertices are ordered such that

$$|w_1| \geq |w_2| \geq \cdots \geq |w_s| \geq |w_{s+1}| = \cdots = |w_N| = 0.$$

In this way, the edge $\overline{\ell(i)\,i}$ can be thought of as the first edge arriving at the right vertex i. Since the vector $\mathbf{w} \in \mathbb{C}^N$ is supported on $S := [s]$, and since $\ell(i) \in S$ whenever $i \in R(S)$, we have

$$(\mathbf{Aw} - \mathbf{w'})_i = \sum_{j=1}^{N} A_{i,j} w_j - w_{\ell(i)} = \sum_{j \in S} 1_{\{\overline{j}i \in E \text{ and } j \neq \ell(i)\}} w_j.$$

Thus, we obtain

$$\|\mathbf{Aw} - \mathbf{w'}\|_1 = \sum_{i=1}^{m} \left| \sum_{j \in S} 1_{\{\overline{j}i \in E \text{ and } j \neq \ell(i)\}} w_j \right| \leq \sum_{i=1}^{m} \sum_{j \in S} 1_{\{\overline{j}i \in E \text{ and } j \neq \ell(i)\}} |w_j|$$

$$= \sum_{j \in S} \left(\sum_{i=1}^{m} 1_{\{\overline{j}i \in E \text{ and } j \neq \ell(i)\}} \right) |w_j| = \sum_{j=1}^{s} c_j |w_j|,$$

where $c_j := \sum_{i=1}^{m} 1_{\{\overline{j}i \in E \text{ and } j \neq \ell(i)\}}$. For all $k \in [s]$, we observe that

$$C_k := \sum_{j=1}^{k} c_j = \sum_{j=1}^{k} \sum_{i=1}^{m} 1_{\{\overline{j}i \in E \text{ and } j \neq \ell(i)\}} = \mathrm{card}(\{\overline{j}i \in E([k]), j \neq \ell(i)\})$$

$$\leq \theta_s \, d \, k, \tag{13.5}$$

where the last inequality was derived from Lemma 13.4. Setting $C_0 = 0$ and performing a *summation by parts*, we have

$$\sum_{j=1}^{s} c_j |w_j| = \sum_{j=1}^{s} (C_j - C_{j-1}) |w_j| = \sum_{j=1}^{s} C_j |w_j| - \sum_{j=1}^{s} C_{j-1} |w_j|$$

$$= \sum_{j=1}^{s} C_j |w_j| - \sum_{j=0}^{s-1} C_j |w_{j+1}| = \sum_{j=1}^{s-1} C_j (|w_j| - |w_{j+1}|) + C_s |w_s|.$$

Since $|w_j| - |w_{j+1}| \geq 0$, the bound (13.5) yields

$$\sum_{j=1}^{s} c_j |w_j| \leq \sum_{j=0}^{s-1} \theta_s \, d \, j (|w_j| - |w_{j+1}|) + \theta_s \, d \, s \, |w_s| = \sum_{j=1}^{s} \theta_s \, d \, |w_j|, \tag{13.6}$$

where the last equality was derived by reversing the summation by parts process after replacing c_j by $\theta_s d$. The result is proved. $\qquad\square$

We are now ready to prove the key result of this section.

Proof (of Theorem 13.11). Let $\mathbf{v} \in \mathbb{C}^N$ be a fixed vector, and let S_0 be an index set of s largest absolute entries of $\mathbf{v}$, S_1 an index set of next s largest absolute entries, etc. It is enough to establish (13.4) for $S = S_0$. We start by writing

$$d\,\|\mathbf{v}_{S_0}\|_1 = d \sum_{j\in S_0} |v_j| = \sum_{\substack{\overline{ji}\in E(S_0)}} |v_j| = \sum_{i\in R(S_0)} \sum_{\substack{j\in S_0 \\ \overline{ji}\in E}} |v_j|$$

$$= \sum_{i\in R(S_0)} |v_{\ell(i)}| + \sum_{i\in R(S_0)} \sum_{\substack{j\in S_0\setminus\{\ell(i)\} \\ \overline{ji}\in E}} |v_j|, \tag{13.7}$$

where the notation of Lemma 13.13 has been used. We now observe that, for $i \in R(S_0)$,

$$(\mathbf{Av})_i = \sum_{j\in[N]} A_{i,j} v_j = \sum_{\substack{j\in[N] \\ \overline{ji}\in E}} v_j = \sum_{k\geq 0}\sum_{\substack{j\in S_k \\ \overline{ji}\in E}} v_j = v_{\ell(i)} + \sum_{\substack{j\in S_0\setminus\{\ell(i)\} \\ \overline{ji}\in E}} v_j + \sum_{k\geq 1}\sum_{\substack{j\in S_k \\ \overline{ji}\in E}} v_j.$$

It follows that

$$|v_{\ell(i)}| \leq \sum_{\substack{j\in S_0\setminus\{\ell(i)\} \\ \overline{ji}\in E}} |v_j| + \sum_{k\geq 1}\sum_{\substack{j\in S_k \\ \overline{ji}\in E}} |v_j| + |(\mathbf{Av})_i|.$$

Summing over all $i \in R(S_0)$ and substituting into (13.7), we obtain

$$d\,\|\mathbf{v}_{S_0}\|_1 \leq 2 \sum_{i\in R(S_0)}\sum_{\substack{j\in S_0\setminus\{\ell(i)\} \\ \overline{ji}\in E}} |v_j| + \sum_{k\geq 1}\sum_{i\in R(S_0)}\sum_{\substack{j\in S_k \\ \overline{ji}\in E}} |v_j| + \|\mathbf{Av}\|_1. \tag{13.8}$$

For the first term in the right-hand side of (13.8), we apply Lemma 13.13 to $\mathbf{w} = |\mathbf{v}_{S_0}|$ (i.e., $w_j = |v_j|$ if $j \in S_0$ and $w_j = 0$ otherwise) to obtain

$$\sum_{i\in R(S_0)}\sum_{\substack{j\in S_0\setminus\{\ell(i)\} \\ \overline{ji}\in E}} |v_j| = \|\mathbf{Aw} - \mathbf{w}'\|_1 \leq \theta_s\, d\,\|\mathbf{w}\|_1 = \theta_s\, d\,\|\mathbf{v}_{S_0}\|_1. \tag{13.9}$$

For the second term in the right-hand side of (13.8), we apply Lemmas 13.12 and 6.10 to obtain

$$\sum_{k\geq 1}\sum_{i\in R(S_0)}\sum_{\substack{j\in S_k \\ \overline{ji}\in E}} |v_j| = \sum_{k\geq 1} \|(\mathbf{A}|\mathbf{v}_{S_k}|)_{R(S_0)}\|_1 \leq \sum_{k\geq 1} \theta_{2s}\, d\, 2s\, \|\mathbf{v}_{S_k}\|_\infty$$

$$\leq 2\theta_{2s}\, d \sum_{k\geq 1} \|\mathbf{v}_{S_{k-1}}\|_1 \leq 2\theta_{2s}\, d\,\|\mathbf{v}\|_1. \tag{13.10}$$

Finally, substituting (13.9) and (13.10) into (13.8), we deduce

$$d\,\|\mathbf{v}_{S_0}\|_1 \leq 2\,\theta_s\,d\,\|\mathbf{v}_{S_0}\|_1 + 2\,\theta_{2s}\,d\,\|\mathbf{v}\|_1 + \|\mathbf{A}\mathbf{v}\|_1$$
$$\leq 4\,\theta_{2s}\,d\,\|\mathbf{v}_{S_0}\|_1 + 2\,\theta_{2s}\,d\,\|\mathbf{v}_{\overline{S_0}}\|_1 + \|\mathbf{A}\mathbf{v}\|_1.$$

Rearranging the latter leads to the desired inequality (13.4). □

To close this section, we highlight that the exact s-sparse recovery by basis pursuit using lossless expanders provides, in retrospect, a lower bound for the number of right vertices in a lossless expander.

Corollary 13.14. *For $s \geq 2$ and $\theta < 1/25$, an (s, d, θ)-lossless expander with N left vertices must have a number m of right vertices bounded from below by*

$$m \geq \frac{c_1}{\theta}\,s\,\ln\left(\frac{c_2\theta N}{s}\right)$$

for some absolute constants $c_1, c_2 > 0$.

Proof. Let us consider $k := \lfloor 1/(25\theta) \rfloor \geq 1$ and $s' := \lfloor s/2 \rfloor \geq 1$. According to Proposition 13.2, we have

$$\theta_{4ks'} \leq 4k\,\theta_{2s'} \leq 4k\,\theta \leq \frac{4}{25} < \frac{1}{6}.$$

Therefore, Theorem 13.10 implies that every $2ks'$ sparse vector $\mathbf{x} \in \mathbb{R}^N$ is recovered from $\mathbf{y} = \mathbf{A}\mathbf{x} \in \mathbb{R}^m$ via ℓ_1-minimization. Theorem 10.11 then implies that, with $c = 1/\ln 9$,

$$m \geq c\,k\,s'\,\ln\left(\frac{N}{4ks'}\right).$$

In view of $1/(50\theta) \leq k \leq 1/(25\theta)$ and of $2s/3 \leq s' \leq s/2$, we conclude that

$$m \geq \frac{c}{75\,\theta}\,s\,\ln\left(\frac{25\theta N}{2s}\right),$$

which is the desired result with $c_1 = c/75$ and $c_2 = 25/2$. □

13.4 Analysis of an Iterative Thresholding Algorithm

In this section, we prove that lossless expanders provide suitable measurement matrices for other algorithms besides basis pursuit. First, in the real setting, we consider a variation of the iterative hard thresholding algorithm. Precisely, starting with an initial s-sparse vector $\mathbf{x}^0 \in \mathbb{R}^N$, typically $\mathbf{x}^0 = \mathbf{0}$, we iterate the scheme

$$\mathbf{x}^{n+1} = H_s(\mathbf{x}^n + \mathcal{M}(\mathbf{y} - \mathbf{A}\mathbf{x}^n)). \tag{13.11}$$

The nonlinear operator $\mathcal{M} = \mathcal{M}_{\mathbf{A}}$ is the *median operator*, which is defined componentwise by

$$(\mathcal{M}(\mathbf{z}))_j := \text{median}[z_i, i \in R(j)] \quad \text{for } \mathbf{z} \in \mathbb{C}^m \text{ and } j \in [N].$$

Here, $R(j) = R(\{j\})$ denotes the set of right vertices connected to j, and the median of the d numbers z_i, $i \in R(j)$, is defined to be the $\lceil d/2 \rceil$th largest of these numbers. The properties of the algorithm (13.11) are very similar to the properties established in Sect. 6.3 for the iterative hard thresholding algorithm.

Theorem 13.15. *Suppose that* $\mathbf{A} \in \{0,1\}^{m \times N}$ *is the adjacency matrix of a left* d-*regular bipartite graph satisfying*

$$\theta_{3s} < \frac{1}{12}.$$

Then, for $\mathbf{x} \in \mathbb{R}^N$, $\mathbf{e} \in \mathbb{R}^m$, *and* $S \subset [N]$ *with* $\text{card}(S) = s$, *the sequence* $(\mathbf{x}^n)$ *defined by* (13.11) *with* $\mathbf{y} = \mathbf{A}\mathbf{x} + \mathbf{e}$ *satisfies, for any* $n \geq 0$,

$$\|\mathbf{x}^n - \mathbf{x}_S\|_1 \leq \rho^n \|\mathbf{x}^0 - \mathbf{x}_S\|_1 + \frac{\tau}{d}\|\mathbf{A}\mathbf{x}_{\overline{S}} + \mathbf{e}\|_1, \quad\quad (13.12)$$

where $\rho < 1$ *and* τ *depend only on* θ_{3s}. *In particular, if the sequence* $(\mathbf{x}^n)$ *clusters around some* $\mathbf{x}^\sharp \in \mathbb{R}^N$, *then*

$$\|\mathbf{x} - \mathbf{x}^\sharp\|_1 \leq C\sigma_s(\mathbf{x})_1 + \frac{D}{d}\|\mathbf{e}\|_1$$

for some constants $C, D > 0$ *depending only on* θ_{3s}.

The proof relies on the fact that the median operator approximately inverts the action of $\mathbf{A}$ on sparse vectors. We state this as a lemma involving the slightly more general *quantile operators* $\mathcal{Q}_k$ in place of $\mathcal{M} = \mathcal{Q}_{\lceil d/2 \rceil}$. It is defined componentwise by

$$(\mathcal{Q}_k(\mathbf{z}))_j := q_k[z_i, i \in R(j)] \quad \text{for } \mathbf{z} \in \mathbb{C}^m \text{ and } j \in [N],$$

where the quantile q_k denotes the kth largest element, i.e.,

$$q_k[a_1, \ldots, a_d] = a_{\pi(k)}$$

if $\pi : [d] \to [d]$ is a permutation for which $a_{\pi(1)} \geq a_{\pi(2)} \geq \cdots \geq a_{\pi(d)}$. We will use the following observations, to be established in Exercise 13.10:

$$|q_k[a_1, \ldots, a_d]| \leq q_k[|a_1|, \ldots, |a_d|], \quad\quad \text{if } 2k \leq d+1, \quad\quad (13.13)$$

$$q_k[b_1, \ldots, b_d] \leq \frac{b_1 + \cdots + b_d}{k} \quad\quad \text{if } b_j \geq 0 \text{ for all } j. \quad\quad (13.14)$$

Lemma 13.16. *Let* $\mathbf{A} \in \{0,1\}^{m \times N}$ *be the adjacency matrix of a left d-regular bipartite graph and let k be an integer satisfying $2\theta_s d < k \leq (d+1)/2$. If S is a subset of $[N]$ with size s, then*

$$\| (\mathcal{Q}_k(\mathbf{A}\mathbf{x}_S + \mathbf{e}) - \mathbf{x})_S \|_1 \leq \frac{2\theta_s d}{k - 2\theta_s d} \|\mathbf{x}_S\|_1 + \frac{1}{k - 2\theta_s d} \|\mathbf{e}_{R(S)}\|_1 \quad (13.15)$$

for all $\mathbf{x} \in \mathbb{R}^N$ and all $\mathbf{e} \in \mathbb{R}^m$.

Proof. According to the definition of $\mathcal{Q}_k$ and to (13.13), we have

$$\| (\mathcal{Q}_k(\mathbf{A}\mathbf{x}_S + \mathbf{e}) - \mathbf{x})_S \|_1 = \sum_{j \in S} |q_k[(\mathbf{A}\mathbf{x}_S + \mathbf{e})_i, i \in R(j)] - x_j|$$

$$= \sum_{j \in S} |q_k[(\mathbf{A}\mathbf{x}_S)_i + e_i - x_j, i \in R(j)]|$$

$$\leq \sum_{j \in S} q_k[|(\mathbf{A}\mathbf{x}_S)_i - x_j + e_i|, i \in R(j)]$$

$$= \sum_{j \in S} q_k\left[\left| \sum_{\substack{\ell \in S \setminus \{j\} \\ \overline{\ell i} \in E}} x_\ell + e_i \right|, i \in R(j) \right].$$

We now proceed by induction on $s = \mathrm{card}(S)$ to show that the latter is bounded above by the right-hand side of (13.15). If $s = 1$, i.e., if $S = \{j\}$ for some $j \in S$ so that there is no $\ell \in S \setminus \{j\}$, we have the stronger estimate

$$q_k\left[\left| \sum_{\substack{\ell \in S \setminus \{j\} \\ \overline{\ell i} \in E}} x_\ell + e_i \right|, i \in R(j) \right] = q_k[|e_i|, i \in R(j)] \leq \frac{1}{k} \|\mathbf{e}_{R(j)}\|_1,$$

where we have used (13.14). Let us now assume that the induction hypothesis holds up to $s - 1$ for some $s \geq 2$, and let us show that it holds for s, too. For $S \subset [N]$ with $\mathrm{card}(S) = s$ and for $j \in S$, we introduce the set

$$R_1(j, S) := R(j) \setminus \bigcup_{\ell \in S \setminus \{j\}} R(\ell)$$

of right vertices connected only to j in S. We recall from Corollary 13.5 that

$$\sum_{j \in S} \mathrm{card}(R_1(j, S)) = \mathrm{card}(R_1(S)) \geq (1 - 2\theta_s)d\, s. \quad (13.16)$$

Thus, there exists $j^* \in S$ such that $r := \mathrm{card}(R_1(j^*, S)) \geq (1 - 2\theta_s)d$. This means that there are at most $d - r \leq 2\theta_s d$ right vertices in $R(j^*) \setminus R_1(j^*, S)$. By definition of q_k, there exist k distinct $i_1, \dots, i_k \in R(j^*)$ such that, for all $h \in [k]$,

$$q_k \left[\left| \sum_{\substack{\ell \in S \setminus \{j^*\} \\ \overline{\ell i} \in E}} x_\ell + e_i \right|, i \in R(j^*) \right] \leq \left| \sum_{\substack{\ell \in S \setminus \{j^*\} \\ \overline{\ell i_h} \in E}} x_\ell + e_{i_h} \right|. \tag{13.17}$$

At least $k' := k - (d - r) \geq k - 2\theta_s d$ elements among $i_1, \ldots, i_k$ are in $R_1(j^*, S)$. Averaging (13.17) over these elements i_h, keeping in mind that there are no $\ell \in S \setminus \{j^*\}$ with $\overline{\ell i_h} \in E$ in this case, we obtain

$$q_k \left[\left| \sum_{\substack{\ell \in S \setminus \{j^*\} \\ \overline{\ell i} \in E}} x_\ell + e_i \right|, i \in R(j^*) \right] \leq \frac{1}{k - 2\theta_s d} \| \mathbf{e}_{R_1(j^*, S)} \|_1. \tag{13.18}$$

On the other hand, if $T := S \setminus \{j^*\}$ and if $j \in T$, we have

$$\left| \sum_{\substack{\ell \in S \setminus \{j\} \\ \overline{\ell i} \in E}} x_\ell + e_i \right| = \left| \sum_{\substack{\ell \in T \setminus \{j\} \\ \overline{\ell i} \in E}} x_\ell + 1_{\{\overline{j^* i} \in E\}} x_{j^*} + e_i \right|.$$

Applying the induction hypothesis with S replaced by T and e_i replaced by $e'_i = 1_{\{\overline{j^* i} \in E\}} x_{j^*} + e_i$ gives, in view of $\theta_{s-1} \leq \theta_s$,

$$\sum_{j \in T} q_k \left[\left| \sum_{\substack{\ell \in S \setminus \{j\} \\ \overline{\ell i} \in E}} x_\ell + e_i \right|, i \in R(j) \right] \leq \frac{2\theta_s d}{k - 2\theta_s d} \| \mathbf{x}_T \|_1 + \frac{1}{k - 2\theta_s d} \| \mathbf{e}'_{R(T)} \|_1. \tag{13.19}$$

In order to bound $\| \mathbf{e}'_{R(T)} \|_1$, we observe that

$$\sum_{i \in R(T)} 1_{\{\overline{j^* i} \in E\}} = \sum_{i=1}^{m} 1_{\{\overline{j^* i} \in E \text{ and } \overline{j i} \in E \text{ for some } j \in T\}}$$

$$= \sum_{i \in R(j^*)} 1_{\{\overline{j i} \in E \text{ for some } j \in T\}} = \text{card}(R(j^*) \setminus R_1(j^*, S)) \leq 2\theta_s d,$$

which allows us to derive

$$\| \mathbf{e}'_{R(T)} \|_1 \leq \sum_{i \in R(T)} 1_{\{\overline{j^* i} \in E\}} |x_{j^*}| + \| \mathbf{e}_{R(T)} \|_1 \leq 2\theta_s d |x_{j^*}| + \| \mathbf{e}_{R(T)} \|_1.$$

Taking this bound into account in (13.19) and summing with (13.18) give

$$\sum_{j \in S} q_k \left[\left| \sum_{\substack{\ell \in S \setminus \{j\} \\ \overline{\ell i} \in E}} x_\ell + e_i \right|, i \in R(j) \right]$$

$$= \sum_{j \in T} q_k \left[\left| \sum_{\substack{\ell \in S \setminus \{j\} \\ \overline{\ell i} \in E}} x_\ell + e_i \right|, i \in R(j) \right] + q_k \left[\left| \sum_{\substack{\ell \in S \setminus \{j^*\} \\ \overline{\ell i} \in E}} x_\ell + e_i \right|, i \in R(j^*) \right]$$

$$\leq \frac{2\theta_s d}{k - 2\theta_s d}\|\mathbf{x}_T\|_1 + \frac{1}{k - 2\theta_s d}\left(2\theta_s d\,|x_{j^*}| + \|\mathbf{e}_{R(T)}\|_1\right) + \frac{1}{k - 2\theta_s d}\|\mathbf{e}_{R_1(j^*,S)}\|_1$$

$$\leq \frac{2\theta_s d}{k - 2\theta_s d}\|\mathbf{x}_S\|_1 + \frac{1}{k - 2\theta_s d}\|\mathbf{e}_{R(S)}\|_1,$$

where the fact that $R_1(j^*, S)$ and $R(T)$ are disjoint subsets of $R(S)$ was used in the last inequality. This concludes the inductive proof. $\qquad\square$

Proof (of Theorem 13.15). We are going to prove that, for any $n \geq 0$,

$$\|\mathbf{x}^{n+1} - \mathbf{x}_S\|_1 \leq \rho\|\mathbf{x}^n - \mathbf{x}_S\|_1 + \frac{(1 - \rho)\tau}{d}\|\mathbf{A}\mathbf{x}_{\overline{S}} + \mathbf{e}\|_1. \qquad (13.20)$$

We use the triangle inequality and the fact that $\mathbf{x}^{n+1}$ is a better s-term approximation than $\mathbf{x}_S$ to $\mathbf{u}^{n+1} := (\mathbf{x}^n + \mathcal{M}(\mathbf{y} - \mathbf{A}\mathbf{x}^n))_{T^{n+1}}$, where $T^{n+1} := S \cup \operatorname{supp}(\mathbf{x}^n) \cup \operatorname{supp}(\mathbf{x}^{n+1})$, to derive that

$$\|\mathbf{x}^{n+1} - \mathbf{x}_S\|_1 \leq \|\mathbf{x}^{n+1} - \mathbf{u}^{n+1}\|_1 + \|\mathbf{x}_S - \mathbf{u}^{n+1}\|_1 \leq 2\|\mathbf{x}_S - \mathbf{u}^{n+1}\|_1.$$

Since $\mathbf{y} = \mathbf{A}\mathbf{x} + \mathbf{e} = \mathbf{A}\mathbf{x}_S + \mathbf{e}'$ with $\mathbf{e}' := \mathbf{A}\mathbf{x}_{\overline{S}} + \mathbf{e}$, Lemma 13.16 implies that

$$\|\mathbf{x}^{n+1} - \mathbf{x}_S\|_1 \leq 2\|(\mathbf{x}_S - \mathbf{x}^n - \mathcal{M}(\mathbf{A}(\mathbf{x}_S - \mathbf{x}^n) + \mathbf{e}'))_{T^{n+1}}\|_1$$

$$\leq \frac{4\theta_{3s} d}{\lceil d/2\rceil - 2\theta_{3s} d}\|\mathbf{x}_S - \mathbf{x}^n\|_1 + \frac{2}{\lceil d/2\rceil - 2\theta_{3s} d}\|\mathbf{e}'\|_1$$

$$\leq \frac{8\theta_{3s}}{1 - 4\theta_{3s}}\|\mathbf{x}_S - \mathbf{x}^n\|_1 + \frac{4}{(1 - 4\theta_{3s})d}\|\mathbf{e}'\|_1.$$

This is the desired inequality (13.20) with $\rho := 8\theta_{3s}/(1 - 4\theta_{3s}) < 1$ and $\tau := 4/(1 - 12\theta_{3s})$. The estimate (13.12) follows immediately by induction. Next, if $\mathbf{x}^\sharp$ is a cluster point of the sequence $(\mathbf{x}^n)_{n\geq 0}$, we deduce

$$\|\mathbf{x}^\sharp - \mathbf{x}_S\|_1 \leq \frac{\tau}{d}\|\mathbf{A}\mathbf{x}_{\overline{S}} + \mathbf{e}\|_1 \leq \frac{\tau}{d}\|\mathbf{A}\mathbf{x}_{\overline{S}}\|_1 + \frac{\tau}{d}\|\mathbf{e}\|_1,$$

where we choose S as an index set of s largest absolute entries of $\mathbf{x}$. In view of the inequality

$$\|\mathbf{A}\mathbf{v}\|_1 = \sum_{i=1}^{m}\left|\sum_{j=1}^{N} a_{i,j} v_j\right| \leq \sum_{j=1}^{N}\sum_{i=1}^{m} a_{i,j}|v_j| = \sum_{j=1}^{N} d|v_j| = d\|\mathbf{v}\|_1$$

applied to $\mathbf{v} = \mathbf{x}_{\overline{S}}$, it follows that

$$\|\mathbf{x}^\sharp - \mathbf{x}\|_1 \leq \|\mathbf{x}_{\overline{S}}\|_1 + \|\mathbf{x}^\sharp - \mathbf{x}_S\|_1 \leq (1 + \tau)\sigma_s(\mathbf{x})_1 + \frac{\tau}{d}\|\mathbf{e}\|_1.$$

This is the desired estimate with $C = 1 + \tau$ and $D = \tau$. $\qquad\square$

13.5   Analysis of a Simple Sublinear-Time Algorithm

The relative simplicity of the algorithm of Sect. 13.4 is counterbalanced by the nonoptimality of its runtime. Indeed, the dimension N enters at least linearly when forming $\mathbf{x}^n + \mathcal{M}(\mathbf{y} - \mathbf{A}\mathbf{x}^n)$. One aims, however, at exploiting some features of the measurement matrix in order to devise algorithms with a smaller runtime than linear in N, for instance, polylogarithmic in N and polynomial in s. Such algorithms are called *sublinear-time algorithms*. This section illustrates that sublinear-time algorithms are indeed possible, although the most sophisticated ones are not presented. As a first indication of this possibility, we consider the special case of 1-sparse vectors. We introduce the *bit-tester* matrix $\mathbf{B} \in \{0,1\}^{\ell \times N}$, $\ell := \lceil \log_2(N) \rceil$, defined by

$$\mathbf{B} = \begin{bmatrix} b_1(1) & \cdots & b_1(N) \\ \vdots & \ddots & \vdots \\ b_\ell(1) & \cdots & b_\ell(N) \end{bmatrix},$$

where $b_i(j) \in \{0,1\}$ denotes the ith digit in the binary expansion of $j - 1$, i.e.,

$$j - 1 = b_\ell(j)2^{\ell-1} + b_{\ell-1}(j)2^{\ell-2} + \cdots + b_2(j)2 + b_1(j). \tag{13.21}$$

If $\mathbf{x} = \mathbf{e}_j$ is a canonical unit vector, then the value of j is deduced from the measurement $\mathbf{B}\mathbf{e}_j = [b_1(j), \ldots, b_\ell(j)]^\top$ via (13.21). Moreover, if the support of $\mathbf{x} \in \mathbb{C}^N$ is a singleton $\{j\}$ and if we append a row of ones after the last row of $\mathbf{B}$ to form the augmented bit-tester matrix

$$\mathbf{B}' = \begin{bmatrix} b_1(1) & \cdots & b_1(N) \\ \vdots & \ddots & \vdots \\ b_\ell(1) & \cdots & b_\ell(N) \\ 1 & \cdots & 1 \end{bmatrix},$$

then the measurement vector $\mathbf{B}'\mathbf{x} = x_j[b_1(j), \ldots, b_\ell(j), 1]^\top$ allows one to determine both j and x_j using only a number of algebraic operations roughly proportional to $\log_2(N)$. This simple strategy can be extended from 1-sparse vectors to s-sparse vectors with $s > 1$ using lossless expanders. Precisely, given a matrix $\mathbf{A} \in \{0,1\}^{m \times N}$ with d ones per columns, we first construct a matrix $\mathbf{A}' \in \{0,1\}^{m' \times N}$ whose $m' = m(\ell + 1)$ rows are all the pointwise products of rows of $\mathbf{A}$ with rows of $\mathbf{B}'$, precisely

$$A'_{(i-1)(\ell+1)+k,j} = B'_{k,j}A_{i,j}, \qquad i \in [m], \ k \in [\ell + 1], \ j \in [N]. \tag{13.22}$$

Next, given $\mathbf{y} \in \mathbb{C}^m$, we construct a sequence of vectors $(\mathbf{x}^n)$ starting with $\mathbf{x}^0 = \mathbf{0}$ and iterating the instructions:

- For all $i \in [m]$ satisfying $v_i := (\mathbf{y} - \mathbf{A}'\mathbf{x}^n)_{i(\ell+1)} \neq 0$, compute the integer

$$j_i := 1 + \left\lfloor \frac{1}{v_i} \sum_{k=1}^{\ell} (\mathbf{y} - \mathbf{A}'\mathbf{x}^n)_{(i-1)(\ell+1)+k} \, 2^{k-1} \right\rfloor.$$

- If there exist $r \geq d/2$ distinct right vertices $i_1, \cdots, i_r \in [m]$ such that $(j_{i_1}, v_{i_1}) = \cdots = (j_{i_r}, v_{i_r}) =: (j, v)$, set

$$x_j^{n+1} = x_j^n + v.$$

The procedure stops when $v_i = 0$ for all $i \in [m]$, i.e., when $\mathbf{A}'\mathbf{x}^n = \mathbf{y}$. If $\mathbf{A}$ is the adjacency matrix of a lossless expander and (neglecting stability and robustness issues) if $\mathbf{y} = \mathbf{A}'\mathbf{x}$ for some exactly s-sparse $\mathbf{x} \in \mathbb{C}^N$, then the algorithm recovers the vector $\mathbf{x}$ in a finite number of iterations, as shown below. The number of measurements approaches the optimal value $c\,s\log_2(N/s)$ up to the logarithmic factor $\log_2(N)$.

Theorem 13.17. *If $m' \geq c\,s\log_2(N/s)\log_2(N)$, then there is a measurement matrix $\mathbf{A}' \in \{0,1\}^{m' \times N}$ such that the procedure described above reconstructs every s-sparse vector $\mathbf{x} \in \mathbb{C}^N$ from $\mathbf{y} = \mathbf{A}'\mathbf{x}$ with a number of algebraic operations at most proportional to $s^2 \log_2(N) \log_2(N/s) \log_2(s)$.*

Proof. Let $\mathbf{A} \in \{0,1\}^{m \times N}$ be the adjacency matrix of a left regular bipartite graph satisfying $\theta_s < 1/16$, and let d denotes its left degree. According to Theorem 13.6, such a matrix exists provided $m \asymp s\log_2(N/s)$. Let $\mathbf{A}' \in \{0,1\}^{m \times N}$ be the matrix constructed in (13.22). Its number of rows satisfies $m' = m(\ell+1) \asymp s\log_2(N/s)\log_2(N)$. We claim that if $(\mathbf{x}^n)$ is the sequence produced by the algorithm described above and if $S^n := \text{supp}(\mathbf{x} - \mathbf{x}^n)$, then we have $\text{card}(S^{n+1}) < \text{card}(S^n)/2$, so that $\mathbf{x}^{\bar{n}} = \mathbf{x}$ when $\bar{n} = \lceil \log_2(s) \rceil$.

To justify this claim, we observe that elements $i \notin R(S^n)$ do not produce any change from $\mathbf{x}^n$ to $\mathbf{x}^{n+1}$, since

$$v_i = (\mathbf{A}'(\mathbf{x} - \mathbf{x}^n))_{i(\ell+1)} = \sum_{j \in S^n} A'_{i(\ell+1),j}(\mathbf{x} - \mathbf{x}^n)_j = \sum_{j \in S^n} A_{i,j}(\mathbf{x} - \mathbf{x}^n)_j = 0.$$

Next, we prove that elements $i \in R_1(S^n) = \cup_{j \in S^n} R_1(j, S^n)$ create many zero entries in $\mathbf{x} - \mathbf{x}^{n+1}$. Indeed, let $i \in R_1(j^*, S^n)$ for some $j^* \in S^n$, i.e., the right vertex i is connected only to the left vertex j^* in S^n. We have, for any $k \in [\ell + 1]$,

$$(\mathbf{y} - \mathbf{A}'\mathbf{x}^n)_{(i-1)(\ell+1)+k} = (\mathbf{A}'(\mathbf{x} - \mathbf{x}^n))_{(i-1)(\ell+1)+k}$$

$$= \sum_{j \in S^n} A'_{(i-1)(\ell+1)+k,j}(\mathbf{x} - \mathbf{x}^n)_j = \sum_{j \in S^n} B_{k,j} A_{i,j}(\mathbf{x} - \mathbf{x}^n)_j$$

$$= B_{k,j^*}(\mathbf{x} - \mathbf{x}^n)_{j^*}.$$

In particular, since $B_{\ell+1,j^*} = 1$, setting $k = \ell + 1$ yields

$$(\mathbf{y} - \mathbf{A}'\mathbf{x}^n)_{i(\ell+1)} = (\mathbf{x} - \mathbf{x}^n)_{j^*} \neq 0.$$

Furthermore, since $B_{k,j^*} = b_k(j^*)$ for $k \in [\ell]$, we obtain

$$\sum_{k=0}^{\ell} (\mathbf{y} - \mathbf{A}'\mathbf{x}^n)_{(i-1)(\ell+1)+k}\, 2^{k-1} = \sum_{k=0}^{\ell} b_k(j^*)\, 2^{k-1}(\mathbf{x} - \mathbf{x}^n)_{j^*}$$

$$= (j^* - 1)(\mathbf{x} - \mathbf{x}^n)_{j^*}.$$

This means that $v_i = (\mathbf{x} - \mathbf{x}^n)_{j^*}$ and that $j_i = j^*$. Therefore, it follows that $(\mathbf{x} - \mathbf{x}^{n+1})_{j^*} = x_{j^*} - (x^n_{j^*} + v_i) = 0$ provided $\mathrm{card}(R_1(j^*, S^n)) \geq d/2$. If t denotes the number of such j^*, Corollary 13.5 implies that

$$(1 - 2\theta_s)d\,\mathrm{card}(S^n) \leq \mathrm{card}(R_1(S^n)) = \sum_{j \in S^n} \mathrm{card}(R_1(j, S^n))$$

$$\leq t\,d + (\mathrm{card}(S^n) - t)\,d/2,$$

which yields $t \geq (1 - 4\theta_s)\,\mathrm{card}(S^n)$. Therefore, at least $(1 - 4\theta_s)\,\mathrm{card}(S^n)$ zeros entries of $\mathbf{x} - \mathbf{x}^{n+1}$ are created by elements $i \in R_1(S^n)$.

Finally, we take into account that elements $i \in R(S^n) \setminus R_1(S^n)$ may potentially corrupt zero entries of $\mathbf{x} - \mathbf{x}^n$ to nonzero entries of $\mathbf{x} - \mathbf{x}^{n+1}$. For a corruption to occur, we need a group of at least $d/2$ elements in $R(S^n) \setminus R_1(S^n)$, which has size at most $2\theta_s d\,\mathrm{card}(S^n)$; hence, the number of corruptions is at most $4\theta_s\,\mathrm{card}(S^n)$. Putting everything together, we deduce the desired claim from

$$\mathrm{card}(S^{n+1}) \leq \mathrm{card}(S^n) - (1 - 4\theta_s)\,\mathrm{card}(S^n) + 4\theta_s\,\mathrm{card}(S^n) = 8\theta_s\,\mathrm{card}(S^n)$$

$$< \frac{\mathrm{card}(S^n)}{2}.$$

It now remains to count the number of algebraic operations the procedure requires. At each iteration, we notice that the first step requires $\mathcal{O}(m(s + \ell s)) = \mathcal{O}(sm\ell)$ operations, since the sparsity of $\mathbf{x}^n$ ensures that each component of $\mathbf{A}'\mathbf{x}^n$ can be computed in $\mathcal{O}(s)$ operations, and that the second step requires $\mathcal{O}(s)$ operations; since the previous argument ensures that at most $\mathcal{O}(s)$ entries change from $\mathbf{x}^n$ to $\mathbf{x}^{n+1}$. Overall, the total number of algebraic operations is then $\mathcal{O}(\bar{n}sm\ell) = \mathcal{O}(\log_2(s)s^2 \log_2(N/s) \log_2(N))$. $\square$

Notes

Some authors use the terms unbalanced expander or left regular bipartite expander instead of lossless expander. We opted for the terminology of [279]. As already pointed out, a lossless expander is different from an expander. We present here two equivalent definitions of the latter. They both concern an undirected graph $G = (V, E)$, with set of vertices V and set of edges E, which is d-regular in the sense that the number d of edges attached to a vertex is the same for all vertices. For $0 < \mu < 1$, the combinatorial property defining a μ-*edge expander* is $\mathrm{card}(E(S, \overline{S})) \geq \mu d \, \mathrm{card}(S)$ for all $S \subset V$ with $\mathrm{card}(S) \leq \mathrm{card}(V)/2$, where $E(S, \overline{S})$ denotes the set of edges between S and its complement $\overline{S}$. For $0 < \lambda < 1$, the algebraic property defining a λ-expander uses its adjacency matrix $\mathbf{A}$ defined by $A_{i,j} = 1$ if there is an edge connecting i and j and $A_{i,j} = 0$ if there is none. Note the usual identification of V to $[n]$ with $n := \mathrm{card}(V)$. Since the matrix $\mathbf{A}/d$ is symmetric and *stochastic*, i.e., it has nonnegative entries summing to one along each row and along each column, it has n real eigenvalues $\lambda_1 = 1, \lambda_2, \ldots, \lambda_n$ ordered as $|\lambda_1| \geq |\lambda_2| \geq \cdots \geq |\lambda_n|$. The graph G is then called a λ-*expander* if $|\lambda_2| \leq \lambda$ or equivalently if its *spectral gap* $1 - |\lambda_2|$ is at least $1 - \lambda$. The combinatorial and algebraic definitions are equivalent, since a λ-expander is a $(1 - \lambda)/2$-edge expander and a μ-edge expander is a $(1 - \mu^2/2)$-expander. We refer the reader to [19, Chap. 21] for more details on the subject.

The use of probabilistic methods to prove that lossless expanders exist is part of the folklore in theoretical computer science. For instance, the result of Theorem 13.8 is stated in [101]. To date, there are no explicit constructions of lossless expanders with optimal parameters, but there are deterministic (i.e., computable in polynomial time) constructions of (s, d, θ)-lossless expanders with $d \asymp (\log(N) \log(s)/\theta)^{1+1/\alpha}$ and $m \asymp d^2 s^{1+\alpha}$ for any $\alpha > 0$; see [251].

The stable null space property for adjacency matrices of lossless expanders was established by Berinde et al. in [41]. We mainly followed their arguments to prove the robust null space property in Theorem 13.11, except that we did not call upon the ℓ_1-restricted isometry property that they established first—see Exercise 13.5.

The algorithm (13.11) is a modification of the sparse matching pursuit algorithm proposed by Berinde et al. in [42]. The analysis of the latter is also based on Lemma 13.16; see Exercise 13.11. The way we proved Lemma 13.16 differs from the original proof of [42]. There are other iterative algorithms yielding stable and robust reconstruction using adjacency matrices of lossless expanders; see the survey [285] by Indyk and Gilbert. For instance, the expander matching pursuit algorithm of [286] precedes the sparse matching pursuit algorithm and runs in linear time, while the HHS (heavy hitters on steroids) pursuit of [225] runs in sublinear time. The sublinear-time algorithm of Theorem 13.17 is taken from [41] and the one of Exercise 13.12 from [291], but they were not designed with stability in mind. There are also sublinear-time algorithms for matrices other than adjacency matrices of lossless expanders; for instance, [261,262,287] deals with partial Fourier matrices.

Exercises

13.1. Show that the expansion property (13.1) for $\text{card}(S) = s$ does not necessarily imply the expansion property for $\text{card}(S) < s$.

13.2. Let $\mathbf{A} \in \{0,1\}^{m \times N}$ be a matrix with exactly d ones per column. With $\mathbf{a}_1, \ldots, \mathbf{a}_N$ denoting these columns, suppose that $\langle \mathbf{a}_i, \mathbf{a}_j \rangle \leq \mu d$ for all distinct $i, j \in [N]$. Prove that $\mathbf{A}$ is the adjacency matrix of a left d-regular bipartite graph satisfying $\theta_s \leq (s-1)\mu/2$.

13.3. Prove that a left d-regular bipartite graph is an $(s, d, (d-1)/d)$-lossless expander if and only if, for any set S of left vertices with $\text{card}(S) \leq s$, one can find for each $j \in J$ an edge $\overline{j\, i_j}$ in such a way that the right vertices i_j, $j \in S$, are all distinct. You may use *Hall's theorem*: For finite subsets $X_1, X_2, \ldots, X_n$ of a set X, one can find distinct points $x_1 \in X_1, x_2 \in X_2, \ldots, x_n \in X_n$ if and only if $\text{card}(\cup_{k \in K} X_k) \geq \text{card}(K)$ for all $K \subset [n]$.

13.4. Let $R_{\geq k}(S)$ be the set of right vertices connected to at least k left vertices of a set S in a left d-regular bipartite graph. Prove that the graph is an (s, d, θ)-lossless expander if and only if $\sum_{k \geq 2} \text{card}(R_{\geq k}(S)) \leq \theta d \, \text{card}(S)$ for any set S of at most s left vertices. Deduce that $\text{card}(R_{\geq 2}(S)) \leq \theta d \, \text{card}(S)$ for any set S of at most s left vertices if the graph is an (s, d, θ)-lossless expander.

13.5. Prove that the $m \times N$ adjacency matrix $\mathbf{A}$ of an (s, d, θ)-lossless expander satisfies the property that

$$d(1 - 2\theta)\|\mathbf{z}\|_1 \leq \|\mathbf{A}\mathbf{z}\|_1 \leq d\|\mathbf{z}\|_1 \quad \text{for all } s\text{-sparse } \mathbf{z} \in \mathbb{C}^N,$$

which can be interpreted as a scaled restricted isometry property in ℓ_1.

13.6. For a fixed $\delta > 0$, suppose that a measurement matrix $\mathbf{A} \in \{0,1\}^{m \times N}$ satisfies $\delta_s(\gamma \mathbf{A}) \leq \delta$ for some $\gamma > 0$. Let c and r denote the minimal number of ones per columns of $\mathbf{A}$ and the maximal number of ones per row of $\mathbf{A}$. Show that

$$c \leq \frac{r \, m}{N}.$$

Observe also that $c \geq (1 - \delta)/\gamma^2$ by considering a suitable 1-sparse vector. Then, by considering any vector in $\{0,1\}^N$ with exactly s ones, deduce that

$$c \leq \frac{1+\delta}{1-\delta} \frac{m}{s}.$$

Next, by considering a suitable vector in $\{0,1\}^N$ with exactly $t := \min\{r, s\}$ ones, observe that

$$t \leq \frac{1+\delta}{\gamma^2} \leq \frac{1+\delta}{1-\delta} c.$$

Separating the cases $r \geq s$ and $r < s$, conclude that

$$m \geq \min \left\{ \frac{1-\delta}{1+\delta} N, \left(\frac{1-\delta}{1+\delta} \right)^2 s^2 \right\},$$

so that matrices populated with zeros and ones do not satisfy the classical restricted isometry property in the parameter range relevant to compressive sensing.

13.7. Let $\mathbf{A} \in \{0,1\}^{m \times N}$ be the adjacency matrix of a left regular bipartite graph and let $S \subset [N]$ be a fixed index set. Suppose that every nonnegative vector supported on S is uniquely recovered via ℓ_1-minimization using $\mathbf{A}$ as a measurement matrix. Prove that every nonnegative vector $\mathbf{x}$ supported on S is in fact the unique vector in the set $\{\mathbf{z} \in \mathbb{R}^N : \mathbf{z} \geq 0, \mathbf{Az} = \mathbf{Ax}\}$.

13.8. Extend Theorem 13.10 to the case of a measurement error considered in ℓ_p-norms, $p \geq 1$. Precisely, given the adjacency matrix $\mathbf{A}$ of a left d-regular bipartite graph such that $\theta_{2s} < 1/6$, for $\mathbf{x} \in \mathbb{C}^N$ and $\mathbf{y} = \mathbf{Ax} + \mathbf{e}$ with $\|\mathbf{e}\|_p \leq \eta$, prove that a solution $\mathbf{x}^{\sharp}$ of

$$\underset{\mathbf{z} \in \mathbb{C}^N}{\text{minimize}} \ \|\mathbf{z}\|_1 \quad \text{subject to} \ \|\mathbf{Az} - \mathbf{y}\|_p \leq \eta$$

satisfies

$$\|\mathbf{x} - \mathbf{x}^{\sharp}\|_1 \leq \frac{2(1-2\theta)}{(1-6\theta)} \sigma_s(\mathbf{x})_1 + \frac{4}{(1-6\theta)d} \frac{s^{1-1/p}}{d^{1/p}} \eta.$$

13.9. For the adjacency matrix $\mathbf{A} \in \{0,1\}^{m \times N}$ of a left regular bipartite graph, let $\mathbf{A}' \in \{-1,1\}^{m \times N}$ be the matrix obtained from $\mathbf{A}$ by replacing the zeros by negative ones. Given $\mathbf{x} \in \mathbb{C}^N$, prove that the solutions of the two problems

$$\text{minimize} \ \|\mathbf{z}\|_1 \quad \text{subject to} \ \mathbf{Az} = \mathbf{Ax},$$

$$\text{minimize} \ \|\mathbf{z}\|_1 \quad \text{subject to} \ \mathbf{A}'\mathbf{z} = \mathbf{A}'\mathbf{x},$$

are identical.

13.10. For the quantiles q_k, prove the inequalities (13.13), (13.14), as well as

$$q_k[a_1, \ldots, a_d] \leq q_k[b_1, \ldots, b_d] \qquad \qquad \text{if } a_j \leq b_j \text{ for all } j,$$

$$q_{2k}[a_1 + b_1, \ldots, a_d + b_d] \leq q_k[a_1, \ldots, a_d] + q_k[b_1, \ldots, b_d] \ \text{ if } a_j, b_j \geq 0 \text{ for all } j.$$

13.11. Establish an analog of Theorem 13.15 when $\theta_{4s} < 1/20$ for the *sparse matching pursuit* algorithm consisting in the scheme

$$\mathbf{u}^{n+1} := H_{2s}(\mathcal{M}(\mathbf{y} - \mathbf{Ax}^n)), \quad \mathbf{x}^{n+1} := H_s(\mathbf{x}^n + \mathbf{u}^{n+1}).$$

13.12. Let $\mathbf{A} \in \{0,1\}^{m \times N}$ be the adjacency matrix of an (s, d, θ)-lossless expander. If θ is small enough, prove that every s-sparse vector is recovered from $\mathbf{y} = \mathbf{A}\mathbf{x}$ in a finite number of iterations of the algorithm:

- For each $i \in [m]$, compute

$$v_i := (\mathbf{y} - \mathbf{A}\mathbf{x}^n)_i.$$

- If there are $i_1, \ldots, i_r \in R(j)$ with $r \geq d/2$ and $v_{i_1} = \cdots = v_{i_r} =: v \neq 0$, set

$$x_j^{n+1} = x_j^n + v.$$

Chapter 14
Recovery of Random Signals using Deterministic Matrices

In this chapter, we change the point of view and work with a deterministic measurement matrix but treat the sparse signal to be recovered as random. In particular, the support set of the sparse vector (and additionally the signs of the nonzero coefficients) is chosen at random. In this scenario, the coherence of the measurement matrix only needs to obey mild conditions, much weaker than the ones outlined in Chap. 5 for the recovery of all s-sparse vectors. We recall that these conditions together with the lower bound of Theorem 5.7 lead to the quadratic bottleneck: arguments based on the coherence necessarily require a number of measurements $m \geq Cs^2$. In contrast, we will see that, with high probability, recovering a random s-sparse vector using ℓ_1-minimization is possible with $m \geq Cs \ln(N)$ measurements, provided the coherence satisfies $\mu \leq c(\ln N)^{-1}$. The latter condition is satisfied for many deterministic constructions of measurements and is indeed much milder than the optimal achievable bound $\mu \leq cm^{-1/2}$. Moreover, the coherence has the advantage of being easily evaluated for an explicitly given matrix.

The results in this chapter are weaker than the ones of Chap. 5 in the sense that they apply only to *most* signals instead of to *all* signals, but they show that the deterministic bounds using coherence may be somewhat pessimistic even if no bounds on the restricted isometry constants are available. Moreover, our analysis reveals that conclusions from numerical experiments for the performance evaluation of measurement matrices have to be handled with care, as testing with randomly chosen sparse signals does not necessarily give much information about recovery of *all* signals or about the restricted isometry constants of the matrix. Indeed, our results apply also to examples of measurement matrices from Chap. 12, for which there exist s-sparse signals that cannot be recovered from fewer than cs^2 measurements; see, e.g., the discussion after Proposition 12.4. Nevertheless, *most* s-sparse signals can be recovered from far fewer samples.

Recovery results for random signals are especially important in the context of sparse approximation, where the matrix $\mathbf{A}$ takes the role of a redundant dictionary and $\mathbf{y} \in \mathbb{C}^m$ is a signal that has a sparse representation in terms of the columns of

S. Foucart and H. Rauhut, *A Mathematical Introduction to Compressive Sensing*, Applied and Numerical Harmonic Analysis, DOI 10.1007/978-0-8176-4948-7_14, © Springer Science+Business Media New York 2013

$\mathbf{A}$, i.e., $\mathbf{y} = \mathbf{A}\mathbf{x}$. In such context, one cannot design $\mathbf{A}$. Since it is hard to verify the restricted isometry property for deterministic matrices in the optimal range of parameters, recovery conditions for random signals that overcome the limits of their deterministic counterparts are essential. Nevertheless, such types of bounds are also relevant in the context of compressive sensing when $\mathbf{A}$ takes the role of a measurement matrix that can be designed—especially in situations where good bounds for the restricted isometry property are not (yet) available.

In Sect. 14.1, we derive bounds on the conditioning of a random column submatrix of a given matrix. The methods build on moment bounds, decoupling, and matrix deviation inequalities developed in Chap. 8. In Sect. 14.2, we then develop guarantees for ℓ_1-minimization based on Corollary 4.28 for the recovery of individual sparse vectors.

14.1 Conditioning of Random Submatrices

Throughout this chapter, we assume that $\mathbf{A} = [\mathbf{a}_1 | \dots | \mathbf{a}_N] \in \mathbb{C}^{m \times N}$ is a measurement matrix with ℓ_2-normalized columns (i.e., $\|\mathbf{a}_j\|_2 = 1$) and coherence

$$\mu = \max_{k \neq \ell} |\langle \mathbf{a}_k, \mathbf{a}_\ell \rangle|.$$

We will use two probability models for selecting a random support set $S \subset [N]$:

- **Uniform Model**. The set S is selected uniformly at random among all subsets of $[N]$ of cardinality $s \leq N$.
- **Bernoulli Model**. Choose $\delta = s/N$, and introduce independent Bernoulli selectors δ_j, $j \in [N]$, that take the value 1 with probability δ and the value 0 with probability $1 - \delta$. Then define the random set

$$S = \{j \in [N], \delta_j = 1\}.$$

In this model, the cardinality of S is random but its expectation satisfies $\mathbb{E}\,\mathrm{card}(S) = s$ due to the choice $\delta = s/N$. By Bernstein's inequality (Corollary 7.31), the size of S concentrates around s as

$$\mathbb{P}(|\,\mathrm{card}(S) - s| \geq t\sqrt{s}) = \mathbb{P}(|\sum_{j=1}^{N}(\delta_j - \delta)| \geq t\sqrt{s}) \leq 2\exp\left(-\frac{st^2/2}{s + \sqrt{s}t/3}\right)$$

$$\leq 2\exp(-3t^2/8), \quad \text{for } 0 < t \leq \sqrt{s}.$$

To verify that Bernstein inequality applies, note that $\mathbb{E}(\delta_j - \delta) = 0$, $|\delta_j - \delta| \leq 1$, and $\mathbb{E}(\delta_j - \delta)^2 = \delta - \delta^2 \leq \delta = s/N$.

The first probability model may be more intuitive because the cardinality of S is always s, but the second probability model is easier to analyze because of the independence of the Bernoulli selectors δ_j. In any case, both probability models are closely related as we will see below.

We are interested in the conditioning of $\mathbf{A}_S$, i.e., in the operator norm

$$\|\mathbf{A}_S^* \mathbf{A}_S - \mathbf{Id}_S\|_{2\to 2}.$$

We have the following probabilistic bound on this norm.

Theorem 14.1. *Let $\mathbf{A} \in \mathbb{C}^{m\times N}$, $m \leq N$, be a matrix with ℓ_2-normalized columns and coherence μ. Let S be a subset of $[N]$ selected at random according to the uniform model with $\operatorname{card}(S) = s$ or to the Bernoulli model with $\mathbb{E}\operatorname{card}(S) = s$. Assume that, for $\eta, \varepsilon \in (0,1)$,*

$$\mu \leq c\frac{\eta}{\ln(N/\varepsilon)}, \tag{14.1}$$

$$\frac{s}{N}\|\mathbf{A}\|_{2\to 2}^2 \leq c\frac{\eta^2}{\ln(N/\varepsilon)}, \tag{14.2}$$

for an appropriate constant $c > 0$. Then, with probability at least $1 - \varepsilon$,

$$\|\mathbf{A}_S^* \mathbf{A}_S - \mathbf{Id}_S\|_{2\to 2} \leq \eta.$$

Remark 14.2. The proof reveals the more precise estimate

$$\mathbb{P}(\|\mathbf{A}_S^* \mathbf{A}_S - \mathbf{Id}_S\|_{2\to 2} \geq c_1\mu u + c_2\sqrt{\frac{s}{N}\|\mathbf{A}\|_{2\to 2}^2 u} + 2e\frac{s}{N}\|\mathbf{A}\|_{2\to 2}^2)$$

$$\leq c_3 N^4 \exp(-u)$$

with $c_1 \approx 4.8078$, $c_2 \approx 11.21$ and $c_3 \approx 70.15$.

In order for Theorem 14.1 to be valuable, the quantity $\frac{s}{N}\|\mathbf{A}\|_{2\to 2}^2$ should be small. Let us briefly comment on this condition. We note that $\operatorname{tr}(\mathbf{A}^*\mathbf{A}) = \sum_{j=1}^N \|\mathbf{a}_j\|_2^2 = N$ and that

$$\operatorname{tr}(\mathbf{AA}^*) \leq m\lambda_{\max}(\mathbf{AA}^*) = m\|\mathbf{AA}^*\|_{2\to 2} = m\|\mathbf{A}\|_{2\to 2}^2, \tag{14.3}$$

so that taking $\operatorname{tr}(\mathbf{A}^*\mathbf{A}) = \operatorname{tr}(\mathbf{AA}^*)$ into account gives

$$\|\mathbf{A}\|_{2\to 2}^2 \geq \frac{N}{m}. \tag{14.4}$$

Equality holds if and only if equality holds in (14.3), i.e., if and only if all the eigenvalues of $\mathbf{AA}^*$ are equal to N/m. This means $\mathbf{AA}^* = (N/m)\mathbf{Id}_m$; in other words, the columns of $\mathbf{A}$ form a unit norm tight frame.

Unit norm tight frames appear frequently in the context of sparse approximation, for instance, as union of several orthonormal bases. In the important case of unit norm tight frames, we therefore have

$$\frac{s}{N}\|\mathbf{A}\|_{2\to2}^2 = \frac{s}{m}. \tag{14.5}$$

There are also relevant examples where $\mathbf{A}$ does not form a tight frame but where at least $\|\mathbf{A}\|_{2\to2}^2$ is close to its lower bound N/m, so that (14.2) still represents a reasonable condition on m.

Choosing the probability $\varepsilon = N^{-1}$, say, Condition (14.2) then becomes the familiar one

$$m \geq c\eta^{-2}s\ln(N),$$

while (14.1) is only a very mild condition on the coherence of $\mathbf{A}$, namely, $\mu \leq c\eta \ln^{-1}(N)$.

We now return to Theorem 14.1 and its proof, which we develop in several steps. Let us start with some notation. We introduce the hollow Gram matrix

$$\mathbf{H} = \mathbf{A}^*\mathbf{A} - \mathbf{Id},$$

which has zero diagonal because $\mathbf{A}$ has ℓ_2-normalized columns by assumption. Let $\mathbf{P} = \mathbf{P}_S$ be the projection operator onto S, i.e., for $\mathbf{x} \in \mathbb{C}^N$,

$$(\mathbf{Px})_\ell = \begin{cases} x_\ell & \text{if } \ell \in S, \\ 0 & \text{if } \ell \notin S. \end{cases}$$

With this notation we observe that

$$\|\mathbf{A}_S^*\mathbf{A}_S - \mathbf{Id}_S\|_{2\to2} = \|\mathbf{PHP}\|_{2\to2}.$$

We analyze the Bernoulli model and later reduce the uniform model to the Bernoulli model. Note that $\mathbf{P}$ is the random diagonal matrix

$$\mathbf{P} = \text{diag}[\delta_1, \ldots, \delta_N],$$

where the δ_j are independent Bernoulli selectors with $\mathbb{E}\delta_j = \delta = s/N$. We will bound the moments of $\|\mathbf{PHP}\|_{2\to2}$. We use decoupling to overcome difficulties in the analysis which arise from the double appearance of $\mathbf{P}$. Theorem 8.12 implies that, for $p \geq 1$,

$$(\mathbb{E}\|\mathbf{PHP}\|_{2\to2}^p)^{1/p} \leq 2(\mathbb{E}\|\mathbf{P}'\mathbf{HP}\|_{2\to2}^p)^{1/p},$$

where $\mathbf{P}'$ is an independent copy of $\mathbf{P}$. Then the matrix $\mathbf{B} = \mathbf{P}'\mathbf{H}$ is independent of $\mathbf{P}$. We first derive a moment estimate for $\|\mathbf{BP}\|_{2\to2}$ with general fixed $\mathbf{B}$.

Lemma 14.3. *Let* $\mathbf{B} \in \mathbb{C}^{N \times N}$ *and let* $\mathbf{P}$ *be a random diagonal matrix of independent Bernoulli variables with mean* $\delta \in [0, 1]$. *Then, for* $p \geq 2$,

$$(\mathbb{E}\|\mathbf{BP}\|_{2\to2}^p)^{1/p} \leq C(C_2 N)^{2/p} \sqrt{p} (\mathbb{E}\|\mathbf{BP}\|_{1\to2}^p)^{1/p} + \sqrt{\delta}\|\mathbf{B}\|_{2\to2}.$$

The constants satisfy $C \leq 1.0310$ *and* $C_2 \leq 4.1878$.

Proof. With $\mathbf{b}_1, \ldots, \mathbf{b}_N$ denoting the columns of $\mathbf{B}$, we observe that

$$\|\mathbf{BP}\|_{2\to2}^2 = \|\mathbf{BPB}^*\|_{2\to2} = \|\sum_{j=1}^N \delta_j \mathbf{b}_j \mathbf{b}_j^*\|_{2\to2},$$

since $\mathbf{P} = \mathbf{P}^2$. Taking expectation followed by the triangle inequality and symmetrization (Lemma 8.4) yields, for $r \geq 1$,

$$(\mathbb{E}\|\mathbf{BP}\|_{2\to2}^{2r})^{1/r} \leq (\mathbb{E}\|\sum_{j=1}^N (\delta_j - \delta)\mathbf{b}_j \mathbf{b}_j^*\|_{2\to2}^r)^{1/r} + \delta\|\sum_{j=1}^N \mathbf{b}_j \mathbf{b}_j^*\|_{2\to2}$$

$$\leq 2(\mathbb{E}\|\sum_{j=1}^N \epsilon_j \delta_j \mathbf{b}_j \mathbf{b}_j^*\|_{2\to2}^r)^{1/r} + \delta\|\mathbf{BB}^*\|_{2\to2},$$

where ϵ is a Rademacher sequence. The tail inequality for matrix Rademacher sums (Proposition 8.20) states that, conditionally on δ,

$$\mathbb{P}_\epsilon(\|\sum_{j=1}^N \epsilon_j \delta_j \mathbf{b}_j \mathbf{b}_j^*\|_{2\to2} \geq t\sigma) \leq 2N e^{-t^2/2}, \quad t > 0,$$

where

$$\sigma = \|\sum_{j=1}^N (\delta_j \mathbf{b}_j \mathbf{b}_j^*)^2\|_{2\to2}^{1/2} = \|\sum_{j=1}^N \delta_j^2 \|\mathbf{b}_j\|_2^2 \mathbf{b}_j \mathbf{b}_j^*\|_{2\to2}^{1/2}$$

$$\leq \max_{j \in [N]} \{\delta_j \|\mathbf{b}_j\|_2\} \|\sum_{j=1}^N \delta_j \mathbf{b}_j \mathbf{b}_j^*\|_{2\to2}^{1/2} = \|\mathbf{BP}\|_{1\to2}\|\mathbf{BP}\|_{2\to2}.$$

Hereby, we have applied the explicit expression (A.11) of the norm $\|\cdot\|_{1\to2}$. It follows from Proposition 7.13 that, for $r \geq 1$,

$$\left(\mathbb{E}_\epsilon \|\sum_{j=1}^N \epsilon_j \delta_j \mathbf{b}_j \mathbf{b}_j^*\|_{2\to2}^r\right)^{1/r} \leq C_1(C_2 N)^{1/r} \sqrt{r}\|\mathbf{BP}\|_{1\to2}\|\mathbf{BP}\|_{2\to2}$$

with $C_1 = e^{1/(2e)}e^{-1/2} \approx 0.729$ and $C_2 = 2C_{2,2} = 2\sqrt{\pi}e^{1/6} \approx 4.1878$. Taking expectation also with respect to δ and applying the Cauchy–Schwarz inequality yield

$$\mathbb{E}\|\sum_{j=1}^{N} \epsilon_j\delta_j\mathbf{b}_j\mathbf{b}_j^*\|_{2\to2}^r \leq C_2 N \cdot C_1^r r^{r/2}(\mathbb{E}\|\mathbf{BP}\|_{1\to2}^{2r})^{1/2}(\mathbb{E}\|\mathbf{BP}\|_{2\to2}^{2r})^{1/2}.$$

By combining the above estimates and choosing $r = p/2$ we arrive at

$$(\mathbb{E}\|\mathbf{BP}\|_{2\to2}^p)^{2/p} \leq 2(\mathbb{E}\|\sum_{j=1}^{N} \epsilon_j\delta_j\mathbf{b}_j\mathbf{b}_j^*\|_{2\to2}^r)^{1/r} + \delta\|\mathbf{B}\|_{2\to2}^2$$

$$\leq 2(C_2 N)^{1/r}C_1\sqrt{r}(\mathbb{E}\|\mathbf{BP}\|_{1\to2}^{2r})^{1/(2r)}(\mathbb{E}\|\mathbf{BP}\|_{2\to2}^{2r})^{1/(2r)} + \delta\|\mathbf{B}\|_{2\to2}^2$$

$$= 2(C_2 N)^{2/p}C_1\sqrt{p/2}(\mathbb{E}\|\mathbf{BP}\|_{1\to2}^p)^{1/p}(\mathbb{E}\|\mathbf{BP}\|_{2\to2}^p)^{1/p} + \delta\|\mathbf{B}\|_{2\to2}^2.$$

Setting $E := (\mathbb{E}\|\mathbf{BP}\|_{2\to2}^p)^{1/p}$, this inequality takes the form $E^2 \leq \alpha E + \beta$. Completing squares gives $(E - \alpha/2)^2 \leq \alpha^2/4 + \beta$, so that

$$E \leq \alpha/2 + \sqrt{\alpha^2/4 + \beta} \leq \alpha + \sqrt{\beta}. \tag{14.6}$$

We conclude that

$$(\mathbb{E}\|\mathbf{BP}\|_{2\to2}^p)^{1/p} \leq (C_2 N)^{2/p}\sqrt{2}C_1\sqrt{p}(\mathbb{E}\|\mathbf{BP}\|_{1\to2}^p)^{1/p} + \delta^{1/2}\|\mathbf{B}\|_{2\to2}.$$

This finishes the proof.   □

The above lemma requires a moment bound for $\|\mathbf{BP}\|_{1\to2}$. Noting that we will later use $\mathbf{B} = \mathbf{P'H}$, we actually need to estimate $\|\mathbf{P'HP}\|_{1\to2} = \|\mathbf{P'\tilde{B}}\|_{1\to2}$ with $\tilde{\mathbf{B}} = \mathbf{HP}$. The next lemma requires the norm

$$\|\mathbf{B}\|_{\max} := \max_{j,k}|B_{j,k}|,$$

i.e., the ℓ_∞-norm over all matrix entries.

Lemma 14.4. *Let* $\mathbf{B} \in \mathbb{C}^{N\times N}$ *and let* $\mathbf{P}$ *be a random diagonal matrix of independent Bernoulli variables with mean* $\delta \in [0,1]$. *Then, for* $p \geq 2$,

$$(\mathbb{E}\|\mathbf{PB}\|_{1\to2}^p)^{1/p} \leq C_3(2N)^{2/p}\sqrt{p}(\mathbb{E}\|\mathbf{PB}\|_{\max}^p)^{1/p} + \sqrt{\delta}\|\mathbf{B}\|_{1\to2} \tag{14.7}$$

with $C_3 = 2(2e)^{-1/2} \approx 0.8578$. *Moreover, for* $u > 0$,

$$\mathbb{P}(\|\mathbf{PB}\|_{1\to2} \geq \sqrt{2\delta}\|\mathbf{B}\|_{1\to2} + 2\|\mathbf{B}\|_{\max}u) \leq 4N^2 e^{-u^2}. \tag{14.8}$$

Proof. Similarly to the previous proof, we set $E = (\mathbb{E}\|\mathbf{PB}\|_{1\to2}^p)^{1/p}$ and $r = p/2$. Symmetrization (Lemma 8.4), the explicit expression for $\|\cdot\|_{1\to2}$, and the triangle inequality yield

$$E^2 = \left(\mathbb{E}\Big(\max_{k\in[N]}\sum_{j=1}^N \delta_j|B_{jk}|^2\Big)^r\right)^{1/r}$$

$$\leq 2\left(\mathbb{E}_\delta\mathbb{E}_\epsilon \max_{k\in[N]}\Big|\sum_{j=1}^N \epsilon_j\delta_j|B_{jk}|^2\Big|^r\right)^{1/r} + \delta\|\mathbf{B}\|_{1\to2}^2. \qquad (14.9)$$

Estimating the maximum by a sum and using the Khintchine inequality (8.9), we arrive at

$$\left(\mathbb{E}_\epsilon \max_{k\in[N]}\Big|\sum_{j=1}^N \epsilon_j\delta_j|B_{jk}|^2\Big|^r\right)^{1/r} \leq \left(\sum_{k=1}^N \mathbb{E}_\epsilon\left[\Big|\sum_{j=1}^N \epsilon_j\delta_j|B_{jk}|^2\Big|^r\right]\right)^{1/r}$$

$$\leq 2^{1/r}e^{-1/2}\sqrt{r}\left(\sum_{k=1}^N\Big(\sum_{j=1}^N \delta_j|B_{j,k}|^4\Big)^{r/2}\right)^{1/r}$$

$$\leq 2^{1/r}e^{-1/2}\sqrt{r}N^{1/r}\max_{k\in[N]}\sqrt{\Big(\max_{j\in[N]}\delta_j|B_{j,k}|^2\Big)\sum_{j=1}^N \delta_j|B_{j,k}|^2}$$

$$\leq (2N)^{1/r}e^{-1/2}\sqrt{r}\|\mathbf{PB}\|_{\max}\|\mathbf{PB}\|_{1\to2}.$$

By the Cauchy–Schwarz inequality,

$$\mathbb{E}_\delta\mathbb{E}_\epsilon \max_{k\in[N]}\Big|\sum_{j=1}^N \epsilon_j\delta_j|B_{jk}|^2\Big|^r \leq 2Ne^{-r/2}r^{r/2}\left(\mathbb{E}\|\mathbf{PB}\|_{\max}^{2r}\right)^{1/2}\left(\mathbb{E}\|\mathbf{PB}\|_{1\to2}^{2r}\right)^{1/2}.$$

Altogether, we have obtained

$$E^2 \leq 2e^{-1/2}(2N)^{1/r}\sqrt{r}(\mathbb{E}\|\mathbf{PB}\|_{\max}^{2r})^{1/(2r)}(\mathbb{E}\|\mathbf{PB}\|_{1\to2}^{2r})^{1/(2r)} + \delta\|\mathbf{B}\|_{1\to2}^2$$

$$= 2e^{-1/2}(2N)^{2/p}\sqrt{p/2}(\mathbb{E}\|\mathbf{PB}\|_{\max}^p)^{1/p}E + \delta\|\mathbf{B}\|_{1\to2}^2.$$

As above, since solutions to $E^2 \leq \alpha E + \beta$ satisfy (14.6), we reach

$$E \leq 2(2e)^{-1/2}(2N)^{2/p}\sqrt{p}(\mathbb{E}\|\mathbf{PB}\|_{\max}^p)^{1/p} + \delta^{1/2}\|\mathbf{B}\|_{1\to2}.$$

Although the probability bound (14.8) with slightly worse constants can be deduced from (14.7), it is instructive to derive it via moment-generating functions. For $\theta > 0$, we obtain, by applying symmetrization (Lemma 8.4) with the convex function $F(\mathbf{u}) = \exp(\theta \max_k |u_k|)$,

$$\mathbb{E}\exp(\theta(\|\mathbf{PB}\|_{1\to 2}^2 - \delta\|\mathbf{B}\|_{1\to 2}^2)) \le \mathbb{E}\exp\left(2\theta \max_{k\in[N]} \left| \sum_{j=1}^N \epsilon_j \delta_j |B_{jk}|^2 \right|\right)$$

$$\le \sum_{k=1}^N \mathbb{E}\exp\left(2\theta \left| \sum_{j=1}^N \epsilon_j \delta_j |B_{jk}|^2 \right|\right) \le 2N\mathbb{E}_\delta \exp(2\theta^2 \|\mathbf{B}\|_{\max}^2 \|\mathbf{PB}\|_{1\to 2}^2),$$

where in the last step we have used the fact that $\sum_{j=1}^N \epsilon_j |B_{j,k}|^2$ is subgaussian by Theorem 7.27. Assuming that $2\theta\|\mathbf{B}\|_{\max}^2 \le 1/2$, Hölder's (or Jensen's) inequality gives

$$\exp(-\theta\delta\|\mathbf{B}\|_{1\to 2}^2)\mathbb{E}[\exp(\theta\|\mathbf{PB}\|_{1\to 2}^2)] \le 2N\mathbb{E}[\exp(\theta\|\mathbf{PB}\|_{1\to 2}^2/2)]$$

$$\le 2N\left(\mathbb{E}[\exp(\theta\|\mathbf{PB}\|_{1\to 2}^2)]\right)^{1/2}.$$

Rearranging this inequality results in

$$\mathbb{E}\left[\exp(\theta(\|\mathbf{PB}\|_{1\to 2}^2 - 2\delta\|\mathbf{B}\|_{1\to 2}^2))\right] \le 4N^2 \quad \text{for all } 0 < \theta \le \frac{1}{4\|\mathbf{B}\|_{\max}^2}.$$

Markov's inequality together with the choice $\theta = 1/(4\|\mathbf{B}\|_{\max}^2)$ yields

$$\mathbb{P}(\|\mathbf{PB}\|_{1\to 2}^2 - 2\delta\|\mathbf{B}\|_{1\to 2}^2 \ge t) \le 4N^2 e^{-\theta t} = 4N^2 e^{-t/(4\|\mathbf{B}\|_{\max}^2)}.$$

Taking square roots inside the probability above and substituting $u = \sqrt{t}/(2\|\mathbf{B}\|_{\max})$ imply

$$\mathbb{P}(\|\mathbf{PB}\|_{1\to 2} \ge \sqrt{2\delta}\|\mathbf{B}\|_{1\to 2} + 2\|\mathbf{B}\|_{\max} u) \le 4N^2 e^{-u^2}.$$

This completes the proof. ☐

Proof (of Theorem 14.1). We first derive a moment estimate for $\|\mathbf{PHP}\|_{2\to 2}$. Using the decoupling inequality (8.18) (noticing that $\mathbf{H}$ has zero diagonal) and applying Lemma 14.3 twice, we obtain, for $p \ge 2$,

$$(\mathbb{E}\|\mathbf{PHP}\|_{2\to 2}^p)^{1/p} \le 2(\mathbb{E}\|\mathbf{PHP}'\|_{2\to 2}^p)^{1/p}$$

$$\le 2\left(\mathbb{E}_\mathbf{P}(C(C_2 N)^{2/p}\sqrt{p}\mathbb{E}_{\mathbf{P}'}\|\mathbf{PHP}'\|_{1\to 2} + \sqrt{\delta}\|\mathbf{PH}\|_{2\to 2})^p\right)^{1/p}$$

$$\le 2C(C_2 N)^{2/p}\sqrt{p}(\mathbb{E}\|\mathbf{PHP}'\|_{1\to 2}^p)^{1/p} + 2\sqrt{\delta}(\mathbb{E}\|\mathbf{HP}\|_{2\to 2}^p)^{1/p}$$

$$\le 2C(C_2 N)^{2/p}\sqrt{p}(\mathbb{E}\|\mathbf{PHP}'\|_{1\to 2}^p)^{1/p} + 2\sqrt{\delta} \cdot C(C_2 N)^{2/p}\sqrt{p}(\mathbb{E}\|\mathbf{HP}\|_{1\to 2}^p)^{1/p}$$

$$+ 2\delta\|\mathbf{H}\|_{2\to 2}^2.$$

Hereby, we have also used that $\|\mathbf{HP}\|_{2\to2} = \|(\mathbf{HP})^*\|_{2\to2} = \|\mathbf{PH}\|_{2\to2}$ since $\mathbf{H}$ and $\mathbf{P}$ are self-adjoint. Next we exploit the properties of $\mathbf{H}$. Since $\mu = \|\mathbf{H}\|_{\max}$, we have $\|\mathbf{PHP}'\|_{\max} \leq \mu$ for any realization of $\mathbf{P}$ and $\mathbf{P}'$. Moreover,

$$\|\mathbf{H}\|_{1\to2} = \|\mathbf{A}^*\mathbf{A} - \mathbf{Id}\|_{1\to2} \leq \|\mathbf{A}^*\mathbf{A}\|_{1\to2} = \max_{k\in[N]} \|\mathbf{A}^*\mathbf{a}_k\|_2 \leq \|\mathbf{A}\|_{2\to2}$$

because the columns $\mathbf{a}_k$ are ℓ_2-normalized. It follows that

$$\|\mathbf{HP}\|_{1\to2} \leq \|\mathbf{H}\|_{1\to2} \leq \|\mathbf{A}\|_{2\to2} \tag{14.10}$$

for any realization of $\mathbf{P}$. Furthermore,

$$\|\mathbf{H}\|_{2\to2} = \|\mathbf{A}^*\mathbf{A} - \mathbf{Id}\|_{2\to2} = \max\{1, \|\mathbf{A}\|_{2\to2}^2 - 1\} \leq \|\mathbf{A}\|_{2\to2}^2$$

because $\|\mathbf{A}\|_{2\to2}^2 \geq N/m$ by (14.4). An application of Lemma 14.4 conditionally on $\mathbf{P}'$ leads to

$$(\mathbb{E}\|\mathbf{PHP}'\|_{1\to2}^p)^{1/p} = (\mathbb{E}_{\mathbf{P}'}\mathbb{E}_{\mathbf{P}}\|\mathbf{PHP}'\|_{1\to2}^p)^{1/p}$$

$$\leq \left(\mathbb{E}_{\mathbf{P}'}\left(C_3(2N)^{2/p}\sqrt{p}(\mathbb{E}_{\mathbf{P}}\|\mathbf{PHP}'\|_{\max}^p)^{1/p} + \sqrt{\delta}\|\mathbf{HP}'\|_{1\to2}\right)^p\right)^{1/p}$$

$$\leq C_3(2N)^{2/p}\sqrt{p}\|\mathbf{H}\|_{\max} + \sqrt{\delta}\|\mathbf{H}\|_{1\to2}.$$

Combining the previous estimates, we arrive at

$$(\mathbb{E}\|\mathbf{PHP}\|_{2\to2}^p)^{1/p}$$

$$\leq 2C(C_2N)^{2/p}\sqrt{p}\left(C_3(2N)^{2/p}\sqrt{p}\|\mathbf{H}\|_{\max} + \sqrt{\delta}\|\mathbf{H}\|_{1\to2}\right)$$

$$+ 2C(C_2N)^{2/p}\sqrt{p\delta}\mathbb{E}\|\mathbf{H}\|_{1\to2} + 2\delta\|\mathbf{H}\|_{2\to2}$$

$$\leq (2C_2N^2)^{2/p}\left(C_4p\mu + C_5\sqrt{p\delta}\|\mathbf{A}\|_{2\to2} + 2\delta\|\mathbf{A}\|_{2\to2}^2\right).$$

with $C_4 = 2CC_3 = 2e^{1/(2e)}\sqrt{2/e} \cdot 2(2e)^{-1/2} = 4e^{1/(2e)-1} \approx 1.7687$ and $C_5 = 4C = 4e^{1/(2e)}\sqrt{2/e} \approx 4.1239$. It follows from Proposition 7.15 that, for $u \geq 2$,

$$\mathbb{P}(\|\mathbf{PHP}\|_{2\to2} \geq eC_4\mu u + eC_5\sqrt{\delta}\|\mathbf{A}\|_{2\to2}\sqrt{u} + 2e\delta\|\mathbf{A}\|_{2\to2}^2) \leq C_6N^4\exp(-u)$$

with $C_6 = (2C_2)^2 \approx 70.15$. This implies that

$$\|\mathbf{PHP}\|_{2\to2} \leq \eta$$

with probability at least $1 - \varepsilon$ provided

$$eC_4\mu\ln(C_6N^4/\varepsilon) \leq \eta/6, \quad eC_5\sqrt{\delta}\|\mathbf{A}\|_{2\to2}\sqrt{\ln(C_6N^4/\epsilon)} \leq 4\eta/5,$$

$$2e\delta\|\mathbf{A}\|_{2\to2}^2 \leq \eta/30.$$

The first two relations are equivalent to

$$\mu \leq \frac{\eta}{C_7 \ln(C_6 N^4/\varepsilon)},$$

$$\delta \|\mathbf{A}\|_{2\to2}^2 \leq \frac{\eta^2}{C_8 \ln(C_6 N^4/\varepsilon)}, \tag{14.11}$$

with $C_7 = 6eC_4 \approx 28.85$, $C_8 = 25e^2C_5^2/16 \approx 196.35$. Then (14.11) also implies $2e\delta\|\mathbf{A}\|_{2\to2}^2 \leq \eta/30$. Noting that $\delta = s/N$ finishes the proof for the Bernoulli model.

For the uniform model, we proceed similarly to the proof of Corollary 12.38 to bound the probability by the one for the Bernoulli model. Let $\mathbb{P}_B$ denote the probability in the Bernoulli model and $\mathbb{P}_{U,r}$ the one in the uniform model, where S is selected uniformly at random among all subsets of cardinality r. Then, for $t > 0$,

$$\mathbb{P}_B(\|\mathbf{PHP}\|_{2\to2} \geq t)$$

$$= \sum_{r=0}^{N} \mathbb{P}_B(\|\mathbf{P}_S\mathbf{HP}_S\|_{2\to2} \geq t \,|\, \mathrm{card}(S) = r)\,\mathbb{P}_B(\mathrm{card}(S) = r)$$

$$\geq \sum_{r=s}^{N} \mathbb{P}_B(\|\mathbf{P}_S\mathbf{HP}_S\|_{2\to2} \geq t \,|\, \mathrm{card}(S) = r)\,\mathbb{P}_B(\mathrm{card}(S) = r)$$

$$= \sum_{r=s}^{N} \mathbb{P}_{U,r}(\|\mathbf{P}_S\mathbf{HP}_S\|_{2\to2} \geq t)\,\mathbb{P}_B(\mathrm{card}(S) = r). \tag{14.12}$$

Since the norm of a submatrix does not exceed the norm of the full matrix (see Lemma A.9), we have $\|\mathbf{P}_S\mathbf{HP}_S\|_{2\to2} \leq \|\mathbf{P}_{S'}\mathbf{HP}_{S'}\|_{2\to2}$ whenever $S \subset S' \subset [N]$. This implies that

$$\mathbb{P}_{U,r+1}(\|\mathbf{P}_S\mathbf{HP}_S\|_{2\to2} \geq t) \geq \mathbb{P}_{U,r}(\|\mathbf{P}_S\mathbf{HP}_S\|_{2\to2} \geq t).$$

Moreover, since s is an integer, it is the median of the binomial distribution (see (7.6)), so that

$$\sum_{r=s}^{N} \mathbb{P}_B(\mathrm{card}(S) = r) = \mathbb{P}_B(\mathrm{card}(S) \geq s) \geq 1/2.$$

It follows that

$$\mathbb{P}_B(\|\mathbf{PHP}\|_{2\to2} \geq t) \geq \frac{1}{2}\mathbb{P}_{U,s}(\|\mathbf{P}_S\mathbf{HP}_S\|_{2\to2} \geq t).$$

This implies the claim for the uniform model.   □

14.2 Sparse Recovery via ℓ_1-Minimization

Based on the previous result on the conditioning of random submatrices, we derive recovery guarantees for random sparse vectors via ℓ_1-minimization. Here, we choose both the support of the vector and the signs of its nonzero coefficients at random.

Theorem 14.5. *Let* $\mathbf{A} \in \mathbb{C}^{m \times N}$, $m \leq N$, *be a matrix with* ℓ_2-*normalized columns and coherence* μ. *Let* S *be a subset of* $[N]$ *selected at random according to the uniform model with* $\mathrm{card}(S) = s$ *or to the Bernoulli model with* $\mathbb{E}\,\mathrm{card}(S) = s$. *Let* $\mathbf{x} \in \mathbb{C}^N$ *be a vector supported on* S *for which* $\mathrm{sgn}(\mathbf{x}_S)$ *is a Steinhaus or Rademacher vector independent of* S. *Assume that, for* $\eta, \varepsilon \in (0, 1)$,

$$\mu \leq \frac{c}{\ln(N/\varepsilon)}, \tag{14.13}$$

$$\frac{s}{N}\|\mathbf{A}\|_{2 \to 2}^2 \leq \frac{c}{\ln(N/\varepsilon)}, \tag{14.14}$$

for an appropriate constant $c > 0$. *Then, with probability at least* $1 - \varepsilon$, *the vector* $\mathbf{x}$ *is the unique minimizer of* $\|\mathbf{z}\|_1$ *subject to* $\mathbf{A}\mathbf{z} = \mathbf{A}\mathbf{x}$.

Explicit constants can be found in the proof. We recall from (14.5) that for a unit norm tight frame, relation (14.14) is satisfied under the familiar condition

$$m \geq Cs \ln(N/\varepsilon).$$

Only the mild condition (14.13) is imposed on the coherence.

Proof. The proof relies on the recovery result for vectors with random signs in Proposition 12.15, which in turn builds on the recovery conditions for individual vectors in Corollary 4.28. We are hence led to bound the term

$$\max_{\ell \in \overline{S}} \|\mathbf{A}_S^\dagger \mathbf{a}_\ell\|_2 = \max_{\ell \in \overline{S}} \|(\mathbf{A}_S^* \mathbf{A}_S)^{-1} \mathbf{A}_S^* \mathbf{a}_\ell\|_2$$

for a random choice of S. If $\|\mathbf{A}_S^* \mathbf{A}_S - \mathbf{Id}_S\|_{2 \to 2} \leq \eta$, as analyzed in the proof of Theorem 14.1, then $\|(\mathbf{A}_S^* \mathbf{A}_S)^{-1}\|_{2 \to 2} \leq (1 - \eta)^{-1}$, and we obtain the bound

$$\max_{\ell \in \overline{S}} \|\mathbf{A}_S^\dagger \mathbf{a}_\ell\|_2 \leq (1 - \eta)^{-1} \max_{\ell \notin S} \|\mathbf{A}_S^* \mathbf{a}_\ell\|_2.$$

Using $\mathbf{H} = \mathbf{A}^* \mathbf{A} - \mathbf{Id}$ and the projection $\mathbf{P} = \mathbf{P}_S$ as in the previous section, we realize that

$$\max_{\ell \in \overline{S}} \|\mathbf{A}_S^* \mathbf{a}_\ell\|_2 = \|\mathbf{P}\mathbf{H}(\mathbf{Id} - \mathbf{P})\|_{1 \to 2} \leq \|\mathbf{P}\mathbf{H}\|_{1 \to 2}.$$

Assuming the Bernoulli model with $\delta = s/N$ for now, Lemma 14.4 implies that

$$\mathbb{P}(\|\mathbf{PH}\|_{1 \to 2} \geq \sqrt{2\delta}\|\mathbf{H}\|_{1 \to 2} + 2\|\mathbf{H}\|_{\max}u) \leq 4N^2 e^{-u^2}$$

Since $\|\mathbf{H}\|_{\max} = \mu$ and $\|\mathbf{H}\|_{1 \to 2} \leq \|\mathbf{A}\|_{2 \to 2}$ (see (14.10)), we therefore obtain

$$\mathbb{P}(\|\mathbf{PH}\|_{1 \to 2} \geq \sqrt{2\delta}\|\mathbf{A}\|_{2 \to 2} + 2\mu u) \leq 4N^2 e^{-u^2}. \qquad (14.15)$$

It follows from Proposition 12.15 that, for any $\alpha > 0$, the probability of failure of reconstruction via ℓ_1-minimization can be bounded by

$$\mathbb{P}(\max_{\ell \in \overline{S}} \|\mathbf{A}_S^\dagger \mathbf{a}_\ell\| \geq \alpha) + 2Ne^{-\alpha^{-2}/2}$$

$$\leq 2Ne^{-\alpha^{-2}/2} + \mathbb{P}(\|\mathbf{A}_S^* \mathbf{A}_S - \mathbf{Id}_S\|_{2 \to 2} \geq 3/4) + \mathbb{P}(\|\mathbf{PH}\|_{1 \to 2} \geq \alpha/4),$$
$$(14.16)$$

where we have set $\eta = 3/4$ in the inequalities in the beginning of this proof. Let $C_6 \approx 70.15$, $C_7 \approx 28.85$, $C_8 \approx 196.35$ be the constants from the proof of Theorem 14.1; see (14.11). With $\alpha = 1/\sqrt{2\ln(2C_6 N^4/\varepsilon)}$, the first term in (14.16) is bounded by ε/C_6. Assume further that

$$\mu \leq \frac{3/4}{C_7 \ln(2C_6 N^4/\varepsilon)}, \quad \text{and} \quad \sqrt{\delta}\|\mathbf{A}\|_{2 \to 2} \leq \frac{3/4}{\sqrt{C_8 \ln(2C_6 N^4/\varepsilon)}}. \qquad (14.17)$$

Then, by Theorem 14.1,

$$\mathbb{P}(\|\mathbf{A}_S^* \mathbf{A}_S - \mathbf{Id}_S\|_{2 \to 2} \geq 3/4) \leq \varepsilon/2.$$

The second inequality in (14.17) also implies

$$\sqrt{2\delta}\|\mathbf{A}\|_{2 \to 2} \leq c_1 \alpha$$

with $c_1 = 3/(2C_8^{1/2}) \approx 0.107$. Furthermore, with $c_2 = 0.14$ and $u = C_9 \sqrt{\ln(2C_6 N^4/\varepsilon)}$ for $C_9 = \sqrt{2}c_2 C_7/3 \approx 1.904$, we have $2\mu u \leq c_2 \alpha$, so that

$$\sqrt{2\delta}\|\mathbf{A}\|_{2 \to 2} + 2\mu u \leq (c_1 + c_2)\alpha \leq \alpha/4.$$

Therefore, by (14.15), we deduce

$$\mathbb{P}(\|\mathbf{PH}\|_{1 \to 2} \geq \alpha/4) \leq \mathbb{P}(\|\mathbf{PH}\|_{1 \to 2} \geq \sqrt{2\delta}\|\mathbf{A}\|_{2 \to 2} + 2\mu u) \leq 4N^2 \exp(-u^2)$$

$$= 4N^2 \exp(-C_9^2 \ln(2C_6 N^4/\varepsilon)) \leq 4N^2 \frac{\varepsilon}{2C_6 N^4} = 2\varepsilon/C_6.$$

Altogether the failure probability is bounded by $\varepsilon/C_6 + \varepsilon/2 + 2\varepsilon/C_6 \leq \varepsilon$. This completes the proof for the Bernoulli model.

For the uniform model, we proceed similarly to the end of the proof of Theorem 14.1 and show that

$$\mathbb{P}_{U,s}(\|\mathbf{PH}\|_{1\rightarrow 2} \geq t) \leq 2\mathbb{P}_B(\|\mathbf{PH}\|_{1\rightarrow 2} \geq t), \tag{14.18}$$

where again $\mathbb{P}_{U,s}$ denotes the probability under the uniform model, for which subsets S are selected uniformly at random among all subsets of cardinality s, while $\mathbb{P}_B$ denotes the probability under the Bernoulli model. With this observation, the proof is concluded in the same way as above. $\qquad\square$

Theorem 14.5 explains why one can expect recovery of s-sparse signals from $m \geq Cs\log(N)$ measurements under coherence conditions much milder than the ones of Chap. 5, even in situations when estimates on the restricted isometry constants are unavailable or known to fail.

Usual numerical performance tests take the support set of the signal and the nonzero coefficients at random, so that the results of this chapter explain the high success rate of these experiments. However, conclusions for the recovery of "real-world" signals from such numerical experiments should be handled with care. Certainly, Theorem 14.5 still indicates that recovery is possible under mild conditions, but it is often hard to argue rigorously that the support set of a "natural" signal is random. For instance, the wavelet coefficients of a natural image follow the edges of an image, so that the nonzero (large) coefficients are rather organized in trees. Such tree structure is definitely not random—at least the support set does not follow a *uniform* distribution. Therefore, the results of the preceding chapters holding for *all* sparse signals remain very important. Moreover, currently available stability results for random signals are weaker than the ones based on the restricted isometry property.

Notes

Conditioning of random submatrices (subdictionaries) based on coherence was first studied by Tropp in [481], where he derived slightly weaker estimates. Indeed, the bounds in [481] require in addition to (14.2) that $\mu^2 s \ln(s) \leq c$, which is harder to satisfy than (14.1) (unless s is tiny, in which case the "quadratic" bounds of Chap. 5 would also be fine). Tropp refined his estimates later in [480] to the ones presented in this chapter. Using more sophisticated decoupling techniques together with the matrix Chernoff inequality [486], Chrétien and Darses [118] obtained slightly better constants than the ones stated in Theorem 14.1 on the conditioning of random submatrices. Candès and Plan applied Tropp's result in the context of statistical sparse estimation using the Dantzig selector, where they also allowed random noise on the measurements [87]. Tropp's paper [481] also contains refined results for the case where the matrix $\mathbf{A}$ is the concatenation of two orthonormal bases. Candès and Romberg [92] treated the special case of the concatenation of the canonical and the Fourier basis; see also [482].

Tropp's original methods in [480, 481] use the noncommutative Khintchine inequalities (8.114) instead of the tail inequality (8.36) for matrix-valued Rademacher sums. The "random compression bound" of Lemma 14.3 goes back to Rudelson and Vershynin [432]; see also [483, Proposition 12].

Analyses of sparse recovery algorithms for random choices of signals have been carried out as well in the context of multichannel sparse recovery or multiple measurement vectors [184, 241], where the measurement matrix $\mathbf{A}$ is applied to a collection of sparse signals $\mathbf{x}^{(1)}, \ldots, \mathbf{x}^{(L)} \in \mathbb{C}^N$ with common support, i.e.,

$$\left[\mathbf{y}^{(1)} | \cdots | \mathbf{y}^{(L)}\right] = \mathbf{A}\left[\mathbf{x}^{(1)} | \cdots | \mathbf{x}^{(L)}\right]$$

and $\mathrm{supp}(\mathbf{x}^{(\ell)}) = S$ for all $\ell \in [L]$. In this context a nonzero coefficient is actually a vector $\mathbf{x}_k = (x_k^{(1)}, \ldots, x_k^{(L)}) \in \mathbb{C}^L$ chosen at random (for instance, according to a multivariate Gaussian distribution or the uniform distribution on the sphere). The results in [184, 241] apply to multichannel variants of ℓ_1-minimization and greedy algorithms and predict that the probability of failure decreases exponentially in L provided that a very mild condition on the number of samples hold. The estimates outlined in this chapter are partly used in these contributions.

The bound on the conditioning of random matrices of Theorem 14.1 is somewhat related to the Bourgain–Tzafriri restricted invertibility theorem [66, 67]. We state a strengthened version due to Spielman and Srivastava [449].

Theorem 14.6. *Let* $\mathbf{A} \in \mathbb{C}^{m \times N}$ *with* ℓ_2-*normalized columns and* $\alpha \in (0, 1)$ *be a prescribed parameter. There exists a subset* $S \subset [N]$ *with*

$$\mathrm{card}(S) \geq \frac{\alpha^2 N}{\|\mathbf{A}\|_{2 \to 2}^2}$$

such that

$$(1 - \alpha)^2 \|\mathbf{x}\|_2^2 \leq \|\mathbf{A}_S \mathbf{x}\|_2^2$$

for all $\mathbf{x} \in \mathbb{C}^S$.

The assumptions in this theorem are weaker than the one of Theorem 14.1; in particular, no reference to the coherence or a similar quantity is made. But the statement is only about *existence* of a submatrix with controlled smallest singular value and not about properties of *most* (i.e., random) submatrices. Indeed, one cannot expect Theorem 14.1 to hold without any assumption on the coherence because a random submatrix of a matrix which consists of a duplicated orthonormal basis (hence, $\mu = 1$) will contain a duplicated column with high probability, so that the singular value will be zero. Nevertheless, well-conditioned submatrices certainly exist in this case, such as the submatrix consisting of one copy of the orthonormal basis. Further information on the restricted invertibility theorem can be found, for instance, in [102, 484, 499].

There is also a relation between Theorem 14.1 and another theorem of Bourgain and Tzafriri [67] stated next.

Theorem 14.7. *Let* $\mathbf{H} \in \mathbb{C}^{N \times N}$ *with* $\|\mathbf{H}\|_{2 \to 2} \leq 1$ *whose entries satisfy*

$$|H_{j,k}| \leq \frac{1}{\ln^2 N}.$$

Let $c \in (0, 1)$ *and let* $S \subset [N]$ *of size* cN *be selected uniformly at random. Then* $\|\mathbf{P}_S \mathbf{H} \mathbf{P}_S\|_{2 \to 2} \leq 1/2$ *with probability at least* $1 - N^{-c}$.

In contrast to Theorem 14.1, this result does not only apply to matrices of the form $\mathbf{H} = \mathbf{A}^* \mathbf{A} - \mathbf{Id}$ (i.e., with zero diagonal), but it requires a condition on the size of the matrix entries slightly stronger than (14.1).

Exercises

14.1. Let $\mathbf{B} \in \mathbb{C}^{N \times N}$ and $\mathbf{P}$ be a random diagonal matrix of independent Bernoulli variables with mean $\delta \in [0, 1]$. Show that

$$\mathbb{E}\|\mathbf{BP}\|_{2 \to 2} \leq \sqrt{8 \ln(2N)}(\mathbb{E}\|\mathbf{BP}\|_{1 \to 2}^2)^{1/2} + \sqrt{\delta}\|\mathbf{B}\|_{2 \to 2}.$$

Hint: Use (8.117).

14.2. Let $\mathbf{B} \in \mathbb{C}^{N \times N}$ and $\mathbf{P}$ be a random diagonal matrix of independent Bernoulli variables with mean $\delta \in [0, 1]$. Show that

$$\mathbb{E}\|\mathbf{PB}\|_{1 \to 2} \leq \sqrt{8 \ln(2N)}(\mathbb{E}\|\mathbf{PB}\|_{\max}^2)^{1/2} + \sqrt{\delta}\|\mathbf{B}\|_{1 \to 2}.$$

14.3. Let $\mathbf{A} \in \mathbb{C}^{m \times N}$, $m \leq N$, be a matrix with ℓ_2-normalized columns and coherence μ. Let S be a subset of $[N]$ selected at random according to the Bernoulli model with $\mathbb{E} \operatorname{card}(S) = s$. Show that

$$\mathbb{E}\|\mathbf{A}_S^* \mathbf{A}_S - \mathbf{Id}_S\|_{2 \to 2} \leq 16 \ln(2N)\mu + \sqrt{128 \ln(2N)\frac{s}{N}\|\mathbf{A}\|_{2 \to 2}^2} + 2\frac{s}{N}\|\mathbf{A}\|_{2 \to 2}^2.$$

14.4. Verify (14.18) in detail.

Chapter 15
Algorithms for ℓ_1-Minimization

Throughout this book, ℓ_1-minimization plays a central role as a recovery method for compressive sensing. So far, however, no *algorithm* for this minimization problem was introduced. In Chap. 3, we mentioned that ℓ_1-minimization can be recast as a linear program in the real case (see (P'_1)) and as a second-order cone program in the complex case (see $(\mathrm{P}'_{1,\eta})$). In these situations, standard software is available, which is based on interior point methods or the older simplex method for linear programs. Despite the ease of use and the reliability of such standard software, it is developed for general linear and second-order cone problems. Algorithms designed specifically for ℓ_1-minimization may be faster than general purpose methods. This chapter introduces and analyzes several of these algorithms. However, we do not make an attempt to give a complete picture of all available algorithms. The present choice was made for the sake of simplicity of exposition and in order to give different flavors of possible approaches. Nevertheless, we give a brief overview on further optimization methods in the Notes section.

The homotopy method introduced in Sect. 15.1, which is restricted to the real case, has similarities with the orthogonal matching pursuit, but is guaranteed to always provide an ℓ_1-minimizer. In Sect. 15.2, we present an algorithm due to Chambolle and Pock. It applies to a whole class of optimization problems including ℓ_1-minimization. Our third algorithm, iteratively reweighted least squares, is only a proxy for ℓ_1-minimization, and its output may be different from the ℓ_1-minimizer. But its formulation is motivated by ℓ_1-minimization, and in certain cases, it indeed provides the ℓ_1-minimizer. Under the stable null space property (equivalent to exact and approximate sparse recovery via ℓ_1-minimization), we will show error guarantees similar to ℓ_1-minimization.

S. Foucart and H. Rauhut, *A Mathematical Introduction to Compressive Sensing*,
Applied and Numerical Harmonic Analysis, DOI 10.1007/978-0-8176-4948-7_15,
© Springer Science+Business Media New York 2013

15.1   The Homotopy Method

The homotopy method computes a minimizer of the ℓ_1-minimization problem

$$\min_{\mathbf{x} \in \mathbb{R}^N} \|\mathbf{x}\|_1 \quad \text{subject to } \mathbf{Ax} = \mathbf{y} \tag{15.1}$$

in the real case, that is, for $\mathbf{A} \in \mathbb{R}^{m \times N}$ and $\mathbf{y} \in \mathbb{R}^m$. Moreover, a slight variant solves the quadratically constrained ℓ_1-minimization problem

$$\min_{\mathbf{x} \in \mathbb{R}^N} \|\mathbf{x}\|_1 \quad \text{subject to } \|\mathbf{Ax} - \mathbf{y}\|_2 \leq \eta. \tag{15.2}$$

For $\lambda > 0$, we consider the ℓ_1-regularized least squares functional

$$F_\lambda(\mathbf{x}) = \frac{1}{2}\|\mathbf{Ax} - \mathbf{y}\|_2^2 + \lambda\|\mathbf{x}\|_1, \quad \mathbf{x} \in \mathbb{R}^N, \tag{15.3}$$

and denote by $\mathbf{x}_\lambda$ a minimizer of F_λ. We will see below that if $\lambda = \hat{\lambda}$ is large enough, then $\mathbf{x}_{\hat{\lambda}} = \mathbf{0}$. Furthermore, we essentially have $\lim_{\lambda \to 0} \mathbf{x}_\lambda = \mathbf{x}^\sharp$, where $\mathbf{x}^\sharp$ is a minimizer of (15.1). A precise statement is contained in the next result.

Proposition 15.1. *Assume that* $\mathbf{Ax} = \mathbf{y}$ *has a solution. If the minimizer* $\mathbf{x}^\sharp$ *of* (15.1) *is unique, then*

$$\lim_{\lambda \to 0^+} \mathbf{x}_\lambda = \mathbf{x}^\sharp.$$

More generally, the $\mathbf{x}_\lambda$ *are bounded and any cluster point of* $(\mathbf{x}_{\lambda_n})$, *where* (λ_n) *is a positive sequence such that* $\lim_{n \to \infty} \lambda_n = 0^+$, *is a minimizer of* (15.1).

Proof. For $\lambda > 0$, let $\mathbf{x}_\lambda$ be a minimizer of F_λ and $\mathbf{x}^\sharp$ be a minimizer of $\|\mathbf{z}\|_1$ subject to $\mathbf{Az} = \mathbf{y}$. Then

$$\frac{1}{2}\|\mathbf{Ax}_\lambda - \mathbf{y}\|_2^2 + \lambda\|\mathbf{x}_\lambda\|_1 = F_\lambda(\mathbf{x}_\lambda) \leq F_\lambda(\mathbf{x}^\sharp) = \lambda\|\mathbf{x}^\sharp\|_1. \tag{15.4}$$

This implies that

$$\|\mathbf{x}_\lambda\|_1 \leq \|\mathbf{x}^\sharp\|_1, \tag{15.5}$$

so that $(\mathbf{x}_\lambda)$ is bounded and $(\mathbf{x}_{\lambda_n})$ possesses a cluster point $\mathbf{x}'$. Inequality (15.4) also implies that

$$\frac{1}{2}\|\mathbf{Ax}_\lambda - \mathbf{y}\|_2^2 \leq \lambda\|\mathbf{x}^\sharp\|_1.$$

Letting $\lambda \to 0$ shows that $\|\mathbf{Ax}' - \mathbf{y}\|_2 = 0$, that is, $\mathbf{Ax}' = \mathbf{y}$. Exploiting (15.5) yields $\|\mathbf{x}'\|_1 \leq \|\mathbf{x}^\sharp\|_1$ and, by definition of $\mathbf{x}^\sharp$, this means that $\|\mathbf{x}'\|_1 = \|\mathbf{x}^\sharp\|_1$. Therefore, also $\mathbf{x}'$ minimizes $\|\mathbf{z}\|_1$ subject to $\mathbf{Az} = \mathbf{y}$. If the ℓ_1-minimizer is unique, then $\mathbf{x}' = \mathbf{x}^\sharp$, and this argument shows that any subsequence of $(\mathbf{x}_{\lambda_n})$ converges to $\mathbf{x}^\sharp$; hence, the whole sequence $(\mathbf{x}_{\lambda_n})$ converges to $\mathbf{x}^\sharp$. □

The basic idea of the homotopy method is to follow the solution $\mathbf{x}_\lambda$ from $\mathbf{x}_{\hat\lambda} = \mathbf{0}$ to $\mathbf{x}^\sharp$. As we will show below, the solution path $\lambda \mapsto \mathbf{x}_\lambda$ is piecewise linear, and it is enough to trace the endpoints of the linear pieces.

By Theorem B.21, the minimizer of (15.3) can be characterized using the subdifferential defined in (B.11). The subdifferential of F_λ is given by

$$\partial F_\lambda(\mathbf{x}) = \mathbf{A}^*(\mathbf{A}\mathbf{x} - \mathbf{y}) + \lambda \partial \|\mathbf{x}\|_1,$$

where the subdifferential of the ℓ_1-norm is given by

$$\partial \|\mathbf{x}\|_1 = \{\mathbf{v} \in \mathbb{R}^N : v_\ell \in \partial |x_\ell|, \ell \in [N]\}.$$

Hereby, the subdifferential of the absolute value is given by

$$\partial |z| = \begin{cases} \{\operatorname{sgn}(z)\} & \text{if } z \neq 0, \\ [-1, 1] & \text{if } z = 0. \end{cases}$$

A vector $\mathbf{x}$ is a minimizer of F_λ if and only if $\mathbf{0} \in \partial F_\lambda(\mathbf{x})$; see Theorem B.21. By the above considerations, this is equivalent to

$$(\mathbf{A}^*(\mathbf{A}\mathbf{x} - \mathbf{y}))_\ell = -\lambda \operatorname{sgn}(x_\ell) \qquad \text{if } x_\ell \neq 0, \qquad (15.6)$$

$$|(\mathbf{A}^*(\mathbf{A}\mathbf{x} - \mathbf{y}))_\ell| \leq \lambda \qquad \text{if } x_\ell = 0, \qquad (15.7)$$

for all $\ell \in [N]$.

The homotopy method starts with $\mathbf{x}^{(0)} = \mathbf{x}_\lambda = \mathbf{0}$. By condition (15.7), the corresponding λ is chosen as $\lambda = \lambda^{(0)} = \|\mathbf{A}^*\mathbf{y}\|_\infty$.

In the further steps $j = 1, 2, \ldots$, the algorithm varies λ, computes corresponding minimizers $\mathbf{x}^{(1)}, \mathbf{x}^{(2)}, \ldots$, and maintains an active (support) set S_j. Denote by

$$\mathbf{c}^{(j)} = \mathbf{A}^*(\mathbf{A}\mathbf{x}^{(j-1)} - \mathbf{y})$$

the current residual vector and, as usual, by $\mathbf{V}_1, \ldots, \mathbf{V}_N$ the columns of $\mathbf{A}$.

Step $j = 1$: Let

$$\ell^{(1)} := \underset{\ell \in [N]}{\operatorname{argmax}} |(\mathbf{A}^*\mathbf{y})_\ell| = \underset{\ell \in [N]}{\operatorname{argmax}} |c_\ell^{(1)}|. \qquad (15.8)$$

One assumes here and also in the further steps that the maximum is attained at only one index ℓ. This is the generic situation—at least if $\mathbf{A}$ and $\mathbf{y}$ are not integer valued, say. We will comment later on the case where the maximum is simultaneously attained at two or more indices ℓ in (15.8) or in (15.9), (15.11), (15.12) below.

Now set $S_1 = \{\ell^{(1)}\}$. We introduce the vector $\mathbf{d}^{(1)} \in \mathbb{R}^N$ describing the direction of the solution (homotopy) path with entries

$$d^{(1)}_{\ell^{(1)}} = \|\mathbf{a}_{\ell^{(1)}}\|_2^{-2}\mathrm{sgn}((\mathbf{A}^*\mathbf{y})_{\ell^{(1)}}) \quad \text{and} \quad d^{(1)}_\ell = 0, \quad \ell \neq \ell^{(1)}.$$

The first linear piece of the solution path then takes the form

$$\mathbf{x} = \mathbf{x}(\gamma) = \mathbf{x}^{(0)} + \gamma \mathbf{d}^{(1)} = \gamma \mathbf{d}^{(1)}, \quad \gamma \in [0, \gamma^{(1)}],$$

with some $\gamma^{(1)}$ to be determined below. One verifies with the definition of $\mathbf{d}^{(1)}$ that (15.6) is always satisfied for $\mathbf{x} = \mathbf{x}(\gamma)$ and $\lambda = \lambda(\gamma) = \lambda^{(0)} - \gamma, \gamma \in [0, \lambda^{(0)}]$. The next breakpoint is found by determining the maximal $\gamma = \gamma^{(1)} > 0$ for which (15.7) is still satisfied, that is,

$$\gamma^{(1)} = \min_{\ell \neq \ell^{(1)}}^{+} \left\{ \frac{\lambda^{(0)} + c_\ell^{(1)}}{1 - (\mathbf{A}^*\mathbf{A}\mathbf{d}^{(1)})_\ell}, \frac{\lambda^{(0)} - c_\ell^{(1)}}{1 + (\mathbf{A}^*\mathbf{A}\mathbf{d}^{(1)})_\ell} \right\}, \quad (15.9)$$

where the symbol $\overset{+}{\min}$ indicates that the minimum is taken only over positive arguments. (There will always be at least one positive argument due to the assumption that the maximum in (15.8) is taken at only one index.) Then $\mathbf{x}^{(1)} = \mathbf{x}(\gamma^{(1)}) = \gamma^{(1)}\mathbf{d}^{(1)}$ is the next minimizer of F_λ for $\lambda = \lambda^{(1)} := \lambda^{(0)} - \gamma^{(1)}$. This $\lambda^{(1)}$ satisfies $\lambda^{(1)} = \|\mathbf{c}^{(2)}\|_\infty$. Let $\ell^{(2)}$ be the index where the minimum in (15.9) is attained (where we again assume that the minimum is attained only at one index), and set $S_2 = \{\ell^{(1)}, \ell^{(2)}\}$.

Step $j \geq 2$: The new direction $\mathbf{d}^{(j)}$ of the homotopy path is determined by

$$\mathbf{A}^*_{S_j}\mathbf{A}_{S_j}\mathbf{d}^{(j)}_{S_j} = -\mathrm{sgn}(\mathbf{c}^{(j)}_{S_j}). \quad (15.10)$$

This amounts to solving a linear system of equations of size $|S_j| \times |S_j|$, where $|S_j| \leq j$. Outside the components in S_j, we set $d^{(j)}_\ell = 0, \ell \notin S_j$. The next linear piece of the path is given by

$$\mathbf{x}(\gamma) = \mathbf{x}^{(j-1)} + \gamma \mathbf{d}^{(j)}, \quad \gamma \in [0, \gamma^{(j)}].$$

To verify this, we note that (15.6) for $\mathbf{x} = \mathbf{x}^{(j-1)}$ implies that $\mathrm{sgn}(\mathbf{c}^{(j)}_{S_{j-1}}) = -\mathrm{sgn}(\mathbf{x}^{(j-1)}_{S_{j-1}})$ and that (15.7) for $\mathbf{x} = \mathbf{x}^{(j-1)}$ implies that $|c^{(j)}_\ell| = \lambda^{(j-1)}$ for an index ℓ that has possibly been added to S_{j-1} to obtain S_j; see below. Hence, $(\mathbf{A}^*(\mathbf{A}\mathbf{x}^{(j-1)} - \mathbf{y}))_\ell = \lambda^{(j-1)}\mathrm{sgn}(c^{(j)}_\ell)$ for all $\ell \in S_j$ (which clearly also holds if an index has been removed from S_{j-1} to obtain S_j; see again below). By continuity together with (15.7) and (15.6), this implies that necessary and sufficient conditions for $\mathbf{x}(\gamma)$ to be on the solution path are $|\mathbf{A}^*(\mathbf{A}(\mathbf{x}(\gamma) - \mathbf{y}))_\ell| \leq \lambda^{(j-1)} - \gamma$ for all $\ell \notin S_j$ and

$$(\mathbf{A}^*(\mathbf{A}\mathbf{x}(\gamma) - \mathbf{y}))_\ell = (\lambda^{(j-1)} - \gamma)\mathrm{sgn}(c^{(j)}_\ell) \quad \text{for all } \ell \in S_j.$$

One easily verifies these equations with the definition of $\mathbf{x}(\gamma)$ and $\mathbf{d}^{(j)}$. The maximal γ such that $\mathbf{x}(\gamma)$ satisfies (15.7) is

$$\gamma_+^{(j)} = \min_{\ell \notin S_j}{}^+ \left\{ \frac{\lambda^{(j-1)} + c_\ell^{(j)}}{1 - (\mathbf{A}^* \mathbf{A} \mathbf{d}^{(j)})_\ell}, \frac{\lambda^{(j-1)} - c_\ell^{(j)}}{1 + (\mathbf{A}^* \mathbf{A} \mathbf{d}^{(j)})_\ell} \right\}. \tag{15.11}$$

The maximal γ such that $\mathbf{x}(\gamma)$ obeys (15.6), that is, $\mathrm{sgn}(c_{S^j}^{(j)}) = -\mathrm{sgn}(\mathbf{x}(\gamma)_{S^j})$, is given by

$$\gamma_-^{(j)} = \min_{\ell \in S_j, d_\ell^{(j)} \neq 0}{}^+ \left\{ -x_\ell^{(j-1)} / d_\ell^{(j)} \right\}. \tag{15.12}$$

If all arguments in the minimum are nonpositive, this means that $\mathrm{sgn}(x_{S^{(j)}}^{(j-1)}) = \mathrm{sgn}(x_{S^{(j)}}^{(j-1)} + \gamma d_{S^{(j)}}^{(j)})$ for all positive γ and we set $\gamma_-^{(j)} = \infty$. The next breakpoint is given by $\mathbf{x}^{(j)} = \mathbf{x}(\gamma^{(j)})$ with $\gamma^{(j)} = \min\{\gamma_+^{(j)}, \gamma_-^{(j)}\}$. If $\gamma_+^{(j)}$ determines the minimum, then the index $\ell_+^{(j)} \notin S_j$ providing the minimum in (15.11) is added to the active set, that is, $S_{j+1} = S_j \cup \{\ell_+^{(j)}\}$. If $\gamma^{(j)} = \gamma_-^{(j)}$, then the index $\ell_-^{(j)} \in S_j$ at which the minimum in (15.12) is attained is removed from the active set (because $\mathbf{x}(\gamma_-^{(j)})_{\ell_-^{(j)}} = 0$ by definition of $\gamma_-^{(j)}$), that is, $S_{j+1} = S_j \setminus \{\ell_-^{(j)}\}$. We update $\lambda^{(j)} = \lambda^{(j-1)} - \gamma^{(j)}$ so that $\lambda^{(j)} = \|\mathbf{c}^{(j+1)}\|_\infty$.

We note that, by construction of the algorithm, the linear system (15.10) always has a (generically unique) solution $\mathbf{d}_{S_j}^{(j)}$. It may in principle happen that $\mathbf{d}_{S_j}^{(j)}$ is not unique (if the minimizer of F_λ for the corresponding λ is not unique), but then any choice of $\mathbf{d}_{S_j}^{(j)}$ satisfying (15.10) is suitable.

The algorithm stops when $\lambda^{(j)} = \|\mathbf{c}^{(j+1)}\|_\infty = 0$, i.e., when the residual vanishes, and it outputs $\mathbf{x}^\sharp = \mathbf{x}^{(j)}$.

We say that the minimizer $\ell^{(j)} \in [N]$ is unique at step j if it is the unique minimizer of

$$\gamma^{(j)} = \min\{\gamma_+^{(j)}, \gamma_-^{(j)}\}$$

$$= \min \left\{ \min_{\ell \notin S_j}{}^+ \left\{ \frac{\lambda^{(j-1)} + c_\ell^{(j)}}{1 - (\mathbf{A}^* \mathbf{A} \mathbf{d}^{(j)})_\ell}, \frac{\lambda^{(j-1)} - c_\ell^{(j)}}{1 + (\mathbf{A}^* \mathbf{A} \mathbf{d}^{(j)})_\ell} \right\}, \min_{\ell \in S_j, d_\ell^{(j)} \neq 0}{}^+ \left\{ -x_\ell^{(j-1)} / d_\ell^{(j)} \right\} \right\}.$$

The following result about the homotopy method holds.

Theorem 15.2. *Assume that the ℓ_1-minimizer $\mathbf{x}^\sharp$ of (15.1) is unique. If the minimizer $\ell^{(j)}$ is unique at each step, then the homotopy algorithm outputs $\mathbf{x}^\sharp$.*

Proof. Following the description of the algorithm above, it only remains to show that the algorithm eventually stops. To this end, we note that the sign patterns $\mathrm{sgn}(\mathbf{x}_{\lambda^{(j)}})$, $j = 1, 2, \ldots$, are pairwise different. Indeed, if $\mathrm{sgn}(\mathbf{x}_{\lambda_1}) = \mathrm{sgn}(\mathbf{x}_{\lambda_2})$ for some $\lambda_1 > \lambda_2 > 0$, then it is a straightforward consequence of (15.6) and (15.7) that every point on the line segment connecting $\mathbf{x}_{\lambda_1}$ and $\mathbf{x}_{\lambda_2}$ is a minimizer $\mathbf{x}_\lambda$ for some $\lambda \in [\lambda_1, \lambda_2]$. By construction, the points $\mathbf{x}_{\lambda^{(j)}}$ are endpoints of such line segments, and since the algorithm adds or removes an element of the support in each step, the vectors $\text{sgn}(\mathbf{x}_{\lambda_1})$ and $\text{sgn}(\mathbf{x}_{\lambda_2})$ have to be different. There are only a finite number of possible sign patterns, and hence, the algorithm eventually stops and the final parameter is $\lambda^{(n)} = 0$. Since $\mathbf{x}_\lambda \to \mathbf{x}^\sharp$ as $\lambda \to 0$ by Proposition 15.1, the final point $\mathbf{x}_{\lambda^{(n)}} = \mathbf{x}_0$ equals $\mathbf{x}^\sharp$. This completes the proof. $\square$

Remark 15.3. (a) The theorem still holds if $\mathbf{Ax} = \mathbf{y}$ has a solution but the ℓ_1-minimizer is not unique: The homotopy algorithm then outputs one of these ℓ_1-minimizers.

(b) If the minimum in (15.11) or (15.12) is attained in more than one index, then additional effort is required to compute the next index set S_{j+1}. Denote by $T \subset [N] \setminus S_j$ the set of indices ℓ for which the minimum $\gamma^{(j)}$ is attained in (15.11) and (15.12) (we only have to consider both (15.11) and (15.12) if $\gamma_+^{(j)} = \gamma_-^{(j)}$). We go through all possible subsets $L \subset T$ and check whether $S_{j+1} = S_j \cup L$ is valid as next support set. This amounts to solving (15.10) for the corresponding next direction $\mathbf{d}^{(j)}$ of the solution path and checking whether (15.6) and (15.7) are satisfied for the potential solution $\mathbf{x} = \mathbf{x}(\gamma) = \mathbf{x}^{(j-1)} + \gamma \mathbf{d}^{(j)}$ and $\lambda = \lambda^{(j-1)} - \gamma$ for some suitable $\gamma > 0$. When such a valid S_{j+1} is found, we continue as before.

If the algorithm is stopped before the residual vanishes, say at some iteration j, then it yields the minimizer of $F_\lambda = F_{\lambda^{(j)}}$. In particular, obvious stopping rules may also be used to solve the problems

$$\min \|\mathbf{x}\|_1 \quad \text{subject to } \|\mathbf{Ax} - \mathbf{y}\|_2 \leq \eta, \tag{15.13}$$

$$\min \|\mathbf{Ax} - \mathbf{y}\|_2 \quad \text{subject to } \|\mathbf{x}\|_1 \leq \tau. \tag{15.14}$$

The first of these appears in (15.2), and the second is called the LASSO (least absolute shrinkage and selection operator); see Chap. 3.

The LARS (least angle regression) algorithm is a simple modification of the homotopy method, which only adds elements to the active set at each step. Thus, $\gamma_-^{(j)}$ in (15.12) is not considered. (Sometimes the homotopy method is therefore also called modified LARS.) LARS is not guaranteed to yield the solution of (15.1) anymore. However, it is observed empirically in sparse recovery problems that the homotopy method merely removes elements from the active set, so that in this case LARS and homotopy perform the same steps. If the solution of (15.1) is s-sparse and the homotopy method never removes elements, then the solution is obtained after precisely s steps. Furthermore, the most demanding computational part at step j is then the solution of the $j \times j$ linear system of equations (15.10).

In conclusion, the homotopy and LARS methods are very efficient when the solution is very sparse. For only moderately sparse solutions, the methods in the next sections may be better suited.

15.2 Chambolle and Pock's Primal-Dual Algorithm

This section presents an iterative primal–dual algorithm for the numerical solution of general optimization problems including the various ℓ_1-minimization problems appearing in this book. We require some background on convex analysis and optimization as covered in Appendix B.

Remark 15.4. We formulate everything below in the complex setting of $\mathbb{C}^N$, although the material in Appendix B is treated only for the real case. As noted there, everything carries over to the complex case by identifying $\mathbb{C}^N$ with $\mathbb{R}^{2N}$. The only formal difference when making this identification concrete is that complex inner products have to be replaced by real inner products $\mathrm{Re}\langle \cdot, \cdot \rangle$. Reversely, everything below holds also if $\mathbb{C}^N$ is replaced by $\mathbb{R}^N$, of course.

We consider a general optimization problem of the form

$$\min_{\mathbf{x} \in \mathbb{C}^N} F(\mathbf{Ax}) + G(\mathbf{x}), \tag{15.15}$$

where $\mathbf{A} \in \mathbb{C}^{m \times N}$ and where $F : \mathbb{C}^m \to (-\infty, \infty]$, $G : \mathbb{C}^N \to (-\infty, \infty]$ are extended real-valued lower semicontinuous convex functions; see Definition B.13 for the notion of lower semicontinuity. (Note that the functions are allowed to take the value ∞ so that the requirement of continuity would be too strong.) We will explain in detail below how ℓ_1-minimization fits into this framework.

The dual problem of (15.15) is given by

$$\max_{\boldsymbol{\xi} \in \mathbb{C}^m} -F^*(\boldsymbol{\xi}) - G^*(-\mathbf{A}^*\boldsymbol{\xi}); \tag{15.16}$$

see (B.49). Here F^* and G^* are the convex conjugate functions of F and G (Definition B.17).

Theorem B.30 states that strong duality holds for the pair (15.15) and (15.16) under mild assumptions on F and G, which are always met in the special cases of our interest. Furthermore, by Theorem B.30, the joint primal–dual optimization of (15.15) and (15.16) is equivalent to solving the saddle-point problem

$$\min_{\mathbf{x} \in \mathbb{C}^N} \max_{\boldsymbol{\xi} \in \mathbb{C}^m} \mathrm{Re}\langle \mathbf{Ax}, \boldsymbol{\xi} \rangle + G(\mathbf{x}) - F^*(\boldsymbol{\xi}). \tag{15.17}$$

The algorithm we describe below uses the proximal mappings (B.13) of F^* and G. It will be convenient to introduce another parameter $\tau > 0$ into these mappings by setting, for $\mathbf{z} \in \mathbb{C}^N$,

$$P_G(\tau; \mathbf{z}) := P_{\tau G}(\mathbf{z}) = \underset{\mathbf{x} \in \mathbb{C}^N}{\operatorname{argmin}} \left\{ \tau G(\mathbf{x}) + \frac{1}{2}\|\mathbf{x} - \mathbf{z}\|_2^2 \right\}, \tag{15.18}$$

and $P_{F^*}(\tau; \mathbf{z})$ is defined in the same way.

The algorithm below is efficient if the proximal mappings $P_{F^*}(\tau; \mathbf{z})$ and $P_G(\tau; \mathbf{z})$ are easy to evaluate. Note that by Moreau's identity (B.16), the proximal mapping associated with F^* is easy to compute once the one associated with F is.

Primal–Dual Algorithm

Input: $\mathbf{A} \in \mathbb{C}^{m \times N}$, convex functions F, G.
Parameters: $\theta \in [0, 1], \tau, \sigma > 0$ such that $\tau\sigma\|\mathbf{A}\|_{2 \to 2}^2 < 1$.
Initialization: $\mathbf{x}^0 \in \mathbb{C}^N, \boldsymbol{\xi}^0 \in \mathbb{C}^m, \bar{\mathbf{x}}^0 = \mathbf{x}^0$.
Iteration: repeat until a stopping criterion is met at $n = \bar{n}$:

$$\boldsymbol{\xi}^{n+1} := P_{F^*}(\sigma; \boldsymbol{\xi}^n + \sigma\mathbf{A}\bar{\mathbf{x}}^n), \tag{PD$_1$}$$

$$\mathbf{x}^{n+1} := P_G(\tau; \mathbf{x}^n - \tau\mathbf{A}^*\boldsymbol{\xi}^{n+1}), \tag{PD$_2$}$$

$$\bar{\mathbf{x}}^{n+1} := \mathbf{x}^{n+1} + \theta(\mathbf{x}^{n+1} - \mathbf{x}^n). \tag{PD$_3$}$$

Output: Approximation $\boldsymbol{\xi}^\sharp = \boldsymbol{\xi}^{\bar{n}}$ to a solution of the dual problem (15.16),
Approximation $\mathbf{x}^\sharp = \mathbf{x}^{\bar{n}}$ to a solution of the primal problem (15.15).

We will analyze this algorithm for the parameter choice $\theta = 1$. In the case where F^* or G are uniformly convex, an acceleration can be achieved by varying the parameters θ, τ, σ during the iterations; see Notes section.

A possible stopping criterion is based on the primal–dual gap (B.30), which in our case reads

$$E(\mathbf{x}, \boldsymbol{\xi}) = F(\mathbf{A}\mathbf{x}) + G(\mathbf{x}) + F^*(\boldsymbol{\xi}) + G^*(-\mathbf{A}^*\boldsymbol{\xi}) \geq 0.$$

For the primal–dual optimum $(\mathbf{x}^*, \boldsymbol{\xi}^*)$, we have $E(\mathbf{x}^*, \boldsymbol{\xi}^*) = 0$. The condition $E(\mathbf{x}^n, \boldsymbol{\xi}^n) \leq \eta$ for some prescribed tolerance $\eta > 0$ can be taken as a criterion to stop the iterations at n.

Remark 15.5. In Examples 15.7(a) and (b) below, F can take the value ∞, so that E may also be infinite during the iterations and gives only limited information about the quality of the approximation of the iterates to the optimal solution. In this case, one may modify the primal–dual gap so that the value ∞ does not occur. Empirically, a modified primal–dual gap still provides a good stopping criterion.

Note that fast matrix multiplication routines for $\mathbf{A}$ and $\mathbf{A}^*$ can easily be exploited to speed up the primal–dual algorithm.

A variant of the algorithm is obtained by interchanging the updates for $\boldsymbol{\xi}^{n+1}$ and $\mathbf{x}^{n+1}$ and carrying along an auxiliary variable $\bar{\boldsymbol{\xi}}^n$, that is,

$$\mathbf{x}^{n+1} = P_G(\tau; \mathbf{x}^n - \tau\mathbf{A}^*\bar{\boldsymbol{\xi}}^n),$$

$$\boldsymbol{\xi}^{n+1} = P_{F^*}(\sigma; \boldsymbol{\xi}^n + \sigma\mathbf{A}\mathbf{x}^{n+1}),$$

$$\bar{\boldsymbol{\xi}}^{n+1} = \boldsymbol{\xi}^{n+1} + \theta(\boldsymbol{\xi}^{n+1} - \boldsymbol{\xi}^n).$$

The algorithm can be interpreted as a fixed-point iteration.

Proposition 15.6. *A point* $(\mathbf{x}^\sharp, \boldsymbol{\xi}^\sharp)$ *is a fixed point of the iterations* (PD$_1$)–(PD$_3$) *(for any choice of θ) if and only if* $(\mathbf{x}^\sharp, \boldsymbol{\xi}^\sharp)$ *is a saddle point of* (15.17)*, that is, a primal–dual optimal point for the problems* (15.15) *and* (15.16)*.*

Proof. It follows from the characterization of the proximal mapping in Proposition B.23 that a fixed point $(\mathbf{x}^\sharp, \boldsymbol{\xi}^\sharp)$ satisfies

$$\boldsymbol{\xi}^\sharp + \sigma \mathbf{A}\mathbf{x}^\sharp \in \boldsymbol{\xi}^\sharp + \sigma \partial F^*(\boldsymbol{\xi}^\sharp),$$

$$\mathbf{x}^\sharp - \tau \mathbf{A}^* \boldsymbol{\xi}^\sharp \in \mathbf{x}^\sharp + \tau \partial G(\mathbf{x}^\sharp),$$

where ∂F^* and ∂G are the subdifferentials of F^* and G; see Definition B.20. Equivalently,

$$\mathbf{0} \in -\mathbf{A}\mathbf{x}^\sharp + \partial F^*(\boldsymbol{\xi}^\sharp) \quad \text{and} \quad \mathbf{0} \in \mathbf{A}^*\boldsymbol{\xi}^\sharp + \partial G(\mathbf{x}^\sharp).$$

By Theorem B.21, these relations are equivalent to $\boldsymbol{\xi}^\sharp$ being a maximizer of the function $\boldsymbol{\xi} \mapsto \text{Re}\langle \mathbf{A}\mathbf{x}^\sharp, \boldsymbol{\xi} \rangle + G(\mathbf{x}^\sharp) - F^*(\boldsymbol{\xi})$ and $\mathbf{x}^\sharp$ being a minimizer of the function $\mathbf{x} \mapsto \text{Re}\langle \mathbf{x}, \mathbf{A}^*\boldsymbol{\xi}^\sharp \rangle + G(\mathbf{x}) - F^*(\boldsymbol{\xi}^\sharp)$. This is equivalent to $(\mathbf{x}^\sharp, \boldsymbol{\xi}^\sharp)$ being a saddle point of (15.17).

These arguments show as well the converse that a saddle point of (15.17) is a fixed point of the primal–dual algorithm. $\square$

Before continuing with the analysis of this algorithm, let us illustrate the setup for various ℓ_1-minimization problems.

Example 15.7. (a) The ℓ_1-minimization problem

$$\min_{\mathbf{x} \in \mathbb{C}^N} \|\mathbf{x}\|_1 \quad \text{subject to } \mathbf{A}\mathbf{x} = \mathbf{y} \tag{15.19}$$

is equivalent to (15.15) where $G(\mathbf{x}) = \|\mathbf{x}\|_1$ and where

$$F(\mathbf{z}) = \chi_{\{\mathbf{y}\}}(\mathbf{z}) = \begin{cases} 0 & \text{if } \mathbf{z} = \mathbf{y}, \\ \infty & \text{if } \mathbf{z} \neq \mathbf{y}, \end{cases}$$

is the characteristic function of the singleton $\{\mathbf{y}\}$. Note that F is trivially lower semicontinuous. By Example B.19, the convex conjugates are given by

$$F^*(\boldsymbol{\xi}) = \text{Re}\langle \mathbf{y}, \boldsymbol{\xi} \rangle,$$

$$G^*(\boldsymbol{\zeta}) = \chi_{B^N_{\|\cdot\|_\infty}}(\boldsymbol{\zeta}) = \begin{cases} 0 & \text{if } \|\boldsymbol{\zeta}\|_\infty \leq 1, \\ \infty & \text{otherwise.} \end{cases} \tag{15.20}$$

Since points where the objective function takes the value $-\infty$ can be discarded when maximizing, we can make such a constraint explicit so that the dual program (15.16) becomes

$$\max_{\boldsymbol{\xi}\in\mathbb{C}^m} -\text{Re}\langle \mathbf{y}, \boldsymbol{\xi}\rangle \quad \text{subject to } \|\mathbf{A}^*\boldsymbol{\xi}\|_\infty \leq 1.$$

(Note that in Appendix B.5 the dual of the ℓ_1-minimization problem is derived in a slightly different way; see (B.32) and (B.33).) The saddle-point problem (15.17) reads

$$\min_{\mathbf{x}\in\mathbb{C}^N} \max_{\boldsymbol{\xi}\in\mathbb{C}^m} \text{Re}\langle \mathbf{A}\mathbf{x} - \mathbf{y}, \boldsymbol{\xi}\rangle + \|\mathbf{x}\|_1. \tag{15.21}$$

The proximal mapping of F is the projection onto $\{\mathbf{y}\}$, that is, the constant map

$$P_F(\sigma; \boldsymbol{\xi}) = \mathbf{y} \quad \text{for all } \boldsymbol{\xi} \in \mathbb{C}^m.$$

By Moreau's identity (B.16) (or by direct computation), the proximal mapping of F^* is therefore

$$P_{F^*}(\sigma; \boldsymbol{\xi}) = \boldsymbol{\xi} - \sigma\mathbf{y}.$$

For the proximal mapping of $G(\mathbf{x}) = \|\mathbf{x}\|_1$, we first observe by a direct computation that the proximal mapping of the modulus function satisfies, for $z \in \mathbb{C}$,

$$P_{|\cdot|}(\tau; z) = \underset{x\in\mathbb{C}}{\text{argmin}} \left\{ \frac{1}{2}|x - z|^2 + \tau|x| \right\} = \begin{cases} \text{sgn}(z)(|z| - \tau) & \text{if } |z| \geq \tau, \\ 0 & \text{otherwise,} \end{cases}$$

$$=: S_\tau(z), \tag{15.22}$$

where the sign function is given by $\text{sgn}(z) = z/|z|$ for $z \neq 0$, as usual. The function $S_\tau(z)$ is called (complex) soft thresholding operator. (In the real case it is computed in (B.18).) Since the optimization problem defining the proximal mapping of $\|\cdot\|_1$ decouples, $P_G(\tau; \mathbf{z}) =: S_\tau(\mathbf{z})$ is given componentwise by

$$P_G(\tau; \mathbf{z})_\ell = S_\tau(z_\ell), \quad \ell \in [N]. \tag{15.23}$$

The primal–dual algorithm for the ℓ_1-minimization problem (15.19) reads

$$\boldsymbol{\xi}^{n+1} = \boldsymbol{\xi}^n + \sigma(\mathbf{A}\bar{\mathbf{x}}^n - \mathbf{y}),$$

$$\mathbf{x}^{n+1} = S_\tau(\mathbf{x}^n - \tau\mathbf{A}^*\boldsymbol{\xi}^{n+1}),$$

$$\bar{\mathbf{x}}^{n+1} = \mathbf{x}^{n+1} + \theta(\mathbf{x}^{n+1} - \mathbf{x}^n).$$

(b) The quadratically constrained ℓ_1-minimization problem

$$\min_{\mathbf{x}\in\mathbb{C}^N} \|\mathbf{x}\|_1 \quad \text{subject to } \|\mathbf{A}\mathbf{x} - \mathbf{y}\|_2 \leq \eta \tag{15.24}$$

takes the form (15.15) with $G(\mathbf{x}) = \|\mathbf{x}\|_1$ and

$$F(\mathbf{z}) = \chi_{B(\mathbf{y},\eta)}(\mathbf{z}) = \begin{cases} 0 & \text{if } \|\mathbf{z} - \mathbf{y}\|_2 \leq \eta, \\ \infty & \text{otherwise.} \end{cases}$$

The function F is lower semicontinuous because the set $B(\mathbf{y}, \eta)$ is closed. Example B.19(d) shows that its convex conjugate is given by

$$F^*(\boldsymbol{\xi}) = \sup_{\mathbf{z}:\|\mathbf{z}-\mathbf{y}\|_2\leq\eta} \mathrm{Re}\langle \mathbf{z}, \boldsymbol{\xi} \rangle = \mathrm{Re}\langle \mathbf{y}, \boldsymbol{\xi} \rangle + \eta\|\boldsymbol{\xi}\|_2.$$

The convex conjugate of G is given by (15.20). The dual problem to (15.24) is therefore

$$\max_{\boldsymbol{\xi}\in\mathbb{C}^m} - \mathrm{Re}\langle \mathbf{y}, \boldsymbol{\xi} \rangle - \eta\|\boldsymbol{\xi}\|_2 \quad \text{subject to } \|\mathbf{A}^*\boldsymbol{\xi}\|_\infty \leq 1,$$

while the associated saddle-point problem is given by

$$\min_{\mathbf{x}\in\mathbb{C}^N} \max_{\boldsymbol{\xi}\in\mathbb{C}^m} \mathrm{Re}\langle \mathbf{A}\mathbf{x} - \mathbf{y}, \boldsymbol{\xi} \rangle - \eta\|\boldsymbol{\xi}\|_2 + \|\mathbf{x}\|_1. \tag{15.25}$$

The proximal mapping of F is the orthogonal projection onto the ball $B(\mathbf{y}, \eta)$, i.e.,

$$P_F(\sigma; \boldsymbol{\xi}) = \operatorname*{argmin}_{\boldsymbol{\zeta}\in\mathbb{C}^m:\|\boldsymbol{\zeta}-\mathbf{y}\|_2\leq\eta} \|\boldsymbol{\zeta} - \boldsymbol{\xi}\|_2$$

$$= \begin{cases} \boldsymbol{\xi} & \text{if } \|\boldsymbol{\xi} - \mathbf{y}\|_2 \leq \eta, \\ \mathbf{y} + \dfrac{\eta}{\|\boldsymbol{\xi} - \mathbf{y}\|_2}(\boldsymbol{\xi} - \mathbf{y}) & \text{otherwise.} \end{cases}$$

By Moreau's identity (B.16), the proximal mapping of F^* is given by

$$P_{F^*}(\sigma; \boldsymbol{\xi}) = \begin{cases} \mathbf{0} & \text{if } \|\boldsymbol{\xi} - \sigma\mathbf{y}\|_2 \leq \eta\sigma, \\ \left(1 - \dfrac{\eta\sigma}{\|\boldsymbol{\xi} - \sigma\mathbf{y}\|_2}\right)(\boldsymbol{\xi} - \sigma\mathbf{y}) & \text{otherwise.} \end{cases}$$

After these computations, our primal–dual algorithm for (15.24) reads

$$\boldsymbol{\xi}^{n+1} = P_{F^*}(\sigma; \boldsymbol{\xi}^n + \sigma\mathbf{A}\bar{\mathbf{x}}^n)$$

$$= \begin{cases} \mathbf{0} & \text{if } \|\sigma^{-1}\boldsymbol{\xi}^n + \mathbf{A}\bar{\mathbf{x}}^n - \mathbf{y}\|_2 \leq \eta, \\ \left(1 - \dfrac{\eta\sigma}{\|\boldsymbol{\xi}^n + \sigma(\mathbf{A}\bar{\mathbf{x}}^n - \mathbf{y})\|_2}\right)(\boldsymbol{\xi}^n + \sigma(\mathbf{A}\bar{\mathbf{x}}^n - \mathbf{y})) & \text{otherwise,} \end{cases}$$

$$\mathbf{x}^{n+1} = S_\tau(\mathbf{x}^n - \tau\mathbf{A}^*\boldsymbol{\xi}^{n+1}),$$

$$\bar{\mathbf{x}}^{n+1} = \mathbf{x}^{n+1} + \theta(\mathbf{x}^{n+1} - \mathbf{x}^n).$$

(c) Consider the ℓ_1-regularized least squares problem

$$\min_{\mathbf{x} \in \mathbb{C}^N} \|\mathbf{x}\|_1 + \frac{\gamma}{2} \|\mathbf{A}\mathbf{x} - \mathbf{y}\|_2^2, \qquad (15.26)$$

with some regularization parameter $\gamma > 0$. This problem is equivalent to (15.3) after the parameter change $\lambda = \gamma^{-1}$. It can be written in the form (15.15) with $G(\mathbf{x}) = \|\mathbf{x}\|_1$ and

$$F(\mathbf{z}) = \frac{\gamma}{2} \|\mathbf{z} - \mathbf{y}\|_2^2.$$

The function F is continuous in this case. It follows either from a direct computation or from Proposition B.18(d) and (e) together with Example B.19(a) that

$$F^*(\boldsymbol{\xi}) = \operatorname{Re}\langle \mathbf{y}, \boldsymbol{\xi} \rangle + \frac{1}{2\gamma} \|\boldsymbol{\xi}\|_2^2.$$

The dual to (15.26) is the optimization problem

$$\max_{\boldsymbol{\xi} \in \mathbb{C}^m} - \operatorname{Re}\langle \mathbf{y}, \boldsymbol{\xi} \rangle - \frac{1}{2\gamma} \|\boldsymbol{\xi}\|_2^2 \quad \text{subject to } \|\mathbf{A}^*\boldsymbol{\xi}\|_\infty \leq 1,$$

and the associated saddle-point problem reads

$$\min_{\mathbf{x} \in \mathbb{C}^N} \max_{\boldsymbol{\xi} \in \mathbb{C}^m} \operatorname{Re}\langle \mathbf{A}\mathbf{x} - \mathbf{y}, \boldsymbol{\xi} \rangle - \frac{1}{2\gamma} \|\boldsymbol{\xi}\|_2^2 + \|\mathbf{x}\|_1.$$

A direct calculation gives

$$P_F(\sigma; \boldsymbol{\xi}) = \frac{\sigma\gamma}{\sigma\gamma + 1}\mathbf{y} + \frac{1}{\sigma\gamma + 1}\boldsymbol{\xi}.$$

By Moreau's identity (B.16),

$$P_{F^*}(\sigma; \boldsymbol{\xi}) = \frac{\gamma}{\gamma + \sigma}(\boldsymbol{\xi} - \sigma\mathbf{y}).$$

With these relations, our primal–dual algorithm for the numerical solution of (15.26) is given by

$$\boldsymbol{\xi}^{n+1} = \frac{\gamma}{\gamma + \sigma}(\boldsymbol{\xi}^n + \sigma(\mathbf{A}\bar{\mathbf{x}}^n - \mathbf{y})),$$

$$\mathbf{x}^{n+1} = \mathcal{S}_\tau(\mathbf{x}^n - \tau\mathbf{A}^*\boldsymbol{\xi}^{n+1}),$$

$$\bar{\mathbf{x}}^{n+1} = \mathbf{x}^{n+1} + \theta(\mathbf{x}^{n+1} - \mathbf{x}^n).$$

In the above examples, $\mathbf{x}^0 = \bar{\mathbf{x}}^0 = \mathbf{A}^*\mathbf{y}$ and $\boldsymbol{\xi}^0 = \mathbf{0}$ are reasonable starting points, and if $\mathbf{A}$ is normalized, i.e., $\|\mathbf{A}\|_{2\to 2} = 1$, then suitable parameters are $\sigma = \tau = 1/2$. (Note that we can always renormalize the functional so that

$\|\mathbf{A}\|_{2\to2} = 1$: For instance, with $\tilde{\tau} = \tau\|\mathbf{A}\|_{2\to2}^2$, $\tilde{\mathbf{A}} = \|\mathbf{A}\|_{2\to2}^{-1}\mathbf{A}$, and $\tilde{\mathbf{y}} = \|\mathbf{A}\|_{2\to2}^{-1}\mathbf{y}$, we have $\tau\|\mathbf{A}\mathbf{x} - \mathbf{y}\|_2^2 = \tilde{\tau}\|\tilde{\mathbf{A}}\mathbf{x} - \tilde{\mathbf{y}}\|_2^2$.) Determining the best parameters in a specific situation is a matter of experiment.

The convergence of the primal–dual algorithm is settled by the following theorem.

Theorem 15.8. *Assume that the problem* (15.17) *has a saddle point. Choose* $\theta = 1$ *and* $\sigma, \tau > 0$ *such that* $\tau\sigma\|\mathbf{A}\|_{2\to2}^2 < 1$. *Let* $(\mathbf{x}^n, \bar{\mathbf{x}}^n, \boldsymbol{\xi}^n)$ *be the sequence generated by* (PD_1)–(PD_3). *Then the sequence* $(\mathbf{x}^n, \boldsymbol{\xi}^n)$ *converges to a saddle-point* $(\mathbf{x}^\sharp, \boldsymbol{\xi}^\sharp)$ *of* (15.17). *In particular,* $(\mathbf{x}^n)$ *converges to a minimizer of* (15.15).

We develop the proof in several steps. We will require the Lagrangian

$$L(\mathbf{x}, \boldsymbol{\xi}) := \mathrm{Re}\langle\mathbf{A}\mathbf{x}, \boldsymbol{\xi}\rangle + G(\mathbf{x}) - F^*(\boldsymbol{\xi}).$$

In order to simplify notation, we introduce, for a sequence $(\mathbf{u}^n)_{n\geq0}$ (of scalars or vectors), the divided difference

$$\Delta_\tau\mathbf{u}^n := \frac{\mathbf{u}^n - \mathbf{u}^{n-1}}{\tau}, \quad n \in \mathbb{N}.$$

This term can be interpreted as a discrete derivative with step size τ. We also use Δ_τ for related expressions such as

$$\Delta_\tau\|\mathbf{u}^{n+1}\|_2^2 = \frac{\|\mathbf{u}^{n+1}\|_2^2 - \|\mathbf{u}^n\|_2^2}{\tau}.$$

We have the following identities which closely resemble corresponding relations for the usual (continuous) derivative.

Lemma 15.9. *Let* $\mathbf{u}, \mathbf{u}^n \in \mathbb{C}^N$, $n \geq 0$. *Then*

$$2\,\mathrm{Re}\langle\Delta_\tau\mathbf{u}^n, \mathbf{u}^n - \mathbf{u}\rangle = \Delta_\tau\|\mathbf{u} - \mathbf{u}^n\|_2^2 + \tau\|\Delta_\tau\mathbf{u}^n\|_2^2. \tag{15.27}$$

Moreover, if $(\mathbf{v}^n)_{n\geq0}$ *is another sequence of vectors, then a discrete integration by parts formula holds, namely, for* $M \in \mathbb{N}$,

$$\tau\sum_{n=1}^M \left(\langle\Delta_\tau\mathbf{u}^n, \mathbf{v}^n\rangle + \langle\mathbf{u}^{n-1}, \Delta_\tau\mathbf{v}^n\rangle\right) = \langle\mathbf{u}^M, \mathbf{v}^M\rangle - \langle\mathbf{u}^0, \mathbf{v}^0\rangle. \tag{15.28}$$

Proof. Set $\tilde{\mathbf{u}}^n = \mathbf{u}^n - \mathbf{u}$. Then $\Delta_\tau\tilde{\mathbf{u}}^n = \Delta_\tau\mathbf{u}^n$ and

$$2\tau\langle\Delta_\tau\mathbf{u}^n, \mathbf{u}^n - \mathbf{u}\rangle = 2\tau\langle\Delta_\tau\tilde{\mathbf{u}}^n, \tilde{\mathbf{u}}^n\rangle = 2\langle\tilde{\mathbf{u}}^n - \tilde{\mathbf{u}}^{n-1}, \tilde{\mathbf{u}}^n\rangle$$

$$= \langle\tilde{\mathbf{u}}^n - \tilde{\mathbf{u}}^{n-1}, \tilde{\mathbf{u}}^n + \tilde{\mathbf{u}}^{n-1}\rangle + \langle\tilde{\mathbf{u}}^n - \tilde{\mathbf{u}}^{n-1}, \tilde{\mathbf{u}}^n - \tilde{\mathbf{u}}^{n-1}\rangle$$

$$= \langle\tilde{\mathbf{u}}^n - \tilde{\mathbf{u}}^{n-1}, \tilde{\mathbf{u}}^n + \tilde{\mathbf{u}}^{n-1}\rangle + \tau^2\|\Delta_\tau\mathbf{u}^n\|_2^2.$$

Noting that

$$\mathrm{Re}\langle \tilde{\mathbf{u}}^n - \tilde{\mathbf{u}}^{n-1}, \tilde{\mathbf{u}}^n + \tilde{\mathbf{u}}^{n-1} \rangle = \|\tilde{\mathbf{u}}^n\|_2^2 - \|\tilde{\mathbf{u}}^{n-1}\|_2^2 = \tau \Delta_\tau \|\tilde{\mathbf{u}}^n\|_2^2$$

completes the proof of the first statement. Next observe that

$$\Delta_\tau \langle \mathbf{u}^n, \mathbf{v}^n \rangle = \frac{\langle \mathbf{u}^n, \mathbf{v}^n \rangle - \langle \mathbf{u}^{n-1}, \mathbf{v}^{n-1} \rangle}{\tau} = \langle \Delta_\tau \mathbf{u}^n, \mathbf{v}^n \rangle + \langle \mathbf{u}^{n-1}, \Delta_\tau \mathbf{v}^n \rangle.$$

Summing this identity over all $n \in [M]$ and using the telescoping identity $\tau \sum_{n=1}^{M} \Delta_\tau \langle \mathbf{u}^n, \mathbf{v}^n \rangle = \langle \mathbf{u}^M, \mathbf{v}^M \rangle - \langle \mathbf{u}^0, \mathbf{v}^0 \rangle$ gives the second statement. □

Lemma 15.10. *Let* $(\mathbf{x}^n, \bar{\mathbf{x}}^n, \boldsymbol{\xi}^n)_{n \geq 0}$ *be the sequence generated by* (PD$_1$)–(PD$_3$), *and let* $\mathbf{x} \in \mathbb{C}^N$, $\boldsymbol{\xi} \in \mathbb{C}^m$ *be arbitrary. Then, for any* $n \geq 1$,

$$\frac{1}{2} \Delta_\sigma \|\boldsymbol{\xi} - \boldsymbol{\xi}^n\|_2^2 + \frac{1}{2} \Delta_\tau \|\mathbf{x} - \mathbf{x}^n\|_2^2 + \frac{\sigma}{2} \|\Delta_\sigma \boldsymbol{\xi}^n\|_2^2 + \frac{\tau}{2} \|\Delta_\tau \mathbf{x}^n\|_2^2$$

$$\leq L(\mathbf{x}, \boldsymbol{\xi}^n) - L(\mathbf{x}^n, \boldsymbol{\xi}) + \mathrm{Re}\langle \mathbf{A}(\mathbf{x}^n - \bar{\mathbf{x}}^{n-1}), \boldsymbol{\xi} - \boldsymbol{\xi}^n \rangle. \tag{15.29}$$

Proof. It follows from the characterization of the proximal mapping in Proposition B.23 that the iterates satisfy the relations (replacing $n + 1$ by n)

$$\boldsymbol{\xi}^{n-1} + \sigma \mathbf{A}\bar{\mathbf{x}}^{n-1} \in \boldsymbol{\xi}^n + \sigma \partial F^*(\boldsymbol{\xi}^n),$$

$$\mathbf{x}^{n-1} - \tau \mathbf{A}^* \boldsymbol{\xi}^n \in \mathbf{x}^n + \tau \partial G(\mathbf{x}^n),$$

where ∂F^* and ∂G are the subdifferentials of F^* and G. By the definition (B.11) of the subdifferential (and recalling that inner products have to be replaced by $\mathrm{Re}\langle \cdot, \cdot \rangle$ when passing from the real to the complex case), this implies, for all $\boldsymbol{\xi} \in \mathbb{C}^m$ and $\mathbf{z} \in \mathbb{C}^N$,

$$\mathrm{Re}\langle -\boldsymbol{\xi}^n + \boldsymbol{\xi}^{n-1} + \sigma \mathbf{A}\bar{\mathbf{x}}^{n-1}, \boldsymbol{\xi} - \boldsymbol{\xi}^n \rangle \leq \sigma F^*(\boldsymbol{\xi}) - \sigma F^*(\boldsymbol{\xi}^n),$$

$$\mathrm{Re}\langle -\mathbf{x}^n + \mathbf{x}^{n-1} - \tau \mathbf{A}^* \boldsymbol{\xi}^n, \mathbf{x} - \mathbf{x}^n \rangle \leq \tau G(\mathbf{x}) - \tau G(\mathbf{x}^n),$$

or, with our definition of the divided difference,

$$\mathrm{Re}\langle \Delta_\sigma \boldsymbol{\xi}^n, \boldsymbol{\xi}^n - \boldsymbol{\xi} \rangle + \mathrm{Re}\langle \mathbf{A}\bar{\mathbf{x}}^{n-1}, \boldsymbol{\xi} - \boldsymbol{\xi}^n \rangle \leq F^*(\boldsymbol{\xi}) - F^*(\boldsymbol{\xi}^n),$$

$$\mathrm{Re}\langle \Delta_\tau \mathbf{x}^n, \mathbf{x}^n - \mathbf{x} \rangle - \mathrm{Re}\langle \mathbf{A}(\mathbf{x} - \mathbf{x}^n), \boldsymbol{\xi}^n \rangle \leq G(\mathbf{x}) - G(\mathbf{x}^n).$$

Summing both inequalities and exploiting (15.27) yields

$$\frac{1}{2} \Delta_\sigma \|\boldsymbol{\xi} - \boldsymbol{\xi}^n\|_2^2 + \frac{1}{2} \Delta_\tau \|\mathbf{x} - \mathbf{x}^n\|_2^2 + \frac{\sigma}{2} \|\Delta_\sigma \boldsymbol{\xi}^n\|_2^2 + \frac{\tau}{2} \|\Delta_\tau \mathbf{x}^n\|_2^2$$

$$\leq F^*(\boldsymbol{\xi}) - F^*(\boldsymbol{\xi}^n) + G(\mathbf{x}) - G(\mathbf{x}^n)$$

$$+ \mathrm{Re}\langle \mathbf{A}(\mathbf{x} - \mathbf{x}^n), \boldsymbol{\xi}^n \rangle - \mathrm{Re}\langle \mathbf{A}\bar{\mathbf{x}}^{n-1}, \boldsymbol{\xi} - \boldsymbol{\xi}^n \rangle$$

$$= (\mathrm{Re}\langle \mathbf{A}\mathbf{x}, \boldsymbol{\xi}^n \rangle + G(\mathbf{x}) - F^*(\boldsymbol{\xi}^n)) - (\mathrm{Re}\langle \mathbf{A}\mathbf{x}^n, \boldsymbol{\xi} \rangle + G(\mathbf{x}^n) - F^*(\boldsymbol{\xi}))$$
$$+ \mathrm{Re}\langle \mathbf{A}(\mathbf{x}^n - \bar{\mathbf{x}}^{n-1}), \boldsymbol{\xi} - \boldsymbol{\xi}^n \rangle.$$

This finishes the proof. □

Remark 15.11. Inequality (15.29) suggests to ideally set $\bar{\mathbf{x}}^{n-1} = \mathbf{x}^n$. However, this leads to an implicit scheme, where the accordingly modified equations (PD$_1$) and (PD$_2$) defining the iterations become as hard to solve as the original problem.

Lemma 15.12. *Let* $(\mathbf{x}^n, \bar{\mathbf{x}}^n, \boldsymbol{\xi}^n)_{n \geq 0}$ *be the sequence generated by* (PD$_1$)–(PD$_3$) *with the parameter choice* $\theta = 1$, *and let* $\mathbf{x} \in \mathbb{C}^N$, $\boldsymbol{\xi} \in \mathbb{C}^m$ *be arbitrary. Then, for* $M \geq 1$,

$$\sum_{n=1}^{M} (L(\mathbf{x}^n, \boldsymbol{\xi}) - L(\mathbf{x}, \boldsymbol{\xi}^n)) + \frac{1}{2\tau}\|\mathbf{x} - \mathbf{x}^M\|_2^2 + \frac{1 - \sigma\tau\|\mathbf{A}\|_{2\to 2}^2}{2\sigma}\|\boldsymbol{\xi} - \boldsymbol{\xi}^M\|_2^2$$

$$+ \frac{1 - \sqrt{\sigma\tau}\|\mathbf{A}\|_{2\to 2}}{2\tau} \sum_{n=1}^{M-1} \|\mathbf{x}^n - \mathbf{x}^{n-1}\|_2^2 + \frac{1 - \sqrt{\sigma\tau}\|\mathbf{A}\|_{2\to 2}}{2\sigma} \sum_{n=1}^{M} \|\boldsymbol{\xi}^n - \boldsymbol{\xi}^{n-1}\|_2^2$$

$$\leq \frac{1}{2\tau}\|\mathbf{x} - \mathbf{x}^0\|_2^2 + \frac{1}{2\sigma}\|\boldsymbol{\xi} - \boldsymbol{\xi}^0\|_2^2. \tag{15.30}$$

Proof. Note that $\mathbf{x}^n - \bar{\mathbf{x}}^{n-1} = \mathbf{x}^n - \mathbf{x}^{n-1} - (\mathbf{x}^{n-1} - \mathbf{x}^{n-2}) = \tau^2 \Delta_\tau \Delta_\tau \mathbf{x}^n =: \tau^2 \Delta_\tau^2 \mathbf{x}^n$ for $n \geq 2$ and that the formula extends to $n = 1$ when setting $\mathbf{x}^{-1} = \mathbf{x}^0$ because by definition $\bar{\mathbf{x}}^0 = \mathbf{x}^0$. Summing inequality (15.29) from $n = 1$ to $n = M$ gives

$$\frac{1}{2\sigma}(\|\boldsymbol{\xi} - \boldsymbol{\xi}^M\|_2^2 - \|\boldsymbol{\xi} - \boldsymbol{\xi}^0\|_2^2) + \frac{1}{2\tau}(\|\mathbf{x} - \mathbf{x}^M\|_2^2 - \|\mathbf{x} - \mathbf{x}^0\|_2^2)$$

$$+ \frac{1}{2\sigma}\sum_{n=1}^{M}\|\boldsymbol{\xi}^n - \boldsymbol{\xi}^{n-1}\|_2^2 + \frac{1}{2\tau}\sum_{n=1}^{M}\|\mathbf{x}^n - \mathbf{x}^{n-1}\|_2^2$$

$$\leq \sum_{n=1}^{M}(L(\mathbf{x}, \boldsymbol{\xi}^n) - L(\mathbf{x}^n, \boldsymbol{\xi})) + \tau^2 \sum_{n=1}^{M} \mathrm{Re}\langle \mathbf{A}\Delta_\tau^2 \mathbf{x}^n, \boldsymbol{\xi} - \boldsymbol{\xi}^n \rangle. \tag{15.31}$$

Next we exploit the discrete integration by parts formula (15.28) and $\Delta_\tau \mathbf{x}^0 = \mathbf{0}$ to reach

$$\tau^2 \sum_{n=1}^{M} \mathrm{Re} \langle \mathbf{A} \Delta_\tau^2 \mathbf{x}^n, \boldsymbol{\xi} - \boldsymbol{\xi}^n \rangle$$

$$= \tau^2 \sum_{n=1}^{M} \mathrm{Re} \langle \mathbf{A} \Delta_\tau \mathbf{x}^{n-1}, \Delta_\tau \boldsymbol{\xi}^n \rangle + \tau \, \mathrm{Re} \langle \mathbf{A} \Delta_\tau \mathbf{x}^M, \boldsymbol{\xi} - \boldsymbol{\xi}^M \rangle$$

$$= \sigma\tau \sum_{n=1}^{M} \mathrm{Re} \langle \mathbf{A} \Delta_\tau \mathbf{x}^{n-1}, \Delta_\sigma \boldsymbol{\xi}^n \rangle + \tau \, \mathrm{Re} \langle \Delta_\tau \mathbf{x}^M, \mathbf{A}^*(\boldsymbol{\xi} - \boldsymbol{\xi}^M) \rangle.$$

Since $2ab \leq \alpha a^2 + \alpha^{-1} b^2$ for positive a, b, α, we have

$$\sigma\tau \, \mathrm{Re} \langle \mathbf{A} \Delta_\tau \mathbf{x}^{n-1}, \Delta_\sigma \boldsymbol{\xi}^n \rangle \leq \sigma\tau \|\mathbf{A}\|_{2\to2} \|\Delta_\tau \mathbf{x}^{n-1}\|_2 \|\Delta_\sigma \boldsymbol{\xi}^n\|_2$$

$$\leq \frac{\sigma\tau \|\mathbf{A}\|_{2\to2}}{2} \left(\alpha \|\Delta_\tau \mathbf{x}^{n-1}\|_2^2 + \alpha^{-1} \|\Delta_\sigma \boldsymbol{\xi}^n\|_2^2 \right)$$

$$\leq \frac{\sigma\alpha \|\mathbf{A}\|_{2\to2}}{2\tau} \|\mathbf{x}^{n-1} - \mathbf{x}^{n-2}\|_2^2 + \frac{\tau \|\mathbf{A}\|_{2\to2}}{2\alpha\sigma} \|\boldsymbol{\xi}^n - \boldsymbol{\xi}^{n-1}\|_2^2.$$

We choose $\alpha = \sqrt{\tau/\sigma}$ to obtain

$$\sigma\tau \, \mathrm{Re} \langle \mathbf{A} \Delta_\tau \mathbf{x}^{n-1}, \Delta_\sigma \boldsymbol{\xi}^n \rangle$$

$$\leq \frac{\sqrt{\sigma\tau} \|\mathbf{A}\|_{2\to2}}{2\tau} \|\mathbf{x}^{n-1} - \mathbf{x}^{n-2}\|_2^2 + \frac{\sqrt{\sigma\tau} \|\mathbf{A}\|_{2\to2}}{2\sigma} \|\boldsymbol{\xi}^n - \boldsymbol{\xi}^{n-1}\|_2^2. \qquad (15.32)$$

Furthermore, we have

$$\tau \, \mathrm{Re} \langle \Delta_\tau \mathbf{x}^M, \mathbf{A}^*(\boldsymbol{\xi} - \boldsymbol{\xi}^M) \rangle \leq \frac{\tau}{2} \left(\|\Delta_\tau \mathbf{x}^M\|_2^2 + \|\mathbf{A}\|_{2\to2}^2 \|\boldsymbol{\xi} - \boldsymbol{\xi}^M\|_2^2 \right)$$

$$= \frac{1}{2\tau} \|\mathbf{x}^M - \mathbf{x}^{M-1}\|_2^2 + \frac{\sigma\tau \|\mathbf{A}\|_{2\to2}^2}{2\sigma} \|\boldsymbol{\xi} - \boldsymbol{\xi}^M\|_2^2.$$

Substituting these estimates into the second sum in (15.31) and using the fact that $\mathbf{x}^{-1} = \mathbf{x}^0$ yields

$$\tau^2 \sum_{n=1}^{M} \mathrm{Re} \langle \mathbf{A} \Delta_\tau^2 \mathbf{x}^n, \boldsymbol{\xi} - \boldsymbol{\xi}^n \rangle$$

$$\leq \frac{\sqrt{\sigma\tau} \|\mathbf{A}\|_{2\to2}}{2\tau} \sum_{n=1}^{M-1} \|\mathbf{x}^n - \mathbf{x}^{n-1}\|_2^2 + \frac{\sqrt{\sigma\tau} \|\mathbf{A}\|_{2\to2}}{2\sigma} \sum_{n=1}^{M} \|\boldsymbol{\xi}^n - \boldsymbol{\xi}^{n-1}\|_2^2$$

$$+ \frac{1}{2\tau} \|\mathbf{x}^M - \mathbf{x}^{M-1}\|_2^2 + \frac{\sigma\tau \|\mathbf{A}\|_{2\to2}^2}{2\sigma} \|\boldsymbol{\xi} - \boldsymbol{\xi}^M\|_2^2.$$

Together with inequality (15.31), we arrive at the claim. $\square$

Corollary 15.13. *Let* $(\mathbf{x}^\sharp, \boldsymbol{\xi}^\sharp)$ *be a primal–dual optimum, that is, a saddle point of* (15.17). *Then the iterates of the primal–dual algorithm with* $\theta = 1$ *and* $\sigma\tau\|\mathbf{A}\|_{2\to 2} < 1$ *satisfy*

$$\frac{1}{2\sigma}\|\boldsymbol{\xi}^\sharp - \boldsymbol{\xi}^M\|_2^2 + \frac{1}{2\tau}\|\mathbf{x}^\sharp - \mathbf{x}^M\|_2^2 \leq C\left(\frac{1}{2\sigma}\|\boldsymbol{\xi}^\sharp - \boldsymbol{\xi}^0\|_2^2 + \frac{1}{2\tau}\|\mathbf{x}^\sharp - \mathbf{x}^0\|_2^2\right),$$

where $C = (1 - \sigma\tau\|\mathbf{A}\|_{2\to 2}^2)^{-1}$. *In particular, the iterates* $(\mathbf{x}^n, \boldsymbol{\xi}^n)$ *are bounded.*

Proof. For a saddle-point $(\mathbf{x}^\sharp, \boldsymbol{\xi}^\sharp)$, the summands $L(\mathbf{x}^n, \boldsymbol{\xi}^\sharp) - L(\mathbf{x}^\sharp, \boldsymbol{\xi}^n)$ are all nonnegative, so that every term on the left-hand side of (15.30) is nonnegative. We obtain in particular

$$\frac{1}{2\tau}\|\mathbf{x}^\sharp - \mathbf{x}^M\|_2^2 + \frac{1 - \sigma\tau\|\mathbf{A}\|_{2\to 2}^2}{2\sigma}\|\boldsymbol{\xi}^\sharp - \boldsymbol{\xi}^M\|_2^2 \leq \frac{1}{2\tau}\|\mathbf{x}^\sharp - \mathbf{x}^0\|_2^2 + \frac{1}{2\sigma}\|\boldsymbol{\xi}^\sharp - \boldsymbol{\xi}^0\|_2^2.$$

This yields the claim. $\qquad\square$

We are now in a position to complete the convergence proof for the primal–dual algorithm.

Proof (of Theorem 15.8). We first note that the boundedness of the sequence $(\mathbf{x}^n, \boldsymbol{\xi}^n)$ established in Corollary 15.13 implies the existence of a convergent subsequence, say $(\mathbf{x}^{n_k}, \boldsymbol{\xi}^{n_k}) \to (\mathbf{x}^\circ, \boldsymbol{\xi}^\circ)$ as $k \to \infty$. Choosing $(\mathbf{x}, \boldsymbol{\xi})$ to be a saddle-point $(\mathbf{x}^\sharp, \boldsymbol{\xi}^\sharp)$ in (15.30) makes all terms nonnegative, and we derive in particular that

$$\frac{1 - \sqrt{\sigma\tau}\|\mathbf{A}\|_{2\to 2}}{2\sigma}\sum_{n=1}^{M-1}\|\mathbf{x}^n - \mathbf{x}^{n-1}\|_2^2 \leq \frac{1}{2\tau}\|\mathbf{x} - \mathbf{x}^0\|_2^2 + \frac{1}{2\sigma}\|\boldsymbol{\xi} - \boldsymbol{\xi}^0\|_2^2.$$

Since the right-hand side is independent of M, and since $\sqrt{\sigma\tau}\|\mathbf{A}\|_{2\to 2} < 1$, we deduce that $\|\mathbf{x}^n - \mathbf{x}^{n-1}\|_2 \to 0$ as $n \to \infty$. Similarly, $\|\boldsymbol{\xi}^n - \boldsymbol{\xi}^{n-1}\|_2 \to 0$. In particular, the subsequence $(\mathbf{x}^{n_k-1}, \boldsymbol{\xi}^{n_k-1})$ converges to $(\mathbf{x}^\circ, \boldsymbol{\xi}^\circ)$, too. It follows that $(\mathbf{x}^\circ, \boldsymbol{\xi}^\circ)$ is a fixed point of the primal–dual algorithm, so that by Proposition 15.6, it is a primal–dual optimal point (or saddle point).

We choose $(\mathbf{x}, \boldsymbol{\xi}) = (\mathbf{x}^\circ, \boldsymbol{\xi}^\circ)$ in (15.29), so that $L(\mathbf{x}^n, \boldsymbol{\xi}^\circ) - L(\mathbf{x}^\circ, \boldsymbol{\xi}^n) \geq 0$. We now proceed similarly to the proof of Lemma 15.12. Summing (15.29) from $n = n_k$ to $n = M > n_k$ results in

$$\frac{1}{2\sigma}\left(\|\boldsymbol{\xi}^\circ - \boldsymbol{\xi}^M\|_2^2 - \|\boldsymbol{\xi}^\circ - \boldsymbol{\xi}^{n_k}\|_2^2\right) + \frac{1}{2\tau}\left(\|\mathbf{x}^\circ - \mathbf{x}^M\|_2^2 - \|\mathbf{x}^\circ - \mathbf{x}^{n_k}\|_2^2\right)$$

$$+ \frac{1}{2\sigma}\sum_{n=n_k}^{M}\|\boldsymbol{\xi}^n - \boldsymbol{\xi}^{n-1}\|_2^2 + \frac{1}{2\tau}\sum_{n=n_k}^{M}\|\mathbf{x}^n - \mathbf{x}^{n-1}\|_2^2$$

$$\leq \tau^2 \sum_{n=n_k}^{M} \mathrm{Re}\langle \mathbf{A}\Delta_\tau^2\mathbf{x}^n, \boldsymbol{\xi}^\circ - \boldsymbol{\xi}^n\rangle. \tag{15.33}$$

Discrete integration by parts (15.28) yields

$$\tau^2 \sum_{n=n_k}^{M} \mathrm{Re}\langle \mathbf{A}\Delta_\tau^2 \mathbf{x}^n, \boldsymbol{\xi}^\circ - \boldsymbol{\xi}^n \rangle$$

$$= \sigma\tau \sum_{n=n_k}^{M} \mathrm{Re}\langle \mathbf{A}\Delta_\tau \mathbf{x}^{n-1}, \Delta_\sigma \boldsymbol{\xi}^n \rangle + \tau \, \mathrm{Re}\langle \mathbf{A}\Delta_\tau \mathbf{x}^M, \boldsymbol{\xi}^\circ - \boldsymbol{\xi}^M \rangle$$

$$- \tau \, \mathrm{Re}\langle \mathbf{A}\Delta_\tau \mathbf{x}^{n_k-1}, \boldsymbol{\xi}^\circ - \boldsymbol{\xi}^{n_k} \rangle.$$

Inequality (15.32) therefore implies

$$\frac{1}{2\sigma}\|\boldsymbol{\xi}^\circ - \boldsymbol{\xi}^M\|_2^2 + \frac{1}{2\tau}\|\mathbf{x}^\circ - \mathbf{x}^M\|_2^2 + \frac{1 - \sqrt{\sigma\tau}\|\mathbf{A}\|_{2\to2}}{2\sigma} \sum_{n=n_k}^{M} \|\boldsymbol{\xi}^n - \boldsymbol{\xi}^{n-1}\|_2^2$$

$$+ \frac{1 - \sqrt{\sigma\tau}\|\mathbf{A}\|_{2\to2}}{2\tau} \sum_{n=n_k}^{M-1} \|\mathbf{x}^n - \mathbf{x}^{n-1}\|_2^2$$

$$+ \frac{1}{2\tau}\left(\|\mathbf{x}^M - \mathbf{x}^{M-1}\|_2^2 - \sqrt{\sigma\tau}\|\mathbf{A}\|_{2\to2}\|\mathbf{x}^{n_k-1} - \mathbf{x}^{n_k-2}\|_2^2\right)$$

$$- \mathrm{Re}\langle \mathbf{A}(\mathbf{x}^M - \mathbf{x}^{M-1}), \boldsymbol{\xi}^\circ - \boldsymbol{\xi}^M \rangle + \mathrm{Re}\langle \mathbf{A}(\mathbf{x}^{n_k-1} - \mathbf{x}^{n_k-2}), \boldsymbol{\xi}^\circ - \boldsymbol{\xi}^{n_k} \rangle$$

$$\leq \frac{1}{2\sigma}\|\boldsymbol{\xi}^\circ - \boldsymbol{\xi}^{n_k}\|_2^2 + \frac{1}{2\tau}\|\mathbf{x}^\circ - \mathbf{x}^{n_k}\|_2^2.$$

In view of $\lim_{n\to\infty}\|\mathbf{x}^n - \mathbf{x}^{n-1}\|_2 = \lim_{n\to\infty}\|\boldsymbol{\xi}^n - \boldsymbol{\xi}^{n-1}\|_2 = 0$ and of $\lim_{k\to\infty}\|\mathbf{x}^\circ - \mathbf{x}^{n_k}\|_2 = \lim_{k\to\infty}\|\boldsymbol{\xi}^\circ - \boldsymbol{\xi}^{n_k}\|_2 = 0$, it finally follows that $\lim_{M\to\infty}\|\mathbf{x}^\circ - \mathbf{x}^M\|_2 = \lim_{M\to\infty}\|\boldsymbol{\xi}^\circ - \boldsymbol{\xi}^M\|_2 = 0$. We have established the claim. □

15.3 Iteratively Reweighted Least Squares

We now turn to an iterative algorithm that serves as a proxy for ℓ_1-minimization. It does not always compute an ℓ_1-minimizer, but provides similar error estimates under the null space property.

The starting point is the trivial observation that $|t| = |t|^2/|t|$ for $t \neq 0$. Therefore, an ℓ_1-minimization can be recast as a weighted ℓ_2-minimization in the following sense: For $\mathbf{A} \in \mathbb{C}^{m\times N}$ with $m \leq N$, if $\mathbf{x}^\sharp$ is a minimizer of

$$\min_{\mathbf{x}\in\mathbb{C}^N} \|\mathbf{x}\|_1 \quad \text{subject to} \quad \mathbf{A}\mathbf{x} = \mathbf{y} \tag{15.34}$$

and if $x_j^\sharp \neq 0$ for all $j \in [N]$, then $\mathbf{x}^\sharp$ is also a minimizer of the weighted ℓ_2-problem

$$\min_{\mathbf{x} \in \mathbb{C}^N} \sum_{j=1}^{N} |x_j|^2 |x_j^\sharp|^{-1} \quad \text{subject to } \mathbf{A}\mathbf{x} = \mathbf{y}.$$

This reformulation is advantageous because minimizing the smooth quadratic function $|t|^2$ is an easier task than minimizing the nonsmooth function $|t|$. However, as obvious drawbacks, we neither know $\mathbf{x}^\sharp$ a priori (this is the vector we would like to compute!) nor can we expect that $x_j^\sharp \neq 0$ for all $j \in [N]$, since one targets sparse solutions. In fact, by Theorem 3.1, the ℓ_1-minimizer is always m-sparse in the real case provided it is unique.

Nevertheless, the above observation motivates us to iteratively solve weighted ℓ_1-minimization problems, where the weight in the current iterate is computed from the solution of the weighted least squares problem of the previous iterate.

Key to the formulation and analysis of the algorithm is the functional

$$\mathcal{J}(\mathbf{x}, \mathbf{w}, \varepsilon) = \frac{1}{2}\left[\sum_{j=1}^{N} |x_j|^2 w_j + \sum_{j=1}^{N}(\varepsilon^2 w_j + w_j^{-1})\right], \tag{15.35}$$

where $\mathbf{x} \in \mathbb{C}^N$, $\varepsilon \geq 0$, and $\mathbf{w} \in \mathbb{R}^N$ is a positive weight vector, i.e., $w_j > 0$ for all $j \in [N]$. The formulation of the algorithm below uses the nonincreasing rearrangement $(\mathbf{x}^n)^* \in \mathbb{R}^N$ of the iterate $\mathbf{x}^n \in \mathbb{C}^N$; see Definition 2.4.

Iteratively reweighted least squares (IRLS)

Input: $\mathbf{A} \in \mathbb{C}^{m \times N}, \mathbf{y} \in \mathbb{C}^m$.
Parameter: $\gamma > 0$, $s \in [N]$.
Initialization: $\mathbf{w}^0 = [1, 1, \ldots, 1]^\top \in \mathbb{R}^N$, $\varepsilon_0 = 1$.
Iteration: repeat until $\varepsilon_n = 0$ or a stopping criterion is met at $n = \bar{n}$:

$$\mathbf{x}^{n+1} := \operatorname*{argmin}_{\mathbf{z} \in \mathbb{C}^N} \mathcal{J}(\mathbf{z}, \mathbf{w}^n, \varepsilon_n) \quad \text{subject to } \mathbf{A}\mathbf{z} = \mathbf{y}, \tag{IRLS$_1$}$$

$$\varepsilon_{n+1} := \min\{\varepsilon_n, \gamma\,(\mathbf{x}^{n+1})_{s+1}^*\}, \tag{IRLS$_2$}$$

$$\mathbf{w}^{n+1} := \operatorname*{argmin}_{\mathbf{w} > 0} \mathcal{J}(\mathbf{x}^{n+1}, \mathbf{w}, \varepsilon_{n+1}). \tag{IRLS$_3$}$$

Output: A solution $\mathbf{x}^\sharp = \mathbf{x}^{\bar{n}}$ of $\mathbf{A}\mathbf{x} = \mathbf{y}$, approximating the sparsest one.

Since $\mathbf{w}^n$ and ε_n are fixed in the minimization problem (IRLS$_1$), the second sum in the definition (15.35) of $\mathcal{J}$ is constant, so that $\mathbf{x}^{n+1}$ is the minimizer of the weighted least squares problem

$$\min_{\mathbf{z} \in \mathbb{C}^N} \|\mathbf{z}\|_{2, \mathbf{w}^n} \quad \text{subject to } \mathbf{A}\mathbf{z} = \mathbf{y},$$

where $\|\mathbf{z}\|_{2,\mathbf{w}^n} = \left(\sum_{j=1}^N |z_j|^2 w_j^n\right)^{1/2}$. By (A.35), the minimizer $\mathbf{x}^{n+1}$ is given explicitly by the formula $\mathbf{x}^{n+1} = \mathbf{D}_{\mathbf{w}^n}^{-1/2}(\mathbf{AD}_{\mathbf{w}^n}^{-1/2})^\dagger\mathbf{y}$, where $(\mathbf{AD}_{\mathbf{w}^n}^{-1/2})^\dagger$ denotes the Moore–Penrose pseudo-inverse of $\mathbf{AD}_{\mathbf{w}^n}^{-1/2}$ (see Definition A.19) and where $\mathbf{D}_{\mathbf{w}^n} = \mathrm{diag}\,[w_1^n, \ldots, w_N^n]$ is the diagonal matrix determined by the weight $\mathbf{w}^n$. If $\mathbf{A}$ has full rank (which is usually the case in the setting of compressive sensing), then (A.36) yields

$$\mathbf{x}^{n+1} = \mathbf{D}_{\mathbf{w}^n}^{-1}\mathbf{A}^*(\mathbf{AD}_{\mathbf{w}^n}^{-1}\mathbf{A}^*)^{-1}\mathbf{y},$$

where $\mathbf{D}_{\mathbf{w}^n}^{-1} = \mathrm{diag}\,[1/w_1^n, \ldots, 1/w_N^n]$. In particular, we can write

$$\mathbf{x}^{n+1} = \mathbf{D}_{\mathbf{w}^n}^{-1}\mathbf{A}^*\mathbf{v} \quad \text{where} \quad \mathbf{AD}_{\mathbf{w}^n}^{-1}\mathbf{A}^*\mathbf{v} = \mathbf{y}, \tag{15.36}$$

so that computing $\mathbf{x}^{n+1}$ involves solving the above linear system for the vector $\mathbf{v}$. We refer to Appendix A.3 for more information on least squares and weighted least squares problems.

Remark 15.14. If $\mathbf{A}$ provides a fast matrix multiplication algorithm as in situations described in Chap. 12, then it is not advisable to solve the linear system of equations in (15.36) by a direct method such as Gaussian elimination because it does not exploit fast forward transforms. Instead, one preferably works with iterative methods such as conjugate gradients, which use only forward applications of $\mathbf{A}$ and $\mathbf{A}^*$ in order to approximately solve for $\mathbf{x}^{n+1}$. However, determining the accuracy required in each step to ensure overall convergence is a subtle problem; see also the Notes section.

The minimization in (IRLS$_3$) can be performed explicitly, namely,

$$w_j^{n+1} = \frac{1}{\sqrt{|x_j^{n+1}|^2 + \varepsilon_{n+1}^2}}, \quad j \in [N]. \tag{15.37}$$

This formula also illustrates the role of ε_n. While for the naive approach above, the suggested weight $w_j^{n+1} = |x_j^{n+1}|^{-1}$ may grow unboundedly when x_j^{n+1} approaches zero, the introduction of ε_{n+1} regularizes $\mathbf{w}^{n+1}$—in particular, we have $\|\mathbf{w}^{n+1}\|_\infty \leq \varepsilon_{n+1}^{-1}$. Nevertheless, during the iterations we aim at approaching the ℓ_1-minimizer, which requires ε_n to decrease with n. The choice (IRLS$_2$) indeed ensures that ε_n does not grow. When $\mathbf{x}^n$ tends to an s-sparse vector, then ε_n tends to zero. In particular, the parameter s of the algorithm controls the desired sparsity.

We point out that other update rules for ε_{n+1} or for the weight $\mathbf{w}^{n+1}$ are possible; see also the Notes section.

The formulation of the main convergence result for the algorithm requires to introduce, for $\varepsilon > 0$, the auxiliary functional

$$F_\varepsilon(\mathbf{x}) := \sum_{j=1}^{N} \sqrt{|x_j|^2 + \varepsilon^2} \qquad (15.38)$$

and the optimization problem

$$\min_{\mathbf{z} \in \mathbb{C}^N} F_\varepsilon(\mathbf{z}) \quad \text{subject to } \mathbf{Az} = \mathbf{y}. \qquad (15.39)$$

We denote by $\mathbf{x}^{(\varepsilon)}$ its minimizer, which is unique by strict convexity of F_ε.

The recovery theorem for iteratively reweighted least squares below is based on the stable null space property (Definition 4.11) and closely resembles the corresponding statements for ℓ_1-minimization. Recall that it was shown in Sect. 9.4 that Gaussian random matrices satisfy the null space property with high probability under appropriate conditions. Moreover, the stable null space property is a consequence of the restricted isometry property (Theorem 6.13), so that the various random matrices considered in this book also satisfy the stable null space property under appropriate conditions.

Theorem 15.15. *Assume that* $\mathbf{A} \in \mathbb{C}^{m \times N}$ *satisfies the stable null space property of order s with constant $\rho < 1$. For $\mathbf{x} \in \mathbb{C}^N$ and $\mathbf{y} = \mathbf{Ax} \in \mathbb{C}^m$, the sequence $(\mathbf{x}^n)$ generated by the IRLS algorithm with parameters γ and s has a limit $\mathbf{x}^\sharp \in \mathbb{C}^N$. Moreover:*

(a) *If* $\lim_{n \to \infty} \varepsilon_n = 0$, *then* $\mathbf{x}^\sharp$ *is an s-sparse ℓ_1-minimizer of (15.34) and*

$$\|\mathbf{x} - \mathbf{x}^\sharp\|_1 \le \frac{2(1+\rho)}{1-\rho} \sigma_s(\mathbf{x})_1, \qquad (15.40)$$

so that exact recovery occurs if $\mathbf{x}$ is s-sparse.

(b) *If* $\varepsilon := \lim_{n \to \infty} \varepsilon_n > 0$, *then* $\mathbf{x}^\sharp$ *is a minimizer of (15.39) and*

$$\|\mathbf{x} - \mathbf{x}^\sharp\|_1 \le \frac{3+\rho}{(1-\rho) - (1+\rho)N\gamma/(s+1-\tilde{s})} \sigma_{\tilde{s}}(\mathbf{x})_1 \qquad (15.41)$$

whenever $\tilde{s} < s + 1 - (1 + \rho)N\gamma/(1-\rho)$. *For instance, if $\rho < 1/3$ and $\gamma = 1/(2N)$, then*

$$\|\mathbf{x} - \mathbf{x}^\sharp\|_1 \le \frac{2(3+\rho)}{1-3\rho} \sigma_s(\mathbf{x})_1.$$

(c) *With $\rho < 1/3$ and $\gamma = 1/(2N)$, if $\mathbf{x}$ is s-sparse, then necessarily $\mathbf{x}^\sharp = \mathbf{x}$ and* $\lim_{n \to \infty} \varepsilon_n = 0$.

Remark 15.16. Other versions of part (c) with different values of γ and ρ are valid when making the stronger assumption that $\mathbf{x}$ is $\tilde{s}$-sparse with $\tilde{s} < s + 1 - (1 + \rho)N\gamma/(1-\rho)$.

We develop the proof of this theorem in several steps. We start with some properties of the iterates.

Lemma 15.17. *Let* $\mathbf{x}^n$ *and* $\mathbf{w}^n$ *be the iterates of the IRLS algorithm. Then, for* $n \geq 1$,

$$J(\mathbf{x}^n, \mathbf{w}^n, \varepsilon_n) = \sum_{j=1}^{N} \sqrt{|x_j^n|^2 + \varepsilon_n^2} = F_{\varepsilon_n}(\mathbf{x}^n) \tag{15.42}$$

and

$$J(\mathbf{x}^n, \mathbf{w}^n, \varepsilon_n) \leq J(\mathbf{x}^n, \mathbf{w}^{n-1}, \varepsilon_n) \leq J(\mathbf{x}^n, \mathbf{w}^{n-1}, \varepsilon_{n-1}) \tag{15.43}$$

$$\leq J(\mathbf{x}^{n-1}, \mathbf{w}^{n-1}, \varepsilon_{n-1}). \tag{15.44}$$

Moreover, the sequence $(\mathbf{x}^n)$ *is bounded, namely,*

$$\|\mathbf{x}^n\|_1 \leq J(\mathbf{x}^1, \mathbf{w}^0, \varepsilon_0) =: B, \quad n \geq 1, \tag{15.45}$$

and the weights $\mathbf{w}^n$ *are bounded from below, namely,*

$$w_j^n \geq B^{-1}, \quad j \in [N], \ n \geq 1. \tag{15.46}$$

Proof. The relation (15.42) is derived from (15.35) and (15.37) by an easy calculation.

The first inequality in (15.43) follows from the minimization property defining $\mathbf{w}^n$, the second from $\varepsilon_n \leq \varepsilon_{n-1}$, and the inequality (15.44) is a consequence of the minimization property defining $\mathbf{x}^n$.

It now follows from (15.42) that

$$\|\mathbf{x}^n\|_1 \leq \sum_{j=1}^{N} \sqrt{|x_j^n|^2 + \varepsilon_n^2} = J(\mathbf{x}^n, \mathbf{w}^n, \varepsilon_n) \leq J(\mathbf{x}^1, \mathbf{w}^0, \varepsilon_0) = B,$$

where the last inequality uses (15.43). This establishes (15.45). Finally,

$$(w_j^n)^{-1} = \sqrt{|x_j^n|^2 + \varepsilon_n^2} \leq J(\mathbf{x}^n, \mathbf{w}^n, \varepsilon_n) \leq B, \quad j \in [N],$$

yields (15.46). □

Note that (15.44) says that each iteration decreases the value of the functional J. As the next step, we establish that the difference of successive iterates converges to zero.

Lemma 15.18. *The iterates of the IRLS algorithm satisfy*

$$\sum_{n=1}^{\infty} \|\mathbf{x}^{n+1} - \mathbf{x}^n\|_2^2 \leq 2B^2,$$

where B *is the constant in* (15.45). *Consequently,* $\lim_{n \to \infty}(\mathbf{x}^{n+1} - \mathbf{x}^n) = \mathbf{0}$.

Proof. The monotonicity property in (15.43) implies

$$2\left(\mathcal{J}(\mathbf{x}^n, \mathbf{w}^n, \varepsilon_n) - \mathcal{J}(\mathbf{x}^{n+1}, \mathbf{w}^{n+1}, \varepsilon_{n+1})\right)$$

$$\geq 2\left(\mathcal{J}(\mathbf{x}^n, \mathbf{w}^n, \varepsilon_n) - \mathcal{J}(\mathbf{x}^{n+1}, \mathbf{w}^n, \varepsilon_n)\right)$$

$$= \sum_{j=1}^{N}(|x_j^n|^2 - |x_j^{n+1}|^2)w_j^n = \operatorname{Re}\langle \mathbf{x}^n + \mathbf{x}^{n+1}, \mathbf{x}^n - \mathbf{x}^{n+1}\rangle_{\mathbf{w}^n},$$

where we have used the weighted inner product $\langle \mathbf{x}, \mathbf{z}\rangle_{\mathbf{w}} = \sum_{j=1}^{N} x_j \bar{z}_j w_j$. By their definitions in (IRLS$_1$), both $\mathbf{x}^n$ and $\mathbf{x}^{n+1}$ satisfy $\mathbf{A}\mathbf{x}^n = \mathbf{y} = \mathbf{A}\mathbf{x}^{n+1}$, so that $\mathbf{x}^n - \mathbf{x}^{n+1} \in \ker \mathbf{A}$. The characterization (A.37) of the minimizer of a weighted least squares problem implies that $\operatorname{Re}\langle \mathbf{x}^{n+1}, \mathbf{x}^n - \mathbf{x}^{n+1}\rangle_{\mathbf{w}^n} = 0$. Therefore, from the above inequality,

$$2\left(\mathcal{J}(\mathbf{x}^n, \mathbf{w}^n, \varepsilon_n) - \mathcal{J}(\mathbf{x}^{n+1}, \mathbf{w}^{n+1}, \varepsilon_{n+1})\right)$$

$$\geq \operatorname{Re}\langle \mathbf{x}^n - \mathbf{x}^{n+1}, \mathbf{x}^n - \mathbf{x}^{n+1}\rangle_{\mathbf{w}^n} = \|\mathbf{x}^n - \mathbf{x}^{n+1}\|_{2,\mathbf{w}^n}^2$$

$$= \sum_{j=1}^{N} w_j^n |x_j^n - x_j^{n+1}|^2 \geq B^{-1}\|\mathbf{x}^n - \mathbf{x}^{n+1}\|_2^2,$$

where we have used (15.46) in the last step. Summing these inequalities over n shows that

$$\sum_{n=1}^{\infty}\|\mathbf{x}^n - \mathbf{x}^{n+1}\|_2^2 \leq 2B\sum_{n=1}^{\infty}\left(\mathcal{J}(\mathbf{x}^n, \mathbf{w}^n, \varepsilon_n) - \mathcal{J}(\mathbf{x}^{n+1}, \mathbf{w}^{n+1}, \varepsilon_{n+1})\right)$$

$$\leq 2B\mathcal{J}(\mathbf{x}^1, \mathbf{w}^1, \varepsilon_1) \leq 2B^2,$$

by Lemma 15.17. □

We now characterize the minimizer $\mathbf{x}^{(\varepsilon)}$ of F_ε; see (15.38) and (15.39). We recall that $\mathbf{x}^{(\varepsilon)}$ is unique by strict convexity of F_ε.

Lemma 15.19. *Given $\varepsilon > 0$, the minimizer $\mathbf{x}^{(\varepsilon)}$ of F_ε is characterized by*

$$\operatorname{Re}\langle \mathbf{x}^{(\varepsilon)}, \mathbf{v}\rangle_{\mathbf{w}_{\mathbf{z},\varepsilon}} = 0 \quad \text{for all } \mathbf{v} \in \ker \mathbf{A},$$

where $(\mathbf{w}_{\mathbf{z},\varepsilon})_j = (|z_j|^2 + \varepsilon^2)^{-1/2}$.

Proof. First assume that $\mathbf{z} = \mathbf{x}^{(\varepsilon)}$ is the minimizer of (15.39). For an arbitrary $\mathbf{v} \in \ker \mathbf{A}$, consider the differentiable function

$$G(t) = F_\varepsilon(\mathbf{z} + t\mathbf{v}) - F_\varepsilon(\mathbf{z}), \quad t \in \mathbb{R}.$$

By the minimizing property and the fact that $A(z + tv) = y$ for all $t \in \mathbb{R}$, we have $G(t) \geq G(0) = 0$ for all $t \in \mathbb{R}$, so that $G'(0) = 0$. By a direct calculation,

$$G'(0) = \sum_{j=1}^{N} \frac{\mathrm{Re}(z_j \overline{v_j})}{\sqrt{|z_j|^2 + \varepsilon^2}} = \mathrm{Re}\langle z, v \rangle_{\mathbf{w}_{z,\varepsilon}},$$

and consequently $\mathrm{Re}\langle z, v \rangle_{\mathbf{w}_{z,\varepsilon}} = 0$ for all $v \in \ker A$.

Conversely, assume that z satisfies $Az = y$ and $\mathrm{Re}\langle z, v \rangle_{\mathbf{w}_{z,\varepsilon}} = 0$ for all $v \in \ker A$. By convexity of the function $f(u) := \sqrt{|u|^2 + \varepsilon^2}, u \in \mathbb{C}$, and Proposition B.11(a), we have for any $u, u_0 \in \mathbb{C}$,

$$\sqrt{|u|^2 + \varepsilon^2} \geq \sqrt{|u_0|^2 + \varepsilon^2} + \frac{\mathrm{Re}(u_0(u - u_0))}{\sqrt{|u_0|^2 + \varepsilon^2}}.$$

Therefore, for any $v \in \ker A$, we have

$$F_\varepsilon(z + v) \geq F_\varepsilon(z) + \sum_{j=1}^{N} \frac{\mathrm{Re}(z_j \overline{v_j})}{\sqrt{|z_j|^2 + \varepsilon^2}} = F_\varepsilon(z) + \mathrm{Re}\langle z, v \rangle_{\mathbf{w}_{z,\varepsilon}} = F_\varepsilon(z).$$

Since $v \in \ker A$ is arbitrary, it follows that z is a minimizer of (15.39). $\qquad \square$

Now we are in a position to prove Theorem 15.15 on the convergence of the iteratively reweighted least squares algorithm.

Proof (of Theorem 15.15). We first note that the sequence (ε_n) always converges since it is nonincreasing and bounded from below. We denote by ε its limit.

(a) Case $\varepsilon = 0$: First assume that $\varepsilon_{n_0} = 0$ for some $n_0 \geq 1$. Then the algorithm stops and we can set $\mathbf{x}^n = \mathbf{x}^{n_0}$ for $n \geq n_0$, so that $\lim_{n \to \infty} \mathbf{x}^n = \mathbf{x}^{n_0} = \mathbf{x}^\sharp$. By definition of ε_{n_0}, the nonincreasing rearrangement of the current iterate satisfies $(\mathbf{x}^{n_0})^*_{s+1} = 0$, so that $\mathbf{x}^\sharp = \mathbf{x}^{n_0}$ is s-sparse. The null space property of order s guarantees that $\mathbf{x}^\sharp$ is the unique ℓ_1-minimizer of (15.34). In addition, if $\mathbf{x}$ is s-sparse, then $\mathbf{x}$ is also the unique ℓ_1-minimizer, so that $\mathbf{x} = \mathbf{x}^\sharp$. For a general $\mathbf{x} \in \mathbb{C}^N$, not necessarily s-sparse, it follows from Theorem 4.12 that

$$\|\mathbf{x} - \mathbf{x}^\sharp\|_1 \leq \frac{2(1 + \rho)}{1 - \rho} \sigma_s(\mathbf{x})_1.$$

Now assume that $\varepsilon_n > 0$ for all $n \geq 1$. Since $\lim_{n \to \infty} \varepsilon_n = 0$, there exists an increasing sequence of indices (n_j) such that $\varepsilon_{n_j} < \varepsilon_{n_j - 1}$ for all $j \geq 1$. By definition (IRLS$_2$) of ε_n, this implies that the nonincreasing rearrangement of $\mathbf{x}^{n_j}$ satisfies

$$(\mathbf{x}^{n_j})^*_{s+1} < \gamma^{-1} \varepsilon_{n_j - 1}, \quad j \geq 1.$$

Since (15.45) shows that the sequence $(\mathbf{x}^n)$ is bounded, there exists a subsequence (n_{j_ℓ}) of (n_j) such that $(\mathbf{x}^{n_{j_\ell}})$ converges to some $\mathbf{x}^\sharp$ satisfying $\mathbf{A}\mathbf{x}^\sharp = \mathbf{y}$. It follows from the Lipschitz property (2.1) of the nonincreasing rearrangement that also $(\mathbf{x}^{n_{j_\ell}})^*$ converges to $(\mathbf{x}^\sharp)^*$, so that

$$(\mathbf{x}^\sharp)^*_{s+1} = \lim_{\ell \to \infty} (\mathbf{x}^{n_{j_\ell}})^*_{s+1} \le \lim_{\ell \to \infty} \gamma^{-1} \varepsilon_{n_{j_\ell}} = 0.$$

This shows that $\mathbf{x}^\sharp$ is s-sparse. As above the null space property of order s guarantees that $\mathbf{x}^\sharp$ is the unique ℓ_1-minimizer. We still need to prove that the whole sequence $(\mathbf{x}^n)$ converges to $\mathbf{x}^\sharp$. Since $\mathbf{x}^{n_{j_\ell}} \to \mathbf{x}^\sharp$ and $\varepsilon_{n_{j_\ell}} \to 0$ as $\ell \to \infty$, the identity (15.42) implies that

$$\lim_{\ell \to \infty} \mathcal{J}(\mathbf{x}^{n_{j_\ell}}, \mathbf{w}^{n_{j_\ell}}, \varepsilon_{n_{j_\ell}}) = \|\mathbf{x}^\sharp\|_1.$$

It follows from the monotonicity properties in (15.43) and (15.44) that $\lim_{n \to \infty} \mathcal{J}(\mathbf{x}^n, \mathbf{w}^n, \varepsilon_n) = \|\mathbf{x}^\sharp\|_1$. From (15.42), we deduce that

$$\mathcal{J}(\mathbf{x}^n, \mathbf{w}^n, \varepsilon_n) - N\varepsilon_n \le \|\mathbf{x}^n\|_1 \le \mathcal{J}(\mathbf{x}^n, \mathbf{w}^n, \varepsilon_n),$$

so that $\lim_{n \to \infty} \|\mathbf{x}^n\|_1 = \|\mathbf{x}^\sharp\|_1$. By the stable null space property and Theorem 4.14, we finally obtain

$$\|\mathbf{x}^n - \mathbf{x}^\sharp\|_1 \le \frac{1+\rho}{1-\rho} \left(\|\mathbf{x}^n\|_1 - \|\mathbf{x}^\sharp\|_1 \right).$$

Since the right-hand side converges to zero, this shows that $\mathbf{x}^n \to \mathbf{x}^\sharp$. The error estimate (15.40) follows from Theorem 4.12 as above.

(b) Case $\varepsilon > 0$: We first show that $\mathbf{x}^n \to \mathbf{x}^{(\varepsilon)}$, where $\mathbf{x}^{(\varepsilon)}$ is the minimizer of (15.39). By Lemma 15.17 the sequence $(\mathbf{x}^n)$ is bounded, so that it has cluster points. Let $(\mathbf{x}^{n_\ell})$ be a convergent subsequence with limit $\mathbf{x}^\sharp$. We claim that $\mathbf{x}^\sharp = \mathbf{x}^{(\varepsilon)}$, which by uniqueness of $\mathbf{x}^{(\varepsilon)}$ implies that every convergent subsequence converges to $\mathbf{x}^\sharp$, therefore the whole sequence $(\mathbf{x}^n)$ converges to $\mathbf{x}^\sharp$ as $n \to \infty$.

From $w_j^n = (|x_j^n|^2 + \varepsilon^2)^{-1/2} \le \varepsilon^{-1}$, we obtain

$$\lim_{\ell \to \infty} w_j^{n_\ell} = (|x_j^\sharp|^2 + \varepsilon^2)^{-1/2} = (w_{\mathbf{x}^\sharp, \varepsilon})_j =: w_j^\sharp, \quad j \in [N],$$

where we have used the same notation as in Lemma 15.19. Lemma 15.18 ensures that also $\mathbf{x}^{n_\ell+1}$ converges to $\mathbf{x}^\sharp$. By the characterization (A.37) of $\mathbf{x}^{n_\ell+1}$ as the minimizer of (IRLS$_1$), we have, for every $\mathbf{v} \in \ker \mathbf{A}$,

$$\mathrm{Re}\langle \mathbf{x}^\sharp, \mathbf{v} \rangle_{\mathbf{w}^\sharp} = \lim_{\ell \to \infty} \mathrm{Re}\langle \mathbf{x}^{n_\ell+1}, \mathbf{v} \rangle_{\mathbf{w}^{n_\ell}} = 0.$$

The characterization in Lemma 15.19 implies that $\mathbf{x}^\sharp = \mathbf{x}^{(\varepsilon)}$.

Now we show the error estimate (15.41). For our $\mathbf{x} \in \mathbb{C}^N$ with $\mathbf{Ax} = \mathbf{y}$, the minimizing property of $\mathbf{x}^{(\varepsilon)}$ yields

$$\|\mathbf{x}^{(\varepsilon)}\|_1 \leq F_\varepsilon(\mathbf{x}^{(\varepsilon)}) \leq F_\varepsilon(\mathbf{x}) = \sum_{j=1}^N \sqrt{|x_j|^2 + \varepsilon^2} \leq N\varepsilon + \sum_{j=1}^N |x_j| = N\varepsilon + \|\mathbf{x}\|_1.$$

It follows from the stable null space property of order $\tilde{s} \leq s$ and Theorem 4.14 that

$$\|\mathbf{x}^{(\varepsilon)} - \mathbf{x}\|_1 \leq \frac{1+\rho}{1-\rho}\left(\|\mathbf{x}^{(\varepsilon)}\|_1 - \|\mathbf{x}\|_1 + 2\sigma_{\tilde{s}}(\mathbf{x})_1\right) \leq \frac{1+\rho}{1-\rho}\left(N\varepsilon + 2\sigma_{\tilde{s}}(\mathbf{x})_1\right).$$

$$(15.47)$$

We have, by (IRLS$_2$),

$$\varepsilon \leq \gamma(\mathbf{x}^{(\varepsilon)})_{s+1}^*$$

and, according to (2.3),

$$(s + 1 - \tilde{s})(\mathbf{x}^\varepsilon)_{s+1}^* \leq \|\mathbf{x}^{(\varepsilon)} - \mathbf{x}\|_1 + \sigma_{\tilde{s}}(\mathbf{x})_1.$$

Hence, we deduce that

$$\varepsilon \leq \frac{\gamma}{s+1-\tilde{s}}\left(\|\mathbf{x}^{(\varepsilon)} - \mathbf{x}\|_1 + \sigma_{\tilde{s}}(\mathbf{x})_1\right).$$

We substitute this inequality into (15.47) to obtain

$$\|\mathbf{x}^{(\varepsilon)} - \mathbf{x}\|_1 \leq \frac{1+\rho}{1-\rho}\left(\frac{N\gamma}{s+1-\tilde{s}}\|\mathbf{x}^{(\varepsilon)} - \mathbf{x}\|_1 + \left(\frac{N\gamma}{s+1-\tilde{s}} + 2\right)\sigma_{\tilde{s}}(\mathbf{x})_1\right).$$

Rearranging and using the assumption that $(1 + \rho)N\gamma/(s + 1 - \tilde{s}) < 1 - \rho$ yields

$$\left((1-\rho) - \frac{(1+\rho)N\gamma}{s+1-\tilde{s}}\right)\|\mathbf{x}^{(\varepsilon)} - \mathbf{x}\|_1 \leq (1+\rho)\left(\frac{N\gamma}{s+1-\tilde{s}} + 2\right)\sigma_{\tilde{s}}(\mathbf{x})_1$$

$$\leq ((1-\rho) + 2(1+\rho))\,\sigma_{\tilde{s}}(\mathbf{x})_1 = (3+\rho)\sigma_{\tilde{s}}(\mathbf{x})_1.$$

This is the desired result. The conclusion for $\rho < 1/3$ and $\gamma = 1/(2N)$ is then immediate.

(c) For $\rho < 1/3$ and $\gamma = 1/(2N)$, we observe that $\mathbf{x}^\sharp = \mathbf{x}$ by (15.41), so that also $\mathbf{x}^\sharp$ is s-sparse. But this implies that $(\mathbf{x}^\sharp)_{s+1}^* = 0$ and therefore $\varepsilon = \lim_{n\to\infty} \varepsilon_n = 0$ by definition (IRLS$_2$) of ε_n. $\square$

We conclude this section with an estimate on the rate of convergence of the iteratively reweighted least squares algorithm. The estimated rate starts only when the iterates are close enough to the limit. Nothing is said about the initial phase, although experience reveals that the initial phase does not take overly long. The estimate for the exactly sparse case below shows linear convergence in the ℓ_1-norm.

Theorem 15.20. *Assume that* $\mathbf{A} \in \mathbb{C}^{m \times N}$ *satisfies the stable null space property of order* s *with constant* $\rho < 1/3$. *Let* $\kappa \in (0,1)$ *be such that*

$$\mu := \frac{3\rho(1+\rho)}{2(1-\kappa)} < 1.$$

For an s-*sparse* $\mathbf{x} \in \mathbb{C}^N$ *and for* $\mathbf{y} = \mathbf{A}\mathbf{x} \in \mathbb{C}^m$, *the sequence* $(\mathbf{x}^n)$ *produced by the IRLS algorithm with parameters* $\gamma = 1/(2N)$ *and* s *converges to* $\mathbf{x}$.
If $n_0 \geq 1$ *is such that*

$$\|\mathbf{x} - \mathbf{x}^{n_0}\|_1 \leq R := \kappa \min_{j \in \mathrm{supp}(\mathbf{x})} |x_j|,$$

then, for all $n \geq n_0$, *we have*

$$\|\mathbf{x} - \mathbf{x}^{n+1}\|_1 \leq \mu \|\mathbf{x} - \mathbf{x}^n\|_1, \tag{15.48}$$

hence, $\|\mathbf{x}^n - \mathbf{x}\|_1 \leq \mu^{n-n_0} \|\mathbf{x}^{n_0} - \mathbf{x}\|_1$ *for all* $n \geq n_0$.

Proof. First notice that $\mathbf{x}^n \to \mathbf{x}$ as $n \to \infty$ by Theorem 15.15(c). Let us denote $\mathbf{v}^n = \mathbf{x}^n - \mathbf{x} \in \ker \mathbf{A}$. By the minimizing property (IRLS$_1$) of $\mathbf{x}^{n+1}$ and the characterization of the minimizer in (A.37), we have

$$0 = \mathrm{Re}\langle \mathbf{x}^{n+1}, \mathbf{v}^{n+1}\rangle_{\mathbf{w}^n} = \mathrm{Re}\langle \mathbf{x} + \mathbf{v}^{n+1}, \mathbf{v}^{n+1}\rangle_{\mathbf{w}^n}.$$

Denoting $S = \mathrm{supp}(\mathbf{x})$ and rearranging terms gives

$$\sum_{j=1}^{N} |v_j^{n+1}|^2 w_j^n = -\mathrm{Re}\left(\sum_{j \in S} x_j \overline{v_j^{n+1}} w_j^n\right) = -\mathrm{Re}\left(\sum_{j \in S} \frac{x_j}{\sqrt{|x_j^n|^2 + \varepsilon_n^2}} \overline{v_j^{n+1}}\right).$$

$$\tag{15.49}$$

Now let $n \geq n_0$, so that $E_n := \|\mathbf{x} - \mathbf{x}^n\|_1 \leq R$. Then, for $j \in S$,

$$|v_j^n| \leq \|\mathbf{v}^n\|_1 = E_n \leq \kappa |x_j|,$$

so that

$$\frac{|x_j|}{\sqrt{|x_j^n|^2 + \varepsilon_n^2}} \le \frac{|x_j|}{|x_j^n|} = \frac{|x_j|}{|x_j + v_j^n|} \le \frac{1}{1 - \kappa}. \tag{15.50}$$

By combining (15.49) and (15.50) with the stable null space property, we reach

$$\sum_{j=1}^N |v_j^{n+1}|^2 w_j^n \le \frac{1}{1 - \kappa} \|v_S^{n+1}\|_1 \le \frac{\rho}{1 - \kappa} \|v_{\overline{S}}^{n+1}\|_1.$$

The Cauchy–Schwarz inequality yields

$$\begin{aligned}
\|v_{\overline{S}}^{n+1}\|_1^2 &\le \left(\sum_{j \in \overline{S}} |v_j^{n+1}|^2 w_j^n \right) \left(\sum_{j \in \overline{S}} \sqrt{|x_j^n|^2 + \varepsilon_n^2} \right) \\
&\le \left(\sum_{j=1}^N |v_j^{n+1}|^2 w_j^n \right) (\|\mathbf{v}^n\|_1 + N\varepsilon_n) \\
&\le \frac{\rho}{1 - \kappa} \|v_{\overline{S}}^{n+1}\|_1 (\|\mathbf{v}^n\|_1 + N\varepsilon_n). \tag{15.51}
\end{aligned}$$

This implies that

$$\|v_{\overline{S}}^{n+1}\|_1 \le \frac{\rho}{1 - \kappa} (\|\mathbf{v}^n\|_1 + N\varepsilon_n).$$

Furthermore, by definition of ε_n and due to (2.1), we have

$$\varepsilon_n \le \gamma(\mathbf{x}^n)_{s+1}^* = \frac{1}{2N} ((\mathbf{x}^n)_{s+1}^* - (\mathbf{x})_{s+1}^*) \le \frac{1}{2N} \|\mathbf{x}^n - \mathbf{x}\|_\infty \le \frac{1}{2N} \|\mathbf{v}^n\|_1.$$

Thus,

$$\|v_{\overline{S}}^{n+1}\|_1 \le \frac{3\rho}{2(1 - \kappa)} \|\mathbf{v}^n\|_1,$$

and finally, using the stable null space property, we arrive at

$$\|\mathbf{v}^{n+1}\|_1 \le (1 + \rho) \|v_{\overline{S}}^{n+1}\|_1 \le \frac{3\rho(1 + \rho)}{2(1 - \kappa)} \|\mathbf{v}^n\|_1,$$

as announced.   □

Remark 15.21. The precise update rule (IRLS$_2$) for ε_n is not essential for the analysis. When $\|\mathbf{x} - \mathbf{x}^{n_0}\|_1 \le R$, the inequality $\|\mathbf{x} - \mathbf{x}^{n+1}\|_1 \le \mu_0(\|\mathbf{x} - \mathbf{x}^n\|_1 + N\varepsilon_n)$ with $\mu_0 = \rho(1 + \rho)/(1 - \kappa)$ is still valid. The rule (IRLS$_2$) only guarantees that $\|\mathbf{x} - \mathbf{x}^{n_0}\|_1 \le R$ is indeed satisfied for some n_0 by Theorem 15.15.

Notes

As outlined in Chap. 3, the basic ℓ_1-minimization problem (BP) is equivalent to the linear optimization problem (P_1') in the real case and to the second-order cone problem $(\text{P}_{1,\eta}')$ (with $\eta = 0$) in the complex case. For such problems, general purpose optimization algorithms apply, including the well-known simplex method [369] and the more recent interior point methods [70, 369].

The homotopy method—or modified LARS—was introduced and analyzed in [169, 177, 376, 377]. Theorem 15.2 was shown in [177]. A Mathematica implementation is described in [335]. It is also able to deal with the weighted case $\min_{\mathbf{x} \in \mathbb{R}^N} \sum_\ell |x_\ell| w_\ell$ subject to $\mathbf{Ax} = \mathbf{y}$ with nonnegative weights $w_\ell \geq 0$ (the case $w_\ell = 0$ needs special treatment).

The adaptive inverse scale space method [79] is another fast ℓ_1-minimization algorithm, which resembles the homotopy method. It builds up the support iteratively. At each step, however, one solves a least squares problem with a nonnegativity constraint instead of a system of linear equations. Like the homotopy method, the inverse scale space method seems to apply only for the real-valued case.

The primal–dual algorithm of Sect. 15.2 was first introduced for a special case in [396]. In full generality, it was presented and analyzed by Chambolle and Pock in [107]. They showed that the algorithm converges with the rate $\mathcal{O}(1/n)$; see Exercise 15.7 for details. For the case where either F^* or G is strongly convex with known strong convexity constant γ in (B.6), a modification of the algorithm where θ, τ, σ are varying throughout the iterations is introduced in [107]. This variant has an improved convergence rate $\mathcal{O}(1/n^2)$. On the other hand, it was proved by Nesterov [367] that the convergence rate $\mathcal{O}(1/n)$ cannot be improved for general convex functions F, G, so that in this sense the rate of Theorem 15.8 is optimal. Also note that for the basis pursuit problems in Examples 15.7(a) and (b), the strong convexity assumptions fail and only the described basic primal–dual algorithm applies. But for Example (15.7)(c) the modified algorithm does apply. A preconditioned version of the primal–dual algorithm is described in [395]. It may be faster than the original method for certain problems and has the additional advantage that the parameters σ and τ are set automatically.

The proof technique involving the discrete derivative (see Lemma 15.9) was introduced by Bartels in [32]. The main motivations of Chambolle and Pock for their algorithm were total variation and related minimization problems appearing in imaging applications [107]. The parameter choice $\theta = 0$ in (PD3) yields the Arrow–Hurwicz method [20, 518]. Chambolle and Pock's algorithm is also related to Douglas–Rachford splitting methods [328]; see below and [107] for more details on this relation. In [264], it is shown that the primal–dual algorithm, modified with a correction step, converges also for values of θ in $[-1, 1]$ for suitable choices of σ and τ. Further primal–dual algorithms for ℓ_1-minimization based on the so-called Bregman iterations, including the inexact Uzawa algorithm, are discussed in [517, Sect. 5]; see also [516].

The iteratively reweighted least squares algorithm of Sect. 15.3 was introduced and analyzed in [139]. The result on the convergence rate of Theorem 15.20 can be extended to approximately sparse vectors; see [139, Theorem 6.4]. Moreover, it is shown in [139] that a variant of the algorithm where the update rule for the weight is motivated by ℓ_p-minimization with $p < 1$ converges superlinearly in a neighborhood of the limit (provided it converges at all). Approximating ℓ_p-minimizers via iteratively reweighted least squares has also been proposed earlier by Chartrand and coauthors; see the empirical analyses in [109–111].

In order to speed up the intermediate least squares steps—in particular, when a fast matrix–vector multiplication is available—one may use the conjugate gradient method in [303] for computing an approximate least squares solution. Denoting $\tau_n = \|\mathbf{x}^n - \hat{\mathbf{x}}^n\|_{\mathbf{w}^n}^2$ with $\mathbf{x}^n$ being the computed iterate and $\hat{\mathbf{x}}^n$ the true weighted ℓ_2-minimizer, it is shown in [511] that a sufficient condition on the accuracy τ_n to ensure overall convergence is

$$\tau_n \leq \frac{a_n^2 \varepsilon_n^2}{\|\mathbf{x}^{n-1}\|_{\mathbf{w}^n}^2 (\|\mathbf{x}^{n-1}\|_\infty + \varepsilon_{n-1})^2}$$

for a positive sequence (a_n) satisfying $\sum_{n=1}^\infty a_n < \infty$.

A variant of iteratively reweighted least squares for low-rank matrix recovery is contained in [205]. Translating the corresponding algorithm back to the vector case, this paper considers a slightly different update rule for the weight, namely,

$$w_j^{n+1} = \min\{|x_j^{n+1}|^{-1}, \varepsilon_n^{-1}\}.$$

Convergence results can also be shown for this variant; see [205] for precise statements. Versions of iteratively reweighted least squares methods appeared also earlier in [121, 318, 375].

Further Algorithms for ℓ_1-Minimization. In order to broaden the picture on algorithms for ℓ_1-minimization (and more general optimization problems), let us briefly describe several other approaches. More information can be found, for instance, in [70, 128, 369], [451, Chap. 7].

Probing the Pareto curve. The so-called SPGL1 method [493] solves the ℓ_1-minimization problem

$$\min_{\mathbf{x}} \|\mathbf{x}\|_1 \quad \text{subject to} \quad \|\mathbf{A}\mathbf{x} - \mathbf{y}\|_2 \leq \sigma \qquad (15.52)$$

for a given value of $\sigma \geq 0$ (so that $\sigma = 0$ also covers the equality-constrained ℓ_1-minimization problem). Both the real case $\mathbf{x} \in \mathbb{R}^N$ and the complex case $\mathbf{x} \in \mathbb{C}^N$ can be treated by this method. It proceeds by solving a sequence of instances of the LASSO problem

$$\min_{\mathbf{x}} \|\mathbf{A}\mathbf{x} - \mathbf{y}\|_2 \quad \text{subject to} \quad \|\mathbf{x}\|_1 \leq \tau \qquad (15.53)$$

for varying values of the parameter τ (recall Proposition 3.2 on the relation of the problems (15.52) and (15.53)). Denoting by $\mathbf{x}_\tau$ the optimal solution of (15.53) and by $\mathbf{r}_\tau = \mathbf{y} - \mathbf{A}\mathbf{x}_\tau$ the corresponding residual, the function

$$\phi(\tau) = \|\mathbf{r}_\tau\|_2$$

gives the optimal value of (15.53) for $\tau \geq 0$. It provides a parametrization of the so-called Pareto curve, which defines the optimal tradeoff between the residual term $\|\mathbf{A}\mathbf{x} - \mathbf{y}\|_2$ and $\|\mathbf{x}\|_1$. We now search for a solution τ_σ of

$$\phi(\tau) = \sigma$$

via Newton's method. The corresponding minimizer $\mathbf{x}_{\tau_\sigma}$ of (15.53) for $\tau = \tau_\sigma$ yields a minimizer of (15.52). It is shown in [493] that the function ϕ is differentiable on $(0, \tau_0)$, where τ_0 is the minimal value of τ such that $\phi(\tau) = 0$ (assuming that $\mathbf{A}$ has full rank). Its derivative satisfies $\phi'(\tau) = -\lambda_\tau$, where λ_τ is the solution of the dual problem of (15.53). With this information, Newton's method $\tau_{k+1} = \tau_k + \Delta\tau_k$ with $\Delta\tau_k := (\sigma - \phi(\tau_k))/\phi'(\tau_k)$ can be invoked.

The intermediate LASSO problem (15.53) for $\tau = \tau_k$ is solved via the spectral projected gradient (SPG) method [49]. This iterative method starts with an initial point $\mathbf{x}_0$ and in each step updates the current iterate $\mathbf{x}^n$ by selecting an appropriate point $\mathbf{x}^{n+1}$ on the projected gradient path $\alpha \mapsto P_\tau(\mathbf{x}^n - \alpha\nabla f(\mathbf{x}^n))$, where $\nabla f(\mathbf{x}) = \mathbf{A}^*(\mathbf{A}\mathbf{x} - \mathbf{y})$ is the gradient of $f(\mathbf{x}) = \frac{1}{2}\|\mathbf{A}\mathbf{x} - \mathbf{y}\|_2^2$ (which is a multiple of the squared objective function) and where $P_\tau(\mathbf{x})$ is the orthogonal projection onto the scaled ℓ_1-unit ball, i.e.,

$$P_\tau(\mathbf{x}) = \operatorname{argmin}\{\|\mathbf{z} - \mathbf{x}\|_2 : \|\mathbf{z}\|_1 \leq \tau\}.$$

This orthogonal projection can be computed efficiently; see [493] or [140] for details. Note also the relation to the proximal mapping of the ℓ_∞-norm (Exercise 15.4). The rule for selecting the update $\mathbf{x}^{n+1}$ on the projected gradient path described in [49, 493] ensures that $\mathbf{x}^n$ converges to a minimizer of (15.53). A similar method for computing the solution of (15.53) is contained in [140].

For more details on the SPGL1 method, the reader should consult [493].

Forward–backward splitting methods. Consider a general optimization problem of the form

$$\min_{\mathbf{x}} F(\mathbf{x}) + G(\mathbf{x}) \tag{15.54}$$

for a differentiable convex function $F : \mathbb{C}^N \to \mathbb{R}$ and a lower semicontinuous convex function $G : \mathbb{C}^N \to (-\infty, \infty]$ satisfying $F(\mathbf{x}) + G(\mathbf{x}) \to \infty$ as $\|\mathbf{x}\|_2 \to \infty$. Then a minimizer of (15.54) exists. We assume that the gradient ∇F of F is L-Lipschitz, i.e.,

$$\|\nabla F(\mathbf{x}) - \nabla F(\mathbf{y})\|_2 \leq L\|\mathbf{x} - \mathbf{y}\|_2 \qquad \text{for all } \mathbf{x}, \mathbf{y} \in \mathbb{C}^N.$$

By Theorem B.21, a vector $\mathbf{x}$ is a minimizer of (15.54) if and only if $\mathbf{0} \in \nabla F(\mathbf{x}) + \partial G(\mathbf{x})$. For any $\tau > 0$, this is equivalent to

$$\mathbf{x} - \tau \nabla F(\mathbf{x}) \in \mathbf{x} + \tau \partial G(\mathbf{x}).$$

By Proposition B.23, a minimizer of (15.54) therefore satisfies the fixed-point equation

$$\mathbf{x} = P_G(\tau; \mathbf{x} - \tau \nabla F(\mathbf{x})),$$

where $P_G(\tau; \cdot) = P_{\tau G}$ denotes the proximal mapping associated with G. Motivated by this relation, the basic forward–backward splitting method starts with some point $\mathbf{x}^0$ and performs the iteration

$$\mathbf{x}^{n+1} = P_G(\tau; \mathbf{x}^n - \tau \nabla F(\mathbf{x}^n)). \tag{15.55}$$

The computation of $\mathbf{z}^n = \mathbf{x} - \tau \nabla F(\mathbf{x}^n)$ is called a gradient or forward step. By Proposition B.23, the iterate $\mathbf{x}^{n+1} = P_G(\tau; \mathbf{z}^n)$ satisfies

$$\frac{\mathbf{x}^{n+1} - \mathbf{z}^n}{\tau} \in \partial G(\mathbf{x}^{n+1}),$$

so that this step can be viewed as an implicit subgradient step, also called backward step. (The iteration scheme (15.55) is referred to as splitting method because the objective function $H(\mathbf{x}) = F(\mathbf{x}) + G(\mathbf{x})$ is split into its components F and G in order to define the algorithm.) We remark that, on the one hand, if $G = 0$, then (15.55) reduces to the gradient method $\mathbf{x}^{n+1} = \mathbf{x}^n - \tau F(\mathbf{x}^n)$ [44]. On the other hand, the choice $F = 0$ yields the proximal point algorithm $\mathbf{x}^{n+1} = P_G(\tau, \mathbf{x}^n)$ for minimizing a nondifferentiable function [128,324,345,423]. The forward–backward algorithm (15.55) can be interpreted as a combination of these two schemes.

It is shown in [129] that the sequence $(\mathbf{x}^n)$ generated by (15.55) converges to a minimizer of (15.54) if $\tau < 2/L$. A modified version of (15.55) uses varying parameters τ_n and introduces additional relaxation parameters λ_n, resulting in the scheme

$$\mathbf{x}^{n+1} = \mathbf{x}^n + \lambda_n(P_G(\tau_n; \mathbf{x}^n - \tau_n \nabla F(\mathbf{x}^n)) - \mathbf{x}^n).$$

If $\lambda_n \in [\varepsilon, 1]$ and $\tau_n \in [\varepsilon/2, 2/L - \varepsilon]$ for some $\varepsilon \in (0, \min\{1, 1/L\})$, then again $(\mathbf{x}^n)$ converges to a minimizer of (15.54); see [129]. For another variant of the forward–backward splitting, we refer to [128, Algorithm 10.5] and [34].

In general, the forward–backward method (15.55) may be slow, although linear convergence rates can be shown under additional assumptions; see [71, 112] and references in [128]. Following ideas of Nesterov [366], Beck and Teboulle [36, 37] introduced the following *accelerated proximal gradient method* which starts with some $\mathbf{x}^0 = \mathbf{z}^0$ and a parameter $t_0 = 1$ and performs the iterations

$$\mathbf{x}^{n+1} = P_G(L^{-1}; \mathbf{z}^n - L^{-1}\nabla F(\mathbf{z}^n)), \tag{15.56}$$

$$t_{n+1} = \frac{1 + \sqrt{4t_n^2 + 1}}{2}, \qquad \lambda_n = 1 + \frac{t_n - 1}{t_{n+1}}, \tag{15.57}$$

$$\mathbf{z}^{n+1} = \mathbf{x}^n + \lambda_n(\mathbf{x}^{n+1} - \mathbf{x}^n). \tag{15.58}$$

This algorithm achieves convergence rate $\mathcal{O}(1/n^2)$, which is optimal in general [365].

The ℓ_1-regularized least squares problem (basis pursuit denoising)

$$\min_{\mathbf{x}} \frac{1}{2}\|\mathbf{A}\mathbf{x} - \mathbf{y}\|_2^2 + \lambda\|\mathbf{x}\|_1 \tag{15.59}$$

is of the form (15.54) with the particular functions $F(\mathbf{x}) = \frac{1}{2}\|\mathbf{A}\mathbf{x} - \mathbf{y}\|_2^2$ and $G(\mathbf{x}) = \lambda\|\mathbf{x}\|_1$. The gradient of F is given by $\nabla F(\mathbf{x}) = \mathbf{A}^*(\mathbf{A}\mathbf{x} - \mathbf{y})$ and satisfies $\|\nabla F(\mathbf{x}) - \nabla F(\mathbf{z})\|_2 = \|\mathbf{A}^*\mathbf{A}(\mathbf{x} - \mathbf{z})\|_2 \le \|\mathbf{A}\|_{2\to 2}^2\|\mathbf{x} - \mathbf{z}\|_2$, so that ∇F is Lipschitz continuous with constant $L \le \|\mathbf{A}\|_{2\to 2}^2$. The proximal mapping for G is the soft thresholding operator S_λ in (15.23). Therefore, the forward–backward algorithm (15.55) for (15.59) reads

$$\mathbf{x}^{n+1} = S_{\tau\lambda}(\mathbf{x}^n + \tau\mathbf{A}^*(\mathbf{y} - \mathbf{A}\mathbf{x}^n)). \tag{15.60}$$

This scheme is also called iterative shrinkage-thresholding algorithm (ISTA) or simply iterative soft thresholding. It was introduced and analyzed in [138, 196]; see also [202, 203]. Convergence is guaranteed if $\tau < 2/\|\mathbf{A}\|_{2\to 2}^2$ [129, 138]. Without the soft thresholding operator S_λ, (15.60) reduces to the so-called Landweber iteration (see e.g. [186]) and therefore (15.60) is sometimes also called thresholded Landweber iteration. The specialization of the accelerated scheme (15.56)–(15.58) to the ℓ_1-minimization problem (15.59) is called fast iterative shrinkage-thresholding algorithm (FISTA) [36].

The fixed-point continuation (FPC) algorithm introduced in [257] (see also [229] for a version adapted to low-rank matrix recovery) solves (15.59) by the forward–backward iteration (15.55), where the parameter τ is changed in a suitable way throughout the iterations.

Generalizations of the backward–forward algorithm to the minimization of functions $H(\mathbf{x}) = F(\mathbf{x}) + \sum_{j=1}^n G_j(\mathbf{x})$, where F is convex and differentiable and the G_j are convex but not necessarily differentiable, are contained in [405]. Taking $F = 0$, these algorithms include the Douglas–Rachford splitting method discussed below.

We remark that (PD$_1$) of Chambolle and Pock's primal dual algorithm may be viewed as a forward–backward step on the dual variable and (PD$_2$) as a forward–backward step on the primal variable.

Douglas–Rachford splitting. The forward–backward splitting method requires F to be differentiable, therefore it is not applicable to constrained

ℓ_1-minimization problems. Douglas–Rachford splitting methods provide an alternative for solving the minimization problem (15.54), where both convex functions $F, G : \mathbb{C}^N \to (-\infty, \infty]$ are not necessarily differentiable. If $\mathbf{x}$ is a minimizer of (15.54), then $\mathbf{x} = P_F(\tau; \mathbf{z})$ for some $\mathbf{z}$ satisfying

$$P_F(\tau; \mathbf{z}) = P_G(\tau; 2P_F(\tau; \mathbf{z}) - \mathbf{z});$$

see Exercise 15.8. This motivates the Douglas–Rachford algorithm which starts with some $\mathbf{z}^0$ and, for $n \geq 0$, performs the iterations

$$\mathbf{x}^n = P_F(\tau; \mathbf{z}^n),$$
$$\mathbf{z}^{n+1} = P_G(\tau; 2\mathbf{x}^n - \mathbf{z}^n) + \mathbf{z}^n - \mathbf{x}^n.$$

The convergence of the sequence $(\mathbf{x}^n)$ to a minimizer of (15.54) has been established in [127]. The method was originally introduced by Douglas and Rachford in [172] for the solution of linear operator equations; see also [328].

The equality-constrained ℓ_1-minimization problem

$$\min \|\mathbf{x}\|_1 \quad \text{subject to } \mathbf{Ax} = \mathbf{y} \tag{15.61}$$

takes the form (15.54) with $G(\mathbf{x}) = \|\mathbf{x}\|_1$ and $F = \chi_{\mathcal{F}_{\mathbf{A},\mathbf{y}}}$, the characteristic function of the affine set $\mathcal{F}_{\mathbf{A},\mathbf{y}} = \{\mathbf{x} : \mathbf{Ax} = \mathbf{y}\}$. We have already seen that the proximal mapping of G is given by the soft thresholding operator $P_G(\tau; \mathbf{x}) = S_\tau(\mathbf{x})$. The proximal mapping of F is given by

$$P_F(\tau; \mathbf{x}) = \operatorname{argmin} \{\|\mathbf{z} - \mathbf{x}\|_2 : A\mathbf{z} = \mathbf{y}\} = \mathbf{x} + \mathbf{A}^\dagger(\mathbf{y} - \mathbf{Ax}),$$

where $\mathbf{A}^\dagger$ is the Moore–Penrose pseudo-inverse of $\mathbf{A}$. With this information, the Douglas–Rachford algorithm for (15.61) can be implemented. In general, the application of $\mathbf{A}^\dagger$ involves the solution of a linear system, which may be computationally expensive (the conjugate gradient method can be applied for speed up). However, in the important special case where $\mathbf{AA}^* = \lambda \mathbf{Id}$, we have $\mathbf{A}^\dagger = \mathbf{A}^*(\mathbf{AA}^*)^{-1} = \lambda^{-1}\mathbf{A}^*$, so that the computation of $P_F(\tau; \mathbf{x})$ greatly simplifies and the iterations of the Douglas–Rachford algorithm become efficient. The case $\mathbf{AA}^* = \lambda \mathbf{Id}$ occurs, for instance, for random partial Fourier matrices; see Example 4 in Sect. 12.1 and more generally the situation of Example 6 in Sect. 12.1.

In [446], it is shown that the alternating split Bregman algorithm introduced in [230] is equivalent to a Douglas–Rachford algorithm. We refer to [127, 128, 451] for more information on Douglas–Rachford splittings.

Alternating Direction Method of Multipliers (ADMM). For convex lower semi-continuous functions $F : \mathbb{C}^N \to (-\infty, \infty]$, $G : \mathbb{C}^k \to (-\infty, \infty]$, and a matrix $\mathbf{B} \in \mathbb{C}^{k \times N}$, we consider the optimization problem

$$\min_{\mathbf{x}} F(\mathbf{x}) + G(\mathbf{Bx}), \tag{15.62}$$

which is equivalent to

$$\min_{\mathbf{x},\mathbf{y}} F(\mathbf{x}) + G(\mathbf{y}) \quad \text{subject to } \mathbf{y} = \mathbf{Bx}.$$

Its so-called augmented Lagrangian of index $\tau > 0$ is given by

$$L_\tau(\mathbf{x}, \mathbf{y}, \boldsymbol{\xi}) = F(\mathbf{x}) + G(\mathbf{y}) + \frac{1}{\tau} \operatorname{Re}\langle \boldsymbol{\xi}, \mathbf{Bx} - \mathbf{y} \rangle + \frac{1}{2\tau} \|\mathbf{Bx} - \mathbf{y}\|_2^2.$$

A step of the ADMM minimizes L_τ over $\mathbf{x}$, then over $\mathbf{y}$, and finally applies a proximal (backward) step over the dual variable (Lagrange multiplier) $\boldsymbol{\xi}$. We assume that $\mathbf{B}^*\mathbf{B}$ is invertible and introduce the modified proximal mapping

$$P_F^{\mathbf{B}}(\tau; \mathbf{y}) = \operatorname*{argmin}_{\mathbf{z} \in \mathbb{C}^N} \left\{ \tau F(\mathbf{z}) + \frac{1}{2} \|\mathbf{Bz} - \mathbf{y}\|_2^2 \right\}.$$

Formally, for a fixed parameter $\tau > 0$, the ADMM algorithm starts with vectors $\mathbf{u}^0, \mathbf{z}^0 \in \mathbb{C}^k$ and, for $n \geq 0$, performs the iterations

$$\mathbf{x}^n = P_F^{\mathbf{B}}(\tau; \mathbf{u}^n - \mathbf{z}^n),$$

$$\mathbf{v}^n = \mathbf{Bx}^n,$$

$$\mathbf{u}^{n+1} = P_G(\tau; \mathbf{v}^n + \mathbf{z}^n),$$

$$\mathbf{z}^{n+1} = \mathbf{z}^n + \mathbf{v}^n - \mathbf{u}^{n+1}.$$

For the convergence of the sequence $(\mathbf{x}^n)$ to a minimizer of (15.62), we refer to [217, 226]. A connection to Douglas–Rachford splittings was observed in [216].

Let us consider the special case of basis pursuit denoising

$$\min_{\mathbf{x} \in \mathbb{C}^N} \frac{1}{2} \|\mathbf{Ax} - \mathbf{y}\|_2^2 + \lambda \|\mathbf{x}\|_1,$$

which is (15.62) for $F(\mathbf{x}) = \|\mathbf{Ax} - \mathbf{y}\|_2^2/2$, $G(\mathbf{x}) = \lambda \|\mathbf{x}\|_1$, and $\mathbf{B} = \mathbf{Id}$, that is, $k = N$. Then the ADMM algorithm reads

$$\mathbf{x}^n = (\mathbf{A}^*\mathbf{A} + \tau^{-1}\mathbf{Id})^{-1}(\mathbf{A}^*\mathbf{y} + \tau^{-1}(\mathbf{u}^n - \mathbf{z}^n)),$$

$$\mathbf{u}^{n+1} = S_{\tau\lambda}(\mathbf{x}^n + \mathbf{z}^n),$$

$$\mathbf{z}^{n+1} = \mathbf{z}^n + \mathbf{x}^n - \mathbf{u}^{n+1}. \tag{15.63}$$

For further information about ADMM, we refer to [4, 128, 188].

Interior point methods. Consider a constrained convex optimization problem of the form

$$\min_{\mathbf{x} \in \mathbb{R}^N} F(\mathbf{x}) \quad \text{subject to } \mathbf{A}\mathbf{x} = \mathbf{y}, \tag{15.64}$$

$$\text{and } F_j(\mathbf{x}) \le 0, \quad j \in [M], \tag{15.65}$$

where $F, F_1, \ldots, F_M : \mathbb{R}^N \to \mathbb{R}$ are convex functions, $\mathbf{A} \in \mathbb{R}^{m \times N}$, and $\mathbf{y} \in \mathbb{R}^m$. As discussed in Chap. 3, the basis pursuit problem is equivalent to a problem of this form in the real setting. (The principle discussed below can be adapted to conic optimization problems, so that the complex case, quadratically constrained ℓ_1-minimization, and nuclear norm minimization are also covered.) Denoting by $\chi_{\{F_j \le 0\}}$ the characteristic function of the set $\{\mathbf{x} : F_j(\mathbf{x}) \le 0\}$, the optimization problem (15.65) is equivalent to

$$\min_{\mathbf{x} \in \mathbb{R}^N} F(\mathbf{x}) + \sum_{j=1}^{n} \chi_{\{F_j \le 0\}}(\mathbf{x}) \quad \text{subject to } \mathbf{A}\mathbf{x} = \mathbf{y}.$$

The idea of a basic interior point method, called the log-barrier method, is to replace, for a parameter $t > 0$, the characteristic function $\chi_{\{F_j \le 0\}}(\mathbf{x})$, by the convex function $-t^{-1} \ln(-F_j(\mathbf{x}))$, which is twice differentiable on $\{\mathbf{x} : F_j(\mathbf{x}) < 0\}$ provided F_j is. This yields the equality-constrained optimization problem

$$\min_{\mathbf{x} \in \mathbb{R}^N} F(\mathbf{x}) - t^{-1} \sum_{j=1}^{n} \ln(-F_j(\mathbf{x})) \quad \text{subject to } \mathbf{A}\mathbf{x} = \mathbf{y}.$$

The log-barrier method consists in solving a sequence of these equality-constrained minimization problems for increasing values of t via Newton's method. In the limit $t \to \infty$, the corresponding iterates converge to a solution of (15.65) under suitable hypotheses. We refer to [70, Chap. 11] for details.

Primal–dual versions of interior point methods have also been developed; see e.g. [70, 369]. In [302], an interior point method specialized to ℓ_1-minimization is described.

Further information on numerical methods for sparse recovery can be found in [201]. Other optimization methods specialized to ℓ_1-minimization are described in [38, 197, 256, 323]. Many implementations and toolboxes are available freely on the Internet. Numerical comparisons of several algorithms/implementations can be found, for instance, in [334, 405].

Exercises

15.1. Apply the homotopy method by hand to the ℓ_1-minimization problem $\min_{x \in \mathbb{R}^N} \|x\|_1$ subject to $Ax = y$ with A and y given by

(a)

$$A = Id, \quad y = [12, -3, 6, 8, 3]^\top,$$

(b)

$$A = \begin{bmatrix} -3 & 4 & 4 \\ -5 & 1 & 4 \\ 5 & 1 & -4 \end{bmatrix}, \quad y = \begin{bmatrix} 24 \\ 17 \\ -7 \end{bmatrix}.$$

15.2. Verify formula (B.18) for the soft thresholding operator. Show that

$$S_\tau(y)^2 = \min_{|x| \le \tau} (x - y)^2.$$

15.3. Show that the minimizer $x^\sharp$ of F_λ in (15.3) satisfies

$$x^\sharp = S_\lambda(x^\sharp + A^*(y - Ax^\sharp)).$$

15.4. Let $\| \cdot \|$ be some norm and let $\| \cdot \|_*$ denote its dual norm. Show that the proximal mapping satisfies

$$P_{\|\cdot\|}(\tau; x) = x - P_{\tau B_{\|\cdot\|_*}}(x),$$

where $P_{\tau B_{\|\cdot\|_*}}(x) := \operatorname*{argmin}_{z \in \tau B_{\|\cdot\|_*}} \|x - z\|_2$ is the orthogonal projection onto the scaled dual unit ball $\tau B_{\|\cdot\|_*} = \{z : \|z\|_* \le \tau\}$; see (B.14).

15.5. Let $\| \cdot \|_*$ be the nuclear norm (A.25).

(a) For a matrix $X \in \mathbb{C}^{n_1 \times n_2}$ whose singular value decomposition has the form $X = U \operatorname{diag}[\sigma_1, \ldots, \sigma_n] V^*$, $n = \min\{n_1, n_2\}$, show that the proximal mapping for the nuclear norm satisfies

$$P_{\|\cdot\|_*}(\tau; X) = U \operatorname{diag}\left[S_\tau(\sigma_1), \ldots, S_\tau(\sigma_n)\right] V^*,$$

where S_τ is the soft thresholding operator.

(b) For a linear map $\mathcal{A} : \mathbb{C}^{n_1 \times n_2} \to \mathbb{C}^m$ and $y \in \mathbb{C}^m$, formulate Chambolle and Pock's primal–dual algorithm for the nuclear norm minimization problems

$$\min_{Z \in \mathbb{C}^{n_1 \times n_2}} \|Z\|_* \quad \text{subject to } \mathcal{A}(Z) = y$$

and, for $\eta > 0$,

$$\min_{\mathbf{Z} \in \mathbb{C}^{n_1 \times n_2}} \|\mathbf{Z}\|_* \quad \text{subject to } \|\mathcal{A}(\mathbf{Z}) - \mathbf{y}\|_F^2 \le \eta.$$

15.6. Formulate Chambolle and Pock's primal–dual algorithm for the solution of the Dantzig selector

$$\min_{\mathbf{x} \in \mathbb{C}^N} \|\mathbf{x}\|_1 \quad \text{subject to } \|\mathbf{A}^*(\mathbf{A}\mathbf{x} - \mathbf{y})\|_\infty \le \tau,$$

where $\mathbf{A} \in \mathbb{C}^{m \times N}$, $\mathbf{y} \in \mathbb{C}^m$, and $\tau \ge 0$.

15.7. Consider the optimization problem (15.15), where $\mathbf{A} \in \mathbb{C}^{m \times N}$ and where $F : \mathbb{C}^m \to (-\infty, \infty]$, $G : \mathbb{C}^N \to (-\infty, \infty]$ are extended real-valued lower semicontinuous convex functions.

(a) For subsets $B_1 \subset \mathbb{C}^N$ and $B_2 \subset \mathbb{C}^m$, let

$$\mathcal{G}_{B_1, B_2}(\mathbf{x}, \boldsymbol{\xi}) := \sup_{\boldsymbol{\xi}' \in B_2} L(\mathbf{x}, \boldsymbol{\xi}') - \inf_{\mathbf{x}' \in B_1} L(\mathbf{x}', \boldsymbol{\xi}) \tag{15.66}$$

be the so-called *partial primal–dual gap*. Show that if $B_1 \times B_2$ contains a primal–dual optimal point $(\hat{\mathbf{x}}, \hat{\boldsymbol{\xi}})$, that is, a saddle point of (15.17), then $\mathcal{G}_{B_1, B_2}(\mathbf{x}, \boldsymbol{\xi}) \ge 0$ for all $(\mathbf{x}, \boldsymbol{\xi}) \in \mathbb{C}^N \times \mathbb{C}^m$ and $\mathcal{G}(\mathbf{x}, \boldsymbol{\xi}) = 0$ if and only if $(\mathbf{x}, \boldsymbol{\xi})$ is a saddle point of $\mathcal{G}_{B_1, B_2}$. (Therefore, $\mathcal{G}_{B_1, B_2}(\mathbf{x}, \boldsymbol{\xi})$ can be taken as a measure of the distance of the pair $(\mathbf{x}, \boldsymbol{\xi})$ to the primal–dual optimizer.)

(b) For the sequence $(\mathbf{x}^n, \bar{\mathbf{x}}^n, \boldsymbol{\xi}^n)_{n \ge 0}$ generated by Chambolle and Pock's primal–dual algorithm (PD$_1$)–(PD$_3$), define $\mathbf{x}_M := M^{-1} \sum_{n=1}^M \mathbf{x}^n$ and $\boldsymbol{\xi}_M := M^{-1} \sum_{n=1}^M \boldsymbol{\xi}^n$. If $B_1 \subset \mathbb{C}^N$ and $B_2 \subset \mathbb{C}^m$ are bounded sets, show that

$$\mathcal{G}_{B_1, B_2}(\mathbf{x}_M, \boldsymbol{\xi}_M) \le \frac{D(B_1, B_2)}{M}, \tag{15.67}$$

where $D(B_1, B_2) := (2\tau)^{-1} \sup_{\mathbf{x} \in B_1} \|\mathbf{x} - \mathbf{x}^0\|_2^2 + (2\sigma)^{-1} \sup_{\boldsymbol{\xi} \in B_2} \|\boldsymbol{\xi} - \boldsymbol{\xi}^0\|_2^2$. (Therefore, we say that Chambolle and Pock's algorithm converges at the rate $\mathcal{O}(M^{-1})$.)

15.8. Let $F, G : \mathbb{C}^N \to (-\infty, \infty]$ be convex lower semicontinuous functions. Suppose that the optimization problem

$$\min_{\mathbf{x} \in \mathbb{C}^N} F(\mathbf{x}) + G(\mathbf{x})$$

possesses a minimizer $\mathbf{x}^\sharp$. For an arbitrary parameter $\tau > 0$, show that $\mathbf{x}^\sharp = P_F(\tau; \mathbf{z})$ for some vector $\mathbf{z}$ satisfying

$$P_F(\tau; \mathbf{z}) = P_G(\tau; 2P_F(\tau; \mathbf{z}) - \mathbf{z}),$$

where P_F and P_G are the proximal mappings associated to F and G.

15.9. Implement some of the algorithms of this chapter. Choose $\mathbf{A} \in \mathbb{R}^{m \times N}$ as a Gaussian random matrix or $\mathbf{A} \in \mathbb{C}^{m \times N}$ as a partial random Fourier matrix. In the latter case, exploit the fast Fourier transform. Test the algorithm on randomly generated s-sparse signals, where first the support is chosen at random and then the nonzero coefficients. By varying m, s, N, evaluate the empirical success probability of recovery. Compare the runtime of the algorithms for small and medium sparsity s.

Appendix A
Matrix Analysis

This appendix collects useful background from linear algebra and matrix analysis, such as vector and matrix norms, singular value decompositions, Gershgorin's disk theorem, and matrix functions. Much more material can be found in various books on the subject including [47, 51, 231, 273, 280, 281, 475].

A.1 Vector and Matrix Norms

We work with real or complex vector spaces X, usually $X = \mathbb{R}^n$ or $X = \mathbb{C}^n$. We usually write the vectors in $\mathbb{C}^n$ in boldface, $\mathbf{x}$, while their entries are denoted x_j, $j \in [n]$, where $[n] := \{1, \ldots, n\}$. The canonical unit vectors in $\mathbb{R}^n$ are denoted by $\mathbf{e}_1, \ldots, \mathbf{e}_n$. They have entries

$$(\mathbf{e}_\ell)_j = \delta_{\ell,j} = \begin{cases} 1 & \text{if } j = \ell, \\ 0 & \text{otherwise.} \end{cases}$$

Definition A.1. A nonnegative function $\| \cdot \| : X \to [0, \infty)$ is called a norm if

(a) $\|\mathbf{x}\| = 0$ if and only if $\mathbf{x} = \mathbf{0}$ (definiteness).
(b) $\|\lambda \mathbf{x}\| = |\lambda| \|\mathbf{x}\|$ for all scalars λ and all vectors $\mathbf{x} \in X$ (homogeneity).
(c) $\|\mathbf{x} + \mathbf{y}\| \leq \|\mathbf{x}\| + \|\mathbf{y}\|$ for all vectors $\mathbf{x}, \mathbf{y} \in X$ (triangle inequality).

If only (b) and (c) hold, so that $\|\mathbf{x}\| = 0$ does not necessarily imply $\mathbf{x} = \mathbf{0}$, then $\| \cdot \|$ is called a seminorm.

If (a) and (b) hold, but (c) is replaced by the weaker quasitriangle inequality

$$\|\mathbf{x} + \mathbf{y}\| \leq C(\|\mathbf{x}\| + \|\mathbf{y}\|)$$

for some constant $C \geq 1$, then $\| \cdot \|$ is called a quasinorm. The smallest constant C is called its quasinorm constant.

S. Foucart and H. Rauhut, *A Mathematical Introduction to Compressive Sensing*, Applied and Numerical Harmonic Analysis, DOI 10.1007/978-0-8176-4948-7, © Springer Science+Business Media New York 2013

A space X endowed with a norm $\| \cdot \|$ is called a normed space.

Definition A.2. Let X be a set. A function $d : X \times X \rightarrow [0, \infty)$ is called a metric if

(a) $d(x, y) = 0$ if and only if $x = y$.
(b) $d(x, y) = d(y, x)$ for all $x, y \in X$.
(c) $d(x, z) \leq d(x, y) + d(y, z)$ for all $x, y, z \in X$.

If only (b) and (c) hold, then d is called a pseudometric.

The set X endowed with a metric d is called a metric space. A norm $\| \cdot \|$ on X induces a metric on X by

$$d(\mathbf{x}, \mathbf{y}) = \|\mathbf{x} - \mathbf{y}\|.$$

A seminorm induces a pseudometric in the same way.

The ℓ_p-norm (or simply p-norm) on $\mathbb{R}^n$ or $\mathbb{C}^n$ is defined for $1 \leq p < \infty$ as

$$\|\mathbf{x}\|_p := \left(\sum_{j=1}^n |x_j|^p \right)^{1/p}, \tag{A.1}$$

and for $p = \infty$ as

$$\|\mathbf{x}\|_\infty := \max_{j \in [n]} |x_j|.$$

For $0 < p < 1$, the expression (A.1) only defines a quasinorm with quasinorm constant $C = 2^{1/p-1}$. This is derived from the p-triangle inequality

$$\|\mathbf{x} + \mathbf{y}\|_p^p \leq \|\mathbf{x}\|_p^p + \|\mathbf{y}\|_p^p.$$

Therefore, the ℓ_p-quasinorm induces a metric via $d(\mathbf{x}, \mathbf{y}) = \|\mathbf{x} - \mathbf{y}\|_p^p$ for $0 < p < 1$.

We define a ball of radius $t \geq 0$ around a point $\mathbf{x}$ in a metric space (X, d) by

$$B(\mathbf{x}, t) = B_d(\mathbf{x}, t) = \{\mathbf{z} \in X : d(\mathbf{x}, \mathbf{z}) \leq t\}.$$

If the metric is induced by a norm $\| \cdot \|$ on a vector space, then we also write

$$B_{\|\cdot\|}(\mathbf{x}, t) = \{\mathbf{z} \in X : \|\mathbf{x} - \mathbf{z}\| \leq t\}. \tag{A.2}$$

If $\mathbf{x} = \mathbf{0}$ and $t = 1$, then $B = B_{\|\cdot\|} = B_{\|\cdot\|}(\mathbf{0}, 1)$ is called unit ball.

The canonical inner product on $\mathbb{C}^n$ is defined by

$$\langle \mathbf{x}, \mathbf{y} \rangle = \sum_{j=1}^n x_j \overline{y_j}, \qquad \mathbf{x}, \mathbf{y} \in \mathbb{C}^n.$$

On $\mathbb{R}^n$, it is given by $\langle \mathbf{x}, \mathbf{y} \rangle = \sum_{j=1}^n x_j y_j$, $\mathbf{x}, \mathbf{y} \in \mathbb{R}^n$. The ℓ_2-norm is related to the canonical inner product by

$$\|\mathbf{x}\|_2 = \sqrt{\langle \mathbf{x}, \mathbf{x} \rangle}.$$

The Cauchy–Schwarz inequality states that

$$|\langle \mathbf{x}, \mathbf{y} \rangle| \leq \|\mathbf{x}\|_2 \|\mathbf{y}\|_2 \quad \text{for all } \mathbf{x}, \mathbf{y} \in \mathbb{C}^n.$$

More generally, for $p, q \in [1, \infty]$ such that $1/p + 1/q = 1$ (with the convention that $1/\infty = 0$ and $1/0 = \infty$), Hölder's inequality states that

$$|\langle \mathbf{x}, \mathbf{y} \rangle| \leq \|\mathbf{x}\|_p \|\mathbf{y}\|_q \quad \text{for all } \mathbf{x}, \mathbf{y} \in \mathbb{C}^n.$$

Given $0 < p < q \leq \infty$, applying the latter with $\mathbf{x}$, $\mathbf{y}$, p, and q replaced by $[|x_1|^p, \ldots, |x_n|^p]^\top$, $[1, \ldots, 1]^\top$, q/p, and $q/(q-p)$ gives $\|\mathbf{x}\|_p^p \leq \|\mathbf{x}\|_q^p n^{(q-p)/p}$, i.e.,

$$\|\mathbf{x}\|_p \leq n^{1/p - 1/q} \|\mathbf{x}\|_q. \tag{A.3}$$

We point out the important special cases

$$\|\mathbf{x}\|_1 \leq \sqrt{n} \|\mathbf{x}\|_2 \quad \text{and} \quad \|\mathbf{x}\|_2 \leq \sqrt{n} \|\mathbf{x}\|_\infty.$$

If $\mathbf{x}$ has at most s nonzero entries, that is, $\|\mathbf{x}\|_0 = \mathrm{card}(\{\ell : x_\ell \neq 0\}) \leq s$, then the above inequalities become $\|\mathbf{x}\|_p \leq s^{1/p - 1/q} \|\mathbf{x}\|_q$, and in particular,

$$\|\mathbf{x}\|_1 \leq \sqrt{s} \|\mathbf{x}\|_2 \leq s \|\mathbf{x}\|_\infty.$$

For $0 < p < q \leq \infty$, we also have the reverse inequalities

$$\|\mathbf{x}\|_q \leq \|\mathbf{x}\|_p, \tag{A.4}$$

and in particular, $\|\mathbf{x}\|_\infty \leq \|\mathbf{x}\|_2 \leq \|\mathbf{x}\|_1$. Indeed, the bound $\|\mathbf{x}\|_\infty \leq \|\mathbf{x}\|_p$ is obvious, and for $p < q < \infty$,

$$\|\mathbf{x}\|_q^q = \sum_{j=1}^n |x_j|^q = \sum_{j=1}^n |x_j|^{q-p} |x_j|^p \leq \|\mathbf{x}\|_\infty^{q-p} \sum_{j=1}^n |x_j|^p \leq \|\mathbf{x}\|_p^{q-p} \|\mathbf{x}\|_p^p = \|\mathbf{x}\|_p^q.$$

Both bounds (A.3) and (A.4) are sharp in general. Indeed, equality holds in (A.3) for vectors with constant absolute entries, while equality holds in (A.4) for scalar multiples of a canonical unit vector.

Definition A.3. Let $\| \cdot \|$ be a norm on $\mathbb{R}^n$ or $\mathbb{C}^n$. Its dual norm $\| \cdot \|_*$ is defined by

$$\|\mathbf{x}\|_* := \sup_{\|\mathbf{y}\| \leq 1} |\langle \mathbf{y}, \mathbf{x} \rangle|.$$

In the real case, the dual norm may equivalently be defined via

$$\|\mathbf{x}\|_* = \sup_{\mathbf{y} \in \mathbb{R}^n, \|\mathbf{y}\| \leq 1} \langle \mathbf{y}, \mathbf{x} \rangle,$$

while in the complex case

$$\|\mathbf{x}\|_* = \sup_{\mathbf{y} \in \mathbb{C}^n, \|\mathbf{y}\| \leq 1} \text{Re} \langle \mathbf{y}, \mathbf{x} \rangle.$$

The dual of the dual norm $\| \cdot \|_*$ is the norm $\| \cdot \|$ itself. In particular, we have

$$\|\mathbf{x}\| = \sup_{\|\mathbf{y}\|_* \leq 1} |\langle \mathbf{x}, \mathbf{y} \rangle| = \sup_{\|\mathbf{y}\|_* \leq 1} \text{Re} \langle \mathbf{x}, \mathbf{y} \rangle. \tag{A.5}$$

The dual of $\| \cdot \|_p$ is $\| \cdot \|_q$ with $1/p + 1/q = 1$. In particular, $\| \cdot \|_2$ is self-dual, i.e.,

$$\|\mathbf{x}\|_2 = \sup_{\|\mathbf{y}\|_2 \leq 1} |\langle \mathbf{y}, \mathbf{x} \rangle|, \tag{A.6}$$

while $\| \cdot \|_\infty$ is the dual norm of $\| \cdot \|_1$ and vice versa.

Given a subspace W of a vector space X, the *quotient space* X/W consists of the residue classes

$$[\mathbf{x}] := \mathbf{x} + W = \{\mathbf{x} + \mathbf{w}, \mathbf{w} \in W\}, \qquad \mathbf{x} \in X.$$

The *quotient map* is the surjective linear map $\mathbf{x} \mapsto [\mathbf{x}] = \mathbf{x} + W \in X/W$. The *quotient norm* on X/W is defined by

$$\|[\mathbf{x}]\|_{X/W} := \inf\{\|\mathbf{v}\|, \mathbf{v} \in [\mathbf{x}] = \mathbf{x} + W\}, \qquad [\mathbf{x}] \in X/W.$$

Let us now turn to norms of matrices (or more generally, linear mappings between normed spaces). The entries of $\mathbf{A} \in \mathbb{C}^{m \times n}$ are denoted $A_{j,k}, j \in [m], k \in [n]$. The columns of $\mathbf{A}$ are denoted $\mathbf{a}_k$, so that $\mathbf{A} = [\mathbf{a}_1 | \cdots | \mathbf{a}_n]$. The transpose of $\mathbf{A} \in \mathbb{C}^{m \times n}$ is the matrix $\mathbf{A}^\top \in \mathbb{C}^{n \times m}$ with entries $(\mathbf{A}^\top)_{k,j} = A_{j,k}$. A matrix $\mathbf{B} \in \mathbb{C}^{n \times n}$ is called symmetric if $\mathbf{B}^\top = \mathbf{B}$. The adjoint (or Hermitian transpose) of $\mathbf{A} \in \mathbb{C}^{m \times n}$ is the matrix $\mathbf{A}^* \in \mathbb{C}^{n \times m}$ with entries $(\mathbf{A}^*)_{k,j} = \overline{A_{j,k}}$. For $\mathbf{x} \in \mathbb{C}^n$ and $\mathbf{y} \in \mathbb{C}^m$, we have $\langle \mathbf{Ax}, \mathbf{y} \rangle = \langle \mathbf{x}, \mathbf{A}^*\mathbf{y} \rangle$. A matrix $\mathbf{B} \in \mathbb{C}^{n \times n}$ is called self-adjoint (or Hermitian) if $\mathbf{B}^* = \mathbf{B}$. The identity matrix on $\mathbb{C}^n$ is denoted $\mathbf{Id}$ or $\mathbf{Id}_n$. A matrix $\mathbf{U} \in \mathbb{C}^{n \times n}$ is called unitary if $\mathbf{U}^*\mathbf{U} = \mathbf{U}\mathbf{U}^* = \mathbf{Id}$. A self-adjoint matrix $\mathbf{B} \in \mathbb{C}^{n \times n}$ possesses an eigenvalue decomposition of the form $\mathbf{B} = \mathbf{U}^*\mathbf{DU}$, where $\mathbf{U}$ is a unitary matrix $\mathbf{U} \in \mathbb{C}^{n \times n}$ and $\mathbf{D} = \text{diag}[\lambda_1, \ldots, \lambda_n]$ is a diagonal matrix containing the real eigenvalues $\lambda_1, \ldots, \lambda_n$ of $\mathbf{B}$.

Definition A.4. Let $\mathbf{A} : X \to Y$ be a linear map between two normed spaces $(X, \| \cdot \|)$ and $(Y, \|\cdot\|)$. The operator norm of $\mathbf{A}$ is defined as

$$\|\mathbf{A}\| := \sup_{\|\mathbf{x}\| \leq 1} \|\mathbf{A}\mathbf{x}\| = \sup_{\|\mathbf{x}\|=1} \|\mathbf{A}\mathbf{x}\|. \tag{A.7}$$

In particular, for a matrix $\mathbf{A} \in \mathbb{C}^{m \times n}$ and $1 \leq p, q \leq \infty$, we define the matrix norm (or operator norm) between ℓ_p and ℓ_q as

$$\|\mathbf{A}\|_{p \to q} := \sup_{\|\mathbf{x}\|_p \leq 1} \|\mathbf{A}\mathbf{x}\|_q = \sup_{\|\mathbf{x}\|_p=1} \|\mathbf{A}\mathbf{x}\|_q. \tag{A.8}$$

For $\mathbf{A} : X \to Y$ and $\mathbf{x} \in X$, the definition implies that $\|\mathbf{A}\mathbf{x}\| \leq \|\mathbf{A}\|\|\mathbf{x}\|$. It is an easy consequence of the definition that the norm of the product of two matrices $\mathbf{A} \in \mathbb{C}^{m \times n}$ and $\mathbf{B} \in \mathbb{C}^{n \times k}$ satisfies

$$\|\mathbf{A}\mathbf{B}\|_{p \to r} \leq \|\mathbf{A}\|_{q \to r}\|\mathbf{B}\|_{p \to q}, \quad 1 \leq p, q, r \leq \infty. \tag{A.9}$$

We summarize formulas for the matrix norms $\|\mathbf{A}\|_{p \to q}$ for some special choices of p and q. The lemma below also refers to the singular values of a matrix, which will be covered in the next section.

Lemma A.5. *Let* $\mathbf{A} \in \mathbb{C}^{m \times n}$.

(a) We have

$$\|\mathbf{A}\|_{2 \to 2} = \sqrt{\lambda_{\max}(\mathbf{A}^*\mathbf{A})} = \sigma_{\max}(\mathbf{A}),$$

where $\lambda_{\max}(\mathbf{A}^*\mathbf{A})$ *is the largest eigenvalue of* $\mathbf{A}^*\mathbf{A}$ *and* $\sigma_{\max}(\mathbf{A})$ *the largest singular value of* $\mathbf{A}$.

In particular, if $\mathbf{B} \in \mathbb{C}^{n \times n}$ *is self-adjoint, then* $\|\mathbf{B}\|_{2 \to 2} = \max_{j \in [n]} |\lambda_j(\mathbf{B})|$, *where the* $\lambda_j(\mathbf{B})$ *denote the eigenvalues of* $\mathbf{B}$.

(b) For $1 \leq p \leq \infty$, *we have* $\|\mathbf{A}\|_{1 \to p} = \max_{k \in [n]} \|\mathbf{a}_k\|_p$. *In particular,*

$$\|\mathbf{A}\|_{1 \to 1} = \max_{k \in [n]} \sum_{j=1}^{m} |A_{j,k}|, \tag{A.10}$$

$$\|\mathbf{A}\|_{1 \to 2} = \max_{k \in [n]} \|\mathbf{a}_k\|_2. \tag{A.11}$$

(c) We have

$$\|\mathbf{A}\|_{\infty \to \infty} = \max_{j \in [m]} \sum_{k=1}^{n} |A_{j,k}|.$$

Proof. (a) Since $\mathbf{A}^*\mathbf{A} \in \mathbb{C}^{n \times n}$ is self-adjoint, it can be diagonalized as $\mathbf{A}^*\mathbf{A} = \mathbf{U}^*\mathbf{D}\mathbf{U}$ with a unitary $\mathbf{U}$ and a diagonal $\mathbf{D}$ containing the eigenvalues λ_ℓ of $\mathbf{A}^*\mathbf{A}$ on the diagonal. For $\mathbf{x} \in \mathbb{C}^n$ with $\|\mathbf{x}\|_2 = 1$,

$$\|\mathbf{A}\mathbf{x}\|_2^2 = \langle \mathbf{A}\mathbf{x}, \mathbf{A}\mathbf{x} \rangle = \langle \mathbf{A}^*\mathbf{A}\mathbf{x}, \mathbf{x} \rangle = \langle \mathbf{U}^*\mathbf{D}\mathbf{U}\mathbf{x}, \mathbf{x} \rangle = \langle \mathbf{D}\mathbf{U}\mathbf{x}, \mathbf{U}\mathbf{x} \rangle.$$

Since $\mathbf{U}$ is unitary, we have $\|\mathbf{U}\mathbf{x}\|_2^2 = \langle \mathbf{U}\mathbf{x}, \mathbf{U}\mathbf{x}\rangle = \langle \mathbf{x}, \mathbf{U}^*\mathbf{U}\mathbf{x}\rangle = \|\mathbf{x}\|_2^2 = 1$. Moreover, for an arbitrary vector $\mathbf{z} \in \mathbb{C}^n$,

$$\langle \mathbf{D}\mathbf{z}, \mathbf{z}\rangle = \sum_{j=1}^n \lambda_j |z_j|^2 \leq \max_{j\in[n]} \lambda_j \sum_{j=1}^n |z_j|^2 = \lambda_{\max}(\mathbf{A}^*\mathbf{A})\|\mathbf{z}\|_2^2.$$

Combining these facts establishes the inequality $\|\mathbf{A}\|_{2\to 2} \leq \sqrt{\lambda_{\max}(\mathbf{A}^*\mathbf{A})}$. Now choose $\mathbf{x}$ to be an eigenvector corresponding to the largest eigenvalue of $\mathbf{A}^*\mathbf{A}$, that is, $\mathbf{A}^*\mathbf{A}\mathbf{x} = \lambda_{\max}(\mathbf{A}^*\mathbf{A})\mathbf{x}$. Then

$$\|\mathbf{A}\mathbf{x}\|_2^2 = \langle \mathbf{A}\mathbf{x}, \mathbf{A}\mathbf{x}\rangle = \langle \mathbf{A}^*\mathbf{A}\mathbf{x}, \mathbf{x}\rangle = \lambda_{\max}(\mathbf{A}^*\mathbf{A})\langle \mathbf{x}, \mathbf{x}\rangle = \lambda_{\max}(\mathbf{A}^*\mathbf{A})\|\mathbf{x}\|_2^2.$$

This proves the reverse inequality $\|\mathbf{A}\|_{2\to 2} \geq \sqrt{\lambda_{\max}(\mathbf{A}^*\mathbf{A})}$. It is shown in Sect. A.2 that $\sigma_{\max}(\mathbf{A}) = \sqrt{\lambda_{\max}(\mathbf{A}^*\mathbf{A})}$, see (A.18). If $\mathbf{B}$ is self-adjoint, then its singular values satisfy $\{\sigma_j(\mathbf{B}), j \in [n]\} = \{|\lambda_j(\mathbf{B})|, j \in [n]\}$, so that $\|\mathbf{B}\|_{2\to 2} = \max_{j\in[n]} |\lambda_j(\mathbf{B})|$ as claimed.

(b) For $\mathbf{x} \in \mathbb{C}^n$ with $\|\mathbf{x}\|_1 = 1$, the triangle inequality gives

$$\|\mathbf{A}\mathbf{x}\|_p = \|\sum_{j=1}^n x_j \mathbf{a}_k\|_p \leq \sum_{k=1}^n |x_k| \|\mathbf{a}_k\|_p \leq \|\mathbf{x}\|_1 \max_{k\in[n]} \|\mathbf{a}_k\|_p. \qquad (A.12)$$

This shows that $\|\mathbf{A}\|_{1\to p} \leq \max_{k\in[n]} \|\mathbf{a}_k\|_p$. For the reverse inequality, we choose the canonical unit vector $\mathbf{x} = \mathbf{e}_{k_0}$ with k_0 being the index that realizes the previous maximum. Then $\|\mathbf{A}\mathbf{x}\|_p = \|\mathbf{a}_{k_0}\|_p = \max_{k\in[n]} \|\mathbf{a}_k\|_p$. This establishes the statement (b).

(c) For $\mathbf{x} \in \mathbb{C}^n$ with $\|\mathbf{x}\|_\infty = 1$, we have

$$\|\mathbf{A}\mathbf{x}\|_\infty = \max_{j\in[m]} |\sum_{k=1}^n A_{j,k} x_k| \leq \max_{j\in[m]} \sum_{k=1}^n |A_{j,k}||x_k| \leq \|\mathbf{x}\|_\infty \max_{j\in[m]} \sum_{k=1}^n |A_{j,k}|.$$

To see that this inequality is sharp in general, we choose an index $j \in [m]$ that realizes the maximum in the previous expression and set $x_k = \mathrm{sgn}(\overline{A_{j,k}}) = \overline{A_{j,k}}$ $/|A_{j,k}|$ if $A_{j,k} \neq 0$ and $x_k = 0$ if $A_{j,k} = 0$. Then $\|\mathbf{x}\|_\infty = 1$ (unless $\mathbf{A} = \mathbf{0}$, in which case the statement is trivial), and

$$(\mathbf{A}\mathbf{x})_j = \sum_{k=1}^n A_{j,k} x_k = \sum_{k=1}^n |A_{j,k}| = \max_{j\in[m]} \sum_{k=1}^n |A_{j,k}|.$$

Together with the inequality established above, this shows the claim. $\square$

Remark A.6. (a) The general identity

$$\|\mathbf{A}\|_{p \to q} = \|\mathbf{A}^*\|_{q' \to p'}$$

where $1/p + 1/p' = 1 = 1/q + 1/q'$ shows that (c) (and more general statements) can be deduced from (b).

(b) Computing the operator norms $\|\mathbf{A}\|_{\infty \to 1}$, $\|\mathbf{A}\|_{2 \to 1}$, and $\|\mathbf{A}\|_{\infty \to 2}$ is an NP-hard problem; see [426]. (Although the case $\|\mathbf{A}\|_{\infty \to 2} = \|\mathbf{A}\|_{2 \to 1}$ is not treated explicitly in [426], it follows from similar considerations as for $\|\mathbf{A}\|_{\infty \to 1}$.)

Lemma A.7. *For* $\mathbf{A} \in \mathbb{C}^{m \times n}$,

$$\|\mathbf{A}\|_{2 \to 2} = \sup_{\|\mathbf{y}\|_2 \leq 1} \sup_{\|\mathbf{x}\|_2 \leq 1} |\langle \mathbf{A}\mathbf{x}, \mathbf{y} \rangle| = \sup_{\|\mathbf{y}\|_2 \leq 1} \sup_{\|\mathbf{x}\|_2 \leq 1} \mathrm{Re}\,\langle \mathbf{A}\mathbf{x}, \mathbf{y} \rangle. \qquad (A.13)$$

If $\mathbf{B} \in \mathbb{C}^{n \times n}$ *is self-adjoint, then*

$$\|\mathbf{B}\|_{2 \to 2} = \sup_{\|\mathbf{x}\|_2 \leq 1} |\langle \mathbf{B}\mathbf{x}, \mathbf{x} \rangle|.$$

Proof. The first statement follows from (A.8) with $p = q = 2$ and (A.6). For the second statement, let $\mathbf{B} = \mathbf{U}^*\mathbf{D}\mathbf{U}$ be the eigenvalue decomposition of $\mathbf{B}$ with a unitary matrix $\mathbf{U}$ and a diagonal matrix $\mathbf{D}$ with the eigenvalues λ_j of $\mathbf{B}$ on the diagonal. Then

$$\sup_{\|\mathbf{x}\|_2 \leq 1} |\langle \mathbf{B}\mathbf{x}, \mathbf{x} \rangle| = \sup_{\|\mathbf{x}\|_2 \leq 1} |\langle \mathbf{U}^*\mathbf{D}\mathbf{U}\mathbf{x}, \mathbf{x} \rangle| = \sup_{\|\mathbf{x}\|_2 \leq 1} |\langle \mathbf{D}\mathbf{U}\mathbf{x}, \mathbf{U}\mathbf{x} \rangle|$$

$$= \sup_{\|\mathbf{x}\|_2 \leq 1} |\langle \mathbf{D}\mathbf{x}, \mathbf{x} \rangle| = \sup_{\|\mathbf{x}\|_2 \leq 1} \left| \sum_{j=1}^{n} \lambda_j |x_j|^2 \right| = \max_{j \in [n]} |\lambda_j|$$

$$= \|\mathbf{B}\|_{2 \to 2},$$

where the last step used the second part of Lemma A.5(a). For the identity $\sup_{\|\mathbf{x}\|_2 \leq 1} |\sum_{j=1}^{n} \lambda_j |x_j|^2| = \max_{j \in [n]} |\lambda_j|$ above, we observe on the one hand

$$\left| \sum_{j=1}^{n} \lambda_j |x_j|^2 \right| \leq \max_{j \in [n]} |\lambda_j| \sum_{j=1}^{n} |x_j|^2 \leq \max_{j \in [n]} |\lambda_j|.$$

One the other hand, if $\mathbf{x} = \mathbf{e}_{j_0}$ is the canonical unit vector corresponding to the index j_0 where $|\lambda_j|$ is maximal, we have $|\sum_{j=1}^{n} \lambda_j |x_j|^2| = |\lambda_{j_0}| = \max_{j \in [n]} |\lambda_j|$. This point completes the proof. □

Specializing the above identity to the rank-one matrix $\mathbf{B} = \mathbf{u}\mathbf{u}^*$ yields, for any vector $\mathbf{u} \in \mathbb{C}^n$,

$$\|\mathbf{u}\mathbf{u}^*\|_{2\to 2} = \sup_{\|\mathbf{x}\|_2 \le 1} |\langle \mathbf{u}\mathbf{u}^*\mathbf{x}, \mathbf{x}\rangle| = \sup_{\|\mathbf{x}\|_2 \le 1} |\langle \mathbf{u}^*\mathbf{x}, \mathbf{u}^*\mathbf{x}\rangle|$$

$$= \sup_{\|\mathbf{x}\|_2 \le 1} |\langle \mathbf{x}, \mathbf{u}\rangle|^2 = \|\mathbf{u}\|_2^2, \tag{A.14}$$

where we also applied (A.6).

The next statement is known as Schur test.

Lemma A.8. *For* $\mathbf{A} \in \mathbb{C}^{m\times n}$,

$$\|\mathbf{A}\|_{2\to 2} \le \sqrt{\|\mathbf{A}\|_{1\to 1}\|\mathbf{A}\|_{\infty\to\infty}}.$$

In particular, for a self-adjoint matrix $\mathbf{B} \in \mathbb{C}^{n\times n}$,

$$\|\mathbf{B}\|_{2\to 2} \le \|\mathbf{B}\|_{1\to 1}.$$

Proof. The statement follows immediately from the Riesz–Thorin interpolation theorem. For readers not familiar with interpolation theory, we give a more elementary proof. By the Cauchy–Schwarz inequality, the jth entry of $\mathbf{A}\mathbf{x}$ satisfies

$$|(\mathbf{A}\mathbf{x})_j| \le \sum_{k=1}^{n} |x_k||A_{j,k}| \le \Big(\sum_{k=1}^{n} |x_k|^2 |A_{j,k}|\Big)^{1/2} \Big(\sum_{\ell=1}^{n} |A_{j,\ell}|\Big)^{1/2}.$$

Squaring and summing this inequality yields

$$\|\mathbf{A}\mathbf{x}\|_2^2 = \sum_{j=1}^{m} |(\mathbf{A}\mathbf{x})_j|^2 \le \sum_{j=1}^{m} \Big(\sum_{k=1}^{n} |x_k|^2 |A_{j,k}|\Big)\Big(\sum_{\ell=1}^{n} |A_{j,\ell}|\Big)$$

$$\le \max_{j\in[n]} \sum_{\ell=1}^{n} |A_{j,\ell}| \sum_{j=1}^{m} \sum_{k=1}^{n} |x_k|^2 |A_{j,k}|$$

$$\le \max_{j\in[n]} \sum_{\ell=1}^{n} |A_{j,\ell}| \max_{k\in[m]} \sum_{j=1}^{m} |A_{j,k}| \sum_{k=1}^{n} |x_k|^2$$

$$= \|\mathbf{A}\|_{\infty\to\infty}\|\mathbf{A}\|_{1\to 1}\|\mathbf{x}\|_2^2.$$

This establishes the first claim. If $\mathbf{B}$ is self-adjoint, then $\|\mathbf{B}\|_{1\to 1} = \|\mathbf{B}\|_{\infty\to\infty}$ by Lemma A.5(b)–(c) or Remark A.6(a). This implies the second claim. $\square$

We note that the above inequality may be crude for some important matrices encountered in this book. For a general matrix, however, it cannot be improved.

Lemma A.9. *The operator norm* $\|\cdot\|_{2\to 2}$ *of a submatrix is bounded by one of the whole matrix. More precisely, if* $\mathbf{A} \in \mathbb{C}^{m\times n}$ *has the form*

$$\mathbf{A} = \begin{bmatrix} \mathbf{A}^{(1)} & \mathbf{A}^{(2)} \\ \mathbf{A}^{(3)} & \mathbf{A}^{(4)} \end{bmatrix}$$

for matrices $\mathbf{A}^{(\ell)}$, *then* $\|\mathbf{A}^{(\ell)}\|_{2\to 2} \le \|\mathbf{A}\|_{2\to 2}$ *for* $\ell = 1, 2, 3, 4$. *In particular, any entry of* $\mathbf{A}$ *satisfies* $|A_{j,k}| \le \|\mathbf{A}\|_{2\to 2}$.

Proof. We give the proof for $\mathbf{A}^{(1)}$. The other cases are analogous. Let $\mathbf{A}^{(1)}$ be of size $m_1 \times n_1$. Then for $\mathbf{x}^{(1)} \in \mathbb{C}^{n_1}$, we have

$$\|\mathbf{A}^{(1)}\mathbf{x}^{(1)}\|_2^2 \le \|\mathbf{A}^{(1)}\mathbf{x}^{(1)}\|_2^2 + \|\mathbf{A}^{(3)}\mathbf{x}^{(1)}\|_2^2 = \left\| \begin{pmatrix} \mathbf{A}^{(1)} \\ \mathbf{A}^{(3)} \end{pmatrix} \mathbf{x}^{(1)} \right\|_2^2 = \left\| \mathbf{A} \begin{pmatrix} \mathbf{x}^{(1)} \\ \mathbf{0} \end{pmatrix} \right\|_2^2.$$

The set T_1 of vectors $\begin{pmatrix} \mathbf{x}^{(1)} \\ \mathbf{0} \end{pmatrix} \in \mathbb{C}^n$ with $\|\mathbf{x}^{(1)}\|_2 \le 1$ is contained in the set $T := \{\mathbf{x} \in \mathbb{C}^n, \|\mathbf{x}\|_2 \le 1\}$. Therefore, the supremum over $\mathbf{x}^{(1)} \in T_1$ above is bounded by $\sup_{\mathbf{x} \in T} \|\mathbf{A}\mathbf{x}\|_2^2 = \|\mathbf{A}\|_{2\to 2}^2$. This concludes the proof.   □

Remark A.10. The same result and proof hold for the operator norms $\|\cdot\|_{p\to q}$ with $1 \le p, q \le \infty$.

Gershgorin's disk theorem stated next provides information about the locations of the eigenvalues of a square matrix.

Theorem A.11. *Let* λ *be an eigenvalue of a square matrix* $\mathbf{A} \in \mathbb{C}^{n \times n}$. *There exists an index* $j \in [n]$ *such that*

$$|\lambda - A_{j,j}| \le \sum_{\ell \in [n] \setminus \{j\}} |A_{j,\ell}|.$$

Proof. Let $\mathbf{u} \in \mathbb{C}^n \setminus \{\mathbf{0}\}$ be an eigenvector associated with λ, and let $j \in [n]$ such that $|u_j|$ is maximal, i.e., $|u_j| = \|\mathbf{u}\|_\infty$. Then $\sum_{\ell \in [n]} A_{j,\ell} u_\ell = \lambda u_j$, and a rearrangement gives $\sum_{\ell \in [n] \setminus \{j\}} A_{j,\ell} u_\ell = \lambda u_j - A_{j,j} u_j$. The triangle inequality yields

$$|\lambda - A_{j,j}| |u_j| \le \sum_{\ell \in [n] \setminus \{j\}} |A_{j,\ell}| |u_\ell| \le \|\mathbf{u}\|_\infty \sum_{\ell \in [n] \setminus \{j\}} |A_{j,\ell}| = |u_j| \sum_{\ell \in [n] \setminus \{j\}} |A_{j,\ell}|.$$

Dividing by $|u_j| > 0$ yields the desired statement.   □

More information on Gershgorin's theorem and its variations can be found, for instance, in the monograph [496].

The trace of a square matrix $\mathbf{B} \in \mathbb{C}^{n \times n}$ is the sum of its diagonal elements, i.e.,

$$\operatorname{tr}(\mathbf{B}) = \sum_{j=1}^{n} B_{jj}.$$

The trace is cyclic, i.e., $\operatorname{tr}(\mathbf{A}\mathbf{B}) = \operatorname{tr}(\mathbf{B}\mathbf{A})$ for all $\mathbf{A} \in \mathbb{C}^{m \times n}$ and $\mathbf{B} \in \mathbb{C}^{n \times m}$. It induces an inner product defined on $\mathbb{C}^{m \times n}$ by

$$\langle \mathbf{A}, \mathbf{B} \rangle_F := \operatorname{tr}(\mathbf{A}\mathbf{B}^*). \tag{A.15}$$

The Frobenius norm of a matrix $\mathbf{A} \in \mathbb{C}^{m \times n}$ is then

$$\|\mathbf{A}\|_F := \sqrt{\operatorname{tr}(\mathbf{A}\mathbf{A}^*)} = \sqrt{\operatorname{tr}(\mathbf{A}^*\mathbf{A})} = \left(\sum_{j \in [m], k \in [n]} |A_{j,k}|^2 \right)^{1/2}. \tag{A.16}$$

After identifying matrices on $\mathbb{C}^{m \times n}$ with vectors in $\mathbb{C}^{mn}$, the Frobenius norm can be interpreted as an ℓ_2-norm on $\mathbb{C}^{mn}$.

The operator norm on ℓ_2 is bounded by the Frobenius norm, i.e.,

$$\|\mathbf{A}\|_{2 \to 2} \leq \|\mathbf{A}\|_F. \tag{A.17}$$

Indeed, for $\mathbf{x} \in \mathbb{C}^n$, the Cauchy–Schwarz inequality yields

$$\|\mathbf{A}\mathbf{x}\|_2^2 = \sum_{j=1}^m \left(\sum_{k=1}^n A_{j,k} x_j \right)^2 \leq \sum_{j=1}^m \left(\sum_{k=1}^n |x_j|^2 \right) \left(\sum_{\ell=1}^n |A_{j,\ell}|^2 \right) = \|\mathbf{A}\|_F^2 \|\mathbf{x}\|_2^2.$$

Next we state a bound for the operator norm of the inverse of a square matrix.

Lemma A.12. *Suppose that* $\mathbf{B} \in \mathbb{C}^{n \times n}$ *satisfies*

$$\|\mathbf{B} - \mathbf{Id}\|_{2 \to 2} \leq \eta$$

for some $\eta \in [0, 1)$. *Then* $\mathbf{B}$ *is invertible and* $\|\mathbf{B}^{-1}\|_{2 \to 2} \leq (1 - \eta)^{-1}$.

Proof. We first note that, with $\mathbf{H} = \mathbf{Id} - \mathbf{B}$, the Neumann series $\sum_{k=0}^{\infty} \mathbf{H}^k$ converges. Indeed, by the triangle inequality and the fact that $\|\mathbf{H}^k\|_{2 \to 2} \leq \|\mathbf{H}\|_{2 \to 2}^k$ derived from (A.9), we have

$$\left\| \sum_{k=0}^{\infty} \mathbf{H}^k \right\|_{2 \to 2} \leq \sum_{k=0}^{\infty} \|\mathbf{H}\|_{2 \to 2}^k \leq \sum_{k=0}^{\infty} \eta^k = \frac{1}{1 - \eta}.$$

Now we observe that

$$(\mathbf{Id} - \mathbf{H}) \sum_{k=0}^{\infty} \mathbf{H}^k = \sum_{k=0}^{\infty} \mathbf{H}^k - \sum_{k=1}^{\infty} \mathbf{H}^k = \mathbf{Id}$$

by convergence of the Neumann series, and similarly $\sum_{k=0}^{\infty} \mathbf{H}^k(\mathbf{Id} - \mathbf{H}) = \mathbf{Id}$. Therefore, the matrix $\mathbf{Id} - \mathbf{H} = \mathbf{B}$ is invertible and

$$\mathbf{B}^{-1} = (\mathbf{Id} - \mathbf{H})^{-1} = \sum_{k=0}^{\infty} \mathbf{H}^k.$$

This establishes the claim. $\qquad\qquad\qquad\qquad\qquad\qquad\qquad\qquad\qquad\qquad\qquad\square$

A.2 The Singular Value Decomposition

While the concept of eigenvalues and eigenvectors applies only to square matrices, every (possibly rectangular) matrix possesses a singular value decomposition.

Proposition A.13. *For* $\mathbf{A} \in \mathbb{C}^{m \times n}$, *there exist unitary matrices* $\mathbf{U} \in \mathbb{C}^{m \times m}$, $\mathbf{V} \in \mathbb{C}^{n \times n}$, *and uniquely defined nonnegative numbers* $\sigma_1 \geq \sigma_2 \geq \cdots \geq \sigma_{\min\{m,n\}} \geq 0$, *called singular values of* $\mathbf{A}$, *such that*

$$\mathbf{A} = \mathbf{U}\boldsymbol{\Sigma}\mathbf{V}^*, \qquad \boldsymbol{\Sigma} = \mathrm{diag}[\sigma_1, \ldots, \sigma_{\min\{m,n\}}] \in \mathbb{R}^{m \times n}.$$

Remark A.14. Writing $\mathbf{U} = [\mathbf{u}_1 | \cdots | \mathbf{u}_m]$ and $\mathbf{V} = [\mathbf{v}_1 | \cdots | \mathbf{v}_n]$, the vectors $\mathbf{u}_\ell$ are called left singular vectors, while the $\mathbf{v}_\ell$ are called right singular vectors.

Proof. Let $\mathbf{v}_1 \in \mathbb{C}^n$ be a vector with $\|\mathbf{v}_1\|_2 = 1$ that realizes the maximum in the definition (A.8) of the operator norm $\|\mathbf{A}\|_{2\to 2}$, and set

$$\sigma_1 = \|\mathbf{A}\mathbf{v}_1\|_2 = \|\mathbf{A}\mathbf{v}_1\|_2.$$

By compactness of the sphere $S^{n-1} = \{\mathbf{x} \in \mathbb{C}^n, \|\mathbf{x}\|_2 = 1\}$, such a vector $\mathbf{v}_1$ always exists. If $\sigma_1 = 0$, then $\mathbf{A} = \mathbf{0}$, and we set $\sigma_\ell = 0$ for all $\ell \in [\min\{m,n\}]$, and $\mathbf{U}, \mathbf{V}$ are arbitrary unitary matrices. Therefore, we assume $\sigma_1 > 0$ and set

$$\mathbf{u}_1 = \sigma_1^{-1}\mathbf{A}\mathbf{v}_1.$$

We can extend $\mathbf{u}_1, \mathbf{v}_1$ to orthonormal bases in order to form unitary matrices $\mathbf{U}_1 = [\mathbf{u}_1|\tilde{\mathbf{U}}_1]$, $\mathbf{V}_1 = [\mathbf{v}_1|\tilde{\mathbf{V}}_1]$. Since $\tilde{\mathbf{U}}_1^*\mathbf{A}\mathbf{v}_1 = \sigma_1\tilde{\mathbf{U}}_1^*\mathbf{u}_1 = \mathbf{0}$, the matrix $\mathbf{A}_1 = \mathbf{U}_1^*\mathbf{A}\mathbf{V}_1$ takes the form

$$\mathbf{A}_1 = \begin{bmatrix} \sigma_1 & \mathbf{b}^* \\ \mathbf{0} & \mathbf{B} \end{bmatrix},$$

where $\mathbf{b}^* = \mathbf{u}_1^*\mathbf{A}\tilde{\mathbf{V}}_1$ and $\mathbf{B} = \tilde{\mathbf{U}}_1^*\mathbf{A}\tilde{\mathbf{V}}_1 \in \mathbb{C}^{(m-1)\times(n-1)}$. It follows from

$$\|\mathbf{A}_1\|_{2\to 2}\sqrt{\sigma_1^2 + \|\mathbf{b}\|_2^2} \geq \left\|\mathbf{A}_1\begin{pmatrix}\sigma_1 \\ \mathbf{b}\end{pmatrix}\right\|_2 = \left\|\begin{pmatrix}\sigma_1^2 + \|\mathbf{b}\|_2^2 \\ \mathbf{Bb}\end{pmatrix}\right\|_2 \geq \sigma_1^2 + \|\mathbf{b}\|_2^2$$

that $\|\mathbf{A}_1\|_{2\to 2} \geq \sqrt{\sigma_1^2 + \|\mathbf{b}\|_2^2}$. But since $\mathbf{U}_1, \mathbf{V}_1$ are unitary, we have $\|\mathbf{A}_1\|_{2\to 2} = \|\mathbf{A}\|_{2\to 2} = \sigma_1$, and therefore, $\mathbf{b} = \mathbf{0}$. In conclusion,

$$\mathbf{A}_1 = \mathbf{U}_1^*\mathbf{A}\mathbf{V}_1 = \begin{pmatrix} \sigma_1 & \mathbf{0} \\ \mathbf{0} & \mathbf{B} \end{pmatrix}.$$

With the same arguments, we can further decompose $\mathbf{B} \in \mathbb{C}^{(m-1)\times(n-1)}$, and by induction we arrive at the stated singular value decomposition. □

The previous proposition reveals that the largest and smallest singular values satisfy

$$\sigma_{\max}(\mathbf{A}) = \sigma_1(\mathbf{A}) = \|\mathbf{A}\|_{2\to 2} = \max_{\|\mathbf{x}\|_2=1} \|\mathbf{A}\mathbf{x}\|_2\,,$$

$$\sigma_{\min}(\mathbf{A}) = \sigma_{\min\{m,n\}}(\mathbf{A}) = \min_{\|\mathbf{x}\|_2=1} \|\mathbf{A}\mathbf{x}\|_2.$$

If $\mathbf{A}$ has rank r, then its r largest singular values $\sigma_1 \geq \cdots \geq \sigma_r$ are positive, while $\sigma_{r+1} = \sigma_{r+2} = \cdots = 0$. Sometimes it is more convenient to work with the reduced singular value decomposition. For $\mathbf{A}$ of rank r with (full) singular value decomposition $\mathbf{A} = \mathbf{U}\mathbf{\Sigma}\mathbf{V}^*$, we consider $\tilde{\mathbf{\Sigma}} = \text{diag}[\sigma_1,\ldots,\sigma_r] \in \mathbb{R}^{r\times r}$ and the submatrices $\tilde{\mathbf{U}} = [\mathbf{u}_1|\cdots|\mathbf{u}_r] \in \mathbb{C}^{m\times r}$, $\tilde{\mathbf{V}} = [\mathbf{v}_1|\cdots|\mathbf{v}_r] \in \mathbb{C}^{n\times r}$ of $\mathbf{U} = [\mathbf{u}_1|\cdots|\mathbf{u}_m]$, $\mathbf{V} = [\mathbf{v}_1|\cdots|\mathbf{v}_n]$. We have

$$\mathbf{A} = \tilde{\mathbf{U}}\tilde{\mathbf{\Sigma}}\tilde{\mathbf{V}}^* = \sum_{j=1}^{r} \sigma_j \mathbf{u}_j \mathbf{v}_j^*.$$

Given $\mathbf{A} \in \mathbb{C}^{m\times n}$ with reduced singular value decomposition $\mathbf{A} = \tilde{\mathbf{U}}\tilde{\mathbf{\Sigma}}\tilde{\mathbf{V}}^*$, we observe that

$$\mathbf{A}^*\mathbf{A} = \tilde{\mathbf{V}}\tilde{\mathbf{\Sigma}}\tilde{\mathbf{U}}^*\tilde{\mathbf{U}}\tilde{\mathbf{\Sigma}}\tilde{\mathbf{V}}^* = \tilde{\mathbf{V}}\tilde{\mathbf{\Sigma}}^2\tilde{\mathbf{V}}^*,$$

$$\mathbf{A}\mathbf{A}^* = \tilde{\mathbf{U}}\tilde{\mathbf{\Sigma}}\tilde{\mathbf{V}}^*\tilde{\mathbf{V}}\tilde{\mathbf{\Sigma}}\tilde{\mathbf{U}}^* = \tilde{\mathbf{U}}\tilde{\mathbf{\Sigma}}^2\tilde{\mathbf{U}}^*.$$

Thus, we obtain the (reduced) eigenvalue decompositions of $\mathbf{A}^*\mathbf{A}$ and $\mathbf{A}\mathbf{A}^*$. In particular, the singular values $\sigma_j = \sigma_j(\mathbf{A})$ satisfy

$$\sigma_j(\mathbf{A}) = \sqrt{\lambda_j(\mathbf{A}^*\mathbf{A})} = \sqrt{\lambda_j(\mathbf{A}\mathbf{A}^*)}, \qquad j \in [\min\{m,n\}], \qquad (\text{A.18})$$

where $\lambda_1(\mathbf{A}^*\mathbf{A}) \geq \lambda_2(\mathbf{A}^*\mathbf{A}) \geq \cdots$ are the eigenvalues of $\mathbf{A}^*\mathbf{A}$ arranged in nonincreasing order. Moreover, the left and right singular vectors listed in $\mathbf{U}, \mathbf{V}$ can be obtained from the eigenvalue decomposition of the positive semidefinite matrices $\mathbf{A}^*\mathbf{A}$ and $\mathbf{A}\mathbf{A}^*$. (One can also prove the existence of the singular value decomposition via the eigenvalue decompositions of $\mathbf{A}^*\mathbf{A}$ and $\mathbf{A}\mathbf{A}^*$.)

For the purpose of this book, the following observation is very useful.

Proposition A.15. *Let* $\mathbf{A} \in \mathbb{C}^{m \times n}$, $m \geq n$. *If*

$$\|\mathbf{A}^*\mathbf{A} - \mathbf{Id}\|_{2 \to 2} \leq \delta \tag{A.19}$$

for some $\delta \in [0, 1]$, *then the largest and smallest singular values of* $\mathbf{A}$ *satisfy*

$$\sigma_{\max}(\mathbf{A}) \leq \sqrt{1 + \delta}, \qquad \sigma_{\min}(\mathbf{A}) \geq \sqrt{1 - \delta}. \tag{A.20}$$

Conversely, if both inequalities in (A.20) *hold, then* (A.19) *follows.*

Proof. By (A.18) the eigenvalues $\lambda_j(\mathbf{A}^*\mathbf{A})$ of $\mathbf{A}^*\mathbf{A}$ are the squared singular values $\sigma_j^2(\mathbf{A})$ of $\mathbf{A}$, $j \in [n]$. Thus, the eigenvalues of $\mathbf{A}^*\mathbf{A} - \mathbf{Id}$ are given by $\sigma_j^2(\mathbf{A}) - 1$, $j \in [n]$, and Lemma A.5(a) yields

$$\max\{\sigma_{\max}^2(\mathbf{A}) - 1, 1 - \sigma_{\min}^2(\mathbf{A})\} = \|\mathbf{A}^*\mathbf{A} - \mathbf{Id}\|_{2 \to 2}.$$

This establishes the claim. □

The largest and smallest singular values are 1-Lipschitz functions with respect to the operator norm and the Frobenius norm.

Proposition A.16. *The smallest and largest singular values* $\sigma_{\min}$ *and* $\sigma_{\max}$ *satisfy, for all matrices* $\mathbf{A}, \mathbf{B}$ *of equal dimensions,*

$$|\sigma_{\max}(\mathbf{A}) - \sigma_{\max}(\mathbf{B})| \leq \|\mathbf{A} - \mathbf{B}\|_{2 \to 2} \leq \|\mathbf{A} - \mathbf{B}\|_F, \tag{A.21}$$

$$|\sigma_{\min}(\mathbf{A}) - \sigma_{\min}(\mathbf{B})| \leq \|\mathbf{A} - \mathbf{B}\|_{2 \to 2} \leq \|\mathbf{A} - \mathbf{B}\|_F. \tag{A.22}$$

Proof. By the identification of the largest singular value with the operator norm, we have

$$|\sigma_{\max}(\mathbf{A}) - \sigma_{\max}(\mathbf{B})| = |\|\mathbf{A}\|_{2 \to 2} - \|\mathbf{B}\|_{2 \to 2}| \leq \|\mathbf{A} - \mathbf{B}\|_{2 \to 2}.$$

The inequality for the smallest singular is deduced as follows:

$$\sigma_{\min}(\mathbf{A}) = \inf_{\|\mathbf{x}\|_2 = 1} \|\mathbf{A}\mathbf{x}\|_2 \leq \inf_{\|\mathbf{x}\|_2 = 1} (\|\mathbf{B}\mathbf{x}\|_2 + \|(\mathbf{A} - \mathbf{B})\mathbf{x}\|_2)$$

$$\leq \inf_{\|\mathbf{x}\|_2 = 1} (\|\mathbf{B}\mathbf{x}\|_2 + \|\mathbf{A} - \mathbf{B}\|_{2 \to 2}) = \sigma_{\min}(\mathbf{B}) + \|\mathbf{A} - \mathbf{B}\|_{2 \to 2}.$$

Therefore, we have $\sigma_{\min}(\mathbf{A}) - \sigma_{\min}(\mathbf{B}) \leq \|\mathbf{A} - \mathbf{B}\|_{2 \to 2}$, and (A.22) follows by symmetry. The estimates by the Frobenius norm in (A.21) and (A.22) follow from the domination (A.17) of the operator norm by the Frobenius norm. □

The singular values $\sigma_1(\mathbf{A}) \geq \cdots \geq \sigma_{\min\{m,n\}}(\mathbf{A}) \geq 0$ of a matrix $\mathbf{A} \in \mathbb{C}^{m \times n}$ obey the useful variational characterization

$$\sigma_k(\mathbf{A}) = \max_{\substack{\mathcal{M} \subset \mathbb{C}^n \\ \dim \mathcal{M} = k}} \min_{\substack{\mathbf{x} \in \mathcal{M} \\ \|\mathbf{x}\|_2 = 1}} \|\mathbf{A}\mathbf{x}\|_2.$$

This follows from a characterization of the eigenvalues $\lambda_1(\mathbf{A}) \geq \cdots \geq \lambda_n(\mathbf{A})$ of a self-adjoint matrix $\mathbf{A} \in \mathbb{C}^{n \times n}$ often referred to as *Courant–Fischer minimax theorem* or simply *minimax principle*, namely

$$\lambda_k(\mathbf{A}) = \max_{\substack{\mathcal{M} \subset \mathbb{C}^n \\ \dim \mathcal{M} = k}} \min_{\substack{\mathbf{x} \in \mathcal{M} \\ \|\mathbf{x}\|_2 = 1}} \langle \mathbf{A}\mathbf{x}, \mathbf{x} \rangle. \tag{A.23}$$

With $(\mathbf{u}_1, \ldots, \mathbf{u}_n)$ denoting an orthonormal basis of eigenvectors for the eigenvalues $\lambda_1(\mathbf{A}) \geq \cdots \geq \lambda_n(\mathbf{A})$, the fact that $\lambda_k(\mathbf{A})$ is not larger than the right-hand side of (A.23) follows by taking $\mathcal{M} = \mathrm{span}(\mathbf{u}_1, \ldots, \mathbf{u}_k)$, so that, if $\mathbf{x} = \sum_{j=1}^{k} c_j \mathbf{u}_j \in \mathcal{M}$ has unit norm, then

$$\langle \mathbf{A}\mathbf{x}, \mathbf{x} \rangle = \sum_{j=1}^{k} \lambda_j(\mathbf{A}) c_j^2 \geq \lambda_k(\mathbf{A}) \sum_{j=1}^{k} c_j^2 = \lambda_k(\mathbf{A}) \|\mathbf{x}\|_2^2 = \lambda_k(\mathbf{A}).$$

For the fact that $\lambda_k(\mathbf{A})$ is not smaller than the right-hand side of (A.23), given a k-dimensional subspace $\mathcal{M}$ of $\mathbb{C}^n$, we choose a unit-normed vector $\mathbf{x} \in \mathcal{M} \cap \mathrm{span}(\mathbf{u}_k, \ldots, \mathbf{u}_n)$, so that, with $\mathbf{x} = \sum_{j=1}^{k} c_j \mathbf{u}_j$,

$$\langle \mathbf{A}\mathbf{x}, \mathbf{x} \rangle = \sum_{j=k}^{n} \lambda_j(\mathbf{A}) c_j^2 \leq \lambda_k(\mathbf{A}) \sum_{j=k}^{n} c_j^2 = \lambda_k(\mathbf{A}) \|\mathbf{x}\|_2^2 = \lambda_k(\mathbf{A}).$$

The characterization (A.23) generalizes to *Wielandt's minimax principle* for sums of eigenvalues. The latter states that, for any $1 \leq i_1 < \cdots < i_k \leq n$,

$$\sum_{j=1}^{k} \lambda_{i_j}(\mathbf{A}) = \max_{\substack{\mathcal{M}_1 \subset \cdots \subset \mathcal{M}_k \subset \mathbb{C}^n \\ \dim \mathcal{M}_j = i_j}} \min_{\substack{(\mathbf{x}_1, \ldots, \mathbf{x}_k) \text{ orthonormal} \\ \mathbf{x}_j \in \mathcal{M}_j}} \sum_{j=1}^{k} \langle \mathbf{A}\mathbf{x}_j, \mathbf{x}_j \rangle.$$

We refer to [47] for a proof. Next we state *Lidskii's inequality* (which reduces to the so-called *Weyl's inequality* in the case $k = 1$). It can be proved using Wielandt's minimax principle, but we prefer to give a direct proof below.

Proposition A.17. *Let* $\mathbf{A}, \mathbf{B} \in \mathbb{C}^{n \times n}$ *be two self-adjoint matrices, and let* $(\lambda_j(\mathbf{A}))_{j \in [n]}, (\lambda_j(\mathbf{B}))_{j \in [n]}, (\lambda_j(\mathbf{A} + \mathbf{B}))_{j \in [n]}$ *denote the eigenvalues of* $\mathbf{A}$, $\mathbf{B}$, *and* $\mathbf{A} + \mathbf{B}$ *arranged in nonincreasing order. For any* $1 \leq i_1 < \cdots < i_k \leq n$,

$$\sum_{j=1}^{k} \lambda_{i_j}(\mathbf{A} + \mathbf{B}) \leq \sum_{j=1}^{k} \lambda_{i_j}(\mathbf{A}) + \sum_{i=1}^{k} \lambda_i(\mathbf{B}). \tag{A.24}$$

Proof. Since the prospective inequality (A.24) is invariant under the change $\mathbf{B} \leftrightarrow \mathbf{B} - c\,\mathrm{Id}$, we may and do assume that $\lambda_{k+1}(\mathbf{B}) = 0$. Then, from the unitarily diagonalized form of $\mathbf{B}$, namely,

$$\mathbf{B} = \mathbf{U}\,\mathrm{diag}\left[\lambda_1(\mathbf{B}), \ldots, \lambda_k(\mathbf{B}), \lambda_{k+1}(\mathbf{B}), \ldots, \lambda_n(\mathbf{B})\right]\mathbf{U}^*,$$

we define the positive semidefinite matrix $\mathbf{B}^+ \in \mathbb{C}^{n \times n}$ as

$$\mathbf{B}^+ := \mathbf{U}\,\mathrm{diag}\left[\lambda_1(\mathbf{B}), \ldots, \lambda_k(\mathbf{B}), 0, \ldots, 0\right]\mathbf{U}^*.$$

We notice that $\mathbf{A} + \mathbf{B} \preccurlyeq \mathbf{A} + \mathbf{B}^+$ and $\mathbf{A} \preccurlyeq \mathbf{A} + \mathbf{B}^+$. Hence, the minimax characterization (A.23) implies that, for all $i \in [n]$,

$$\lambda_i(\mathbf{A} + \mathbf{B}) \leq \lambda_i(\mathbf{A} + \mathbf{B}^+) \quad \text{and} \quad \lambda_i(\mathbf{A}) \leq \lambda_i(\mathbf{A} + \mathbf{B}^+).$$

It follows that

$$\sum_{j=1}^{k}\left(\lambda_{i_j}(\mathbf{A} + \mathbf{B}) - \lambda_{i_j}(\mathbf{A})\right) \leq \sum_{j=1}^{k}\left(\lambda_{i_j}(\mathbf{A} + \mathbf{B}^+) - \lambda_{i_j}(\mathbf{A})\right)$$

$$\leq \sum_{i=1}^{n}\left(\lambda_i(\mathbf{A} + \mathbf{B}^+) - \lambda_i(\mathbf{A})\right) = \mathrm{tr}\,(\mathbf{A} + \mathbf{B}^+) - \mathrm{tr}\,(\mathbf{A}) = \mathrm{tr}\,(\mathbf{B}^+) = \sum_{i=1}^{k}\lambda_i(\mathbf{B}).$$

This is the desired inequality. $\square$

As a consequence of Proposition A.17, we establish the following lemma.

Lemma A.18. *If the matrices* $\mathbf{X} \in \mathbb{C}^{m \times n}$ *and* $\mathbf{Y} \in \mathbb{C}^{m \times n}$ *have the singular values* $\sigma_1(\mathbf{X}) \geq \cdots \geq \sigma_\ell(\mathbf{X}) \geq 0$ *and* $\sigma_1(\mathbf{Y}) \geq \cdots \geq \sigma_\ell(\mathbf{Y}) \geq 0$, *where* $\ell := \min\{m, n\}$, *then, for any* $k \in [\ell]$,

$$\sum_{j=1}^{k}|\sigma_j(\mathbf{X}) - \sigma_j(\mathbf{Y})| \leq \sum_{j=1}^{k}\sigma_j(\mathbf{X} - \mathbf{Y}).$$

Proof. The *self-adjoint dilations* $S(\mathbf{X}), S(\mathbf{Y}) \in \mathbb{C}^{(m+n) \times (m+n)}$ defined by

$$S(\mathbf{X}) = \begin{bmatrix} \mathbf{0} & \mathbf{X} \\ \mathbf{X}^* & \mathbf{0} \end{bmatrix} \quad \text{and} \quad S(\mathbf{Y}) = \begin{bmatrix} \mathbf{0} & \mathbf{Y} \\ \mathbf{Y}^* & \mathbf{0} \end{bmatrix}$$

have eigenvalues

$$\sigma_1(\mathbf{X}) \geq \cdots \geq \sigma_\ell(\mathbf{X}) \geq 0 = \cdots = 0 \geq -\sigma_\ell(\mathbf{X}) \geq \cdots \geq -\sigma_1(\mathbf{X}),$$

$$\sigma_1(\mathbf{Y}) \geq \cdots \geq \sigma_\ell(\mathbf{Y}) \geq 0 = \cdots = 0 \geq -\sigma_\ell(\mathbf{Y}) \geq \cdots \geq -\sigma_1(\mathbf{Y}).$$

Therefore, given $k \in [\ell]$, there exists a subset I_k of $[m + n]$ with size k such that

$$\sum_{j=1}^{k} |\sigma_j(\mathbf{X}) - \sigma_j(\mathbf{Y})| = \sum_{j \in I_k} (\lambda_j(S(\mathbf{X})) - \lambda_j(S(\mathbf{Y}))).$$

Using (A.24) with $\mathbf{A} = S(\mathbf{Y})$, $\mathbf{B} = S(\mathbf{X} - \mathbf{Y})$, and $\mathbf{A} + \mathbf{B} = S(\mathbf{X})$ yields

$$\sum_{j=1}^{k} |\sigma_j(\mathbf{X}) - \sigma_j(\mathbf{Y})| \leq \sum_{j=1}^{k} \lambda_j(S(\mathbf{X} - \mathbf{Y})) = \sum_{j=1}^{k} \sigma_j(\mathbf{X} - \mathbf{Y}).$$

The proof is complete. $\square$

Lemma A.18 implies in particular the triangle inequality

$$\sum_{j=1}^{\ell} \sigma_j(\mathbf{A} + \mathbf{B}) \leq \sum_{j=1}^{\ell} \sigma_j(\mathbf{A}) + \sum_{j=1}^{\ell} \sigma_j(\mathbf{B}), \qquad \mathbf{A}, \mathbf{B} \in \mathbb{C}^{m \times n},$$

where $\ell = \min\{m, n\}$. Moreover, it is easy to verify that $\sum_{j=1}^{\ell} \sigma_j(\mathbf{A}) = 0$ if and only if $\mathbf{A} = \mathbf{0}$ and that $\sum_{j=1}^{\ell} \sigma_j(\lambda \mathbf{A}) = |\lambda| \sum_{j=1}^{\ell} \sigma_j(\mathbf{A})$. These three properties show that the expression

$$\|\mathbf{A}\|_* := \sum_{j=1}^{\min\{m,n\}} \sigma_j(\mathbf{A}), \qquad \mathbf{A} \in \mathbb{C}^{m \times n}, \tag{A.25}$$

defines a norm on $\mathbb{C}^{m \times n}$, called the *nuclear norm*. It is also referred to as the Schatten 1-norm, in view of the fact that, for all $1 \leq p \leq \infty$, the expression

$$\|\mathbf{A}\|_{S_p} := \left[\sum_{j=1}^{\min\{m,n\}} \sigma_j(\mathbf{A})^p \right]^{1/p}, \qquad \mathbf{A} \in \mathbb{C}^{m \times n},$$

defines a norm on $\mathbb{C}^{m \times n}$, called the Schatten p-norm. We note that it reduces to the Frobenius norm for $p = 2$ and to the operator norm for $p = \infty$.

Next we introduce the Moore–Penrose pseudo-inverse, which generalizes the usual inverse of a square matrix, but exists for any (possibly rectangular) matrix.

Definition A.19. Let $\mathbf{A} \in \mathbb{C}^{m \times n}$ of rank r with reduced singular value decomposition

$$\mathbf{A} = \tilde{\mathbf{U}} \tilde{\mathbf{\Sigma}} \tilde{\mathbf{V}}^* = \sum_{j=1}^{r} \sigma_j(\mathbf{A}) \mathbf{u}_j \mathbf{v}_j^*.$$

Then its *Moore–Penrose pseudo-inverse* $\mathbf{A}^\dagger \in \mathbb{C}^{n \times m}$ is defined as

$$\mathbf{A}^\dagger = \tilde{\mathbf{V}} \tilde{\mathbf{\Sigma}}^{-1} \tilde{\mathbf{U}}^* = \sum_{j=1}^{r} \sigma_j^{-1}(\mathbf{A}) \mathbf{v}_j \mathbf{u}_j^*.$$

Note that the singular values satisfy $\sigma_j(\mathbf{A}) > 0$ for all $j \in [r]$, $r = \text{rank}(\mathbf{A})$, so that $\mathbf{A}^\dagger$ is well defined. If $\mathbf{A}$ is an invertible square matrix, then one easily verifies that $\mathbf{A}^\dagger = \mathbf{A}^{-1}$. It follows immediately from the definition that $\mathbf{A}^\dagger$ has the same rank r as $\mathbf{A}$ and that

$$\sigma_{\max}(\mathbf{A}^\dagger) = \|\mathbf{A}^\dagger\|_{2 \to 2} = \sigma_r^{-1}(\mathbf{A}).$$

In particular, if $\mathbf{A}$ has full rank, then

$$\|\mathbf{A}^\dagger\|_{2 \to 2} = \sigma_{\min}^{-1}(\mathbf{A}). \tag{A.26}$$

If $\mathbf{A}^*\mathbf{A} \in \mathbb{C}^{n \times n}$ is invertible (implying $m \geq n$), then

$$\mathbf{A}^\dagger = (\mathbf{A}^*\mathbf{A})^{-1}\mathbf{A}^*. \tag{A.27}$$

Indeed,

$$(\mathbf{A}^*\mathbf{A})^{-1}\mathbf{A}^* = (\tilde{\mathbf{V}}\tilde{\mathbf{\Sigma}}^2\tilde{\mathbf{V}}^*)^{-1}\tilde{\mathbf{V}}\tilde{\mathbf{\Sigma}}\tilde{\mathbf{U}}^* = \tilde{\mathbf{V}}\tilde{\mathbf{\Sigma}}^{-2}\tilde{\mathbf{V}}^*\tilde{\mathbf{V}}\tilde{\mathbf{\Sigma}}\tilde{\mathbf{U}}^* = \tilde{\mathbf{V}}\tilde{\mathbf{\Sigma}}^{-1}\tilde{\mathbf{U}}^* = \mathbf{A}^\dagger.$$

Similarly, if $\mathbf{A}\mathbf{A}^* \in \mathbb{C}^{m \times m}$ is invertible (implying $n \geq m$), then

$$\mathbf{A}^\dagger = \mathbf{A}^*(\mathbf{A}\mathbf{A}^*)^{-1}. \tag{A.28}$$

The Moore–Penrose pseudo-inverse is closely connected to least squares problems considered next.

A.3 Least Squares Problems

Let us first connect least squares problems with the Moore–Penrose pseudo-inverse introduced above.

Proposition A.20. *Let* $\mathbf{A} \in \mathbb{C}^{m \times n}$ *and* $\mathbf{y} \in \mathbb{C}^m$. *Define* $\mathcal{M} \subset \mathbb{C}^n$ *to be the set of minimizers of* $\mathbf{x} \mapsto \|\mathbf{A}\mathbf{x} - \mathbf{y}\|_2$. *The optimization problem*

$$\underset{\mathbf{x} \in \mathcal{M}}{\text{minimize}} \ \|\mathbf{x}\|_2 \tag{A.29}$$

has the unique solution $\mathbf{x}^\sharp = \mathbf{A}^\dagger \mathbf{y}$.

Proof. The (full) singular value decomposition of $\mathbf{A}$ can be written $\mathbf{A} = \mathbf{U}\boldsymbol{\Sigma}\mathbf{V}^*$ with

$$\boldsymbol{\Sigma} = \begin{bmatrix} \tilde{\boldsymbol{\Sigma}} & \mathbf{0} \\ \mathbf{0} & \mathbf{0} \end{bmatrix} \in \mathbb{R}^{m \times n},$$

where $\tilde{\boldsymbol{\Sigma}} \in \mathbb{R}^{r \times r}$, $r = \mathrm{rank}(\mathbf{A})$, is the diagonal matrix containing the nonzero singular values $\sigma_1(\mathbf{A}), \ldots, \sigma_r(\mathbf{A})$ on its diagonal. We introduce the vectors

$$\mathbf{z} = \begin{pmatrix} \mathbf{z}_1 \\ \mathbf{z}_2 \end{pmatrix} = \mathbf{V}^*\mathbf{x}, \quad \mathbf{z}_1 \in \mathbb{C}^r,$$

$$\mathbf{b} = \begin{pmatrix} \mathbf{b}_1 \\ \mathbf{b}_2 \end{pmatrix} = \mathbf{U}^*\mathbf{y}, \quad \mathbf{b}_1 \in \mathbb{C}^r.$$

Since the ℓ_2-norm is invariant under orthogonal transformations, we have

$$\|\mathbf{A}\mathbf{x} - \mathbf{y}\|_2 = \|\mathbf{U}^*(\mathbf{A}\mathbf{x} - \mathbf{y})\|_2 = \|\boldsymbol{\Sigma}\mathbf{V}^*\mathbf{x} - \mathbf{b}\|_2 = \left\| \begin{pmatrix} \tilde{\boldsymbol{\Sigma}}\mathbf{z}_1 - \mathbf{b}_1 \\ -\mathbf{b}_2 \end{pmatrix} \right\|_2.$$

This ℓ_2-norm is minimized for $\mathbf{z}_1 = \tilde{\boldsymbol{\Sigma}}^{-1}\mathbf{b}_1$ and arbitrary $\mathbf{z}_2$. Fixing $\mathbf{z}_1$, we notice that $\|\mathbf{x}\|_2^2 = \|\mathbf{V}^*\mathbf{x}\|_2^2 = \|\mathbf{z}\|_2^2 = \|\mathbf{z}_1\|_2^2 + \|\mathbf{z}_2\|_2^2$ is minimized for $\mathbf{z}_2 = \mathbf{0}$. Altogether, the minimizer $\mathbf{x}^\sharp$ of (A.29) is given by

$$\mathbf{x} = \mathbf{V} \begin{pmatrix} \mathbf{z}_1 \\ \mathbf{0} \end{pmatrix} = \mathbf{V} \begin{bmatrix} \tilde{\boldsymbol{\Sigma}}^{-1} & \mathbf{0} \\ \mathbf{0} & \mathbf{0} \end{bmatrix} \mathbf{U}^*\mathbf{y} = \mathbf{A}^\dagger\mathbf{y}$$

by definition of the Moore–Penrose pseudo-inverse.   □

Let us highlight two special cases.

Corollary A.21. *Let $\mathbf{A} \in \mathbb{C}^{m \times n}$, $m \geq n$, be of full rank n, and let $\mathbf{y} \in \mathbb{C}^m$. Then the least squares problem*

$$\underset{\mathbf{x} \in \mathbb{C}^n}{\text{minimize}} \ \|\mathbf{A}\mathbf{x} - \mathbf{y}\|_2 \tag{A.30}$$

has the unique solution $\mathbf{x}^\sharp = \mathbf{A}^\dagger\mathbf{y}$.

This follows from Proposition A.20 because a minimizer $\mathbf{x}^\sharp$ of $\|\mathbf{A}\mathbf{x} - \mathbf{y}\|_2$ is unique: Indeed, $\mathbf{A}\mathbf{x}^\sharp$ is the orthogonal projection of $\mathbf{y}$ onto the range of $\mathbf{A}$, and this completely determines $\mathbf{x}^\sharp$ because $\mathbf{A}$ has full rank. Notice then that $\mathbf{A}\mathbf{A}^\dagger$ is the orthogonal projection onto the range of $\mathbf{A}$. Since $\mathbf{A}$ is assumed to have full rank and $m \geq n$, the matrix $\mathbf{A}^*\mathbf{A}$ is invertible, so that (A.27) yields $\mathbf{A}^\dagger = (\mathbf{A}^*\mathbf{A})^{-1}\mathbf{A}^*$. Therefore, $\mathbf{x}^\sharp = \mathbf{A}^\dagger\mathbf{y}$ is equivalent to the normal equation

$$\mathbf{A}^*\mathbf{A}\mathbf{x}^\sharp = \mathbf{A}^*\mathbf{y}. \tag{A.31}$$

Corollary A.22. *Let* $\mathbf{A} \in \mathbb{C}^{m \times n}$, $n \geq m$, *be of full rank* m, *and let* $\mathbf{y} \in \mathbb{C}^m$. *Then the least squares problem*

$$\underset{\mathbf{x} \in \mathbb{C}^n}{\text{minimize}} \; \|\mathbf{x}\|_2 \quad \text{subject to } \mathbf{A}\mathbf{x} = \mathbf{y} \tag{A.32}$$

has the unique solution $\mathbf{x}^\sharp = \mathbf{A}^\dagger \mathbf{y}$.

Since $\mathbf{A}$ is of full rank $m \leq n$, (A.28) yields $\mathbf{A}^\dagger = \mathbf{A}^*(\mathbf{A}\mathbf{A}^*)^{-1}$. Therefore, the minimizer $\mathbf{x}^\sharp$ of (A.32) satisfies the normal equation of the second kind

$$\mathbf{x}^\sharp = \mathbf{A}^*\mathbf{b}, \quad \text{where } \mathbf{A}\mathbf{A}^*\mathbf{b} = \mathbf{y}. \tag{A.33}$$

We can also treat the weighted ℓ_2-minimization problem

$$\underset{\mathbf{z} \in \mathbb{C}^n}{\text{minimize}} \; \|\mathbf{z}\|_{2,\mathbf{w}} = \left(\sum_{j=1}^n |z_j|^2 w_j \right)^{1/2} \quad \text{subject to } \mathbf{A}\mathbf{z} = \mathbf{y}, \tag{A.34}$$

where $\mathbf{w}$ is a sequence of weights $w_j > 0$. Introducing the diagonal matrix $\mathbf{D}_{\mathbf{w}} = \text{diag}[w_1, \ldots, w_n] \in \mathbb{R}^{n \times n}$ and making the substitution $\mathbf{x} = \mathbf{D}_{\mathbf{w}}^{1/2}\mathbf{z}$, the minimizer $\mathbf{z}^\sharp$ of (A.34) is deduced from the minimizer $\mathbf{x}^\sharp$ of

$$\underset{\mathbf{x} \in \mathbb{C}^n}{\text{minimize}} \; \|\mathbf{x}\|_2 \quad \text{subject to } \mathbf{A}\mathbf{D}_{\mathbf{w}}^{-1/2}\mathbf{x} = \mathbf{y}$$

via

$$\mathbf{z}^\sharp = \mathbf{D}_{\mathbf{w}}^{-1/2}\mathbf{x}^\sharp = \mathbf{D}_{\mathbf{w}}^{-1/2}(\mathbf{A}\mathbf{D}_{\mathbf{w}}^{-1/2})^\dagger \mathbf{y}. \tag{A.35}$$

In particular, if $n \geq m$ and $\mathbf{A}$ has full rank, then

$$\mathbf{z}^\sharp = \mathbf{D}_{\mathbf{w}}^{-1}\mathbf{A}^*(\mathbf{A}\mathbf{D}_{\mathbf{w}}^{-1}\mathbf{A}^*)^{-1}\mathbf{y}. \tag{A.36}$$

Instead of (A.35), the following characterization in terms of the inner product $\langle \mathbf{x}, \mathbf{x}' \rangle_{\mathbf{w}} = \sum_{j=1}^n x_j \overline{x'}_j w_j$ can be useful.

Proposition A.23. *A vector* $\mathbf{z}^\sharp \in \mathbb{C}^n$ *is a minimizer of* (A.34) *if and only if* $\mathbf{A}\mathbf{z}^\sharp = \mathbf{y}$ *and*

$$\text{Re} \, \langle \mathbf{z}^\sharp, \mathbf{v} \rangle_{\mathbf{w}} = 0 \quad \text{for all } \mathbf{v} \in \ker \mathbf{A}. \tag{A.37}$$

Proof. Given $\mathbf{z}^\sharp$ with $\mathbf{A}\mathbf{z}^\sharp = \mathbf{y}$, a vector $\mathbf{z} \in \mathbb{C}^n$ is feasible for (A.34) if and only if it can be written as $\mathbf{z} = \mathbf{x} + t\mathbf{v}$ with $t \in \mathbb{R}$ and $\mathbf{v} \in \ker \mathbf{A}$. Observe that

$$\|\mathbf{x} + t\mathbf{v}\|_{2,\mathbf{w}}^2 = \|\mathbf{x}\|_{2,\mathbf{w}}^2 + t^2 \|\mathbf{v}\|_{2,\mathbf{w}}^2 + 2t \, \text{Re} \, \langle \mathbf{x}, \mathbf{v} \rangle_{\mathbf{w}}. \tag{A.38}$$

Therefore, if $\operatorname{Re} \langle \mathbf{x}, \mathbf{v} \rangle_{\mathbf{w}} = 0$, then $t = 0$ is the minimizer of $t \mapsto \|\mathbf{x} + t\mathbf{v}\|_{2,\mathbf{w}}$, and in turn $\mathbf{z}^{\sharp}$ is a minimizer of (A.34). Conversely, if $\mathbf{z}^{\sharp}$ is a minimizer of (A.34), then $t = 0$ is a minimizer of $t \mapsto \|\mathbf{x} + t\mathbf{v}\|_{2,\mathbf{w}}$ for all $\mathbf{v} \in \ker \mathbf{A}$. However, if $\operatorname{Re} \langle \mathbf{x}, \mathbf{v} \rangle_{\mathbf{w}} \neq 0$, then (A.38) with a nonzero t sufficiently small and of opposite sign to $\operatorname{Re} \langle \mathbf{x}, \mathbf{v} \rangle_{\mathbf{w}}$ reads $\|\mathbf{x} + t\mathbf{v}\|_{2,\mathbf{w}} < \|\mathbf{x}\|_2$, which is a contradiction. $\square$

Although (A.31) and (A.33) suggest to solve the normal equations via a method such as Gauss elimination in order to obtain the solution of least squares problems, the use of specialized methods presents some numerical advantages. An overview of various approaches can be found in [51]. We briefly mention the prominent method of solving least squares problems via the QR decomposition.

For any matrix $\mathbf{A} \in \mathbb{C}^{m \times n}$ with $m \geq n$, there exists a unitary matrix $\mathbf{Q} \in \mathbb{C}^{m \times m}$ and an upper triangular matrix $\mathbf{R} \in \mathbb{C}^{n \times n}$ with nonnegative diagonal entries such that

$$\mathbf{A} = \mathbf{Q} \begin{pmatrix} \mathbf{R} \\ \mathbf{0} \end{pmatrix}.$$

We refer to [51] for the existence (see [51, Theorem 1.3.1]) and for methods to compute this QR decomposition. Now consider the least squares problem (A.30). Since $\mathbf{Q}$ is unitary, we have

$$\|\mathbf{A}\mathbf{x} - \mathbf{y}\|_2 = \|\mathbf{Q}^*\mathbf{A}\mathbf{x} - \mathbf{Q}^*\mathbf{y}\|_2 = \left\| \begin{pmatrix} \mathbf{R} \\ \mathbf{0} \end{pmatrix} \mathbf{x} - \mathbf{Q}^*\mathbf{y} \right\|_2.$$

Partitioning $\mathbf{b} = \begin{pmatrix} \mathbf{b}_1 \\ \mathbf{b}_2 \end{pmatrix} = \mathbf{Q}^*\mathbf{y}$ with $\mathbf{b}_1 \in \mathbb{C}^n$, this is minimized by solving the triangular system $\mathbf{R}\mathbf{x} = \mathbf{b}_1$ with a simple backward elimination. (If $\mathbf{R}$ has some zeros on the diagonal, a slight variation of this procedure can be applied via partitioning of $\mathbf{R}$.)

The orthogonal matching pursuit algorithm involves successive optimization problems of the type (A.30) where a new column is added to $\mathbf{A}$ at each step. In such a situation, it is beneficial to work with the QR decomposition of $\mathbf{A}$, as it is numerically cheap to update the QR decomposition when a new column is added; see [51, Sect. 3.2.4] for details.

The least squares problem (A.32) may also be solved using the QR decomposition of $\mathbf{A}^*$; see [51, Theorem 1.3.3] for details.

If fast matrix–vector multiplication algorithms are available for $\mathbf{A}$ and $\mathbf{A}^*$ (for instance, via the fast Fourier transform or if $\mathbf{A}$ is sparse), then iterative algorithms for least squares problems are fast alternatives to QR decompositions. Conjugate gradients [51, 231] and especially the variant in [303] belong to this class of algorithms.

A.4 Vandermonde Matrices

The *Vandermonde matrix* associated with $x_0, x_1, \ldots, x_n \in \mathbb{C}$ is defined as

$$
\mathbf{V} := \mathbf{V}(x_0, x_1, \ldots, x_n) := \begin{bmatrix} 1 & x_0 & x_0^2 & \cdots & x_0^n \\ 1 & x_1 & x_1^2 & \cdots & x_1^n \\ \vdots & \vdots & \vdots & \cdots & \vdots \\ 1 & x_n & x_n^2 & \cdots & x_n^n \end{bmatrix}. \tag{A.39}
$$

Theorem A.24. *The determinant of the Vandermonde matrix (A.39) equals*

$$
\det \mathbf{V} = \prod_{0 \leq k < \ell \leq n} (x_\ell - x_k).
$$

Proof. The proof can be done by induction on $n \geq 1$. For $n = 1$, the result is clear. For $n \geq 2$, we remark that $\det \mathbf{V}(x_0, x_1, \ldots, x_n)$ is a polynomial in x_n, has degree at most n, and vanishes at $x_0, \ldots, x_{n-1}$. Therefore,

$$
\det \mathbf{V}(x_0, x_1, \ldots, x_n) = c \prod_{0 \leq k < n} (x_n - x_k) \tag{A.40}
$$

for some constant c depending on $x_1, \ldots, x_{n-1}$. We notice that the constant c is the coefficient of x_n^n in $\det \mathbf{V}(x_0, x_1, \ldots, x_n)$. By expanding the determinant of the matrix $\mathbf{V}(x_0, x_1, \ldots, x_n)$ along its last row, we observe that $c = \det \mathbf{V}(x_0, x_1, \ldots, x_{n-1})$. Using the induction hypothesis to substitute the value of c in (A.40) concludes the proof. $\square$

We now establish a more involved result on the total positivity of Vandermonde matrices.

Theorem A.25. *If $x_n > \cdots x_1 > x_0 > 0$, then the Vandermonde matrix (A.39) is totally positive, i.e., for any sets $I, J \subset [n+1]$ of equal size,*

$$
\det \mathbf{V}_{I,J} > 0,
$$

where $\mathbf{V}_{I,J}$ is the submatrix of $\mathbf{V}$ with rows and columns indexed by I and J.

We start with the following lemma, known as *Descartes' rule of signs*.

Lemma A.26. *For a polynomial $p(x) = a_n x^n + \cdots + a_1 x + a_0 \neq 0$, the number $Z(p)$ of positive zeros of p and the number $S(a) := \mathrm{card}(\{i \in [n] : a_{i-1} a_i < 0\})$ of sign changes of $a = (a_0, a_1, \ldots, a_n)$ satisfy*

$$
Z(p) \leq S(a).
$$

Proof. We proceed by induction on $n \geq 1$. For $n = 1$, the desired result is clear. Let us now assume that the result holds up to an integer $n - 1$, $n \geq 2$. We want to establish that, given $p(x) = a_n x^n + \cdots + a_1 x + a_0 \neq 0$, we have $Z(p) \leq S(a)$. We suppose that $a_0 \neq 0$; otherwise, the result is clear from the induction hypothesis. Changing p in $-p$ if necessary, we may assume $a_0 > 0$, and we consider the smallest positive integer k such that $a_k \neq 0$—the result is clear of no such k exists. We separate two cases.

1. $a_0 > 0$ and $a_k < 0$.
   The result follows from Rolle's theorem and the induction hypothesis via

$$Z(p) \leq Z(p') + 1 \leq S(a_1, \ldots, a_n) + 1 = S(a_0, a_1, \ldots, a_n).$$

2. $a_0 > 0$ and $a_k > 0$.
   Let t denote the smallest positive zero of p—again the result is clear if no such t exists. Let us assume that p' does not vanish on $(0, t)$. Since $p'(0) = ka_k > 0$, we derive that $p'(x) > 0$ for all $x \in (0, t)$. If follows that $a_0 = p(0) < p(t) = 0$, which is absurd. Therefore, there must be a zero of p' in $(0, t)$. Taking into account the zeros of p' guaranteed by Rolle's theorem, the result follows from the induction hypothesis via

$$Z(p) \leq Z(p') \leq S(a_1, \ldots, a_n) = S(a_0, a_1, \ldots, a_n).$$

This concludes the inductive proof.   □

Proof (of Theorem A.25). We will prove by induction on $k \in [n]$ that

$$\det \begin{bmatrix} x_{i_1}^{j_1} & x_{i_1}^{j_2} & \cdots & x_{i_1}^{j_k} \\ x_{i_2}^{j_1} & x_{i_2}^{j_2} & \cdots & x_{i_2}^{j_k} \\ \vdots & \vdots & \ddots & \vdots \\ x_{i_k}^{j_1} & x_{i_k}^{j_2} & \cdots & x_{i_k}^{j_k} \end{bmatrix} > 0$$

for all $0 < x_0 < x_1 < \cdots < x_n$ and for all $0 \leq i_1 < i_2 < \cdots < i_k \leq n$ and $0 \leq j_1 < j_2 < \cdots < j_k \leq n$. For $k = 1$, this is nothing else than the positivity of the x_i's. Let us now suppose that the result holds up to an integer $k - 1$, $2 \leq k \leq n$, and assume that

$$\det \begin{bmatrix} x_{i_1}^{j_1} & x_{i_1}^{j_2} & \cdots & x_{i_1}^{j_k} \\ x_{i_2}^{j_1} & x_{i_2}^{j_2} & \cdots & x_{i_2}^{j_k} \\ \vdots & \vdots & \ddots & \vdots \\ x_{i_k}^{j_1} & x_{i_k}^{j_2} & \cdots & x_{i_k}^{j_k} \end{bmatrix} \leq 0 \qquad (A.41)$$

for some $0 < x_0 < x_1 < \cdots < x_n$ and for some $0 \leq i_1 < i_2 < \cdots < i_k \leq n$ and $0 \leq j_1 < j_2 < \cdots < j_k \leq n$. We introduce the polynomial p defined by

$$p(x) := \det \begin{bmatrix} x_{i_1}^{j_1} & x_{i_1}^{j_2} & \cdots & x_{i_1}^{j_k} \\ x_{i_2}^{j_1} & x_{i_2}^{j_2} & \cdots & x_{i_2}^{j_k} \\ \vdots & \vdots & \ddots & \vdots \\ x^{j_1} & x^{j_2} & \cdots & x^{j_k} \end{bmatrix}.$$

Expanding with respect to the last row and invoking Descartes' rule of signs, we observe that $Z(p) \leq k - 1$. Since the polynomial p vanishes at the positive points $x_{i_1}, \ldots, x_{i_{k-1}}$, it cannot vanish elsewhere. The assumption (A.41) then implies that $p(x) < 0$ for all $x > x_{i_{k-1}}$. But this contradicts the induction hypothesis, because

$$\lim_{x \to +\infty} \frac{p(x)}{x^{j_k}} = \det \begin{bmatrix} x_{i_1}^{j_1} & x_{i_1}^{j_2} & \cdots & x_{i_1}^{j_{k-1}} \\ x_{i_2}^{j_1} & x_{i_2}^{j_2} & \cdots & x_{i_2}^{j_{k-1}} \\ \vdots & \vdots & \ddots & \vdots \\ x_{i_{k-1}}^{j_1} & x_{i_{k-1}}^{j_2} & \cdots & x_{i_{k-1}}^{j_{k-1}} \end{bmatrix} > 0.$$

Thus, we have shown that the desired result holds for the integer k, and this concludes the inductive proof. □

A.5 Matrix Functions

In this section, we consider functions of self-adjoint matrices and some of their basic properties. We recall that a matrix $\mathbf{A} \in \mathbb{C}^{n \times n}$ is called self-adjoint if $\mathbf{A} = \mathbf{A}^*$. It is called positive semidefinite if additionally $\langle \mathbf{A}\mathbf{x}, \mathbf{x} \rangle \geq 0$ for all $\mathbf{x} \in \mathbb{C}^n$ and positive definite if $\langle \mathbf{A}\mathbf{x}, \mathbf{x} \rangle > 0$ for all $\mathbf{x} \neq \mathbf{0}$. For two self-adjoint matrices $\mathbf{A}, \mathbf{B} \in \mathbb{C}^{n \times n}$, we write $\mathbf{A} \preccurlyeq \mathbf{B}$ or $\mathbf{B} \succcurlyeq \mathbf{A}$ if $\mathbf{B} - \mathbf{A}$ is positive semidefinite and $\mathbf{A} \prec \mathbf{B}$ or $\mathbf{B} \succ \mathbf{A}$ if $\mathbf{B} - \mathbf{A}$ is positive definite.

A self-adjoint matrix $\mathbf{A}$ possesses an eigenvalue decomposition of the form

$$\mathbf{A} = \mathbf{U}\mathbf{D}\mathbf{U}^*,$$

where $\mathbf{U} \in \mathbb{C}^{n \times n}$ is a unitary matrix and $\mathbf{D} = \text{diag}[\lambda_1, \ldots, \lambda_n]$ is a diagonal matrix with the eigenvalues of $\mathbf{A}$ (repeated according to their multiplicities) on the diagonal. For a function $f : I \to \mathbb{R}$ with $I \subset \mathbb{R}$ containing the eigenvalues of $\mathbf{A}$, we define the self-adjoint matrix $f(\mathbf{A}) \in \mathbb{C}^{n \times n}$ via the spectral mapping

$$f(\mathbf{A}) = \mathbf{U}f(\mathbf{D})\mathbf{U}^*, \quad f(\mathbf{D}) = \text{diag}\left[f(\lambda_1), \ldots, f(\lambda_n)\right]. \tag{A.42}$$

This definition does not depend on the particular eigenvalue decomposition. It is easy to verify that for polynomials f, the definition coincides with the natural one. For instance, if $f(t) = t^2$, then, because $\mathbf{U}$ is unitary,

$$f(\mathbf{A}) = \mathbf{U}\mathbf{D}^2\mathbf{U}^* = \mathbf{U}\mathbf{D}\mathbf{U}^*\mathbf{U}\mathbf{D}\mathbf{U}^* = \mathbf{A}^2.$$

Note that, if $f(x) \leq g(x)$, respectively $f(x) < g(x)$, for all $x \in [a, b]$, then

$$f(\mathbf{A}) \preccurlyeq g(\mathbf{A}), \qquad \text{respectively } f(\mathbf{A}) \prec g(\mathbf{A}), \tag{A.43}$$

for all $\mathbf{A}$ with eigenvalues contained in $[a, b]$. It is a simple consequence of the definition that for a block-diagonal matrix with self-adjoint blocks $\mathbf{A}_1, \ldots, \mathbf{A}_L$ on the diagonal,

$$f\left(\begin{bmatrix} \mathbf{A}_1 & 0 & \cdots & 0 \\ 0 & \mathbf{A}_2 & \ddots & \vdots \\ \vdots & \ddots & \ddots & 0 \\ 0 & \cdots & 0 & \mathbf{A}_L \end{bmatrix}\right) = \begin{bmatrix} f(\mathbf{A}_1) & 0 & \cdots & 0 \\ 0 & f(\mathbf{A}_2) & \ddots & \vdots \\ \vdots & \ddots & \ddots & 0 \\ 0 & \cdots & 0 & f(\mathbf{A}_L) \end{bmatrix}. \tag{A.44}$$

Moreover, if $\mathbf{A}$ commutes with $\mathbf{B}$, i.e., $\mathbf{AB} = \mathbf{BA}$, then also $f(\mathbf{A})$ commutes with $\mathbf{B}$, i.e., $f(\mathbf{A})\mathbf{B} = \mathbf{B}f(\mathbf{A})$.

The *matrix exponential function* of a self-adjoint matrix $\mathbf{A}$ may be defined by applying (A.42) with the function $f(x) = e^x$, or equivalently via the power series

$$e^{\mathbf{A}} := \exp(\mathbf{A}) := \mathbf{Id} + \sum_{k=1}^{\infty} \frac{1}{k!}\mathbf{A}^k. \tag{A.45}$$

(The power series definition actually applies to any square, not necessarily self-adjoint, matrix.) The matrix exponential of a self-adjoint matrix is always positive definite by (A.43). Moreover, it follows from $1 + x \leq e^x$ for all $x \in \mathbb{R}$ and from (A.43) again that, for any self-adjoint matrix $\mathbf{A}$,

$$\mathbf{Id} + \mathbf{A} \preccurlyeq \exp(\mathbf{A}). \tag{A.46}$$

Lemma A.27. *If* $\mathbf{A}$ *and* $\mathbf{B}$ *commute, i.e.,* $\mathbf{AB} = \mathbf{BA}$*, then*

$$\exp(\mathbf{A} + \mathbf{B}) = \exp(\mathbf{A}) \exp(\mathbf{B}).$$

Proof. If $\mathbf{A}$ and $\mathbf{B}$ commute, then

$$\frac{1}{k!}(\mathbf{A} + \mathbf{B})^k = \frac{1}{k!} \sum_{j=0}^{k} \binom{k}{j} \mathbf{A}^j \mathbf{B}^{k-j} = \sum_{j=0}^{k} \frac{\mathbf{A}^j}{j!} \frac{\mathbf{B}^{k-j}}{(k-j)!}.$$

Therefore,

$$\exp(\mathbf{A} + \mathbf{B}) = \sum_{k=0}^{\infty} \sum_{j=0}^{k} \frac{\mathbf{A}^j}{j!} \frac{\mathbf{B}^{k-j}}{(k-j)!} = \sum_{j=0}^{\infty} \sum_{k=j}^{\infty} \frac{\mathbf{A}^j}{j!} \frac{\mathbf{B}^{k-j}}{(k-j)!} = \sum_{j=0}^{\infty} \frac{1}{j!} \mathbf{A}^j \sum_{\ell=0}^{\infty} \frac{1}{\ell!} \mathbf{B}^\ell$$

$$= \exp(\mathbf{A}) \exp(\mathbf{B}).$$

This yields the claim. $\square$

This lemma fails in the general case where $\mathbf{A}$ and $\mathbf{B}$ do not commute.

Corollary A.28. *For any square matrix* $\mathbf{A}$, *the matrix exponential* $\exp(\mathbf{A})$ *is invertible and*

$$\exp(\mathbf{A})^{-1} = \exp(-\mathbf{A}).$$

Proof. Since $\mathbf{A}$ and $-\mathbf{A}$ commute, the previous lemma yields

$$\exp(\mathbf{A}) \exp(-\mathbf{A}) = \exp(\mathbf{A} - \mathbf{A}) = \exp(\mathbf{0}) = \mathbf{Id},$$

and similarly $\exp(-\mathbf{A}) \exp(\mathbf{A}) = \mathbf{Id}$. $\square$

Of special interest is the *trace exponential*

$$\operatorname{tr} \exp : \mathbf{A} \mapsto \operatorname{tr} \exp(\mathbf{A}). \tag{A.47}$$

The trace exponential is monotone with respect to the semidefinite order. Indeed, for self-adjoint matrices $\mathbf{A}, \mathbf{B}$, we have

$$\operatorname{tr} \exp \mathbf{A} \leq \operatorname{tr} \exp \mathbf{B} \quad \text{whenever } \mathbf{A} \preccurlyeq \mathbf{B}. \tag{A.48}$$

This fact follows from the more general statement below.

Proposition A.29. *Let* $f : \mathbb{R} \to \mathbb{R}$ *be a nondecreasing function, and let* $\mathbf{A}, \mathbf{B}$ *be self-adjoint matrices. Then* $\mathbf{A} \preccurlyeq \mathbf{B}$ *implies*

$$\operatorname{tr} f(\mathbf{A}) \leq \operatorname{tr} f(\mathbf{B}).$$

Proof. The minimax principle in (A.23) and the assumption $\mathbf{A} \preccurlyeq \mathbf{B}$ imply that the eigenvalues $\lambda_1(\mathbf{A}) \geq \lambda_2(\mathbf{A}) \geq \cdots$ and $\lambda_1(\mathbf{B}) \geq \lambda_2(\mathbf{B}) \geq \cdots$ of $\mathbf{A}$ and $\mathbf{B}$ satisfy

$$\lambda_k(\mathbf{A}) = \max_{\substack{\mathcal{M} \subset \mathbb{C}^n \\ \dim \mathcal{M} = k}} \min_{\substack{\mathbf{x} \in \mathcal{M} \\ \|\mathbf{x}\|_2 = 1}} \langle \mathbf{A}\mathbf{x}, \mathbf{x} \rangle \leq \max_{\substack{\mathcal{M} \subset \mathbb{C}^n \\ \dim \mathcal{M} = k}} \min_{\substack{\mathbf{x} \in \mathcal{M} \\ \|\mathbf{x}\|_2 = 1}} \langle \mathbf{B}\mathbf{x}, \mathbf{x} \rangle = \lambda_k(\mathbf{B}).$$

Since f is nondecreasing, it follows that

$$\operatorname{tr} f(\mathbf{A}) = \sum_{k=1}^{n} f(\lambda_k(\mathbf{A})) \leq \sum_{k=1}^{n} f(\lambda_k(\mathbf{B})) = \operatorname{tr} f(\mathbf{B}).$$

This completes the proof. $\square$

Next we show that certain inequalities for scalar functions extend to traces of matrix-valued functions; see [382] for more details.

Theorem A.30. *Given* $f_\ell, g_\ell : [a, b] \to \mathbb{R}$ *and* $c_\ell \in \mathbb{R}$ *for* $\ell \in [M]$, *if*

$$\sum_{\ell=1}^{M} c_\ell f_\ell(x) g_\ell(y) \geq 0 \qquad \text{for all } x, y \in [a, b],$$

then, for all self-adjoint matrices $\mathbf{A}, \mathbf{B}$ *with eigenvalues in* $[a, b]$,

$$\operatorname{tr} \left(\sum_{\ell=1}^{M} c_\ell f_\ell(\mathbf{A}) g_\ell(\mathbf{B}) \right) \geq 0.$$

Proof. Let $\mathbf{A} = \sum_{j=1}^{n} \lambda_j \mathbf{u}_j \mathbf{u}_j^*$ and $\mathbf{B} = \sum_{k=1}^{n} \eta_k \mathbf{v}_k \mathbf{v}_k^*$ be the eigenvalue decompositions of $\mathbf{A}$ and $\mathbf{B}$. Then

$$\operatorname{tr} \left(\sum_{\ell=1}^{M} c_\ell f_\ell(\mathbf{A}) g_\ell(\mathbf{B}) \right) = \operatorname{tr} \left(\sum_{\ell=1}^{M} c_\ell \sum_{j,k=1}^{n} f_\ell(\lambda_j) g_\ell(\eta_k) \mathbf{u}_j \mathbf{u}_j^* \mathbf{v}_k \mathbf{v}_k^* \right)$$

$$= \sum_{j,k=1}^{n} \sum_{\ell=1}^{M} c_\ell f_\ell(\lambda_j) g_\ell(\eta_k) \operatorname{tr} \left(\mathbf{u}_j \mathbf{u}_j^* \mathbf{v}_k \mathbf{v}_k^* \right) = \sum_{j,k=1}^{n} \sum_{\ell=1}^{M} c_\ell f_\ell(\lambda_j) g_\ell(\eta_k) |\langle \mathbf{u}_j, \mathbf{v}_k \rangle|^2 \geq 0.$$

Hereby, we have used the cyclicity of the trace in the second-to-last step.   □

A function f is called *matrix monotone* (or operator monotone) if $\mathbf{A} \preccurlyeq \mathbf{B}$ implies

$$f(\mathbf{A}) \preccurlyeq f(\mathbf{B}). \tag{A.49}$$

Somewhat surprisingly, not every nondecreasing function $f : \mathbb{R} \to \mathbb{R}$ extends to a matrix monotone function via (A.42). A simple example is provided by $f(t) = t^2$.

In order to study matrix monotonicity for some specific functions below, we first make an easy observation.

Lemma A.31. *If* $\mathbf{A} \preccurlyeq \mathbf{B}$, *then* $\mathbf{Y}^* \mathbf{A} \mathbf{Y} \preccurlyeq \mathbf{Y}^* \mathbf{B} \mathbf{Y}$ *for all matrices* $\mathbf{Y}$ *with appropriate dimensions. In addition, if* $\mathbf{Y}$ *is invertible and* $\mathbf{A} \prec \mathbf{B}$, *then* $\mathbf{Y}^* \mathbf{A} \mathbf{Y} \prec \mathbf{Y}^* \mathbf{B} \mathbf{Y}$.

Proof. For every vector $\mathbf{x}$, we have

$$\langle \mathbf{Y}^* \mathbf{A} \mathbf{Y} \mathbf{x}, \mathbf{x} \rangle = \langle \mathbf{A} \mathbf{Y} \mathbf{x}, \mathbf{Y} \mathbf{x} \rangle \leq \langle \mathbf{B} \mathbf{Y} \mathbf{x}, \mathbf{Y} \mathbf{x} \rangle = \langle \mathbf{Y}^* \mathbf{B} \mathbf{Y} \mathbf{x}, \mathbf{x} \rangle,$$

which shows the first part. The second part follows with minor changes.   □

Next, we establish the matrix monotonicity of the negative inverse map.

Proposition A.32. *The matrix function* $f(\mathbf{A}) = -\mathbf{A}^{-1}$ *is matrix monotone on the set of positive definite matrices.*

Proof. Given $0 \prec \mathbf{A} \preccurlyeq \mathbf{B}$, the matrix $\mathbf{B}^{-1/2}$ exists (and may be defined via (A.42)). According to Lemma A.31,

$$\mathbf{B}^{-1/2}\mathbf{A}\mathbf{B}^{-1/2} \preccurlyeq \mathbf{B}^{-1/2}\mathbf{B}\mathbf{B}^{-1/2} = \mathrm{Id}.$$

The matrix $\mathbf{C} = \mathbf{B}^{-1/2}\mathbf{A}\mathbf{B}^{-1/2}$ has an eigenvalue decomposition $\mathbf{C} = \mathbf{U}\mathbf{D}\mathbf{U}^*$ with a unitary matrix $\mathbf{U}$ and a diagonal matrix $\mathbf{D}$. By Lemma A.31, the above relation implies $0 \prec \mathbf{D} \preccurlyeq \mathrm{Id}$. It follows that $\mathrm{Id} \preccurlyeq \mathbf{D}^{-1}$. Then Lemma A.31 yields

$$\mathrm{Id} \preccurlyeq \mathbf{U}\mathbf{D}^{-1}\mathbf{U}^* = \mathbf{C}^{-1} = (\mathbf{B}^{-1/2}\mathbf{A}\mathbf{B}^{-1/2})^{-1} = \mathbf{B}^{1/2}\mathbf{A}^{-1}\mathbf{B}^{1/2}.$$

Applying Lemma A.31 again shows that $\mathbf{B}^{-1} = \mathbf{B}^{-1/2}\mathrm{Id}\mathbf{B}^{-1/2} \preccurlyeq \mathbf{A}^{-1}$. □

The *matrix logarithm* can be defined for positive definite matrices via the spectral mapping formula (A.42) with $f(x) = \ln(x)$. It is the inverse of the matrix exponential, i.e.,

$$\exp(\ln(\mathbf{A})) = \mathbf{A}. \qquad (A.50)$$

Remark A.33. The definition of the matrix logarithm can be extended to invertible, not necessarily self-adjoint, matrices, just like the matrix exponential extends to all square matrices via the power series expansion (A.45). Similarly to the extension of the logarithm to the complex numbers, one encounters the nonuniqueness of the logarithm defined via (A.50). One usually chooses the principal branch, thus restricting the domain to matrices with eigenvalues outside of the negative real line.

Unlike the matrix exponential, the matrix logarithm is matrix monotone.

Proposition A.34. *Let* $\mathbf{A}, \mathbf{B}$ *be positive definite matrices. Then*

$$\ln(\mathbf{A}) \preccurlyeq \ln(\mathbf{B}) \quad \text{whenever } \mathbf{A} \preccurlyeq \mathbf{B}.$$

Proof. We first observe that the (scalar) logarithm satisfies

$$\ln(x) = \int_0^\infty \left(\frac{1}{t+1} - \frac{1}{t+x} \right) dt, \qquad x > 0. \qquad (A.51)$$

Indeed, a simple integral transformation shows that, for $R > 0$,

$$\int_0^R \left(\frac{1}{t+1} - \frac{1}{t+x} \right) dt = \ln\left(\frac{t+1}{t+x} \right)\Big|_0^R = \ln\left(\frac{R+1}{R+x} \right) - \ln\left(\frac{1}{x} \right) \xrightarrow{R\to\infty} \ln(x).$$

It now follows from Proposition A.32 that, for $t \geq 0$, the matrix function

$$g_t(\mathbf{A}) := \frac{1}{t+1}\mathbf{Id} - (t\mathbf{Id} + \mathbf{A})^{-1}$$

is matrix monotone on the set of positive definite matrices. By (A.51) and by the definition of the matrix logarithm via (A.42), we derive that, for a positive definite matrix $\mathbf{A}$,

$$\ln(\mathbf{A}) = \int_0^\infty g_t(\mathbf{A})dt.$$

Therefore, the matrix logarithm is matrix monotone, since integrals preserve the semidefinite ordering.   □

The square-root function $\mathbf{A} \mapsto \mathbf{A}^{1/2}$ and all the power functions $\mathbf{A} \mapsto \mathbf{A}^p$ with $0 < p \leq 1$ are also matrix monotone on the set of positive semidefinite matrices; see [47] or [379] for a simple proof.

We continue the discussion on matrix function in Sect. B.6, where we treat convexity issues.

Appendix B
Convex Analysis

This appendix provides an overview of convex analysis and convex optimization. Much more information can be found in various books on the subject such as [70, 178, 275, 293, 424, 425].

For the purpose of this exposition on convexity, we work with real vector spaces $\mathbb{R}^N$ and treat sets in and functions on $\mathbb{C}^N$ by identifying $\mathbb{C}^N$ with $\mathbb{R}^{2N}$. In order to reverse this identification in some of the statements and definitions below, one needs to replace the inner product $\langle \mathbf{x}, \mathbf{z} \rangle$ by $\mathrm{Re}\langle \mathbf{x}, \mathbf{z} \rangle$ for $\mathbf{x}, \mathbf{z} \in \mathbb{C}^N$.

B.1 Convex Sets

Let us start with some basic definitions.

Definition B.1. A subset $K \subset \mathbb{R}^N$ is called convex, if for all $\mathbf{x}, \mathbf{z} \in K$, the line segment connecting $\mathbf{x}$ and $\mathbf{z}$ is entirely contained in K, that is,

$$t\mathbf{x} + (1-t)\mathbf{z} \in K \quad \text{for all } t \in [0, 1].$$

It is straightforward to verify that a set $K \in \mathbb{R}^N$ is convex if and only if, for all $\mathbf{x}_1, \ldots, \mathbf{x}_n \in K$ and $t_1, \ldots, t_n \geq 0$ such that $\sum_{j=1}^{n} t_j = 1$, the convex combination $\sum_{j=1}^{n} t_j \mathbf{x}_j$ is also contained in K.

Definition B.2. The convex hull $\mathrm{conv}(T)$ of a set $T \subset \mathbb{R}^N$ is the smallest convex set containing T.

It is well known [424, Theorem 2.3] that the convex hull of T consists of the finite convex combinations of T, i.e.,

$$\mathrm{conv}(T) = \left\{ \sum_{j=1}^{n} t_j \mathbf{x}_j : n \geq 1, t_1, \ldots, t_n \geq 0, \sum_{j=1}^{n} t_j = 1, \mathbf{x}_1, \ldots, \mathbf{x}_n \in T \right\}.$$

S. Foucart and H. Rauhut, *A Mathematical Introduction to Compressive Sensing*, Applied and Numerical Harmonic Analysis, DOI 10.1007/978-0-8176-4948-7, © Springer Science+Business Media New York 2013

Simple examples of convex sets include subspaces, affine spaces, half spaces, polygons, or norm balls $B(\mathbf{x}, t)$; see (A.2). The intersection of convex sets is again a convex set.

Definition B.3. A set $K \subset \mathbb{R}^N$ is called a cone if, for all $\mathbf{x} \in K$ and all $t \geq 0$, $t\mathbf{x}$ also belongs to K. In addition, if K is convex, then K is called a convex cone.

Obviously, the zero vector is contained in every cone. A set K is a convex cone if, for all $\mathbf{x}, \mathbf{z} \in K$ and $t, s \geq 0$, $s\mathbf{x} + t\mathbf{z}$ also belongs to K.

Simple examples of convex cones include subspaces, half spaces, the positive orthant $\mathbb{R}_+^N = \{\mathbf{x} \in \mathbb{R}^N : x_i \geq 0 \text{ for all } i \in [N]\}$, or the set of positive semidefinite matrices in $\mathbb{R}^{N \times N}$. Another important example of a convex cone is the second-order cone

$$\left\{ \mathbf{x} \in \mathbb{R}^{N+1} : \sqrt{\sum_{j=1}^{N} x_j^2} \leq x_{N+1} \right\}. \tag{B.1}$$

For a cone $K \subset \mathbb{R}^N$, its dual cone K^* is defined via

$$K^* := \left\{ \mathbf{z} \in \mathbb{R}^N : \langle \mathbf{x}, \mathbf{z} \rangle \geq 0 \text{ for all } \mathbf{x} \in K \right\}. \tag{B.2}$$

As the intersection of half spaces, K^* is closed and convex, and it is straightforward to verify that K^* is again a cone. If K is a closed nonempty cone, then $K^{**} = K$. Moreover, if $H, K \subset \mathbb{R}^N$ are cones such that $H \subset K$, then $K^* \subset H^*$. As an example, the dual cone of the positive orthant $\mathbb{R}_+^N$ is $\mathbb{R}_+^N$ itself—in other words, $\mathbb{R}_+^N$ is self-dual. Note that the dual cone is closely related to the polar cone, which is defined by

$$K^\circ := \left\{ \mathbf{z} \in \mathbb{R}^N : \langle \mathbf{x}, \mathbf{z} \rangle \leq 0 \text{ for all } \mathbf{x} \in K \right\} = -K^*. \tag{B.3}$$

The conic hull $\text{cone}(T)$ of a set $T \subset \mathbb{R}^N$ is the smallest convex cone containing T. It can be described as

$$\text{cone}(T) = \left\{ \sum_{j=1}^{n} t_j \mathbf{x}_j : n \geq 1, t_1, \dots, t_n \geq 0, \mathbf{x}_1, \dots, \mathbf{x}_n \in T \right\}. \tag{B.4}$$

Convex sets can be separated by hyperplanes as stated next.

Theorem B.4. Let $K_1, K_2 \subset \mathbb{R}^N$ be convex sets whose interiors have empty intersection. Then there exist a vector $\mathbf{w} \in \mathbb{R}^N$ and a scalar λ such that

$$K_1 \subset \{\mathbf{x} \in \mathbb{R}^N : \langle \mathbf{x}, \mathbf{w} \rangle \leq \lambda\},$$

$$K_2 \subset \{\mathbf{x} \in \mathbb{R}^N : \langle \mathbf{x}, \mathbf{w} \rangle \geq \lambda\}.$$

Remark B.5. The theorem applies in particular when $K_1 \cap K_2 = \emptyset$ or when K_1, K_2 intersect in only one point, i.e., $K_1 \cap K_2 = \{x_0\}$. In the latter case, one chooses $\lambda = \langle x_0, w \rangle$. If K_2 is a subset of a hyperplane, then one can choose w and λ such that $K_2 \subset \{x \in \mathbb{R}^N : \langle x, w \rangle = \lambda\}$.

Next we consider the notion of extreme points.

Definition B.6. Let $K \subset \mathbb{R}^N$ be a convex set. A point $x \in K$ is called an extreme point of K if $x = ty + (1 - t)z$ for $y, z \in K$ and $t \in (0, 1)$ implies $x = y = z$.

Compact convex sets can be described via their extreme points as stated next (see, for instance, [424, Corollary 18.5.1] or [275, Theorem 2.3.4]).

Theorem B.7. *A compact convex set is the convex hull of its extreme points.*

If K is a polygon, then its extreme points are the zero-dimensional faces of K, and the above statement is rather intuitive.

B.2 Convex Functions

We work with extended-valued functions $F : \mathbb{R}^N \to (-\infty, \infty] = \mathbb{R} \cup \{\infty\}$. Sometimes we also consider an additional extension of the values to $-\infty$. Addition, multiplication, and inequalities in $(-\infty, \infty]$ are understood in the "natural" sense— for instance, $x + \infty = \infty$ for all $x \in \mathbb{R}$, $\lambda \cdot \infty = \infty$ for $\lambda > 0$, $x < \infty$ for all $x \in \mathbb{R}$. The domain of an extended-valued function F is defined as

$$\mathrm{dom}(F) = \{x \in \mathbb{R}^N, F(x) \neq \infty\}.$$

A function with $\mathrm{dom}(F) \neq \emptyset$ is called proper. A function $F : K \to \mathbb{R}$ on a subset $K \subset \mathbb{R}^N$ can be converted to an extended-valued function by setting $F(x) = \infty$ for $x \notin K$. Then $\mathrm{dom}(F) = K$, and we call this extension the canonical one.

Definition B.8. An extended-valued function $F : \mathbb{R}^N \to (-\infty, \infty]$ is called convex if, for all $x, z \in \mathbb{R}^N$ and $t \in [0, 1]$,

$$F(tx + (1 - t)z) \leq tF(x) + (1 - t)F(z). \tag{B.5}$$

F is called strictly convex if, for all $x \neq z$ and all $t \in (0, 1)$,

$$F(tx + (1 - t)z) < tF(x) + (1 - t)F(z).$$

F is called strongly convex with parameter $\gamma > 0$ if, for all $x, z \in \mathbb{R}^N$ and $t \in [0, 1]$,

$$F(tx + (1 - t)z) \leq tF(x) + (1 - t)F(z) - \frac{\gamma}{2}t(1 - t)\|x - z\|_2^2. \tag{B.6}$$

A function $F : \mathbb{R}^N \to [-\infty, \infty)$ is called (strictly, strongly) concave if $-F$ is (strictly, strongly) convex.

Obviously, a strongly convex function is strictly convex.

The domain of a convex function is convex, and a function $F : K \to \mathbb{R}^N$ on a convex subset $K \subset \mathbb{R}^N$ is called convex if its canonical extension to $\mathbb{R}^N$ is convex (or alternatively if $\mathbf{x}, \mathbf{z}$ in the definition (B.5) are assumed to be in K). A function F is convex if and only if its epigraph

$$\mathrm{epi}(F) = \{(\mathbf{x}, r) : \mathbf{x} \in \mathbb{R}^N, r \geq F(\mathbf{x})\} \subset \mathbb{R}^N \times \mathbb{R}$$

is a convex set.

Convexity can also be characterized using general convex combinations: A function $F : \mathbb{R}^N \to (-\infty, \infty]$ is convex if and only if for all $\mathbf{x}_1, \ldots, \mathbf{x}_n \in \mathbb{R}^N$ and $t_1, \ldots, t_n \geq 0$ such that $\sum_{j=1}^n t_j = 1$,

$$F\left(\sum_{j=1}^n t_j \mathbf{x}_j\right) \leq \sum_{j=1}^n t_j F(\mathbf{x}_j).$$

Let us summarize some results on the composition of convex functions.

Proposition B.9. (a) Let F, G be convex functions on $\mathbb{R}^N$. For $\alpha, \beta \geq 0$, the function $\alpha F + \beta G$ is convex.
(b) Let $F : \mathbb{R} \to \mathbb{R}$ be convex and nondecreasing and $G : \mathbb{R}^N \to \mathbb{R}$ be convex. Then the function $H(\mathbf{x}) = F(G(\mathbf{x}))$ is convex.

Proof. Verifying (a) is straightforward. For (b), given $\mathbf{x}, \mathbf{y} \in \mathbb{R}^N$ and $t \in [0, 1]$, we have

$$H(t\mathbf{x} + (1 - t)\mathbf{y}) = F(G(t\mathbf{x} + (1 - t)\mathbf{y})) \leq F(tG(\mathbf{x}) + (1 - t)G(\mathbf{y}))$$
$$\leq tF(G(\mathbf{x})) + (1 - t)F(G(\mathbf{y})) = tH(\mathbf{x}) + (1 - t)H(\mathbf{y}),$$

where we have used convexity of G and monotonicity of F in the first inequality and convexity of F in the second inequality. $\qquad\Box$

Let us give some classical examples of convex functions.

Example B.10. (a) For $p \geq 1$, the function $F(x) = |x|^p$, $x \in \mathbb{R}$, is convex.
(b) Every norm $\| \cdot \|$ on $\mathbb{R}^N$ is a convex function. This follows from the triangle inequality and homogeneity.
(c) The ℓ_p-norms $\| \cdot \|_p$ are strictly convex for $1 < p < \infty$, but they are not strictly convex for $p = 1$ or $p = \infty$.
(d) For a nondecreasing convex function $F : \mathbb{R} \to (-\infty, \infty]$ and a norm $\| \cdot \|$ on $\mathbb{R}^N$, the function $H(\mathbf{x}) = F(\|\mathbf{x}\|)$ is convex. This follows from (a) and Proposition B.9(b). In particular, the function $\mathbf{x} \mapsto \|\mathbf{x}\|^p$ is convex provided that $p \geq 1$.

(e) For a positive semidefinite matrix $\mathbf{A} \in \mathbb{R}^{N \times N}$, the function $F(\mathbf{x}) = \mathbf{x}^* \mathbf{A} \mathbf{x}$ is convex. If $\mathbf{A}$ is positive definite, then F is strongly convex.

(f) For a convex set K, the characteristic function

$$\chi_K(\mathbf{x}) = \begin{cases} 0 & \text{if } \mathbf{x} \in K, \\ \infty & \text{if } \mathbf{x} \notin K \end{cases} \qquad (\text{B.7})$$

is convex.

For differentiable functions, the following characterizations of convexity holds.

Proposition B.11. *Let* $F : \mathbb{R}^N \to \mathbb{R}$ *be a differentiable function.*

(a) F is convex if and only if, for all $\mathbf{x}, \mathbf{y} \in \mathbb{R}^N$,

$$F(\mathbf{x}) \geq F(\mathbf{y}) + \langle \nabla F(\mathbf{y}), \mathbf{x} - \mathbf{y} \rangle,$$

where $\nabla F(\mathbf{y}) := \left[\frac{\partial F}{\partial y_1}(\mathbf{y}), \ldots, \frac{\partial F}{\partial y_N}(\mathbf{y}) \right]^\top$ is the gradient of F at $\mathbf{y}$.

(b) F is strongly convex with parameter $\gamma > 0$ if and only if, for all $\mathbf{x}, \mathbf{y} \in \mathbb{R}^N$,

$$F(\mathbf{x}) \geq F(\mathbf{y}) + \langle \nabla F(\mathbf{y}), \mathbf{x} - \mathbf{y} \rangle + \frac{\gamma}{2} \| \mathbf{x} - \mathbf{y} \|_2^2.$$

(c) If F is twice differentiable, then it is convex if and only if, for all $\mathbf{x} \in \mathbb{R}^N$,

$$\nabla^2 F(\mathbf{x}) \succcurlyeq 0,$$

where $\nabla^2 F(\mathbf{x}) := \left[\frac{\partial^2 F}{\partial x_i \partial x_j}(\mathbf{x}) \right]_{i,j=1}^N$ is the Hessian of F at $\mathbf{x}$.

We continue with a discussion about continuity properties.

Proposition B.12. *A convex function $F : \mathbb{R}^N \to \mathbb{R}$ is continuous on $\mathbb{R}^N$.*

The treatment of extended-valued functions requires the notion of lower semicontinuity (which is a particularly useful notion in infinite-dimensional Hilbert spaces, where, for instance, the norm $\|\cdot\|$ is not continuous but lower semicontinuous with respect to the weak topology; see, for instance, [178]).

Definition B.13. *A function $F : \mathbb{R}^N \to (-\infty, \infty]$ is called lower semicontinuous if, for every $\mathbf{x} \in \mathbb{R}^N$ and every sequence $(\mathbf{x}_j)_{j \geq 1} \subset \mathbb{R}$ converging to $\mathbf{x}$,*

$$\liminf_{j \to \infty} F(\mathbf{x}_j) \geq F(\mathbf{x}).$$

A continuous function $F : \mathbb{R}^N \to \mathbb{R}$ is lower semicontinuous. A nontrivial example of lower semicontinuity is provided by the characteristic function χ_K of a proper subset $K \subset \mathbb{R}^N$ defined in (B.7). Indeed, χ_K is not continuous, but it is lower semicontinuous if and only if K is closed.

A function is lower semicontinuous if and only if its epigraph is closed.

Convex functions have nice properties related to minimization. A (global) minimizer of a function $F : \mathbb{R}^N \rightarrow (-\infty, \infty]$ is a point $\mathbf{x} \in \mathbb{R}^N$ satisfying $F(\mathbf{x}) \leq F(\mathbf{y})$ for all $\mathbf{y} \in \mathbb{R}^N$. A local minimizer of F is a point $\mathbf{x} \in \mathbb{R}^N$ such that there exists $\varepsilon > 0$ with $F(\mathbf{x}) \leq F(\mathbf{y})$ for all $\mathbf{y} \in \mathbb{R}^N$ satisfying $\|\mathbf{x} - \mathbf{y}\|_2 \leq \varepsilon$. (The Euclidean norm $\| \cdot \|_2$ can be replaced by any norm $\| \cdot \|$ in this definition.)

Proposition B.14. *Let* $F : \mathbb{R}^N \rightarrow (-\infty, \infty]$ *be a convex function.*

(a) *A local minimizer of F is a global minimizer.*
(b) *The set of minimizers of F is convex.*
(c) *If F is strictly convex, then the minimizer of F is unique.*

Proof. (a) Let $\mathbf{x}$ be a local minimizer and $\mathbf{z} \in \mathbb{R}^N$ be arbitrary. Let $\varepsilon > 0$ be the parameter appearing in the definition of a local minimizer. There exists $t \in (0, 1)$ such that $\mathbf{y} := t\mathbf{x} + (1-t)\mathbf{z}$ satisfies $\|\mathbf{x} - \mathbf{y}\|_2 \leq \varepsilon$. Then $F(\mathbf{x}) \leq F(\mathbf{y})$ and by convexity $F(\mathbf{y}) = F(t\mathbf{x} + (1-t)\mathbf{z}) \leq tF(\mathbf{x}) + (1-t)F(\mathbf{z})$. It follows that $(1-t)F(\mathbf{x}) \leq (1-t)F(\mathbf{z})$; hence, $F(\mathbf{x}) \leq F(\mathbf{z})$ because $1 - t > 0$. Since $\mathbf{z}$ was arbitrary, this shows that $\mathbf{x}$ is a global minimizer.

(b) Let $\mathbf{x}, \mathbf{y} \in \mathbb{R}^N$ be two minimizers, i.e., $F(\mathbf{x}) = F(\mathbf{y}) = \inf_\mathbf{z} F(\mathbf{z})$. Then, for $t \in [0, 1]$,

$$F(t\mathbf{x} + (1 - t)\mathbf{y}) \leq tF(\mathbf{x}) + (1 - t)F(\mathbf{y}) = \inf_\mathbf{z} F(\mathbf{z}),$$

so that $t\mathbf{x} + (1 - t)\mathbf{y}$ is a minimizer as well.

(c) Suppose that $\mathbf{x} \neq \mathbf{y}$ are both minimizers of F. Then, for $t \in (0, 1)$,

$$F(t\mathbf{x} + (1 - t)\mathbf{y}) < tF(\mathbf{x}) + (1 - t)F(\mathbf{y}) = \inf_\mathbf{z} F(\mathbf{z}),$$

which is a contradiction.   □

The fact that local minimizers of convex functions are automatically global minimizers essentially explains the availability of efficient methods for convex optimization problems.

We say that a function $f(\mathbf{x}, \mathbf{y})$ of two arguments $\mathbf{x} \in \mathbb{R}^n, \mathbf{y} \in \mathbb{R}^m$ is jointly convex if it is convex as a function of the variable $\mathbf{z} = (\mathbf{x}, \mathbf{y})$. Partial minimization of a jointly convex function in one variable gives rise to a new convex function as stated next.

Theorem B.15. *Let* $f : \mathbb{R}^n \times \mathbb{R}^m \rightarrow (-\infty, \infty]$ *be a jointly convex function. Then* $g(\mathbf{x}) := \inf_{\mathbf{y} \in \mathbb{R}^m} f(\mathbf{x}, \mathbf{y})$, *if well defined, is a convex function of $\mathbf{x} \in \mathbb{R}^n$.*

Proof. For simplicity, we assume that the infimum is always attained. The general case has to be treated with an ε-argument.

Given $\mathbf{x}_1, \mathbf{x}_2 \in \mathbb{R}^n$, there exist $\mathbf{y}_1, \mathbf{y}_2 \in \mathbb{R}^m$ such that

$$g(\mathbf{x}_1) = f(\mathbf{x}_1, \mathbf{y}_1) = \min_{\mathbf{y} \in \mathbb{R}^m} f(\mathbf{x}_1, \mathbf{y}), \qquad g(\mathbf{x}_2) = f(\mathbf{x}_2, \mathbf{y}_2) = \min_{\mathbf{y} \in \mathbb{R}^m} f(\mathbf{x}_2, \mathbf{y}).$$

For $t \in [0, 1]$, the joint convexity implies that

$$g(t\mathbf{x}_1 + (1 - t)\mathbf{x}_2) \leq f(t\mathbf{x}_1 + (1 - t)\mathbf{x}_2, t\mathbf{y}_1 + (1 - t)\mathbf{y}_2)$$
$$\leq tf(\mathbf{x}_1, \mathbf{y}_1) + (1 - t)f(\mathbf{x}_2, \mathbf{y}_2) = tg(\mathbf{x}_1) + (1 - t)g(\mathbf{x}_2).$$

This point finishes the argument.   □

The previous theorem shows as well that partial maximization of a jointly concave function gives rise to a concave function.

Next we consider the maximum of a convex function over a convex set.

Theorem B.16. *Let $K \subset \mathbb{R}^N$ be a compact convex set and let $F : K \to \mathbb{R}$ be a convex function. Then F attains its maximum at an extreme point of K.*

Proof. Let $\mathbf{x} \in K$ such that $F(\mathbf{x}) \geq F(\mathbf{z})$ for all $\mathbf{z} \in K$. By Theorem B.7, the set K is the convex hull of its extreme points, so we can write $\mathbf{x} = \sum_{j=1}^m t_j \mathbf{x}_j$ for some integer $m \geq 1, t_1, \ldots, t_m > 0$ with $\sum_{j=1}^m t_j = 1$, and extreme points $\mathbf{x}_1, \ldots, \mathbf{x}_m$ of K. By convexity,

$$F(\mathbf{x}) = F\left(\sum_{j=1}^m t_j \mathbf{x}_j\right) \leq \sum_{j=1}^m t_j F(\mathbf{x}_j) \leq \sum_{j=1}^m t_j F(\mathbf{x}) = F(\mathbf{x})$$

because $F(\mathbf{x}_j) \leq F(\mathbf{x})$ by definition of $\mathbf{x}$. Thus, all inequalities hold with equality, which is only possible if $F(\mathbf{x}_j) = F(\mathbf{x})$ for all $j \in [m]$. Therefore, the maximum of F is attained at an extreme point of K.   □

B.3 The Convex Conjugate

The convex conjugate is a very useful concept in convex analysis and optimization.

Definition B.17. Given a function $F : \mathbb{R}^N \to (-\infty, \infty]$, the convex conjugate (or Fenchel dual) function of F is the function $F^* : \mathbb{R}^N \to (-\infty, \infty]$ defined by

$$F^*(\mathbf{y}) := \sup_{\mathbf{x} \in \mathbb{R}^N} \{\langle \mathbf{x}, \mathbf{y} \rangle - F(\mathbf{x})\}.$$

The convex conjugate F^* is always a convex function, no matter whether the function F is convex or not. The definition of F^* immediately gives the Fenchel (or Young, or Fenchel–Young) inequality

$$\langle \mathbf{x}, \mathbf{y} \rangle \leq F(\mathbf{x}) + F^*(\mathbf{y}) \qquad \text{for all } \mathbf{x}, \mathbf{y} \in \mathbb{R}^N. \tag{B.8}$$

Let us summarize some properties of convex conjugate functions.

Proposition B.18. *Let $F : \mathbb{R}^N \to (-\infty, \infty]$.*

(a) The convex conjugate F^* is lower semicontinuous.
(b) The biconjugate F^{**} is the largest lower semicontinuous convex function satisfying $F^{**}(\mathbf{x}) \le F(\mathbf{x})$ for all $\mathbf{x} \in \mathbb{R}^N$. In particular, if F is convex and lower semicontinuous, then $F^{**} = F$.
(c) For $\tau \ne 0$, if $F_\tau(\mathbf{x}) := F(\tau\mathbf{x})$, then $(F_\tau)^*(\mathbf{y}) = F^*(\mathbf{y}/\tau)$.
(d) For $\tau > 0$, $(\tau F)^*(\mathbf{y}) = \tau F^*(\mathbf{y}/\tau)$.
(e) For $\mathbf{z} \in \mathbb{R}^N$, if $F^{(\mathbf{z})} := F(\mathbf{x} - \mathbf{z})$, then $(F^{(\mathbf{z})})^*(\mathbf{y}) = \langle \mathbf{z}, \mathbf{y} \rangle + F^*(\mathbf{y})$.

Proof. For (a) and (b), we refer to [424, Corollary 12.1.1 and Theorem 12.2]. For (d), a substitution gives

$$(\tau F)^*(\mathbf{y}) = \sup_{\mathbf{x} \in \mathbb{R}^N} \{\langle \mathbf{x}, \mathbf{y} \rangle - \tau F(\mathbf{x})\} = \tau \sup_{\mathbf{x} \in \mathbb{R}^N} \{\langle \mathbf{x}, \mathbf{y}/\tau \rangle - F(\mathbf{x})\} = \tau F^*(\mathbf{y}/\tau).$$

The statements (c) and (e) are obtained from simple calculations. □

The biconjugate F^{**} is sometimes called the convex relaxation of F because of (b).
    Let us compute the convex conjugate for some examples.

Example B.19. (a) Let $F(\mathbf{x}) = \frac{1}{2}\|\mathbf{x}\|_2^2$, $\mathbf{x} \in \mathbb{R}^N$. Then $F^*(\mathbf{y}) = \frac{1}{2}\|\mathbf{y}\|_2^2 = F(\mathbf{y})$, $\mathbf{y} \in \mathbb{R}^N$. Indeed, since

$$\langle \mathbf{x}, \mathbf{y} \rangle \le \frac{1}{2}\|\mathbf{x}\|_2^2 + \frac{1}{2}\|\mathbf{y}\|_2^2, \tag{B.9}$$

we have

$$F^*(\mathbf{y}) = \sup_{\mathbf{x} \in \mathbb{R}^N} \{\langle \mathbf{x}, \mathbf{y} \rangle - F(\mathbf{x})\} \le \frac{1}{2}\|\mathbf{y}\|_2^2.$$

For the converse inequality, we just set $\mathbf{x} = \mathbf{y}$ in the definition of the convex conjugate to obtain

$$F^*(\mathbf{y}) \ge \|\mathbf{y}\|_2^2 - \frac{1}{2}\|\mathbf{y}\|_2^2 = \frac{1}{2}\|\mathbf{y}\|_2^2.$$

This is the only example of a function F on $\mathbb{R}^N$ satisfying $F^* = F$.
(b) Let $F(x) = \exp(x)$, $x \in \mathbb{R}$. The function $x \mapsto xy - \exp(x)$ takes its maximum at $x = \ln y$ if $y > 0$, so that

$$F^*(y) = \sup_{x \in \mathbb{R}}\{xy - e^x\} = \begin{cases} y \ln y - y & \text{if } y > 0, \\ 0 & \text{if } y = 0, \\ \infty & \text{if } y < 0. \end{cases}$$

The Fenchel inequality for this particular pair reads

$$xy \leq e^x + y\ln(y) - y \quad \text{for all } x \in \mathbb{R}, y > 0. \tag{B.10}$$

(c) Let $F(\mathbf{x}) = \|\mathbf{x}\|$ for some norm on $\mathbb{R}^N$ and let $\|\cdot\|_*$ be its dual norm; see Definition A.3. Then the convex conjugate of F is the characteristic function of the dual norm ball, that is,

$$F^*(\mathbf{y}) = \chi_{B_{\|\cdot\|_*}}(\mathbf{y}) = \begin{cases} 0 & \text{if } \|\mathbf{y}\|_* \leq 1, \\ \infty & \text{otherwise.} \end{cases}$$

Indeed, by the definition of the dual norm, we have $\langle \mathbf{x}, \mathbf{y} \rangle \leq \|\mathbf{x}\|\|\mathbf{y}\|_*$. It follows that

$$F^*(\mathbf{y}) = \sup_{\mathbf{x} \in \mathbb{R}^N} \{\langle \mathbf{x}, \mathbf{y} \rangle - \|\mathbf{x}\|\} \leq \sup_{\mathbf{x} \in \mathbb{R}^N} \{(\|\mathbf{y}\|_* - 1)\|\mathbf{x}\|\},$$

so that $F^*(\mathbf{y}) \leq 0$ if $\|\mathbf{y}\|_* \leq 1$. The choice $\mathbf{x} = \mathbf{0}$ shows that $F^*(\mathbf{y}) = 0$ in this case. If $\|\mathbf{y}\|_* > 1$, then there exists $\mathbf{x}$ such that $\langle \mathbf{x}, \mathbf{y} \rangle > \|\mathbf{x}\|$. Replacing $\mathbf{x}$ by $\lambda\mathbf{x}$ and letting $\lambda \to \infty$ show that $F^*(\mathbf{y}) = \infty$ in this case.

(d) Let $F = \chi_K$ be the characteristic function of a convex set K; see (B.7). Its convex conjugate is given by

$$F^*(\mathbf{y}) = \sup_{\mathbf{x} \in K} \langle \mathbf{x}, \mathbf{y} \rangle.$$

B.4 The Subdifferential

The subdifferential generalizes the gradient to nondifferentiable functions.

Definition B.20. The subdifferential of a convex function $F : \mathbb{R}^N \to (-\infty, \infty]$ at a point $\mathbf{x} \in \mathbb{R}^N$ is defined by

$$\partial F(\mathbf{x}) = \{\mathbf{v} \in \mathbb{R}^N : F(\mathbf{z}) \geq F(\mathbf{x}) + \langle \mathbf{v}, \mathbf{z} - \mathbf{x} \rangle \text{ for all } \mathbf{z} \in \mathbb{R}^N\}. \tag{B.11}$$

The elements of $\partial F(\mathbf{x})$ are called subgradients of F at $\mathbf{x}$.

The subdifferential $\partial F(\mathbf{x})$ of a convex function $F : \mathbb{R}^N \to \mathbb{R}$ is always nonempty. If F is differentiable at $\mathbf{x}$, then $\partial F(\mathbf{x})$ contains only the gradient, i.e.,

$$\partial F(\mathbf{x}) = \{\nabla F(\mathbf{x})\},$$

see Proposition B.11(a) for one direction. A simple example of a function with a nontrivial subdifferential is the absolute value function $F(x) = |x|$, for which

$$\partial F(x) = \begin{cases} \{\operatorname{sgn}(x)\} & \text{if } x \neq 0, \\ [-1,1] & \text{if } x = 0, \end{cases}$$

where $\operatorname{sgn}(x) = +1$ for $x > 0$ and $\operatorname{sgn}(x) = -1$ for $x < 0$ as usual.

The subdifferential allows a simple characterization of minimizers of convex functions.

Theorem B.21. *A vector $\mathbf{x}$ is a minimum of a convex function F if and only if $0 \in \partial F(\mathbf{x})$.*

Proof. This is obvious from the definition of the subdifferential. □

Convex conjugates and subdifferentials are related in the following way.

Theorem B.22. *Let $F : \mathbb{R}^N \to (-\infty, \infty]$ be a convex function and let $\mathbf{x}, \mathbf{y} \in \mathbb{R}^N$. The following conditions are equivalent:*

(a) $\mathbf{y} \in \partial F(\mathbf{x})$.
(b) $F(\mathbf{x}) + F^*(\mathbf{y}) = \langle \mathbf{x}, \mathbf{y} \rangle$.

Additionally, if F is lower semicontinuous, then (a) and (b) are equivalent to

(c) $\mathbf{x} \in \partial F^*(\mathbf{y})$.

Proof. By definition of the subgradient, condition (a) reads

$$\langle \mathbf{x}, \mathbf{y} \rangle - F(\mathbf{x}) \geq \langle \mathbf{z}, \mathbf{y} \rangle - F(\mathbf{z}) \quad \text{for all } \mathbf{z} \in \mathbb{R}^N. \tag{B.12}$$

Therefore, the function $\mathbf{z} \mapsto \langle \mathbf{z}, \mathbf{y} \rangle - F(\mathbf{z})$ attains its maximum at $\mathbf{x}$. By definition of the convex conjugate, this implies that $F^*(\mathbf{y}) = \langle \mathbf{x}, \mathbf{y} \rangle - F(\mathbf{x})$, i.e., $F^*(\mathbf{y}) + F(\mathbf{x}) = \langle \mathbf{x}, \mathbf{y} \rangle$. This shows that (a) implies (b). Conversely, condition (b) implies by the Fenchel inequality and the definition of the convex conjugate that the function $\mathbf{z} \mapsto \langle \mathbf{z}, \mathbf{y} \rangle - F(\mathbf{z})$ attains its maximum in $\mathbf{x}$, which is nothing else than (B.12). It follows from the definition of the subdifferential that $\mathbf{y} \in \partial F(\mathbf{x})$.

Now if F is lower semicontinuous, then $F^{**} = F$ by Proposition B.18(b) so that (b) is equivalent to $F^{**}(\mathbf{x}) + F^*(\mathbf{y}) = \langle \mathbf{x}, \mathbf{y} \rangle$. Using the equivalence of (a) and (b) with F replaced by F^* concludes the proof. □

As a consequence, if F is a convex lower semicontinuous function, then ∂F is the inverse of ∂F^* in the sense that $\mathbf{x} \in \partial F^*(\mathbf{y})$ if and only if $\mathbf{y} \in \partial F(\mathbf{x})$.

Next we consider the so-called proximal mapping (also known as proximation, resolvent operator, or proximity operator). Let $F : \mathbb{R}^N \to (-\infty, \infty]$ be a convex function. Then, for $\mathbf{z} \in \mathbb{R}^N$, the function

$$\mathbf{x} \mapsto F(\mathbf{x}) + \frac{1}{2}\|\mathbf{x} - \mathbf{z}\|_2^2$$

is strictly convex due to the strict convexity of $\mathbf{x} \mapsto \|\mathbf{x}\|_2^2$. By Proposition B.14(c), its minimizer is unique. The mapping

$$P_F(\mathbf{z}) := \operatorname*{argmin}_{\mathbf{x} \in \mathbb{R}^N} F(\mathbf{x}) + \frac{1}{2}\|\mathbf{x} - \mathbf{z}\|_2^2 \tag{B.13}$$

is called the proximal mapping associated with F. In the special case where $F = \chi_K$ is the characteristic function of a convex set K defined in (B.7), then $P_K := P_{\chi_K}$ is the orthogonal projection onto K, that is,

$$P_K(\mathbf{z}) = \operatorname*{argmin}_{\mathbf{x} \in K} \|\mathbf{x} - \mathbf{z}\|_2. \tag{B.14}$$

If K is a subspace of $\mathbb{R}^N$, then this is the usual orthogonal projection onto K, which in particular is a linear map.

The proximal mapping can be expressed via subdifferentials as shown in the next statement.

Proposition B.23. *Let* $F : \mathbb{R}^N \to (-\infty, \infty]$ *be a convex function. Then* $\mathbf{x} = P_F(\mathbf{z})$ *if and only if* $\mathbf{z} \in \mathbf{x} + \partial F(\mathbf{x})$.

Proof. By Theorem B.21, we have $\mathbf{x} = P_F(\mathbf{z})$ if and only if

$$\mathbf{0} \in \partial\left(\frac{1}{2}\| \cdot -\mathbf{z}\|_2^2 + F\right)(\mathbf{x}).$$

The function $\mathbf{x} \mapsto \frac{1}{2}\|\mathbf{x} - \mathbf{z}\|_2^2$ is differentiable with gradient $\nabla\left(\frac{1}{2}\| \cdot -\mathbf{z}\|_2^2\right)(\mathbf{x}) = \mathbf{x} - \mathbf{z}$, so that the above condition reads $\mathbf{0} \in \mathbf{x} - \mathbf{z} + \partial F(\mathbf{x})$, which is equivalent to $\mathbf{z} \in \mathbf{x} + \partial F(\mathbf{x})$. $\square$

The previous proposition justifies the writing

$$P_F = (\mathbf{Id} + \partial F)^{-1}.$$

Moreau's identity stated next relates the proximal mappings of F and F^*.

Theorem B.24. *Let* $F : \mathbb{R}^N \to (-\infty, \infty]$ *be a lower semicontinuous convex function. Then, for all* $\mathbf{z} \in \mathbb{R}^N$,

$$P_F(\mathbf{z}) + P_{F^*}(\mathbf{z}) = \mathbf{z}.$$

Proof. Let $\mathbf{x} := P_F(\mathbf{z})$, and set $\mathbf{y} := \mathbf{z} - \mathbf{x}$. By Proposition B.23, we have $\mathbf{z} \in \mathbf{x} + \partial F(\mathbf{x})$, i.e., $\mathbf{y} = \mathbf{z} - \mathbf{x} \in \partial F(\mathbf{x})$. Since F is lower semicontinuous, it follows from Theorem B.22 that $\mathbf{x} \in \partial F^*(\mathbf{y})$, i.e., $\mathbf{z} \in \mathbf{y} + \partial F^*(\mathbf{y})$. By Proposition B.23 again, we have $\mathbf{y} = P_{F^*}(\mathbf{z})$. In particular, we have shown that $P_F(\mathbf{z}) + P_{F^*}(\mathbf{z}) = \mathbf{x} + \mathbf{y} = \mathbf{z}$ by definition of $\mathbf{y}$. $\square$

If P_F is easy to compute, then the previous result shows that $P_{F^*}(\mathbf{z}) = \mathbf{z} - P_F(\mathbf{z})$ is also easy to compute. It is useful to note that applying Moreau's identity to the function τF for some $\tau > 0$ shows that

$$P_{\tau F}(\mathbf{z}) + \tau P_{\tau^{-1}F^*}(\mathbf{z}/\tau) = \mathbf{z}. \tag{B.15}$$

Indeed, we have $P_{\tau F}(\mathbf{z}) + P_{(\tau F)^*}(\mathbf{z}) = \mathbf{z}$, so it remains to show that $P_{(\tau F)^*}(\mathbf{z}) = \tau P_{\tau^{-1}F}(\mathbf{z}/\tau)$. This follows from Proposition B.18(d), since

$$P_{(\tau F)^*}(\mathbf{z}) = \underset{\mathbf{x} \in \mathbb{R}^N}{\operatorname{argmin}} \left\{ \frac{1}{2}\|\mathbf{x} - \mathbf{z}\|_2^2 + (\tau F)^*(\mathbf{x}) \right\}$$

$$= \underset{\mathbf{x} \in \mathbb{R}^N}{\operatorname{argmin}} \left\{ \frac{1}{2}\|\mathbf{x} - \mathbf{z}\|_2^2 + \tau F^*(\mathbf{x}/\tau) \right\}$$

$$= \underset{\mathbf{x} \in \mathbb{R}^N}{\operatorname{argmin}} \left\{ \tau^2 \left(\frac{1}{2}\|\mathbf{x}/\tau - \mathbf{z}/\tau\|_2^2 + \tau^{-1}F^*(\mathbf{x}/\tau) \right) \right\} = \tau P_{\tau^{-1}F^*}(\mathbf{z}/\tau).$$

Since F is a lower semicontinuous convex function, we have $F^{**} = F$, so applying (B.15) to F^* in place of F gives

$$P_{\tau F^*}(\mathbf{z}) + \tau P_{\tau^{-1}F}(\mathbf{z}/\tau) = \mathbf{z}. \tag{B.16}$$

Theorem B.25. *For a convex function $F : \mathbb{R}^N \to (-\infty, \infty]$, the proximal mapping P_F is nonexpansive, i.e.,*

$$\|P_F(\mathbf{z}) - P_F(\mathbf{z}')\|_2 \le \|\mathbf{z} - \mathbf{z}'\|_2 \quad \text{for all } \mathbf{z}, \mathbf{z}' \in \mathbb{R}^N.$$

Proof. Set $\mathbf{x} = P_F(\mathbf{z})$ and $\mathbf{x}' = P_F(\mathbf{z}')$. By Proposition B.23, we have $\mathbf{z} \in \mathbf{x} + \partial F(\mathbf{x})$, so that $\mathbf{y} := \mathbf{z} - \mathbf{x} \in \partial F(\mathbf{x})$. Theorem B.22 shows that $F(\mathbf{x}) + F^*(\mathbf{y}) = \langle \mathbf{x}, \mathbf{y} \rangle$. Similarly, we can find $\mathbf{y}'$ such that $\mathbf{z}' = \mathbf{x}' + \mathbf{y}'$ and $F(\mathbf{x}') + F^*(\mathbf{y}') = \langle \mathbf{x}', \mathbf{y}' \rangle$. It follows that

$$\|\mathbf{z} - \mathbf{z}'\|_2^2 = \|\mathbf{x} - \mathbf{x}'\|_2^2 + \|\mathbf{y} - \mathbf{y}'\|_2^2 + 2\langle \mathbf{x} - \mathbf{x}', \mathbf{y} - \mathbf{y}' \rangle. \tag{B.17}$$

Note that, by the Fenchel inequality (B.8), we have

$$\langle \mathbf{x}', \mathbf{y} \rangle \le F(\mathbf{x}') + F^*(\mathbf{y}) \quad \text{and} \quad \langle \mathbf{x}, \mathbf{y}' \rangle \le F(\mathbf{x}) + F^*(\mathbf{y}').$$

Therefore,

$$\langle \mathbf{x} - \mathbf{x}', \mathbf{y} - \mathbf{y}' \rangle = \langle \mathbf{x}, \mathbf{y} \rangle + \langle \mathbf{x}', \mathbf{y}' \rangle - \langle \mathbf{x}', \mathbf{y} \rangle - \langle \mathbf{x}, \mathbf{y}' \rangle$$

$$= F(\mathbf{x}) + F^*(\mathbf{y}) + F(\mathbf{x}') + F^*(\mathbf{y}') - \langle \mathbf{x}', \mathbf{y} \rangle - \langle \mathbf{x}, \mathbf{y}' \rangle \ge 0.$$

Together with (B.17), this shows that $\|\mathbf{x} - \mathbf{x}'\|_2^2 \le \|\mathbf{z} - \mathbf{z}'\|_2^2$. $\square$

Let us conclude with an important example of a proximal mapping. Let $F(x) = |x|$, $x \in \mathbb{R}$, be the absolute value function. A straightforward computation shows that, for $\tau > 0$,

$$P_{\tau F}(y) = \operatorname*{argmin}_{x \in \mathbb{R}} \left\{ \frac{1}{2}(x-y)^2 + \tau |x| \right\} = \begin{cases} y - \tau & \text{if } y \geq \tau, \\ 0 & \text{if } |y| \leq \tau, \\ y + \tau & \text{if } y \leq -\tau, \end{cases}$$

$$=: S_\tau(y). \tag{B.18}$$

The function S_τ is called soft thresholding or shrinkage operator. More generally, if $F(\mathbf{x}) = \|\mathbf{x}\|_1$ is the ℓ_1-norm on $\mathbb{R}^N$, then the minimization problem defining the proximal operator decouples and $P_{\tau F}(\mathbf{y})$ is given entrywise for $\mathbf{y} \in \mathbb{R}^N$ by

$$P_{\tau F}(\mathbf{y})_\ell = S_\tau(y_\ell), \quad \ell \in [N]. \tag{B.19}$$

B.5 Convex Optimization Problems

An optimization problem takes the form

$$\min_{\mathbf{x} \in \mathbb{R}^N} F_0(\mathbf{x}) \quad \text{subject to } \mathbf{A}\mathbf{x} = \mathbf{y}, \tag{B.20}$$

$$\text{and } F_j(\mathbf{x}) \leq b_j, \quad j \in [M], \tag{B.21}$$

where the function $F_0 : \mathbb{R}^N \to (-\infty, \infty]$ is called *objective function*, the functions $F_1, \ldots, F_M : \mathbb{R}^N \to (-\infty, \infty]$ are called *constraint functions*, and $\mathbf{A} \in \mathbb{R}^{m \times N}, \mathbf{y} \in \mathbb{R}^m$ provide the equality constraint. A point $\mathbf{x} \in \mathbb{R}^N$ satisfying the constraints is called *feasible* and (B.20) is called feasible if there exists a feasible point. A feasible point $\mathbf{x}^\sharp$ for which the minimum is attained, that is, $F_0(\mathbf{x}^\sharp) \leq F_0(\mathbf{x})$ for all feasible $\mathbf{x}$ is called a minimizer or optimal point, and $F_0(\mathbf{x}^\sharp)$ is the optimal value.

We note that the equality constraint may be removed and represented by inequality constraints of the form $F_j(\mathbf{x}) \leq y_j$ and $-F_j(\mathbf{x}) \leq -y_j$ with $F_j(\mathbf{x}) := \langle \mathbf{A}_j, \mathbf{x} \rangle$ where $\mathbf{A}_j \in \mathbb{R}^N$ is the jth row of $\mathbf{A}$.

The set of feasible points described by the constraints is given by

$$K = \{\mathbf{x} \in \mathbb{R}^N : \mathbf{A}\mathbf{x} = \mathbf{y}, F_j(\mathbf{x}) \leq b_j, j \in [M]\}. \tag{B.22}$$

Two optimization problems are said to be equivalent if, given the solution of one problem, the solution of the other problem can be "easily" computed. For the purpose of this exposition, we settle for this vague definition of equivalence which will hopefully be clear in concrete situations.

The optimization problem (B.20) is equivalent to the problem of minimizing F_0 over K, i.e.,

$$\min_{\mathbf{x} \in K} F_0(\mathbf{x}). \tag{B.23}$$

Recalling that the characteristic function of K is

$$\chi_K(\mathbf{x}) = \begin{cases} 0 & \text{if } \mathbf{x} \in K, \\ \infty & \text{if } \mathbf{x} \notin K, \end{cases}$$

the optimization problem becomes as well equivalent to the unconstrained optimization problem

$$\min_{\mathbf{x} \in \mathbb{R}^N} F_0(\mathbf{x}) + \chi_K(\mathbf{x}).$$

A *convex optimization problem* (or *convex program*) is a problem of the form (B.20), in which the objective function F_0 and the constraint functions F_j are convex. In this case, the set of feasible points K defined in (B.22) is convex. The convex optimization problem is equivalent to the unconstrained optimization problem of minimizing the convex function

$$F(\mathbf{x}) = F_0(\mathbf{x}) + \chi_K(\mathbf{x}).$$

Due to this equivalence, we may freely switch between constrained and unconstrained optimization problems. Clearly, the statements of Proposition B.14 carry over to constrained optimization problems. We only note that, for constrained optimization problems, the function F_0 is usually taken to be finite, i.e., $\mathrm{dom}(F_0) = \mathbb{R}^N$.

In a *linear optimization problem* (or *linear program*), the objective function F_0 and all the constraint functions $F_1, \ldots, F_M$ are linear. This is a special case of a convex optimization problem.

The *Lagrange function* of an optimization problem of the form (B.20) is defined for $\mathbf{x} \in \mathbb{R}^N, \boldsymbol{\xi} \in \mathbb{R}^m, \boldsymbol{\nu} \in \mathbb{R}^M$ with $\nu_\ell \geq 0$ for all $\ell \in [M]$, by

$$L(\mathbf{x}, \boldsymbol{\xi}, \boldsymbol{\nu}) := F_0(\mathbf{x}) + \langle \boldsymbol{\xi}, \mathbf{A}\mathbf{x} - \mathbf{y} \rangle + \sum_{\ell=1}^{M} \nu_\ell(F_\ell(\mathbf{x}) - b_\ell). \tag{B.24}$$

For an optimization problem without inequality constraints, we just set

$$L(\mathbf{x}, \boldsymbol{\xi}) := F_0(\mathbf{x}) + \langle \boldsymbol{\xi}, \mathbf{A}\mathbf{x} - \mathbf{y} \rangle. \tag{B.25}$$

The variables $\boldsymbol{\xi}$ and $\boldsymbol{\nu}$ are called *Lagrange multipliers*. For ease of notation, we write $\boldsymbol{\nu} \succcurlyeq \mathbf{0}$ if $\nu_\ell \geq 0$ for all $\ell \in [M]$. The *Lagrange dual function* is defined by

$$H(\boldsymbol{\xi}, \boldsymbol{\nu}) := \inf_{\mathbf{x} \in \mathbb{R}^N} L(\mathbf{x}, \boldsymbol{\xi}, \boldsymbol{\nu}), \quad \boldsymbol{\xi} \in \mathbb{R}^m, \boldsymbol{\nu} \in \mathbb{R}^M, \boldsymbol{\nu} \succcurlyeq \mathbf{0}.$$

If $\mathbf{x} \mapsto L(\mathbf{x}, \boldsymbol{\xi}, \boldsymbol{\nu})$ is unbounded from below, then we set $H(\boldsymbol{\xi}, \boldsymbol{\nu}) = -\infty$. Again, if there are no inequality constraints, then

$$H(\boldsymbol{\xi}) := \inf_{\mathbf{x} \in \mathbb{R}^N} L(\mathbf{x}, \boldsymbol{\xi}) = \inf_{\mathbf{x} \in \mathbb{R}^N} \left\{ F_0(\mathbf{x}) + \langle \boldsymbol{\xi}, \mathbf{A}\mathbf{x} - \mathbf{y} \rangle \right\}, \quad \boldsymbol{\xi} \in \mathbb{R}^m.$$

The dual function is always concave (even if the original problem (B.20) is not convex) because it is the pointwise infimum of a family of affine functions. The dual function provides a bound on the optimal value of $F_0(\mathbf{x}^\sharp)$ for the minimization problem (B.20), namely,

$$H(\boldsymbol{\xi}, \boldsymbol{\nu}) \le F_0(\mathbf{x}^\sharp) \quad \text{for all } \boldsymbol{\xi} \in \mathbb{R}^m, \boldsymbol{\nu} \in \mathbb{R}^M, \boldsymbol{\nu} \succcurlyeq \mathbf{0}. \tag{B.26}$$

Indeed, if $\mathbf{x}$ is a feasible point for (B.20), then $\mathbf{A}\mathbf{x} - \mathbf{y} = \mathbf{0}$ and $F_\ell(\mathbf{x}) - b_\ell \le 0$ for all $\ell \in [M]$, so that, for all $\boldsymbol{\xi} \in \mathbb{R}^m$ and $\boldsymbol{\nu} \succcurlyeq \mathbf{0}$,

$$\langle \boldsymbol{\xi}, \mathbf{A}\mathbf{x} - \mathbf{y} \rangle + \sum_{\ell=1}^{M} \nu_\ell (F_\ell(\mathbf{x}) - b_\ell) \le 0.$$

Therefore,

$$L(\mathbf{x}, \boldsymbol{\xi}, \boldsymbol{\nu}) = F_0(\mathbf{x}) + \langle \boldsymbol{\xi}, \mathbf{A}\mathbf{x} - \mathbf{y} \rangle + \sum_{\ell=1}^{M} \nu_\ell (F_\ell(\mathbf{x}) - b_\ell) \le F_0(\mathbf{x}).$$

Using $H(\boldsymbol{\xi}, \boldsymbol{\nu}) \le L(\mathbf{x}, \boldsymbol{\xi}, \boldsymbol{\nu})$ and taking the infimum over all feasible $\mathbf{x} \in \mathbb{R}^N$ in the right-hand side yields (B.26). We would like this lower bound to be as tight as possible. This motivates us to consider the optimization problem

$$\max_{\boldsymbol{\xi} \in \mathbb{R}^m, \boldsymbol{\nu} \in \mathbb{R}^M} H(\boldsymbol{\xi}, \boldsymbol{\nu}) \quad \text{subject to } \boldsymbol{\nu} \succcurlyeq \mathbf{0}. \tag{B.27}$$

This optimization problem is called the *dual problem* to (B.20), which in this context is sometimes called the primal problem. Since H is concave, this problem is equivalent to the convex optimization problem of minimizing the convex function $-H$ subject to the positivity constraint $\boldsymbol{\nu} \succcurlyeq \mathbf{0}$. A pair $(\boldsymbol{\xi}, \boldsymbol{\nu})$ with $\boldsymbol{\xi} \in \mathbb{R}^m$ and $\boldsymbol{\nu} \in \mathbb{R}^M$ such that $\boldsymbol{\nu} \succcurlyeq \mathbf{0}$ is called dual feasible. A (feasible) maximizer $(\boldsymbol{\xi}^\sharp, \boldsymbol{\nu}^\sharp)$ of (B.27) is referred to as *dual optimal* or optimal Lagrange multipliers. If $\mathbf{x}^\sharp$ is optimal for the primal problem (B.20), then the triple $(\mathbf{x}^\sharp, \boldsymbol{\xi}^\sharp, \boldsymbol{\nu}^\sharp)$ is called primal–dual optimal. Inequality (B.26) shows that we always have

$$H(\boldsymbol{\xi}^\sharp, \boldsymbol{\nu}^\sharp) \le F(\mathbf{x}^\sharp). \tag{B.28}$$

This inequality is called *weak duality*. For most (but not all) convex optimization problems, strong duality holds, meaning that

$$H(\boldsymbol{\xi}^\sharp, \boldsymbol{\nu}^\sharp) = F(\mathbf{x}^\sharp). \tag{B.29}$$

Slater's constraint qualification, stated in a simplified form below, provides a condition ensuring strong duality.

Theorem B.26. *Assume that* $F_0, F_1, \ldots, F_M$ *are convex functions with* dom $(F_0) = \mathbb{R}^N$. *If there exists* $\mathbf{x} \in \mathbb{R}^N$ *such that* $\mathbf{A}\mathbf{x} = \mathbf{y}$ *and* $F_\ell(\mathbf{x}) < b_\ell$ *for all* $\ell \in [M]$, *then strong duality holds for the optimization problem* (B.20).

In the absence of inequality constraints, strong duality holds if there exists $\mathbf{x} \in \mathbb{R}^N$ *with* $\mathbf{A}\mathbf{x} = \mathbf{y}$, *i.e., if* (B.20) *is feasible.*

Proof. See, for instance, [70, Sect. 5.3.2] or [293, Satz 8.1.7]. □

Given a primal–dual feasible point $(\mathbf{x}, \boldsymbol{\xi}, \boldsymbol{\nu})$, that is, $\mathbf{x} \in \mathbb{R}^N$ is feasible for (B.20) and $\boldsymbol{\xi} \in \mathbb{R}^m$, $\boldsymbol{\nu} \in \mathbb{R}^M$ with $\boldsymbol{\nu} \succcurlyeq 0$, the primal–dual gap

$$E(\mathbf{x}, \boldsymbol{\xi}, \boldsymbol{\nu}) = F(\mathbf{x}) - H(\boldsymbol{\xi}, \boldsymbol{\nu}) \tag{B.30}$$

can be used to quantify how close $\mathbf{x}$ is to the minimizer $\mathbf{x}^\sharp$ of the primal problem (B.20) and how close $(\boldsymbol{\xi}, \boldsymbol{\nu})$ is to the maximizer of the dual problem (B.27). If $(\mathbf{x}^\sharp, \boldsymbol{\xi}^\sharp, \boldsymbol{\nu}^\sharp)$ is primal–dual optimal and strong duality holds, then $E(\mathbf{x}^\sharp, \boldsymbol{\xi}^\sharp, \boldsymbol{\nu}^\sharp) = 0$. The primal dual gap is often taken as a stopping criterion in iterative optimization methods.

For illustration, let us compute the dual problem of the ℓ_1-minimization problem

$$\min_{\mathbf{x} \in \mathbb{R}^N} \|\mathbf{x}\|_1 \quad \text{subject to } \mathbf{A}\mathbf{x} = \mathbf{y}. \tag{B.31}$$

The Lagrange function for this problem takes the form

$$L(\mathbf{x}, \boldsymbol{\xi}) = \|\mathbf{x}\|_1 + \langle \boldsymbol{\xi}, \mathbf{A}\mathbf{x} - \mathbf{y} \rangle.$$

The Lagrange dual function is

$$H(\boldsymbol{\xi}) = \inf_{\mathbf{x} \in \mathbb{R}^N} \{ \|\mathbf{x}\|_1 + \langle \mathbf{A}^*\boldsymbol{\xi}, \mathbf{x} \rangle - \langle \boldsymbol{\xi}, \mathbf{y} \rangle \}.$$

If $\|\mathbf{A}^*\boldsymbol{\xi}\|_\infty > 1$, then there exists $\mathbf{x} \in \mathbb{R}^N$ such that $\langle \mathbf{A}^*\boldsymbol{\xi}, \mathbf{x} \rangle < -\|\mathbf{x}\|_1$. Replacing $\mathbf{x}$ by $\lambda\mathbf{x}$ and letting $\lambda \to \infty$ show that $H(\boldsymbol{\xi}) = -\infty$ in this case. If $\|\mathbf{A}^*\boldsymbol{\xi}\|_\infty \leq 1$, then $\|\mathbf{x}\|_1 + \langle \mathbf{A}^*\boldsymbol{\xi}, \mathbf{x} \rangle \geq 0$. The choice $\mathbf{x} = \mathbf{0}$ therefore yields the infimum, and $H(\boldsymbol{\xi}) = -\langle \boldsymbol{\xi}, \mathbf{y} \rangle$. In conclusion,

$$H(\boldsymbol{\xi}) = \begin{cases} -\langle \boldsymbol{\xi}, \mathbf{y} \rangle & \text{if } \|\mathbf{A}^*\boldsymbol{\xi}\|_\infty \leq 1, \\ -\infty & \text{otherwise.} \end{cases}$$

Clearly, it is enough to maximize over the points $\boldsymbol{\xi}$ for which $H(\boldsymbol{\xi}) > -\infty$. Making this constraint explicit, the dual program to (B.31) is given by

$$\min_{\boldsymbol{\xi} \in \mathbb{R}^m} - \langle \boldsymbol{\xi}, \mathbf{y} \rangle \quad \text{subject to } \|\mathbf{A}^* \boldsymbol{\xi}\|_\infty \leq 1. \tag{B.32}$$

By Theorem B.26, strong duality holds for this pair of primal and dual optimization problems provided the primal problem (B.31) is feasible.

Remark B.27. In the complex case where $\mathbf{A} \in \mathbb{C}^{m \times N}$ and $\mathbf{y} \in \mathbb{C}^m$, the inner product has to be replaced by the real inner product $\mathrm{Re}\langle \mathbf{x}, \mathbf{y} \rangle$ as noted in the beginning of this chapter. Following the derivation above, we see that the dual program of (B.31), where the minimum now ranges over $\mathbf{x} \in \mathbb{C}^N$, is given by

$$\max_{\boldsymbol{\xi} \in \mathbb{C}^m} - \mathrm{Re}\langle \boldsymbol{\xi}, \mathbf{y} \rangle \quad \text{subject to } \|\mathbf{A}^* \boldsymbol{\xi}\|_\infty \leq 1. \tag{B.33}$$

A *conic optimization problem* is of the form

$$\min_{\mathbf{x} \in \mathbb{R}^N} F_0(\mathbf{x}) \quad \text{subject to } \mathbf{x} \in K, \tag{B.34}$$

$$\text{and } F_\ell(\mathbf{x}) \leq b_\ell, \quad \ell \in [M],$$

where K is a convex cone and the F_ℓ are convex functions. If K is a second-order cone, see (B.1) (possibly in a subset of variables, or the intersection of second-order cones in different variables), then the above problem is called a second-order cone problem. If K is the cone of positive semidefinite matrices, then the above optimization problem is called a semidefinite program.

Conic programs have their duality theory, too. The Lagrange function of a conic program of the form (B.34) is defined, for $\mathbf{x} \in \mathbb{R}^N$, $\boldsymbol{\xi} \in K^*$, and $\boldsymbol{\nu} \in \mathbb{R}^M$ with $\nu_\ell \geq 0$ for all $\ell \in [M]$, by

$$L(\mathbf{x}, \boldsymbol{\xi}, \boldsymbol{\nu}) := F_0(\mathbf{x}) - \langle \mathbf{x}, \boldsymbol{\xi} \rangle + \sum_{\ell=1}^{M} \nu_\ell (F_\ell(\mathbf{x}) - b_\ell),$$

where K^* is the dual cone of K defined in (B.2). (If there are no inequality constraints, then the last term above is omitted, of course.) The Lagrange dual function is then defined as

$$H(\boldsymbol{\xi}, \boldsymbol{\nu}) := \min_{\mathbf{x} \in \mathbb{R}^N} L(\mathbf{x}, \boldsymbol{\xi}, \boldsymbol{\nu}), \quad \boldsymbol{\xi} \in K^*, \boldsymbol{\nu} \in \mathbb{R}^M, \boldsymbol{\nu} \succcurlyeq 0.$$

Similarly to (B.26), the minimizer $\mathbf{x}^\sharp$ of (B.26) satisfies the lower bound

$$H(\boldsymbol{\xi}, \boldsymbol{\nu}) \leq F_0(\mathbf{x}^\sharp), \quad \text{for all } \boldsymbol{\xi} \in K^*, \boldsymbol{\nu} \in \mathbb{R}^M, \boldsymbol{\nu} \succcurlyeq 0. \tag{B.35}$$

Indeed, if $\mathbf{x} \in K$ and $F_\ell(\mathbf{x}) \leq b_\ell$ for all $\ell \in [M]$, then $\langle \mathbf{x}, \boldsymbol{\xi} \rangle \geq 0$ for all $\boldsymbol{\xi} \in K^*$ by definition (B.2) of the dual cone, and with $\boldsymbol{\nu} \succcurlyeq \mathbf{0}$,

$$-\langle \mathbf{x}, \boldsymbol{\xi} \rangle + \sum_{\ell=1}^{M} \nu_\ell (F_\ell(\mathbf{x}) - b_\ell) \leq 0.$$

Therefore,

$$L(\mathbf{x}, \boldsymbol{\xi}, \boldsymbol{\nu}) = F_0(\mathbf{x}) - \langle \mathbf{x}, \boldsymbol{\xi} \rangle + \sum_{\ell=1}^{M} \nu_\ell (F_\ell(\mathbf{x}) - b_\ell) \leq F_0(\mathbf{x}).$$

This point establishes (B.35). The dual program of (B.34) is then defined as

$$\max_{\boldsymbol{\xi} \in \mathbb{R}^m, \boldsymbol{\nu} \in \mathbb{R}^M} H(\boldsymbol{\xi}, \boldsymbol{\nu}) \quad \text{subject to } \boldsymbol{\xi} \in K^* \text{ and } \boldsymbol{\nu} \succcurlyeq \mathbf{0}. \tag{B.36}$$

Denoting by $(\boldsymbol{\xi}^\sharp, \boldsymbol{\nu}^\sharp)$ a dual optimum, that is, a maximizer of the program (B.36), and by $\mathbf{x}^\sharp$ a minimizer of the primal program (B.34), the triple $(\mathbf{x}^\sharp, \boldsymbol{\xi}^\sharp, \boldsymbol{\nu}^\sharp)$ is called a primal–dual optimum. The above arguments establish the weak duality

$$H(\boldsymbol{\xi}^\sharp, \boldsymbol{\nu}^\sharp) \leq F_0(\mathbf{x}^\sharp). \tag{B.37}$$

If this is an equality, then we say that strong duality holds. Similar conditions to Slater's constraint qualification (Theorem B.26) ensure strong duality for conic programs—for instance, if there exists a point in the interior of K such that all inequality constraints hold strictly; see, e.g., [70, Sect. 5.9].

Let us illustrate duality for conic programs with an example relevant to Sect. 9.2. For a convex cone K and a vector $\mathbf{g} \in \mathbb{R}^N$, we consider the optimization problem

$$\min_{\mathbf{x} \in \mathbb{R}^N} \langle \mathbf{x}, \mathbf{g} \rangle \quad \text{subject to } \mathbf{x} \in K \text{ and } \|\mathbf{x}\|_2^2 \leq 1.$$

Its Lagrange function is given by

$$L(\mathbf{x}, \boldsymbol{\xi}, \nu) = \langle \mathbf{x}, \mathbf{g} \rangle - \langle \boldsymbol{\xi}, \mathbf{x} \rangle + \nu(\|\mathbf{x}\|_2^2 - 1), \quad \boldsymbol{\xi} \in K^*, \nu \geq 0.$$

The minimum with respect to $\mathbf{x}$ of the Lagrange function is attained at $\mathbf{x} = (2\nu)^{-1}(\boldsymbol{\xi} - \mathbf{g})$. By substituting this value into L, the Lagrange dual function turns out to be

$$H(\boldsymbol{\xi}, \nu) = -\frac{1}{4\nu} \|\mathbf{g} - \boldsymbol{\xi}\|_2^2 - \nu.$$

This leads to the dual program

$$\max_{\boldsymbol{\xi}, \nu} \left(-\nu - \frac{1}{4\nu} \|\mathbf{g} - \boldsymbol{\xi}\|_2^2 \right) \quad \text{subject to } \boldsymbol{\xi} \in K^* \text{ and } \nu \geq 0.$$

Solving this optimization program with respect to ν gives $\nu = \frac{1}{2}\|\mathbf{g} - \boldsymbol{\xi}\|_2$, so we arrive at the dual program

$$\max_{\boldsymbol{\xi}} \; - \|\mathbf{g} - \boldsymbol{\xi}\|_2 \quad \text{subject to } \boldsymbol{\xi} \in K^*. \tag{B.38}$$

Note that the maximizer is the orthogonal projection of $\mathbf{g}$ onto the dual cone K^*, which always exists since K^* is convex and closed. Weak duality for this case reads $\max_{\boldsymbol{\xi} \in K^*} -\|\mathbf{g} - \boldsymbol{\xi}\|_2 \leq \min_{\mathbf{x} \in K, \|\mathbf{x}\|_2 \leq 1} \langle \mathbf{g}, \mathbf{x} \rangle$, or

$$\max_{\mathbf{x} \in K, \|\mathbf{x}\|_2 \leq 1} \langle -\mathbf{g}, \mathbf{x} \rangle \leq \min_{\boldsymbol{\xi} \in K^*} \|\mathbf{g} - \boldsymbol{\xi}\|_2. \tag{B.39}$$

In fact, strong duality—that is, equality above—often holds, for instance, if K has nonempty interior. Note that the inequality (B.39) for $-\mathbf{g}$ instead of $\mathbf{g}$ can be rewritten in terms of the polar cone $K^\circ = -K^*$ introduced in (B.3) as

$$\max_{\mathbf{x} \in K, \|\mathbf{x}\|_2 \leq 1} \langle \mathbf{g}, \mathbf{x} \rangle \leq \min_{\boldsymbol{\xi} \in K^\circ} \|\mathbf{g} - \boldsymbol{\xi}\|_2. \tag{B.40}$$

Lagrange duality has a saddle-point interpretation. For ease of exposition, we consider (B.20) without inequality constraints, but extensions that include inequality constraints or conic programs are derived in the same way.

Let $(\mathbf{x}^\sharp, \boldsymbol{\xi}^\sharp)$ be a primal–dual optimal point. Recalling the definition of the Lagrange function L, we have

$$\sup_{\boldsymbol{\xi} \in \mathbb{R}^m} L(\mathbf{x}, \boldsymbol{\xi}) = \sup_{\boldsymbol{\xi} \in \mathbb{R}^m} F_0(\mathbf{x}) + \langle \boldsymbol{\xi}, \mathbf{A}\mathbf{x} - \mathbf{y} \rangle$$

$$= \begin{cases} F_0(\mathbf{x}) & \text{if } \mathbf{A}\mathbf{x} = \mathbf{y}, \\ \infty & \text{otherwise.} \end{cases} \tag{B.41}$$

In other words, the above supremum is infinite if $\mathbf{x}$ is not feasible. A (feasible) minimizer $\mathbf{x}^\sharp$ of the primal problem (B.20) therefore satisfies

$$F_0(\mathbf{x}^\sharp) = \inf_{\mathbf{x} \in \mathbb{R}^N} \sup_{\boldsymbol{\xi} \in \mathbb{R}^m} L(\mathbf{x}, \boldsymbol{\xi}).$$

On the other hand, a dual optimal vector $\boldsymbol{\xi}^\sharp$ satisfies

$$H(\boldsymbol{\xi}^\sharp) = \sup_{\boldsymbol{\xi} \in \mathbb{R}^m} \inf_{\mathbf{x} \in \mathbb{R}^N} L(\mathbf{x}, \boldsymbol{\xi})$$

by definition of the Lagrange dual function. Thus, weak duality implies

$$\sup_{\boldsymbol{\xi} \in \mathbb{R}^m} \inf_{\mathbf{x} \in \mathbb{R}^N} L(\mathbf{x}, \boldsymbol{\xi}) \leq \inf_{\mathbf{x} \in \mathbb{R}^N} \sup_{\boldsymbol{\xi} \in \mathbb{R}^m} L(\mathbf{x}, \boldsymbol{\xi})$$

(which can be derived directly), while strong inequality reads

$$\sup_{\boldsymbol{\xi}\in\mathbb{R}^m}\inf_{\mathbf{x}\in\mathbb{R}^N} L(\mathbf{x},\boldsymbol{\xi}) = \inf_{\mathbf{x}\in\mathbb{R}^N}\sup_{\boldsymbol{\xi}\in\mathbb{R}^m} L(\mathbf{x},\boldsymbol{\xi}).$$

In other words, the order of minimization and maximization can be interchanged in the case of strong duality. This property is called the strong *max–min property* or *saddle-point property*. Indeed, in this case, a primal–dual optimal $(\mathbf{x}^\sharp, \boldsymbol{\xi}^\sharp)$ is a saddle point of the Lagrange function, that is,

$$L(\mathbf{x}^\sharp, \boldsymbol{\xi}) \le L(\mathbf{x}^\sharp, \boldsymbol{\xi}^\sharp) \le L(\mathbf{x}, \boldsymbol{\xi}^\sharp) \qquad \text{for all } \mathbf{x} \in \mathbb{R}^N, \boldsymbol{\xi} \in \mathbb{R}^m. \tag{B.42}$$

Jointly optimizing the primal and dual problem is therefore equivalent to finding a saddle point of the Lagrange function provided that strong duality holds.

Based on these findings, we establish below a theorem relating some convex optimization problems relevant to this book. The theorem also holds in the complex setting by interpreting $\mathbb{C}^N$ as $\mathbb{R}^{2N}$.

Theorem B.28. *Let* $\| \cdot \|$ *be a norm on* $\mathbb{R}^m$ *and* $\|\cdot\|$ *be a norm on* $\mathbb{R}^N$. *For* $\mathbf{A} \in \mathbb{R}^{m\times N}$, $\mathbf{y} \in \mathbb{R}^m$, *and* $\tau > 0$, *consider the optimization problem*

$$\min_{\mathbf{x}\in\mathbb{R}^N} \|\mathbf{A}\mathbf{x} - \mathbf{y}\| \quad \text{subject to } \|\mathbf{x}\| \le \tau. \tag{B.43}$$

If $\mathbf{x}^\sharp$ *is a minimizer of* (B.43), *then there exists a parameter* $\lambda \ge 0$ *such that* $\mathbf{x}^\sharp$ *is also a minimizer of the optimization problem*

$$\min_{\mathbf{x}\in\mathbb{R}^N} \lambda\|\mathbf{x}\| + \|\mathbf{A}\mathbf{x} - \mathbf{y}\|^2. \tag{B.44}$$

Conversely, for $\lambda \ge 0$, *if* $\mathbf{x}^\sharp$ *is a minimizer of* (B.44), *then there exists* $\tau \ge 0$ *such that* $\mathbf{x}^\sharp$ *is a minimizer of* (B.43).

Proof. The optimization problem (B.43) is equivalent to

$$\min_{\mathbf{x}\in\mathbb{R}^N} \|\mathbf{A}\mathbf{x} - \mathbf{y}\|^2 \quad \text{subject to } \|\mathbf{x}\| \le \tau.$$

The Lagrange function of this minimization problem is given by

$$L(\mathbf{x}, \xi) = \|\mathbf{A}\mathbf{x} - \mathbf{y}\|^2 + \xi(\|\mathbf{x}\| - \tau).$$

Since for $\tau > 0$, there exist vectors $\mathbf{x}$ with $\|\mathbf{x}\| < \tau$, so that Theorem B.26 implies strong duality for (B.43). Therefore, there exists a dual optimal $\xi^\sharp \ge 0$. The saddle-point property (B.42) implies that $L(\mathbf{x}^\sharp, \xi^\sharp) \le L(\mathbf{x}, \xi^\sharp)$ for all $\mathbf{x} \in \mathbb{R}^N$. Therefore, $\mathbf{x}^\sharp$ is also a minimizer of $\mathbf{x} \mapsto L(\mathbf{x}, \xi^\sharp)$. Since the constant term $-\xi^\sharp\tau$ does not affect the minimizer, the conclusion follows with $\lambda = \xi^\sharp$.

For the converse statement, one can proceed directly as in the proof of Proposition 3.2(a). □

Remark B.29. (a) The same type of statement and proof is valid for the pair of optimization problems (B.44) and

$$\min_{\mathbf{x} \in \mathbb{R}^N} \|\mathbf{x}\| \quad \text{subject to } \|\mathbf{A}\mathbf{x} - \mathbf{y}\| \leq \eta \tag{B.45}$$

under a strict feasibility assumption for (B.45).

(b) The parameter transformations between τ, λ, η depend on the minimizers and can only be performed after solving the optimization problems. Thus, the theoretical equivalence of the problems (B.43)–(B.45) is too implicit for practical purposes.

For the remainder of this section, we consider a convex optimization problem of the form

$$\min_{\mathbf{x} \in \mathbb{R}^N} F(\mathbf{A}\mathbf{x}) + G(\mathbf{x}), \tag{B.46}$$

where $\mathbf{A} \in \mathbb{R}^{m \times N}$ and where $F : \mathbb{R}^m \to (-\infty, \infty]$, $G : \mathbb{R}^N \to (-\infty, \infty]$ are convex functions. All the relevant optimization problems appearing in this book belong to this class; see Sect. 15.2. For instance, the choice $G(\mathbf{x}) = \|\mathbf{x}\|_1$ and $F = \chi_{\{\mathbf{y}\}}$, the characteristic function (B.7) of the singleton $\{\mathbf{y}\}$, leads to the ℓ_1-minimization problem (B.31).

The substitution $\mathbf{z} = \mathbf{A}\mathbf{x}$ yields the equivalent problem

$$\min_{\mathbf{x} \in \mathbb{R}^N, \mathbf{z} \in \mathbb{R}^m} F(\mathbf{z}) + G(\mathbf{x}) \quad \text{subject to } \mathbf{A}\mathbf{x} - \mathbf{z} = \mathbf{0}. \tag{B.47}$$

The Lagrange dual function to this problem is given by

$$H(\boldsymbol{\xi}) = \inf_{\mathbf{x}, \mathbf{z}} \{F(\mathbf{z}) + G(\mathbf{x}) + \langle \mathbf{A}^*\boldsymbol{\xi}, \mathbf{x} \rangle - \langle \boldsymbol{\xi}, \mathbf{z} \rangle\}$$

$$= -\sup_{\mathbf{z} \in \mathbb{R}^m} \{\langle \boldsymbol{\xi}, \mathbf{z} \rangle - F(\mathbf{z})\} - \sup_{\mathbf{x} \in \mathbb{R}^N} \{\langle \mathbf{x}, -\mathbf{A}^*\boldsymbol{\xi} \rangle - G(\mathbf{x})\}$$

$$= -F^*(\boldsymbol{\xi}) - G^*(-\mathbf{A}^*\boldsymbol{\xi}), \tag{B.48}$$

where F^* and G^* are the convex conjugate functions of F and G, respectively. Therefore, the dual problem of (B.46) is

$$\max_{\boldsymbol{\xi} \in \mathbb{R}^m} \left(-F^*(\boldsymbol{\xi}) - G^*(-\mathbf{A}^*\boldsymbol{\xi}) \right). \tag{B.49}$$

Since the minimal values of (B.46) and (B.47) coincide, we refer to (B.49) also as the dual problem of (B.46)—although strictly speaking (B.49) is not the dual to (B.46) in the sense described above. (Indeed, an unconstrained optimization problem does not introduce dual variables in the Lagrange function. In general, equivalent problems may have nonequivalent duals.)

The following theorem states strong duality of the problems (B.46) and (B.49).

Theorem B.30. *Let* $\mathbf{A} \in \mathbb{R}^{m \times N}$ *and* $F : \mathbb{R}^m \to (-\infty, \infty]$, $G : \mathbb{R}^N \to (-\infty, \infty]$ *be proper convex functions with either* $\mathrm{dom}(F) = \mathbb{R}^m$ *or* $\mathrm{dom}(G) = \mathbb{R}^N$ *and such that there exists* $\mathbf{x} \in \mathbb{R}^N$ *such that* $\mathbf{Ax} \in \mathrm{dom}(F)$. *Assume that the optima in* (B.46) *and* (B.49) *are attained. Then strong duality holds in the form*

$$\min_{\mathbf{x} \in \mathbb{R}^N} \left(F(\mathbf{Ax}) + G(\mathbf{x}) \right) = \max_{\boldsymbol{\xi} \in \mathbb{R}^m} \left(-F^*(\boldsymbol{\xi}) - G^*(-\mathbf{A}^*\boldsymbol{\xi}) \right).$$

Furthermore, a primal–dual optimum $(\mathbf{x}^\sharp, \boldsymbol{\xi}^\sharp)$ *is a solution to the saddle-point problem*

$$\min_{\mathbf{x} \in \mathbb{R}^N} \max_{\boldsymbol{\xi} \in \mathbb{R}^m} \langle \mathbf{Ax}, \boldsymbol{\xi} \rangle + G(\mathbf{x}) - F^*(\boldsymbol{\xi}), \tag{B.50}$$

where F^* *is the convex conjugate of* F.

Proof. The first statement follows from the Fenchel duality theorem; see, e.g., [424, Theorem 31.1]. Strong duality implies the saddle-point property (B.42) of the Lagrange function. By (B.48), the value of the Lagrange function at the primal–dual optimal point is the optimal value of the min–max problem, i.e.,

$$\min_{\mathbf{x}, \mathbf{z} \in \mathbb{R}^N} \max_{\boldsymbol{\xi} \in \mathbb{R}^m} F(\mathbf{z}) + G(\mathbf{x}) + \langle \mathbf{A}^*\boldsymbol{\xi}, \mathbf{x} \rangle - \langle \boldsymbol{\xi}, \mathbf{z} \rangle$$

$$= \min_{\mathbf{x} \in \mathbb{R}^N} \max_{\boldsymbol{\xi} \in \mathbb{R}^m} - \left(\min_{\mathbf{z} \in \mathbb{R}^m} \langle \boldsymbol{\xi}, \mathbf{z} \rangle - F(\mathbf{z}) \right) + \langle \mathbf{A}^*\boldsymbol{\xi}, \mathbf{x} \rangle + G(\mathbf{x})$$

$$= \min_{\mathbf{x} \in \mathbb{R}^N} \max_{\boldsymbol{\xi} \in \mathbb{R}^m} \langle \mathbf{Ax}, \boldsymbol{\xi} \rangle + G(\mathbf{x}) - F^*(\boldsymbol{\xi})$$

by definition of the convex conjugate function. The interchange of the minimum and maximum above is justified by the fact that if $((\mathbf{x}^\sharp, \mathbf{z}^\sharp), \boldsymbol{\xi}^\sharp)$ is a saddle point of $L((\mathbf{x}, \mathbf{z}), \boldsymbol{\xi})$, then $(\mathbf{x}^\sharp, \boldsymbol{\xi}^\sharp)$ is a saddle point of $H(\mathbf{x}, \boldsymbol{\xi}) = \min_{\mathbf{z}} L((\mathbf{x}, \mathbf{z}), \boldsymbol{\xi})$. $\quad\square$

The condition $\mathrm{dom}(F) = \mathbb{R}^m$ or $\mathrm{dom}(G) = \mathbb{R}^N$ above may be relaxed; see, e.g., [424, Theorem 31.1].

B.6   Matrix Convexity

This section uses the notion of matrix functions introduced in Sect. A.5, in particular the matrix exponential and the matrix logarithm. The main goal is to show the following concavity theorem due to Lieb [325], which is a key ingredient in the proof of the noncommutative Bernstein inequality in Sect. 8.5.

Theorem B.31. *For* $\mathbf{H}$ *is a self-adjoint matrix, the function*

$$\mathbf{X} \mapsto \mathrm{tr}\ \exp(\mathbf{H} + \ln(\mathbf{X}))$$

is concave on the set of positive definite matrices.

While the original proof [325] and its variants [187, 439, 440] rely on complex analysis, we proceed as in [485]; see also [50]. This requires us to introduce some background from matrix convexity and some concepts from quantum information theory.

Given a function $f : I \to \mathbb{R}$ defined on an interval $I \subset \mathbb{R}$, we recall that f is extended to self-adjoint matrices $\mathbf{A}$ with eigenvalues contained in I by (A.42). Similarly to the definition (A.49) of operator monotonicity, we say that f is *matrix convex* (or operator convex) if, for all integers $n \geq 1$, all self-adjoint matrices $\mathbf{A}, \mathbf{B} \in \mathbb{C}^{n \times n}$ with eigenvalues in I, and all $t \in [0, 1]$,

$$f(t\mathbf{A} + (1 - t)\mathbf{B}) \preccurlyeq tf(\mathbf{A}) + (1 - t)f(\mathbf{B}). \tag{B.51}$$

Equivalently, f is matrix convex if, for all $n \geq 1$ and all $\mathbf{x} \in \mathbb{C}^n$, the scalar-valued function $\mathbf{A} \mapsto \langle f(\mathbf{A})\mathbf{x}, \mathbf{x} \rangle$ is convex on the set of self-adjoint matrices in $\mathbb{C}^{n \times n}$ with eigenvalues in I. Like matrix monotonicity, matrix convexity is a much stronger property than the usual scalar convexity.

We start with a simple characterization in terms of orthogonal projections, i.e., of self-adjoint matrices $\mathbf{P}$ that satisfy $\mathbf{P}^2 = \mathbf{P}$. Here and in the following, when matrix dimensions are not specified, they are arbitrary, but the matrices are assumed to have matching dimensions so that matrix multiplication is well defined.

Theorem B.32. *Let* $I \subset \mathbb{R}$ *be an interval containing* 0 *and let* $f : I \to \mathbb{R}$. *Then* f *is matrix convex and* $f(0) \leq 0$ *if and only if* $f(\mathbf{PAP}) \preccurlyeq \mathbf{P}f(\mathbf{A})\mathbf{P}$ *for all orthogonal projections* $\mathbf{P}$ *and all self-adjoint matrices* $\mathbf{A}$ *with eigenvalues in* I.

Proof. We prove matrix convexity based on the given condition, since only this direction is needed later. A proof of the converse direction and of further equivalences can be found in [47, Theorem V.2.3] or [258, Theorem 2.1].

Let $\mathbf{A}, \mathbf{B}$ be self-adjoint matrices with eigenvalues in I and let $t \in [0, 1]$. Define $\mathbf{T}, \mathbf{P}, \mathbf{V}_t$ to be the block matrices

$$\mathbf{T} = \begin{bmatrix} \mathbf{A} & 0 \\ 0 & \mathbf{B} \end{bmatrix}, \quad \mathbf{P} = \begin{bmatrix} \mathrm{Id} & 0 \\ 0 & 0 \end{bmatrix}, \quad \mathbf{V}_t = \begin{bmatrix} \sqrt{t}\,\mathrm{Id} & -\sqrt{1-t}\,\mathrm{Id} \\ \sqrt{1-t}\,\mathrm{Id} & \sqrt{t}\,\mathrm{Id} \end{bmatrix}.$$

The matrix $\mathbf{V}_t$ is unitary and the matrix $\mathbf{P}$ is an orthogonal projection. We observe that

$$\mathbf{PV}_t^*\mathbf{TV}_t\mathbf{P} = \begin{bmatrix} t\mathbf{A} + (1-t)\mathbf{B} & 0 \\ 0 & 0 \end{bmatrix}.$$

Therefore, by (A.44) and by the hypothesis, we have

$$\begin{bmatrix} f(t\mathbf{A} + (1-t)\mathbf{B}) & \mathbf{0} \\ \mathbf{0} & f(0) \end{bmatrix} = f(\mathbf{P}\mathbf{V}_t^*\mathbf{T}\mathbf{V}_t\mathbf{P}) \preccurlyeq \mathbf{P}f(\mathbf{V}_t^*\mathbf{T}\mathbf{V}_t)\mathbf{P}$$

$$= \mathbf{P}\mathbf{V}_t^* f(\mathbf{T})\mathbf{V}_t\mathbf{P} = \begin{bmatrix} tf(\mathbf{A}) + (1-t)f(\mathbf{B}) & \mathbf{0} \\ \mathbf{0} & \mathbf{0} \end{bmatrix}.$$

The first equality in the second line is valid since $\mathbf{V}_t$ is unitary. This shows that f is matrix convex and that $f(0) \le 0$.   □

Theorem B.33. *Let f be a continuous function on $[0, \infty)$. Then f is matrix convex and $f(0) \le 0$ if and only if the function $g(t) = f(t)/t$ is matrix monotone on $(0, \infty)$.*

Proof. We prove that f is matrix convex with $f(0) \le 0$ if g is matrix monotone, since only this direction is needed later. For the converse direction, we refer to [47, Theorem V.2.9] or [258, Theorem 2.4].

First, we note that the (scalar) monotonicity of $f(t)/t$ on $(0, \infty)$ implies that $f(t)/t \le f(1)/1$, i.e., $f(t) \le tf(1)$, for all $0 < t \le 1$. Letting $t \to 0$ shows that $f(0) \le 0$.

Let now $\mathbf{A}$ be an arbitrary self-adjoint matrix with eigenvalues in $(0, \infty)$ and $\mathbf{P}$ be an arbitrary orthogonal projection (of the same dimension as $\mathbf{A}$). According to Theorem B.32, to prove that f is matrix convex, we need to show that $f(\mathbf{P}\mathbf{A}\mathbf{P}) \preccurlyeq \mathbf{P}f(\mathbf{A})\mathbf{P}$. For $\varepsilon > 0$, the fact that $\mathbf{P} + \varepsilon\mathbf{Id} \preccurlyeq (1 + \varepsilon)\mathbf{Id}$ implies that $\mathbf{A}^{1/2}(\mathbf{P} + \varepsilon\mathbf{Id})\mathbf{A}^{1/2} \preccurlyeq (1 + \varepsilon)\mathbf{A}$ by Lemma A.31. The matrix monotonicity of g then yields

$$f(\mathbf{A}^{1/2}(\mathbf{P} + \varepsilon\mathbf{Id})\mathbf{A}^{1/2})\mathbf{A}^{-1/2}(\mathbf{P} + \varepsilon\mathbf{Id})^{-1}\mathbf{A}^{-1/2} \preccurlyeq f((1+\varepsilon)\mathbf{A})(1+\varepsilon)^{-1}\mathbf{A}^{-1}.$$

Multiplying on the left by $(\mathbf{P} + \varepsilon\mathbf{Id})\mathbf{A}^{1/2}$ and on the right by $\mathbf{A}^{1/2}(\mathbf{P} + \varepsilon\mathbf{Id})$ and using Lemma A.31, we arrive at

$$(\mathbf{P} + \varepsilon\mathbf{Id})\mathbf{A}^{1/2}f(\mathbf{A}^{1/2}(\mathbf{P} + \varepsilon\mathbf{Id})\mathbf{A}^{1/2})\mathbf{A}^{-1/2}$$

$$\preccurlyeq (\mathbf{P} + \varepsilon\mathbf{Id})\mathbf{A}^{1/2}f((1+\varepsilon)\mathbf{A})(1+\varepsilon)^{-1}\mathbf{A}^{-1/2}(\mathbf{P} + \varepsilon\mathbf{Id}).$$

By continuity of f, letting $\varepsilon \to 0$ gives

$$\mathbf{P}\mathbf{A}^{1/2}f(\mathbf{A}^{1/2}\mathbf{P}\mathbf{A}^{1/2})\mathbf{A}^{-1/2} \preccurlyeq \mathbf{P}\mathbf{A}^{1/2}f(\mathbf{A})\mathbf{A}^{-1/2}\mathbf{P}. \qquad (\text{B.52})$$

The right-hand side of (B.52) equals $\mathbf{P}f(\mathbf{A})\mathbf{P}$ because $\mathbf{A}^{1/2}$ and $f(\mathbf{A})$ commute. Thus, it remains to show that the left-hand side of (B.52) equals $f(\mathbf{P}\mathbf{A}\mathbf{P})$. Such an identity holds for monomials $h(t) = t^n$, since

$$\mathbf{PA}^{1/2}h(\mathbf{A}^{1/2}\mathbf{PA}^{1/2})\mathbf{A}^{-1/2}$$

$$= \mathbf{PA}^{1/2}(\mathbf{A}^{1/2}\mathbf{PA}^{1/2})(\mathbf{A}^{1/2}\mathbf{PA}^{1/2})\cdots(\mathbf{A}^{1/2}\mathbf{PA}^{1/2})\mathbf{A}^{-1/2}$$

$$= \mathbf{PAPA}\cdots\mathbf{PAP} = (\mathbf{PAP})(\mathbf{PAP})\cdots(\mathbf{PAP}) = h(\mathbf{PAP}).$$

By linearity, the identity $\mathbf{PA}^{1/2}h(\mathbf{A}^{1/2}\mathbf{PA}^{1/2})\mathbf{A}^{-1/2} = h(\mathbf{PAP})$ holds for all polynomials h. We now consider an interpolating polynomial h satisfying $h(\lambda) = f(\lambda)$ for all λ in the set of eigenvalues of $\mathbf{A}^{1/2}\mathbf{PA}^{1/2}$—which equals the set of eigenvalues of $\mathbf{PAP}$ since $\mathbf{A}^{1/2}\mathbf{PA}^{1/2} = (\mathbf{A}^{1/2}\mathbf{P})(\mathbf{PA}^{1/2})$ and $\mathbf{PAP} = (\mathbf{PA}^{1/2})(\mathbf{A}^{1/2}\mathbf{P})$. This yields $h(\mathbf{A}^{1/2}\mathbf{PA}^{1/2}) = f(\mathbf{A}^{1/2}\mathbf{PA}^{1/2})$ and $h(\mathbf{PAP}) = f(\mathbf{PAP})$ by the definition (A.42) of matrix functions. It follows that $\mathbf{PA}^{1/2}f(\mathbf{A}^{1/2}\mathbf{PA}^{1/2})\mathbf{A}^{-1/2} = f(\mathbf{PAP})$, as desired. By continuity of f, the relation $f(\mathbf{PAP}) \preccurlyeq \mathbf{P}f(\mathbf{A})\mathbf{P}$ extends to all self-adjoint matrices $\mathbf{A}$ with eigenvalues in $[0, \infty)$. The claim follows therefore from Theorem B.32. □

We are particularly interested in the following special case.

Corollary B.34. *The continuous function defined by $\phi(x) = x\ln(x)$ for $x > 0$ and by $\phi(0) = 0$ is matrix convex on $[0, \infty)$.*

Proof. Combine Proposition A.34 with Theorem B.33.

Next we state the affine version of the Hansen–Pedersen–Jensen inequality.

Theorem B.35. *Let f be a matrix convex function on some interval $I \subset \mathbb{R}$ and let $\mathbf{X}_1, \ldots, \mathbf{X}_n \in \mathbb{C}^{d\times d}$ be square matrices such that $\sum_{j=1}^{n}\mathbf{X}_j^*\mathbf{X}_j = \mathbf{Id}$. Then, for all self-adjoint matrices $\mathbf{A}_1, \ldots, \mathbf{A}_n \in \mathbb{C}^{d\times d}$ with eigenvalues in I,*

$$f\left(\sum_{j=1}^{n}\mathbf{X}_j^*\mathbf{A}_j\mathbf{X}_j\right) \preccurlyeq \sum_{j=1}^{n}\mathbf{X}_j^*f(\mathbf{A}_j)\mathbf{X}_j. \tag{B.53}$$

There is a converse to this theorem, in the sense that if (B.53) holds for arbitrary choices of $\mathbf{A}_j$ and $\mathbf{X}_j$, then f is matrix convex; see [259, Theorem 2.1]. The proof requires the following auxiliary lemma.

Lemma B.36. *Let $\mathbf{B}_{j,\ell} \in \mathbb{C}^{m\times m}$, $j, \ell \in [n]$, be a double sequence of square matrices, and let $\mathbf{B}$ be the block matrix $\mathbf{B} = (\mathbf{B}_{j,\ell}) \in \mathbb{C}^{mn\times mn}$. With $\omega = e^{2\pi i/n}$, let $\mathbf{E} \in \mathbb{C}^{mn\times mn}$ be the unitary block-diagonal matrix $\mathbf{E} = \mathrm{diag}[\omega\mathbf{Id}, \ldots, \omega^n\mathbf{Id}]$. Then*

$$\frac{1}{n}\sum_{k=1}^{n}\mathbf{E}^{-k}\mathbf{BE}^k = \mathrm{diag}[\mathbf{B}_{1,1}, \mathbf{B}_{2,2}, \ldots, \mathbf{B}_{n,n}].$$

Proof. A direct computation shows that

$$(\mathbf{E}^{-k}\mathbf{BE}^k)_{j,\ell} = \omega^{k(\ell-j)}\mathbf{B}_{j,\ell}.$$

The formula for geometric sums implies, for $j, \ell \in [n]$,

$$\sum_{k=1}^{n} \omega^{k(\ell-j)} = \sum_{k=1}^{n} e^{2\pi i(\ell-j)k/n} = \begin{cases} n & \text{if } \ell = j, \\ 0 & \text{otherwise.} \end{cases}$$

The claim follows directly. $\square$

Proof (of Theorem B.35). Define the block matrices

$$\mathbf{X} = \begin{bmatrix} \mathbf{X}_1 \\ \mathbf{X}_2 \\ \vdots \\ \mathbf{X}_n \end{bmatrix} \in \mathbb{R}^{nd \times d} \quad \text{and} \quad \mathbf{U} = \begin{bmatrix} \mathbf{Id} - \mathbf{X}\mathbf{X}^* & \mathbf{X} \\ -\mathbf{X}^* & \mathbf{0} \end{bmatrix} \in \mathbb{R}^{(n+1)d \times (n+1)d}.$$

We observe that $\mathbf{X}^*\mathbf{X} = \mathbf{Id}$ and in turn that

$$\mathbf{U}\mathbf{U}^* = \begin{bmatrix} (\mathbf{Id} - \mathbf{X}\mathbf{X}^*)^2 + \mathbf{X}\mathbf{X}^* & -(\mathbf{Id} - \mathbf{X}\mathbf{X}^*)\mathbf{X} \\ -\mathbf{X}^*(\mathbf{Id} - \mathbf{X}\mathbf{X}^*) & \mathbf{X}^*\mathbf{X} \end{bmatrix} = \mathbf{Id}.$$

Similarly, we compute $\mathbf{U}^*\mathbf{U} = \mathbf{Id}$. Therefore, $\mathbf{U}$ is a unitary matrix (called the unitary dilation of $\mathbf{X}$). We partition $\mathbf{U}$ into $d \times d$ blocks $(\mathbf{U}_{j,\ell})_{j,\ell \in [n+1]}$. Note that $\mathbf{U}_{k,n+1} = \mathbf{X}_k$ for $k \in [n]$ and that $\mathbf{U}_{n+1,n+1} = \mathbf{0}$. Furthermore, let $\mathbf{A}$ be the block-diagonal matrix $\mathbf{A} = \text{diag}[\mathbf{A}_1, \mathbf{A}_2, \ldots, \mathbf{A}_n, \mathbf{0}]$. Using Lemma B.36 with n replaced by $n+1$ together with the matrix convexity of f, we obtain

$$f\left(\sum_{j=1}^{n} \mathbf{X}_j^* \mathbf{A}_j \mathbf{X}_j\right) = f\left((\mathbf{U}^*\mathbf{A}\mathbf{U})_{n+1,n+1}\right)$$

$$= f\left(\left(\frac{1}{n+1}\sum_{k=1}^{n+1} \mathbf{E}^{-k}\mathbf{U}^*\mathbf{A}\mathbf{U}\mathbf{E}^k\right)_{n+1,n+1}\right)$$

$$= \left(f\left(\frac{1}{n+1}\sum_{k=1}^{n+1} \mathbf{E}^{-k}\mathbf{U}^*\mathbf{A}\mathbf{U}\mathbf{E}^k\right)\right)_{n+1,n+1}$$

$$\preccurlyeq \left(\frac{1}{n+1}\sum_{k=1}^{n+1} f\left(\mathbf{E}^{-k}\mathbf{U}^*\mathbf{A}\mathbf{U}\mathbf{E}^k\right)\right)_{n+1,n+1}.$$

The third equality is due to the fact that the matrix in the argument of f in the second line is block-diagonal by Lemma B.36. Since $\mathbf{E}$ and $\mathbf{U}$ are unitary, the definition of f on self-adjoint matrices implies that the last term equals

$$
\left(\frac{1}{n+1} \sum_{k=1}^{n+1} \mathbf{E}^{-k} \mathbf{U}^* f(\mathbf{A}) \mathbf{U} \mathbf{E}^k \right)_{n+1,n+1} = (\mathbf{U}^* f(\mathbf{A}) \mathbf{U})_{n+1,n+1}
$$

$$
= \sum_{j=1}^{n} \mathbf{X}_j^* f(\mathbf{A}_j) \mathbf{X}_j.
$$

This completes the proof. $\square$

Our next tool is the *perspective*. In the scalar case, for a convex function f defined on some convex set $K \subset \mathbb{R}^n$, it is given by $g(x,t) = tf(x/t)$ whenever $t > 0$ and $x/t \in K$. It is straightforward to verify that g is jointly convex in (x,t), i.e., that g is a convex function in the variable $\mathbf{y} = (x,t)$. As an important example, the perspective of the convex function $f(x) = x \ln x$, $x \geq 0$, is the jointly convex function

$$
g(x,t) = x \ln x - x \ln t, \tag{B.54}
$$

which is known as relative entropy.

Now for a matrix convex function $f : (0,\infty) \to \mathbb{R}$, we define its perspective on positive definite matrices $\mathbf{A}, \mathbf{B}$ via

$$
g(\mathbf{A}, \mathbf{B}) = \mathbf{B}^{1/2} f(\mathbf{B}^{-1/2} \mathbf{A} \mathbf{B}^{-1/2}) \mathbf{B}^{1/2}. \tag{B.55}
$$

By the next theorem of Effros [176], it is jointly matrix convex in $(\mathbf{A}, \mathbf{B})$.

Theorem B.37. *For a matrix convex function $f : (0,\infty) \to \mathbb{R}$, the perspective g defined by* (B.55) *is jointly matrix convex in the sense that, for all positive definite* $\mathbf{A}_1, \mathbf{A}_2, \mathbf{B}_1, \mathbf{B}_2$ *of matching dimension and all* $t \in [0,1]$,

$$
g(t\mathbf{A}_1 + (1-t)\mathbf{A}_2, t\mathbf{B}_1 + (1-t)\mathbf{B}_2) \preccurlyeq tg(\mathbf{A}_1, \mathbf{B}_1) + (1-t)g(\mathbf{A}_2, \mathbf{B}_2).
$$

Proof. Let $\mathbf{A} := t\mathbf{A}_1 + (1-t)\mathbf{A}_2$ and $\mathbf{B} := t\mathbf{B}_1 + (1-t)\mathbf{B}_2$. The matrices $\mathbf{X}_1 := (t\mathbf{B}_1)^{1/2}\mathbf{B}^{-1/2}$ and $\mathbf{X}_2 := ((1-t)\mathbf{B}_2)^{1/2}\mathbf{B}^{-1/2}$ satisfy

$$
\mathbf{X}_1^* \mathbf{X}_1 + \mathbf{X}_2^* \mathbf{X}_2 = t\mathbf{B}^{-1/2}\mathbf{B}_1\mathbf{B}^{-1/2} + (1-t)\mathbf{B}^{-1/2}\mathbf{B}_2\mathbf{B}^{-1/2} = \mathbf{Id}.
$$

Theorem B.35 together with Lemma A.31 then implies that

$$g(\mathbf{A}, \mathbf{B}) = \mathbf{B}^{1/2} f\left(\mathbf{B}^{-1/2}\mathbf{A}\mathbf{B}^{-1/2}\right)\mathbf{B}^{1/2}$$

$$= \mathbf{B}^{1/2} f\left(\mathbf{X}_1^*\mathbf{B}_1^{-1/2}\mathbf{A}_1\mathbf{B}_1^{-1/2}\mathbf{X}_1 + \mathbf{X}_2^*\mathbf{B}_2^{-1/2}\mathbf{A}_2\mathbf{B}_2^{-1/2}\mathbf{X}_2\right)\mathbf{B}^{1/2}$$

$$\preccurlyeq \mathbf{B}^{1/2}\left(\mathbf{X}_1^* f(\mathbf{B}_1^{-1/2}\mathbf{A}_1\mathbf{B}_1^{-1/2})\mathbf{X}_1 + \mathbf{X}_2^* f(\mathbf{B}_2^{-1/2}\mathbf{A}_2\mathbf{B}_2^{-1/2})\mathbf{X}_2\right)\mathbf{B}^{1/2}$$

$$= t\mathbf{B}_1^{1/2} f(\mathbf{B}_1^{-1/2}\mathbf{A}_1\mathbf{B}_1^{-1/2})\mathbf{B}_1^{1/2} + (1-t)\mathbf{B}_2^{1/2} f(\mathbf{B}_2^{-1/2}\mathbf{A}_2\mathbf{B}_2^{-1/2})\mathbf{B}_2^{1/2}$$

$$= tg(\mathbf{A}_1, \mathbf{B}_1) + (1-t)g(\mathbf{A}_2, \mathbf{B}_2).$$

This concludes the proof.   □

Next we introduce a concept from quantum information theory [372, 383].

Definition B.38. For two positive definite matrices $\mathbf{A}$ and $\mathbf{B}$, the *quantum relative entropy* is defined as

$$\mathscr{D}(\mathbf{A}, \mathbf{B}) := \operatorname{tr}\left(\mathbf{A}\ln\mathbf{A} - \mathbf{A}\ln\mathbf{B} - (\mathbf{A} - \mathbf{B})\right).$$

If $\mathbf{A}$ and $\mathbf{B}$ are scalars, then the above definition reduces to the scalar relative entropy (B.54) (up to the term $\mathbf{A} - \mathbf{B}$).

The quantum relative entropy is nonnegative, a fact that is also known as *Klein's inequality*.

Theorem B.39. *If $\mathbf{A}$ and $\mathbf{B}$ are positive definite matrices, then*

$$\mathscr{D}(\mathbf{A}, \mathbf{B}) \geq 0.$$

Proof. The scalar function $\phi(x) = x\ln x$, $x > 0$, is convex (and even matrix convex by Corollary B.34). It follows from Proposition B.11 that

$$x\ln x = \phi(x) \geq \phi(y) + \phi'(y)(x-y) = y\ln y + (1+\ln y)(x-y) = x\ln y + (x-y),$$

so that $x\ln x - x\ln y - (x - y) \geq 0$. Theorem A.30 shows that $\mathscr{D}(\mathbf{A}, \mathbf{B}) \geq 0$.   □

As a consequence, we obtain a variational formula for the trace.

Corollary B.40. *If B is a positive definite matrix, then*

$$\operatorname{tr}\mathbf{B} = \max_{\mathbf{A} \succ \mathbf{0}} \operatorname{tr}\left(\mathbf{A}\ln\mathbf{B} - \mathbf{A}\ln\mathbf{A} + \mathbf{A}\right).$$

Proof. By definition of the quantum relative entropy and Theorem B.39,

$$\operatorname{tr}\mathbf{B} \geq \operatorname{tr}\left(\mathbf{A}\ln\mathbf{B} - \mathbf{A}\ln\mathbf{A} + \mathbf{A}\right).$$

Choosing $\mathbf{A} = \mathbf{B}$ yields equality above and establishes the claim.   □

Generalizing the matrix convexity of the standard relative entropy (B.54) (or Kullback–Leibler divergence), the quantum relative entropy is also jointly matrix convex. This fact goes back to Lindblad [327]; see also [403,492]. Our proof based on the perspective was proposed by Effros in [176].

Theorem B.41. *The quantum relative entropy $\mathscr{D}$ is jointly convex on pairs of positive definite matrices.*

Proof. Let $\mathbf{A}, \mathbf{B} \in \mathbb{C}^{n \times n}$ be positive definite matrices. To these matrices, we associate operators acting on the space $\mathbb{C}^{n \times n}$ endowed with the Frobenius inner product $\langle \mathbf{A}, \mathbf{B} \rangle_F = \operatorname{tr}(\mathbf{AB}^*)$; see (A.15). Precisely, we set

$$\mathbf{L_A X} := \mathbf{AX} \quad \text{and} \quad \mathbf{R_B X} := \mathbf{XB}, \qquad \mathbf{X} \in \mathbb{C}^{n \times n}.$$

By associativity of matrix multiplication, the operators $\mathbf{L_A}$ and $\mathbf{R_B}$ commute, and by positivity of $\mathbf{A}$ and $\mathbf{B}$, they are positive operators. Indeed, $\langle \mathbf{L_A}(\mathbf{X}), \mathbf{X} \rangle_F = \operatorname{tr}(\mathbf{AXX}^*) = \|\mathbf{A}^{1/2}\mathbf{X}\|_F^2 > 0$ for all nonzero $\mathbf{X} \in \mathbb{C}^{n \times n}$. The function $\phi(x) = x \ln x$, $x > 0$, is operator convex by Corollary B.34, and its perspective g is given by

$$g(\mathbf{L_A}, \mathbf{R_B}) = \mathbf{R_B} \phi(\mathbf{R_B^{-1} L_A}) = \mathbf{R_B}(\mathbf{R_B^{-1} L_A}) \ln(\mathbf{R_B^{-1} L_A})$$
$$= \mathbf{L_A}(\ln \mathbf{L_A} - \ln \mathbf{R_B}),$$

where these steps hold because $\mathbf{L_A}$ and $\mathbf{R_B}$ commute. By joint matrix convexity of the perspective (Theorem B.37), the scalar-valued function

$$h(\mathbf{A}, \mathbf{B}) := \langle g(\mathbf{L_A}, \mathbf{R_B})(\mathbf{Id}), \mathbf{Id} \rangle_F$$

is jointly convex in $(\mathbf{A}, \mathbf{B})$. Furthermore, $f(\mathbf{L_A})(\mathbf{Id}) = f(\mathbf{A})$ and $f(\mathbf{R_B})(\mathbf{Id}) = f(\mathbf{B})$ for any function f. Indeed, these relations are easily verified for monomials $f(t) = t^n$, then for polynomials, and finally for any f by interpolation. Therefore, the function h takes the form

$$h(\mathbf{A}, \mathbf{B}) = \langle g(\mathbf{L_A}, \mathbf{R_B})(\mathbf{Id}), \mathbf{Id} \rangle_F = \operatorname{tr}(g(\mathbf{L_A}, \mathbf{R_B})(\mathbf{Id}))$$
$$= \operatorname{tr}(\mathbf{L_A}(\ln \mathbf{L_A} - \ln \mathbf{R_B})(\mathbf{Id})) = \operatorname{tr}(\mathbf{A}(\ln \mathbf{A} - \ln \mathbf{B})).$$

We conclude that

$$\mathscr{D}(\mathbf{A}, \mathbf{B}) = h(\mathbf{A}, \mathbf{B}) - \operatorname{tr}(\mathbf{A} - \mathbf{B})$$

is jointly convex in $(\mathbf{A}, \mathbf{B})$. □

We are finally in the position to prove Lieb's concavity theorem.

Proof (of Theorem B.31). Setting $\mathbf{B} = \exp(\mathbf{H} + \ln \mathbf{X})$ in Corollary B.40 yields

$$\operatorname{tr} \exp(\mathbf{H} + \ln \mathbf{X}) = \max_{\mathbf{A} \succ 0} \operatorname{tr} \left(\mathbf{A}(\mathbf{H} + \ln \mathbf{X}) - \mathbf{A} \ln \mathbf{A} + \mathbf{A} \right)$$

$$= \max_{\mathbf{A} \succ 0} \left[\operatorname{tr} (\mathbf{A}\mathbf{H}) + \operatorname{tr} \mathbf{X} - \mathscr{D}(\mathbf{A}, \mathbf{X}) \right].$$

For each self-adjoint matrix $\mathbf{H}$, Theorem B.41 guarantees that the term in square brackets is a jointly concave function of the self-adjoint matrices $\mathbf{X}$ and $\mathbf{A}$. According to Theorem B.15, partial maximization of a jointly concave function gives rise to a concave function; hence, $\mathbf{X} \mapsto \operatorname{tr} \exp(\mathbf{H} + \ln \mathbf{X})$ is concave on the set of positive definite matrices.   □

Appendix C
Miscellanea

C.1 Fourier Analysis

This section recalls some simple facts from Fourier analysis. We cover the finite-dimensional analog of the Shannon sampling theorem mentioned in Sect. 1.2 as well as basic facts on the Fourier matrix and the fast Fourier transform (FFT). More background on Fourier and harmonic analysis can be found in various books on the subject including [39, 198, 236, 271, 272, 390, 400, 452, 505].

Finite-Dimensional Sampling Theorem

We consider trigonometric polynomials of degree at most M, that is, functions of the form

$$f(t) = \sum_{k=-M}^{M} c_k e^{2\pi i k t}, \quad t \in [0,1]. \tag{C.1}$$

The numbers c_k are called Fourier coefficients. They are given in terms of f by

$$c_k = \int_0^1 f(t) e^{-2\pi i k t} dt.$$

The Dirichlet kernel is defined as

$$D_M(t) := \sum_{k=-M}^{M} e^{2\pi i k t} = \begin{cases} \dfrac{\sin(\pi(2M+1)t)}{\sin(\pi t)} & \text{if } t \neq 0, \\ 2M+1 & \text{if } t = 0. \end{cases}$$

S. Foucart and H. Rauhut, *A Mathematical Introduction to Compressive Sensing*, Applied and Numerical Harmonic Analysis, DOI 10.1007/978-0-8176-4948-7, © Springer Science+Business Media New York 2013

The expression for $t \neq 0$ is derived from the geometric sum identity followed by a simplification. The finite-dimensional version of the Shannon sampling theorem reads as follows.

Theorem C.1. *Let f be a trigonometric polynomial of degree at most M. Then, for all $t \in [0, 1]$,*

$$f(t) = \frac{1}{2M+1} \sum_{j=0}^{2M} f\left(\frac{j}{2M+1}\right) D_M\left(t - \frac{j}{2M+1}\right). \qquad (C.2)$$

Proof. Let $f(t) = \sum_{k=-M}^{M} c_k e^{2\pi i k t}$. We evaluate the expression on the right-hand side of (C.2) multiplied by $2M + 1$ as

$$\sum_{j=0}^{2M} f\left(\frac{j}{2M+1}\right) D_M\left(t - \frac{j}{2M+1}\right)$$

$$= \sum_{j=0}^{2M} \sum_{k=-M}^{M} c_k e^{2\pi i k j/(2M+1)} \sum_{\ell=-M}^{M} e^{2\pi i \ell(t - j/(2M+1))}$$

$$= \sum_{k=-M}^{M} c_k \sum_{\ell=-M}^{M} \sum_{j=0}^{2M} e^{2\pi i (k-\ell)j/(2M+1)} e^{2\pi i \ell t}.$$

The identity $\sum_{j=0}^{2M} e^{2\pi i (k-\ell)j/(2M+1)} = (2M+1)\delta_{k,\ell}$ completes the proof. $\qquad \square$

The Fast Fourier Transform

The Fourier matrix $\mathbf{F} \in \mathbb{C}^{N \times N}$ has entries

$$F_{j,k} = \frac{1}{\sqrt{N}} e^{2\pi i (j-1)(k-1)/N}, \qquad j, k \in [N]. \qquad (C.3)$$

The application of $\mathbf{F}$ to a vector $\mathbf{x} \in \mathbb{C}^N$ is called the Fourier transform of $\mathbf{x}$ and denoted by

$$\hat{\mathbf{x}} = \mathbf{F}\mathbf{x}.$$

Intuitively, the coefficient $\hat{x}_j$ reflects the frequency content of $\mathbf{x}$ corresponding to the monomials $j \mapsto e^{2\pi i (j-1)(k-1)/N}$. The Fourier transform arises, for instance, when evaluating a trigonometric polynomial of the form (C.1) at the points $j/(2M+1)$, $j = -M, \ldots, M$.

The Fourier matrix is unitary, i.e., $\mathbf{F}^*\mathbf{F} = \mathbf{Id}$, so that $\mathbf{F}^{-1} = \mathbf{F}^*$; see (12.1). This reflects the fact that its columns form an orthonormal basis of $\mathbb{C}^N$.

A naive implementation of the Fourier transform requires $\mathcal{O}(N^2)$ operations. The fast Fourier transform (FFT) is an algorithm that evaluates the Fourier transform much quicker, namely, in $\mathcal{O}(N \ln N)$ operations. This is what makes the FFT one of the most widely used algorithms. Many devices of modern technology would not work without it.

Let us give the main idea of the FFT algorithm. Assume that N is even. Then the Fourier transform of $\mathbf{x} \in \mathbb{C}^N$ has entries

$$
\hat{x}_j = \frac{1}{\sqrt{N}} \sum_{k=1}^{N} x_k e^{2\pi i(j-1)(k-1)/N}
$$

$$
= \frac{1}{\sqrt{N}} \left(\sum_{\ell=1}^{N/2} x_{2\ell} e^{2\pi i(j-1)(2\ell-1)/N} + \sum_{\ell=1}^{N/2} x_{2\ell-1} e^{2\pi i(j-1)(2\ell-2)/N} \right)
$$

$$
= \frac{1}{\sqrt{N}} \left(e^{2\pi i(j-1)/N} \sum_{\ell=1}^{N/2} x_{2\ell} e^{2\pi i(j-1)(\ell-1)/(N/2)} \right.
$$

$$
\left. + \sum_{\ell=1}^{N/2} x_{2\ell-1} e^{2\pi i(j-1)(\ell-1)/(N/2)} \right).
$$

We have basically reduced the evaluation of $\hat{\mathbf{x}} \in \mathbb{C}^N$ to the evaluation of two Fourier transforms in dimension $N/2$, namely, to the ones of $(x_{2\ell})_{\ell=1}^{N/2}$ and of $(x_{2\ell-1})_{\ell=1}^{N/2}$. If $N = 2^n$, then in this way we can recursively reduce the evaluation of the Fourier transform to the ones of half dimension until we reach the dimension 2. This requires n recursion steps and altogether $\mathcal{O}(n2^n) = \mathcal{O}(N \log N)$ algebraic operations. The resulting algorithm, named after Cooley and Tukey [130], is often simply called the fast Fourier transform. For other composite numbers $N = pq$ similar reduction steps can be made. We refer to [494, 506] for details.

C.2 Covering Numbers

Let T be a subset of a metric space (X, d). For $t > 0$, the covering number $\mathcal{N}(T, d, t)$ is defined as the smallest integer $\mathcal{N}$ such that T can be covered with balls $B(\mathbf{x}_\ell, t) = \{\mathbf{x} \in X, d(\mathbf{x}, \mathbf{x}_\ell) \leq t\}$, $\mathbf{x}_\ell \in T$, $\ell \in [\mathcal{N}]$, i.e.,

$$
T \subset \bigcup_{\ell=1}^{\mathcal{N}} B(\mathbf{x}_\ell, t).
$$

The set of points $\{x_1, \ldots, x_N\}$ is called a t-covering. (Note that some authors only require $x_\ell \in X$, so the points are not necessarily elements of T.)

The packing number $\mathcal{P}(T, d, t)$ is defined, for $t > 0$, as the maximal integer $\mathcal{P}$ such that there are points $x_\ell \in T$, $\ell \in [\mathcal{P}]$, which are t-separated, i.e., $d(x_\ell, x_k) > t$ for all $k, \ell \in [\mathcal{P}]$, $k \neq \ell$.

If $X = \mathbb{R}^n$ is a normed vector space and the metric d is induced by the norm via $d(u, v) = \|u - v\|$, we also write $\mathcal{N}(T, \|\cdot\|, t)$ and $\mathcal{P}(T, \|\cdot\|, t)$.

Let us first state some elementary properties of the covering numbers. The packing numbers satisfy precisely the same properties. For arbitrary sets $S, T \subset X$,

$$\mathcal{N}(S \cup T, d, t) \leq \mathcal{N}(S, d, t) + \mathcal{N}(T, d, t). \tag{C.4}$$

For any $\alpha > 0$,

$$\mathcal{N}(T, \alpha d, t) = \mathcal{N}(T, d, t/\alpha). \tag{C.5}$$

If $X = \mathbb{R}^n$ and d is induced by a norm $\|\cdot\|$, then furthermore

$$\mathcal{N}(\alpha T, d, t) = \mathcal{N}(T, d, t/\alpha). \tag{C.6}$$

Moreover, if d' is another metric on X that satisfies $d'(x, y) \leq d(x, y)$ for all $x, y \in T$, then

$$\mathcal{N}(T, d', t) \leq \mathcal{N}(T, d, t). \tag{C.7}$$

The following simple relations between covering and packing numbers hold.

Lemma C.2. *Let T be a subset of a metric space (X, d) and let $t > 0$. Then*

$$\mathcal{P}(T, d, 2t) \leq \mathcal{N}(T, d, t) \leq \mathcal{P}(T, d, t).$$

Proof. Let $\{x_1, \ldots, x_\mathcal{P}\}$ be a $2t$-separated set and $\{x'_1, \ldots, x'_N\}$ be a t-covering. Then we can assign to each point x_j a point x'_ℓ with $d(x'_\ell, x_j) \leq t$. This assignment is unique since the points x_j are $2t$-separated. Indeed, the assumption that two points x_j, x_k, $j \neq k$, can be assigned the same point x'_ℓ would lead to a contradiction by the triangle inequality $d(x_j, x_k) \leq d(x_j, x'_\ell) + d(x'_\ell, x_k) \leq 2t$. It follows that $\mathcal{P} \leq \mathcal{N}$.

Now let $\{x_1, \ldots, x_N\}$ be a maximal t-packing. Then it is also a t-covering. Indeed, if there were a point x not covered by a ball $B(x_\ell, t)$, $\ell \in [\mathcal{N}]$, then $d(x, x_\ell) > t$ for all $\ell \in [\mathcal{N}]$. This means that we could add x to the t-packing. But this would be a contradiction to the maximality. $\qquad\square$

The following proposition estimates in particular the packing number of a ball or a sphere in a finite-dimensional normed space.

Proposition C.3. *Let* $\| \cdot \|$ *be some norm on* $\mathbb{R}^n$ *and let* U *be a subset of the unit ball* $B = \{\mathbf{x} \in \mathbb{R}^n, \|\mathbf{x}\| \leq 1\}$. *Then the packing and covering numbers satisfy, for* $t > 0$,

$$\mathcal{N}(U, \| \cdot \|, t) \leq \mathcal{P}(U, \| \cdot \|, t) \leq \left(1 + \frac{2}{t}\right)^n . \tag{C.8}$$

Proof. Lemma C.2 shows the first inequality. Let now $\{\mathbf{x}_1, \ldots, \mathbf{x}_\mathcal{P}\} \subset U$ be a maximal t-packing of U. Then the balls $B(\mathbf{x}_\ell, t/2)$ do not intersect and they are contained in the ball $(1 + t/2)B$. By comparing volumes (that is, Lebesgue measures) of the involved balls, we obtain

$$\mathrm{vol}\left(\bigcup_{\ell=1}^{\mathcal{P}} B(\mathbf{x}_\ell, t/2)\right) = \mathcal{P} \, \mathrm{vol}\left((t/2)B\right) \leq \mathrm{vol}\left((1 + t/2)B\right) .$$

On $\mathbb{R}^n$ the volume satisfies the homogeneity relation $\mathrm{vol}\,(tB) = t^n \, \mathrm{vol}\,(B)$; hence, $\mathcal{P}\,(t/2)^n \, \mathrm{vol}\,(B) \leq (1 + t/2)^n \, \mathrm{vol}\,(B)$, i.e., $\mathcal{P} \leq (1 + 2/t)^n$. $\qquad\square$

C.3 The Gamma Function and Stirling's Formula

The Gamma function is defined for $x > 0$ via

$$\Gamma(x) = \int_0^\infty t^{x-1} e^{-t} dt . \tag{C.9}$$

It interpolates the factorial function in the sense that, for positive integers n,

$$\Gamma(n) = (n-1)! . \tag{C.10}$$

In fact, it follows from integration by parts that the Gamma function satisfies the functional equation

$$\Gamma(x+1) = x\Gamma(x) , \quad x > 0 . \tag{C.11}$$

Its value at the point $1/2$ is given by $\Gamma(1/2) = \sqrt{\pi}$.

Stirling's formula states that, for $x > 0$,

$$\Gamma(x) = \sqrt{2\pi} x^{x-1/2} e^{-x} \exp\left(\frac{\theta(x)}{12x}\right) \tag{C.12}$$

with $0 \leq \theta(x) \leq 1$. Using (C.10) and applying the formula (C.12) show that the factorial $n! = n\Gamma(n)$ satisfies

$$n! = \sqrt{2\pi n}\, n^n e^{-n} e^{R_n}, \tag{C.13}$$

with $0 \leq R_n \leq 1/(12n)$.

We also need a technical result about the quantity $\sqrt{2}\dfrac{\Gamma((m+1)/2)}{\Gamma(m/2)}$, which asymptotically behaves like $\sqrt{m}$ as $m \to \infty$.

Lemma C.4. *For integers* $m \geq s \geq 1$,

$$\sqrt{2}\frac{\Gamma((m+1)/2)}{\Gamma(m/2)} - \sqrt{2}\frac{\Gamma((s+1)/2)}{\Gamma(s/2)} \geq \sqrt{m} - \sqrt{s}.$$

Proof. It is sufficient to show that, for any $m \geq 1$,

$$\alpha_{m+1} - \alpha_m \geq \sqrt{m+1} - \sqrt{m}, \qquad \text{where} \quad \alpha_m := \sqrt{2}\frac{\Gamma((m+1)/2)}{\Gamma(m/2)}.$$

It follows from $\Gamma((m+2)/2) = (m/2)\Gamma(m/2)$ that $\alpha_{m+1}\alpha_m = m$. Thus, multiplying the prospective inequality $\alpha_{m+1} - \alpha_m \geq \sqrt{m+1} - \sqrt{m}$ by α_m and rearranging the terms, we need to prove that

$$\alpha_m^2 + \left(\sqrt{m+1} - \sqrt{m}\right)\alpha_m - m \leq 0.$$

In other words, we need to prove that α_m does not exceed the positive root of the quadratic polynomial $z^2 + \left(\sqrt{m+1} - \sqrt{m}\right)z - m$, i.e., that

$$\alpha_m \leq \beta_m := \frac{-\left(\sqrt{m+1} - \sqrt{m}\right) + \sqrt{\left(\sqrt{m+1} - \sqrt{m}\right)^2 + 4m}}{2}.$$

The first step consists in bounding α_m from above. To this end, we use *Gauss' hypergeometric theorem*; see, e.g., [17]. It states that if $\mathrm{Re}(c - a - b) > 0$, then

$$\frac{\Gamma(c)\Gamma(c-a-b)}{\Gamma(c-a)\Gamma(c-b)} = {}_2F_1(a,b;c;1).$$

Here, ${}_2F_1$ denotes *Gauss' hypergeometric function*

$${}_2F_1(a,b;c;z) := \sum_{n=0}^{\infty} \frac{(a)_n(b)_n}{(c)_n}\frac{z^n}{n!},$$

where $(d)_n$ is the *Pochhammer symbol* defined by $(d)_0 = 1$ and $(d)_n := d(d + 1) \cdots (d + n - 1)$ for $n \geq 1$. Therefore, choosing $a = -1/2$, $b = -1/2$, and $c = (m - 1)/2$, we derive

$$\alpha_m^2 = 2 \frac{\Gamma((m+1)/2)^2}{\Gamma(m/2)^2} = (m - 1) \frac{\Gamma((m+1)/2)\,\Gamma((m-1)/2)}{\Gamma(m/2)^2}$$

$$= (m - 1) \sum_{n=0}^{\infty} \frac{[(-1/2)(1/2) \cdots (-1/2 + n - 1)]^2}{((m-1)/2)((m+1)/2) \cdots ((m+2n-3)/2)} \frac{1}{n!}$$

$$= m - \frac{1}{2} + \frac{1}{8} \frac{1}{m+1} + \sum_{n=3}^{\infty} \frac{2^{n-2}[(1/2) \cdots (n - 3/2)]^2}{n!(m+1)(m+3) \cdots (m+2n-3)}.$$

We observe that the quantity $\gamma_m := (m + 1)(\alpha_m^2 - m + 1/2 - 1/(8(m + 1)))$ decreases with m. Thus, for an integer m_0 to be chosen later,

$$\alpha_m^2 \leq m - \frac{1}{2} + \frac{1}{8} \frac{1}{m+1} + \frac{\gamma_{m_0}}{m+1}, \qquad m \geq m_0. \tag{C.14}$$

The next step consists in bounding β_m from below. We start by writing

$$\beta_m \geq \frac{-(\sqrt{m+1} - \sqrt{m}) + \sqrt{4m}}{2} = \frac{3\sqrt{m} - \sqrt{m+1}}{2} =: \delta_m.$$

Then we look for an expansion of $\delta_m^2 = (10m + 1 - 6\sqrt{m(m+1)})/4$. We have

$$\sqrt{m(m+1)} = (m + 1)\sqrt{1 - \frac{1}{m+1}}$$

$$= (m + 1)\left(1 + \sum_{n=1}^{\infty} \frac{(1/2)(-1/2) \cdots (1/2 - n + 1)}{n!}\left(\frac{-1}{m+1}\right)^n\right)$$

$$= m + \frac{1}{2} - \frac{1}{8} \frac{1}{m+1} - \frac{1}{2} \sum_{n=3}^{\infty} \frac{(1/2) \cdots (n - 3/2)}{n!}\left(\frac{1}{m+1}\right)^{n-1}.$$

It now follows that

$$\beta_m^2 \geq \delta_m^2 = m - \frac{1}{2} + \frac{3}{16} \frac{1}{m+1} + \frac{3}{4} \sum_{n=3}^{\infty} \frac{(1/2) \cdots (n - 3/2)}{n!}\left(\frac{1}{m+1}\right)^{n-1}$$

$$\geq m - \frac{1}{2} + \frac{3}{16} \frac{1}{m+1}. \tag{C.15}$$

Subtracting (C.14) from (C.15), we obtain

$$\beta_m^2 - \alpha_m^2 \geq \frac{1}{16} \frac{1}{m+1} - \frac{\gamma_{m_0}}{m+1}, \qquad m \geq m_0.$$

We choose m_0 to be the smallest integer such that $\gamma_{m_0} \leq 1/16$, i.e., $m_0 = 3$, so that $\beta_m^2 \geq \alpha_m^2$ for all $m \geq 3$. The numerical verification that $\beta_m^2 \geq \alpha_m^2$ for $m = 1$ and $m = 2$ is straightforward. This concludes the proof.   □

C.4   The Multinomial Theorem

The multinomial theorem is concerned with the expansion of a power of a sum. It states that, for an integer $n \geq 1$,

$$\left(\sum_{\ell=1}^{m} x_\ell \right)^n = \sum_{k_1+k_2+\ldots+k_m=n} \frac{n!}{k_1! k_2! \cdots k_m!} \prod_{j=1}^{m} x_j^{k_j}.$$

The sum is taken over all possible m-tuples of nonnegative integers $k_1, \ldots, k_m$ that sum up to n. This formula can be proved with the binomial theorem and induction on n.

C.5   Some Elementary Estimates

Lemma C.5. *For integers $n \geq k > 0$,*

$$\left(\frac{n}{k} \right)^k \leq \binom{n}{k} \leq \left(\frac{en}{k} \right)^k.$$

Proof. For the upper bound, we use

$$e^k = \sum_{\ell=0}^{\infty} \frac{k^\ell}{\ell!} \geq \frac{k^k}{k!}$$

to derive the inequality

$$\binom{n}{k} = \frac{n(n-1)\cdots(n-k+1)}{k!} \leq \frac{n^k}{k!} = \frac{k^k}{k!} \frac{n^k}{k^k} \leq e^k \frac{n^k}{k^k}.$$

As for the lower bound, we write

$$\binom{n}{k} = \frac{n(n-1)\cdots(n-k+1)}{k(k-1)\cdots 1} = \prod_{\ell=1}^{k} \frac{n-k+\ell}{\ell} \geq \left(\frac{n}{k}\right)^k,$$

having used that $(n-k+\ell)/\ell = (n-k)/\ell + 1$ decreases with $\ell \geq 1$.   □

Lemma C.6. *Let $N, m, s \geq 1$ with $N \geq s$ be given integers and let $c, d > 0$ be prescribed constants.*

(a) If $m \geq cs \ln(dN/s)$ and $m \geq s$, then $m \geq cs \ln(dN/m)$.
(b) If $m \geq c's \ln(c'N/m)$ with $c' = c(1 + d/e)$, then $m \geq cs \ln(dN/s)$.
(c) If $m \geq cs \ln(dN/m)$, then $m \geq c''s \ln(dN/s)$ with $c'' = ec/(e+c)$ or better $c'' = c/\ln(c)$ provided $c, d \geq e$.

Proof. The statement (a) simply follows from $m \geq s$. For (b) and (c), let us assume that $m \geq \gamma s \ln(\delta N/m)$ for some $\gamma, \delta > 0$. For any $\delta' > 0$, we then have

$$m \geq \gamma s \ln\left(\frac{\delta'N}{s}\right) + \gamma s \ln\left(\frac{\delta s}{\delta'm}\right) = \gamma s \ln\left(\frac{\delta'N}{s}\right) + \frac{\gamma\delta'}{\delta} m \frac{\delta s}{\delta'm} \ln\left(\frac{\delta s}{\delta'm}\right).$$

We notice that the function $f(x) := x \ln(x)$ is decreasing on $(0, 1/e)$ and increasing on $(1/e, +\infty)$, with a minimum value of $-1/e$, to derive

$$m \geq \gamma s \ln\left(\frac{\delta'N}{s}\right) - \frac{\gamma\delta'}{e\delta} m, \quad \text{i.e.,} \quad \left(1 + \frac{\gamma\delta'}{e\delta}\right)m \geq \gamma s \ln\left(\frac{\delta'N}{s}\right).$$

The statement (b) is obtained by taking (among other possible choices) $\gamma = \delta = c(1 + d/e)$ and $\delta' = d$, while the first part of (c) is obtained by taking $\gamma = c$ and $\delta = \delta' = d$. The second part of (c), where $c, d \geq e$, follows from $s/m \leq 1/(c \ln(dN/s)) \leq 1/c$; hence, $f(s/m) \geq f(1/c) = -\ln(c)/c$. The same choice of γ, δ, δ' yields

$$m \geq cs \ln\left(\frac{dN}{s}\right) - \ln(c)m, \quad \text{i.e.,} \quad (1 + \ln(c))m \geq cs \ln\left(\frac{dN}{s}\right),$$

which is a rewriting of the desired conclusion.   □

C.6   Estimates of Some Integrals

Next we provide some useful estimates of certain integrals. The first two lemmas are related to the estimation of the tail of a Gaussian random variable from above and below.

Lemma C.7. *For* $u > 0$,

$$\int_u^\infty e^{-t^2/2} dt \leq \min\left\{\sqrt{\frac{\pi}{2}}, \frac{1}{u}\right\} \exp(-u^2/2).$$

Proof. A change of variables yields

$$\int_u^\infty e^{-t^2/2} dt = \int_0^\infty e^{-(t+u)^2/2} dt = e^{-u^2/2} \int_0^\infty e^{-tu} e^{-t^2/2} dt. \qquad (C.16)$$

On the one hand, using the fact that $e^{-tu} \leq 1$ for $t, u \geq 0$, we have

$$\int_u^\infty e^{-t^2/2} dt \leq e^{-u^2/2} \int_0^\infty e^{-t^2/2} dt = \sqrt{\frac{\pi}{2}} e^{-u^2/2}.$$

On the other hand, the fact that $e^{-t^2/2} \leq 1$ yields

$$\int_u^\infty e^{-t^2/2} dt \leq e^{-u^2/2} \int_0^\infty e^{-tu} dt = \frac{1}{u} e^{-u^2/2}. \qquad (C.17)$$

This shows the desired estimate. $\square$

Lemma C.8. *For* $u > 0$,

$$\int_u^\infty e^{-t^2/2} dt \geq \max\left\{\frac{1}{u} - \frac{1}{u^3}, \sqrt{\frac{\pi}{2}} - u\right\} \exp(-u^2/2).$$

Proof. We use (C.16) together with $e^{-t^2/2} \geq 1 - t^2/2$ to obtain

$$\int_u^\infty e^{-t^2/2} dt \geq e^{-u^2/2} \int_0^\infty \left(1 - \frac{t^2}{2}\right) e^{-tu} dt = e^{-u^2/2} \left(\frac{1}{u} - \frac{1}{u^3}\right).$$

Using instead $e^{-tu} \geq 1 - tu$ in (C.16) yields

$$\int_u^\infty e^{-t^2/2} dt \geq e^{-u^2/2} \int_0^\infty e^{-t^2/2}(1 - ut) dt = e^{-u^2/2}\left(\sqrt{\frac{\pi}{2}} - u\right).$$

This completes the proof. $\square$

Lemma C.9. *For* $\alpha > 0$,

$$\int_0^\alpha \sqrt{\ln(1 + t^{-1})} dt \leq \alpha \sqrt{\ln(e(1 + \alpha^{-1}))}. \qquad (C.18)$$

Proof. First apply the Cauchy–Schwarz inequality to obtain

$$\int_0^\alpha \sqrt{\ln(1+t^{-1})}\,dt \leq \sqrt{\int_0^\alpha 1\,dt \int_0^\alpha \ln(1+t^{-1})\,dt}.$$

A change of variables and integration by parts yield

$$\int_0^\alpha \ln(1+t^{-1})\,dt = \int_{\alpha^{-1}}^\infty u^{-2}\ln(1+u)\,du$$

$$= -u^{-1}\ln(1+u)\Big|_{\alpha^{-1}}^\infty + \int_{\alpha^{-1}}^\infty u^{-1}\frac{1}{1+u}\,du \leq \alpha\ln(1+\alpha^{-1}) + \int_{\alpha^{-1}}^\infty \frac{1}{u^2}\,du$$

$$= \alpha\ln(1+\alpha^{-1}) + \alpha.$$

Combining the above estimates concludes the proof.   □

C.7   Hahn–Banach Theorems

The Hahn–Banach theorems are fundamental results about vector spaces. Although they are only used in a finite-dimensional setting in this book, they remain true in infinite dimensions. Their proof can be found in most textbooks on functional analysis, e.g., [72, 419, 438, 514]. The *Hahn–Banach extension theorem*, stated below, is often used when $\|\cdot\|$ is a norm.

Theorem C.10. *Let X be a vector space equipped with a seminorm $\|\cdot\|$ and let Y be a subspace of X. If λ is a linear functional defined on Y such that $|\lambda(\mathbf{y})| \leq \|\mathbf{y}\|$ for all $\mathbf{y} \in Y$, then there exists a linear functional $\tilde{\lambda}$ defined on X such that $\tilde{\lambda}(\mathbf{y}) = \lambda(\mathbf{y})$ for all $\mathbf{y} \in Y$ and $|\tilde{\lambda}(\mathbf{x})| \leq \|\mathbf{x}\|$ for all $\mathbf{x} \in X$.*

As for the *Hahn–Banach separation theorem*, it reads as follows.

Theorem C.11. *If C and D are two disjoint nonempty convex subsets of a normed space X and if C is open, then there exist a continuous linear functional λ defined on X and a real number t such that*

$$\mathrm{Re}(\lambda(\mathbf{x})) < t \quad \text{for all } \mathbf{x} \in C \qquad and \qquad \mathrm{Re}(\lambda(\mathbf{x})) \geq t \quad \text{for all } \mathbf{x} \in D.$$

C.8   Smoothing Lipschitz Functions

The proof of the concentration of measure results (Theorems 8.34 and 8.40) requires to approximate a Lipschitz function by a smooth Lipschitz function. The following result establishes this rigorously.

Theorem C.12. *Let $f : \mathbb{R}^n \to \mathbb{R}$ be a Lipschitz function with constant $L = 1$ in (8.69). For $\varepsilon > 0$ and $\mathbf{x} \in \mathbb{R}^n$, denote by $B_\varepsilon(\mathbf{x}) = \{\mathbf{y} \in \mathbb{R}^n : \|\mathbf{y} - \mathbf{x}\|_2 \le \varepsilon\}$ the ball of radius ε around $\mathbf{x}$ and by $|B_\varepsilon(\mathbf{x})|$ its volume. Define $g : \mathbb{R}^n \to \mathbb{R}$ by*

$$g(\mathbf{x}) = \frac{1}{|B_\varepsilon(\mathbf{x})|} \int_{B_\varepsilon(\mathbf{x})} f(\mathbf{y}) d\mathbf{y}.$$

Then the function g is differentiable and $\|\nabla g(\mathbf{x})\|_2 \le 1$ for all $\mathbf{x} \in \mathbb{R}^n$ (so that g is Lipschitz with constant $L = 1$). Furthermore,

$$|f(\mathbf{x}) - g(\mathbf{x})| \le \frac{\varepsilon n}{n+1} \le \varepsilon \qquad \text{for all } \mathbf{x} \in \mathbb{R}^n.$$

Proof. We start with the case $n = 1$. Then g is defined via

$$g(x) = \frac{1}{2\varepsilon} \int_{x-\varepsilon}^{x+\varepsilon} f(y) dy.$$

Therefore,

$$g'(x) = \frac{f(x+\varepsilon) - f(x-\varepsilon)}{2\varepsilon},$$

and since f is 1-Lipschitz, it follows that $|g'(x)| \le 1$. Moreover,

$$|f(x) - g(x)| = \left| \frac{1}{2\varepsilon} \int_{x-\varepsilon}^{x+\varepsilon} (f(x) - f(y)) dy \right| \le \frac{1}{2\varepsilon} \int_{x-\varepsilon}^{x+\varepsilon} |f(x) - f(y)| dy$$

$$\le \frac{1}{2\varepsilon} \int_{x-\varepsilon}^{x+\varepsilon} |x - y| dy = \frac{1}{\varepsilon} \int_0^\varepsilon t dt = \frac{\varepsilon^2}{2\varepsilon} = \frac{\varepsilon}{2}.$$

Assume now $n > 1$. We choose a unit vector $\mathbf{u} \in \mathbb{R}^n$ and some $\mathbf{x} \in \mathbb{R}^n$ and show that the function $\psi(t) = g(\mathbf{x} + t\mathbf{u})$ is differentiable with $|\psi'(t)| \le 1$, which is equivalent to $|\langle \nabla g(\mathbf{x}), \mathbf{v} \rangle| \le 1$. As $\mathbf{u}$ and $\mathbf{x}$ are arbitrary, this will establish that g is differentiable with $\|\nabla g(\mathbf{x})\|_2 \le 1$ for all $\mathbf{x} \in \mathbb{R}^n$. Without loss of generality, we assume that $\mathbf{x} = \mathbf{0}$ and that $\mathbf{u} = (0, \ldots, 0, 1)$. Then the orthogonal complement $\mathbf{u}^\perp$ can be identified with $\mathbb{R}^{n-1}$. Let $D_\varepsilon = \{(\mathbf{z}, 0) : \mathbf{z} \in \mathbb{R}^{n-1}, \|\mathbf{z}\|_2 \le \varepsilon\}$. For any $(\mathbf{z}, 0) \in D_\varepsilon$, the intersection of the line through $(\mathbf{z}, 0)$ in the direction of $\mathbf{u}$ with $B_\varepsilon(\mathbf{0})$ is an interval with endpoints of the form $(\mathbf{z}, -a(\mathbf{z}))$ and $(\mathbf{z}, a(\mathbf{z}))$ with $a(\mathbf{z}) > 0$. It follows that

$$|B_\varepsilon(\mathbf{0})| = 2 \int_{D_\varepsilon} a(\mathbf{z}) d\mathbf{z}. \tag{C.19}$$

Now we estimate the derivative of ψ. For $\tau \in \mathbb{R}$, we have

$$\psi(\tau) = g(\tau\mathbf{u}) = \frac{1}{|B_\varepsilon(\tau\mathbf{u})|} \int_{B_\varepsilon(\tau\mathbf{u})} f(\mathbf{y})d\mathbf{y} = \frac{1}{|B_\varepsilon(\mathbf{0})|} \int_{D_\varepsilon} \int_{-a(\mathbf{z})+\tau}^{a(\mathbf{z})+\tau} f(\mathbf{z}, t)dt d\mathbf{z},$$

hence,

$$\psi'(0) = \frac{1}{|B_\varepsilon(\mathbf{0})|} \int_{D_\varepsilon} (f(\mathbf{z}, a(\mathbf{z})) - f(\mathbf{z}, -a(\mathbf{z})))d\mathbf{z}.$$

Since f is 1-Lipschitz, we have $|f(\mathbf{z}, a(\mathbf{z})) - f(\mathbf{z}, -a(\mathbf{z}))| \le 2a(\mathbf{z})$, so by (C.19)

$$|\psi'(0)| = |\langle \nabla g(\mathbf{x}), \mathbf{v} \rangle| \le 1.$$

The approximation property follows similarly to the case $n = 1$, namely,

$$|f(\mathbf{x}) - g(\mathbf{x})| \le \frac{1}{|B_\varepsilon(\mathbf{x})|} \int_{B_\varepsilon(\mathbf{x})} |f(\mathbf{x}) - f(\mathbf{z})|d\mathbf{z} \le \frac{1}{|B_\varepsilon(\mathbf{x})|} \int_{B_\varepsilon(\mathbf{x})} \|\mathbf{x} - \mathbf{z}\|_2 d\mathbf{z}$$

$$= \frac{1}{|B_\varepsilon(\mathbf{0})|} \int_{B_\varepsilon(\mathbf{0})} \|\mathbf{z}\|_2 d\mathbf{z} = \frac{|S^{n-1}|}{|B_\varepsilon(\mathbf{0})|} \int_0^\varepsilon r^n dr = \frac{\varepsilon|S^{n-1}|}{(n+1)|B_1(\mathbf{0})|}, \qquad (C.20)$$

where $|S^{n-1}|$ is the surface area of the sphere $S^{n-1} = \{\mathbf{x} \in \mathbb{R}^n, \|\mathbf{x}\|_2 = 1\}$. Hereby, we used the fact that the volume in $\mathbb{R}^n$ satisfies $|B_\varepsilon(\mathbf{0})| = \varepsilon^n |B_1(\mathbf{0})|$. Denoting by $s_n(r)$ the surface area of the sphere of radius r in $\mathbb{R}^n$ and by $v_n(r)$ the volume of the corresponding ball, we have the relation

$$v_n(r) = \int_0^r s_n(\rho)d\rho. \qquad (C.21)$$

Since $v_n(r) = |B_1(\mathbf{0})|r^n$, differentiating (C.21) shows that $s_n(r) = n|B_1(\mathbf{0})|r^{n-1}$, so that $|S^{n-1}| = s_n(1) = n|B_1(\mathbf{0})|$. Substituting this into (C.20) completes the proof. $\qquad\square$

C.9   Weak and Distributional Derivatives

The concept of weak derivative generalizes the classical derivative. Given a measurable function $f : \mathbb{R} \to \mathbb{R}$, we say that $v : \mathbb{R} \to \mathbb{R}$ is a weak derivative of f if, for all infinitely differentiable functions $\phi : \mathbb{R} \to \mathbb{R}$ with compact support,

$$\int_{-\infty}^{\infty} f(x)\phi'(x)dx = -\int_{-\infty}^{\infty} v(x)\phi(x)dx. \qquad (C.22)$$

In this case, we write $v = f' = \frac{d}{dx}f$. If f is continuously differentiable according to the classical definition, then it follows from integration by parts that the classical derivative f' is a weak derivative. If v and w are weak derivatives of f, then they are equal almost everywhere, and in this sense the weak derivative is unique. If a weak derivative exists, then we say that f is weakly differentiable.

If the function f is defined only on a compact subinterval $[a, b] \subset \mathbb{R}$, then the integrals in (C.22) are only defined on $[a, b]$ and the functions ϕ are assumed to vanish on the boundary, i.e., $\phi(a) = \phi(b) = 0$.

This concept extends to the multivariate case and to higher derivatives in an obvious way. For a function $f : \mathbb{R}^n \to \mathbb{R}$, and a multi-index $\alpha = (\alpha_1, \ldots, \alpha_n)$, $\alpha_j \in \mathbb{N}_0$, we set $|\alpha| = \sum_j \alpha_j$ and $D^\alpha f = \frac{\partial^{\alpha_1}}{\partial x_1^{\alpha_1}} \cdots \frac{\partial^{\alpha_n}}{\partial x_n^{\alpha_n}} f$. Then $v : \mathbb{R}^n \to \mathbb{R}$ is a weak derivative of order α if

$$\int_{\mathbb{R}^d} f(\mathbf{x}) D^\alpha \phi(\mathbf{x}) dx = (-1)^{|\alpha|} \int_{\mathbb{R}^d} v(\mathbf{x}) \phi(\mathbf{x}) dx$$

for all infinitely differentiable functions $\phi : \mathbb{R}^n \to \mathbb{R}$ with compact support. We write $v = D^\alpha f$ in this case.

The concept of weak derivative can be further generalized to distributional derivatives. We denote by $\mathcal{D}$ the space of all infinitely differentiable functions with compact support. A distribution is a functional on $\mathcal{D}$, i.e., a linear mapping from $\mathcal{D}$ into the scalars. A function f on $\mathbb{R}^n$ which is bounded on every compact subset of $\mathbb{R}^n$ (or at least locally integrable) induces a distribution via $f(\phi) = \int_{\mathbb{R}^n} f(\mathbf{x})\phi(\mathbf{x})dx$ for $\phi \in \mathcal{D}$. The distributional derivative of a distribution f is defined via

$$\frac{\partial}{\partial x_j} f(\phi) = -f\left(\frac{\partial}{\partial x_j}\phi\right), \quad \phi \in \mathcal{D}.$$

The distributional derivative always exists. If f can be identified with a function, then it is the functional

$$\frac{\partial}{\partial x_j} f(\phi) = -\int_{\mathbb{R}^n} f(\mathbf{x}) \frac{\partial}{\partial x_j} \phi(\mathbf{x}) dx.$$

If f possesses a weak derivative, then the distributional derivative can be identified with it by (C.22). If f is even differentiable, then both distributional and weak derivative can be identified with the classical derivative.

We say that a distribution f is nonnegative if $f(\phi) \geq 0$ for all nonnegative functions $\phi \in \mathcal{D}$. In this sense, also nonnegativity of distributional derivatives is understood. For instance, we write $\frac{\partial f}{\partial x_j} \geq 0$ for a function f if, for all nonnegative $\phi \in \mathcal{D}$,

$$\int_{\mathbb{R}^n} f(\mathbf{x}) \frac{\partial}{\partial x_j} \phi(\mathbf{x}) dx \geq 0.$$

C.10   Differential Inequalities

The following lemma bounds the solution of a differential inequality by the solution of a corresponding differential equation.

Lemma C.13. *Let $f, g, h : [0, \infty) \to \mathbb{R}$ be continuous functions with $g(x) \geq 0$ and $f(x) > 0$ for all $x \in [0, \infty)$. Assume that $L_0 : [0, \infty) \to \mathbb{R}$ is such that*

$$f(x)L_0'(x) - g(x)L_0(x) = h(x), \quad x \in [0, \infty), \tag{C.23}$$

while L satisfies the differential inequality

$$f(x)L'(x) - g(x)L(x) \leq h(x), \quad x \in [0, \infty). \tag{C.24}$$

If $L(0) \leq L_0(0)$, then $L(x) \leq L_0(x)$ for all $x \in [0, \infty)$.

Proof. We need to prove that $\ell(x) := L(x) - L_0(x) \leq 0$ for all $x \in [0, \infty)$, knowing that the function ℓ satisfies $\ell(0) \leq 0$ and

$$f(x)\ell'(x) - g(x)\ell(x) \leq 0, \tag{C.25}$$

which follows by subtracting (C.23) from (C.24). Let us introduce

$$k(x) := \ell(x) \exp\left(-\int_0^x \frac{g(t)}{f(t)} dt\right).$$

We calculate, for $x \in [0, \infty)$,

$$k'(x) = \left(\ell'(x) - \frac{g(x)}{f(x)}\ell(x)\right) \exp\left(-\int_0^x \frac{g(t)}{f(t)} dt\right) \leq 0.$$

This shows that k is nonincreasing on $[0, \infty)$. As a result, for all $x \in [0, \infty)$, we have $k(x) \leq k(0) \leq 0$; hence, $\ell(x) \leq 0$. $\qquad\square$

C.10. Differential Inequalities

List of Symbols

s	Usually the number of nonzero entries of a vector to be recovered
m	Usually the number of linear measurements
N	Usually the number of entries of a vector to be recovered
$[N]$	The set of the natural numbers not exceeding N, i.e., $\{1, 2, \ldots, N\}$
S	Usually a subset of $[N]$
$\mathrm{card}(S)$	The cardinality of a set S
$\overline{S}$	The complement of a set S, usually $\overline{S} = [N] \setminus S$
$\mathbb{N}$	The set of natural numbers, i.e., $\{1, 2, \ldots\}$
$\mathbb{N}_0$	The set of natural numbers including 0, i.e., $\{0, 1, 2, \ldots\}$
$\mathbb{Z}$	The set of integers, i.e., $\{\ldots, -2, -1, 0, 1, 2, \ldots\}$
$\mathbb{Q}$	The set of rational numbers
$\mathbb{R}$	The set of real numbers
$\mathbb{R}_+$	The subset of $\mathbb{R}$ consisting of the nonnegative real numbers
$\mathbb{C}$	The set of complex numbers
$\mathbb{K}$	The field $\mathbb{R}$ or $\mathbb{C}$
$\mathbb{R}^N$	The N-dimensional real vector space
$\mathbb{C}^N$	The N-dimensional complex vector space
$\mathbb{C}^S$	The space of vectors indexed by the set S, isomorphic to $\mathbb{C}^{\mathrm{card}(S)}$
ℓ_p^N	The space $\mathbb{C}^N$ equipped with the ℓ_p-(quasi)norm
B_p^N	The unit ball of the (quasi)normed space ℓ_p^N
$(\mathbf{e}_1, \ldots, \mathbf{e}_n)$	The canonical basis of $\mathbb{K}^N$
$\mathbf{x}$	Usually the vector in $\mathbb{C}^N$ to be reconstructed
$\mathbf{x}^\sharp$	Usually the vector outputted by a reconstruction algorithm
$\mathbf{x}^*$	The nonincreasing rearrangement of a vector $\mathbf{x}$
$\mathrm{supp}(\mathbf{x})$	The support of a vector $\mathbf{x}$
$\mathbf{x}_S$	Either the vector in $\mathbb{C}^N$ equal to $\mathbf{x}$ on S and to zero on $\overline{S}$ or the vector in $\mathbb{C}^S$ which is the restriction of $\mathbf{x}$ to the entries indexed by S
$L_s(\mathbf{x})$	An index set of s largest absolute entries of a vector $\mathbf{x}$

S. Foucart and H. Rauhut, *A Mathematical Introduction to Compressive Sensing*, Applied and Numerical Harmonic Analysis, DOI 10.1007/978-0-8176-4948-7, © Springer Science+Business Media New York 2013

$H_s(\mathbf{x})$	The hard thresholding operator applied to a vector $\mathbf{x}$, i.e., $H_s(\mathbf{x}) = \mathbf{x}_{L_s(\mathbf{x})}$
$\sigma_s(\mathbf{x})_p$	The error of best s-term approximation to a vector $\mathbf{x}$
$\mathrm{Re}(z), \mathrm{Im}(z)$	The real and imaginary parts of a complex number z
$\mathrm{Re}(\mathbf{x}), \mathrm{Im}(\mathbf{x})$	The real and imaginary part of a vector $\mathbf{x}$ defined componentwise
$\mathrm{sgn}(z)$	The sign of a complex number z
$\mathrm{sgn}(\mathbf{x})$	The sign of a vector $\mathbf{x}$ defined componentwise
$\mathbf{e}$	Usually the vector of measurement errors
$\mathbf{y}$	Usually the measurement vector, i.e., $\mathbf{y} = \mathbf{Ax} + \mathbf{e}$
$\mathbf{Id}$	The identity matrix
$\mathbf{F}$	The Fourier matrix
$\mathbf{H}$	The Hadamard matrix
$\mathbf{J}$	The matrix with all entries equal to one
$\mathbf{A}$	A matrix, usually the measurement matrix
$\mathbf{A}^\top$	The transpose of a matrix $\mathbf{A}$, i.e., $(\mathbf{A}^\top)_{jk} = A_{jk}$
$\mathbf{A}^*$	The adjoint of a matrix $\mathbf{A}$, i.e., $(\mathbf{A}^*)_{jk} = \overline{A_{jk}}$
$\mathbf{A}^\dagger$	The Moore–Penrose pseudo-inverse of the matrix $\mathbf{A}$
$\mathbf{A}_{I,J}$	The submatrix of a matrix $\mathbf{A}$ with rows indexed by I and columns indexed by J
$\mathbf{A}_S$	The submatrix of a matrix $\mathbf{A}$ with columns indexed by S
$\mathbf{a}_j$	The jth column of a matrix $\mathbf{A}$
$\ker \mathbf{A}$	The null space of the matrix $\mathbf{A}$, i.e., $\{\mathbf{x} : \mathbf{Ax} = \mathbf{0}\}$
Δ	Usually a reconstruction map from $\mathbb{C}^m$ to $\mathbb{C}^N$
Δ_1	The reconstruction map associated with equality-constrained ℓ_1-minimization
$\Delta_{1,\eta}$	The reconstruction map associated with quadratically constrained ℓ_1-minimization
$\|\mathbf{x}\|_p$	The ℓ_p-(quasi)norm of a vector $\mathbf{x}$
$\|\mathbf{x}\|_{p,\infty}$	The weak ℓ_p-quasinorm of a vector $\mathbf{x}$
$\|\mathbf{x}\|_0$	The number of nonzero entries of a vector $\mathbf{x}$
$\langle \mathbf{u}, \mathbf{v} \rangle$	The inner product between two vectors $\mathbf{u}$ and $\mathbf{v}$
$\|\mathbf{A}\|_{p \to q}$	The operator norm of a matrix $\mathbf{A}$ from ℓ_p to ℓ_q
$\|\mathbf{A}\|_{2 \to 2}$	The operator norm (largest singular value) of a matrix $\mathbf{A}$ on ℓ_2
$\|\mathbf{A}\|_F$	The Frobenius norm of a matrix $\mathbf{A}$
μ	The coherence of a matrix
μ_1	The ℓ_1-coherence function of a matrix
δ_s	The sth restricted isometry constant of a matrix
θ_s	The sth restricted expansion constant of a left regular bipartite graph
η	Usually an upper bound on the measurement error, i.e., $\|\mathbf{e}\|_2 \le \eta$
ε	Usually a small probability
$\mathbb{E}(X)$	The expectation of a random variable X
$\mathbb{P}(B)$	The probability of an event B
g	A standard Gaussian random variable
$\mathbf{g}$	A standard Gaussian random vector

$\ell(T)$	The Gaussian width of a subset T of $\mathbb{R}^N$
$E^m(K, X)$	The compressive m-width of a subset K of a normed space X
$d^m(K, X)$	The Gelfand m-width of a subset K of a normed space X
$d_m(K, X)$	The Kolmogorov m-width of a subset K of a normed space X
$\mathcal{N}(T, d, t)$	The covering number of a set T by balls of radius t relative to the metric d
$\mathrm{conv}(T)$	The convex hull of a set T
$\mathrm{cone}(T)$	The conic hull of a set T
K^*	The dual cone of a cone $K \subset \mathbb{R}^N$
K°	The polar cone of a cone $K \subset \mathbb{R}^N$
F^*	The convex conjugate of a function F
$\partial F(\mathbf{x})$	The subdifferential of a function F at a point $\mathbf{x}$
$P_F(\tau; \mathbf{x})$	The proximal mapping associated to a convex function F

References

1. P. Abrial, Y. Moudden, J.-L. Starck, J. Fadili, J. Delabrouille, M. Nguyen, CMB data analysis and sparsity. Stat. Meth. **5**, 289–298 (2008) (Cited on p. 428.)
2. R. Adamczak, A tail inequality for suprema of unbounded empirical processes with applications to Markov chains. Electron. J. Probab. **13**(34), 1000–1034 (2008) (Cited on p. 265.)
3. F. Affentranger, R. Schneider, Random projections of regular simplices. Discrete Comput. Geom. **7**(3), 219–226 (1992) (Cited on p. 305.)
4. M. Afonso, J. Bioucas Dias, M. Figueiredo, Fast image recovery using variable splitting and constrained optimization. IEEE Trans. Image Process. **19**(9), 2345–2356 (2010) (Cited on p. 509.)
5. M. Aharon, M. Elad, A. Bruckstein, The K-SVD: An algorithm for designing of overcomplete dictionaries for sparse representation. IEEE Trans. Signal Process. **54**(11), 4311–4322 (2006) (Cited on p. 38.)
6. R. Ahlswede, A. Winter, Strong converse for identification via quantum channels. IEEE Trans. Inform. Theor. **48**(3), 569–579 (2002) (Cited on p. 261.)
7. N. Ailon, B. Chazelle, The fast Johnson-Lindenstrauss transform and approximate nearest neighbors. SIAM J. Comput. **39**(1), 302–322 (2009) (Cited on p. 423.)
8. N. Ailon, E. Liberty, Fast dimension reduction using Rademacher series on dual BCH codes. Discrete Comput. Geom. **42**(4), 615–630 (2009) (Cited on p. 431.)
9. N. Ailon, E. Liberty, Almost optimal unrestricted fast Johnson-Lindenstrauss transform. In *Proceedings of the 22nd Annual ACM-SIAM Symposium on Discrete Algorithms (SODA)*, San Francisco, USA, 22–25 January 2011 (Cited on p. 423.)
10. N. Ailon, H. Rauhut, Fast and RIP-optimal transforms. *Preprint* (2013) (Cited on p. 431.)
11. K. Alexander, Probability inequalities for empirical processes and a law of the iterated logarithm. Ann. Probab. **12**(4), 1041–1067 (1984) (Cited on p. 265.)
12. W.O. Alltop, Complex sequences with low periodic correlations. IEEE Trans. Inform. Theor. **26**(3), 350–354 (1980) (Cited on p. 129.)
13. G. Anderson, A. Guionnet, O. Zeitouni, in *An Introduction to Random Matrices*. Cambridge Studies in Advanced Mathematics, vol. 118 (Cambridge University Press, Cambridge, 2010) (Cited on p. 303.)
14. F. Andersson, M. Carlsson, M.V. de Hoop, Nonlinear approximation of functions in two dimensions by sums of exponentials. Appl. Comput. Harmon. Anal. **29**(2), 198–213 (2010) (Cited on p. 57.)
15. F. Andersson, M. Carlsson, M.V. de Hoop, Sparse approximation of functions using sums of exponentials and AAK theory. J. Approx. Theor. **163**(2), 213–248 (2011) (Cited on p. 57.)
16. J. Andersson, J.-O. Strömberg, On the theorem of uniform recovery of structured random matrices. *Preprint* (2012) (Cited on p. 169.)

S. Foucart and H. Rauhut, *A Mathematical Introduction to Compressive Sensing*,
Applied and Numerical Harmonic Analysis, DOI 10.1007/978-0-8176-4948-7,
© Springer Science+Business Media New York 2013

17. G. Andrews, R. Askey, R. Roy, in *Special Functions*. Encyclopedia of Mathematics and its Applications, vol. 71 (Cambridge University Press, Cambridge, 1999) (Cited on pp. 426, 428, 578.)

18. M. Anthony, P. Bartlett, *Neural Network Learning: Theoretical Foundations* (Cambridge University Press, Cambridge, 1999) (Cited on p. 36.)

19. S. Arora, B. Barak, *Computational Complexity: A Modern Approach* (Cambridge University Press, Cambridge, 2009) (Cited on pp. 57, 455.)

20. K. Arrow, L. Hurwicz, H. Uzawa, *Studies in Linear and Non-linear Programming* (Stanford University Press, Stanford, California, 1958) (Cited on p. 503.)

21. U. Ayaz, H. Rauhut, Nonuniform sparse recovery with subgaussian matrices. *ETNA*, to appear (Cited on p. 303.)

22. J.-M. Azais, M. Wschebor, *Level Sets and Extrema of Random Processes and Fields* (Wiley, Hoboken, NJ, 2009) (Cited on p. 262.)

23. F. Bach, R. Jenatton, J. Mairal, G. Obozinski, Optimization with sparsity-inducing penalties. Found. Trends Mach. Learn. **4**(1), 1–106 (2012) (Cited on p. 36.)

24. B. Bah, J. Tanner, Improved bounds on restricted isometry constants for Gaussian matrices. SIAM J. Matrix Anal. Appl. **31**(5), 2882–2898 (2010) (Cited on p. 306.)

25. Z. Bai, J. Silverstein, *Spectral Analysis of Large Dimensional Random Matrices*. Springer Series in Statistics, 2nd edn. (Springer, New York, 2010) (Cited on p. 303.)

26. W. Bajwa, J. Haupt, G. Raz, S. Wright, R. Nowak, Toeplitz-structured compressed sensing matrices. In *Proc. IEEE Stat. Sig. Proc. Workshop*, pp. 294–298, 2007 (Cited on p. 429.)

27. W. Bajwa, J. Haupt, A.M. Sayeed, R. Nowak, Compressed channel sensing: a new approach to estimating sparse multipath channels. Proc. IEEE **98**(6), 1058–1076 (June 2010) (Cited on p. 37.)

28. A. Bandeira, E. Dobriban, D. Mixon, W. Sawin, Certifying the restricted isometry property is hard. *Preprint* (2012) (Cited on p. 170.)

29. R.G. Baraniuk, Compressive sensing. IEEE Signal Process. Mag. **24**(4), 118–121 (2007) (Cited on p. 34.)

30. R.G. Baraniuk, V. Cevher, M. Duarte, C. Hedge, Model-based compressive sensing. IEEE Trans. Inform. Theor. **56**, 1982–2001 (April 2010) (Cited on p. 38.)

31. R.G. Baraniuk, M. Davenport, R.A. DeVore, M. Wakin, A simple proof of the restricted isometry property for random matrices. Constr. Approx. **28**(3), 253–263 (2008) (Cited on p. 302.)

32. S. Bartels, Total variation minimization with finite elements: convergence and iterative solution. SIAM J. Numer. Anal. **50**(3), 1162–1180 (2012) (Cited on p. 503.)

33. A. Barvinok, Measure concentration. Lecture notes, University of Michigan, Michigan, USA, 2005 (Cited on p. 264.)

34. H. Bauschke, P. Combettes, *Convex Analysis and Monotone Operator Theory in Hilbert Spaces*. CMS Books in Mathematics/Ouvrages de Mathématiques de la SMC (Springer, New York, 2011) (Cited on p. 506.)

35. I. Bechar, A Bernstein-type inequality for stochastic processes of quadratic forms of Gaussian variables. *Preprint* (2009) (Cited on p. 262.)

36. A. Beck, M. Teboulle, A fast iterative shrinkage-thresholding algorithm for linear inverse problems. SIAM J. Imag. Sci. **2**(1), 183–202 (2009) (Cited on pp. 506, 507.)

37. A. Beck, M. Teboulle, Fast gradient-based algorithms for constrained total variation image denoising and deblurring problems. IEEE Trans. Image Process. **18**(11), 2419–2434 (2009) (Cited on p. 506.)

38. S. Becker, J. Bobin, E.J. Candès, NESTA: A fast and accurate first-order method for sparse recovery. SIAM J. Imaging Sci. **4**(1), 1–39 (2011) (Cited on p. 510.)

39. J.J. Benedetto, P.J.S.G. Ferreira (eds.), *Modern Sampling Theory: Mathematics and Applications*. Applied and Numerical Harmonic Analysis (Birkhäuser, Boston, MA, 2001) (Cited on pp. 35, 573.)

40. G. Bennett, Probability inequalities for the sum of independent random variables. J. Amer. Statist. Assoc. **57**, 33–45 (1962) (Cited on p. 199.)

41. R. Berinde, A. Gilbert, P. Indyk, H. Karloff, M. Strauss, Combining geometry and combinatorics: A unified approach to sparse signal recovery. In *Proc. of 46th Annual Allerton Conference on Communication, Control, and Computing*, pp. 798–805, 2008 (Cited on pp. 35, 38, 455.)

42. R. Berinde, P. Indyk, M. Ržić, Practical near-optimal sparse recovery in the L1 norm. In *Proc. Allerton*, 2008 (Cited on p. 455.)

43. S. Bernstein, Sur une modification de l'inégalité de Tchebichef. In *Annals Science Insitute Sav. Ukraine, Sect. Math. I*, 1924 (Cited on p. 199.)

44. D. Bertsekas, J. Tsitsiklis, *Parallel and Distributed Computation: Numerical Methods* (Athena Scientific, Cambridge, MA, 1997) (Cited on p. 506.)

45. G. Beylkin, L. Monzón, On approximation of functions by exponential sums. Appl. Comput. Harmon. Anal. **19**(1), 17–48 (2005) (Cited on p. 57.)

46. G. Beylkin, L. Monzón, Approximation by exponential sums revisited. Appl. Comput. Harmon. Anal. **28**(2), 131–149 (2010) (Cited on p. 57.)

47. R. Bhatia, *Matrix Analysis*. Graduate Texts in Mathematics, vol. 169 (Springer, New York, 1997) (Cited on pp. 515, 528, 542, 565, 566.)

48. P. Bickel, Y. Ritov, A. Tsybakov, Simultaneous analysis of lasso and Dantzig selector. Ann. Stat. **37**(4), 1705–1732 (2009) (Cited on p. 36.)

49. E. Birgin, J. Martínez, M. Raydan, Inexact spectral projected gradient methods on convex sets. IMA J. Numer. Anal. **23**(4), 539–559 (2003) (Cited on p. 505.)

50. I. Bjelakovic, R. Siegmund-Schultze, Quantum Stein's lemma revisited, inequalities for quantum entropies, and a concavity theorem of Lieb. *Preprint* (2012) (Cited on p. 565.)

51. A. Björck, *Numerical Methods for Least Squares Problems* (SIAM, Philadelphia, 1996) (Cited on pp. 515, 534.)

52. R. Blahut, *Algebraic Codes for Data Transmission* (Cambridge University Press, Cambridge, 2003) (Cited on p. 57.)

53. J. Blanchard, C. Cartis, J. Tanner, Compressed sensing: how sharp is the restricted isometry property? SIAM Rev. **53**(1), 105–125 (2011) (Cited on p. 306.)

54. J. Blanchard, A. Thompson, On support sizes of restricted isometry constants. Appl. Comput. Harmon. Anal. **29**(3), 382–390 (2010) (Cited on p. 169.)

55. T. Blu, P. Marziliano, M. Vetterli, Sampling signals with finite rate of innovation. IEEE Trans. Signal Process. **50**(6), 1417–1428 (2002) (Cited on p. 57.)

56. T. Blumensath, M. Davies, Iterative thresholding for sparse approximations. J. Fourier Anal. Appl. **14**, 629–654 (2008) (Cited on p. 169.)

57. T. Blumensath, M. Davies, Iterative hard thresholding for compressed sensing. Appl. Comput. Harmon. Anal. **27**(3), 265–274 (2009) (Cited on p. 169.)

58. T. Blumensath, M. Davies, Normalized iterative hard thresholding: guaranteed stability and performance. IEEE J. Sel. Top. Signal Process. **4**(2), 298–309 (2010) (Cited on p. 169.)

59. S. Boucheron, O. Bousquet, G. Lugosi, P. Massart, Moment inequalities for functions of independent random variables. Ann. Probab. **33**(2), 514–560 (2005) (Cited on p. 265.)

60. S. Boucheron, G. Lugosi, P. Massart, Concentration inequalities using the entropy method. Ann. Probab. **31**(3), 1583–1614 (2003) (Cited on p. 265.)

61. S. Boucheron, G. Lugosi, P. Massart, Concentration inequalities. A nonasymptotic theory of independence. Oxford University Press (2013) (Cited on p. 264.)

62. J. Bourgain, Bounded orthogonal systems and the $\Lambda(p)$-set problem. Acta Math. **162**(3–4), 227–245 (1989) (Cited on p. 423.)

63. J. Bourgain, Λ_p-sets in analysis: results, problems and related aspects. In *Handbook of the Geometry of Banach Spaces*, vol. 1, ed. by W. B. Johnson, J. Lindenstrauss (North-Holland, Amsterdam, 2001), pp. 195–232 (Cited on p. 423.)

64. J. Bourgain, S. Dilworth, K. Ford, S. Konyagin, D. Kutzarova, Breaking the k^2-barrier for explicit RIP matrices. In *STOC'11 Proceedings of the 43rd Annual ACM Symposium on Theory of Computing*, ACM New York, USA, pp. 637–644, 2011 (Cited on p. 170.)

65. J. Bourgain, S. Dilworth, K. Ford, S. Konyagin, D. Kutzarova, Explicit constructions of RIP matrices and related problems. Duke Math. J. **159**(1), 145–185 (2011) (Cited on p. 170.)

66. J. Bourgain, L. Tzafriri, Invertibility of 'large' submatrices with applications to the geometry of Banach spaces and harmonic analysis. Isr. J. Math. **57**(2), 137–224 (1987) (Cited on pp. 262, 472.)

67. J. Bourgain, L. Tzafriri, On a problem of Kadison and Singer. J. Reine Angew. Math. **420**, 1–43 (1991) (Cited on pp. 472, 473.)

68. O. Bousquet, A Bennett concentration inequality and its application to suprema of empirical processes. C. R. Math. Acad. Sci. Paris. **334**(6), 495–500 (2002) (Cited on p. 265.)

69. O. Bousquet, Concentration inequalities for sub-additive functions using the entropy method. In *Stochastic Inequalities and Applications*, ed. by E. Gin, C. Houdr , D. Nualart. Progress in Probability, vol. 56 (Birkhäuser, Basel, 2003), pp. 213–247 (Cited on p. 265.)

70. S. Boyd, L. Vandenberghe, *Convex Optimization* (Cambridge University Press, Cambridge, 2004) (Cited on pp. 71, 503, 504, 510, 543, 558, 560.)

71. K. Bredies, D. Lorenz, Linear convergence of iterative soft-thresholding. J. Fourier Anal. Appl. **14**(5–6), 813–837 (2008) (Cited on p. 506.)

72. H. Brezis, *Functional Analysis, Sobolev Apaces and Partial Differential Equations*. Universitext (Springer, New York, 2011) (Cited on p. 583.)

73. A. Bruckstein, D.L. Donoho, M. Elad, From sparse solutions of systems of equations to sparse modeling of signals and images. SIAM Rev. **51**(1), 34–81 (2009) (Cited on p. 36.)

74. A. Buchholz, Operator Khintchine inequality in non-commutative probability. Math. Ann. **319**, 1–16 (2001) (Cited on p. 261.)

75. A. Buchholz, Optimal constants in Khintchine type inequalities for fermions, Rademachers and q-Gaussian operators. Bull. Pol. Acad. Sci. Math. **53**(3), 315–321 (2005) (Cited on p. 261.)

76. P. Bühlmann, S. van de Geer, *Statistics for High-dimensional Data*. Springer Series in Statistics (Springer, Berlin, 2011) (Cited on p. 36.)

77. H. Buhrman, P. Miltersen, J. Radhakrishnan, S. Venkatesh, Are bitvectors optimal? In *Proceedings of the Thirty-second Annual ACM Symposium on Theory of Computing STOC '00:* , pp. 449–458, ACM, New York, NY, USA, 2000 (Cited on p. 327.)

78. V. Buldygin, Y. Kozachenko, in *Metric Characterization of Random Variables and Random Processes*, Translations of Mathematical Monographs, vol. 188 (American Mathematical Society, Providence, RI, 2000) (Cited on p. 199.)

79. M. Burger, M. Moeller, M. Benning, S. Osher, An adaptive inverse scale space method for compressed sensing. Math. Comp. **82**(281), 269–299 (2013) (Cited on p. 503.)

80. N. Burq, S. Dyatlov, R. Ward, M. Zworski, Weighted eigenfunction estimates with applications to compressed sensing. SIAM J. Math. Anal. **44**(5), 3481–3501 (2012) (Cited on p. 428.)

81. T. Cai, L. Wang, G. Xu, New bounds for restricted isometry constants. IEEE Trans. Inform. Theor. **56**(9), 4388–4394 (2010) (Cited on p. 169.)

82. T. Cai, L. Wang, G. Xu, Shifting inequality and recovery of sparse vectors. IEEE Trans. Signal Process. **58**(3), 1300–1308 (2010) (Cited on p. 169.)

83. T. Cai, A. Zhang, Sharp RIP bound for sparse signal and low-rank matrix recovery, Appl. Comput. Harmon. Anal. **35**(1), 74–93, (2013) (Cited on p. 169.)

84. E.J. Candès, Compressive sampling. In *Proceedings of the International Congress of Mathematicians*, Madrid, Spain, 2006 (Cited on p. 34.)

85. E.J. Candès, The restricted isometry property and its implications for compressed sensing. C. R. Math. Acad. Sci. Paris **346**, 589–592 (2008) (Cited on p. 169.)

86. E.J. Candès, Y.C. Eldar, D. Needell, P. Randall, Compressed sensing with coherent and redundant dictionaries. Appl. Comput. Harmon. Anal. **31**(1), 59–73 (2011) (Cited on p. 302.)

87. E.J. Candès, Y. Plan, Near-ideal model selection by ℓ_1 minimization. Ann. Stat. **37**(5A), 2145–2177 (2009) (Cited on p. 471.)

88. E.J. Candès, Y. Plan, A probabilistic and RIPless theory of compressed sensing. IEEE Trans. Inform. Theor. **57**(11), 7235–7254 (2011) (Cited on p. 421.)

89. E.J. Candès, Y. Plan, Tight oracle bounds for low-rank matrix recovery from a minimal number of random measurements. IEEE Trans. Inform. Theor. **57**(4), 2342–2359 (2011) (Cited on p. 302.)

90. E.J. Candès, B. Recht, Exact matrix completion via convex optimization. Found. Comput. Math. **9**, 717–772 (2009) (Cited on pp. 36, 37.)

91. E. J. Candès, B. Recht Simple bounds for recovering low-complexity models. Math. Program. Springer-Verlag, 1–13 (2012) (Cited on p. 303.)

92. E.J. Candès, J. Romberg, Quantitative robust uncertainty principles and optimally sparse decompositions. Found. Comput. Math. **6**(2), 227–254 (2006) (Cited on pp. 36, 423, 472.)

93. E.J. Candès, J. Romberg, Sparsity and incoherence in compressive sampling. Inverse Probl. **23**(3), 969–985 (2007) (Cited on p. 421.)

94. E.J. Candès, J. Romberg, T. Tao, Robust uncertainty principles: exact signal reconstruction from highly incomplete frequency information. IEEE Trans. Inform. Theor. **52**(2), 489–509 (2006) (Cited on pp. 33, 35, 37, 104, 421, 422.)

95. E.J. Candès, J. Romberg, T. Tao, Stable signal recovery from incomplete and inaccurate measurements. Comm. Pure Appl. Math. **59**(8), 1207–1223 (2006) (Cited on pp. 104, 168.)

96. E.J. Candès, T. Tao, Decoding by linear programming. IEEE Trans. Inform. Theor. **51**(12), 4203–4215 (2005) (Cited on pp. 36, 168.)

97. E.J. Candès, T. Tao, Near optimal signal recovery from random projections: universal encoding strategies? IEEE Trans. Inform. Theor. **52**(12), 5406–5425 (2006) (Cited on pp. 35, 168, 302, 422.)

98. E.J. Candès, T. Tao, The Dantzig selector: statistical estimation when p is much larger than n. Ann. Stat. **35**(6), 2313–2351, (2007) (Cited on pp. 36, 71.)

99. E.J. Candès, T. Tao, The power of convex relaxation: near-optimal matrix completion. IEEE Trans. Inform. Theor. **56**(5), 2053–2080 (2010) (Cited on pp. 36, 37.)

100. E.J. Candès, M. Wakin, An introduction to compressive sampling. IEEE Signal Process. Mag. **25**(2), 21–30 (2008) (Cited on p. 34.)

101. M. Capalbo, O. Reingold, S. Vadhan, A. Wigderson, Randomness conductors and constant-degree lossless expanders. In *Proceedings of the Thirty-Fourth Annual ACM Symposium on Theory of Computing* (electronic) (ACM, New York, 2002), pp. 659–668 (Cited on p. 455.)

102. P. Casazza, J. Tremain, Revisiting the Bourgain-Tzafriri restricted invertibility theorem. Oper. Matrices **3**(1), 97–110 (2009) (Cited on p. 473.)

103. D. Chafaï, O. Guédon, G. Lecué, A. Pajor, *Interactions Between Compressed Sensing, Random Matrices and High-dimensional Geometry*, Panoramas et Synthèses, vol. 37 (Société Mathématique de France, to appear) (Cited on p. 422.)

104. A. Chambolle, V. Caselles, D. Cremers, M. Novaga, T. Pock, An introduction to total variation for image analysis. In *Theoretical Foundations and Numerical Methods for Sparse Recovery*, ed. by M. Fornasier. Radon Series on Computational and Applied Mathematics, vol. 9 (de Gruyter, Berlin, 2010), pp. 263–340 (Cited on p. 37.)

105. A. Chambolle, R.A. DeVore, N.-Y. Lee, B.J. Lucier, Nonlinear wavelet image processing: variational problems, compression, and noise removal through wavelet shrinkage. IEEE Trans. Image Process. **7**(3), 319–335 (1998) (Cited on p. 36.)

106. A. Chambolle, P.-L. Lions, Image recovery via total variation minimization and related problems. Numer. Math. **76**(2), 167–188 (1997) (Cited on p. 37.)

107. A. Chambolle, T. Pock, A first-order primal-dual algorithm for convex problems with applications to imaging. J. Math. Imag. Vis. **40**, 120–145 (2011) (Cited on pp. 37, 503.)

108. V. Chandrasekaran, B. Recht, P. Parrilo, A. Willsky, The convex geometry of linear inverse problems. Found. Comput. Math. **12**(6), 805–849 (2012) (Cited on pp. 104, 303.)

109. R. Chartrand, Exact reconstruction of sparse signals via nonconvex minimization. IEEE Signal Process. Lett. **14**(10), 707–710 (2007) (Cited on p. 504.)

110. R. Chartrand, V. Staneva, Restricted isometry properties and nonconvex compressive sensing. Inverse Probl. **24**(3), 1–14 (2008) (Not cited.)

111. R. Chartrand, W. Yin, Iteratively reweighted algorithms for compressive sensing. In *2008 IEEE International Conference on Acoustics, Speech and Signal Processing*, ICASSP, Las Vegas, Nevada, USA, pp. 3869–3872, 2008 (Cited on p. 504.)

112. G. Chen, R. Rockafellar, Convergence rates in forward-backward splitting. SIAM J. Optim. **7**(2), 421–444 (1997) (Cited on p. 506.)

113. S. Chen, S. Billings, W. Luo, Orthogonal least squares methods and their application to nonlinear system identification. Intl. J. Contr. **50**(5), 1873–1896 (1989) (Cited on p. 71.)

114. S.S. Chen, D.L. Donoho, M.A. Saunders, Atomic decomposition by Basis Pursuit. SIAM J. Sci. Comput. **20**(1), 33–61 (1999) (Cited on pp. 34, 36, 71.)

115. M. Cheraghchi, V. Guruswami, A. Velingker, Restricted isometry of Fourier matrices and list decodability of random linear codes. *Preprint* (2012) (Cited on p. 422.)

116. H. Chernoff, A measure of asymptotic efficiency of tests of a hypothesis based on the sum of observations. Ann. Math. Statist. **23**, 493–507 (1952) (Cited on p. 198.)

117. T. Chihara, *An Introduction to Orthogonal Polynomials* (Gordon and Breach Science Publishers, New York 1978) (Cited on p. 426.)

118. S. Chrétien, S. Darses, Invertibility of random submatrices via tail decoupling and a matrix Chernoff inequality. Stat. Probab. Lett. **82**(7), 1479–1487 (2012) (Cited on p. 471.)

119. O. Christensen, *An Introduction to Frames and Riesz Bases.* Applied and Numerical Harmonic Analysis (Birkhäuser, Boston, 2003) (Cited on pp. 129, 429.)

120. O. Christensen, *Frames and Bases: An Introductory Course.* Applied and Numerical Harmonic Analysis (Birkhäuser, Basel, 2008) (Cited on p. 129.)

121. A. Cline, Rate of convergence of Lawson's algorithm. Math. Commun. **26**, 167–176 (1972) (Cited on p. 504.)

122. A. Cohen, *Numerical Analysis of Wavelet Methods* (North-Holland, Amsterdam, 2003) (Cited on pp. 36, 425.)

123. A. Cohen, W. Dahmen, R.A. DeVore, Compressed sensing and best k-term approximation. J. Amer. Math. Soc. **22**(1), 211–231 (2009) (Cited on pp. 56, 104, 168, 327, 362.)

124. A. Cohen, I. Daubechies, R. DeVore, G. Kerkyacharian, D. Picard, Capturing Ridge Functions in High Dimensions from Point Queries. Constr. Approx. **35**, 225–243 (2012) (Cited on p. 39.)

125. A. Cohen, R. DeVore, S. Foucart, H. Rauhut, Recovery of functions of many variables via compressive sensing. In *Proc. SampTA 2011*, Singapore, 2011 (Cited on p. 39.)

126. R. Coifman, F. Geshwind, Y. Meyer, Noiselets. Appl. Comput. Harmon. Anal. **10**(1), 27–44 (2001) (Cited on pp. 424, 425.)

127. P. Combettes, J.-C. Pesquet, A Douglas-Rachford Splitting Approach to Nonsmooth Convex Variational Signal Recovery. IEEE J. Sel. Top. Signal Process. **1**(4), 564–574 (2007) (Cited on p. 508.)

128. P. Combettes, J.-C. Pesquet, Proximal splitting methods in signal processing. In *Fixed-Point Algorithms for Inverse Problems in Science and Engineering*, ed. by H. Bauschke, R. Burachik, P. Combettes, V. Elser, D. Luke, H. Wolkowicz (Springer, New York, 2011), pp. 185–212 (Cited on pp. 504, 506, 508, 509.)

129. P. Combettes, V. Wajs, Signal recovery by proximal forward-backward splitting. Multiscale Model. Sim. **4**(4), 1168–1200 (electronic) (2005) (Cited on pp. 506, 507.)

130. J. Cooley, J. Tukey, An algorithm for the machine calculation of complex Fourier series. Math. Comp. **19**, 297–301 (1965) (Cited on p. 575.)

131. G. Cormode, S. Muthukrishnan, Combinatorial algorithms for compressed sensing. In *CISS*, Princeton, 2006 (Cited on pp. 35, 38.)

132. H. Cramér, Sur un nouveau théorème-limite de la théorie des probabilités. Actual. Sci. Industr. **736**, 5–23 (1938) (Cited on p. 198.)

133. F. Cucker, S. Smale, On the mathematical foundations of learning. Bull. Am. Math. Soc., New Ser. **39**(1), 1–49 (2002) (Cited on p. 36.)

134. F. Cucker, D.-X. Zhou, *Learning Theory: An Approximation Theory Viewpoint.* Cambridge Monographs on Applied and Computational Mathematics (Cambridge University Press, Cambridge, 2007) (Cited on p. 36.)

135. W. Dai, O. Milenkovic, Subspace Pursuit for Compressive Sensing Signal Reconstruction. IEEE Trans. Inform. Theor. **55**(5), 2230–2249 (2009) (Cited on pp. 72, 170.)

136. S. Dasgupta, A. Gupta, An elementary proof of a theorem of Johnson and Lindenstrauss. Random Struct. Algorithm. **22**(1), 60–65, 2003 (Cited on p. 303.)

137. I. Daubechies, *Ten Lectures on Wavelets*, CBMS-NSF Regional Conference Series in Applied Mathematics, vol. 61 (SIAM, Philadelphia, 1992) (Cited on pp. 36, 425.)

138. I. Daubechies, M. Defrise, C. De Mol, An iterative thresholding algorithm for linear inverse problems with a sparsity constraint. Comm. Pure Appl. Math. **57**(11), 1413–1457 (2004) (Cited on pp. 37, 507.)

139. I. Daubechies, R. DeVore, M. Fornasier, C. Güntürk, Iteratively re-weighted least squares minimization for sparse recovery. Comm. Pure Appl. Math. **63**(1), 1–38 (2010) (Cited on p. 504.)

140. I. Daubechies, M. Fornasier, I. Loris, Accelerated projected gradient methods for linear inverse problems with sparsity constraints. J. Fourier Anal. Appl. **14**(5–6), 764–792 (2008) (Cited on p. 505.)

141. K. Davidson, S. Szarek, in *Local Operator Theory, Random Matrices and Banach Spaces*, ed. by W.B. Johnson, J. Lindenstrauss. Handbook of the Geometry of Banach Spaces, vol. 1 (North-Holland, Amsterdam, 2001), pp. 317–366 (Cited on p. 303.)

142. E. Davies, B. Simon, Ultracontractivity and the heat kernel for Schrödinger operators and Dirichlet Laplacians. J. Funct. Anal. **59**, 335–395 (1984) (Cited on p. 264.)

143. M. Davies, Y. Eldar, Rank awareness in joint sparse recovery. IEEE Trans. Inform. Theor. **58**(2), 1135–1146 (2012) (Cited on p. 38.)

144. M. Davies, R. Gribonval, Restricted isometry constants where ℓ^p sparse recovery can fail for $0 < p \leq 1$. IEEE Trans. Inform. Theor. **55**(5), 2203–2214 (2009) (Cited on p. 169.)

145. G. Davis, S. Mallat, Z. Zhang, Adaptive time-frequency decompositions. Opt. Eng. **33**(7), 2183–2191 (1994) (Cited on p. 71.)

146. V. de la Peña, E. Giné, *Decoupling: From Dependence to Independence*. Probability and its Applications (New York) (Springer, New York, 1999) (Cited on p. 262.)

147. C. De Mol, E. De Vito, L. Rosasco, Elastic-net regularization in learning theory. J. Complex. **25**(2), 201–230 (2009) (Cited on p. 36.)

148. R.A. DeVore, G. Petrova, P. Wojtaszczyk, Instance-optimality in probability with an ℓ_1-minimization decoder. Appl. Comput. Harmon. Anal. **27**(3), 275–288 (2009) (Cited on p. 363.)

149. R.A. DeVore, G. Petrova, P. Wojtaszczyk, Approximation of functions of few variables in high dimensions. Constr. Approx. **33**(1), 125–143 (2011) (Cited on p. 39.)

150. D.L. Donoho, De-noising by soft-thresholding. IEEE Trans. Inform. Theor. **41**(3), 613–627 (1995) (Cited on p. 36.)

151. D.L. Donoho, Neighborly polytopes and sparse solutions of underdetermined linear equations. *Preprint* (2005) (Cited on p. 104.)

152. D.L. Donoho, Compressed sensing. IEEE Trans. Inform. Theor. **52**(4), 1289–1306 (2006) (Cited on pp. 33, 327.)

153. D.L. Donoho, For most large underdetermined systems of linear equations the minimal l^1 solution is also the sparsest solution. Commun. Pure Appl. Anal. **59**(6), 797–829 (2006) (Cited on p. 169.)

154. D.L. Donoho, High-dimensional centrally symmetric polytopes with neighborliness proportional to dimension. Discrete Comput. Geom. **35**(4), 617–652 (2006) (Cited on pp. 38, 303, 304.)

155. D.L. Donoho, M. Elad, Optimally sparse representations in general (non-orthogonal) dictionaries via ℓ^1 minimization. Proc. Nat. Acad. Sci. **100**(5), 2197–2202 (2003) (Cited on pp. 34, 36, 56, 104, 128.)

156. D.L. Donoho, M. Elad, On the stability of the basis pursuit in the presence of noise. Signal Process. **86**(3), 511–532 (2006) (Not cited.)

157. D.L. Donoho, M. Elad, V.N. Temlyakov, Stable recovery of sparse overcomplete representations in the presence of noise. IEEE Trans. Inform. Theor. **52**(1), 6–18 (2006) (Not cited.)

158. D.L. Donoho, X. Huo, Uncertainty principles and ideal atomic decompositions. IEEE Trans. Inform. Theor. **47**(7), 2845–2862 (2001) (Cited on pp. 34, 36, 104, 128, 421.)

159. D.L. Donoho, I.M. Johnstone, Minimax estimation via wavelet shrinkage. Ann. Stat. **26**(3), 879–921 (1998) (Cited on p. 36.)
160. D.L. Donoho, G. Kutyniok, Microlocal analysis of the geometric separation problem. Comm. Pure Appl. Math. **66**(1), 1–47 (2013) (Cited on p. 36.)
161. D.L. Donoho, B. Logan, Signal recovery and the large sieve. SIAM J. Appl. Math. **52**(2), 577–591 (1992) (Cited on p. 34.)
162. D.L. Donoho, A. Maleki, A. Montanari, Message-passing algorithms for compressed sensing. Proc. Natl. Acad. Sci. USA **106**(45), 18914–18919 (2009) (Cited on pp. 72, 306.)
163. D.L. Donoho, P. Stark, Recovery of a sparse signal when the low frequency information is missing. Technical report, Department of Statistics, University of California, Berkeley, June 1989 (Cited on p. 36.)
164. D.L. Donoho, P. Stark, Uncertainty principles and signal recovery. SIAM J. Appl. Math. **48**(3), 906–931 (1989) (Cited on pp. 36, 421, 423.)
165. D.L. Donoho, J. Tanner, Neighborliness of randomly projected simplices in high dimensions. Proc. Natl. Acad. Sci. USA **102**(27), 9452–9457 (2005) (Cited on pp. 38, 303, 304, 305.)
166. D.L. Donoho, J. Tanner, Sparse nonnegative solutions of underdetermined linear equations by linear programming. Proc. Natl. Acad. Sci. **102**(27), 9446–9451 (2005) (Not cited.)
167. D.L. Donoho, J. Tanner, Counting faces of randomly-projected polytopes when the projection radically lowers dimension. J. Am. Math. Soc. **22**(1), 1–53 (2009) (Cited on pp. 36, 38, 303, 304, 305.)
168. D.L. Donoho, J. Tanner, Observed universality of phase transitions in high-dimensional geometry, with implications for modern data analysis and signal processing. Philos. Trans. R. Soc. Lond. Ser. A Math. Phys. Eng. Sci. **367**(1906), 4273–4293 (2009) (Cited on p. 305.)
169. D.L. Donoho, Y. Tsaig, Fast solution of l1-norm minimization problems when the solution may be sparse. IEEE Trans. Inform. Theor. **54**(11), 4789–4812 (2008) (Cited on p. 503.)
170. D.L. Donoho, M. Vetterli, R.A. DeVore, I. Daubechies, Data compression and harmonic analysis. IEEE Trans. Inform. Theor. **44**(6), 2435–2476 (1998) (Cited on p. 36.)
171. R. Dorfman, The detection of defective members of large populations. Ann. Stat. **14**, 436–440 (1943) (Cited on p. 35.)
172. J. Douglas, H. Rachford, On the numerical solution of heat conduction problems in two or three space variables. Trans. Am. Math. Soc. **82**, 421–439 (1956) (Cited on p. 508.)
173. D.-Z. Du, F. Hwang, *Combinatorial Group Testing and Its Applications* (World Scientific, Singapore, 1993) (Cited on p. 35.)
174. M. Duarte, M. Davenport, D. Takhar, J. Laska, S. Ting, K. Kelly, R.G. Baraniuk, Single-Pixel Imaging via Compressive Sampling. IEEE Signal Process. Mag. **25**(2), 83–91 (2008) (Cited on p. 35.)
175. R.M. Dudley, The sizes of compact subsets of Hilbert space and continuity of Gaussian processes. J. Funct. Anal. **1**, 290–330 (1967) (Cited on p. 262.)
176. E. Effros, A matrix convexity approach to some celebrated quantum inequalities. Proc. Natl. Acad. Sci. USA **106**(4), 1006–1008 (2009) (Cited on pp. 569, 571.)
177. B. Efron, T. Hastie, I. Johnstone, R. Tibshirani, Least angle regression. Ann. Stat. **32**(2), 407–499 (2004) (Cited on p. 503.)
178. I. Ekeland, R. Témam, *Convex Analysis and Variational Problems* (SIAM, Philadelphia, 1999) (Cited on pp. 543, 547.)
179. M. Elad, *Sparse and Redundant Representations: From Theory to Applications in Signal and Image Processing* (Springer, New York, 2010) (Cited on p. 36.)
180. M. Elad, M. Aharon, Image denoising via sparse and redundant representations over learned dictionaries. IEEE Trans. Image Process. **15**(12), 3736–3745 (2006) (Cited on p. 36.)
181. M. Elad, A.M. Bruckstein, A generalized uncertainty principle and sparse representation in pairs of bases. IEEE Trans. Inform. Theor. **48**(9), 2558–2567 (2002) (Cited on pp. 34, 36, 104, 421.)
182. Y. Eldar, G. Kutyniok (eds.), *Compressed Sensing: Theory and Applications* (Cambridge University Press, New York, 2012) (Cited on p. 34.)

183. Y. Eldar, M. Mishali, Robust recovery of signals from a structured union of subspaces. IEEE Trans. Inform. Theor. **55**(11), 5302–5316 (2009) (Cited on p. 38.)

184. Y. Eldar, H. Rauhut, Average case analysis of multichannel sparse recovery using convex relaxation. IEEE Trans. Inform. Theor. **56**(1), 505–519 (2010) (Cited on pp. 38, 261, 472.)

185. J. Ender, On compressive sensing applied to radar. Signal Process. **90**(5), 1402–1414 (2010) (Cited on p. 35.)

186. H.W. Engl, M. Hanke, A. Neubauer, *Regularization of Inverse Problems* (Springer, New York, 1996) (Cited on pp. 37, 507.)

187. H. Epstein, Remarks on two theorems of E. Lieb. Comm. Math. Phys. **31**, 317–325 (1973) (Cited on p. 565.)

188. E. Esser, Applications of Lagrangian-based alternating direction methods and connections to split Bregman. *Preprint* (2009) (Cited on p. 509.)

189. A. Fannjiang, P. Yan, T. Strohmer, Compressed remote sensing of sparse objects. SIAM J. Imag. Sci. **3**(3), 596–618 (2010) (Cited on p. 35.)

190. M. Fazel, *Matrix Rank Minimization with Applications*. PhD thesis, 2002 (Cited on p. 37.)

191. H.G. Feichtinger, F. Luef, T. Werther, A guided tour from linear algebra to the foundations of Gabor analysis. In *Gabor and Wavelet Frames*. Lecture Notes Series, Institute for Mathematical Sciences National University of Singapore, vol. 10 (World Sci. Publ., Hackensack, 2007), pp. 1–49 (Cited on p. 429.)

192. H.G. Feichtinger, T. Strohmer, *Gabor Analysis and Algorithms: Theory and Applications* (Birkhäuser, Boston, 1998) (Cited on p. 429.)

193. X. Fernique, Regularité des trajectoires des fonctions aléatoires gaussiennes. In *École d'Été de Probabilités de Saint-Flour, IV-1974*. Lecture Notes in Mathematics, vol. 480 (Springer, Berlin, 1975), pp. 1–96 (Cited on pp. 262, 264.)

194. X. Fernique, *Fonctions Aléatoires Gaussiennes, Vecteurs Aléatoires Gaussiens*. Université de Montréal, Centre de Recherches Mathématiques, 1997 (Cited on p. 262.)

195. P.J.S.G. Ferreira, J.R. Higgins, The establishment of sampling as a scientific principle—a striking case of multiple discovery. Not. AMS **58**(10), 1446–1450 (2011) (Cited on p. 35.)

196. M.A. Figueiredo, R.D. Nowak, An EM algorithm for wavelet-based image restoration. IEEE Trans. Image Process. **12**(8), 906–916 (2003) (Cited on p. 507.)

197. M.A. Figueiredo, R.D. Nowak, S. Wright, Gradient projection for sparse reconstruction: Application to compressed sensing and other inverse problems. IEEE J. Sel. Top. Signal Proces. **1**(4), 586–598 (2007) (Cited on p. 510.)

198. G.B. Folland, *Fourier Analysis and Its Applications* (Wadsworth and Brooks, Pacific Grove, 1992) (Cited on pp. 420, 573.)

199. G.B. Folland, *A Course in Abstract Harmonic Analysis* (CRC Press, Boca Raton, 1995) (Cited on p. 420.)

200. G.B. Folland, A. Sitaram, The uncertainty principle: A mathematical survey. J. Fourier Anal. Appl. **3**(3), 207–238 (1997) (Cited on p. 421.)

201. M. Fornasier, Numerical methods for sparse recovery. In *Theoretical Foundations and Numerical Methods for Sparse Recovery*, ed. by M. Fornasier. Radon Series on Computational and Applied Mathematics, vol. 9 (de Gruyter, Berlin, 2010), pp. 93–200 (Cited on p. 510.)

202. M. Fornasier, H. Rauhut, Iterative thresholding algorithms. Appl. Comput. Harmon. Anal. **25**(2), 187–208 (2008) (Cited on p. 507.)

203. M. Fornasier, H. Rauhut, Recovery algorithms for vector valued data with joint sparsity constraints. SIAM J. Numer. Anal. **46**(2), 577–613 (2008) (Cited on pp. 38, 507.)

204. M. Fornasier, H. Rauhut, Compressive sensing. In *Handbook of Mathematical Methods in Imaging*, ed. by O. Scherzer (Springer, New York, 2011), pp. 187–228 (Cited on p. 34.)

205. M. Fornasier, H. Rauhut, R. Ward, Low-rank matrix recovery via iteratively reweighted least squares minimization. SIAM J. Optim. **21**(4), 1614–1640 (2011) (Cited on p. 504.)

206. M. Fornasier, K. Schnass, J. Vybiral, Learning Functions of Few Arbitrary Linear Parameters in High Dimensions. Found. Comput. Math. **12**, 229–262 (2012) (Cited on p. 39.)

207. S. Foucart, A note on guaranteed sparse recovery via ℓ_1-minimization. Appl. Comput. Harmon. Anal. **29**(1), 97–103 (2010) (Cited on p. 169.)

208. S. Foucart, Hard thresholding pursuit: an algorithm for compressive sensing. SIAM J. Numer. Anal. **49**(6), 2543–2563 (2011) (Cited on p. 169.)

209. S. Foucart, Sparse recovery algorithms: sufficient conditions in terms of restricted isometry constants. In *Approximation Theory XIII: San Antonio 2010*, ed. by M. Neamtu, L. Schumaker. Springer Proceedings in Mathematics, vol. 13 (Springer, New York, 2012), pp. 65–77 (Cited on pp. 56, 169.)

210. S. Foucart, Stability and robustness of ℓ_1-minimizations with Weibull matrices and redundant dictionaries. Lin. Algebra Appl. to appear (Cited on p. 363.)

211. S. Foucart, R. Gribonval, Real vs. complex null space properties for sparse vector recovery. Compt. Rendus Acad. Sci. Math. **348**(15–16), 863–865 (2010) (Cited on p. 104.)

212. S. Foucart, M. Lai, Sparsest solutions of underdetermined linear systems via ℓ_q-minimization for $0 < q \leq 1$. Appl. Comput. Harmon. Anal. **26**(3), 395–407 (2009) (Cited on p. 169.)

213. S. Foucart, A. Pajor, H. Rauhut, T. Ullrich, The Gelfand widths of ℓ_p-balls for $0 < p \leq 1$. J. Complex. **26**(6), 629–640 (2010) (Cited on pp. 327, 362.)

214. J. Friedman, W. Stuetzle, Projection pursuit regressions. J. Am. Stat. Soc. **76**, 817–823 (1981) (Cited on p. 71.)

215. J.J. Fuchs, On sparse representations in arbitrary redundant bases. IEEE Trans. Inform. Theor. **50**(6), 1341–1344 (2004) (Cited on pp. 34, 104.)

216. D. Gabay, Applications of the method of multipliers to variational inequalities. In *Augmented Lagrangian Methods: Applications to the Numerical Solution of Boundary-Value Problems*, ed. by M. Fortin, R. Glowinski (North-Holland, Amsterdam, 1983), pp. 299–331 (Cited on p. 509.)

217. D. Gabay, B. Mercier, A dual algorithm for the solution of nonlinear variational problems via finite elements approximations. Comput. Math. Appl. **2**, 17–40 (1976) (Cited on p. 509.)

218. R. Garg, R. Khandekar, Gradient descent with sparsification: An iterative algorithm for sparse recovery with restricted isometry property. In *Proceedings of the 26th Annual International Conference on Machine Learning*, ICML '09, pp. 337–344, ACM, New York, NY, USA, 2009 (Cited on p. 169.)

219. A. Garnaev, E. Gluskin, On widths of the Euclidean ball. Sov. Math. Dokl. **30**, 200–204 (1984) (Cited on pp. 34, 302, 327.)

220. D. Ge, X. Jiang, Y. Ye, A note on complexity of l_p minimization. Math. Program. **129**, 285–299 (2011) (Cited on p. 57.)

221. Q. Geng, J. Wright, On the local correctness of ℓ^1-minimization for dictionary learning. *Preprint* (2011) (Cited on p. 39.)

222. A. Gilbert, M. Strauss, Analysis of data streams. Technometrics **49**(3), 346–356 (2007) (Cited on p. 38.)

223. A.C. Gilbert, S. Muthukrishnan, S. Guha, P. Indyk, M. Strauss, Near-Optimal Sparse Fourier Representations via Sampling. In *Proceedings of the Thiry-fourth Annual ACM Symposium on Theory of Computing*, STOC '02, pp. 152–161, ACM, New York, NY, USA, 2002 (Cited on pp. 35, 38, 423.)

224. A.C. Gilbert, S. Muthukrishnan, M.J. Strauss, Approximation of functions over redundant dictionaries using coherence. In *Proceedings of the Fourteenth Annual ACM-SIAM Symposium on Discrete Algorithms*, SODA '03, pp. 243–252. SIAM, Philadelphia, PA, 2003 (Cited on pp. 34, 36.)

225. A.C. Gilbert, M. Strauss, J.A. Tropp, R. Vershynin, One sketch for all: fast algorithms for compressed sensing. In *Proceedings of the Thirty-ninth Annual ACM Symposium on Theory of Computing*, STOC '07, pp. 237–246, ACM, New York, NY, USA, 2007 (Cited on pp. 35, 38, 56, 455.)

226. R. Glowinski, T. Le, *Augmented Lagrangian and Operator-Splitting Methods in Nonlinear Mechanics* (SIAM, Philadelphia, 1989) (Cited on p. 509.)

227. E. Gluskin, Norms of random matrices and widths of finite-dimensional sets. Math. USSR-Sb. **48**, 173–182 (1984) (Cited on p. 34.)

228. E. Gluskin, Extremal properties of orthogonal parallelepipeds and their applications to the geometry of Banach spaces. Mat. Sb. (N.S.) **136**(178)(1), 85–96 (1988) (Cited on p. 363.)

229. D. Goldfarb, S. Ma, Convergence of fixed point continuation algorithms for matrix rank minimization. Found. Comput. Math. **11**(2), 183–210 (2011) (Cited on p. 507.)

230. T. Goldstein, S. Osher, The split Bregman method for L1-regularized problems. SIAM J. Imag. Sci. **2**(2), 323–343 (2009) (Cited on p. 508.)

231. G. Golub, C.F. van Loan, *Matrix Computations*, 3rd edn. (The Johns Hopkins University Press, Baltimore, MD, 1996) (Cited on pp. 428, 515, 534.)

232. R.A. Gopinath, Nonlinear recovery of sparse signals from narrowband data. In *Proceedings of the International Conference on Acoustics, Speech, and Signal Processing*. ICASSP '95, vol. 2, pp. 1237–1239, IEEE Computer Society, 1995 (Cited on p. 57.)

233. Y. Gordon, Some inequalities for Gaussian processes and applications. Israel J. Math. **50**(4), 265–289 (1985) (Cited on p. 264.)

234. Y. Gordon, Elliptically contoured distributions. Probab. Theor. Relat. Field. **76**(4), 429–438 (1987) (Cited on p. 264.)

235. Y. Gordon, On Milman's inequality and random subspaces which escape through a mesh in $\mathbb{R}^n$. In *Geometric Aspects of Functional Analysis (1986/87)*, Lecture Notes in Mathematics, vol. 1317 (Springer, Berlin, 1988), pp. 84–106 (Cited on p. 303.)

236. L. Grafakos, *Modern Fourier Analysis*, 2nd edn. Graduate Texts in Mathematics, vol. 250 (Springer, New York, 2009) (Cited on pp. 420, 573.)

237. R. Graham, N. Sloane, Lower bounds for constant weight codes. IEEE Trans. Inform. Theor. **26**(1), 37–43 (1980) (Cited on p. 327.)

238. R. Gribonval, Sparse decomposition of stereo signals with matching pursuit and application to blind separation of more than two sources from a stereo mixture. In *Proceedings of the International Conference on Acoustics, Speech, and Signal Processing (ICASSP'02)*, vol. 3, pp. 3057–3060, 2002 (Cited on p. 36.)

239. R. Gribonval, M. Nielsen, Sparse representations in unions of bases. IEEE Trans. Inform. Theor. **49**(12), 3320–3325 (2003) (Cited on pp. 34, 104, 128.)

240. R. Gribonval, M. Nielsen, Highly sparse representations from dictionaries are unique and independent of the sparseness measure. Appl. Comput. Harmon. Anal. **22**(3), 335–355 (2007) (Cited on p. 104.)

241. R. Gribonval, H. Rauhut, K. Schnass, P. Vandergheynst, Atoms of all channels, unite! Average case analysis of multi-channel sparse recovery using greedy algorithms. J. Fourier Anal. Appl. **14**(5), 655–687 (2008) (Cited on pp. 38, 472.)

242. R. Gribonval, K. Schnass, Dictionary identification—sparse matrix-factorisation via l_1-minimisation. IEEE Trans. Inform. Theor. **56**(7), 3523–3539 (2010) (Cited on pp. 38, 39.)

243. G. Grimmett, D. Stirzaker, *Probability and Random Processes*, 3rd edn. (Oxford University Press, New York, 2001) (Cited on p. 198.)

244. K. Gröchenig, *Foundations of Time-Frequency Analysis*. Applied and Numerical Harmonic Analysis (Birkhäuser, Boston, MA, 2001) (Cited on pp. 36, 429, 430.)

245. D. Gross, Recovering low-rank matrices from few coefficients in any basis. IEEE Trans. Inform. Theor. **57**(3), 1548–1566 (2011) (Cited on pp. 37, 421.)

246. D. Gross, Y.-K. Liu, S.T. Flammia, S. Becker, J. Eisert, Quantum state tomography via compressed sensing. Phys. Rev. Lett. **105**, 150401 (2010) (Cited on p. 37.)

247. L. Gross, Logarithmic Sobolev inequalities. Am. J. Math. **97**(4), 1061–1083 (1975) (Cited on p. 264.)

248. O. Guédon, S. Mendelson, A. Pajor, N. Tomczak-Jaegermann, Majorizing measures and proportional subsets of bounded orthonormal systems. Rev. Mat. Iberoam. **24**(3), 1075–1095 (2008) (Cited on pp. 421, 423.)

249. C. Güntürk, M. Lammers, A. Powell, R. Saab, Ö. Yilmaz, Sobolev duals for random frames and $\Sigma\Delta$ quantization of compressed sensing measurements. Found. Comput. Math. **13**(1), 1–36, Springer-Verlag (2013) (Cited on p. 38.)

250. S. Gurevich, R. Hadani, N. Sochen, On some deterministic dictionaries supporting sparsity. J. Fourier Anal. Appl. **14**, 859–876 (2008) (Cited on p. 129.)

251. V. Guruswani, C. Umans, S. Vadhan, Unbalanced expanders and randomness extractors from Parvaresh-Vardy codes. In *IEEE Conference on Computational Complexity*, pp. 237–246, 2007 (Cited on p. 455.)

252. M. Haacke, R. Brown, M. Thompson, R. Venkatesan, *Magnetic Resonance Imaging: Physical Principles and Sequence Design* (Wiley-Liss, New York, 1999) (Cited on p. 35.)

253. U. Haagerup, The best constants in the Khintchine inequality. Studia Math. **70**(3), 231–283 (1982), 1981 (Cited on p. 260.)

254. T. Hagerup, C. Rüb, A guided tour of Chernoff bounds. Inform. Process. Lett. **33**(6), 305–308 (1990) (Cited on p. 198.)

255. J. Haldar, D. Hernando, Z. Liang, Compressed-sensing MRI with random encoding. IEEE Trans. Med. Imag. **30**(4), 893–903 (2011) (Cited on p. 35.)

256. E. Hale, W. Yin, Y. Zhang, Fixed-point continuation for ℓ_1-minimization: methodology and convergence. SIAM J. Optim. **19**(3), 1107–1130 (2008) (Cited on p. 510.)

257. E. Hale, W. Yin, Y. Zhang, Fixed-point continuation applied to compressed sensing: implementation and numerical experiments. J. Comput. Math. **28**(2), 170–194 (2010) (Cited on p. 507.)

258. F. Hansen, G. Pedersen, Jensen's inequality for operators and Löwner's theorem. Math. Ann. **258**(3), 229–241 (1982) (Cited on pp. 565, 566.)

259. F. Hansen, G. Pedersen, Jensen's operator inequality. Bull. London Math. Soc. **35**(4), 553–564 (2003) (Cited on p. 567.)

260. D. Hanson, F. Wright, A bound on tail probabilities for quadratic forms in independent random variables. Ann. Math. Stat. **42**, 1079–1083 (1971) (Cited on p. 262.)

261. H. Hassanieh, P. Indyk, D. Katabi, E. Price, Nearly optimal sparse Fourier transform. In *Proceedings of the 44th Symposium on Theory of Computing*, STOC '12, pp. 563–578, ACM, New York, NY, USA, 2012 (Cited on pp. 35, 38, 424, 455.)

262. H. Hassanieh, P. Indyk, D. Katabi, E. Price, Simple and practical algorithm for sparse Fourier transform. In *Proceedings of the Twenty-third Annual ACM-SIAM Symposium on Discrete Algorithms*, SODA '12, pp. 1183–1194. SIAM, 2012 (Cited on pp. 35, 455.)

263. J. Haupt, W. Bajwa, G. Raz, R. Nowak, Toeplitz compressed sensing matrices with applications to sparse channel estimation. IEEE Trans. Inform. Theor. **56**(11), 5862–5875 (2010) (Cited on p. 429.)

264. B. He, X. Yuan, Convergence analysis of primal-dual algorithms for a saddle-point problem: from contraction perspective. SIAM J. Imag. Sci. **5**(1), 119–149 (2012) (Cited on p. 503.)

265. D. Healy Jr., D. Rockmore, P. Kostelec, S. Moore, FFTs for the 2-Sphere—Improvements and Variations. J. Fourier Anal. Appl. **9**(4), 341–385 (2003) (Cited on pp. 427, 428.)

266. T. Hemant, V. Cevher, Learning non-parametric basis independent models from point queries via low-rank methods. *Preprint* (2012) (Cited on p. 39.)

267. W. Hendee, C. Morgan, Magnetic resonance imaging Part I—Physical principles. West J. Med. **141**(4), 491–500 (1984) (Cited on p. 35.)

268. M. Herman, T. Strohmer, High-resolution radar via compressed sensing. IEEE Trans. Signal Process. **57**(6), 2275–2284 (2009) (Cited on pp. 35, 430.)

269. F. Herrmann, M. Friedlander, O. Yilmaz, Fighting the curse of dimensionality: compressive sensing in exploration seismology. Signal Process. Mag. IEEE **29**(3), 88–100 (2012) (Cited on p. 37.)

270. F. Herrmann, H. Wason, T. Lin, Compressive sensing in seismic exploration: an outlook on a new paradigm. CSEG Recorder **36**(4), 19–33 (2011) (Cited on p. 37.)

271. J.R. Higgins, *Sampling Theory in Fourier and Signal Analysis: Foundations*, vol. 1 (Clarendon Press, Oxford, 1996) (Cited on pp. 35, 573.)

272. J.R. Higgins, R.L. Stens, *Sampling Theory in Fourier and Signal Analysis: Advanced Topics*, vol. 2 (Oxford University Press, Oxford, 1999) (Cited on pp. 35, 573.)

273. N.J. Higham, *Functions of Matrices. Theory and Computation* (Society for Industrial and Applied Mathematics (SIAM), Philadelphia, PA, 2008) (Cited on p. 515.)

274. A. Hinrichs, J. Vybíral, Johnson-Lindenstrauss lemma for circulant matrices. Random Struct. Algorithm. **39**(3), 391–398 (2011) (Cited on p. 423.)

275. J.-B. Hiriart-Urruty, C. Lemaréchal, *Fundamentals of Convex Analysis*. Grundlehren Text Editions (Springer, Berlin, 2001) (Cited on pp. 543, 545.)

276. W. Hoeffding, Probability inequalities for sums of bounded random variables. J. Am. Stat. Assoc. **58**, 13–30 (1963) (Cited on p. 198.)

277. J. Högborn, Aperture synthesis with a non-regular distribution of interferometer baselines. Astronom. and Astrophys. **15**, 417 (1974) (Cited on p. 71.)

278. D. Holland, M. Bostock, L. Gladden, D. Nietlispach, Fast multidimensional NMR spectroscopy using compressed sensing. Angew. Chem. Int. Ed. **50**(29), 6548–6551 (2011) (Cited on p. 35.)

279. S. Hoory, N. Linial, A. Wigderson, Expander graphs and their applications. Bull. Am. Math. Soc. (N.S.) **43**(4), 439–561 (electronic) (2006) (Cited on p. 455.)

280. R. Horn, C. Johnson, *Matrix Analysis* (Cambridge University Press, Cambridge, 1990) (Cited on p. 515.)

281. R. Horn, C. Johnson, *Topics in Matrix Analysis* (Cambridge University Press, Cambridge, 1994) (Cited on p. 515.)

282. W. Huffman, V. Pless, *Fundamentals of Error-correcting Codes* (Cambridge University Press, Cambridge, 2003) (Cited on p. 36.)

283. M. Hügel, H. Rauhut, T. Strohmer, Remote sensing via ℓ_1-minimization. Found. Comput. Math., to appear. (2012) (Cited on pp. 35, 422.)

284. R. Hunt, On $L(p, q)$ spaces. Enseignement Math. (2) **12**, 249–276 (1966) (Cited on p. 56.)

285. P. Indyk, A. Gilbert, Sparse recovery using sparse matrices. Proc. IEEE **98**(6), 937–947 (2010) (Cited on pp. 38, 455.)

286. P. Indyk, M. Ružić, Near-optimal sparse recovery in the L1 norm. In *Proceedings of the 49th Annual IEEE Symposium on Foundations of Computer Science*, FOCS '08, pp. 199–207 (2008) (Cited on p. 455.)

287. M. Iwen, Combinatorial sublinear-time Fourier algorithms. Found. Comput. Math. **10**(3), 303–338 (2010) (Cited on pp. 35, 423, 455.)

288. M. Iwen, Improved approximation guarantees for sublinear-time Fourier algorithms. Appl. Comput. Harmon. Anal. **34**(1), 57–82 (2013) (Cited on pp. 35, 423.)

289. M. Iwen, A. Gilbert, M. Strauss, Empirical evaluation of a sub-linear time sparse DFT algorithm. Commun. Math. Sci. **5**(4), 981–998 (2007) (Cited on pp. 38, 424.)

290. L. Jacques, J. Laska, P. Boufounos, R. Baraniuk, Robust 1-bit compressive sensing via binary stable embeddings of sparse vectors. IEEE Trans. Inform. Theor. **59**(4), 2082–2102 (2013) (Cited on p. 38.)

291. S. Jafarpour, W. Xu, B. Hassibi, R. Calderbank, Efficient and robust compressed sensing using optimized expander graphs. IEEE Trans. Inform. Theor. **55**(9), 4299–4308 (2009) (Cited on p. 455.)

292. R. James, M. Dennis, N. Daniel, Fast discrete polynomial transforms with applications to data analysis for distance transitive graphs. SIAM J. Comput. **26**(4), 1066–1099 (1997) (Cited on p. 427.)

293. F. Jarre, J. Stoer, *Optimierung* (Springer, Berlin, 2004) (Cited on pp. 543, 558.)

294. A.J. Jerri, The Shannon sampling theorem—its various extensions and applications: A tutorial review. Proc. IEEE. **65**(11), 1565–1596 (1977) (Cited on p. 35.)

295. W.B. Johnson, J. Lindenstrauss, Extensions of Lipschitz mappings into a Hilbert space. In *Conference in Modern Analysis and Probability (New Haven, Conn., 1982)*. Contemporary Mathematics, vol. 26 (American Mathematical Society, Providence, RI, 1984), pp. 189–206 (Cited on p. 303.)

296. W.B. Johnson, J. Lindenstrauss (eds.), *Handbook of the Geometry of Banach Spaces Vol I* (North-Holland Publishing Co., Amsterdam, 2001) (Cited on p. 260.)

297. J.-P. Kahane, Propriétés locales des fonctions à séries de Fourier aléatoires. Studia Math. **19**, 1–25 (1960) (Cited on p. 199.)

298. S. Karlin, *Total Positivity Vol. I* (Stanford University Press, Stanford, 1968) (Cited on p. 57.)

299. B. Kashin, Diameters of some finite-dimensional sets and classes of smooth functions. Math. USSR, Izv. **11**, 317–333 (1977) (Cited on pp. 34, 302, 328.)

300. B.S. Kashin, V.N. Temlyakov, A remark on compressed sensing. Math. Notes **82**(5), 748–755 (2007) (Cited on p. 327.)

301. A. Khintchine, Über dyadische Brüche. Math. Z. **18**(1), 109–116 (1923) (Cited on p. 260.)

302. S. Kim, K. Koh, M. Lustig, S. Boyd, D. Gorinevsky, A method for large-scale l1-regularized least squares problems with applications in signal processing and statistics. IEEE J. Sel. Top. Signal Proces. **4**(1), 606–617 (2007) (Cited on p. 510.)

303. J. King, A minimal error conjugate gradient method for ill-posed problems. J. Optim. Theor. Appl. **60**(2), 297–304 (1989) (Cited on pp. 504, 534.)

304. T. Klein, E. Rio, Concentration around the mean for maxima of empirical processes. Ann. Probab. **33**(3), 1060–1077 (2005) (Cited on p. 265.)

305. H. König, S. Kwapień, Best Khintchine type inequalities for sums of independent, rotationally invariant random vectors. Positivity. **5**(2), 115–152 (2001) (Cited on p. 261.)

306. N. Kôno, Sample path properties of stochastic processes. J. Math. Kyoto Univ. **20**(2), 295–313 (1980) (Cited on p. 263.)

307. F. Krahmer, S. Mendelson, H. Rauhut, Suprema of chaos processes and the restricted isometry property. Comm. Pure Appl. Math. (to appear) (Cited on pp. 263, 264, 429, 430.)

308. F. Krahmer, G.E. Pfander, P. Rashkov, Uncertainty principles for time–frequency representations on finite abelian groups. Appl. Comput. Harmon. Anal. **25**(2), 209–225 (2008) (Cited on p. 429.)

309. F. Krahmer, R. Ward, New and improved Johnson-Lindenstrauss embeddings via the Restricted Isometry Property. SIAM J. Math. Anal. **43**(3), 1269–1281 (2011) (Cited on pp. 303, 423.)

310. F. Krahmer, R. Ward, Beyond incoherence: stable and robust sampling strategies for compressive imaging. *Preprint* (2012) (Cited on p. 427.)

311. I. Krasikov, On the Erdelyi-Magnus-Nevai conjecture for Jacobi polynomials. Constr. Approx. **28**(2), 113–125 (2008) (Cited on p. 428.)

312. M.A. Krasnosel'skij, Y.B. Rutitskij, *Convex Functions and Orlicz Spaces.* (P. Noordhoff Ltd., Groningen, The Netherlands, 1961), p. 249 (Cited on p. 263.)

313. J. Kruskal, Three-way arrays: rank and uniqueness of trilinear decompositions, with application to arithmetic complexity and statistics. Lin. Algebra Appl. **18**(2), 95–138 (1977) (Cited on p. 56.)

314. P. Kuppinger, G. Durisi, H. Bölcskei, Uncertainty relations and sparse signal recovery for pairs of general signal sets. IEEE Trans. Inform. Theor. **58**(1), 263–277 (2012) (Cited on p. 421.)

315. M.-J. Lai, L.Y. Liu, The null space property for sparse recovery from multiple measurement vectors. Appl. Comput. Harmon. Anal. **30**, 402–406 (2011) (Cited on p. 104.)

316. J. Laska, P. Boufounos, M. Davenport, R. Baraniuk, Democracy in action: quantization, saturation, and compressive sensing. Appl. Comput. Harmon. Anal. **31**(3), 429–443 (2011) (Cited on p. 38.)

317. J. Lawrence, G.E. Pfander, D. Walnut, Linear independence of Gabor systems in finite dimensional vector spaces. J. Fourier Anal. Appl. **11**(6), 715–726 (2005) (Cited on p. 429.)

318. C. Lawson, *Contributions to the Theory of Linear Least Maximum Approximation.* PhD thesis, University of California, Los Angeles, 1961 (Cited on p. 504.)

319. J. Lederer, S. van de Geer, The Bernstein-Orlicz norm and deviation inequalities. *Preprint* (2011) (Cited on p. 265.)

320. M. Ledoux, On Talagrand's deviation inequalities for product measures. ESAIM Probab. Stat. **1**, 63–87 (1996) (Cited on p. 265.)

321. M. Ledoux, *The Concentration of Measure Phenomenon.* AMS, 2001. (Cited on pp. 264, 265.)

322. M. Ledoux, M. Talagrand, *Probability in Banach Spaces* (Springer, Berlin, Heidelberg, NewYork, 1991) (Cited on pp. 198, 260, 263, 264, 265.)

323. J. Lee, Y. Sun, M. Saunders, Proximal Newton-type methods for convex optimization. *Preprint* (2012) (Cited on p. 510.)

324. B. Lemaire, The proximal algorithm. In *New Methods in Optimization and Their Industrial Uses (Pau/Paris, 1987)*, Internationale Schriftenreihe Numerischen Mathematik, vol. 87 (Birkhäuser, Basel, 1989), pp. 73–87 (Cited on p. 506.)

325. E. Lieb, Convex trace functions and the Wigner-Yanase-Dyson conjecture. Adv. Math. **11**, 267–288 (1973) (Cited on pp. 564, 565.)

326. M. Lifshits, *Lectures on Gaussian Processes*. Springer Briefs in Mathematics (Springer, New York, 2012) (Cited on p. 262.)

327. G. Lindblad, Expectations and entropy inequalities for finite quantum systems. Comm. Math. Phys. **39**, 111–119 (1974) (Cited on p. 571.)

328. P.-L. Lions, B. Mercier, Splitting algorithms for the sum of two nonlinear operators. SIAM J. Numer. Anal. **16**, 964–979 (1979) (Cited on pp. 503, 508.)

329. A. Litvak, A. Pajor, M. Rudelson, N. Tomczak-Jaegermann, Smallest singular value of random matrices and geometry of random polytopes. Adv. Math. **195**(2), 491–523 (2005) (Cited on p. 363.)

330. Y. Liu, Universal low-rank matrix recovery from Pauli measurements. In *NIPS*, pp. 1638–1646, 2011 (Cited on p. 37.)

331. A. Llagostera Casanovas, G. Monaci, P. Vandergheynst, R. Gribonval, Blind audiovisual source separation based on sparse redundant representations. IEEE Trans. Multimed. **12**(5), 358–371 (August 2010) (Cited on p. 36.)

332. B. Logan, *Properties of High-Pass Signals*. PhD thesis, Columbia University, New York, 1965 (Cited on p. 34.)

333. G. Lorentz, M. von Golitschek, Y. Makovoz, *Constructive Approximation: Advanced Problems* (Springer, New York, 1996) (Cited on p. 327.)

334. D. Lorenz, M. Pfetsch, A. Tillmann, Solving Basis Pursuit: Heuristic optimality check and solver comparison. *Preprint* (2011) (Cited on p. 510.)

335. I. Loris, L1Packv2: a Mathematica package in minimizing an ℓ_1-penalized functional. Comput. Phys. Comm. **179**(12), 895–902 (2008) (Cited on p. 503.)

336. F. Lust-Piquard, Inégalités de Khintchine dans C_p $(1 < p < \infty)$. C. R. Math. Acad. Sci. Paris **303**, 289–292 (1986) (Cited on p. 261.)

337. F. Lust-Piquard, G. Pisier, Noncommutative Khintchine and Paley inequalities. Ark. Mat. **29**(2), 241–260 (1991) (Cited on p. 261.)

338. M. Lustig, D.L. Donoho, J. Pauly, Sparse MRI: The application of compressed sensing for rapid MR imaging. Magn. Reson. Med. **58**(6), 1182–1195 (2007) (Cited on p. 35.)

339. L. Mackey, M. Jordan, R. Chen, B. Farrell, J. Tropp, Matrix concentration inequalities via the method of exchangeable pairs. *Preprint* (2012) (Cited on p. 261.)

340. A. Maleki, Coherence analysis of iterative thresholding algorithms. In *Proc. of 47th Annual Allerton Conference on Communication, Control, and Computing*, pp. 236–243, 2009 (Cited on p. 128.)

341. S. Mallat, *A Wavelet Tour of Signal Processing*. Middleton Academic Press, San Diego, 1998 (Cited on p. 425.)

342. S. Mallat, Z. Zhang, Matching pursuits with time-frequency dictionaries. IEEE Trans. Signal Process. **41**(12), 3397–3415 (1993) (Cited on pp. 34, 36, 71.)

343. M. Marcus, L. Shepp, Sample behavior of Gaussian processes. In *Proceedings of the Sixth Berkeley Symposium on Mathematical Statistics and Probability (Univ. California, Berkeley, Calif., 1970/1971), Vol. II: Probability Theory*, pp. 423–441. University of California Press, Berkeley, 1972 (Cited on p. 264.)

344. S. Marple, *Digital Spectral Analysis with Applications* (Prentice-Hall, Englewood Cliffs, 1987) (Cited on pp. 34, 57.)

345. B. Martinet, Régularisation d'inéquations variationnelles par approximations successives. Rev. Française Informat. Recherche Opérationnelle **4**(Ser. R-3), 154–158 (1970) (Cited on p. 506.)

346. P. Massart, Rates of convergence in the central limit theorem for empirical processes. Ann. Inst. H. Poincaré Probab. Statist. **22**(4), 381–423 (1986) (Cited on p. 265.)

347. P. Massart, About the constants in Talagrand's concentration inequalities for empirical processes. Ann. Probab. **28**(2), 863–884 (2000) (Cited on p. 265.)

348. P. Massart, *Concentration Inequalities and Model Selection.* Lecture Notes in Mathematics, vol. 1896 (Springer, Berlin, 2007) (Cited on pp. 263, 264.)

349. C. McDiarmid, Concentration. In *Probabilistic Methods for Algorithmic Discrete Mathematics.* Algorithms and Combinations, vol. 16 (Springer, Berlin, 1998), pp. 195–248 (Cited on p. 199.)

350. S. Mendelson, A. Pajor, M. Rudelson, The geometry of random $\{-1, 1\}$-polytopes. Discr. Comput. Geom. **34**(3), 365–379 (2005) (Cited on p. 327.)

351. S. Mendelson, A. Pajor, N. Tomczak-Jaegermann, Uniform uncertainty principle for Bernoulli and subgaussian ensembles. Constr. Approx. **28**(3), 277–289 (2009) (Cited on p. 302.)

352. D. Middleton, Channel Modeling and Threshold Signal Processing in Underwater Acoustics: An Analytical Overview. IEEE J. Oceanic Eng. **12**(1), 4–28 (1987) (Cited on p. 430.)

353. M. Mishali, Y.C. Eldar, From theory to practice: Sub-nyquist sampling of sparse wideband analog signals. IEEE J. Sel. Top. Signal Process. **4**(2), 375–391 (April 2010) (Cited on p. 37.)

354. Q. Mo, S. Li, New bounds on the restricted isometry constant δ_{2k}. Appl. Comput. Harmon. Anal. **31**(3), 460–468 (2011) (Cited on p. 169.)

355. Q. Mo, Y. Shen, Remarks on the restricted isometry property in orthogonal matching pursuit algorithm. IEEE Trans. Inform. Theor. **58**(6), 3654–3656 (2012) (Cited on p. 169.)

356. S. Montgomery-Smith, The distribution of Rademacher sums. Proc. Am. Math. Soc. **109**(2), 517–522 (1990) (Cited on p. 363.)

357. T.K. Moon, W.C. Stirling, *Mathematical Methods and Algorithms for Signal Processing.* (Prentice-Hall, Upper Saddle River, NJ, 2000) (Cited on p. 57.)

358. M. Murphy, M. Alley, J. Demmel, K. Keutzer, S. Vasanawala, M. Lustig, Fast ℓ_1-SPIRiT Compressed Sensing Parallel Imaging MRI: Scalable Parallel Implementation and Clinically Feasible Runtime. IEEE Trans. Med. Imag. **31**(6), 1250–1262 (2012) (Cited on p. 35.)

359. B.K. Natarajan, Sparse approximate solutions to linear systems. SIAM J. Comput. **24**, 227–234 (1995) (Cited on pp. 34, 36, 57.)

360. F. Nazarov, A. Podkorytov, Ball, Haagerup, and distribution functions. In *Complex Analysis, Operators, and Related Topics.* Operator Theory: Advances and Applications, vol. 113 (Birkhäuser, Basel, 2000), pp. 247–267 (Cited on p. 260.)

361. D. Needell, J. Tropp, CoSaMP: Iterative signal recovery from incomplete and inaccurate samples. Appl. Comput. Harmon. Anal. **26**(3), 301–321 (2008) (Cited on pp. 72, 169.)

362. D. Needell, R. Vershynin, Uniform uncertainty principle and signal recovery via regularized orthogonal matching pursuit. Found. Comput. Math. **9**(3), 317–334 (2009) (Cited on pp. 72, 170.)

363. D. Needell, R. Vershynin, Signal recovery from incomplete and inaccurate measurements via regularized orthogonal matching pursuit. IEEE J. Sel. Top. Signal Process. **4**(2), 310–316 (April 2010) (Cited on pp. 72, 170.)

364. D. Needell, R. Ward, Stable image reconstruction using total variation minimization. *Preprint* (2012) (Cited on p. 37.)

365. A. Nemirovsky, D. Yudin, *Problem Complexity and Method Efficiency in Optimization.* A Wiley-Interscience Publication. (Wiley, New York, 1983) (Cited on p. 507.)

366. Y. Nesterov, A method for solving the convex programming problem with convergence rate $O(1/k^2)$. Dokl. Akad. Nauk SSSR **269**(3), 543–547 (1983) (Cited on p. 506.)

367. Y. Nesterov, Smooth minimization of non-smooth functions. Math. Program. **103**(1, Ser. A), 127–152 (2005) (Cited on p. 503.)

368. N. Noam, W. Avi, Hardness vs randomness. J. Comput. Syst. Sci. **49**(2), 149–167 (1994) (Cited on p. 327.)

369. J. Nocedal, S. Wright, *Numerical Optimization*, 2nd edn. Springer Series in Operations Research and Financial Engineering (Springer, New York, 2006) (Cited on pp. 71, 503, 504, 510.)

370. E. Novak, Optimal recovery and n-widths for convex classes of functions. J. Approx. Theor. **80**(3), 390–408 (1995) (Cited on pp. 34, 327.)

371. E. Novak, H. Woźniakowski, *Tractability of Multivariate Problems. Vol. 1: Linear Information*. EMS Tracts in Mathematics, vol. 6 (European Mathematical Society (EMS), Zürich, 2008) (Cited on pp. 39, 327.)

372. M. Ohya, D. Petz, *Quantum Entropy and Its Use*. Texts and Monographs in Physics (Springer, New York, 2004) (Cited on p. 570.)

373. R. Oliveira, Concentration of the adjacency matrix and of the Laplacian in random graphs with independent edges. *Preprint* (2009) (Cited on p. 261.)

374. R. Oliveira, Sums of random Hermitian matrices and an inequality by Rudelson. Electron. Commun. Probab. **15**, 203–212 (2010) (Cited on p. 261.)

375. M. Osborne, *Finite Algorithms in Optimization and Data Analysis* (Wiley, New York, 1985) (Cited on p. 504.)

376. M. Osborne, B. Presnell, B. Turlach, A new approach to variable selection in least squares problems. IMA J. Numer. Anal. **20**(3), 389–403 (2000) (Cited on p. 503.)

377. M. Osborne, B. Presnell, B. Turlach, On the LASSO and its dual. J. Comput. Graph. Stat. **9**(2), 319–337 (2000) (Cited on p. 503.)

378. Y.C. Pati, R. Rezaiifar, P.S. Krishnaprasad, Orthogonal Matching Pursuit: Recursive Function Approximation with Applications to Wavelet Decomposition. In *1993 Conference Record of The Twenty-Seventh Asilomar Conference on Signals, Systems and Computers, Nov. 1–3, 1993.*, pp. 40–44, 1993 (Cited on p. 71.)

379. G.K. Pedersen, Some operator monotone functions. Proc. Am. Math. Soc. **36**(1), 309–310 (1972) (Cited on p. 542.)

380. G. Peškir, Best constants in Kahane-Khintchine inequalities for complex Steinhaus functions. Proc. Am. Math. Soc. **123**(10), 3101–3111 (1995) (Cited on p. 261.)

381. G. Peškir, A.N. Shiryaev, The Khintchine inequalities and martingale expanding sphere of their action. Russ. Math. Surv. **50**(5), 849–904 (1995) (Cited on p. 261.)

382. D. Petz, A survey of certain trace inequalities. In *Functional Analysis and Operator Theory (Warsaw, 1992)*. Banach Center Publications, vol. 30 (Polish Academy of Sciences, Warsaw, 1994), pp. 287–298 (Cited on p. 540.)

383. D. Petz, *Quantum Information Theory and Quantum Statistics*. Theoretical and Mathematical Physics (Springer, Berlin, 2008) (Cited on p. 570.)

384. G. Pfander, H. Rauhut, J. Tanner, Identification of matrices having a sparse representation. IEEE Trans. Signal Process. **56**(11), 5376–5388 (2008) (Cited on pp. 35, 430.)

385. G. Pfander, H. Rauhut, J. Tropp, The restricted isometry property for time-frequency structured random matrices. Prob. Theor. Relat. Field. to appear (Cited on pp. 35, 430.)

386. M. Pfetsch, A. Tillmann, The computational complexity of the restricted isometry property, the nullspace property, and related concepts in Compressed Sensing. *Preprint* (2012) (Cited on p. 170.)

387. A. Pinkus, *n-Widths in Approximation Theory* (Springer, Berlin, 1985) (Cited on p. 327.)

388. A. Pinkus, *On L^1-Approximation*. Cambridge Tracts in Mathematics, vol. 93 (Cambridge University Press, Cambridge, 1989) (Cited on p. 104.)

389. A. Pinkus, *Totally Positive Matrices*. Cambridge Tracts in Mathematics, vol. 181 (Cambridge University Press, Cambridge, 2010) (Cited on p. 57.)

390. M. Pinsky, *Introduction to Fourier Analysis and Wavelets*. Graduate Studies in Mathematics, vol. 102 (American Mathematical Society, Providence, RI, 2009) (Cited on pp. 420, 573.)

391. G. Pisier, Conditions d'entropie assurant la continuité de certains processus et applications à l'analyse harmonique. In *Seminar on Functional Analysis, 1979–1980 (French)*, pp. 13–14, 43, École Polytech., 1980 (Cited on p. 263.)

392. G. Pisier, *The Volume of Convex Bodies and Banach Space Geometry*. Cambridge Tracts in Mathematics (Cambridge University Press, Cambridge, 1999) (Cited on p. 262.)

393. Y. Plan, R. Vershynin, One-bit compressed sensing by linear programming. Comm. Pure Appl. Math. **66**(8), 1275–1297 (2013) (Cited on p. 38.)

394. Y. Plan, R. Vershynin, Robust 1-bit compressed sensing and sparse logistic regression: a convex programming approach. IEEE Trans. Inform. Theor. **59**(1), 482–494 (2013) (Cited on p. 38.)

395. T. Pock, A. Chambolle, Diagonal preconditioning for first order primal-dual algorithms in convex optimization. In *IEEE International Conference Computer Vision (ICCV) 2011*, pp. 1762–1769, November 2011 (Cited on p. 503.)

396. T. Pock, D. Cremers, H. Bischof, A. Chambolle, An algorithm for minimizing the Mumford-Shah functional. In *ICCV Proceedings* (Springer, Berlin, 2009) (Cited on p. 503.)

397. L. Potter, E. Ertin, J. Parker, M. Cetin, Sparsity and compressed sensing in radar imaging. Proc. IEEE **98**(6), 1006–1020 (2010) (Cited on p. 35.)

398. D. Potts, Fast algorithms for discrete polynomial transforms on arbitrary grids. Lin. Algebra Appl. **366**, 353–370 (2003) (Cited on p. 427.)

399. D. Potts, G. Steidl, M. Tasche, Fast algorithms for discrete polynomial transforms. Math. Comp. **67**, 1577–1590 (1998) (Cited on p. 427.)

400. D. Potts, G. Steidl, M. Tasche, Fast fourier transforms for nonequispaced data: a tutorial. In *Modern Sampling Theory: Mathematics and Applications* ed. by J. Benedetto, P. Ferreira, Chap. 12 (Birkhäuser, Boston, 2001), pp. 247–270 (Cited on pp. 421, 573.)

401. D. Potts, M. Tasche, Parameter estimation for exponential sums by approximate Prony method. Signal Process. **90**(5), 1631–1642 (2010) (Cited on pp. 34, 57.)

402. R. Prony, Essai expérimental et analytique sur les lois de la Dilatabilité des fluides élastiques et sur celles de la Force expansive de la vapeur de l'eau et de la vapeur de l'alkool, à différentes températures. J. École Polytechnique **1**, 24–76 (1795) (Cited on pp. 34, 57.)

403. W. Pusz, S. Woronowicz, Form convex functions and the WYDL and other inequalities. Lett. Math. Phys. **2**(6), 505–512 (1977/78) (Cited on p. 571.)

404. S. Qian, D. Chen, Signal representation using adaptive normalized Gaussian functions. Signal Process. **36**(1), 1–11 (1994) (Cited on p. 71.)

405. H. Raguet, J. Fadili, G. Peyré, A generalized forward-backward splitting. *Preprint* (2011) (Cited on pp. 507, 510.)

406. R. Ramlau, G. Teschke, Sparse recovery in inverse problems. In *Theoretical Foundations and Numerical Methods for Sparse Recovery*, ed. by M. Fornasier. Radon Series on Computational and Applied Mathematics, vol. 9 (de Gruyter, Berlin, 2010), pp. 201–262 (Cited on p. 37.)

407. M. Raphan, E. Simoncelli, Optimal denoising in redundant representation. IEEE Trans. Image Process. **17**(8), 1342–1352 (2008) (Cited on p. 36.)

408. H. Rauhut, Random sampling of sparse trigonometric polynomials. Appl. Comput. Harmon. Anal. **22**(1), 16–42 (2007) (Cited on pp. 35, 422.)

409. H. Rauhut, On the impossibility of uniform sparse reconstruction using greedy methods. Sampl. Theor. Signal Image Process. **7**(2), 197–215 (2008) (Cited on pp. 35, 169, 422.)

410. H. Rauhut, Circulant and Toeplitz matrices in compressed sensing. In *Proc. SPARS'09* (Saint-Malo, France, 2009) (Cited on p. 429.)

411. H. Rauhut, Compressive sensing and structured random matrices. In *Theoretical Foundations and Numerical Methods for Sparse Recovery*, ed. by M. Fornasier. Radon Series on Computational and Applied Mathematics, vol. 9 (de Gruyter, Berlin, 2010), pp. 1–92 (Cited on pp. 34, 35, 198, 261, 262, 263, 421, 422, 429.)

412. H. Rauhut, G.E. Pfander, Sparsity in time-frequency representations. J. Fourier Anal. Appl. **16**(2), 233–260 (2010) (Cited on pp. 35, 422, 429, 430.)

413. H. Rauhut, J.K. Romberg, J.A. Tropp, Restricted isometries for partial random circulant matrices. Appl. Comput. Harmon. Anal. **32**(2), 242–254 (2012) (Cited on pp. 263, 423, 429.)

414. H. Rauhut, K. Schnass, P. Vandergheynst, Compressed sensing and redundant dictionaries. IEEE Trans. Inform. Theor. **54**(5), 2210–2219 (2008) (Cited on p. 302.)

415. H. Rauhut, R. Ward, Sparse recovery for spherical harmonic expansions. In *Proc. SampTA 2011*, Singapore, 2011 (Cited on p. 428.)

416. H. Rauhut, R. Ward, Sparse Legendre expansions via ℓ_1-minimization. J. Approx. Theor. **164**(5), 517–533 (2012) (Cited on pp. 35, 426, 427.)

417. B. Recht, A simpler approach to matrix completion. J. Mach. Learn. Res. **12**, 3413–3430 (2011) (Cited on pp. 37, 421.)
418. B. Recht, M. Fazel, P. Parrilo, Guaranteed minimum-rank solutions of linear matrix equations via nuclear norm minimization. SIAM Rev. **52**(3), 471–501 (2010) (Cited on pp. 36, 302.)
419. M. Reed, B. Simon, *Methods of Modern Mathematical Physics. I. Functional Analysis* (Academic Press, New York, 1972) (Cited on p. 583.)
420. J. Renes, R. Blume-Kohout, A. Scott, C. Caves, Symmetric informationally complete quantum measurements. J. Math. Phys. **45**(6), 2171–2180 (2004) (Cited on p. 129.)
421. E. Rio, Inégalités de concentration pour les processus empiriques de classes de parties. Probab. Theor. Relat. Field. **119**(2), 163–175 (2001) (Cited on p. 265.)
422. E. Rio, Une inégalité de Bennett pour les maxima de processus empiriques. Ann. Inst. H. Poincaré Probab. Stat. **38**(6), 1053–1057 (2002) (Cited on p. 265.)
423. R.T. Rockafellar, Monotone operators and the proximal point algorithm. SIAM J. Contr. Optim. **14**(5), 877–898 (1976) (Cited on p. 506.)
424. R.T. Rockafellar, *Convex Analysis* (Princeton University Press, reprint edition, 1997) (Cited on pp. 543, 545, 550, 564.)
425. R.T. Rockafellar, R.J.B. Wets, *Variational Analysis*. Grundlehren der Mathematischen Wissenschaften [Fundamental Principles of Mathematical Sciences], vol. 317 (Springer, Berlin, 1998) (Cited on p. 543.)
426. J. Rohn, Computing the norm $\|A\|_{\infty,1}$ is NP-hard. Linear Multilinear Algebra **47**(3), 195–204 (2000) (Cited on p. 521.)
427. J.K. Romberg, Imaging via compressive sampling. IEEE Signal Process. Mag. **25**(2), 14–20 (March, 2008) (Cited on p. 34.)
428. S. Ross, *Introduction to Probability Models*, 9th edn. (Academic Press, San Diego, 2006) (Cited on p. 198.)
429. R. Rubinstein, M. Zibulevsky, M. Elad, Double sparsity: learning sparse dictionaries for sparse signal approximation. IEEE Trans. Signal Process. **58**(3, part 2), 1553–1564 (2010) (Cited on p. 38.)
430. M. Rudelson, Random vectors in the isotropic position. J. Funct. Anal. **164**(1), 60–72 (1999) (Cited on p. 261.)
431. M. Rudelson, R. Vershynin, Geometric approach to error-correcting codes and reconstruction of signals. Int. Math. Res. Not. **64**, 4019–4041 (2005) (Cited on p. 36.)
432. M. Rudelson, R. Vershynin, Sampling from large matrices: an approach through geometric functional analysis. *J. ACM* **54**(4), Art. 21, 19 pp. (electronic), 2007 (Cited on p. 472.)
433. M. Rudelson, R. Vershynin, On sparse reconstruction from Fourier and Gaussian measurements. Comm. Pure Appl. Math. **61**, 1025–1045 (2008) (Cited on pp. 303, 422.)
434. M. Rudelson, R. Vershynin, The Littlewood-Offord problem and invertibility of random matrices. Adv. Math. **218**(2), 600–633 (2008) (Cited on p. 303.)
435. M. Rudelson, R. Vershynin, Non-asymptotic theory of random matrices: extreme singular values. In *Proceedings of the International Congress of Mathematicians*, vol. 3 (Hindustan Book Agency, New Delhi, 2010), pp. 1576–1602 (Cited on p. 303.)
436. L. Rudin, S. Osher, E. Fatemi, Nonlinear total variation based noise removal algorithms. Physica D **60**(1–4), 259–268 (1992) (Cited on pp. 34, 37.)
437. W. Rudin, *Fourier Analysis on Groups* (Interscience Publishers, New York, 1962) (Cited on p. 420.)
438. W. Rudin, *Functional Analysis*, 2nd edn. International Series in Pure and Applied Mathematics (McGraw-Hill Inc., New York, 1991) (Cited on p. 583.)
439. M. Ruskai, Inequalities for quantum entropy: a review with conditions for equality. J. Math. Phys. **43**(9), 4358–4375 (2002) (Cited on p. 565.)
440. M. Ruskai, Erratum: "Inequalities for quantum entropy: a review with conditions for equality". J. Math. Phys. **46**(1), 019901 (2005) (Cited on p. 565.)
441. F. Santosa, W. Symes, Linear inversion of band-limited reflection seismograms. SIAM J. Sci. Stat. Comput. **7**(4), 1307–1330 (1986) (Cited on pp. 34, 37.)

442. G. Schechtman, Special orthogonal splittings of L_1^{2k}. Israel J. Math. **139**, 337–347 (2004) (Cited on p. 328.)

443. K. Schnass, P. Vandergheynst, Dictionary preconditioning for greedy algorithms. IEEE Trans. Signal Process. **56**(5), 1994–2002 (2008) (Cited on p. 129.)

444. B. Schölkopf, A. Smola, *Learning with Kernels* (MIT Press, Cambridge, 2002) (Cited on p. 36.)

445. I. Segal, M. Iwen, Improved sparse Fourier approximation results: Faster implementations and stronger guarantees. Numer. Algorithm. **63**(2), 239–263 (2013) (Cited on p. 424.)

446. S. Setzer, Operator splittings, Bregman methods and frame shrinkage in image processing. Int. J. Comput. Vis. **92**(3), 265–280 (2011) (Cited on p. 508.)

447. Y. Shrot, L. Frydman, Compressed sensing and the reconstruction of ultrafast 2D NMR data: Principles and biomolecular applications. J. Magn. Reson. **209**(2), 352–358 (2011) (Cited on p. 35.)

448. D. Slepian, The one-sided barrier problem for Gaussian noise. Bell System Tech. J. **41**, 463–501 (1962) (Cited on p. 264.)

449. D. Spielman, N. Srivastava, An elementary proof of the restricted invertibility theorem. Israel J. Math. **190**(1), 83–91 (2012) (Cited on p. 472.)

450. J.-L. Starck, E.J. Candès, D.L. Donoho, The curvelet transform for image denoising. IEEE Trans. Image Process. **11**(6), 670–684 (2002) (Cited on p. 36.)

451. J.-L. Starck, F. Murtagh, J. Fadili, *Sparse Image and Signal Processing: Wavelets, Curvelets, Morphological Diversity* (Cambridge University Press, Cambridge, 2010) (Cited on pp. 36, 504, 508.)

452. E.M. Stein, R. Shakarchi, *Functional Analysis: Introduction to Further Topics in Analysis*. Princeton Lectures in Analysis, vol. 4 (Princeton University Press, Princeton, NJ, 2011) (Cited on pp. 420, 573.)

453. M. Stojanovic, *Underwater Acoustic Communications*. In Encyclopedia of Electrical and Electronics Engineering, vol. 22, ed. by M. Stojanovic, J. G. Webster (Wiley, New York, 1999), pp. 688–698 (Cited on p. 430.)

454. M. Stojnic, ℓ_1-optimization and its various thresholds in compressed sensing. In *2010 IEEE International Conference on Acoustics, Speech, and Signal Processing*, ICASSP, pp. 3910–3913, 2010 (Cited on p. 303.)

455. T. Strohmer, B. Friedlander, Analysis of sparse MIMO radar. *Preprint* (2012) (Cited on p. 35.)

456. T. Strohmer, R.W. Heath, Jr, Grassmannian frames with applications to coding and communication. Appl. Comput. Harmon. Anal. **14**(3), 257–275 (2003) (Cited on p. 129.)

457. S. Szarek, On Kashin's almost Euclidean orthogonal decomposition of l_n^1. Bull. Acad. Polon. Sci. Sér. Sci. Math. Astronom. Phys. **26**(8), 691–694 (1978) (Cited on p. 328.)

458. G. Szegő, *Orthogonal Polynomials*, 4th edn. (American Mathematical Society, Providence, 1975) (Cited on pp. 426, 427.)

459. M. Talagrand, Regularity of Gaussian processes. Acta Math. **159**(1–2), 99–149 (1987) (Cited on p. 262.)

460. M. Talagrand, Isoperimetry and integrability of the sum of independent Banach-space valued random variables. Ann. Probab. **17**(4), 1546–1570 (1989) (Cited on p. 265.)

461. M. Talagrand, A new look at independence. Ann. Probab. **24**(1), 1–34 (1996) (Cited on p. 264.)

462. M. Talagrand, Majorizing measures: the generic chaining. Ann. Probab. **24**(3), 1049–1103 (1996) (Cited on p. 262.)

463. M. Talagrand, New concentration inequalities in product spaces. Invent. Math. **126**(3), 505–563 (1996) (Cited on p. 265.)

464. M. Talagrand, Selecting a proportion of characters. Israel J. Math. **108**, 173–191 (1998) (Cited on pp. 421, 422.)

465. M. Talagrand, Majorizing measures without measures. Ann. Probab. **29**(1), 411–417 (2001) (Cited on p. 262.)

466. M. Talagrand, *The Generic Chaining*. Springer Monographs in Mathematics (Springer, Berlin, 2005) (Cited on pp. 262, 263, 264.)

467. M. Talagrand, *Mean Field Models for Spin Glasses. Volume I: Basic Examples* (Springer, Berlin, 2010) (Cited on p. 264.)

468. G. Tauböck, F. Hlawatsch, D. Eiwen, H. Rauhut, Compressive estimation of doubly selective channels in multicarrier systems: leakage effects and sparsity-enhancing processing. IEEE J. Sel. Top. Sig. Process. **4**(2), 255–271 (2010) (Cited on p. 37.)

469. H. Taylor, S. Banks, J. McCoy, Deconvolution with the ℓ_1-norm. Geophysics **44**(1), 39–52 (1979) (Cited on pp. 34, 37.)

470. V. Temlyakov, Nonlinear methods of approximation. Found. Comput. Math. **3**(1), 33–107 (2003) (Cited on p. 71.)

471. V. Temlyakov, Greedy approximation. Act. Num. **17**, 235–409 (2008) (Cited on p. 71.)

472. V. Temlyakov, *Greedy Approximation*. Cambridge Monographs on Applied and Computational Mathematics, vol. 20 (Cambridge University Press, Cambridge, 2011) (Cited on pp. 36, 71, 72, 129.)

473. R. Tibshirani, Regression shrinkage and selection via the lasso. J. Roy. Stat. Soc. B **58**(1), 267–288 (1996) (Cited on pp. 34, 36, 71.)

474. J. Traub, G. Wasilkowski, H. Woźniakowski, *Information-based Complexity*. Computer Science and Scientific Computing (Academic Press Inc., Boston, MA, 1988) With contributions by A.G.Werschulz, T. Boult. (Cited on pp. 34, 39.)

475. L. Trefethen, D. Bau, *Numerical Linear Algebra* (SIAM, Philadelphia, 2000) (Cited on p. 515.)

476. J.A. Tropp, Greed is good: Algorithmic results for sparse approximation. IEEE Trans. Inform. Theor. **50**(10), 2231–2242 (2004) (Cited on pp. 34, 36, 71, 128.)

477. J.A. Tropp, Recovery of short, complex linear combinations via l_1 minimization. IEEE Trans. Inform. Theor. **51**(4), 1568–1570 (2005) (Cited on p. 104.)

478. J.A. Tropp, Algorithms for simultaneous sparse approximation. Part II: Convex relaxation. Signal Process. **86**(3), 589–602 (2006) (Cited on p. 38.)

479. J.A. Tropp, Just relax: Convex programming methods for identifying sparse signals in noise. IEEE Trans. Inform. Theor. **51**(3), 1030–1051 (2006) (Cited on pp. 34, 36.)

480. J.A. Tropp, Norms of random submatrices and sparse approximation. C. R. Math. Acad. Sci. Paris **346**(23–24), 1271–1274 (2008) (Cited on pp. 262, 471, 472.)

481. J.A. Tropp, On the conditioning of random subdictionaries. Appl. Comput. Harmon. Anal. **25**, 1–24 (2008) (Cited on pp. 261, 262, 421, 423, 430, 471, 472.)

482. J.A. Tropp, On the linear independence of spikes and sines. J. Fourier Anal. Appl. **14**(5–6), 838–858 (2008) (Cited on pp. 36, 472.)

483. J.A. Tropp, The random paving property for uniformly bounded matrices. Stud. Math. **185**(1), 67–82 (2008) (Cited on p. 472.)

484. J.A. Tropp, Column subset selection, matrix factorization, and eigenvalue optimization. In *Proceedings of the Twentieth Annual ACM-SIAM Symposium on Discrete Algorithms*, SODA '09, Society for Industrial and Applied Mathematics, pp. 978–986, Philadelphia, PA, USA, 2009. (Cited on p. 473.)

485. J.A. Tropp, From the joint convexity of quantum relative entropy to a concavity theorem of Lieb. Proc. Am. Math. Soc. **140**(5), 1757–1760 (2012) (Cited on p. 565.)

486. J.A. Tropp, User-friendly tail bounds for sums of random matrices. Found. Comput. Math. **12**(4), 389–434 (2012) (Cited on pp. 261, 471.)

487. J.A. Tropp, A.C. Gilbert, M.J. Strauss, Algorithms for simultaneous sparse approximation. Part I: Greedy pursuit. Signal Process. **86**(3), 572–588 (2006) (Cited on p. 38.)

488. J.A. Tropp, J.N. Laska, M.F. Duarte, J.K. Romberg, R.G. Baraniuk, Beyond Nyquist: Efficient sampling of sparse bandlimited signals. IEEE Trans. Inform. Theor. **56**(1), 520–544 (2010) (Cited on pp. 37, 430, 431.)

489. J.A. Tropp, M. Wakin, M. Duarte, D. Baron, R. Baraniuk, Random filters for compressive sampling and reconstruction. *Proc. 2006 IEEE International Conference Acoustics, Speech, and Signal Processing*, vol. 3, pp. 872–875, 2006 (Cited on p. 429.)

490. A. Tsybakov, *Introduction to Nonparametric Estimation*. Springer Series in Statistics (Springer, New York, 2009) (Cited on p. 264.)

491. M. Tygert, Fast algorithms for spherical harmonic expansions, II. J. Comput. Phys. **227**(8), 4260–4279 (2008) (Cited on p. 427.)

492. A. Uhlmann, Relative entropy and the Wigner-Yanase-Dyson-Lieb concavity in an interpolation theory. Comm. Math. Phys. **54**(1), 21–32 (1977) (Cited on p. 571.)

493. E. van den Berg, M. Friedlander, Probing the Pareto frontier for basis pursuit solutions. SIAM J. Sci. Comput. **31**(2), 890–912 (2008) (Cited on pp. 504, 505.)

494. C.F. Van Loan, *Computational Frameworks for the Fast Fourier Transform* (SIAM, Philadelphia, 1992) (Cited on p. 575.)

495. S. Varadhan, Large deviations and applications. In *École d'Été de Probabilités de Saint-Flour XV–XVII, 1985–87*. Lecture Notes in Mathematics, vol. 1362 (Springer, Berlin, 1988), pp. 1–49 (Cited on pp. 198, 199.)

496. R. Varga, *Gershgorin and His Circles*. Springer Series in Computational Mathematics (Springer, Berlin, 2004) (Cited on p. 523.)

497. S. Vasanawala, M. Alley, B. Hargreaves, R. Barth, J. Pauly, M. Lustig, Improved pediatric MR imaging with compressed sensing. Radiology **256**(2), 607–616 (2010) (Cited on p. 35.)

498. A. Vershik, P. Sporyshev, Asymptotic behavior of the number of faces of random polyhedra and the neighborliness problem. Sel. Math. Sov. **11**(2), 181–201 (1992) (Cited on p. 305.)

499. R. Vershynin, John's decompositions: selecting a large part. Israel J. Math. **122**, 253–277 (2001) (Cited on p. 473.)

500. R. Vershynin, Frame expansions with erasures: an approach through the non-commutative operator theory. Appl. Comput. Harmon. Anal. **18**(2), 167–176 (2005) (Cited on p. 262.)

501. R. Vershynin, Introduction to the non-asymptotic analysis of random matrices. In *Compressed Sensing: Theory and Applications*, ed. by Y. Eldar, G. Kutyniok (Cambridge University Press, Cambridge, 2012), pp. xii+544 (Cited on pp. 199, 303.)

502. J. Vybiral, A variant of the Johnson-Lindenstrauss lemma for circulant matrices. J. Funct. Anal. **260**(4), 1096–1105 (2011) (Cited on p. 423.)

503. M. Wakinm, *The Geometry of Low-dimensional Signal Models*, PhD thesis, Rice University, 2006 (Cited on p. 56.)

504. S. Waldron, *An Introduction to Finite Tight Frames*. Applied and Numerical Harmonic Analysis (Birkhäuser, Boston, To appear) (Cited on p. 129.)

505. J. Walker, Fourier analysis and wavelet analysis. Notices Am. Math. Soc. **44**(6), 658–670 (1997) (Cited on pp. 420, 573.)

506. J.S. Walker, *Fast Fourier Transforms* (CRC Press, Littleton, 1991) (Cited on p. 575.)

507. Y. Wiaux, L. Jacques, G. Puy, A. Scaife, P. Vandergheynst, Compressed sensing imaging techniques for radio interferometry. Mon. Not. Roy. Astron. Soc. **395**(3), 1733–1742 (2009) (Cited on p. 37.)

508. P. Wojtaszczyk, *A Mathematical Introduction to Wavelets* (Cambridge University Press, Cambridge, 1997) (Cited on pp. 36, 424, 425.)

509. P. Wojtaszczyk, Stability and instance optimality for Gaussian measurements in compressed sensing. Found. Comput. Math. **10**, 1–13 (2010) (Cited on pp. 362, 363.)

510. P. Wojtaszczyk, ℓ_1 minimisation with noisy data. SIAM J. Numer. Anal. **50**(2), 458–467 (2012) (Cited on p. 363.)

511. S. Worm, Iteratively re-weighted least squares for compressed sensing. Diploma thesis, University of Bonn, 2011 (Cited on p. 504.)

512. G. Wright, Magnetic resonance imaging. IEEE Signal Process. Mag. **14**(1), 56–66 (1997) (Cited on p. 35.)

513. J. Wright, A. Yang, A. Ganesh, S. Sastry, Y. Ma, Robust Face Recognition via Sparse Representation. IEEE Trans. Pattern Anal. Mach. Intell. **31**(2), 210–227 (2009) (Cited on p. 36.)

514. K. Yosida, *Functional Analysis*, 6th edn. Classics in Mathematics (Springer, Berlin, 1995) (Cited on p. 583.)

515. T. Zhang, Sparse recovery with orthogonal matching pursuit under RIP. IEEE Trans. Inform. Theor. **57**(9), 6215–6221 (2011) (Cited on p. 169.)

516. X. Zhang, M. Burger, X. Bresson, S. Osher, Bregmanized nonlocal regularization for deconvolution and sparse reconstruction. SIAM J. Imag. Sci. **3**(3), 253–276 (2010) (Cited on p. 503.)

517. X. Zhang, M. Burger, S. Osher, A unified primal-dual algorithm framework based on Bregman iteration. J. Sci. Comput. **46**, 20–46 (2011) (Cited on p. 503.)

518. M. Zhu, T. Chan, An efficient primal-dual hybrid gradient algorithm for total variation image restoration. Technical report, CAM Report 08–34, UCLA, Los Angeles, CA, 2008 (Cited on p. 503.)

519. J. Zou, A.C. Gilbert, M. Strauss, I. Daubechies, Theoretical and experimental analysis of a randomized algorithm for sparse Fourier transform analysis. J. Comput. Phys. **211**, 572–595 (2005) (Cited on pp. 35, 423.)

520. A. Zymnis, S. Boyd, E.J. Candès, Compressed sensing with quantized measurements. IEEE Signal Process. Lett. **17**(2), 149–152 (2010) (Cited on p. 38.)

Applied and Numerical Harmonic Analysis (63 Volumes)

S. Foucart and H. Rauhut, *A Mathematical Introduction to Compressive Sensing*, Applied and Numerical Harmonic Analysis, DOI 10.1007/978-0-8176-4948-7, © Springer Science+Business Media New York 2013

J.J. Benedetto and P.J.S.G. Ferreira: *Modern Sampling Theory* (ISBN 978-0-8176-4023-1)

D.F. Walnut: *An Introduction to Wavelet Analysis* (ISBN 978-0-8176-3962-4)

A. Abbate, C. DeCusatis, and P.K. Das: *Wavelets and Subbands* (ISBN 978-0-8176-4136-8)

O. Bratteli, P. Jorgensen, and B. Treadway: *Wavelets Through a Looking Glass* (ISBN 978-0-8176-4280-80)

H.G. Feichtinger and T. Strohmer: *Advances in Gabor Analysis* (ISBN 978-0-8176-4239-6)

O. Christensen: *An Introduction to Frames and Riesz Bases* (ISBN 978-0-8176-4295-2)

L. Debnath: *Wavelets and Signal Processing* (ISBN 978-0-8176-4235-8)

G. Bi and Y. Zeng: *Transforms and Fast Algorithms for Signal Analysis and Representations* (ISBN 978-0-8176-4279-2)

J.H. Davis: *Methods of Applied Mathematics with a MATLAB Overview* (ISBN 978-0-8176-4331-7)

J.J. Benedetto and A.I. Zayed: *Modern Sampling Theory* (ISBN 978-0-8176-4023-1)

E. Prestini: *The Evolution of Applied Harmonic Analysis* (ISBN 978-0-8176-4125-2)

L. Brandolini, L. Colzani, A. Iosevich, and G. Travaglini: *Fourier Analysis and Convexity* (ISBN 978-0-8176-3263-2)

W. Freeden and V. Michel: *Multiscale Potential Theory* (ISBN 978-0-8176-4105-4)

O. Christensen and K.L. Christensen: *Approximation Theory* (ISBN 978-0-8176-3600-5)

O. Calin and D.-C. Chang: *Geometric Mechanics on Riemannian Manifolds* (ISBN 978-0-8176-4354-6)

J.A. Hogan: *Time-Frequency and Time-Scale Methods* (ISBN 978-0-8176-4276-1)

C. Heil: *Harmonic Analysis and Applications* (ISBN 978-0-8176-3778-1)

K. Borre, D.M. Akos, N. Bertelsen, P. Rinder, and S.H. Jensen: *A Software-Defined GPS and Galileo Receiver* (ISBN 978-0-8176-4390-4)

T. Qian, M.I. Vai, and Y. Xu: *Wavelet Analysis and Applications* (ISBN 978-3-7643-7777-9)

G.T. Herman and A. Kuba: *Advances in Discrete Tomography and Its Applications* (ISBN 978-0-8176-3614-2)

M.C. Fu, R.A. Jarrow, J.-Y. Yen, and R.J. Elliott: *Advances in Mathematical Finance* (ISBN 978-0-8176-4544-1)

O. Christensen: *Frames and Bases* (ISBN 978-0-8176-4677-6)

P.E.T. Jorgensen, J.D. Merrill, and J.A. Packer: *Representations, Wavelets, and Frames* (ISBN 978-0-8176-4682-0)

M. An, A.K. Brodzik, and R. Tolimieri: *Ideal Sequence Design in Time-Frequency Space* (ISBN 978-0-8176-4737-7)

S.G. Krantz: *Explorations in Harmonic Analysis* (ISBN 978-0-8176-4668-4)

B. Luong: *Fourier Analysis on Finite Abelian Groups* (ISBN 978-0-8176-4915-9)

G.S. Chirikjian: *Stochastic Models, Information Theory, and Lie Groups, Volume 1* (ISBN 978-0-8176-4802-2)

C. Cabrelli and J.L. Torrea: *Recent Developments in Real and Harmonic Analysis* (ISBN 978-0-8176-4531-1)

M.V. Wickerhauser: *Mathematics for Multimedia* (ISBN 978-0-8176-4879-4)

B. Forster, P. Massopust, O. Christensen, K. Gröchenig, D. Labate, P. Vandergheynst, G. Weiss, and Y. Wiaux: *Four Short Courses on Harmonic Analysis* (ISBN 978-0-8176-4890-9)

O. Christensen: *Functions, Spaces, and Expansions* (ISBN 978-0-8176-4979-1)

J. Barral and S. Seuret: *Recent Developments in Fractals and Related Fields* (ISBN 978-0-8176-4887-9)

O. Calin, D.-C. Chang, and K. Furutani, and C. Iwasaki: *Heat Kernels for Elliptic and Sub-elliptic Operators* (ISBN 978-0-8176-4994-4)

C. Heil: *A Basis Theory Primer* (ISBN 978-0-8176-4686-8)

J.R. Klauder: *A Modern Approach to Functional Integration* (ISBN 978-0-8176-4790-2)

J. Cohen and A.I. Zayed: *Wavelets and Multiscale Analysis* (ISBN 978-0-8176-8094-7)

D. Joyner and J.-L. Kim: *Selected Unsolved Problems in Coding Theory* (ISBN 978-0-8176-8255-2)

G.S. Chirikjian: *Stochastic Models, Information Theory, and Lie Groups, Volume 2* (ISBN 978-0-8176-4943-2)

J.A. Hogan and J.D. Lakey: *Duration and Bandwidth Limiting* (ISBN 978-0-8176-8306-1)

G. Kutyniok and D. Labate: *Shearlets* (ISBN 978-0-8176-8315-3)

P.G. Casazza and G. Kutyniok: *Finite Frames* (ISBN 978-0-8176-8372-6)

For an up-to-date list of ANHA titles, please visit **http://www.springer.com/series/4968**

Index

S. Foucart and H. Rauhut, *A Mathematical Introduction to Compressive Sensing*,
Applied and Numerical Harmonic Analysis, DOI 10.1007/978-0-8176-4948-7,
© Springer Science+Business Media New York 2013

Printed in the United States
by Bookmasters

Printed in the United States
By Bookmasters